Study Guide

OHANIAN'S PHYSICS

Second Edition

Study Guide

OHANIAN'S PHYSICS

Second Edition

Van E. Neie
PURDUE UNIVERSITY

Peter Riley
THE UNIVERSITY OF TEXAS AT AUSTIN

W · W · NORTON & COMPANY · NEW YORK · LONDON

Printed in the United States of America.

ISBN 0-393-95752-7

W. W. Norton & Company, Inc., 500 Fifth Avenue, New York, N.Y. 10110
W. W. Norton & Company Ltd., 37 Great Russell Street, London WC1B 3NU

1 2 3 4 5 6 7 8 9 0

Contents

To the Student

Physics is problem solving. Unfortunately, this fact is often taken to mean that one must perform numerical calculations, a process known more colloquially as "plug and chug." In the real sense, however, problem solving means grappling with a new idea or a posed question or problem and reaching a satisfactory solution, be it a numerical value or a verbal statement. The purpose of this guide is to help you improve your problem-solving skills through a variety of tasks designed to foster successful techniques and strategies. To be successful, *you must become actively engaged in the process*. You cannot learn problem solving by watching, any more than you can learn to play the violin by watching a violinist. There is simply no shortcut to proficiency in this skill. Without practice, progress will be minimal.

Each chapter in this guide first provides an overview. Read it before you read the corresponding chapter in the text. It gives you a preview of what is to come in the chapter and connects the chapter to previously covered material. Second, a summary of essential terms is presented. It serves as a reminder of the major topics in the chapter. Next is a section called *Key Concepts*, in which we attempt to expand and clarify ideas discussed in the text. Often these are concepts that traditionally cause difficulty for beginning students. The heart of each study guide chapter is the *Sample Problems* section. There, a worked problem is presented, followed by the authors' solution. Careful attention has been given to explaining the rationale for each step. The solution (most often algebraic rather than numerical) is then discussed in terms of its reasonableness and the way in which the solution changes as the variables take on different values. Then it's *your* turn. A guided problem is presented, followed by a step-by-step scheme for solving the problem. The execution of the various steps is your task, and we present only the final answer (at the end of the chapter). After you are satisfied that you have an acceptable solution, you can check it against the authors' solution. Then go one more step. Examine the solution to see if it fits your expectations, and see how the solution changes with selected variable changes. By looking at known situations (e.g., if one mass is 0, the acceleration is known to be that of free fall), you can determine whether the solution is likely to be correct. It is this kind of critical analysis that will most probably lead to successful problem-solving strategies and to a deeper understanding of the material.

In addition, because the ideas of physics are interrelated, quite often a problem appears similar to one encountered earlier. You should be alert to such occurrences. By seeing these relationships you can begin to build up a set of *schemes* applicable to classes of problems rather than to only individual ones.

We hope that you enjoy the challenge of learning physics and that this guide will offer some assistance toward that goal.

Study Guide

OHANIAN'S PHYSICS

Second Edition

Chapter 1

Measurement of Space, Time, and Mass

I often say that when you can measure what you are speaking about, and express it in numbers, you know something about it; but . . . when you cannot express it in numbers, your knowledge is of a meager and unsatisfactory kind; it may be the beginning of knowledge, but you have scarcely, in your thoughts, advanced to the stage of science, *whatever the matter may be.*

—Lord Kelvin (William Thomson)

Overview

In the preface we were introduced to the considerable range of sizes in the physical world. How is this and other information obtained? All information about our environment involves making comparisons or measurements that must be related in some way to standards that are mutually agreed upon. Measurements are, in a sense, the "scaffolding" needed to erect meaningful models of reality, and Chapter 1 presents and discusses the reference frames and standards associated with the act of measuring.

Essential Terms

ideal particle
event
coordinate systems (rectangular, spherical, and cylindrical)
reference frame
empirical evidence
mass
SI system of units
derived unit
meter
kilogram
second

position
length
time
slug
pound
pound-mass

Key Concepts

1. *Ideal particle*. Many objects and systems are complex in their structure and in their motions, and it often is useful to reduce such objects to simpler conceptual forms to get at certain underlying principles devoid of "distracting appendages." The *ideal particle* does not exist, but by treating objects as if they truly were point masses, we can obtain useful information that might otherwise be buried in the "noise" of secondary characteristics. For example, we might state that a car is located at a position 25 m from some specified origin. But which part of the car? Conceptually, we treat the car as a point particle and eliminate such ambiguity. Should the dimensions of the car become an important consideration, then that assumption may no longer be valid.

You will note a common theme throughout the study of physics: that an object or system behaves *as if* it were a simpler system conceptually. The *as if* idea is not an attempt to change reality; rather, it is a technique for achieving a better understanding of what is happening in a given situation.

2. *Coordinate axes*. Coordinate axes are used for a variety of purposes. There is nothing sacred about the commonly used coordinate systems—rectangular, cylindrical, and spherical; it's just that these most often are the best choice in the traditional classic physics problems. They have the further advantage of being easily visualized because they take advantage of one's spatial perception of the three-dimensional physical environment.

3. *Reference frames*. Reference frames consist of a coordinate grid with suitably adjusted clocks. Knowledge of the *rest frame* of a particle in no way implies any knowledge about the absolute motion of the particle, which is to say that the rest frame does not indicate that the particle is absolutely at rest; rather, the particle's position does not change relative to the origin of that frame. Thus, the particle may be moving relative to one frame while at rest relative to another. For example, we usually say that we are at rest when standing still on the surface of the Earth, as indeed it seems that we are; the Earth rotates on its axis relative to a different reference frame, however, and travels in an orbit about the Sun relative to another. Depending on the choice of reference frame, then, we may be at rest or traveling at a speed of approximately 10,000 km per hour!

4. *Significant figures*. The widespread availability of electronic calculators and computers has opened up new possibilities for performing calculations that previously would have been extremely difficult and time-consuming. But there are drawbacks to this development. All numbers entered into these devices are assumed to be "pure"; for example, 105 is assumed to be *exactly* 105. In physics, however, we often are dealing with measurements

or other data that have limited precision. Thus, 105 cm, unless stated otherwise, means simply that the measurement is closer to 105 cm than it is to 104 cm or 106 cm. We say that the number 105 has *three significant figures*. The purpose of our emphasis on significant figures is to establish a sense of the uncertainty associated with data handling. The calculator sees no difference between 105 and 105.000, but the latter reflects a greater precision because it shows that the three decimal places beyond 105 are in fact zeros and not something else. This idea is very important in reporting results of operations on data. As a general rule, one should report in the result only the number of significant figures present in the datum with the smallest number of significant figures. Consider the following calculations: 2.4 cm $\times$ 4.72 cm $\times$ 10.36 cm. The calculator display reads 117.35808, or 1.1735808×10^2. If we think of each of these measurements as being rounded to the last reported digit, then it is not difficult to show that the result should be expressed as 1.2×10^2 cm^3, or perhaps 1.17×10^2 cm^3 at best. It is tempting to include all digits obtained by the operation, but the extra digits are *meaningless*. Honesty in reporting requires that one not attribute more precision to a number than is justified.

The limitations on precision we have discussed do not apply to *counting numbers*, however. If we require the total mass of 9 marbles, each having a mass of 1.36×10^{-2} kg, the answer is $9 \times 1.36 \times 10^{-2}$ kg $= 1.22 \times 10^{-1}$ kg (three significant figures). The number 9 is an *exact* number, not subject to uncertainty, assuming that one counted properly.

One final note. If you wish to report unambiguously which, if any, zeros in a number are significant, use the power-of-10 notation (express the result as a number between 1 and 10 multiplied by a power of 10). Then any zeros to the right of the decimal point must be significant. For example, the number 13,000 might have 2, 3, 4, or 5 significant figures, depending on the precision. We cannot tell from the way it is written. But if it is expressed as 1.30×10^4, then we know that only one of the zeros is significant, the others simply being placeholders.

5. *Unit of mass*. The international *kilogram* is the standard of mass to which all others are compared, but because certain units in the British engineering system are similar to everyday usage in the United States, there is often confusion as to their meaning. For example, the *pound* is the unit of *force*, but the *pound-mass* is used frequently in everyday language as the unit of mass. And it isn't always clear just what is meant by these terms. For the time being, use the conversion factors given in the text. A clearer picture should emerge when we study Newton's laws in Chapter 5.

Sample Problems

1. *Worked Problem*. A common unit of length (distance) in horse racing is the furlong, defined as one-eighth of a mile. How many meters are equivalent to a furlong?

Solution: From Table 1.3 we know that a mile is equal to 1609.38 m. Thus we have

$$1 \text{ furlong} \times \frac{1 \text{ mi}}{8 \text{ furlongs}} \times \frac{1609.38 \text{ m}}{1 \text{ mi}} = 201.1725 \text{ m}$$
$$= 2.011725 \times 10^2 \text{ m}$$

Note that the answer is expressed to seven figures (read from the calculator). But how many are significant? Although it appears that we should retain only one because the miles and furlongs are expressed to one significant figure, this really is somewhat misleading. Most conversions comparing a standard to some other unit imply that we mean *exactly* one unit of that quantity. For example, 1 mi = 1609.38 m means that exactly 1 mile is equal (to six significant figures) to 1609.38 m. Legitimately we could keep six significant figures in the result. As a practical matter, however, this kind of precision is rarely needed. So a better way to ask the question might be, "Approximately how many meters are in a furlong?" The answer is approximately 200 m.

2. *Guided Problem*. A road sign is to be made that carries dual units for the distance to Middlevale. Suppose the English unit portion reads, "Middlevale 76 mi." How should the distance in kilometers be stated?

Solution scheme

a. Look up the mile-to-kilometer equivalence.
b. How many significant figures does the mile measure in the problem have?
c. Express the kilometer distance to the same number of significant figures (or at most to the nearest kilometer).

3. *Worked Problem*. A dime has a diameter of 18 mm. If it is held about 2 m from the eye, it will just cover the full Moon.

a. What angle (in degrees) does the full Moon subtend at the eye?
b. If the distance to the Moon is 3.9×10^5 km, what is the diameter of the Moon in kilometers?

Solution: Because the dime just covers the Moon at 2 m, at this distance the dime subtends the same angle as the Moon. The geometrical diagram (not to scale) for this situation is shown in Figure 1.1.

a. The circumference of a circle of radius 2 m is approximately 13 m. The dime takes up 0.018 m/13 m of that circumference. But a full circle corresponds to 360°, so that

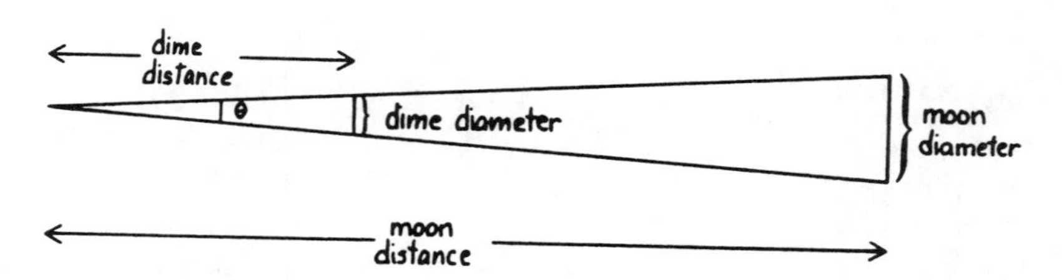

Figure 1.1 Problem 3. Because the dime just covers the Moon, the ratio of diameters is equal to the ratio of distances.

$$\frac{0.018 \text{ m}}{13 \text{ m}} = \frac{x}{360°}$$
$$x = 0.5°$$

b. With reference to the diagram, we see that there are similar triangles, giving

$$\frac{0.018 \text{ m}}{2 \text{ m}} = \frac{D_{\text{m}}}{3.9 \times 10^5 \text{ km}}$$
$$D_{\text{m}} = 3.5 \times 10^3 \text{ km}$$

Comment: It was not necessary to change the distance to the Moon to meters, because the ratio on the left side is dimensionless. Therefore, D_m is expressed in the same units as the Moon distance.

4. *Guided Problem.* Eratosthenes of Egypt was able to deduce the circumference of the Earth (and thus its diameter) by the following method. When the Sun at noon shone directly down a well at Syene, a vertical obelisk in Alexandria, 5000 stadia due north, cast a shadow that indicated that the rays of the Sun make an angle of 7.2° with respect to the vertical, as in Figure 1.2. Determine the diameter of the Earth in stadia. If one stadium (singular form of stadia) is approximately one-tenth of a mile, compare the result with the modern-day value.

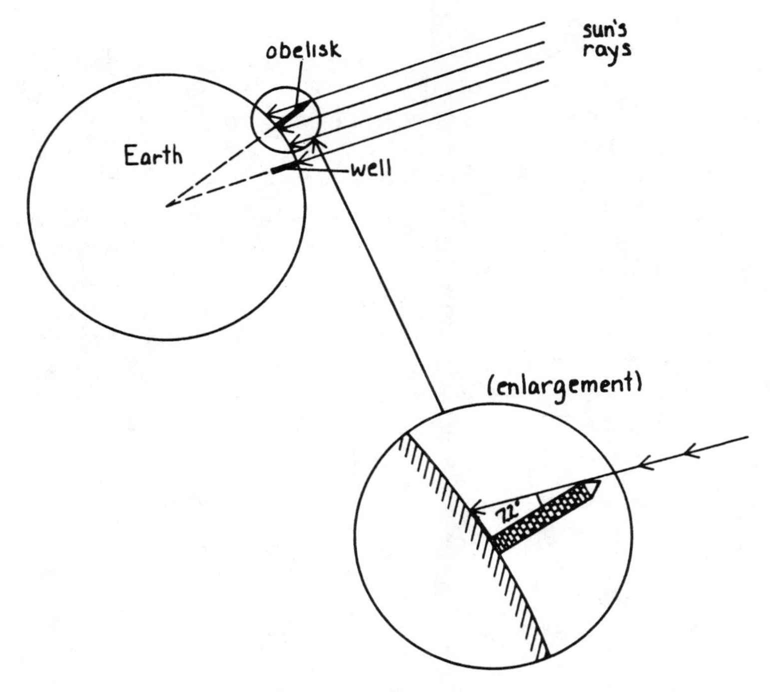

Figure 1.2 Problem 4. The rays of the Sun make an angle of 7.2° with the vertical at the obelisk.

Solution scheme

a. Relate the angle made by the rays of the Sun to the angle subtended at the center of the Earth by the 5000 stadia distance.
b. Note that the angle subtended at the center of the Earth is to 360° as 5000 stadia is to the circumference of the Earth.
c. Use the fact that $C = \pi d$ to find the diameter.

5. *Worked Problem.* In a laboratory experiment, a student measured the dimensions of a rectangular block and obtained the data shown in Figure 1.3. Assume all measurements have been rounded to the nearest millimeter.

a. What volume should be reported for the block?
b. Which measurement contributes most to the uncertainty in the result?

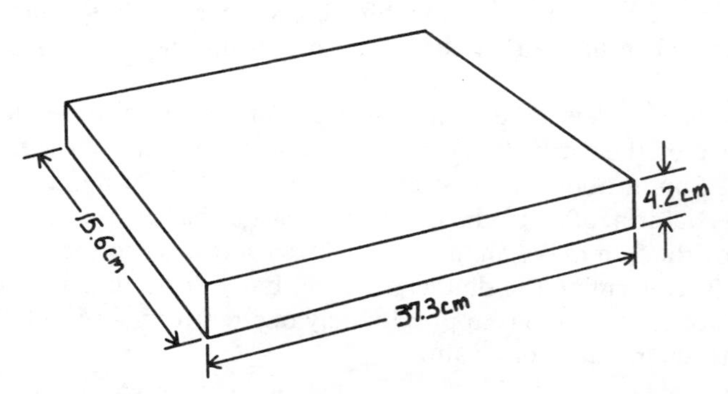

Figure 1.3 Problem 5. The dimensions of a rectangular block.

Solution: Because these are measurements, they are subject to uncertainty; therefore, the volume, obtained by multiplying together the three measurements, should be reported in a manner consistent with the significant-figure concept.

a. $V = 15.6 \text{ cm} \times 37.3 \text{ cm} \times 4.2 \text{ cm} = 2443.896 \text{ cm}^3 = 2.4 \times 10^3 \text{ cm}^3$. Note that the measurement having the least precision is 4.2 cm (two significant figures). Therefore, the volume may be reported only to this precision.
b. It is useful here to consider percentages of change in the result when each measurement, in turn, is changed. We assumed that each of these measurements is accurate to ±0.05 cm, that is, that each has been rounded to the nearest 0.1 cm. For the three measurements, then, the possible percentages of uncertainty are:

$$4.2 \text{ cm:} \quad \frac{0.05 \text{ cm}}{4.2 \text{ cm}} \times 100\% = 1.2\%$$

$$15.6 \text{ cm:} \quad \frac{0.05 \text{ cm}}{15.6 \text{ cm}} \times 100\% = 0.32\%$$

$$37.3 \text{ cm:} \quad \frac{0.05 \text{ cm}}{37.3 \text{ cm}} \times 100\% = 0.13\%$$

We see that an error of ±0.05 cm causes the largest percentage of change in the measurement with the fewest significant figures. This result has important practical implications. If one can measure several quantities to the same absolute degree of precision, then the smallest number has the largest percentage of uncertainty. Therefore, to improve the precision of the result, the smallest dimension should be measured with considerable care, because it contributes most to the uncertainty. To take an extreme case, an uncertainty of ±1 mm in a measurement of 2 mm is 50%, whereas the same uncertainty in a measurement of 1 m (1000 mm) is only 0.1%.

6. *Guided Problem.* A student wishes to determine the height of a building by triangulation. The appropriate measurements taken are shown in Figure 1.4. The angle has an uncertainty of ±1°, and the distance L has an uncertainty of ±0.20 m. What is the percentage of uncertainty in the height measurement?

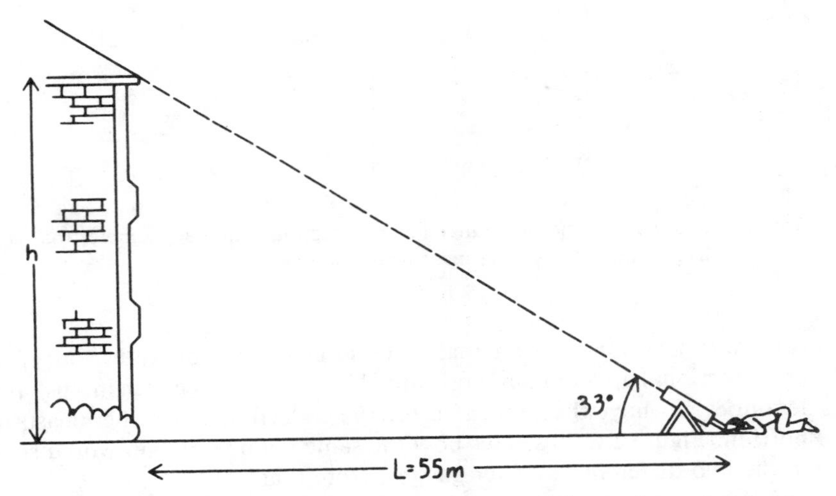

Figure 1.4 Problem 6. By sighting the top of the building to determine the elevation angle, we can determine the height of the building.

Solution scheme

a. Make use of the fact that, in multiplying or dividing data, percentages of uncertainty are *added*; that is, the total percentage of uncertainty is the *sum* of the individual percentages of uncertainty.
b. Calculate the percentages of uncertainty in the length and angle measurements.
c. The height is found by the expression $L \times \tan\theta$. Calculate the uncertainty in $\tan\theta$ by computing $\tan\theta$ for $(\theta + 1)°$ and for $(\theta - 1)°$.
d. Calculate h and add the percentages of uncertainty.

7. *Worked Problem.* A rain gauge is constructed in such a way that the opening at the top has a diameter of 10 cm but the measuring chamber has a diameter of 5 cm, as illustrated in Figure 1.5. How should the measuring chamber be calibrated to read the rainfall in centimeters? In inches?

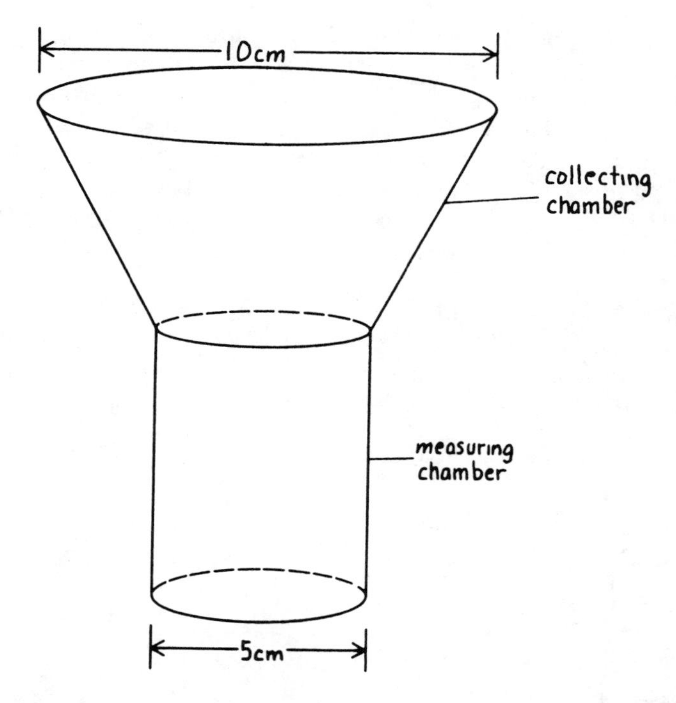

Figure 1.5 Problem 7. The rain gauge has a collecting chamber diameter that is larger than that of the measuring chamber.

Solution: The rain gauge actually measures volume of water, but if the cross-sectional area remains constant, $V \sim h$. The rain gauge in this problem does not have the same area for the collecting as for the measuring portion. Figure 1.6 illustrates how the same volume of rain would stand in the two different diameter gauges. Note that

$$V_c = A_c h_c$$

and

$$V_m = A_m h_m$$

Thus, for a given volume of rain, $V_c = V_m$,

$$A_c h_c = A_m h_m$$
$$\frac{h_c}{h_m} = \frac{A_m}{A_c}$$

Now h_m is the height measured in the measuring tube, but h_c is the "true height," because it represents the height of rain collected in a container of uniform cross-sectional area A_c. Therefore,

$$h_c = \frac{A_m}{A_c} h_m$$

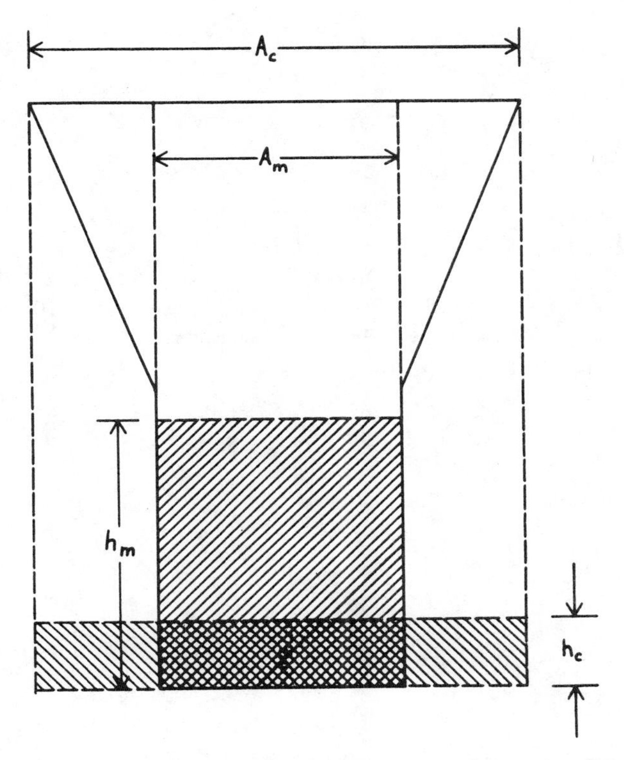

Figure 1.6 Problem 7. If the measuring chamber were of the same diameter as the collecting chamber, the rain would collect to a height of h_c rather than h_m.

is the calibration equation. If the height in the measuring tube is 8.0 cm, the true reading is

$$h_c = \frac{(5\text{ cm})^2}{(10\text{ cm})^2}\,8.0\text{ cm} = 0.40 \times 8.0\text{ cm} = 3.2\text{ cm} = 1.3\text{ in}$$

Thus, at a distance of 8.0 cm above the bottom of the gauge, the gauge should be marked 3.2 cm or 1.3 in.

8. *Guided Problem.* The mole is approximately 6.0×10^{23} particles. To imagine how many particles this represents, suppose you had a mole of tiny glass balls, each 1.0 mm in diameter. Glass has a density of 2.5×10^3 kg/m^3.

a. How much mass would be in a mole of these glass balls?
b. Suppose you could stack them into a cubical pile. What would be the dimensions of such a cube?

Solution scheme

a. Find the volume of one glass ball, given its diameter.
b. Given the density, determine its mass.
c. Each ball, in stacking, occupies a space of 1 mm^3. Use this figure to determine how much volume is occupied by a mole of these balls.

Answers to Guided Problems

2. 122 km
4. 8.0×10^4 stadia, or 8,000 mi. (Compare with 7,936 mi, the present-day value.)
6. tan 33° = 0.65 ± 7.6%
   L = 55 m ± 0.36%
   h = 35.7 m ± 8% = 35.7 m ± 2.9 m
   Almost all the error is attributable to the measurement of θ.
8. a. 1.25×10^{-3} g $\times 6.0 \times 10^{23} = 7.5 \times 10^{20}$ g $= 7.5 \times 10^{17}$ kg, which is about 1/100,000 the mass of the Moon.
   b. A cube 84 km on a side.

Chapter 2

Kinematics in One Dimension

Overview

Now that we have learned to handle combinations of variables, especially the ratio of quantities, the next step is to analyze the phenomenon of motion. In describing how objects undergo changes in position or velocity, we draw upon our knowledge of the basic units of length and time and the ways in which these can be combined to yield derived quantities such as velocity and acceleration. In addition to algebraic techniques, we introduce in this chapter the procedures for analyzing motion graphically; such analysis is a powerful alternative for studying the conceptual aspects of motion.

Essential Terms

kinematics
translational motion
average speed
average velocity
instantaneous velocity
worldline
slope of a line
tangent line
average acceleration
instantaneous acceleration
acceleration of gravity
gee

Key Concepts

1. *Distance, displacement, speed, and worldlines.* These concepts are closely related and sometimes confusing. For example, the symbol Δx cannot literally be called a *distance*, because Δx means $x_2 - x_1$ or $x - x_0$. These variables are positions along a line, and the Δx means "take the difference between x and x_0." This difference may or may not be the actual distance traveled. Suppose you drive in the $+x$ direction 40 km and then turn around

and drive back to a point 20 km from your starting point. $\Delta x = 20$ km, but you have driven a *distance* of 40 km + 20 km = 60 km. To avoid ambiguity, one should use a symbol such as d for distance, where d stands for the number that would be read from, say, the odometer of an automobile. In contrast, *displacement* refers to where you are at a given time relative to where you were at some earlier time, and this difference, Δx, is measured (for one-dimensional motion) along a line, relative to some origin. These differences are spelled out carefully in example 3 on page 29 of the text. Be sure you understand them. The *average speed* is defined as the distance traveled divided by the time. (A closely related concept, instantaneous speed, will be discussed at greater length in Chapter 5.)

The *worldline* is an important idea, both here and within the context of Chapter 41, which deals with special relativity. The *worldline* is a graph of position versus time and is not to be confused with the actual path of the object. Consider a block moving in a single dimension while its motion is recorded on a sheet of graph paper moving at uniform speed perpendicular to the direction of the block's motion as illustrated in Figure 2.1. The motion itself takes place along a line, but it is recorded in two dimensions, one of which is time.

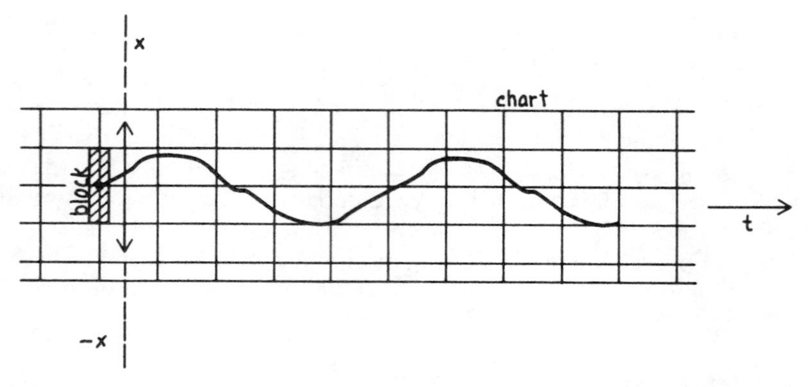

Figure 2.1 The block with ink pen attached moves only along the x axis, while the chart moves uniformly at right angles to the x direction. The trace thus produced is the worldline for the block.

2. *Graphs of motion*. There probably is no better way to understand the phenomenon of motion than to study graphs of a variety of motions. In so doing, one can see the time development of the motion rather than simply the position or velocity or acceleration at a single instant. To help you understand what information a graph conveys, note which *variables* are being plotted. For example, is the graph one of position versus time, or velocity versus time? If it is the former and the position of the object is desired, one simply reads it directly from the graph. If velocity is the quantity being sought, however, one must compute the slope of the curve at the specified time. It is often helpful to remember the following sequence: the slope of the position–time curve gives the velocity, and the slope of the velocity–time curve gives the acceleration.

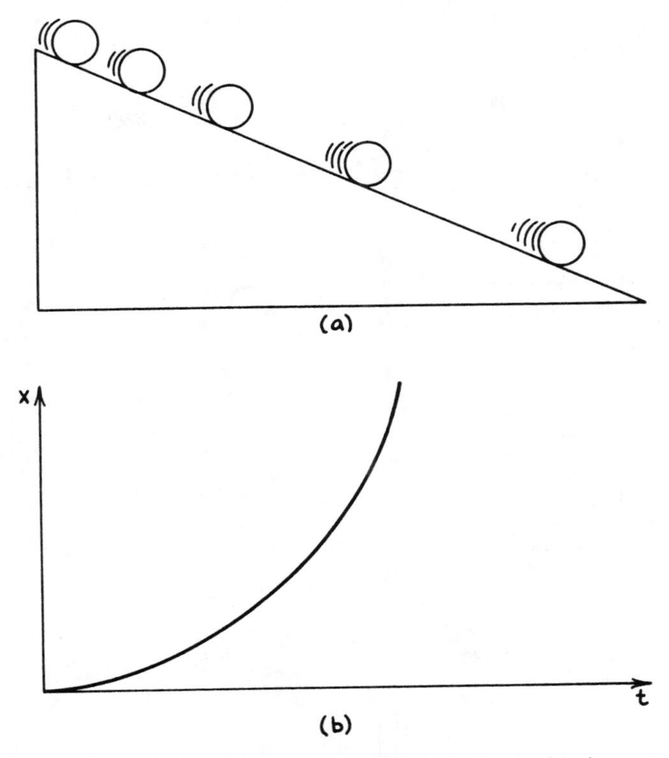

Figure 2.2 (a) A ball rolls down a ramp, uniformly increasing its speed. (b) A position–time graph of the motion.

The text discusses the method by which slopes are determined from the tangent line. The following thought experiment may be of additional help. Suppose a ball is rolled down a straight ramp, as illustrated in Figure 2.2a. We know that it will undergo uniform acceleration and that the worldline will look like that in Figure 2.2b. Now suppose that at some time t' (and associated position x') the ball encounters a straight horizontal shelf in contact with the plane, as shown in Figure 2.3a. As the ball rolls along the shelf, its speed will be constant and equal to its speed when it encountered the shelf. Thus the speed *at* t' can be determined by measuring the constant speed *after* t'. The worldline for this motion is shown in Figure 2.3b. Note that the slope is constant after t'. This slope is easy to determine, as illustrated by the triangle. But if this straight line is extended downward and to the left, as in Figure 2.3c, we generate the *tangent line* at (x',t'). Therefore, the slope of the curve at any point is the slope of the tangent line at that point. These latter graphs should be compared with Figures 2.5(a) and 2.5(b) in the text.

3. *Acceleration*. Just as velocity is the rate of change of position, acceleration is the rate of change of velocity. The acceleration carries no information about the velocity itself; rather, it is a measure of how the velocity is *changing*. The velocity may actually be zero at some instant, but the acceleration can still be non-zero. Moreover, the acceleration and the velocity

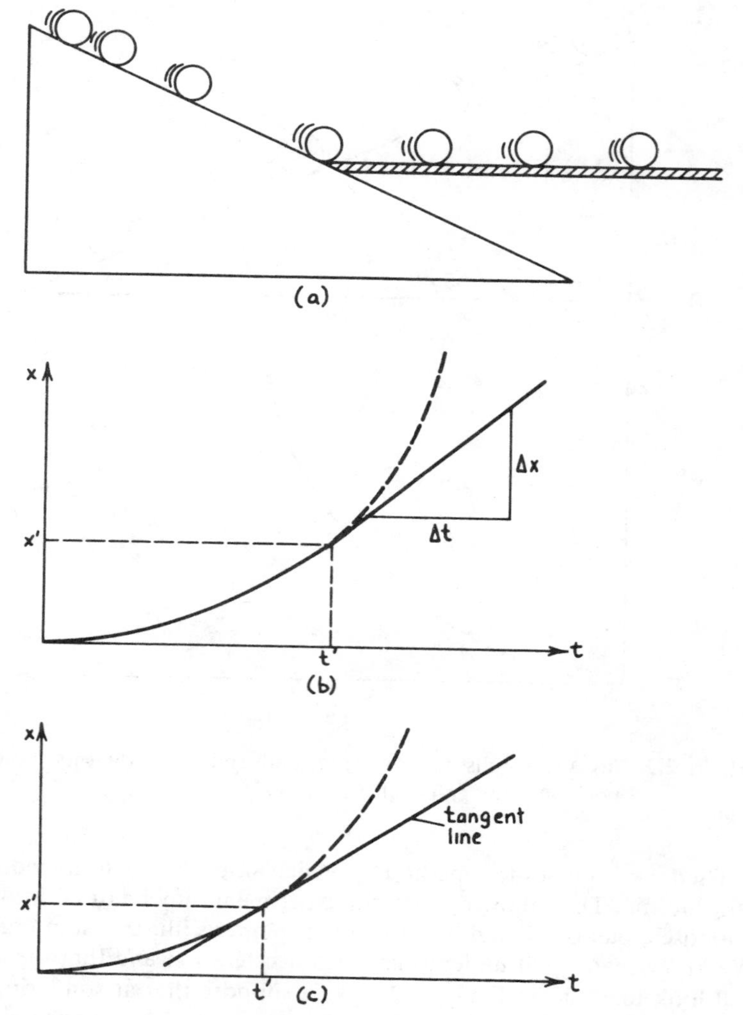

Figure 2.3 (a) A horizontal shelf intercepts the ball. (b) A position–time graph of the situation in (a). (c) The tangent line.

rarely are in the same direction. Only when an object is speeding up along a straight line is the acceleration in the same direction as the velocity. Thus, we should avoid using the *incorrect* definition that acceleration is velocity divided by time. The correct statement is that the acceleration is the *change* in velocity divided by the time interval.

We can see that the units for acceleration must be a velocity unit divided by a time unit. It is not necessary that the unit used for time in the denominator be the same as that used in the velocity unit, but they often are the same, especially when we use the more complex of the motion equations. If so, we get unusual-looking units such as mi/h^2 or m/s^2. This form of the unit is an artifact of the division process. The unit still should be conceptualized as (mi/h)/h or (m/s)/s meaning, for example, in the latter case, that the velocity is changing at the rate of so many meters per second each second.

4. *Free fall*. Because we associate the term *falling* with downward motion, it is easy to forget that the term *free fall*, as used in physics, refers to any motion resulting from only the gravitational force. The object may be moving upward, downward, or, as we shall see in Chapter 5, even at some angle to the horizontal. As long as it is free of all forces other than gravity, we call the motion *free fall*. (A better term might be *free flight*, but *free fall* is the one commonly used.) The *direction* of the acceleration of gravity is always vertically downward, regardless of the direction of motion (as long as the motion is free fall). Whether this direction is labeled positive or negative depends on the choice of coordinate systems. Should the positive direction be taken as upward, then the acceleration, g, is negative. If the downward direction is positive, then g is positive. For example, if a ball is thrown vertically upward, the acceleration of the ball, once it leaves the hand, is downward at all times, even when the ball is at the top of its flight, where the *velocity* is zero. It is common for the upward direction to be taken as positive, and the value of g is then negative. If, however, positive is downward, then g is positive.

5. *Equations of motion*. Although it appears that there are several equations of motion, the two involving t^2 and v^2 are redundant; that is, they are derived equations. One could solve all problems in straight-line kinematics using only the definitions of average velocity and acceleration; the derived expressions are useful, however, because they usually allow us to obtain the solution more quickly and efficiently. But be especially careful in using the expression for average speed:

$$\bar{v} = \frac{v_0 + v}{2}.$$

This expression is valid *only* for uniformly accelerated motion. Let's look at an example for which this relationship is *not* appropriate. Suppose you drive to a city 80 km distant and return by the same route. On the way to the city you drive 80 km/h, but during the return trip you drive only 40 km/h. What is your average speed for the total trip? It is not 60 km/h! To understand why, note that the first part of the trip requires 1 hour, whereas the second part requires 2 hours. Thus, the average speed must be nearer the slower speed, because more time is spent driving at that speed. Because the total distance traveled is 160 km, requiring a time of 3 h, the average speed for the trip is approximately 53 km/h. Moreover, the greater the difference in the two speeds, the closer the average will be to the slower of the two. Now consider a different problem. A car *accelerates uniformly* from 10 km/h to 50 km/h. In this case the average speed is the simple average of the two speeds, 30 km/h.

Sample Problems

1. *Worked Problem*. Many drivers complain that precious time is lost because of the imposition of the 55 mi/h speed limit. Some claim that 65 mi/h is a more reasonable limit. Suppose you drive from Indianapolis to

Chicago, a distance of 200 mi. How much "precious time" do you lose driving at the lower speed?

Solution

$$\text{At 65 mi/h:}\quad t = \frac{d}{\upsilon} = \frac{200\ \text{mi}}{65\ \text{mi/h}} = 3.08\ \text{h} = 184\ \text{min}$$

$$\text{At 55 mi/h:}\quad t = \frac{d}{\upsilon} = \frac{200\ \text{mi}}{55\ \text{mi/h}} = 3.64\ \text{h} = 218\ \text{min}$$

The difference is approximately 34 min. Whether this is "precious" depends on your situation. For a trip of 50 mi, the difference is only 8 min, but for a 1000 mi trip, the difference is 2.8 h. In each case the increase is approximately 18% of the total driving time. For the 1000 mi trip, however, it might be more meaningful to include some stops. Let's assume that the stops in each instance reduce the *average* speed to 45 mi/h and 55 mi/h, respectively, resulting in times of 18.2 h and 22.2 h for the higher and lower speeds. The difference now is 22% of the total driving time.

2. *Guided Problem.* An automobile traveling 50 km/h passes a string of 20 connected stationary railroad cars in 24 s. From this information calculate the length of one car.

Solution scheme

a. Determine the distance that the car travels in 24 s, assuming the speed to remain constant.
b. Divide this distance by 20 to get the length of each car.

3. *Worked Problem.* For the first 2.0 km, a jogger maintains a steady speed of 10 km/h. For the next 2.0 km, she walks at a steady speed of 6.0 km/h.

a. Determine her average speed for the 4.0 km distance.
b. Show that the same answer is obtained independent of the actual distances as long as the distance walked is the same as the distance run.

Solution: This is *not* accelerated motion, so we cannot simply average the two speeds; rather, we must use the definition for average speed given by $\bar{\upsilon} = d/t$, where d is the total distance covered.

a. $d = d_1 + d_2 = 2.0\ \text{km} + 2.0\ \text{km} = 4.0\ \text{km}$

$$t = t_1 + t_2 = \frac{d_1}{\upsilon_1} + \frac{d_2}{\upsilon_2} = \frac{2.0\ \text{km}}{10\ \text{km/h}} + \frac{2.0\ \text{km}}{6.0\ \text{km/h}}$$

$$t = \frac{1}{5}\ \text{h} + \frac{1}{3}\ \text{h} = 0.53\ \text{h}$$

$$\bar{\upsilon} = \frac{4.0\ \text{km}}{0.53\ \text{h}} = 7.5\ \text{km/h}$$

Compare this result with the simple average, which is 8 km/h.

b. Rather than use the specific distances, we will designate them as d_1 and d_2, with $d_1 = d_2 = d'$. Then

$$\bar{\upsilon} = \frac{2d'}{\dfrac{d'}{\upsilon_1} + \dfrac{d'}{\upsilon_2}} = \frac{2}{\dfrac{1}{\upsilon_1} + \dfrac{1}{\upsilon_2}} = \frac{2\upsilon_1\upsilon_2}{\upsilon_1 + \upsilon_2}$$

Note that d' divides out here.

$$\bar{v} = \frac{2 \times 6.0 \text{ km/h} \times 10 \text{ km/h}}{6.0 \text{ km/h} + 10 \text{ km/h}} = 7.5 \text{ km/h}$$

This means, for example, that if one makes a round trip, going and returning by the same route, the average speed will be independent of the distance *if* the speed going and the speed returning, respectively, are the same in each case.

4. *Guided Problem.* A world-class sprinter can run the 100 m dash in 10.0 s. A junior high school student may require 12 s to run the same distance. If the two were to race each other, by how much would the winner be ahead at the conclusion of the race?

Solution scheme

a. Determine the average speed of each runner for the entire 100 m distance.
b. Determine how far the trailing runner has run during the *winner's* time.
c. Subtract this distance from 100 m to determine the lead at the end of the race.

5. *Worked Problem.* Figure 2.4 illustrates the worldlines for two automobiles making the same trip.

a. Which one has the greater average velocity during the trip?
b. Determine the average velocity of each.
c. Which one is traveling faster at $t = 2$ h?
d. At what time(s) do they pass each other?
e. What is the approximate velocity of B at $t = 2$ h?
f. Describe the trip taken by each car.

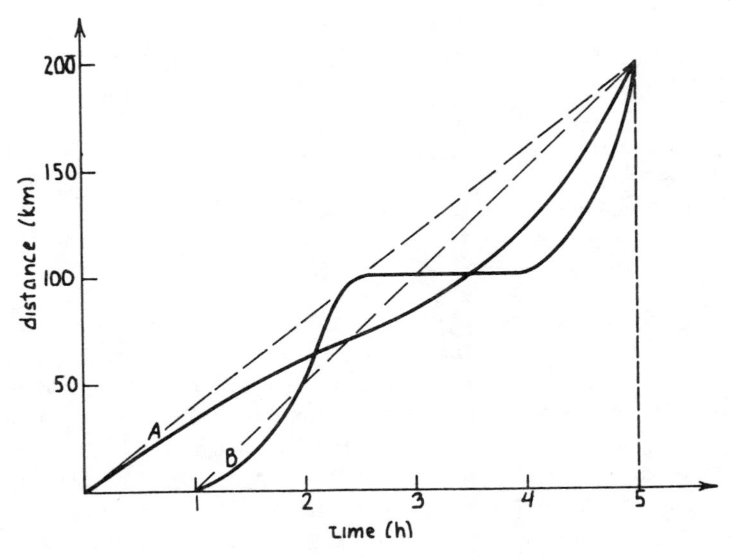

Figure 2.4 Problem 5. Worldlines for automobiles A and B.

Solution: We note first that these are worldlines, that is, *position*–time graphs of the motion. Thus, positions are read directly from the graph but velocities must be found by computing slopes.

a. Average velocity is the total displacement divided by the time. Because each car had the same displacement, B had the greater average velocity: it started later and arrived at the same time.
b. The average velocities are found by computing the slopes of the dashed lines in the figure.

$$\text{A:}\quad \bar{v} = \frac{200\text{ km}}{5\text{ h}} = 40\text{ km/h}$$

$$\text{B:}\quad \bar{v} = \frac{200\text{ km}}{4\text{ h}} = 50\text{ km/h}$$

c. At $t = 2$ h, the worldline of car B has the greater slope, so car B has the greater velocity.
d. To make this determination read the cars' positions from the graph. B is closer to the origin until $t = 2$ h, at which time the positions are the same. From $t = 2$ h to $t = 3.5$ h, B is farther from the origin than A. From $t = 3.5$ h to $t = 5$ h, A is ahead of B. Therefore, B passes A at the 2 h mark and A passes B (who is at rest) at the 3.5 h mark. They meet again at $t = 5$ h.
e. The slope of the worldline for B at $t = 2$ h is approximately 100 km/h; that is the instantaneous velocity of B at that time.
f. Each car starts at the same location, but B starts 1 h later than A. For the first 2.5 h, B travels faster than A. B stops for approximately 1.5 h, however, during which time A not only speeds up but also passes B. Finally, B starts again and travels faster than A, catching A at $t = 5$ h.

6. *Guided Problem.* The following are the positions at various times for a hot-air balloon moving vertically.

Elev. (m)	5	20	30	32	40	45	50	50	38	45
Time (min)	0	1	2	3	4	5	6	7	8	9

a. Draw the worldline for this motion.
b. Calculate the average velocity from $t = 2$ min to $t = 8$ min.
c. Calculate the average speed during this time.
d. Calculate the instantaneous velocity at $t = 1$ min.

Solution scheme

a. Compute the average velocity by determining the *displacement* (net change in position), Δx, and dividing this quantity by the time.
b. The average speed is computed by determining the total *distance* traveled. From the worldline determine how far upward the balloon traveled and add this to the distance it traveled downward, all between the times $t = 2$ min and $t = 8$ min. Signs should be disregarded; you want the total distance here, not the displacement.
c. From the worldline compute the slope of the curve at $t = 1$ min.

7. ***Worked Problem.*** A young man who wishes to catch the bus to work sees it stopped at the curb up the street. If the man runs at a constant speed of 6 m/s and if, when he is 15 m from the bus, the bus begins to accelerate uniformly at 1 m/s^2, will he be able to catch the bus? If so, in how many seconds, and how far will the bus have traveled during this time?

Solution: Our first task is to translate the problem into a statement that can be "understood" by means of the appropriate kinematic expressions. So we ask: At what common time t will the bus and the man have the same position value? We take the origin (arbitrarily) to be the position of the man when the bus starts to accelerate, and $t = 0$ at this instant. So at $t = 0$, $x_{0_b} = 15$ m and $x_{0_m} = 0$. The two motions are not the same. The man moves with constant velocity, but the bus undergoes uniform acceleration. In each case, however, we need to find an expression for x as a function of t. What variables do we know already? For the man, $x_0 = 0$ and $\upsilon = 6$ m/s (constant). For the bus, $x_0 = 15$ m, $\upsilon_0 = 0$, $a = 1$ m/s^2. Looking among the equations governing these motions, we find:

$$\text{Man:} \quad x_m = x_{0_m} + \upsilon_m t$$
$$\text{Bus:} \quad x_b = x_{0_b} + \upsilon_{0_b} t + \tfrac{1}{2}at^2$$

The condition for catching the bus is that $x_m = x_b$ at the common time t.

$$x_{0_m} + \upsilon_m t = x_{0_b} + \upsilon_{0_b} t + \tfrac{1}{2}at^2$$

There is one further condition. The time must have a *real* value, that is, $t^2 \geqslant 0$. Solving for t, noting that υ_{0_b} and x_{0_m} are each zero, we get:

$$t^2 - \frac{2\upsilon_m}{a}t + \frac{2x_{0_b}}{a} = 0$$
$$t = \frac{\upsilon_m}{a}\left[1 \pm \left(1 - \frac{2x_{0_b}a}{\upsilon_m^2}\right)^{1/2}\right]$$

To be a real solution, the quantity in brackets must be $\geqslant 1$. Thus

$$\frac{2x_{0_b}a}{\upsilon_m^2} \leqslant 1$$

For our data,

$$\frac{2x_{0_b}a}{\upsilon_m^2} = \frac{2 \times 15 \text{ m} \times 1 \text{ m/s}^2}{(6 \text{ m/s})^2} = 0.83$$

Because $0.83 < 1$, the man can catch the bus. To find out when, we solve for t:

$$t = \frac{6 \text{ m/s}}{1 \text{ m/s}^2}[1 \pm (1 - 0.83)^{1/2}]$$
$$t = 3.5 \text{ s and } 8.4 \text{ s}$$

Note that we have *two* values for t. What does this mean? We must remember that we asked *mathematically*, "What is t when $x_b = x_m$?" If the man continues to run instead of jumping on the bus, he will pass the bus. But later the bus, because it is accelerating, will catch up to the man and pass him. If we assume that the man's intention is to ride the bus, we can ignore the longer time because it is not relevant to the problem.

Now that we know when the man catches the bus, how far has the bus traveled? It has traveled a distance $x_b - x_{0_b}$:

$$x_b = x_{0_b} + v_{0_b}t + \tfrac{1}{2}a\,t^2$$

$$x_b - x_{0_b} = \tfrac{1}{2}at^2 \text{ (because } v_{0_b} = 0)$$

$$x_b - x_{0_b} = \tfrac{1}{2}(1\ \text{m/s}^2)(3.5\ \text{s})^2$$

$$x_b - x_{0_b} = 6\ \text{m}$$

There is another interesting result. If x_{0_b} is made smaller, that is, if the man is closer to the bus at $t = 0$, he will catch it sooner, as expected. But if he continues to run, the time for the bus to catch up to him again is longer! Do you see why? A position–time graph for this problem is shown in Figure 2.5. Note that if the curve for the bus is moved downward (a smaller x_{0_b}), then the crossing points move farther apart.

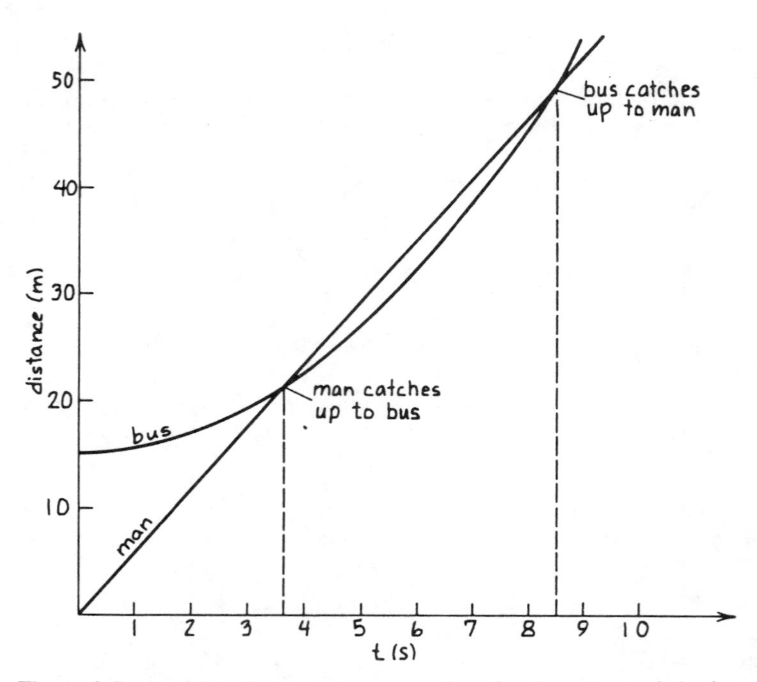

Figure 2.5 Problem 7. Distance versus time for the man and the bus.

8. *Guided Problem.* A speeding car, traveling at a constant speed of 80 km/h, passes a parked police car. At that instant the police car accelerates uniformly at 5 m/s² toward the speeder.

a. If the acceleration remains constant, how much time elapses before the police car catches up to the speeder?
b. How fast is the police car traveling at this time?
c. How far has the police car traveled?
d. In view of your answer to part b, is uniform acceleration of the police car likely in a real situation of this kind?
e. Sketch a velocity–time graph for the problem as stated and for the real situation.

Solution scheme

a. Let $x = 0$ at $t = 0$ for both cars.
b. Convert 80 km/h to meters per second.
c. At the common time t, $x_p = x_s$ (p = police car; s = speeder).
d. Two solutions for t will be obtained. Which do you choose? What is the meaning of the other?
e. Knowing a, t, and v_0 for the police car, calculate v.
f. The distance d (Δx in this case) can be found by using one of the derived equations directly. We can also make use of the fact that because we are dealing with uniformly accelerated motion, the average velocity can be calculated simply.
g. In sketching the graph, note that the police officer must eventually travel at the same speed as the speeder if the officer is to pull the speeder over. (But what will happen if this speed match occurs too soon?)

9. *Worked Problem.* A projectile is fired vertically upward from the edge of a straight-walled cliff, as illustrated in Figure 2.6. The top of the cliff is 118 m above the water. The initial speed of the projectile is 49 m/s.

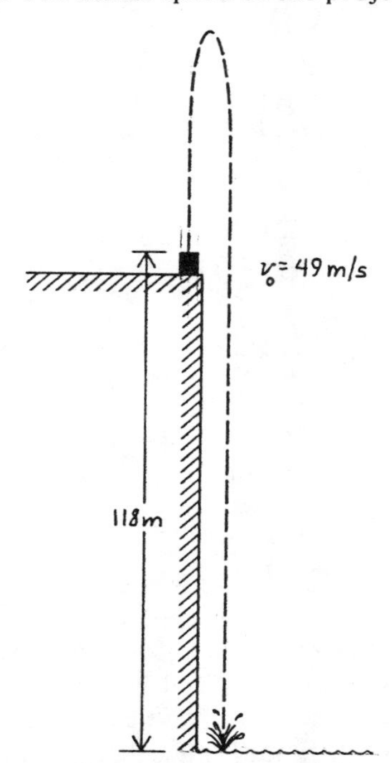

Figure 2.6 Problem 9. A projectile is fired upward and eventually lands in the water.

a. How high will the projectile go?
b. How long will it take to reach this height?
c. At what time will the projectile hit the water?

Solution: We first must establish an origin for our coordinate system. The choice is arbitrary, so we take it to be the top of the cliff. Also, the positive direction is upward. The next step is to determine just what values are given initially. Because the motion is free fall (assuming that friction is negligible), we know that $a = -g = -9.8 \text{ m/s}^2$. The initial velocity, υ_0, is 49 m/s (positive).

a. Translated into mathematics, "How high will the projectile go?" means "What is the position x when $\upsilon = 0$, given the value for a?" This formulation suggests using

$$\upsilon^2 = \upsilon_0^2 + 2a(x - x_0)$$

Solving for x, we get

$$x = x_0 + \frac{\upsilon^2 - \upsilon_0^2}{2a}$$
$$x = 0 + \frac{0 - (49 \text{ m/s})^2}{2(-9.8 \text{ m/s}^2)} = 120 \text{ m}$$

b. In this part we ask, "What is the value for t when $\upsilon = 0$, given a?" We use the equation

$$\upsilon = \upsilon_0 = a\,t$$
$$\upsilon = 49 \text{ m/s} - 9.8 \text{ m/s}^2 \cdot t$$
$$t = 5 \text{ s}$$

c. To determine when the projectile hits the water, we wish to know, "Given a, what is the value for t when $x = -118$ m?" The appropriate equation is

$$x = x_0 + \upsilon_0 t + \tfrac{1}{2}a\,t^2$$
$$-100 \text{ m} = 0 + 49 \text{ m/s} \cdot t - \tfrac{1}{2}(9.8 \text{ m/s}^2)\,t^2$$

Dropping the units temporarily to see better the form of the quadratic equation, we get

$$4.9\,t^2 - 49\,t - 118 = 0$$
$$t^2 - 10\,t - 24 = 0$$
$$(t - 12)(t + 2) = 0$$
$$t = 12 \text{ s}$$
$$t = -2 \text{ s}$$

Although there are two answers, it appears reasonable that the positive solution is the one we want, but what is the meaning of the other value for t? When one "asks" the equation for the value of t when $x = -118$ m, no constraint is placed on the equation. In fact, a projectile thrown upward from the water level at such a speed that it would be traveling at 49 m/s as it passed the top of the cliff would follow exactly the same kind of motion as that described in the original problem. If we start the clock when it passes by, then we can say that it began its motion 2 s prior at a point 118 m below the top of the cliff.

10. *Guided Problem.* A bridge over the Rio Grande Gorge in New Mexico is approximately 200 m above the river below. A girl standing on the bridge drops a rock over the railing. Another girl waits 1 s and throws a rock straight downward with a speed that causes her rock to hit the water at the same time as the one that was dropped. With what speed did she throw the rock?

Solution scheme

a. Given values for υ_0, a, and x, determine the time required for the first stone to hit the water.
b. Use the same equation for the second stone, but for t use $t' - 1$, where t' is the time for the first stone.

Answers to Guided Problems

2. 17 m or 56 ft
4. In 10 s the slower runner runs 85 m, so the difference is 15 m.
6. a. See Figure 2.7.

   b. $\bar{\upsilon} = \dfrac{\Delta x}{\Delta t} = \dfrac{38\text{ m} - 30\text{ m}}{8\text{ min} - 2\text{ min}} = 1.3\text{ m/min}$

   c. The average speed is somewhat uncertain. If we assume that the balloon rises no higher than 50 m, then the distance traveled is (50 m − 32 m) + (50 m − 30 m) = 38 m. The average speed is 38 m/6 min = 6.3 m/min.

   d. The slope of the tangent line at $t = 1$ min is approximately 12 m/min.

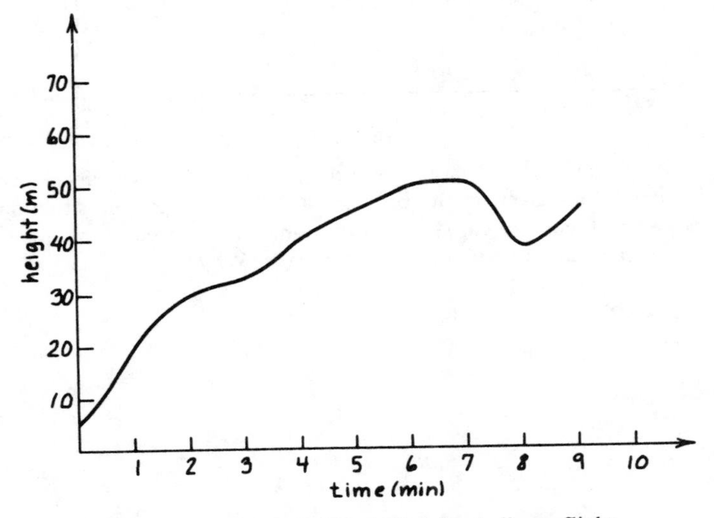

Figure 2.7 Problem 6. Worldline for balloon flight.

8. a. $t = 11$ s (Another solution is $t = 0$. What does that mean?)
   b. $\upsilon = 55$ m/s (200 km/h! Not very reasonable.)
   c. $\Delta x = 300$ m
   d. It would be unwise (and probably not possible) for the police car to accelerate uniformly during the entire time, because it would be traveling twice as fast as the speeder when overtaking. In the more realistic situation, the police car must eventually decelerate to match the other car's speed, but only after catching up to it. Figure 2.8 shows the velocity–time histories for both situations.
10. For the dropped stone, $t = 6.4$ s. Using $t' = 5.4$ s for the thrown stone, the initial velocity must be $\upsilon_0 = 11$ m/s.

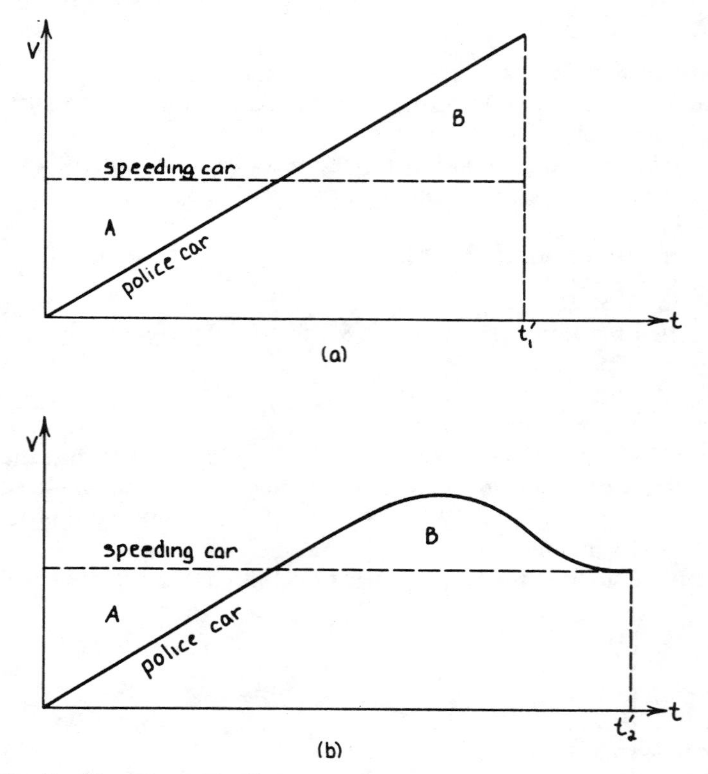

Figure 2.8 Problem 8. (a) The velocity–time graph for the two cars under the conditions given in the problem. (b) A more realistic situation, in which the police car slows down to match the speed of the other car. In each case area A must equal area B. Why?

Chapter 3

Vectors

Overview

To avoid unnecessary complications in our initial study of motion, we chose to limit the analysis to motion along a single dimension. Most real motions, however, take place in two or three dimensions. In preparation for the more complicated analysis in Chapter 4 and in anticipation of the study of forces, moments, and fields, we digress briefly to develop a tool used widely throughout the study of physics: *vector analysis*.

Essential Terms

vector
scalar
displacement vector
resultant
unit vector
vector component
vector addition
vector multiplication

Key Concepts

1. *The displacement and other vectors.* Because it is easily visualized in space, the displacement vector is used to describe the properties of all vectors. The laws governing the addition of displacement vectors, for example, conform to our intuition about how real displacements are carried out. But there are some important differences between the displacement vector and other vectors. For instance, the velocity vector and the force vector, although represented by arrows of given lengths and directions, are specified *at a point*. In other words, the *vector representation* of the velocity is spatial, but we are referring to a magnitude and direction at a point, not along a line. We add the velocity vectors as if they were displacements; in fact, they

are not. This may create problems in the symbolic representation of the vectors associated with such quantities. For example, an airplane flying due north at a speed of 200 km/h may be represented as in Figure 3.1. It is easy to think of the arrow as the actual path along which the plane flies, but in fact all we know is that at the point A (the origin of the vector), the velocity is in the direction specified by the arrow and has a magnitude given by the length of the vector.

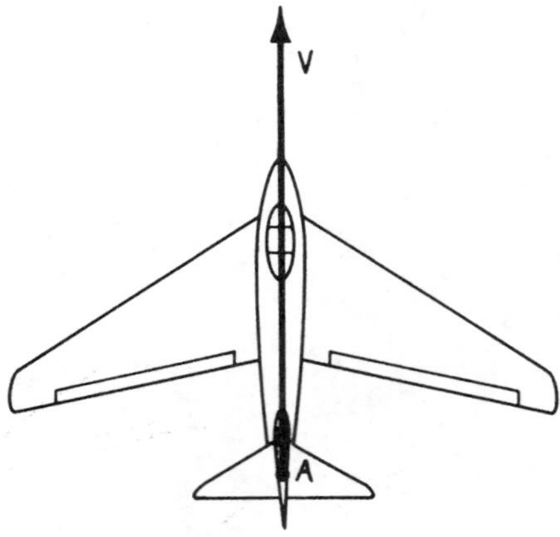

Figure 3.1 The velocity vector $\mathbf{V}$ represents the velocity of the plane only at point A.

2. *Displacement vector versus position vector*. The displacement vector is the directed distance from one point in a coordinate system to another; that is, it is the straight-line distance from one position to another. The positions themselves are designated by vectors, however, because they are specified with respect to a common origin. Thus, the displacement vector is the *vector difference* of two position vectors. Because the position vectors are "anchored" at the origin, they are unique; two objects having the same position vector must occupy the same point in space. Many combinations of position vectors can give the same *displacement*, however. Figure 3.2 shows the same displacement vector for different position vectors. Note that $\Delta\mathbf{d}' = \Delta\mathbf{d}$ even though these vectors do not occupy the same space. They have the same length and point in the same direction, so by definition they are equal vectors. This concept is illustrated in Figure 3.3 of the text.

3. *Vector multiplication*. We have seen that vector addition does not follow the rules of ordinary addition, so it should come as no surprise that vector multiplication is different from ordinary multiplication. Vector multiplication, however, is compounded by having at least two kinds of products —the dot (scalar) product and the cross (vector) product. The former lends itself more readily to a physical interpretation because it gives the "effective product," that is, a product based on how much of one vector lies along the direction of the other. The cross product, in contrast, is a vector that points in a direction perpendicular to the directions of the two vectors being multiplied. Later we will find that this product plays an essential role in the development of some key principles and, if not intuitive, at least operates under a self-consistent set of rules.

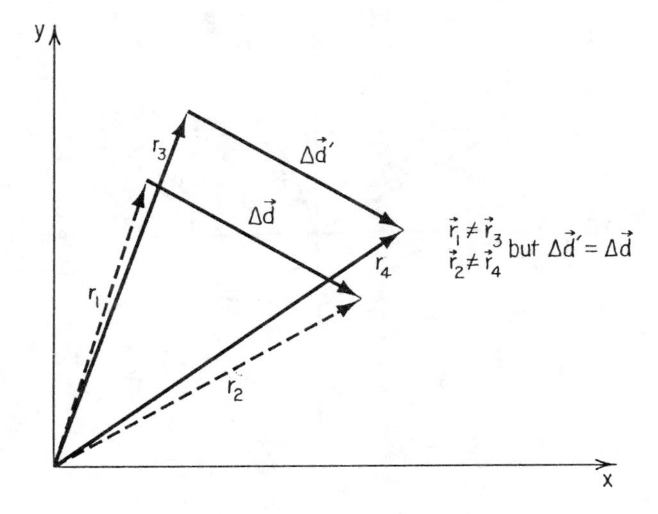

Figure 3.2 Showing the equivalence of displacement vectors.

4. *Right-hand rule.* One of the rules mentioned in this chapter is the "right-hand rule" (see p. 64 of the text). This rule is also called the "right-hand-screw rule" because if one turns a right-hand-thread screw in the direction of the smaller angle from **A** to **B**, then **C** will be in the direction that the screw advances.

Sample Problems

1. *Worked Problem.* Vector **A** has a magnitude of 10 units and is directed due west. Vector **B** has a magnitude of 5 units and is directed 70° N of E. Determine the magnitude and the direction of the resultant vector, using the graphical method.

Solution: Figure 3.3a shows the two vectors drawn to scale, along with the resultant. Figure 3.3b shows the addition of the vectors using the tip-to-tail method. From the first figure we measure directly the magnitude and direction of the resultant by using the scale of the graph. From this we get

$$\mathbf{R} = 9.5 \text{ units, } 152°$$

In the second figure we can use the properties of a triangle to determine these parameters. The law of cosines, for example, gives

$$R^2 = 10^2 + 5^2 - 2 \times 10 \times 5 \times \cos 70° = 90.1$$
$$R = 9.5$$

The law of sines allows us to compute ϕ:

$$\frac{\sin 70°}{9.5} = \frac{\sin \phi}{10} \Rightarrow \sin \phi = \frac{10}{9.5} \sin 70° = 0.989$$
$$\phi = 81.5°$$

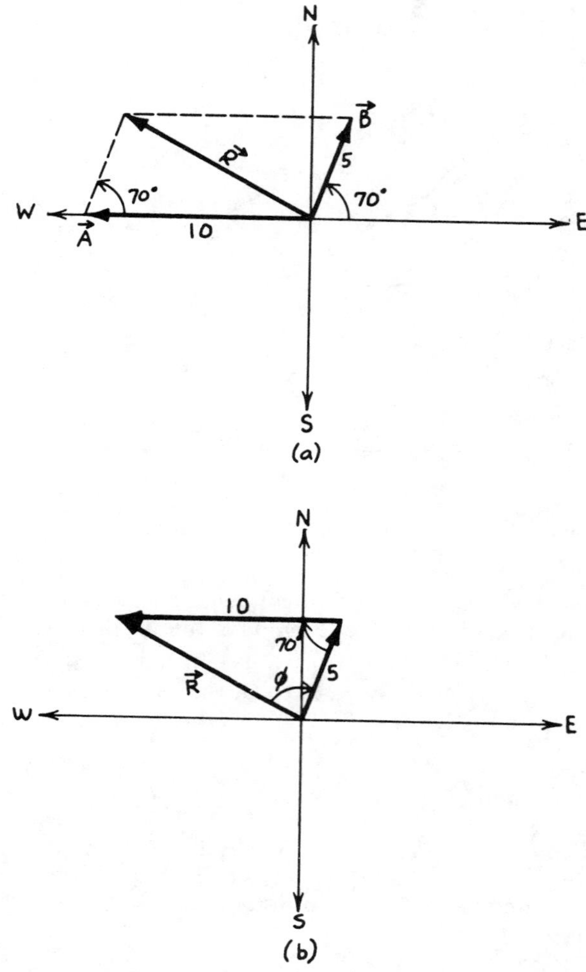

Figure 3.3 Problem 1. a) Solution using the scaled-diagram method. b) Analytic solution using the properties of triangles.

With respect to the $+x$ axis, the angle of the resultant vector is

$$\theta = \phi + 70° = 152°$$

2. *Guided Problem.* A ship sails in a direction NNE for 50 km, then sails NW for 70 km. What is the displacement of the ship at this time?

Solution scheme

a. Convert the directions NNE and NW to angles with respect to some fiducial direction.
b. Draw the vectors to scale on graph paper. Connect them tip to tail, making certain that the length and the direction of each vector remain unchanged. Complete the triangle, and measure the length of the resultant and the angle it makes with respect to the fiducial direction.

or

c. Use the law of cosines and the law of sines to find a more precise solution.

3. *Worked Problem.* From the set of three vectors shown in Figure 3.4, determine

a. the resultant vector
b. the vector that, when added to these three, gives a zero resultant

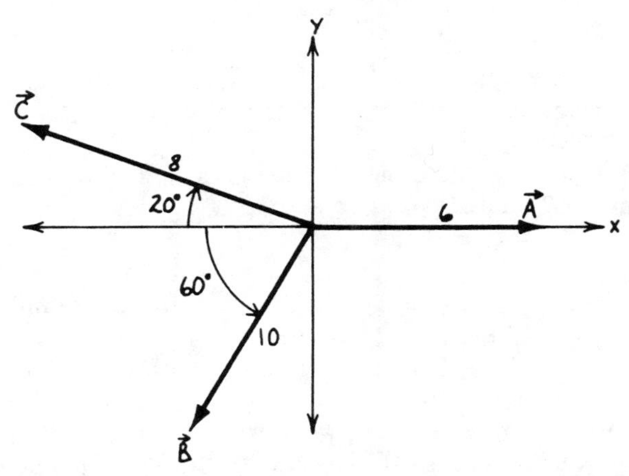

Figure 3.4 Problem 3. The three vectors are to be added to find the resultant.

Solution:

a. By projecting the vectors onto each of the two axes, we can add the components algebraically. The sum of the x components gives the x component of the resultant, whereas the sum of the y components gives the y component of the resultant. Fewer errors are likely if all angles are expressed as the smallest angle between the vector and the x axis. The sign of the component then can be determined by simple inspection.

x *components*	y *components*
$A_x = A \cos \theta$	$A_y = A \sin \theta$
$B_x = B \cos \theta$	$B_y = B \sin \theta$
$C_x = C \cos \theta$	$C_y = C \sin \theta$

Let the resultant **R** equal **A** + **B** + **C**. Then

$$R_x = A_x + B_x + C_x \qquad R_y = A_y + B_y + C_y$$

$$\begin{aligned} R_x &= A \cos \theta + B \cos \theta + C \cos \theta \\ &= 6 \cos 0° - 10 \cos 60° - 8 \cos 20° \\ &= -6.5 \end{aligned}$$

$$\begin{aligned} R_y &= A \sin \theta + B \sin \theta + C \sin \theta \\ &= 6 \sin 0° - 10 \sin 60° + 8 \sin 20° \\ &= -6.0 \end{aligned}$$

The magnitude of the resultant is found from

$$R^2 = R_x^2 + R_y^2 = 78$$
$$R = 8.8$$

The direction is given by

$$\tan \theta_R = \frac{R_y}{R_x} = \frac{-6.0}{-6.5} = 0.92$$
$$\theta_R = 43°$$

So the resultant vector has a magnitude of 8.8 units and is directed 43° from the negative x axis in the third quadrant, or 223° from the $+x$ direction.

With any resultant, the quadrant is easily determined by noting the signs of R_x and R_y:

R_x (+)	; R_y (+)	1st quadrant
R_x (−)	; R_y (+)	2nd quadrant
R_x (−)	; R_y (−)	3rd quadrant
R_x (+)	; R_y (−)	4th quadrant

b. To obtain a zero resultant we must add the resultant from part a to its negative, because $\mathbf{R} + (-\mathbf{R}) = 0$. But the negative of a vector is one having the same magnitude but opposite in direction. Therefore, the required vector is

$$-\mathbf{R} = 8.8 \text{ units, } 43° \quad \text{(quadrant 1)}$$

Comment: If we were to use the unit vector notation, then

$$\mathbf{R} = -6.5\hat{\mathbf{x}} - 6.0\hat{\mathbf{y}}$$
$$-\mathbf{R} = 6.5\hat{\mathbf{x}} + 6.0\hat{\mathbf{y}}$$
$$R = \sqrt{R^2} = \sqrt{\mathbf{R} \cdot \mathbf{R}} = \sqrt{(-6.5)^2 (\hat{\mathbf{x}} \cdot \hat{\mathbf{x}}) + (-6.0)^2 (\hat{\mathbf{y}} \cdot \hat{\mathbf{y}})}$$
$$R = \sqrt{(6.5)^2 + (6.0)^2} = \sqrt{78} = 8.8$$

4. ***Guided Problem.*** With 0° taken to be in the $+x$ direction, two vectors are defined as follows:

$$\mathbf{A} = 8 \text{ units, } 60°$$
$$\mathbf{B} = 5 \text{ units, } 307°$$

a. Determine the x and the y components of the resultant vector $\mathbf{R} = \mathbf{A} + \mathbf{B}$.
b. What is the magnitude and the direction of the resultant vector?
c. Calculate the vector $\mathbf{D} = \mathbf{A} - \mathbf{B}$.

Solution scheme:

a. 1. Sketch the vectors on an x–y coordinate system, labeling the angle between the vector and the x axis. There should be two angles less than 90°.

2. Determine the x and y components of each vector, assigning positive or negative values depending on whether the component is $+x$ or $-x$, $+y$ or $-y$.
3. Add the x components and y components separately, yielding values for R_x and R_y.

b. Use $R^2 = R_x^2 + R_y^2$ to obtain R and $\tan\theta_{\mathbf{R}} = R_y/R_x$ to get the direction for **R**.

c. 1. For $\mathbf{A} - \mathbf{B}$, note that $-\mathbf{B} = -(\mathbf{B})$. Therefore, change $B_x \rightarrow -B_x$ and $B_y \rightarrow -B_y$ and proceed as described.
2. Draw the vectors approximately to scale to see if the result is reasonable.

5. *Worked Problem*. Determine the resultant **R** of the vectors $\mathbf{A} = 2\hat{\mathbf{x}} - 3\hat{\mathbf{y}} + 3\hat{\mathbf{z}}$; $\mathbf{B} = \hat{\mathbf{x}} - 2\hat{\mathbf{y}}$; and $\mathbf{C} = -3\hat{\mathbf{x}} + 4\hat{\mathbf{z}}$.

Solution: According to the rules for vector addition using the unit vector notation,

$$\mathbf{R} = (A_x + B_x + C_x)\hat{\mathbf{x}} + (A_y + B_y + C_y)\hat{\mathbf{y}} + (A_z + B_z + C_z)\hat{\mathbf{z}}$$

where

$$\mathbf{A} = A_x\hat{\mathbf{x}} + A_y\hat{\mathbf{y}} + A_z\hat{\mathbf{z}}, \text{ etc.}$$

Then

$$\mathbf{R} = (2 + 1 - 3)\hat{\mathbf{x}} + (-3 - 1 + 0)\hat{\mathbf{y}} + (3 + 0 + 4)\hat{\mathbf{z}}$$
$$\mathbf{R} = -4\hat{\mathbf{y}} + 7\hat{\mathbf{z}}$$

And the magnitude of **R** is found from

$$R^2 = \mathbf{R} \cdot \mathbf{R} = 4^2 + 7^2 = 65$$
$$R = 8 \text{ units}$$

Because the vector resultant is in three-dimensional space, we must be clear in the way we specify the angle. It is convenient in these cases to express the direction of the vector in terms of two coordinate angles—the *latitude* and the *longitude*, shown in the spherical coordinate system of Chapter 1. We take the $+x$ direction to be 0° for the longitude and define positive angular displacements to conform to the right-hand rule; that is, if the thumb points along the positive z axis, then the fingers curl in the direction of the positive angular displacements. The latitude, the smallest angle from the x–y plane to the vector, is found by forming the ratio of the z component of the vector and the vector's projection onto the x–y plane, the latter being $(A_x^2 + A_y^2)^{1/2}$. The *longitude* angle is found from

$$\tan\theta_{\text{long}} = \frac{A_y}{A_x}$$

where the signs of both A_x and A_y determine the quadrant for this angle, as discussed earlier. For the *latitude* angle,

$$\tan \theta_{\text{lat}} = \frac{A_z}{(A_x^2 + A_y^2)^{1/2}}$$

Note that if A_z is positive, the angle θ_{lat} is between 0° and 90°. If A_z is negative, then θ_{lat} is between 0° and −90° (Figure 3.5). For the problem at

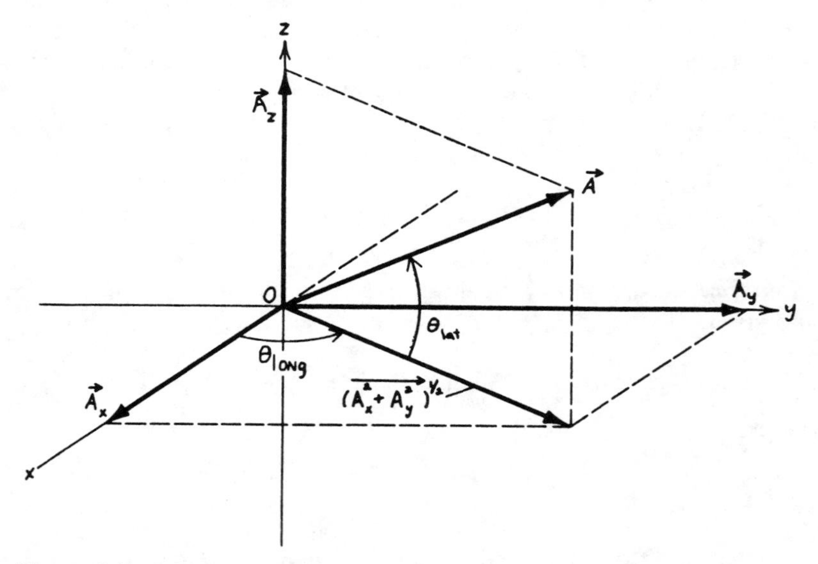

Figure 3.5 Problem 5. The resolution of a vector **A** in three dimensions.

hand, there is no x component and the y component is negative, so the longitude angle must be

$$\theta_{\text{long}} = 270°$$

The latitude angle is found from:

$$\tan \theta_{\text{lat}} = \frac{7}{[0 + (-4)^2]^{1/2}} = +1.75$$

$$\theta_{\text{lat}} = 60°$$

The solution is shown graphically in Figure 3.6.

6. *Guided Problem.* For the vectors $\mathbf{A} = 2\hat{\mathbf{x}} - \hat{\mathbf{y}} + 3\hat{\mathbf{z}}$; $\mathbf{B} = -3\hat{\mathbf{x}} + 2\hat{\mathbf{y}} - 2\hat{\mathbf{z}}$; and $\mathbf{C} = 4\hat{\mathbf{y}} - 4\hat{\mathbf{z}}$, calculate the vectors

a. $\mathbf{D} = 2\mathbf{A} + \mathbf{B} + 3\mathbf{C}$
b. $\mathbf{E} = \mathbf{A} - \mathbf{B} + 2\mathbf{C}$

Solution scheme:

a. As in the previous problems, use the fact that $\mathbf{A} = A_x\hat{\mathbf{x}} + A_y\hat{\mathbf{y}} + A_z\hat{\mathbf{z}}$.
b. Note that $\text{m}\mathbf{A} = \text{m}A_x\hat{\mathbf{x}} + \text{m}A_y\hat{\mathbf{y}} + \text{m}A_z\hat{\mathbf{z}}$.
c. Sum the x, y, and z components separately.
d. Use $R^2 = \mathbf{R} \cdot \mathbf{R} = R_x^2 + R_y^2 + R_z^2$.
e. Determine the longitude and latitude angles as we did previously.

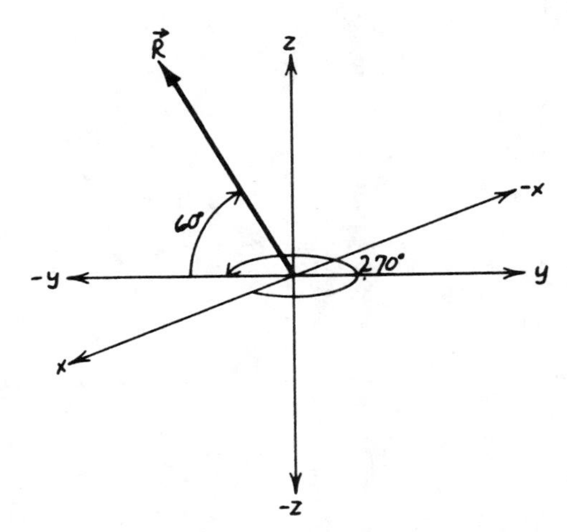

Figure 3.6 Problem 5. The resultant has a magnitude of 8 units, a latitude angle of 60°, and a longitude angle of 270°.

7. *Worked Problem.* For the vectors **A** and **B** in problem 6, find
a. the dot (scalar) product $\mathbf{A} \cdot \mathbf{B}$
b. the cross (vector) product $\mathbf{A} \times \mathbf{B}$

Solution

a. We note that $\mathbf{A} \cdot \mathbf{B} = A_x B_x + A_y B_y + A_z B_z$, so that

$$\mathbf{A} \cdot \mathbf{B} = (2)(-3) + (-1)(2) + (3)(-2) = 14$$

There is no angle to compute, because the dot product is a scalar.

b. $$\begin{aligned} \mathbf{A} \times \mathbf{B} &= (A_y B_z - A_z B_y)\hat{\mathbf{x}} + (A_z B_x - A_x B_z)\hat{\mathbf{y}} + (A_x B_y - A_y B_x)\hat{\mathbf{z}} \\ &= [(-1)(-2) - (3)(2)]\hat{\mathbf{x}} + [(3)(-3) - (2)(-2)]\hat{\mathbf{y}} \\ &\quad + [(2)(2) - (-1)(-3)]\hat{\mathbf{z}} \\ \mathbf{A} \times \mathbf{B} &= \mathbf{R} = -4\hat{\mathbf{x}} - 5\hat{\mathbf{y}} + \hat{\mathbf{z}} \end{aligned}$$

The magnitude of this vector is

$$R = [(A \times B) \cdot (A \times B)]^{1/2} = [(-4)^2 + (-5)^2 + (1)^2]^{1/2}$$
$$R = \sqrt{42} = 6.5 \text{ units}$$

The longitude angle is

$$\tan \theta_{\text{long}} = \frac{-5}{-4} = 1.25$$
$$\theta_{\text{long}} = 231°$$

For the latitude angle we have

$$\tan \theta_{\text{lat}} = \frac{1}{[(-4)^2 + (-5)^2]^{1/2}} = 0.156$$
$$\theta_{\text{lat}} = 9°$$

This result is illustrated in Figure 3.7.

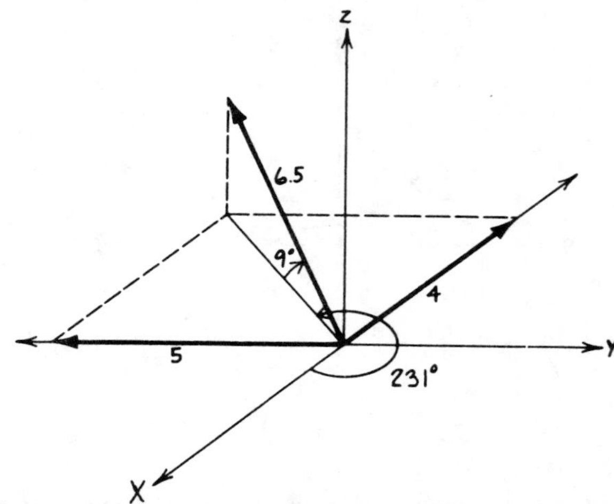

Figure 3.7 Problem 7. The resultant vector has a magnitude of 6.5 units, a latitude angle of 9°, and a longitude angle of 231°.

Comment: Because the vector product is a vector perpendicular to the other two, we can check our result to see if

$$\mathbf{R} \cdot \mathbf{A} = \mathbf{R} \cdot \mathbf{B} = 0$$
$$\mathbf{R} \cdot \mathbf{A} = (-4\hat{\mathbf{x}} - 5\hat{\mathbf{y}} + \hat{\mathbf{z}}) \cdot (2\hat{\mathbf{x}} - \hat{\mathbf{y}} + 3\hat{\mathbf{z}}) = -8 + 5 + 3 = 0$$
$$\mathbf{R} \cdot \mathbf{B} = (-4\hat{\mathbf{x}} - 5\hat{\mathbf{y}} + \hat{\mathbf{z}}) \cdot (-3\hat{\mathbf{x}} + 2\hat{\mathbf{y}} - 2\hat{\mathbf{z}}) = 12 - 10 - 2 = 0$$

8. *Guided Problem.* Prove that the two vectors $\mathbf{A} = \hat{\mathbf{x}} - 3\hat{\mathbf{y}} + 4\hat{\mathbf{z}}$ and $\mathbf{B} = -3\hat{\mathbf{x}} + 9\hat{\mathbf{y}} - 12\hat{\mathbf{z}}$ are parallel.

Solution scheme

a. Because the two vectors have a common origin, to be parallel one must be a multiple of the other; that is, $\mathbf{B} = \pm m\mathbf{A}$, where m is some positive constant.
b. It is also true that $\mathbf{A} \cdot \mathbf{B}$ must be either AB or $-AB$, since $\cos \phi = \pm 1$.

Answers to Guided Problems

2. See Figure 3.8.
   NNE = 67.5° N of E
   NW = 45° N of W = 135°
   $\mathbf{D}$ = 1000 km, 110° (70° N of W)
4. a. $R_x = 7$ units
      $R_y = 2.9$ units
   b. $\mathbf{R} = 7.6$ units, 22.5°
   c. $D_x = 1$ unit
      $D_y = 11$ units
      $\mathbf{D} = 11$ units, 85°
   See Figure 3.9.
6. a. $2\mathbf{A} = 4\hat{\mathbf{x}} - 2\hat{\mathbf{y}} + 6\hat{\mathbf{z}}$
      $3\mathbf{C} = 12\hat{\mathbf{y}} - 12\hat{\mathbf{z}}$

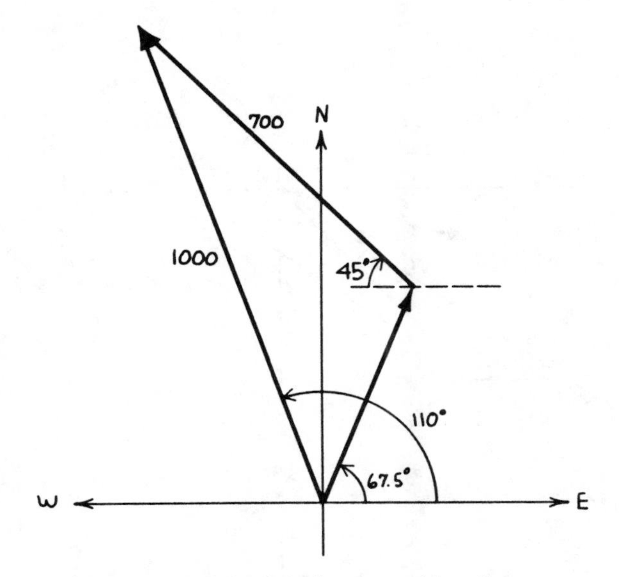

Figure 3.8 Solution to problem 2.

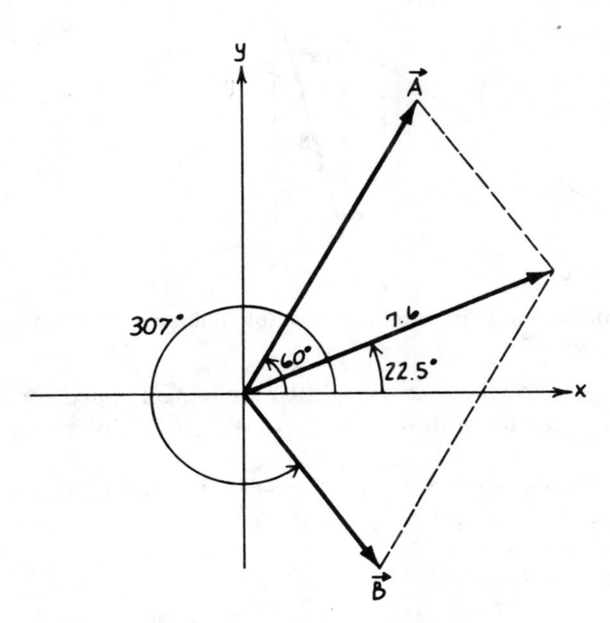

Figure 3.9 Solution to problem 4, part c.

$\mathbf{D} = \hat{\mathbf{x}} - 8\hat{\mathbf{z}}$
$\mathbf{E} = -5\hat{\mathbf{x}} + \hat{\mathbf{y}} - 3\hat{\mathbf{z}}$
$\mathbf{E} = 5.9$ units, $\theta_{\text{long}} = 11°$, $\theta_{\text{lat}} = 30°$
$\mathbf{D} = 8.1$ units, $\theta_{\text{long}} = 0°$, $\theta_{\text{lat}} = -83°$

See Figure 3.10.

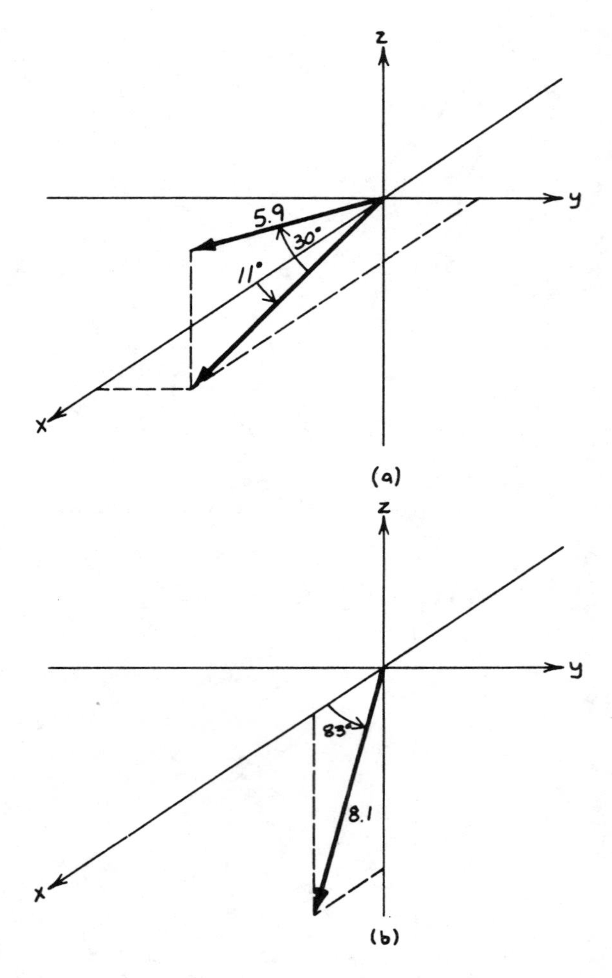

Figure 3.10 (a) Solution to problem 6, part a. (b) Solution to problem 6, part b.

8. Note that $\mathbf{B} = -3\mathbf{A}$, thus satisfying the requirement stated for parallel vectors. Another way to show this relationship is the following:

$$\begin{aligned} \mathbf{A} \cdot \mathbf{B} &= -3\hat{\mathbf{x}} \cdot \hat{\mathbf{x}} - 27\hat{\mathbf{y}} \cdot \hat{\mathbf{y}} - 48\hat{\mathbf{z}} \cdot \hat{\mathbf{z}} = -78 \text{ units} \\ A &= 5.1 \text{ units} \\ B &= 15.3 \text{ units} \\ AB &= 78 \text{ units} \end{aligned}$$

Therefore

$$\mathbf{A} \cdot \mathbf{B} = -AB$$

We note further that

$$\theta^A_{\text{long}} = 288°; \quad \theta^A_{\text{lat}} = 52°$$
$$\theta^B_{\text{long}} = 108°; \quad \theta^B_{\text{lat}} = -52°$$

which shows that the directions of the two vectors are opposite.

Chapter 4

Kinematics in Three Dimensions

Overview

With the techniques for working with vectors at our disposal, we are ready to proceed to the analysis of motion in two and three dimensions. In one-dimensional motion, we handled the directional (vector) properties by using positive and negative signs, because the only possibilities were motions taking place along a single dimension. Now we must take into account motions in any arbitrary direction, the analysis of which we undertake by using the rules for vector operation.

Essential Terms

velocity vector
acceleration vector
projectile motion
range of a projectile
uniform circular motion
centripetal acceleration
centrifugal acceleration
Galilean transformations

Key Concepts

1. *Velocity versus speed revisited*. Chapter 2 distinguished between velocity and speed and noted that sometimes *speed* appears to have two different meanings. We can keep these meanings separate by noting that *average* speed is the number we get when the (finite) *distance* is divided by the associated (finite) time. In contrast, speed as the magnitude of the velocity vector —i.e., the *instantaneous* speed—involves the derivative $|d\mathbf{r}/dt|$. It therefore no longer refers to a finite distance or displacement but, rather, to instantaneous values that result from taking limits as $\Delta t \to 0$. Thus, the instantaneous

speed is the magnitude of the instantaneous velocity and is determined in the same way as the magnitude (length) of a vector. (See the discussion associated with Eqs. 4.10 and 4.11 in the text.)

It is essential that you understand the difference between speed and velocity, especially when you are dealing with changes in these quantities. We found that in one-dimensional motion, a change in velocity always indicates a change in speed, but now we have an alternative way to change the velocity: changing the *direction*. We can and often do have motion with constant speed but with changing velocity. For example, if you maintain your speed at a constant 80 km/h but are traveling on a road that is not straight, your velocity changes every time you change direction. There really is nothing mysterious about this; it simply reflects the way we have defined these quantities. But we will see that such definitions are quite useful in generalizing the one-dimensional results to two or three dimensions.

2. *Projectile motion*. Figure 4.5 of the text shows two balls being released simultaneously, one vertically and the other horizontally. You should note two important outcomes of this experiment. The superimposed horizontal lines show that the two balls accelerate vertically at precisely the same rate. This means that the vertical motion of the projected ball is in no way affected by its horizontal motion. A no less important observation is that the images of the projected ball are equally spaced in the horizontal direction. (You should verify this with a ruler or some other measuring device.) Thus, the vertical motion of this ball has no influence on its horizontal motion. These observations are consistent with Newton's laws of motion, which we study in the next chapter. For now, the importance of the independence of the two motions is that we can apply the equations developed for motion in a single dimension to the two- and three-dimensional case; that is, we can analyze the components of the motion along each of the coordinate axes using the one-dimensional motion equations. These equations then can be combined to provide the solution for the overall motion of the body. Eqs. 4.41, 4.42, and 4.43 are derived in this way. But note that the symbol x_{max} in Eqs. 4.42 and 4.43 does *not* mean maximum range in the sense of the greatest horizontal distance the projectile can cover with a given initial *speed*. Rather, for the given speed and angle, it is the maximum horizontal distance attainable under zero frictional force. If we let the range $R = x_{max}$, then there is an R_{max} that is obtained when θ in the Eqs. 4.42 and 4.43 is 45°. Under these conditions $R_{max} = v_0^2/g$ (see Figure 4.10 in the text). Note also that these derived expressions apply only when the launch and impact elevations are the same. If this is not the case, one should go back to the basic equations from which these were derived.

Projectile motion also provides another opportunity to focus on the independence of velocity and acceleration. During the projectile's flight, the velocity vector continuously changes direction, but the acceleration vector is directed always *downward*. In other words, even though **v** is *not* constant, $d\mathbf{v}/dt$ *is* constant.

3. *Uniform circular motion*. Uniform circular motion also shows the independence of **v** and **a**, in that **v** is directed along the tangent to the object's path, whereas **a** is directed toward the center of the circular path. But unlike

projectile motion, in which **a** is constant both in magnitude and in direction (vertically downward), uniform circular motion has an acceleration constant in magnitude but not in direction. The acceleration (called the *centripetal acceleration*) always is directed toward the center of the circular path, so the vector must change direction continuously to maintain this orientation.

Comment: In Chapter 2 of this guide, we asserted that acceleration carries no information about velocity, only about *changes* in velocity. But since $a = \upsilon^2/r$ in uniform circular motion, it appears that we have a situation in which a knowledge of υ allows us to calculate a. However, υ here is the instantaneous *speed* and carries no information about the *direction* of **v**. Hence, the original statement stands.

Sample Problems

1. *Worked Problem.* At some time t_0 the velocity of a car is given by $\mathbf{v}_0 = 10\hat{\mathbf{x}} + 50\hat{\mathbf{y}}$; 10 s later its velocity is $\mathbf{v} = 30\hat{\mathbf{x}} + 10\hat{\mathbf{y}}$. Measurements for υ_0 and υ are made in kilometers per hour. Determine the average acceleration of the car during the 10 s interval.

Solution: We know that $\mathbf{a}_{avg} = \Delta\mathbf{v}/\Delta t$, so

$$\Delta\mathbf{v} = \mathbf{v} - \mathbf{v}_0 = (30\hat{\mathbf{x}} + 10\hat{\mathbf{y}}) - (10\hat{\mathbf{x}} + 50\hat{\mathbf{y}}) \text{ km/h}$$

$$\Delta\mathbf{v} = 20\hat{\mathbf{x}} - 40\hat{\mathbf{y}} \text{ km/h}$$

$$\mathbf{a}_{avg} = \frac{\Delta\mathbf{v}}{\Delta t} = \frac{(20\hat{\mathbf{x}} - 40\hat{\mathbf{y}}) \text{ km/h}}{10 \text{ s}} = 2\hat{\mathbf{x}} - 4\hat{\mathbf{y}} \text{ km/h/s}$$

$$|\mathbf{a}_{avg}| = \sqrt{(2)^2 + (-4)^2} = 4.5 \text{ km/h/s}$$

$$\tan\theta = \frac{-4}{2}$$

$$\theta = 297°$$

The solution is shown graphically in Figure 4.1.

2. *Guided Problem.* Figure 4.2 represents a driving demonstration in which one car (A) travels at a constant speed of 50 km/h along a straight road, as shown. When car A's position vector makes an angle θ with the x axis, car B starts from rest and then accelerates uniformly at 5.0 m/s^2. Determine θ such that car B just clears the intersection ahead of A. (We assume that this means B arrives at the same time as A, although in reality B must arrive slightly earlier.)

Solution scheme

a. Given the distance from car B's starting position to A's road, and knowing that $\theta_0 = 45°$, calculate how far B must travel to reach the intersection.
b. Use this information, along with knowledge of B's acceleration, to determine the time required for B to reach the intersection.
c. For this time t, determine how far A travels, which will be the distance of A from the intersection when B started.
d. With the origin of coordinates at B's initial position, you can deter-

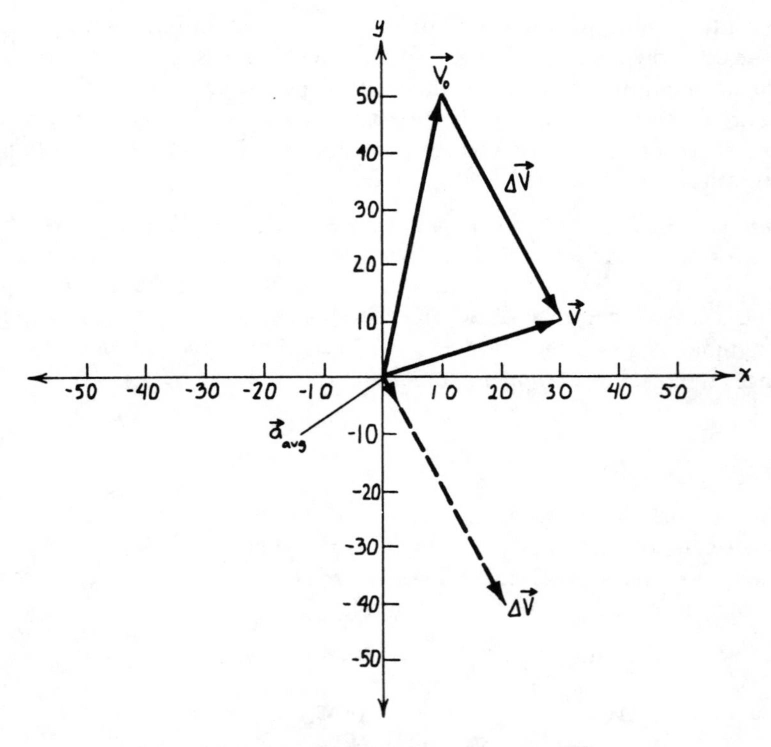

Figure 4.1 Problem 1. The vectors $\Delta\mathbf{v}$ and $\mathbf{a}_{avg}$.

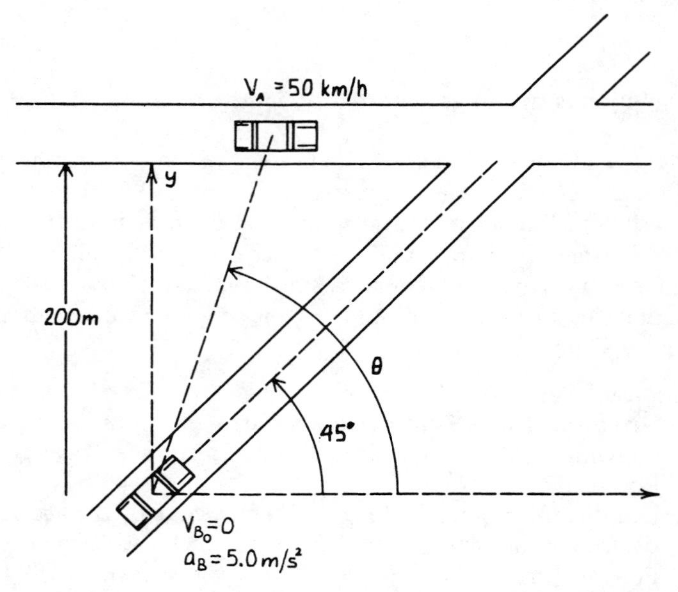

Figure 4.2 Problem 2. Car A travels at a constant speed, while car B accelerates from rest. The two cars meet at the intersection.

mine the x component of A's position when B starts. Since the y component of A's position is constant, the position of A can be calculated.

3. *Worked Problem.* A particle moves in such a way that at time t its position vector is given by

$$\mathbf{r}(t) = (6 + t)\hat{\mathbf{x}} - 4t^2\hat{\mathbf{y}} + (8t - 4t^2)\hat{\mathbf{z}}$$

Determine (a) its velocity and (b) its acceleration as a function of time.

Solution

a. The velocity is the derivative of the position vector:

$$\mathbf{v}(t) = \frac{d\mathbf{r}(t)}{dt} = \hat{\mathbf{x}} - 8t\hat{\mathbf{y}} + (8 - 8t)\hat{\mathbf{z}}$$

b. The acceleration is the derivative of the velocity vector, so

$$\mathbf{a}(t) = \frac{d\mathbf{v}(t)}{dt} = -8\hat{\mathbf{y}} - 8\hat{\mathbf{z}}$$

Comment: Note that the acceleration is constant. In fact, if there are no terms for which t is of higher order than t^2 in the $\mathbf{r}$ equation, then the acceleration must be constant in time. Likewise, if $\mathbf{r}$ contains no terms of higher order than t, then the velocity is constant.

4. *Guided Problem.* The position vector of a particle is given by

$$\mathbf{r}(t) = 10 \cos 2\pi t\ \hat{\mathbf{x}} + 5 \sin 3\pi t\ \hat{\mathbf{y}} \text{ cm}$$

where t is in seconds. Determine the velocity and the acceleration of the particle at $t = 5$ s.

Solution scheme

a. To find $\mathbf{v}(t)$, differentiate $\mathbf{r}(t)$ with respect to time.
b. Differentiate $\mathbf{v}(t)$ to get $\mathbf{a}(t)$.
c. Substitute $t = 5$ s. Note that because the time is a whole number, the arguments of the sine and cosine will be integral multiples of 2π and 3π.
d. Use the methods of Chapter 3 to find $\mathbf{v}$ and $\mathbf{a}$ and their respective directions.

5. *Worked Problem.* A good outfielder can throw a ball with an initial speed of about 40 m/s (90 mph). If the outfielder wishes to throw the ball to the catcher 90 m (300 ft) away without bouncing the ball, at what angle must the ball be thrown?

Solution: We will assume that the catcher catches the ball at the same height as it was thrown. Then we can use the expression

$$x_{max} = \frac{\upsilon_0^2 \sin 2\theta}{g}$$

$$\sin 2\theta = \frac{g\, x_{max}}{\upsilon_0^2} = \frac{(9.8 \text{ m/s}^2)(90 \text{ m})}{(40 \text{ m/s})^2} = 0.55$$

$$2\theta = 33.4°$$

$$\theta = 16.7° \approx 17°$$

Another angle will give the same result. Consider $\theta = 73°$. This gives $2\theta = 146°$. Since sine $146° = 0.55$, the range is the same. So what is the difference in these two situations? Does it mean that the outfielder could throw the ball at a 73° angle and achieve the same result? Not exactly. Consider the times involved in each case. For the 17° angle

$$t_{17} = \frac{2\upsilon_0 \sin \theta}{g} = \frac{2 \times 40 \text{ m/s} \times \sin 17°}{9.8 \text{ m/s}^2} = 2.4 \text{ s}$$

and for the 73° angle

$$t_{73} = \frac{2 \times 40 \text{ m/s} \times \sin 73°}{9.8 \text{ m/s}^2} = 7.8 \text{ s}$$

If the intent is to get the ball to the catcher in a hurry, which it usually is in baseball, the choice clearly is the 17° angle. If one is punting a football, however, the strategy often is to get as much "hang time" as possible to let the tacklers get downfield; in this case one should go for the larger angle. If a "quick kick" is the strategy, then one should choose the lower trajectory.

6. *Guided Problem.* A rock is thrown into the air from ground level. When the rock is 11 m above the ground, its velocity is observed to be $\mathbf{v} = 31\hat{\mathbf{x}} + 21\hat{\mathbf{z}}$ m/s.

a. How high will the rock go?
b. How far will it travel horizontally before hitting the ground?
c. What will be the velocity of the ball just prior to hitting the ground?

Solution scheme

a. Because the x component of the velocity remains constant, the coefficient of $\hat{\mathbf{x}}$ must remain constant. The coefficient of $\hat{\mathbf{z}}$ gives the z component of the velocity at $z = 11$ m.
b. Note that $\upsilon_x = \upsilon_0 \cos \theta$ and $\upsilon_z^2 = (\upsilon_0 \sin \theta)^2 - 2gz$. Thus, you have two equations in the unknowns υ_0 and θ. The unknowns can be found by a straightforward application of equations 4.41, 4.42, and 4.43.

7. *Worked Problem.* A projectile is fired from a cliff 900 m above the plain, as illustrated in Figure 4.3. If the projectile is fired at a 20° angle above the horizontal, at a speed of 300 m/s, how far from the cliff will it land?

Solution: We know that the projectile's components of velocity in the x direction must remain constant. Therefore, we need the total time of flight to get the range. To get the time of flight we use

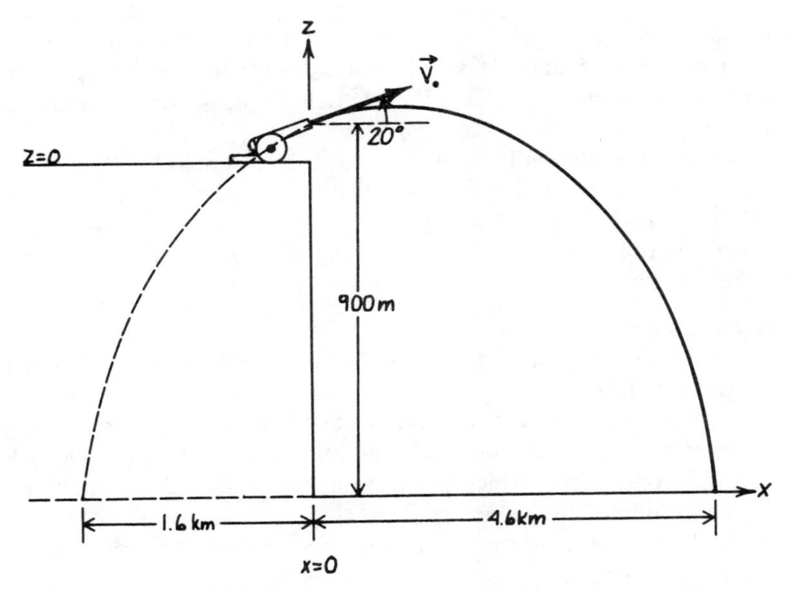

Figure 4.3 Problem 7. A projectile is fired at an angle of 20°.

$$z = z_0 + \upsilon_{0_z}t - \tfrac{1}{2}gt^2 = z_0 + \upsilon_0 \sin \theta t - \tfrac{1}{2}gt^2$$

This gives, substituting $t = \dfrac{x}{\upsilon_0 \cos \theta}$,

$$z = z_0 + (\upsilon_0 \sin \theta) \frac{x}{\upsilon_0 \cos \theta} - \tfrac{1}{2}g \frac{x^2}{\upsilon_0^2 \cos^2 \theta}$$

$$z = z_0 + (\tan \theta)x = \frac{g\, x^2}{\upsilon_0^2 \cos^2 \theta}$$

With the data given, $z = -900$ m, $\theta = 20°$, and $\upsilon_0 = 300$ m/s.

$$-900 \text{ m} = (\tan 20°)x - \frac{9.8 \text{ m/s}^2}{(300 \text{ m/s})^2(\cos 20°)^2}\, x^2$$

$$-900 = 0.36x - 1.2 \times 10^{-4}x^2$$

$$x^2 - 3 \times 10^3 x - 7.5 \times 10^6 = 0$$

$$x = \frac{3 \times 10^3 \pm \sqrt{9 \times 10^6 + 3 \times 10^7}}{2} = \frac{9.2 \times 10^3}{2} \text{ and } \frac{-3.2 \times 10^3}{2}$$

$$x = 4.6 \times 10^3 \text{ m} = 4.6 \text{ km}$$

Comment: Note that we chose the positive solution. The negative distance corresponds to a negative time and is one of the answers to the *mathematical* question "What is the value of x when $z = -900$ m?" In other words, if the projectile obeyed this equation for all time, it would have been at $x = -1.6$ km when $z = -900$ m the first time and then at $x = 4.6$ km when $z = -900$ m the second time.

8. *Guided Problem.* Assume that a baseball batter hitting a home run hits the ball in such a way that it leaves the bat at a 45° angle with respect to the (level) ground. The ball just clears the left field fence, which is 6 m high and 110 m away.

a. How fast was the ball traveling when it left the bat?
b. How fast was the ball traveling when it cleared the fence?
c. If the fence had been 1 m higher, how much faster would the batter have had to hit the ball? Ignore the height of the ball above the ground when it is hit.

Solution scheme

a. Using the expression developed in problem 7, let $\theta = 45°$, $z = 6$ m and $x = 110$ m.
b. To find out how fast the ball is traveling when it clears the fence, determine the z component of the velocity at that point. Along with the x component, which is constant, this allows us to calculate the magnitude of the velocity.
c. Add 1 m to the height of the fence to see what difference in initial speed is required.

9. *Worked Problem.* A ship moving 20° N of E at 20 knots relative to the water (ocean) has smoke coming out of one of its smokestacks. If the smoke leaves the stack at an angle of 30° W of S (relative to the ship) at an apparent speed (relative to the ship) of 15 knots, determine the speed and direction of the wind relative to the water.

Solution: First we must understand that the smoke particles, upon leaving the stack, immediately take on the velocity of the wind. Thus, the velocity of the smoke relative to the ship means also the velocity of the wind relative to the ship. Eq. 4.58 is the basic expression for this motion. First we will rewrite the equation using subscripts. Let *w, o,* and *s* represent the *wind, water* (ocean), and *ship*, respectively. Then Eq. 4.58 can be written

$$\mathbf{v}_{wo} = \mathbf{v}_{ws} + \mathbf{v}_{so}$$

which says that the velocity of the wind relative to the ocean is the vector sum of the velocity of the wind relative to the ship and the velocity of the ship relative to the ocean. The graphical solution to the problem is given in Figure 4.4. The measurements from the graph give us the value $\mathbf{v} = 13$ knots, 331° (29° S of E). Now let's work the problem using the unit vector notation. Let the directions for x and y be due east and due north, respectively. Then

$$\begin{aligned}
\mathbf{v}_{so} &= v_{so} \cos \theta_{so}\hat{\mathbf{x}} + v_{so} \sin \theta_{so}\hat{\mathbf{y}} \\
\mathbf{v}_{ws} &= v_{ws} \cos \theta_{ws}\hat{\mathbf{x}} + v_{ws} \sin \theta_{ws}\hat{\mathbf{y}} \\
\mathbf{v}_{so} &= (20)(0.94)\hat{\mathbf{x}} + (20)(0.34)\hat{\mathbf{y}} \text{ knots} \\
\mathbf{v}_{ws} &= (15)(-0.5)\hat{\mathbf{x}} + (15)(-0.87)\hat{\mathbf{y}} \text{ knots} \\
\mathbf{v}_{so} &= 18.8\hat{\mathbf{x}} + 6.8\hat{\mathbf{y}} \text{ knots} \\
\mathbf{v}_{ws} &= -7.5\hat{\mathbf{x}} - 13\hat{\mathbf{y}} \text{ knots} \\
\mathbf{v}_{wo} &= \mathbf{v}_{ws} + \mathbf{v}_{so} = 11.3\hat{\mathbf{x}} - 6.2\hat{\mathbf{y}} \text{ knots}
\end{aligned}$$

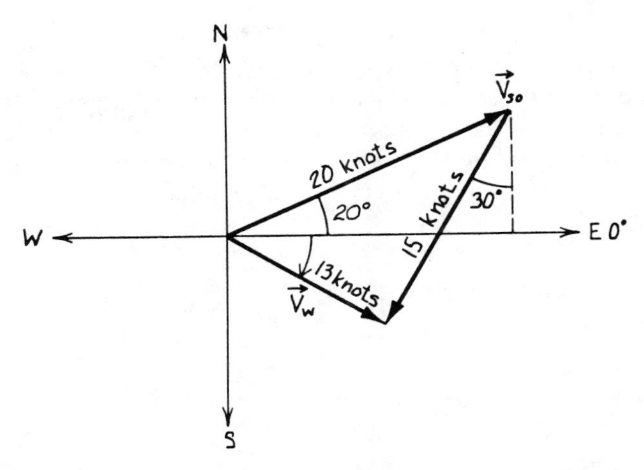

Figure 4.4 Problem 9. The velocity of the wind relative to the ocean is the vector sum of the velocity of the ship relative to the ocean and the velocity of the wind relative to the ship.

$$v_{wo} = \sqrt{(11.3)^2 + (-6.2)^2} = 13 \text{ knots}$$

$$\tan \theta_{wo} = \frac{-6.2}{11.3} = -0.55$$

$$\theta_{wo} = 29° \text{ S of E } (331°)$$

10. *Guided Problem.* A pilot flying cross country wishes to reach a destination 27° W of N relative to his present position. The airplane is capable of flying at a speed *relative to the air* of 150 km/h. Weather data indicate a steady wind blowing 25 km/h from the west. In what compass direction should he head the plane to arrive at his destination? If he is 50 km from his destination, how long will the flight take?

Solution scheme: To find the *heading*, you want the direction of the vector $\mathbf{v}_{PA}$, the velocity of the plane relative to the air, not the ground. The plane actually travels relative to the ground in a different direction. The magnitude of the vector $\mathbf{v}_{PA}$ is given. To find the time required to make the trip, you need the velocity of the plane relative to the ground. You are given the direction, so you must find the magnitude.

a. The simplest method for solving this problem is graphic. First construct the vector $\mathbf{v}_{AG}$, the velocity of the wind relative to the ground. Next, starting from the origin, draw a line along the *direction* 117° (27° N of W). This represents the *direction* of $\mathbf{v}_{PG}$. The length is arbitrary at this point, because you have yet to solve for it.
b. Here is the key part. With a drawing compass (or some similar device), strike an arc of radius 150 km/h with the center at the tip of $\mathbf{v}_{AG}$. Where this arc intercepts the 117° line defines the location of the tip of the vector $\mathbf{v}_{PG}$.
c. Measure the angle between $\mathbf{v}_{PA}$ and the x axis. This gives the heading. By measuring the length of the vector $\mathbf{v}_{PG}$, you can determine the ground speed.
d. Knowing the ground speed and the distance, you can calculate the required time.

Alternate solution: If you write each of the vectors in unit vector notation, you will have two equations and two unknowns. Unfortunately, the solution is a bit messy, because it involves using trigonometric identities and a quadratic equation. The result is the same, however. Regardless of your choice of solution, you should at least sketch the vectors approximately to scale and in the proper direction to check the reasonableness of the result.

Answers to Guided Problems

2. $\theta = 75°$ (Car A is 147 m from the intersection when B starts.)
4. $\mathbf{v} = 15\pi$ cm/s, 270° or $\mathbf{v} = -15\pi\hat{\mathbf{y}}$
   $\mathbf{a} = 40\pi^2$ cm/s², 180° or $\mathbf{a} = -40\pi^2\hat{\mathbf{x}}$
6. $\mathbf{v}_0 = 31\hat{\mathbf{x}} + 26\hat{\mathbf{z}} = 40$ m/s, 40°
   a. 34 m
   b. 253 m
   c. $\mathbf{v} = 31\hat{\mathbf{x}} - 26\hat{\mathbf{z}}$ (compare with $\mathbf{v}_0$)
   $\mathbf{v} = 40$ m/s, 320° (40° below x axis)
8. a. 47.8 m/s
   b. 46.6 m/s
   c. Difference in speed is approximately 0.2 m/s.
10. See Figure 4.5.

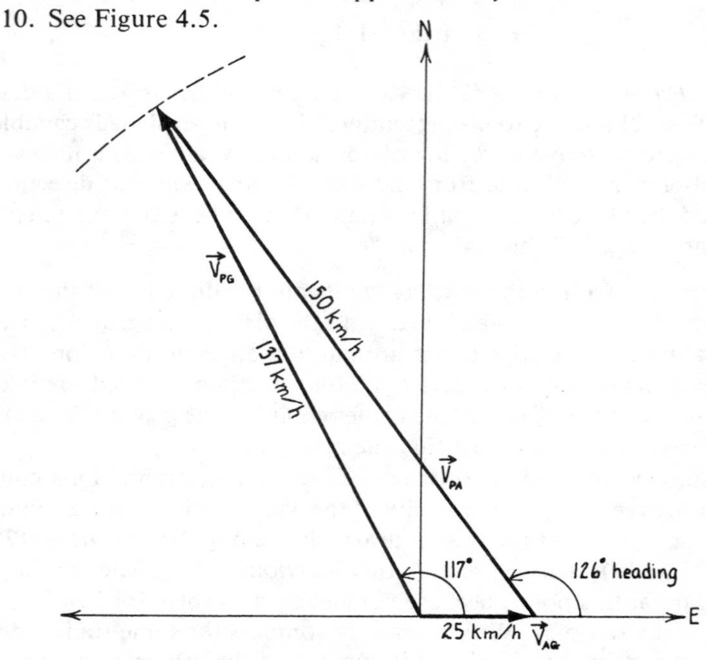

Figure 4.5 The graphic solution to problem 10.

For the unit vector notation, the solution is found in the following way:

a. $\mathbf{v}_{PA} = 150 \cos \theta_{PA}\hat{\mathbf{x}} + 150 \sin \theta_{PA}\hat{\mathbf{y}}$

$\mathbf{v}_{PA} = \upsilon_{PG} - \upsilon_{AG} = (\upsilon_{PG} \cos 117°\hat{\mathbf{x}} + v_{PG} \sin 117°\hat{\mathbf{y}}) - (25\hat{\mathbf{x}} + 0\hat{\mathbf{y}})$

$= (\upsilon_{PG} \cos 117° - 25°)\hat{\mathbf{x}} - \upsilon_{PG} \sin 117°\hat{\mathbf{y}}$

Equate a and b. You get two expressions, one in $\cos \theta_{PA}$ and one in $\sin \Theta_{PA}$. Use $1 - \cos^2 \Theta = \sin^2 \Theta$. Eventually Θ_{PA} is eliminated, and υ_{PA} can be calculated.

Chapter 5

Dynamics—Newton's Laws

Overview

Thus far we have analyzed the behavior of particles by examining their motions from a descriptive point of view. Little or nothing has been said about the origins of these motions. The study of the latter is called *dynamics* and concerns itself with the mechanisms causing changes in motion. We call these mechanisms *forces*, and it is to the properties of forces that we now direct our attention.

Essential Terms

force
Newton's First Law
inertia
inertial reference frame
Newton's Second Law
newton (unit of force)
superposition of forces
Newton's Third Law
linear momentum
angular momentum

Key Concepts

1. *Newton's First Law.* The law states that objects tend to remain at rest or in uniform motion in a straight line, unless acted upon by an external force. This seems at once both to violate common experience and to be obvious. But further examination reveals that the law serves as a starting point for a definition of force and is a postulate concerning the way a particle should behave under conditions not precisely testable. The law of inertia, as the First Law is often called, has to do with free bodies. Free bodies are defined as bodies that are free from any interactions, but this is an ideal definition. We cannot find a situation, even in deep space, where interac-

tions with the body in question truly are nonexistent. We can carry out some simple experiments, however, that show consistency with the First Law. For example, a block sliding without friction on a horizontal surface does have forces acting on it, but only in the vertical direction. And we observe in this situation that the block moves with zero acceleration in the horizontal direction, as predicted by Newton's First Law. Thus, we can infer that the block, in the absence of forces, would move with constant velocity. It is then but a short leap conceptually to show that the same thing occurs when the forces are not zero but add to zero.

2. *Inertial frames of reference.* The inertial frame of reference is a concept that is used extensively in physics. In some sense the definition of an inertial frame appears to be part of a circular definition—an inertial frame is one that obeys Newton's First Law, and if the object obeys the First Law, then the frame is inertial! What we seek, however, is a self-consistent set of formulations, and Newton's laws are precisely that. If all motions of a particular particle within a reference frame are consistent with the law of inertia, then we can define that frame as being inertial.

But there are important restrictions on the frame. For example, if a particle is at rest in a particular frame, the frame may or may not be inertial, depending on whatever forces have to be invented to explain why the object is at rest. Let's look at a particular situation. Suppose you are seated in a train while the train is rounding a curve. In the rest frame of the car, a tray that previously had been at rest appears to accelerate. There is no real force that you can identify causing the plate to accelerate, so you invent one. Such forces as *centrifugal* and *Coriolis* are fictitious forces arising from attempts to apply Newton's laws in noninertial frames. But these forces were invented to keep Newton's laws valid, when in fact the laws do not apply in this accelerated (noninertial) reference frame. Often a system's motion can be conceptualized more simply in the noninertial frame, but in general it is better to stay with inertial frames and real forces.

3. *Acceleration due to gravity.* As noted in the text, the noninertial character of the rest frame of the Earth is hidden in the *measured* values of the acceleration due to gravity. In fact, in nearly all instances, when we speak of g, we mean the acceleration due to gravity taking into account the centripetal acceleration correction. The possibility for confusion comes from how we define the *weight* of an object, mg. If we mean by weight the *pull of gravity* on the object, then g is the value without the correction. If, however, we are concerned with the weight of the body as it affects the tension in a rope from which the body is suspended, then g must contain the correction, assuming that we are treating the system as being at rest in the Earth's frame. Fortunately, the correction is so small that for most purposes we can take the values to be the same. More will be said about true weight versus apparent weight in the next chapter.

4. *Forces and Newton's Second Law.* The Second Law states that the acceleration is directly proportional to the net force and inversely proportional to the mass. Like that of inertial reference frames, the concept of a force appears to be circular. Defining a force as a push or pull is satisfactory

for the first step, but ultimately we must appeal to the Second Law for an unambiguous definition. The Second Law in fact provides an *operational* definition for force; that is, it permits force to be defined in terms of a specified set of operations. The difficulty is in defining a force for the situation in which the object is not accelerating. Here is another instance, however, in which we can extrapolate from the operational definition. Experience tells us that when only one force acts on an object, the object will accelerate according to Newton's Second Law. Therefore, even if the object is at rest, we can speak of a particular force as having a magnitude such that *if* that force were to act alone, it would cause a given mass to have a certain acceleration. For example, this idea allows us to state that a block of mass m, suspended from a rope at rest, has a downward force mg resulting from the pull of gravity. We know this because if we break the rope so that the block is allowed to fall freely, the only force on the block is the gravitational force and the block accelerates at a rate g. Thus, $\mathbf{F} = m\mathbf{a}$ translates to $\mathbf{F}_g = w = mg$. This idea is developed further in the next chapter.

5. *Inertia and mass.* Newton's Second Law tells us that the rate at which an object changes its velocity under the influence of a given force has much to do with how "heavy" the object is. For example, we know that a bowling ball will accelerate more slowly than a marble if each is subjected to the same force. Quantitatively we are talking about the mass of the object, and, indeed, we define the mass in this case to be the ratio F/a. In a qualitative sense, however, we are looking at a *tendency* for objects to remain at rest or in a state of uniform motion in a straight line. This tendency we give the name *inertia.* Because this property manifests itself only when the object is being accelerated, we call the mass defined by F/a the *inertial mass.* In Chapter 9 we shall see how this property is to be distinguished conceptually (but not quantitatively) from the *gravitational mass*. We need not make that distinction here, so we shall speak only of mass, without a qualifier.

6. *Units for force and mass.* In the SI system, the units for force and mass are the newton (N) and the kilogram (kg), respectively. Although the newton is somewhat unfamiliar, the kilogram is quite familiar and often is mistaken for a force unit. There probably is no real harm in this misconception in everyday dealings in the marketplace. For example, to say that a car weighs 2000 kg technically is incorrect, but the usage is widespread and will very likely cause little or no confusion. In working with Newton's Second Law, however, we must clearly distinguish between mass and weight. The unit *kilogram* is a *mass* unit, and because we use the term *weight* to mean the *force* of gravity, the distinction must be maintained. Further confusion often arises over the commercial use of the *pound* as a unit of mass. A 1 lb can of coffee has a mass of 1 lb-mass. The purchaser is more interested in how much coffee (mass) is in the can than in how heavy it is to lift (weight). On the other hand, if a 1000 lb box is to be moved, the weight may be an important consideration. In practical matters this distinction is not very important, because the weight varies so little from place to place on the Earth. But to use Newton's laws, we must have correct and consistent force and mass units. For example, if the force is expressed in pounds (British engineering system), the mass must be in *slugs* to give the proper unit for acceleration,

ft/s^2. Remember that mass is an invariant property of an object, independent of location or of any forces that might be acting on the object. The weight (which is described more fully in section 6.2) is a force and varies according to location.

7. *Superposition of forces*. The technique of adding several forces is merely an extension of the procedures for vector addition. We need only remember that although forces add like displacements, they are not displacements. The arrows we draw representing the force vectors do not tell us anything about how the object is moving or, indeed, whether it is moving at all. The length of the vector tells us the size of the force at the point of application and at that instant. The *resultant* vector dictates only what the *instantaneous* acceleration will be. If we wish to know the displacement of the object, we need further information about the time variation of the vector.

8. *Tensions in wires, cables, and the like*. The concept of *tension* is used frequently in solving physics problems. Tension is the force exerted by stretched ropes, wires, and other similar stretchable objects. (There is an apparent paradox here. A rope cannot be under tension unless it stretches a bit, but we often assume that the rope is inextensible. We assume that it is stretched a small amount initially and then maintains this stretch, keeping the length constant. For nearly all strings, wires, or cables the stretch is negligible compared with the total length.) *The tension force at a given point in a rope is defined as the force the rope exerts on that to which it is attached*—more rope, a box, or whatever. If the system is at rest or moving with constant velocity, then the tension must be the same throughout the rope; otherwise, if we consider a small element somewhere in the rope for which the tension at each end of the element is different, according to Newton's Second Law, that element must accelerate.

Now let's look at the case in which the rope *is* accelerating, shown in Figure 5.1. We again consider a small element of the rope, m_r. If T and T' are the magnitudes of the tensions at each end of this element, respectively, then Newton's Second Law gives us $T - T' = m_r a$. So if $a \neq 0$, then $T \neq T'$. If m_r is very small and a is not too large, then T is very nearly equal to T'. Indeed, in many cases the mass of the rope or string is so small compared with the other masses in the system that we can ignore it. In that case we set T equal to T'. In other words, the tension is the same throughout the entire length of the rope. This is what we mean when we say that a massless rope transmits the force undiminished from one part of the system to another. It must be understood, however, that what we are saying is true only if there are no other external forces acting on the rope. Suppose we attach a weight

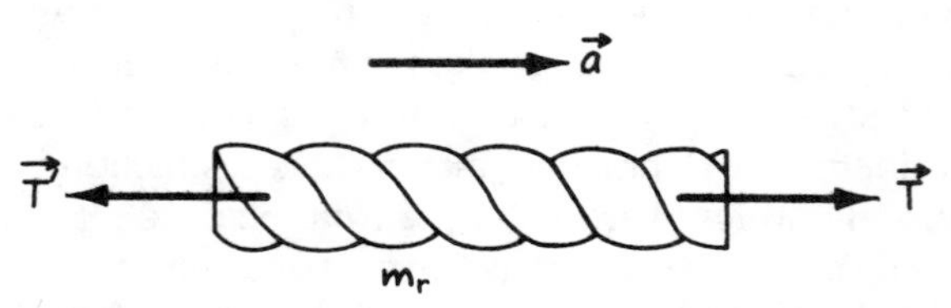

Figure 5.1 A small element of rope and the tension forces causing acceleration.

to the rope and let is slide slowly at constant speed over the edge of a table as in Figure 5.2. Because of the friction force acting on the rope where it passes over the edge, the tension T' will be less than T because if υ is constant, $T' + f = T$. In fact, if the friction force is very large, then T' can be quite small compared with T. If the table edge is very smooth, the friction force is approximately zero and the tension is the same throughout the rope.

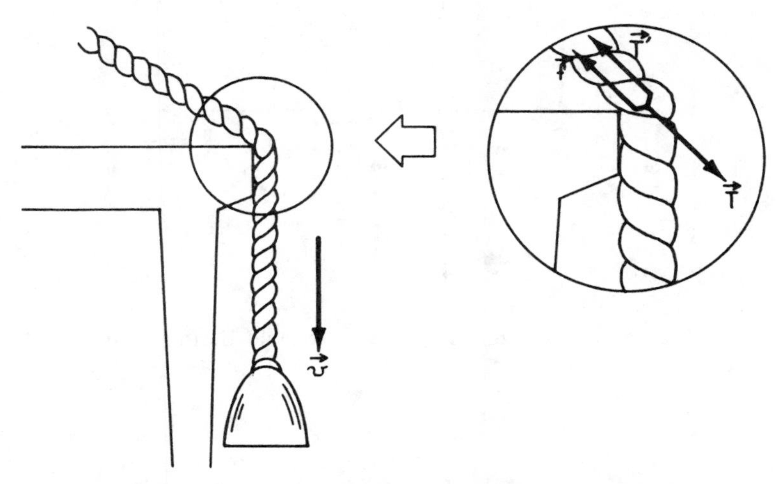

Figure 5.2 The tension is not as great in the portion of the rope above the table, because of the friction force **f**.

9. *Newton's Third Law.* For every action (force), there exists an equal and opposite reaction. This is perhaps the most misunderstood of the three laws. It is deceptive in that one must be certain to give proper consideration to the bodies that receive these so-called *action–reaction* pairs of forces. It might be helpful to think of the action–reaction pair as a single *interaction* of two objects, each being the recipient of this interaction (force). For example, when two bodies collide, there is an interaction that imparts a force to each body. There are several ways to identify correctly the action–reaction force pairs. One is to consider the verbal statement regarding the forces acting on each body. We merely interchange the subject and the object: For example, if we say that a bat (subject) exerts a force on a ball (object), then the reaction is the ball (subject) exerting a force on the bat (object). Another approach is to use subscripts in labeling the forces. Let $\mathbf{F}_{Bb}$ be the force that the bat (B) exerts on the ball (b), and $\mathbf{F}_{bB}$ the force that the ball exerts on the bat. According to the Third Law, $F_{Bb} = F_{bB}$, or, in vector form, $\mathbf{F}_{Bb} = -\mathbf{F}_{bB}$. With this notation we see that the Third Law force pairs are identified by ***reversed subscripts***.

Let's look at an example in which this notation may be helpful. In Figure 5.3 we have a book at rest on a horizontal table. The forces acting on the system are as indicated. Using the subscript notation, we see that the action–reaction pairs are

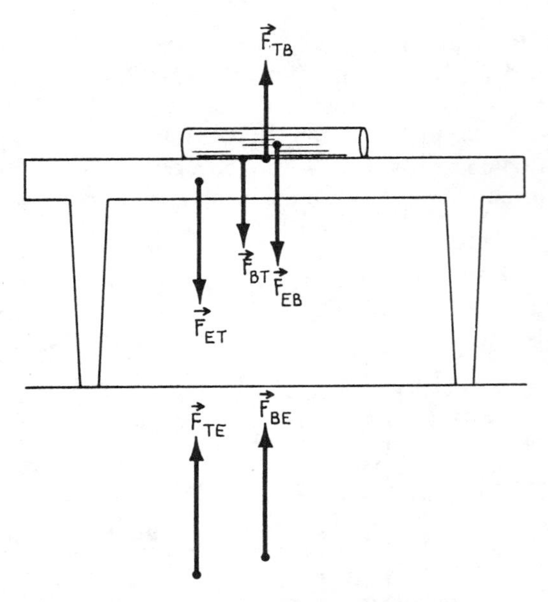

Figure 5.3 The forces on the book–table–Earth system.

$$\mathbf{F}_{BE} = -\mathbf{F}_{EB}$$
$$\mathbf{F}_{BT} = -\mathbf{F}_{TB}$$

and

$$\mathbf{F}_{TE} = -\mathbf{F}_{ET}$$

It is true also that $\mathbf{F}_{EB} = -\mathbf{F}_{TB}$, that is, that the Earth exerts a force on the book equal and opposite to the force the table exerts on the book. These two forces, however, are *not* an action–reaction pair, because they act on the same body and originate from different sources. They are equal and opposite in *this* situation only because the system is in equilibrium. Indeed, if the book and table were in an elevator accelerating upward, then $\mathbf{F}_{TB} > \mathbf{F}_{EB}$, the difference being what gives rise to the acceleration of the book. Thus, it is not sufficient that forces be equal and opposite to be action–reaction pairs; they must act on different bodies and be part of a mutual interaction between two objects.

10. *Linear momentum.* The linear momentum of a particle, **p**, is defined as the product of mass and velocity. It is an important concept, especially when a system of two or more particles is considered (see Chapter 10). We observe first of all that momentum is a *product*. This means that a given momentum can be obtained by many different combinations of mass and velocity. A 10,000 kg freight car traveling with a velocity of 2 m/s has the same momentum as a 5000 kg truck traveling at 4 m/s in the same direction. To change the momentum of a particle requires an external force, and the amount of force is equal to the rate at which the momentum is changed. This idea fits our intuitive notions about forces; in fact, it is an alternative expression for Newton's Second Law. The importance of the momentum

form of the law becomes apparent when the system consists of more than a single particle or when the mass of the system is not constant over time.

Another important attribute of momentum is its *vector* property. Failure to observe this property can lead to erroneous conclusions when computing the momentum change of a particle in such situations as collisions. For example, in Figure 5.4 a 0.5 kg ball travels with a speed of 5 m/s toward the

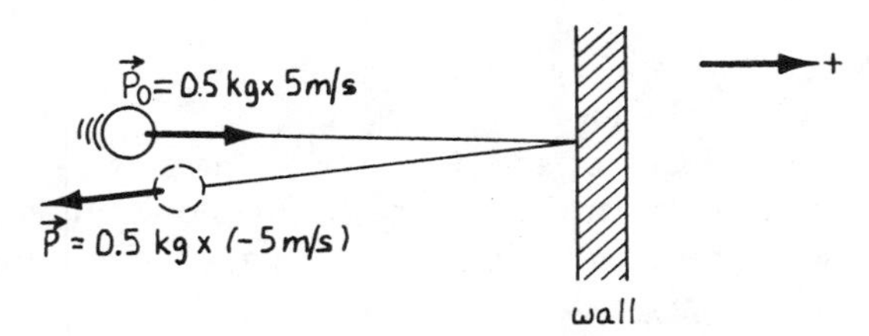

Figure 5.4 The *change* in momentum of the ball is in the *negative* direction.

right, hits a wall, and rebounds in the opposite direction with the same speed. The momentum change, however, is not zero. Rather, the change is $\Delta\mathbf{p} = 0.5\text{ kg} \times (-5\text{ m/s} - 5\text{ m/s}) = -5\text{ kg-m/s}$. Note that $\Delta\mathbf{p} = \mathbf{p} - \mathbf{p}_0$ has one negative sign "built in" because it represents the *difference* in two vectors. The vectors themselves also have signs associated with their respective directions. To avoid errors in this regard, always write first the expression $\Delta\mathbf{p} = \mathbf{p} - \mathbf{p}_0$, then substitute the individual values of the momenta, keeping the signs intact. As a check on the result, bear in mind that the resultant force on the particle must be in the same direction as the change in momentum. If you consider the change in direction of the particle's motion, it usually will be obvious in what direction the force must act to produce such a change. The sign associated with the momentum change must be consistent with the direction of that force.

You may have observed that units consisting of combinations of other units often are given special names. For example, the newton is the same as kg-m/s^2. Others have no special name. For instance, the unit for momentum is simply kg-m/s.

11. *Angular momentum.* In Chapter 3 we encountered a mathematical entity called the *cross (vector) product*, defined as $\mathbf{A} = \mathbf{B} \times \mathbf{C}$, where $|\mathbf{A}| = |\mathbf{B}||\mathbf{C}| \sin\theta$ and the vector $\mathbf{A}$ is perpendicular to the plane formed by $\mathbf{B}$ and $\mathbf{C}$. We come now to our first use of this concept: we will define the *angular momentum* to be the cross product of $\mathbf{r}$ and $\mathbf{p}$, where $\mathbf{r}$ is the position vector of the particle and $\mathbf{p}$ is the linear momentum. So if we let $\mathbf{L}$ be the angular momentum, then $\mathbf{L} = \mathbf{r} \times \mathbf{p}$. Although linear momentum is independent of the location of the origin in a particular reference frame, the angular momentum is not. For example, if at some instant the momentum vector $\mathbf{p}$ points directly toward or away from the origin, the angular momentum will be zero, because under these circumstances $\mathbf{r}$ and $\mathbf{p}$ are parallel vectors and $\mathbf{r} \times \mathbf{p} = 0$. If, however, the origin of the frame is moved to another location, so that $\mathbf{r}$ and $\mathbf{p}$ are no longer parallel vectors, $\mathbf{L}$ no longer equals zero.

Note that the definition of *torque*, $\tau = \mathbf{r} \times \mathbf{F}$, is developed in an analogous way such that the relationship $\tau = d\mathbf{L}/dt$ carries a kind of internal consistency with respect to these definitions. The consistent nature of these relationships will become clearer when we examine the dynamics of rigid bodies under rotation in Chapter 13.

Sample Problems

1. *Worked Problem.* A 250 g particle moves along the x axis according to the equation $x(t) = (2t^3 - t^2 + 5t)m$.

a. Calculate the velocity of and the force on the particle.

b. Plot a graph of these two variables as a function of time from $t = 0$ to $t = 5$ s, using the information from part a. Do the graphs agree with what we would expect using the slope concept?

Solution: We know that the velocity is related to the position by $\upsilon(t) = dx(t)/dt$. To obtain the force, we must know the value of the mass and the acceleration. The acceleration is defined as $a(t) = d\upsilon(t)/dt$.

a. $\upsilon(t) = (6t^2 - 2t + 5)$ m/s

$a(t) = (12t - 2)$ m/s^2

$F(t) = m\,a(t) = 0.25$ kg $\times$ $(12t - 2)$ m/s^2 $= (3t - 0.5)$ N

b. The graphs may be plotted by setting up a table of values for $\upsilon(t)$ and $F(t)$ from $t = 0$ to $t = 5$s. (See Tables 5.1 and 5.2, respectively.) These data are plotted in Figures 5.5 and 5.6.

Table 5.1 Values of $\upsilon(t)$ and t for the equation in problem 1

t (s)	$\upsilon(t)$ (m/s)
0	5
1	9
2	25
3	54
4	93
5	145

Table 5.2 Values of $F(t)$ and t for the equation in problem 1

t (s)	$F(t)$ (N)
0	−0.5
1	2.5
2	5.5
3	8.5
4	11.5
5	14.5

Comment: The *slope* of the $\upsilon(t)$ curve at each instant is proportional to $F(t)$ because $a(t) = F(t)/m$. Thus, the force curve should reflect the steadily increasing slope[1] of the $\upsilon(t)$ curve, which it does. You may have observed

[1]The slope is almost steadily increasing; there is a slight dip around $t = 0.17$ s.

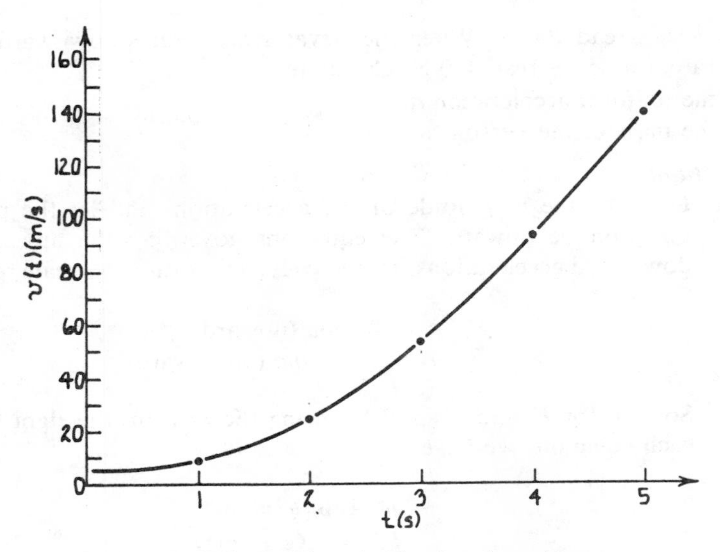

Figure 5.5 Problem 1. The function $\upsilon(t)$.

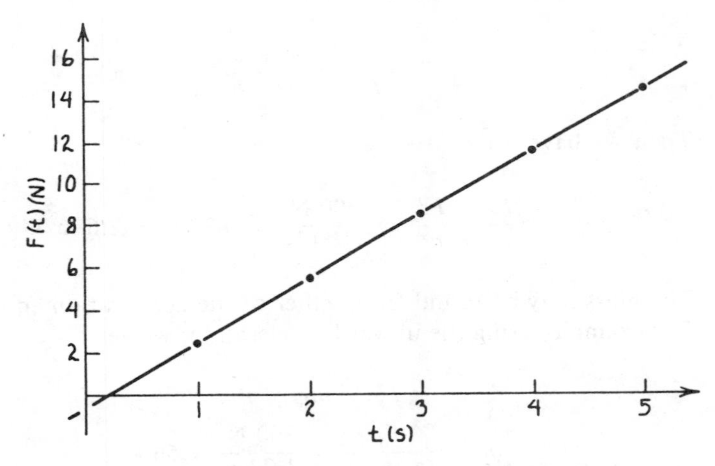

Figure 5.6 Problem 1. The function $F(t)$.

that $\upsilon(t)$ is in the mathematical form of a parabola and $F(t)$ is the slope-intercept form of the equation for a straight line.

2. *Guided Problem*. A 1.0 kg particle moves along the x axis in such a way that it obeys the equation $\upsilon(t) = 5t^2 - 3t + 4$ m/s. Calculate the force as a function of time, and graph the values of the force from 0 to 10 s.

Solution scheme

a. How is the acceleration related to the velocity equation?
b. What is the relationship among the force, mass, and acceleration?

3. *Worked Problem*. A person of mass m is standing on scales while riding in an elevator. When the elevator accelerates upward uniformly at a rate

a, the scales read 700 N. When the elevator accelerates downward at the same rate, the scales read 400 N. Calculate

a. the uniform acceleration a
b. the mass of the person

Solution:

a. Let a be the magnitude of the acceleration, and let the positive direction be upward. The equations governing the upward and downward accelerations, respectively, are (with F_s = scale reading)

$$F_s - mg = ma \text{ (upward)}$$
$$F_s' - mg = -ma \text{ (downward)}$$

Solving for F_s and F_s' and factoring the m from the right side of each equation, we have

$$F_s = m(g + a)$$
$$F_s' = m(g - a)$$

Dividing the two equations results in

$$\frac{F_s}{F_s'} = \frac{g + a}{g - a}$$

Then we have for a

$$a = g\frac{F_s - F_s'}{F_s + F_s'} = \frac{300 \text{ N}}{1100 \text{ N}} = 0.27\ g = 2.6 \text{ m/s}^2$$

b. The mass may be found from either of the acceleration equations. For example, using the upward acceleration, we get

$$F_s - mg = ma$$

$$m = \frac{F_s}{(g + a)} = \frac{700 \text{ N}}{1.27\ g} = 56 \text{ kg}$$

4. *Guided Problem.* A helium balloon with a total mass of 100 kg descends with a uniform acceleration of 2.0 m/s^2. If the balloon is to accelerate upward at 1.0 m/s^2, how much ballast must be jettisoned? Assume that the upward force is the same whether the balloon is ascending or descending.

Solution scheme

a. What is the net force on the balloon while it is descending? Set this equal to the product of the mass of the balloon and the downward acceleration.
b. What is the net force on the balloon while it is rising? (Remember, the total mass is now smaller.) Set this force equal to the product of the new mass and the upward acceleration.
c. Eliminate the constant upward force from the two equations, and solve for the mass of the ballast.

5. *Worked Problem.* An object on a frictionless horizontal surface is subjected to a constant horizontal force F. The speed of the object changes from 0.30 m/s to 0.60 m/s in 0.50 s. When a different force F' replaces F, the object changes its speed from 0.60 m/s to 0.80 m/s in the same time interval. Calculate the ratio of the two forces, F'/F.

Solution: The mass of the object is constant, so if we form the ratio of the two forces, the mass divides out and the ratio is simply $F'/F = a'/a$. But $a = \Delta \upsilon/\Delta t$ for constant a, and because the time intervals are the same, $a'/a = \Delta \upsilon'/\Delta \upsilon$. Therefore

$$\frac{F'}{F} = \frac{\Delta \upsilon'}{\Delta \upsilon} = \frac{(0.60 - 0.30)\ \text{m/s}}{(0.80 - 0.60)\ \text{m/s}} = \frac{0.30}{0.20} = 1.5$$

6. *Guided Problem.* Figure 5.7 shows the velocity, along a straight line, of a 2.0 kg particle as a function of time.

a. Plot the force as a function of time.

b. Suppose we have a different velocity–time graph, identical in shape to the first, but with $\upsilon_0 = 0$. What changes, if any, would there be in the force–time graph?

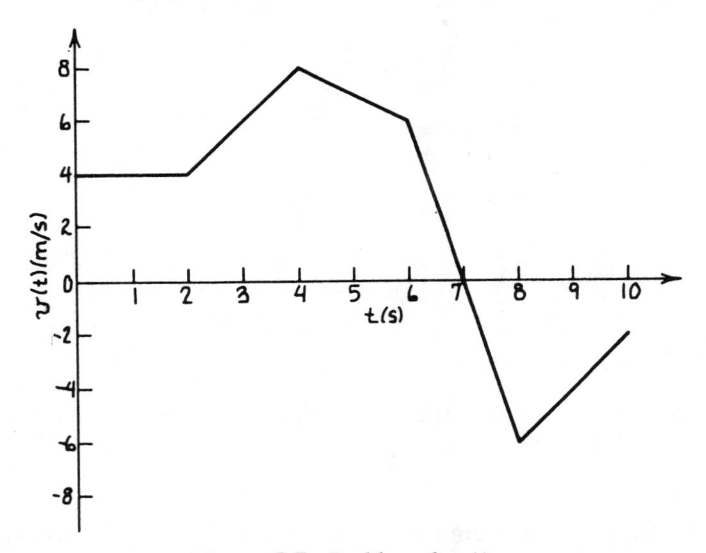

Figure 5.7 Problem 6. $\upsilon(t)$ versus t.

Solution scheme

a. To obtain $F(t)$ when m is given, you need to know $a(t)$. How is $a(t)$ related to $\upsilon(t)$ mathematically?

b. Because $\upsilon(t)$ is not expressed mathematically in this problem, you will have to use graphic data. What operation do you perform on the $\upsilon(t)$ graph to find $a(t)$?

7. *Worked Problem.* Because a fire in a building has blocked all normal exits, a man decides to slide down a rope that he has found in a closet. The rope has a breaking strength of 400 N, but the man weighs 735 N.

a. Is there a way he can slide down the rope without breaking it?
b. If the maximum speed at which he can hit the ground below without sustaining fatal injuries is 30 km/h, what is the maximum height from which he can escape safely using this technique?

Solution: The man certainly cannot change his weight significantly, so some method must be found for keeping the tension in the rope at 400 N or less. If he accelerates while sliding downward on the rope, the tension in the rope could vary from almost zero (essentially free fall) to 735 N (momentarily) if he slides at constant speed. Thus, the plan is to choose the minimum acceleration that will not break the rope but will enable him to avoid fatal injuries. Figure 5.8 shows the forces acting at the point of contact between the body and the rope.

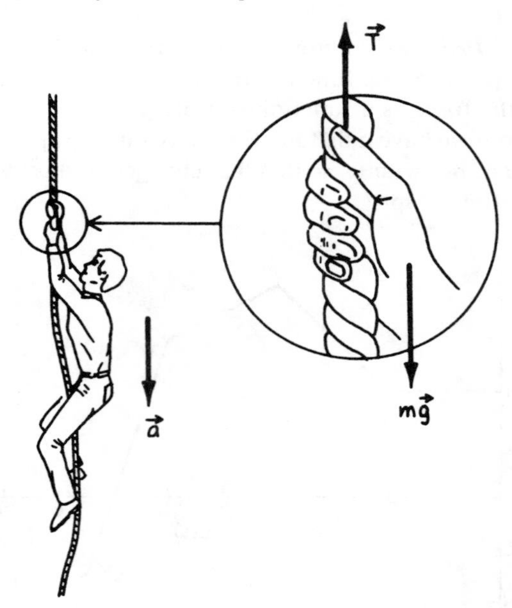

Figure 5.8 Problem 7. As the man accelerates downward, the tension in the rope is less than the man's weight.

a. We find the minimum acceleration possible by summing the forces on the rope, taking the downward direction as positive.

$$mg - T = ma$$
$$a = g - T/m$$
$$a = 9.8 \text{ m/s}^2 - \frac{400 \text{ N}}{75 \text{ kg}}$$
$$a = 5.3 \text{ m/s}^2$$

b. The problem now is to find the initial height such that when the man reaches the ground, he will be traveling 30 km/h. We have 30 km/h = 8.3 m/s, so that with the variables given, the appropriate equation is

$$\upsilon^2 = \upsilon_0^2 + 2ay$$
$$(8.3\ \text{m/s}^2)^2 = 0 + 2 \times 5.3\ \text{m/s}^2\ y$$
$$y = 6.5\ \text{m}$$

8. *Guided Problem.* A 510 kg rocket accelerates straight upward at 20 m/s^2. We will assume the fuel loss is negligible compared with the rocket mass, so that a remains constant. The rocket engine shuts down after 5.0 s and the rocket continues in free flight, maintaining its vertical orientation at all times. On the rocket's way down, the engine is restarted to ensure that the rocket will land with a speed of no greater than 1.0 m/s.

a. How much thrust is developed by the engine?
b. At what height above the ground must the engine be restarted?

Solution scheme

a. The acceleration of the rocket is upward, so the net force must be upward also. What is the net force on the rocket?
b. At engine shutdown how fast is the rocket traveling and how high is it above the ground?
c. Take these values as initial conditions for the free-fall portion of the flight, and determine how high the rocket goes before starting back down.
d. If h is the total height, the rocket falls freely for a distance $h - y$, then decelerates at 20 m/s^2 for a distance y, where y is the height at which the engine is restarted. You know the speed at which the rocket hits the ground, so you can solve for y.

9. *Worked Problem.* The Russell traction apparatus, shown in Figure 5.9, serves two functions: one is to support the weight of the broken leg, and the other is to maintain the proper tension in the leg for the bone to set properly.

a. Explain how the system can provide a variable tension for the leg while maintaining a constant support for the leg.
b. If the mass of the leg is 4.5 kg, how much weight should be placed on the end of the cord?
c. If the tension in the leg is to be 60 N, what is the required value for the angle θ?

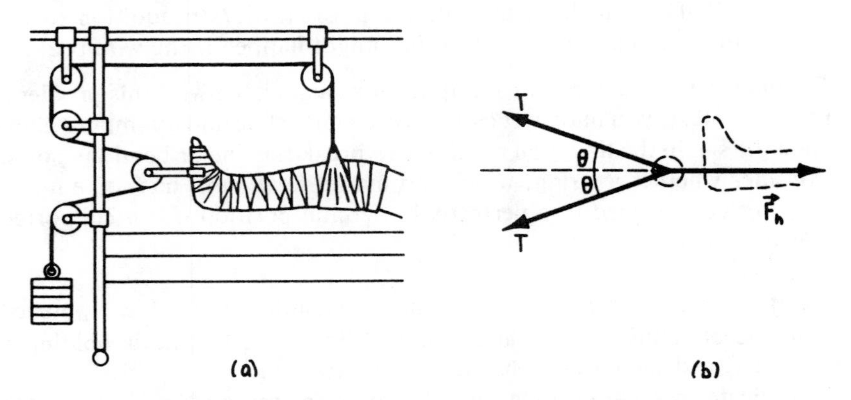

Figure 5.9 Problem 9. (a) Russell traction apparatus. (b) Tension forces at the foot connector.

Solution

a. We assume that the friction in the pulleys is negligible and that the cord is massless. Therefore, the tension is the same everywhere throughout the cord. To support the 4.5 kg leg a 4.5 kg mass must be hung from the end of the cord. This is true regardless of the angle θ. If, however, we consider the pull at the foot connection, clearly if the cord were vertical at this point, no horizontal pull would result. But if θ were zero, we would expect the maximum pull, equal to $2T$. So the angle adjustment provides for a variable force on the leg horizontally while keeping the support force constant.

b. We have noted already that the weight must be 44 N. To find the angle, we use the fact that the forces are in equilibrium at the foot connector. The forces acting at this point are shown in Figure 5.9b. We have for the horizontal forces

$$F_h = 2T \cos \theta$$
$$\cos \theta = \frac{F_h}{2T} = \frac{60 \text{ N}}{2 \times 44 \text{ N}} = 0.68$$
$$\theta = \cos^{-1} 0.68$$
$$\theta = 47°$$

10. *Guided Problem.* In a circus performance, a 50 kg man walks along a wire suspended from two rigging posts 10 m apart. If the wire is tightened so that the center sags no more than 15 cm when the man is in that position, how much tension is in the wire when the man is at the center? Assume that the mass of the wire itself is negligible.

Solution scheme

a. Use the approximation for small angles, $\tan \theta \approx \sin \theta$.

b. Sum the forces acting at the point at which the man is standing on the wire. Resolve them into horizontal and vertical components. What force must the vertical components of the tension force balance? (See the previous problem for the method of dealing with the tension components.)

c. Your answer should be a fairly large number, because the angle is so small. Do you think this situation is realistic? In addition to the man's falling, what do you think might happen if the wire broke?

Comment: There are practical implications associated with this problem. For example, if you hang wet clothes on a clothesline and attempt to eliminate the sag in the line, you are likely to break the line or bend the poles. Consider whether a string, which, no matter how light, has some mass, can ever be stretched to a perfectly horizontal position if it is supported only at its ends.

11. *Worked Problem.* A 50 kg woman is standing in a cable-supported elevator accelerating upward at 2.0 m/s^2 (Figure 5.10). She is holding a 2.0 kg box, and the elevator has a mass of 1000 kg.

a. Indicate in separate diagrams the forces acting on the elevator, the woman, and the box, respectively.

Figure 5.10 Problem 11. A woman holds a box as she accelerates upward in an elevator.

b. Identify the reaction forces associated with each force described in part a.
c. Write separately the equation of motion for each object.
d. By combining these three equations algebraically, obtain the equation of motion for the *system*, consisting of the three objects together.
e. Show that the same equation is obtained if you consider the system all together and sum the forces acting on it.
f. Calculate the tension in the cable for the conditions stated.

Solution: For each of the objects we shall consider the forces to act at a point, although for clarity the force vectors may have to be displaced laterally to avoid overlap in the diagram.

a. The forces acting on each object in the system are shown in Figure 5.11. The subscripts are identified in the figure. Note that $\mathbf{F}_{EL}$, $\mathbf{F}_{EW}$, and $\mathbf{F}_{EB}$ are the *weights* of the three objects, respectively.

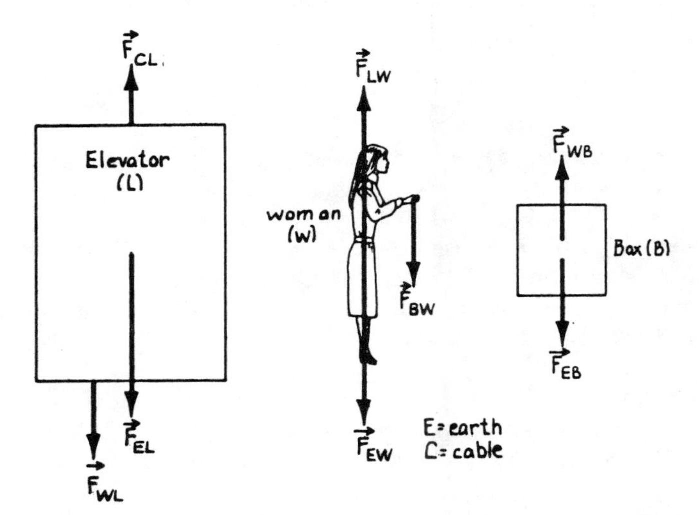

Figure 5.11 The forces acting on the objects in Figure 5.10.

b. The reaction forces are

$$\mathbf{F}_{CL} = -\mathbf{F}_{LC} \qquad \mathbf{F}_{EW} = -\mathbf{F}_{WE}$$
$$\mathbf{F}_{EL} = -\mathbf{F}_{LE} \qquad \mathbf{F}_{BW} = -\mathbf{F}_{WB}$$
$$\mathbf{F}_{WL} = -\mathbf{F}_{LW} \qquad \mathbf{F}_{EB} = -\mathbf{F}_{BE}$$

c. Take the upward direction to be positive. Then

$$\text{Elevator:} \quad F_{CL} - F_{WL} - F_{EL} = m_L a$$
$$\text{Woman:} \quad F_{LW} - F_{BW} - F_{EW} = m_w a$$
$$\text{Box:} \quad F_{WB} - F_{EB} = m_B a$$

These can be rewritten as

$$F_{CL} - F_{WL} - m_L g = m_L a \qquad (1)$$
$$F_{LW} - F_{BW} - m_w g = m_W a \qquad (2)$$
$$F_{WB} - m_B g = m_B a \qquad (3)$$

d. If we substitute Eq. (3) into Eq. (2) and make use of the fact that $F_{BW} = F_{WB}$ (magnitudes), we get

$$F_{LW} - (m_B g + m_B a) - m_W g = m_W a$$

or

$$F_{LW} - (m_B + m_W)g = (m_B + m_W)a$$

Then substituting the latter equation into Eq. (1) gives

$$F_{CL} - (m_B + m_W + m_L)g = (m_B + m_W + m_L)a$$
$$F_{CL} = (m_B + m_W + m_L)(g + a)$$

e. If we consider all of the forces acting on the entire system of mass $(m_W + m_B + m_L)$, we get

$$F_{CL} - F_{EL} - \cancel{F_{WL}} + \cancel{F_{LW}} - F_{EW} - \cancel{F_{BW}} + \cancel{F_{WB}} - F_{EB}$$
$$= (m_W + m_B + m_L)a$$
$$F_{CL} - m_L g - m_W g - m_B g = (m_L + m_W + m_B)a$$
$$F_{CL} = (m_L + m_W + m_B)(g + a)$$

where we once again used the equality of the magnitudes of Newton's Third Law action–reaction force pairs.

f. The tension in the cable is the force that the cable exerts on the elevator. (We assume the mass of the cable to be negligible.) This force is F_{CL}.

$$F_{CL} = T = (m_L + m_W + m_B)(g + a)$$
$$= (1000\text{ kg} + 50\text{ kg} + 2\text{ kg})(9.8\text{ m/s}^2 + 2.0\text{ m/s}^2)$$
$$= 1052\text{ kg} \times 11.2\text{ m/s}^2$$
$$T = 1.2 \times 10^4\text{ N} \quad (2.6 \times 10^3\text{ lb})$$

12. *Guided Problem.* A 225 kg horse pulls a sled loaded with logs across a field. The total mass of the sled and logs is 90 kg. A friction force of 400 N must be overcome to start the sled moving.

a. With how much force must the horse push horizontally against the ground to get the sled moving?
b. What is the *net* force exerted by the horse on the ground?
c. How much force does the horse exert on the sled? Assume the pull of the horse on the sled is horizontal.
d. With how much force does the sled pull on the horse?
e. In view of your answers to parts c and d, how can the horse ever get the sled moving?

Solution scheme

a. Both the horse and the sled must be moved, so the force needed to move them is the net force on the *system* (horse plus sled).
b. The forward force on the system is the *reaction by* the ground to the push of the horse horizontally *on* the ground.
c. From the horse the ground experiences not only the horizontal push from the horse, but also the horse's weight. What is the resultant of these two forces?
d. With the horse and sled taken together as a single system, the force exerted by the horse on the sled and the force exerted by the sled on the horse constitute an internal action–reaction pair of forces. Can these equal and opposite forces prevent motion of the system?

13. *Worked Problem.* A 0.15 kg baseball traveling 40 m/s (90 mph) is struck squarely by a bat, causing the ball to rebound in the opposite direction at 60 m/s (135 mph). The time of contact between the bat and the ball has been measured in experiments to be approximately 1.3×10^{-3} s.

a. Calculate the average force exerted on the ball by the bat.
b. If we assume that the contact force increases linearly from zero to some maximum value, F_{max}, and decreases the same way, compute the value for F_{max}. What is happening to the ball during this time?

Solution: This problem can be approached using the concept of either force–momentum or force–acceleration. For a single particle and constant mass, there is no difference. We shall use the momentum concept.

a. To find $\mathbf{F}_{avg}$, we must compute $\Delta\mathbf{p}$. Assume that the initial direction of the ball is positive. Then

$$\Delta\mathbf{p} = \mathbf{p} - \mathbf{p}_0 = m\mathbf{v} - m\mathbf{v}_0 = m(\mathbf{v} - \mathbf{v}_0)$$
$$\Delta\mathbf{p} = 0.15 \text{ kg} \times (-60 \text{ m/s} - 40 \text{ m/s})$$
$$\Delta\mathbf{p} = -15 \text{ kg-m/s}$$
$$F_{avg} = \frac{\Delta\mathbf{p}}{\Delta t} = \frac{-15 \text{ kg-m/s}}{1.3 \times 10^{-3} \text{ s}} = -1.5 \times 10^4 \text{ N} \quad (-2600 \text{ lb})$$

The negative sign in the answer means that the force is in the direction of the final velocity of the ball.

b. If the force increases and decreases linearly, then

$$F_{avg} = \frac{0 + F_{max}}{2}$$

$$F_{max} = 2F_{avg} = -2.3 \times 10^4 \text{ N} \quad (-5200 \text{ lb})$$

Stop-action photographs of a ball being struck by a bat show large deformations of the ball.

14. *Guided Problem.* The human tibia (leg bone) will fracture when the compressional force exceeds about 2×10^5 N (4500 lb). Suppose an 80 kg man were to land squarely on his feet with knees locked so that the stopping force is absorbed by the bone structure as a compressional force.

a. If the body is brought to rest in 0.003 s, from what maximum height could he jump without sustaining a tibial fracture? Assume that the force is distributed evenly to both legs and the force builds up linearly, as in the previous problem.
b. If the man jumps from a height of 2 m and, upon making contact with the ground, bends his knees so that the stopping time is increased to 0.15 s, what is the maximum force sustained, assuming a linear increase as before? What force mechanisms are brought into play in this situation?

Solution scheme: This is another momentum change problem. Because there are two tibiae to absorb the shock, what is the maximum total force allowed? What is the *average* force permissible?

a. Given the force and the interaction time, what is the change in momentum?
b. What landing speed will result in this momentum change?
c. From what height must the man fall freely to achieve this speed?
d. Change the interaction time. How does that affect the force?

15. *Worked Problem.* A 0.50 kg particle is constrained to move in a circular path of radius 1.5 m in the x–y plane. Starting from rest, it moves with a tangential acceleration of 0.20 m/s^2 in the counterclockwise direction (looking down from above the plane).

a. Determine the angular momentum (with respect to the center of the circle) of the particle as a function of time.
b. Compute the torque on the particle relative to the same origin.

Solution

a. Because $\mathbf{v}$ is always perpendicular to $\mathbf{r}$ when the origin is at the center of the circle, $\mathbf{L} = \mathbf{r} \times \mathbf{p} = rp\,\hat{\mathbf{z}} = rmv\,\hat{\mathbf{z}}$. But since the particle is accelerating in the direction of $\mathbf{v}$, $|\mathbf{v}| = v = at$ where a is the tangential acceleration. Thus

$$\mathbf{L} = rmv\,\hat{\mathbf{z}} = rmat\,\hat{\mathbf{z}}$$

b. The torque is defined as $\tau = \mathbf{v} \times \mathbf{F}$, but $\mathbf{F} = m\mathbf{a}$, and since $\mathbf{F}$ is in the direction of $\mathbf{v}$ (because the acceleration is tangential only), we have

$$\tau = \mathbf{r} \times \mathbf{F} = rF\,\hat{\mathbf{z}} = rma\,\hat{\mathbf{z}}$$

16. *Guided Problem.* A 0.05 kg particle has a velocity given by $\mathbf{v} = (3\hat{\mathbf{x}} + 4\hat{\mathbf{y}})$ m/s when it is at the position $\mathbf{r} = (2\hat{\mathbf{x}} - 4\hat{\mathbf{y}})$ *m*. Calculate the angular momentum of the particle about the origin of **r** when the particle is at the stated position.

Solution scheme

a. Write down the definition of **L** in the cross-product notation.
b. You are given **v**. What is **p**?
c. Carry out the cross-product operation.
d. In what direction is the vector **L**? If you got a negative sign for **L**, what would that mean?

Answers to Guided Problems

2. $a = 10t - 3$. See Figure 5.12.

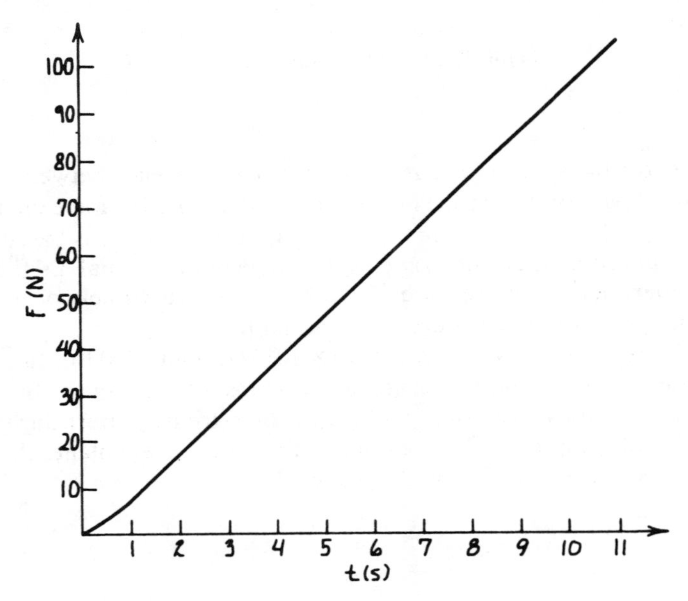

Figure 5.12 F versus t for problem 2.

4. $m_b = 28$ kg
6. a. See Figure 5.13.
   b. Slope is the same for both curves, so $F(t)$ remains the same also.
8. a. Thrust $= 1.5 \times 10^4$ N
   b. $y = 56$ m
10. $T = 8 \times 10^3$ N
12. a. Just over 400 N
    b. 400 N rearward, 2200 N downward (weight)
    c. 400 N
    d. Same as in part c but in opposite direction
    e. These forces are action–reaction pairs, and if they were the only forces acting on the system, no movement would be possible. What

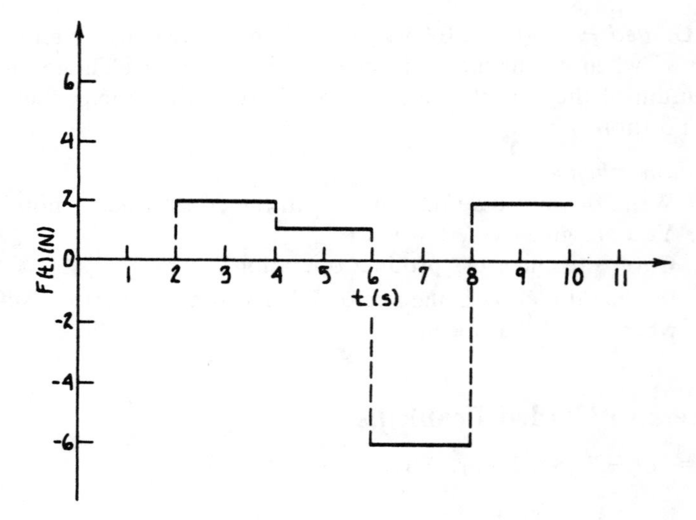

Figure 5.13 $F(t)$ versus t for problem 6.

causes the system to start to move is the difference between the forward push of the ground on the horse and the friction force on the sled.

14. a. $h_{max} = 11$ m. This value may seem a bit high, but it is based on the assumption that the force is strictly compressional in the tibia. Other, perhaps more serious, injuries would very likely occur if one jumped from heights even lower than this.
    b. $F_{avg} = 3.3 \times 10^3$ N, so $F_{max} = 6.6 \times 10^3$ N (about 1500 lb). In this case the leg muscles under tension absorb most of the impact. But the leg muscles are very strong, and F_{max} is of extremely short duration.
16. $\mathbf{L} = (-0.2\ \hat{\mathbf{z}})$ kg-m^2/s. Since $\hat{\mathbf{z}}$ points up out of the x–y plane, the vector $\mathbf{L}$ points downward, below the plane.

Chapter 6

Dynamics—Forces and the Solution of the Equation of Motion

Overview

In this chapter we continue to discuss the concepts encountered in Chapter 5 by focusing on the equation of motion (Newton's Second Law) in more detail. Of particular interest will be the analysis of the motion of an object or system of objects subjected to a constant friction force in addition to other possible forces. We will also look at another special case of motion under a constant force, uniform circular motion.

Essential Terms

weight
apparent weight
normal force
"free-body" diagram
friction (kinetic and static)
coefficient of friction
elastic force
Hooke's Law
spring constant
centripetal force

Key Concepts

1. *Weight*. The term *weight* has come to have many different connotations, so we must be sure to use it in a correct and consistent way. Although weight is defined as the force of gravity, the *perception* of weight is the result of an interaction with another object. For example, when you stand on a bathroom scale, the scale is not recording your weight directly but,

rather, is recording the force that you exert on the scale. The reaction (the scale pushing up on you) is what we call the *normal* force. But because you are in equilibrium, the normal force must be equal in magnitude to your weight. If you were to stand on the same scale in an elevator accelerating upward, however, the normal force would be greater than your actual weight, thereby causing you to feel "heavier." As you accelerate upward, the scales push with an upward force greater than the downward pull the Earth is exerting on you. Reacting to this push, you exert an equal and opposite force on the scale, and it registers a greater "weight." In other words, your actual weight has not changed, but your perceived, or *apparent*, weight has. Your arm, for example, appears to have weight because when you lift it, various body muscles are under tension, producing a force to overcome the downward pull of gravity. But in free fall, all parts of the body accelerate downward at the same rate, so that no muscle forces are required to "hold up" the arm or any other part of the body. In your reference frame, the body behaves in the same way it would if there were *no* gravitational force acting on it. Your *apparent* weight is therefore zero, but your *true* weight is still mg.

We must also remember that mass and weight are very distinct quantities. Whereas mass is an invariant, *scalar*, property of matter, related to inertia, weight is a *force*, equal in magnitude to mg, where g is the acceleration due to gravity. An object does not have to accelerate to have weight, however. By allowing an object to fall freely, we can establish that the weight must be equal to the product of two quantities, m and g, but because the gravitational force must exist on the object independent of its state of motion, the weight will be equal to mg at all times.

2. *"Free-body" diagrams*. To solve the equations of motion successfully, it is essential that one be able to construct accurate and useful "free-body" diagrams (FBDs). As noted in your text, one constructs an FBD by isolating the object (or system of objects) and drawing vectors representing the forces acting *only on that object* (or system). The key idea is that only those forces in *direct contact* with the object are to be included in the FBD, with one exception: if there are any "action-at-a-distance" forces (e.g., gravity) acting on the object, these should be included. Until we get to the chapters dealing with electricity and magnetism, we need be concerned only with gravity as an action-at-a-distance force. So our rule for deciding what forces to include in the FBD is as follows:

a. First draw in the gravitational force vector (weight). This force always points downward (toward the center of the Earth).
b. Next draw all other vectors that represent forces in *direct contact* with the object. These are called *contact forces*.

For example, consider the situation in which three blocks are tied together and accelerating under the action of an applied force, as shown in Figure 6.1. To construct an FBD for block m_2, we first isolate it from its environment, then proceed to draw all the forces acting *on* it, as in Figure 6.2. Note that there are four forces acting on the block that added *vectorially* must equal $m_2\mathbf{a}$: $\mathbf{F}_{net} = \mathbf{T}_1 + \mathbf{T}_2 + \mathbf{N}_2 + \mathbf{W}_2 = m_2\mathbf{a}$. A common error made by beginning students is to attempt to include in FBDs a "force due to the acceleration." In inertial reference frames (those in which Newton's laws are

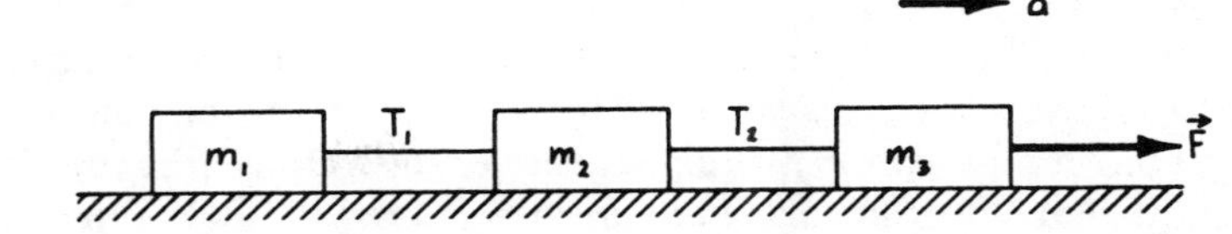

Figure 6.1 Three blocks, tied together by massless cords, are accelerated along a frictionless surface by the externally applied force **F**.

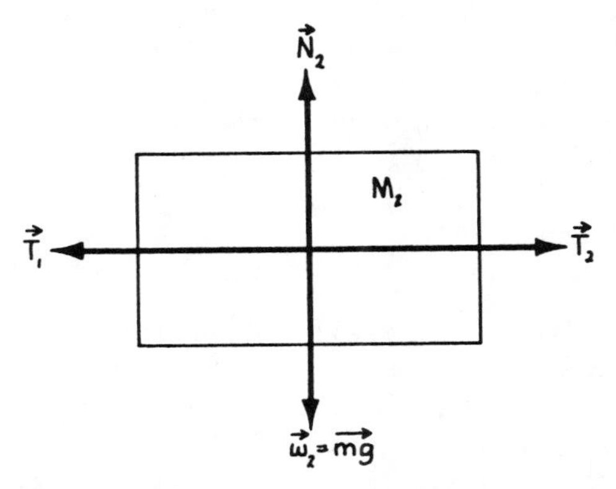

Figure 6.2 The "free-body" diagram for mass m_2 in Figure 6.1.

valid), no such force exists. Newton's Second Law states that the vector sum of all forces acting on a body gives rise to an acceleration of that body. In other words, forces cause acceleration—accelerations do not cause forces!

3. *Friction*. For our purposes we can describe two kinds of friction forces, *static* and *kinetic*. The latter rarely causes much conceptual difficulty, but the static friction force is often misunderstood and therefore incorrectly used. The static friction force, like the normal force, is a *passive* force; that is, it merely responds to some other force. Thus, if a block is at rest on a horizontal table and no forces with horizontal components are applied to the block, there is no static friction force. If there were, an unbalanced force would exist and the block would be accelerated. Now suppose some small horizontal force is applied to the block but the block remains at rest. Then we must conclude that an equal and opposite force acts on the block to prevent acceleration. This is the static friction force. If the horizontally applied force increases, the friction force will increase by the same amount until its maximum value is reached. So the static friction force is *not* a fixed value but, rather, is always equal to the applied force, unless the maximum value of the friction force has been exceeded. This is the meaning of $f_s \leqslant \mu_s N$. For example, suppose that the coefficient of static friction between the block and the table is 0.4 and that the normal force is 30 N. Then the *maximum* value of the friction force is 12 N. If a force of only 8 N pulls laterally on the block, however, the static friction force also will be only 8 N, just balancing the lateral pulling force.

4. *Centripetal force*. We often hear it said that for an object moving in a circular path, "there is a force directed toward the center of the circle called

the centripetal force." The problem with this statement is that it makes it appear that there is a special kind of force, the centripetal force, that acts in addition to the other forces. *Centripetal* means "center-seeking" and merely defines the direction of the *net* force. It is better to say that "the real forces acting on the body must have a *resultant* directed toward the center of the circular path and equal in magnitude to mv^2/R." There may, in fact, be no actual force directed toward the center. For example, in the case of the banked curve (see Example 13, p. 148 in the text), the forces acting on the automobile are the *normal* force, the *weight*, and, possibly, *friction*. None of these is directed toward the center of the curved path; the *resultant* of these forces must be so directed, however, if the car is to negotiate the curve at constant speed.

Sample Problems

1. *Worked Problem.* Two blocks having masses of 3 kg and 5 kg, respectively, are connected by a massless string that passes over a massless, frictionless pulley, as shown in Figure 6.3. The blocks are in contact with frictionless planes of angles 37° and 53°. Determine (a) the acceleration of each block, (b) the tension in the string, and (c) the normal force on each block.

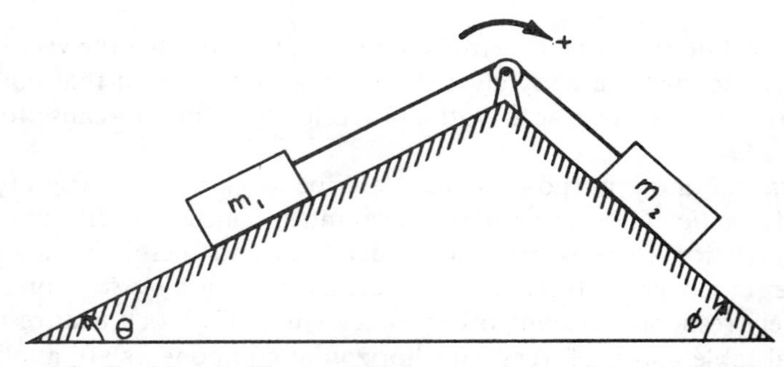

Figure 6.3 Problem 1. The two blocks are connected by a string passing over a frictionless pulley.

Solution: In Figure 6.3 we have assumed that the acceleration is to the right (more precisely, up the plane for m_1 and down the plane for m_2). We begin by drawing FBDs for blocks m_1 and m_2 separately, as in Figure 6.4. Because the string is assumed to be inextensible, $a_1 = a_2 = a$, and because it is massless, $T_1 = T_2 = T$. With our coordinate axes chosen so that $+x$ is in the direction of the assumed acceleration, the equations of motion for m_1 are

$$\Sigma F_x = T - m_1 g \sin\theta = m_1 a$$
$$\Sigma F_y = N_1 - m_1 g \cos\theta = 0$$

For m_2

$$\Sigma F_x = m_2 g \sin\phi - T = m_2 a$$
$$\Sigma F_y = N_2 - m_2 g \cos\phi = 0$$

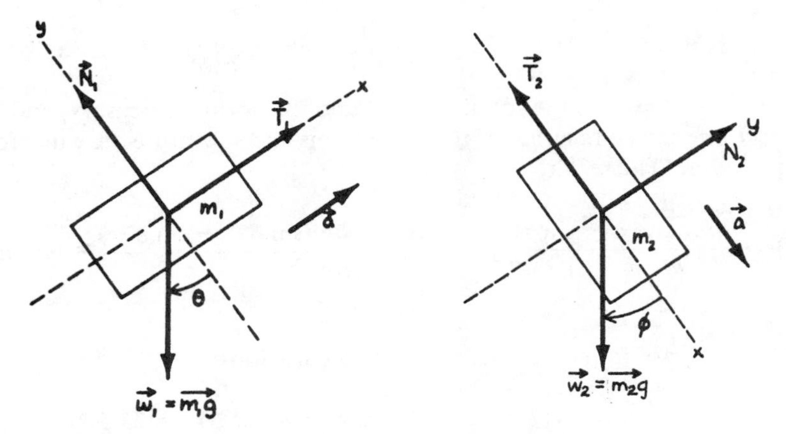

Figure 6.4 Problem 1. "Free-body" diagrams for the two masses in Figure 6.3.

By adding the two x equations, we can eliminate the variable T. After rearranging terms, we get

$$a = \frac{(m_2 \sin \phi - m_1 \sin \theta)}{(m_1 + m_2)} g$$

or

$$a = \frac{5 \text{ kg} \sin 53° - 3 \text{ kg} \sin 37°}{3 \text{ kg} + 5 \text{ kg}} g = 2.7 \text{ m/s}^2$$

Before solving for T, let's examine this solution. First note that if $m_1 \sin \theta > m_2 \sin \phi$, the system will accelerate to the "left," that is, opposite to the direction we assumed to be positive. If these two terms are equal, the acceleration will be zero, which means that the system is either at rest or moving at constant velocity. The solution includes two interesting special cases. If $\phi = 0°$ and $\theta = 90°$, we have the system shown in Figure 6.5 (for which we get an acceleration).

$$a = \frac{m_2}{(m_l + m_2)} g$$

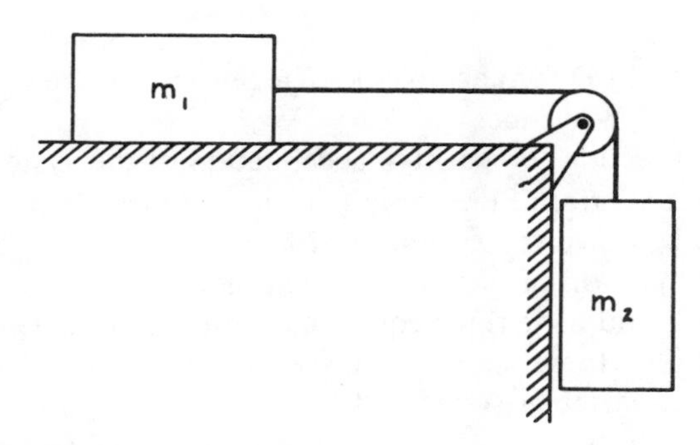

Figure 6.5 Problem 1. The special case for $\phi = 0°$ and $\theta = 90°$.

What is the acceleration if θ and ϕ are each equal to 90°? Describe the system.

Now let's look at the solution for T and the normal forces, N_1 and N_2. For T one can go to either of the x equations and substitute the value for a just found. The result is

$$T = \frac{m_1 m_2 (\sin \phi - \sin \theta)}{(m_1 + m_2)} g = \frac{3 \text{ kg} \times 5 \text{ kg} (\sin 53° - \sin 37°)}{8 \text{ kg}} \times 9.8 \text{ m/s}^2$$
$$= 3.6 \text{ N}$$

N_1 and N_2 are found directly from the y equations:

$$N_1 = m_1 g \cos \theta = 3 \text{ kg} \times 9.8 \text{ m/s}^2 \times \cos 37° = 23.5 \text{ N}$$
$$N_2 = m_2 g \cos \phi = 5 \text{ kg} \times 9.8 \text{ m/s}^2 \times \cos 53° = 29.5 \text{ N}$$

Comment: If we were given no numerical data, we would have no way of knowing whether the acceleration is zero or to the right or the left. In this case we would simply choose a positive direction for acceleration and let the results tell us the proper direction based on assumptions about the data.

2. *Guided Problem.* Figure 6.6 shows surfaces and pulleys that are frictionless. Determine (a) the acceleration of the system and (b) the tension in each cord.

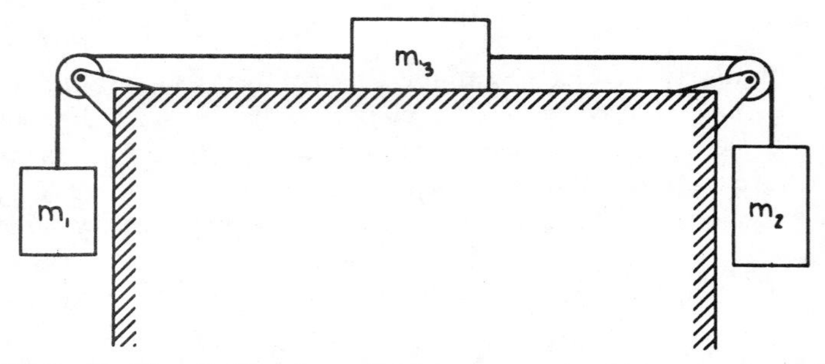

Figure 6.6 Problem 2. The three masses are connected by a massless string passing over frictionless pulleys.

Solution scheme

a. Construct an FBD for each block. The tensions in the two cords are not the same. Why not?
b. Choose a direction for positive acceleration of the system, and call that the $+x$ direction. For example, if this is to the right for m_3, then $+x$ should be down for m_2 and up for m_1.
c. Write the equation of motion for each block.
d. You now should have three equations with a and the two Ts as unknowns. Solve these simultaneously to obtain algebraic solutions for the three variables.

Hint: If $m_3 = 0$, then $T_{13} = T_{23}$. Does this happen with your solution?

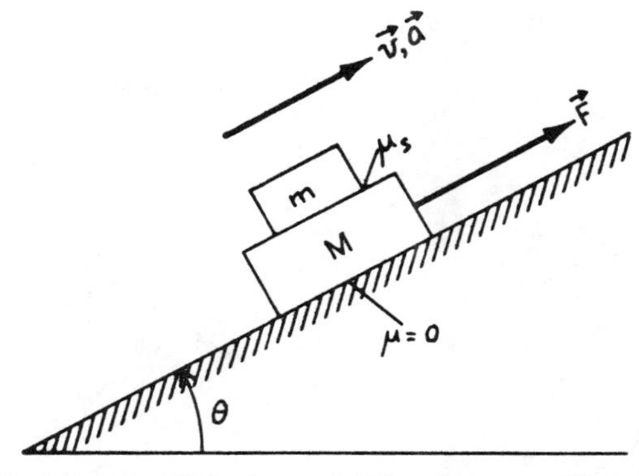

Figure 6.7 Problem 3. There is no friction between M and the surface of the incline. The coefficient of static friction between m and M is μ_s.

3. *Worked Problem.* Figure 6.7 shows a set up that at first looks quite complicated, but we will show that a systematic application of the equation of motion to the various parts of the system makes the problem tractable.

The two blocks are to be accelerated up the incline by means of an applied force F without having the small block slip. The coefficient of static friction between the two blocks is μ_s, but there is no friction between the larger block and the plane. We wish to find the maximum value of the force F that will maintain the nonslipping condition.

Solution: We know that for the smaller block (m) to be accelerated, there must be a net force on it directed up the incline. But the only force on this block that can possibly have a component in that direction is the friction force f_s, and because f_s has a maximum value, $f_{s_{max}}$, the block m has a maximum acceleration a_{max}. Therefore, if the larger block is given an acceleration greater than this value, slipping will occur between the two blocks. So our solution scheme is this:

1. Find the maximum acceleration for m.
2. Let this be the maximum acceleration for M also, and compute the force F necessary to produce this acceleration.

We first construct an FBD for the small block m. To do this we have to decide in which direction the friction force acts. Note that the static friction force always *opposes the tendency to slip*. In the present configuration, the small block is tending to slip backward relative to the larger block. Thus, the friction force f_s will be directed up the plane to oppose this. So the FBD for m is that shown in Figure 6.8a. We first compute the sum of the forces along the two independent axes, x and y:

$$\Sigma F_x = f_s - mg \sin \theta = ma$$
$$\Sigma F_y = N_1 - mg \cos \theta = 0$$

At this point we have two independent equations but three unknowns. We are looking for the *maximum* value of the acceleration, however, and this occurs when f_s is at its maximum value: $f_{s_{max}} = \mu_s N_1$. Therefore, the x equation becomes

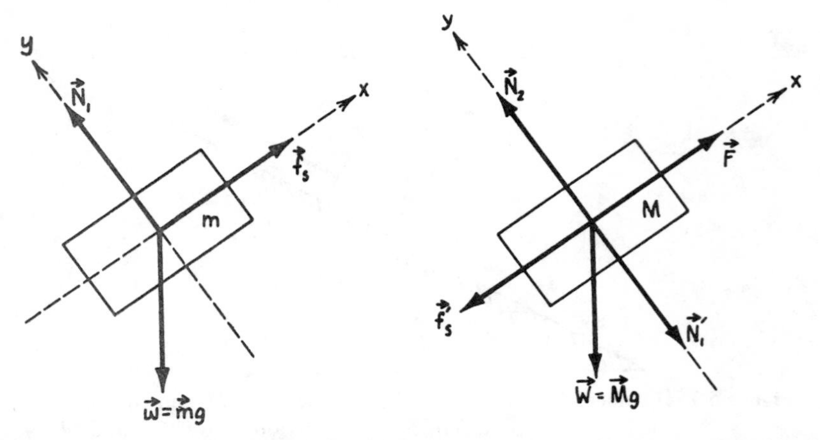

Figure 6.8 Problem 3. The "free-body" diagrams for (a) block m and (b) block M.

$$\mu_s N_1 - mg \sin \theta = ma_{max}$$

Next we solve the y equation for N_1 and substitute this value into the x equation:

$$\mu_s mg \cos \theta - mg \sin \theta = ma_{max}$$
$$a_{max} = g(\mu_s \cos \theta - \sin \theta)$$

Next we construct an FBD for M, as shown in Figure 6.8b. The force N_1' is the reaction force to N_1, and f_s' is the reaction force to f_s. Note that these are action–reaction pairs, as defined by Newton's Third Law, and that they act on different bodies. The force N_2 is the normal force on the block exerted by the surface of the inclined plane. Once again we write the equations of motion for the directions x and y:

$$\Sigma F_x = F - f_s' - Mg \sin \theta = Ma$$
$$\Sigma F_y = N_2 - N_1' - Mg \cos \theta = 0$$

From Newton's Third Law we know that $f_s' = f_s$ and $N_1' = N_1$, and since we are looking at the condition $a = a_{max}$, $f_s' = f_{s_{max}}' = \mu_s N_1'$. Therefore,

$$\Sigma F_x = F_{max} - \mu_s mg \cos \theta - Mg \sin \theta = Ma_{max}$$

and

$$\Sigma F_y = N_2 - mg \cos \theta - Mg \cos \theta = 0$$

We can see that the x equation is sufficient for us to solve for the unknown F_{max}:

$$F_{max} = \mu_s mg \cos \theta + Mg \sin \theta + Mg(\mu_s \cos \theta - \sin \theta)$$
$$F_{max} = \mu_s(M + m)g \cos \theta$$

An alternative method is to draw an FBD for the two blocks as a single system after solving for a_{max}, as shown in Figure 6.9. Note that now the

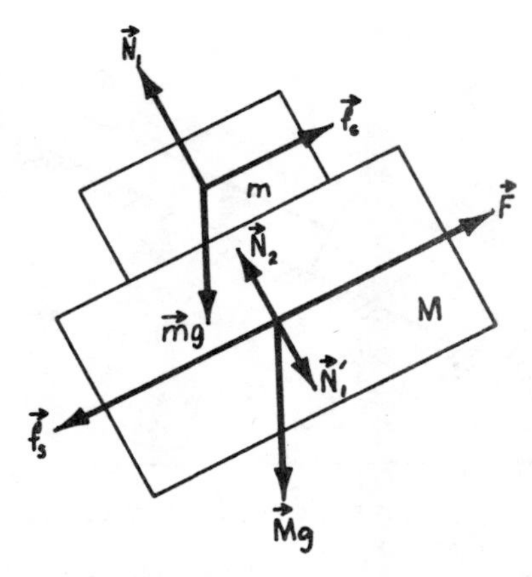

Figure 6.9 The "free-body" diagram for m and M taken as a system.

forces f_s and f_s', as well as N_1 and N_1', cancel because they are internal to the system. Thus,

$$\Sigma F_x = F_{\text{max}} + \cancel{f_s} - \cancel{f_s'} - (M + m)g \sin \theta = (M + m)a_{\text{max}}$$
$$\Sigma F_y = \cancel{N_1} - \cancel{N_1'} + N_2 - (M + m)g \cos \theta = 0$$

which leads to the same result as before.

Now let's examine the result. First we note that the maximum possible force becomes larger as θ approaches 0°. This is as expected, because there is a smaller component of the gravitational force down the incline tending to make the block slip. In addition, there is a larger normal force, so that $f_{s_{\text{max}}}$ is larger. We cannot extrapolate the result to 90°, however, because somewhere in the range from 0° to 90° the block will slip even if there is no acceleration. This occurs when $\tan \theta = \mu_s$ (compare Example 7, p. 139 in the text). If we call this angle the critical angle, θ_c, then

$$F_{\text{max}} = \frac{\sin \theta_c}{\cos \theta_c} (M + m)g \cos \theta$$

If $\theta = \theta_c$, then

$$F_{\text{max}} = (M + m)g \sin \theta_c$$

In other words, F_{max} must just balance the component of the total weight down the incline. At this angle, the block is already on the verge of slipping, so any acceleration at all will cause slipping. If $\theta > \theta_c$, the blocks no longer move as a unit, so we must alter our original equation.

Once again we note the power of the algebraic versus the numerical solution. The former provides a vehicle for exploring a wide range of conditions in the problem and is likely to result in deeper insights into just what is happening physically.

4. *Guided Problem.* A block of mass M is at rest on an incline that makes an angle θ with respect to the horizontal, as shown in Figure 6.10. A string

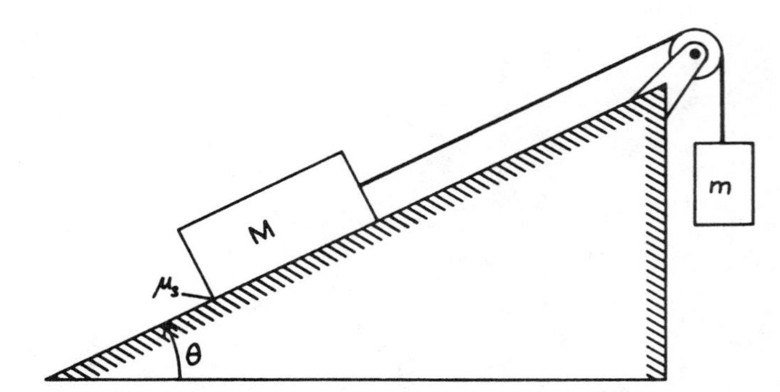

Figure 6.10 Problem 4. There is static friction between the block M and the surface of the plane.

attached to this block passes over a massless, frictionless pulley and is connected to a block of mass m, which is suspended freely. The coefficient of static friction between the block M and the surface of the plane is μ_s. Find the maximum and minimum values of m for which the block M will remain at rest.

Solution scheme

a. First explain why there should be a maximum and a minimum value for m—that is, why there should be a *range* of values.
b. For the situations in which m is minimum and maximum, respectively, determine the direction of the static friction force.
c. Draw an FBD for the block M, labeling all the forces acting on it. Do the same for m.
d. Write the equations of motion for each block. For M you will need both x and y equations.
e. The friction force has its maximum value under the conditions of the problem. Thus, it can be written in terms of the normal force.
f. Eliminate T from the equations, and solve for m.

Hint: If $\mu_s = 0$ you should have a single value for the required mass m. Does your answer reflect this?

5. *Worked Problem.* Figure 6.11 shows two ideal springs of unstretched lengths l_1 and l_2, respectively, attached to rigid walls. The walls are a distance L apart, where $L > l_1 + l_2$. The springs are connected to each other by a small ring. Find the distance from the left wall to the ring when the system is in equilibrium. Assume negligible sag in the system.

Solution: The ring is in equilibrium when the pull from each spring is the same. But the pull of the spring is just the spring constant k multiplied by the amount the spring is stretched. Thus

$$k_1x_1 = k_2x_2$$

where the x's represent the magnitude of the displacements from equilibrium. From Figure 6.11 we see that

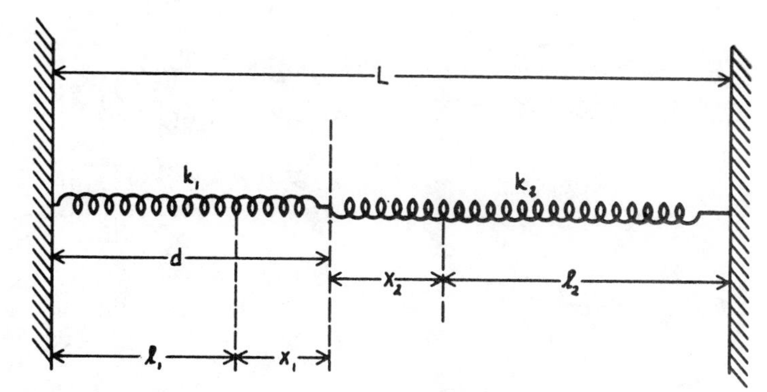

Figure 6.11 Problem 5. The springs are stretched from their equilibrium positions by x_1 and x_2, respectively.

$$x_1 = d - l_1 \quad \text{and} \quad x_2 = L - d - l_2$$

so that

$$k_1(d - l_1) = k_2(L - d - l_2)$$
$$d = \frac{k_2(L - l_2) + k_1 l_1}{k_1 + k_2}$$

Let's look at some special cases:

a. $l_1 = l_2 = l$ and $k_1 = k_2 = k$
This means that the springs are identical, so we should expect the ring to settle at a point midway between the walls.

$$d = \frac{k(L - l) + kl}{2k} = \frac{L}{2}$$

b. Suppose $k_1 \gg k_2$, which means that the left-hand spring is very much stronger. The ring should settle close to the relaxed position of this spring.

$$d = \frac{\frac{k_2}{k_1}(L - l_1) + l_1}{1 + \frac{k_2}{k_1}}$$

If $k_1 \gg k_2$ we see that $d \approx l_1$.

6. *Guided Problem.* A 2 kg block is accelerated to the right, as shown in Figure 6.12. The coefficient of kinetic friction between the block and the surface is 0.30. The pulley is frictionless. A massless spring of spring constant 200 N/m and a massless string connect the 2 kg block to a 3 kg block suspended freely. When the system has settled down so that the two blocks have the same acceleration, how much is the spring stretched from its equilibrium length?

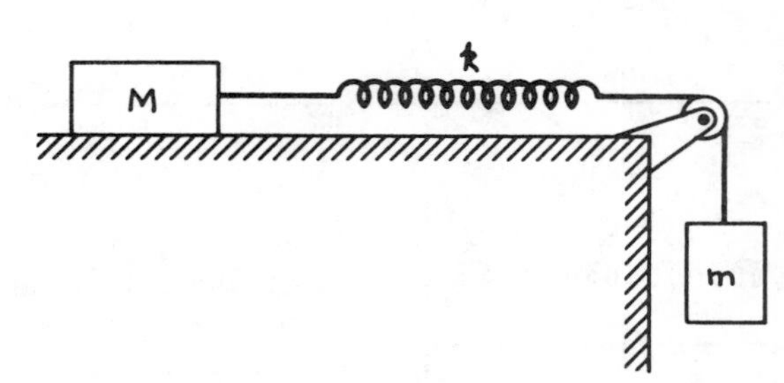

Figure 6.12 Problem 6. The two masses eventually acquire the same acceleration. What is the amount of stretch in the spring when this happens?

Solution scheme

a. Draw the FBDs for each object, including the spring.
b. Write down the equation of motion for each object.
c. If the spring mass is zero, what is the tension in the string at each end of the spring? What does this tell you about the tension in the spring itself?
d. Now solve each motion equation for T, and eliminate this variable to find a. Then use *one* of the motion equations to find T.
e. How much must the spring be stretched to produce this value of T?

7. *Worked Problem.* A model airplane is attached to a line that makes an angle θ with respect to the vertical, as shown in Figure 6.13. The plane is flying in a horizontal circle of radius R at speed υ. The lift force on the plane is perpendicular to the wings and makes an angle ϕ, as illustrated in the figure. Determine (a) the tension in the line and (b) the lift force.

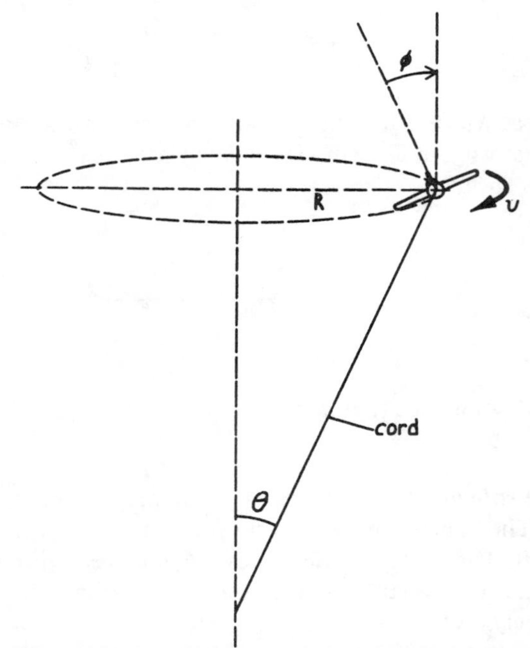

Figure 6.13 Problem 7. A model airplane flies in a circle of radius R with constant speed υ. The wings are at an angle ϕ with respect to the horizontal.

Solution: Because the plane is in uniform circular motion, the net force must be directed toward the center of the circular path and must have a magnitude equal to $m\upsilon^2/R$. Two forces have components centrally directed: the lift force L and the tension T. Because L and T are both unknown, we need two independent equations to solve for them. Therefore, our scheme in solving the problem is to sum the forces along both the x and the y directions, where $+x$ is radially inward and $+y$ is vertically upward. The FBD for the plane is shown in Figure 6.14. Writing the equation of motion for each coordinate, we have

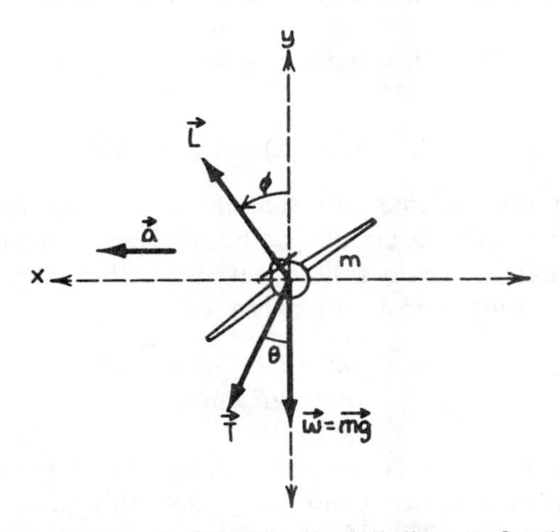

Figure 6.14 Problem 7. The "free-body" diagram for the airplane.

$$\Sigma F_x = L \sin \phi + T \sin \theta = ma = m\upsilon^2/R$$
$$\Sigma F_y = L \cos \phi - T \cos \theta - Mg = 0$$

or

$$L \sin \phi = m\upsilon^2/R - T \sin \theta$$
$$L \cos \phi = T \cos \theta + Mg$$

Dividing these two equations gives us

$$\tan \phi = \frac{m\upsilon^2/R - T \sin \theta}{T \cos \theta + Mg}$$

This equation can be solved for T:

$$T = \frac{m\upsilon^2/R - mg \tan \phi}{\cos \theta \tan \phi + \sin \theta}$$

Then the expression for L becomes

$$L = \left[\frac{(m\upsilon^2/R - mg \tan \phi) \cos \theta}{\cos \theta \tan \phi + \sin \theta} + Mg\right] \Big/ \cos \theta$$

These general results appear quite complicated, but let's examine them for some special cases to check the reasonableness of the solutions.

a. Suppose $\phi = 0°$. Then the expression for T becomes

$$T = (mv^2/R)/\sin\theta$$

or

$$T\sin\theta = mv^2/R$$

which is the same expression one gets for the conical pendulum. (Why should this be so?) Under these same conditions, we have

$$L = T\cos\theta + Mg$$

or

$$L = (mv^2/R)\cot\theta = Mg$$

b. Another situation of interest is that in which the tension no longer is needed to maintain the circular path, that is, the lift force L is sufficient. At what angle θ does this occur? From the expression for T we see that the numerator must be zero, or

$$0 = mv^2/R - mg\tan\phi$$
$$v^2 = gR\tan\phi$$

This result should be compared with that for the car rounding a banked curve (or conical pendulum). Does this make sense?

c. Now suppose $\phi = 90°$. This is an interesting case, because with L pointing toward the center of the circular path and T directed below the horizontal, there is no upward force component available to support the weight of the plane and the downward component of T. Let's substitute $\phi = 90°$ into the expression for T anyway and see what happens. First we rewrite the expression in a more convenient form:

$$T = \frac{\dfrac{mv^2/R}{\tan\phi} - mg}{\cos\theta - \dfrac{\sin\theta}{\tan\phi}}$$

If $\phi = 90°$

$$T = -\frac{mg}{\cos\theta} \quad \text{or} \quad -T\cos\theta = mg$$

In other words, T must be pointing in the opposite direction and have a vertical component equal to mg to balance the weight of the plane. If T is in fact in this direction, then

$$L = mv^2/R + mg\tan\theta$$

which means that L not only must provide the force necessary to maintain the circular path but must in addition overcome the outward component of T.

8. *Guided Problem.* A block of mass M is at rest on the surface of an incline at a distance of 10 cm above the base (Figure 6.15). The coefficient of static friction between the block and the plane is 0.85, and $\theta = 15°$. If the whole assembly is rotated about an axis AA' at n rev/s, what is the maximum value of n at which the block will not slip?

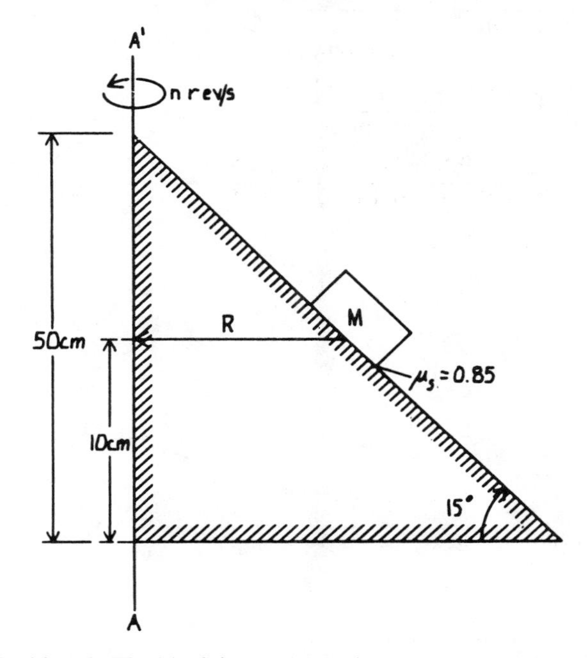

Figure 6.15 Problem 8. The block is at rest relative to the plane, which rotates at n rev/s about AA'.

Solution scheme

a. Set up a coordinate system such that the acceleration (radial) is along one of the axes.
b. Construct an FBD for the block. What direction is f_s?
c. Resolve forces along the x and y directions. Write the equation of motion for each coordinate axis.
d. Note that the object is on the verge of slipping when the static friction force has its maximum value, $f_{s_{max}}$. This gives f_s in terms of the normal force, N.
e. You should have two equations in N and v. Solve for v.
f. The number n is the inverse of the time taken for the block to complete one revolution. One revolution is $2\pi R$, and R can be found from the geometry of the situation.

Answers to Guided Problems

2. $$a = \frac{m_2 - m_1}{m_1 + m_2 + m_3} g$$

$$T_{13} = \frac{m_1(2m_2 + m_3)}{m_1 + m_2 + m_3} g$$

$$T_{23} = \frac{m_2(2m_1 + m_3)}{m_1 + m_2 + m_3} g$$

4. $M(\sin\theta - \mu_s \cos\theta) \leqslant m \leqslant M(\sin\theta + \mu_s \cos\theta) \rightarrow 2.9 \text{ kg} \leqslant m \leqslant 4.8 \text{ kg}$

6. $$x = \frac{(\mu_k + 1)Mm}{M + m}\frac{g}{k} = \frac{(0.30 + 1)(2 \text{ kg} \times 3 \text{ kg})}{(5 \text{ kg})} \times \frac{9.8 \text{ m/s}^2}{200 \text{ N/m}}$$
$$= 0.076 \text{ m} = 7.6 \text{ cm}$$

8. $$n = \frac{1}{2\pi}\left[\frac{g}{H - h}\,\frac{\mu_s - \tan\theta}{\mu_s + \cot\theta}\right]^{1/2}$$
$$= \frac{1}{2\pi}\left[\frac{9.8 \text{ m/s}^2}{(0.50 \text{ m} - 0.10 \text{ m})} \times \frac{0.85 - \tan 15^\circ}{0.85 + \cot 15^\circ}\right]^{1/2} = 0.28 \text{ rev/s}$$

Chapter 7

Work and Energy

Overview

The concept of a force developed in the preceding chapter allows us to examine the causes of motion and analyze motions resulting from known forces. We can gain further insights by examining the relationship between the forces and their displacements. We introduce the concept *work* to tie these variables together, and we shall see that work is connected in an intimate way to another concept, *energy*.

Essential Terms

work
energy
kinetic energy
potential energy
mechanical energy
work–energy theorem
joule
foot-pound
law of conservation of mechanical energy

Key Concepts

1. *Work* (*general definition*). Because the term *work* has so many colloquial meanings, it is important to keep in mind that *work* in physics (sometimes called *physical work*) is unambiguously defined as the product of *force* and *displacement*. The product may be a simple algebraic product or the more general dot (scalar) product. Because the latter includes the former (when the angle between the force and the displacement is zero), we can state generally that the work is the dot product of the force and the displacement. Here we see another instance of a physical quantity involving the product of two variables. Just as a particular value of momentum can be

obtained by a variety of combinations of mass and velocity, the same amount of work can be done by different combinations of force and displacement. This is the principle behind many "labor-saving devices," which combine a smaller force and a greater displacement to accomplish the same amounts of work. Unlike momentum, which is the product of a scalar and a vector, work is the product of two vectors. More specifically, work is the dot product of two vectors, which means that work is always a *scalar* quantity. Moreover, it means that we must consider not only the force and the displacement but also the angle between these quantities. Whereas a given mass and velocity fix the value of the momentum, a force and displacement of given magnitude may result in different amounts of work, depending on the relative orientation of the two vectors.

2. *Positive and negative work*. The definition of *work* as the dot product allows also for negative values of work. But you should not view the positive and negative signs as indicators of the direction of the work, because work is a scalar and has no directional properties. The force and displacement vectors themselves have directions, however, and the way in which these vectors are oriented relative to each other determines the sign associated with the work. A simple way to think of this situation is to consider the work to be the magnitude of the displacement multiplied by the *component* of the force along the line of the displacement. If that component is in the same direction as the displacement, the work is positive. If the component is opposite to the displacement, then the work is negative. Figure 7.1 illustrates this idea.

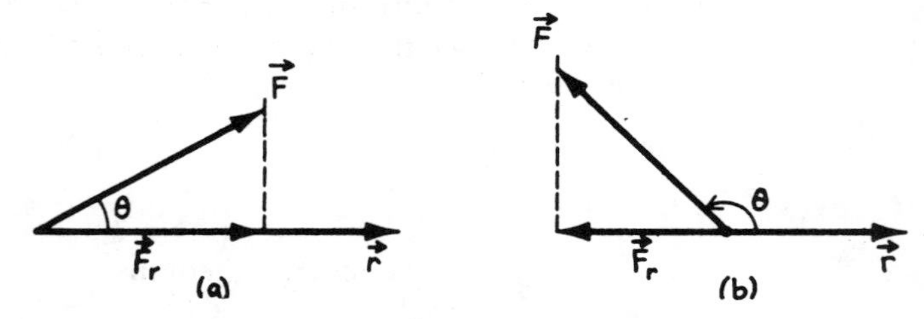

Figure 7.1 (a) The work done by the force is positive. (b) The work done by **F** is negative.

3. *Forces that do work and forces that do no work*. When the angle between the force and the displacement is 90°, the dot product is zero; that is, the force has no component along the direction of the displacement, and the work done is zero. There is another situation in which forces do no work. If the force causes zero displacement—for example, when we lean against a wall—no physical work is done. These are fairly straightforward consequences of applying the dot product rule. There are other situations, however, in which the results are not quite so clear. For example, when you walk you are propelled forward and slightly upward by the reaction force to your foot pushing against the ground. But there is no relative motion between your foot and the ground, so how can any work be done? Actually, no work is done by this force. Through a series of actions and reactions, this force is

transmitted to muscles that stretch and contract, and these forces do work on your body. This is sometimes called "internal work." Thus, although the reaction force at the ground is necessary for you to walk, this force itself does no work. These so-called workless forces exist in many situations. For example, in the acceleration of an automobile, the contact force between the tire and the road does no work, but this force is transmitted ultimately through a system of gears that produces the motion of the car.

4. *Units for work.* The SI unit for work (and for energy) is the *joule* (J), defined to be the work done by a 1 N force during a displacement of 1 m, if the force and the displacement are in the same direction. Otherwise, we must consider the component of the force along the direction of the displacement. Regardless, in the expression $W = \mathbf{F} \cdot \mathbf{x}$, if the force is in newtons and the displacement is in meters, the work will be in joules. In the British engineering system, the corresponding unit is the *foot-pound* (ft-lb).

5. *Work done by a variable force and the concept of average force.* In evaluating the work done by a force that does not remain constant throughout the displacement, we usually resort to the methods of the integral calculus. Actually we can mimic the procedures employed in the integral calculus by computing areas under curves, which really is what integral calculus is all about. For example, in Chapter 2 one way we could have determined the displacement would have been to compute the area under the velocity–time curve, because $\Delta x = \int_{t_1}^{t_2} \upsilon(t)dt$. It is generally true that if we plot a pair of variables, one a function of the other, on a graph, the area under the curve will represent a quantity whose unit is the product of the two separate units. Thus, a graph of $\upsilon(t)$ versus t gives displacement (velocity times time), $\mathbf{F}(x)$ versus $\mathbf{x}$ gives work (force times displacement), and so forth.

Let's look at the work idea a little more closely. We begin with the definition for work: $W = \int_{x_1}^{x_2} F(x)dx$. From the calculus we know that the average value of a function $f(x)$ is defined as

$$f_{avg}(x) = \int_{x_1}^{x2} f(x)dx \,/\, (x_2 - x_1)$$

or

$$f_{avg}\Delta x = \int_{x_1}^{x_2} f(x)dx$$

Comparing this with the work definition, we can see that the *average force* over the interval $\Delta x = x_2 - x_1$ is just

$$F_{avg} = \int_{x_1}^{x_2} F(x)dx/(x_2 - x_1)$$

But the integral is the area under the $F(x)$ versus x curve over the interval Δx, so that

$$F_{avg} = \frac{\text{area under } F(x) \text{ versus } x \text{ curve}}{x_2 - x_1}$$

In Figure 7.2 we show a variable force $F(x)$ as a function of x. The work done by this force over the interval x_1 to x_2 is indicated by the cross-hatched area. But note that the same *area* can be obtained from a rectangle of width x_2-x_1 and height $F'(x)$, that is, $F'(x)\Delta x =$ area. According to our earlier discussion, however, $F'(x)$ is just F_{avg}. Thus, the average force is that *constant* force which, acting over the same interval as the variable force, does the same amount of work. To point up the similarity between this and the velocity concept, we can define in the same way an *average velocity* as that *constant* velocity acting over the same time interval as the variable velocity that produces the same displacement (area under the velocity–time curve). Note, however, that we are dealing here with a *time-averaged* velocity, whereas the average force obtained by means of the work concept is a *displacement-averaged* force. We will see in Chapter 11 that we can define a *time-averaged* force that is related to the momentum change.

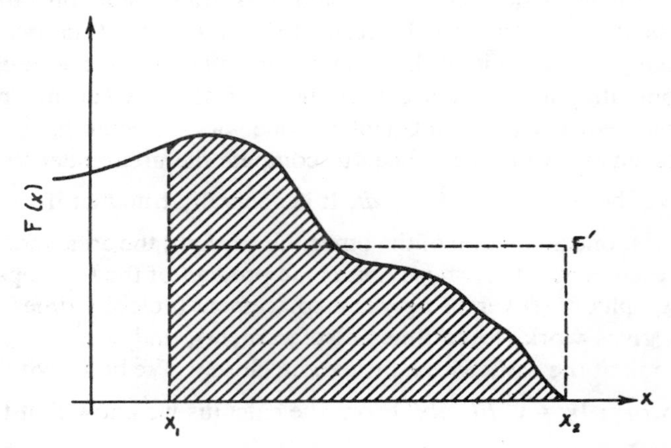

Figure 7.2 The cross-hatched area is equal to the area bounded by the $F(x)$ versus x curve and is the same as the area enclosed by the rectangle, which is $F'(x_2 - x_1)$. F' is the average force.

6. *Kinetic energy and the work–energy theorem*. The work done by the *net* (resultant) force on a particle is equal to the change in kinetic energy of the particle. Note the separation of extrinsic and intrinsic variables. In the expression $\int \mathbf{F}_{net} \cdot d\mathbf{x} = \Delta(\frac{1}{2}mv^2)$, we have on the left-hand side variables *extrinsic* to the particle—the externally applied force and the displacement. On the right-hand side we have the mass and the velocity—variables that are *intrinsic* to the particle. Furthermore, we see that the *change* in kinetic energy is a function only of the initial and the final speeds. It does not depend on the detailed way in which the force and the velocity behave during the acceleration, nor does it depend on the time. This result—that the change in kinetic energy is equal to the work done by the net force on the particle—is called the *work–energy theorem*. The theorem is applicable only to work done by the *net force*. If the net force is composed of more than one force, *the work–energy theorem cannot be applied separately to each force*.

What does it mean to say that the object possesses kinetic energy? Because reducing the speed of the particle decreases its kinetic energy, a deceleration will cause a negative change in the kinetic energy. In other words, negative work is done *on* the particle *by* the net force. As a consequence of Newton's Third Law, the particle must do positive work on its surroundings. Thus, the kinetic energy is equal to the work that the particle is capable of doing on its surroundings as it is brought to rest.

The units for kinetic energy (indeed, all energies) and work are the same, as they must be if the above discussion is valid: In the SI system, the unit is the *joule*, whereas in the British system it is the *foot-pound*.

7. *Potential energy*. We have shown that energy results from the performance of work; work also may result from an object's changing its position or orientation. A hammer dropping onto a nail, water running downhill, and a stretched spring snapping back all result in work being done. But this work results from the configuration or the position of the object. For example, the spring is doing no work *while* it is stretched, but we know that it can do work *because* it is stretched. Therefore, we say that the spring has *potential energy*. Similarly, if we raise an object to a higher elevation, we have done work on the object in lifting it against the gravity force. The object now has gravitational potential energy, which can be converted back to work as the object is allowed to return to its initial position. Because the gravitational force is always downward, it does negative work on the object as it is displaced upward. But displacing the object upward gives it more gravitational potential energy. Therefore, we define the *change* in the gravitational potential energy as the negative of the work done by the gravitational force in the displacement of the object.

It is not always necessary to identify the work that was done to get the object to a certain position. For example, a boulder sitting at the edge of a high cliff has potential energy relative to the ground below, but was it actually lifted to that position? The answer is that gravitational potential energy may be equated to the work that would have to be done *if* the object were lifted to that height. Thus, if 50,000 J of work would be required to lift the boulder to that height, then the boulder has a gravitational potential energy of 50,000 J relative to the ground.

This discussion brings us to another crucial point: the zero of potential energy is purely arbitrary, because we relate any work done to energy *changes*. It often is convenient to let the lowest position that an object will attain be the zero of the gravitational potential energy, but it is not necessary. Also, there is no unique value for the gravitational potential energy of an object at a given position. A book lying on a desktop has zero potential energy relative to the desktop, because no work can be performed by the book in being allowed to fall to the desk—it is already there. In contrast, if the book is allowed to fall to the floor, it can do work as it comes to this new position. So it has a non-zero potential energy relative to the floor. Therefore, we must always specify the potential energy with respect to some (arbitrary) location.

We have said that the rock, or the book, has gravitational potential energy. Actually, the energy is in the rock–Earth or book–Earth system. For example, when we lift the rock, we are separating the rock–Earth system and storing potential energy in the gravitational field, much as stretching stores energy in a spring. In most cases, however, we are considering masses that are very small compared with the mass of the Earth, so the two approaches are essentially equivalent.

8. *Mechanical energy and the conservation of energy.* If we combine the potential energy and the kinetic energy, we have what is called the *mechanical energy, E,* of the system. Furthermore, if the only force doing work on the system is the gravitational force, then the total mechanical energy is conserved. In other words, the energy E remains constant although both the kinetic and the potential energies may vary separately. In Chapter 8 we shall see that gravity is but one of a class of forces that result in the conservation of mechanical energy.

Sample Problems

1. *Worked Problem.* Figure 7.3 shows a force-versus-displacement curve. Determine the work done by this force in the interval from 0 to 7 m.

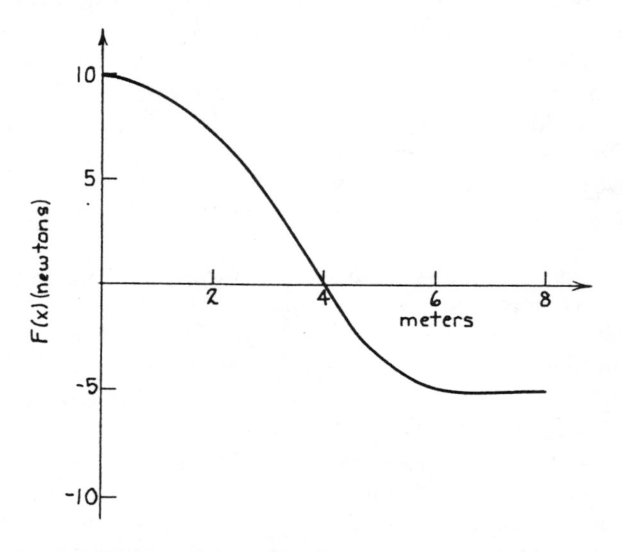

Figure 7.3 Problem 1. Force-versus-displacement curve.

Solution: Because this curve has no obvious mathematical form, we must resort to calculating the area under the curve. We can do this quite readily by drawing the curve on graph paper and counting squares. Note that the area corresponding to $F > 0$ represents positive work, whereas the area for $F < 0$ represents negative work. In this case the answer is approximately as follows:

Positive area ≈ 25 J
Negative area ≈ 10 J
Work done = 25 J − 10 J = 15 J

2. *Guided Problem.* Consider the force given by the equation $F(x) = 6$ N $+ 5x$ N. Calculate the work done by this force over the interval $x = 1$ m to $x = 5$ m.

Solution scheme: How is the work done by a variable force computed by means of the techniques of integral calculus?

3. *Worked Problem.* A one-dimensional force acting along the x axis has the form $F_x = 4(x^2 + 1)$ N, where x is the distance in meters from the origin. This force acts on a particle whose mass is 2 kg.

a. How much work does the force do in the interval from $x = 1$ m to $x = 3$ m?
b. If the particle is traveling 1.5 m/s at $x = 1$ m, what is its kinetic energy at $x = 3$ m?

Solution

a. $W = 4\int_1^3 x^2 dx + 4\int_1^3 dx = 4\left[\frac{x^3}{3}\right]_1^3 + 4[x]_1^3$

$W = 41$ J

b. Using the work–energy theorem, we get

$$\Delta K = 41 \text{ J}$$
$$K_0 = 2 \text{ kg} \cdot (1.5 \text{ m/s})^2 = 2.2 \text{ J}$$
$$K = K_0 + W = 41 \text{ J} + 2.2 \text{ J}$$
$$K = 43 \text{ J}$$

4. *Guided Problem.* A rope of mass per unit length μ and total length L is coiled on the floor. If one end of the rope is pulled slowly upward at constant speed, how much work is required to lift the end of the rope through a vertical height l, where $l < L$?

Solution scheme

a. Let the height variable be z. What is the weight of a piece of rope of length z, where z is the length off the floor?
b. How much force is required to lift this much rope?
c. The force will change as more rope is lifted off the floor. How do you compute the work done by a variable force over a specified distance?

5. *Worked Problem.* Figure 7.4 shows two projectiles of equal mass being fired with the same initial speed but at different angles with respect to the horizontal. Consider the kinetic energies of the projectiles at the positions marked *A, B,* and *C*. How do these kinetic energies compare with one another?

Solution: Each projectile is fired with the same kinetic energy, because kinetic energy is a function only of mass and speed, not of direction. The gravitational potential energy at z' is the same for each, because it is a function only of z. Therefore, if air friction is assumed negligible, total mechanical energy is conserved and the kinetic energies at *A, B,* and *C* will be the same.

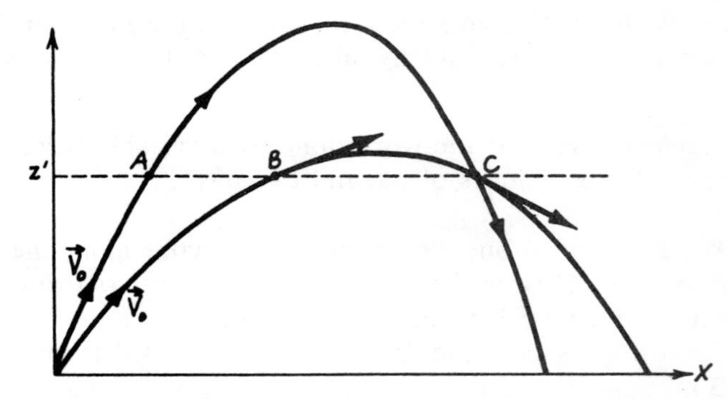

Figure 7.4 Problem 5. Equal-mass projectiles fired at different angles but with the same initial speed.

6. ***Guided Problem.*** A 2 kg mass hangs from a massless, ideal spring, as shown in Figure 7.5. The spring constant $k = 100$ N/m. How much work is required to pull the spring downward an *additional* distance of 15 cm?

Solution scheme

a. Determine the *initial* stretch of the spring when the mass is suspended from it.
b. How much work must be done to stretch the spring an *additional* 15 cm?
c. Does the weight of the mass suspended from the spring have anything to do with the work done in stretching the spring the additional amount?

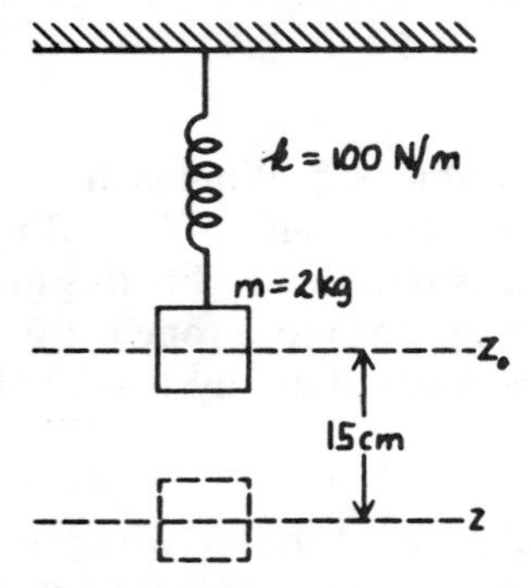

Figure 7.5 Problem 6. A mass suspended from an ideal spring at $z = z_0$ is pulled slowly downward an additional 15 cm.

7. ***Worked Problem.*** A 5 kg and a 3 kg mass are connected by a massless cord passing over a massless and frictionless pulley (Figure 7.6). The 3 kg mass is held to the floor and then released.

a. How fast is the 5 kg mass traveling when it hits the floor?
b. How high will the 3 kg mass go, assuming that the pulley does not interfere with its upward flight?

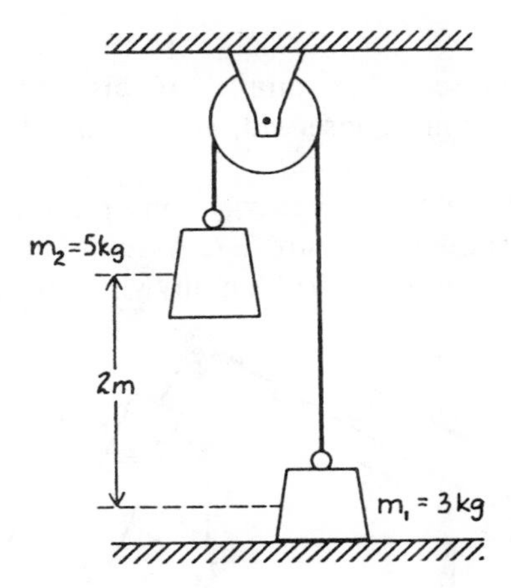

Figure 7.6 Problem 7. The mass m_1 is held to the floor, then released.

Solution: The only force doing work on this system is the gravitational force, so mechanical energy is conserved. Thus, our solution is as follows:

a. Let $z = 0$ and $U = 0$ at floor. $\Delta E = 0$, so

$$U_0 + K_0 = U + K$$
$$m_2gz_0 = m_1gz_0 + \tfrac{1}{2}(m_1 + m_2)\upsilon^2$$
$$2(m_2 - m_1)gz_0 = (m_1 + m_2)\upsilon^2$$
$$\upsilon^2 = \frac{2(m_2 - m_1)}{(m_1 + m_2)}gz_0 = \frac{2(2\text{ kg})(9.8\text{ m/s}^2)(2\text{ m})}{8\text{ kg}} = 9.8\text{ m}^2/\text{s}^2$$
$$\upsilon = 3.1\text{ m/s}$$

b. We now focus our attention on the 3 kg mass alone, after the 5 kg mass hits the floor. Why? Because now the 3 kg mass is the only one that has *only* a gravitational force acting on it. The other mass is at rest and is no longer undergoing any energy changes. Our solution is then to look at the energy changes of the 3 kg mass. Note that this mass is moving upward at 0.1 m/s when the other mass hits the floor. So

$$\not{m}_1gz_0 + \tfrac{1}{2}\not{m}_1\upsilon^2 = \not{m}_1gz$$
$$z = z_0 + \frac{\upsilon^2}{2g}$$
$$z = 2\text{ m} + \frac{(3.1\text{ m/s})^2}{2 \times 9.8\text{ m/s}^2}$$
$$z = 2\text{ m} + 0.5\text{ m}$$
$$z = 2.5\text{ m}$$

Comment: It might have been tempting to set the initial potential energy, at the beginning of the problem, equal to the final potential energy, when the 3 kg block is at its highest point. During this time the 5 kg block hit the floor, however, and work was done by forces other than gravity in bringing this block to rest. Thus, mechanical energy of the *total* system (the two blocks) is not conserved. Examples of this type of problem, in which mechanical energy is not conserved, are presented in Chapter 8.

8. *Guided Problem.* Calculate, using energy conservation principles, the acceleration of the system shown in Figure 7.7. Assume that the pulley and cord are massless and that there is no friction anywhere in the system.

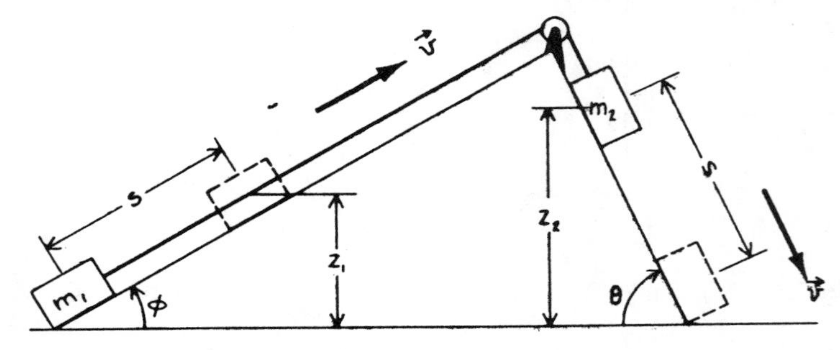

Figure 7.7 Problem 8. The two blocks slide without friction on the two surfaces as shown. Using the principle of the conservation of energy is one way to solve this problem.

Solution scheme

a. Establish a zero level for gravitational potential energy.
b. For simplicity, allow the system to start from rest. What is the total initial mechanical energy?
c. Assume that the cord cannot stretch. How does the speed of m_1 compare with that of m_2 at any time?
d. Let the blocks move through a distance s. How is s related to the angles θ and φ and the heights z_1 and z_2?
e. Now solve for υ. From your knowledge of kinematics, how are υ, a, and s related?
f. This problem was solved using Newton's Second Law in Chapter 6 (problem 6.1).

9. *Worked Problem.* A 10 kg block rests on a table. A strong but very light cord attached to the block passes over two pulleys, as illustrated in Figure 7.8. A 6 kg mass is attached to the other end of the cord 1.5 m below the pulley. From what angle (measured from the vertical) must the latter block be released so that the heavier block will just begin to lift off the floor?

Solution: Why would the heavier block be lifted from the floor by the lighter block? Recall that a force is required to keep an object moving in a circle, so that at the bottom of the path, the tension in the cord must not only support the weight but also provide an additional force (tension) for the circular motion. This tension is transmitted through the cord to the other block. To find the additional tension, we need to compute the centripetal force. To do so, we must know the speed, which we can derive by applying the principle of the conservation of mechanical energy.

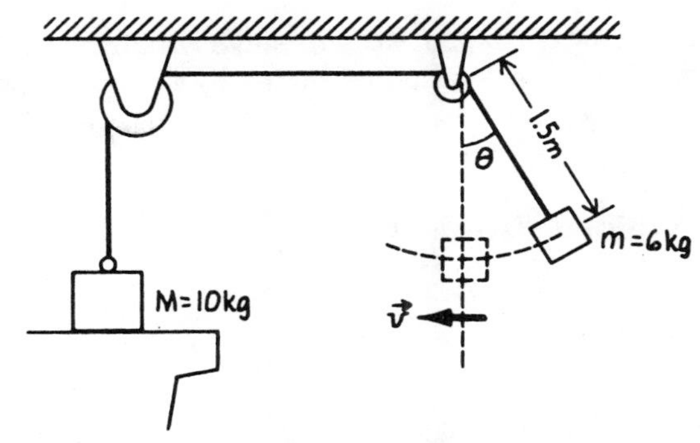

Figure 7.8 Problem 9. By allowing the smaller mass to swing in an arc, one can lift the larger mass from the table.

$$\Delta E = 0$$
$$U_0 + K_0 = U + K$$
$$mgz_0 = \tfrac{1}{2}m\,v^2$$

But

$$z_0 = l(1 - \cos\theta)$$
$$\not{m}gl(1 - \cos\theta) = \tfrac{1}{2}\not{m}\,v^2$$
$$v^2 = 2gl(1 - \cos\theta)$$

The tension T is

$$T = mg + mv^2/l = Mg$$
$$mg + m\,\frac{2g(1 - \cos\theta)l}{l} = Mg$$

A rearranging of terms gives

$$\cos\theta = \frac{3m - M}{2m} = \frac{18\text{ kg} - 10\text{ kg}}{12\text{ kg}} = \frac{2}{3}$$
$$\theta = 48°$$

10. *Guided Problem.* A 3 kg block slides without friction along a horizontal surface. Its speed is 5 m/s. It strikes an ideal massless spring (k = 400 N/m) and compresses the spring a distance x when the speed of the block is reduced to 2 m/s.

a. How much work is done on the spring by the block?
b. From your answer in part a, determine x, the compression of the spring.

Solution scheme

a. By how much is the kinetic energy of the block reduced? How does this change relate to the work done on the block by the spring? What is the work done on the spring by the block?
b. What is the work done by an ideal spring when it is compressed a distance x?

Comment: In the next chapter we will solve problems of this type using the principle of the conservation of energy by defining a potential energy for the ideal spring.

Answers to Guided Problems

2. $W = \int_1^5 F(x)dx = 84 \text{ J}$

4. $W = \int F(z)dz = \int_0^l \mu gz dz = \mu g l^2/2$

6. $x_0 = \dfrac{mg}{k} = 0.2$ m, so the range of x is from 0.2 m to (0.2 m + 0.15 m).

   $W = \int_{0.2 \text{ m}}^{0.35 \text{ m}} kx dx = 4$ J (The mass is needed only to compute x_0.)

8. $a = \dfrac{m_2 \sin \varphi - m_1 \sin \theta}{m_1 + m_2} g$

10. a. $W = \Delta K = K - K_0 = \frac{1}{2}m(v^2 - v_0^2) = -31.5 \text{ J}$ = work done by the spring on the block. Work done by the block on the spring is +31.5 J.

    b. $W = \frac{1}{2}kx^2$ for the spring, so $x = \sqrt{\dfrac{2W}{k}} = 0.4$ m.

Chapter 8

Conservation of Energy

Overview

The concepts of work and energy developed in the last chapter were treated within the limited context of gravitational potential energy and frictionless systems. The ideas presented there are only special cases involving energy conservation and the work–energy theorem. This chapter develops the more general approach.

Essential Terms

conservative force
potential-energy function
elastic potential energy
mass–energy equivalence
power
watt
horsepower
kilowatt-hour

Key Concepts

1. *Conservative force.* If a given force is a function of *position*, we say the force is *conservative*. The choice of this term is neither coincidental nor arbitrary. We see that when only conservative forces are acting (or when only conservative forces are doing work), the mechanical energy is *conserved*; that is, it remains constant. Another property of a conservative force is that it does zero work in a round trip (closed path) and it is derivable from a potential-energy function. All of these are equivalent definitions. Examples of conservative forces that we will be using quite extensively are the gravitational force and the elastic (Hooke's Law) force. In later chapters we will encounter other conservative forces, ones that are characteristically "field" forces, like the electrical force and the magnetic force.

2. *Potential energy (a second look)*. In Chapter 7 we showed that the work done by the gravitational force is equal to the negative of the change in gravitational potential energy. We now find that this result is part of a more general relationship, namely, that the work done by *any* conservative force is equal to the negative of the potential-energy change. If the conservative force obeys Hooke's Law, we have an associated *elastic potential energy*. In Chapter 22 we shall see that there exists an electrical potential energy associated with the electrical force.

3. *Potential energy of a spring (elastic potential energy)*. We saw in Chapter 7 that the work done by an ideal spring when it is stretched or compressed a distance Δx is $\frac{1}{2}k(x^2 - x_0^2)$. This expression is therefore the *change* in the elastic potential energy. If we arbitrarily set the potential energy equal to zero at $x = 0$, then the potential energy of the spring (or elastic system) is $U_{\text{elas}}(x) = \frac{1}{2}kx^2$. Note carefully the x^2 dependence. It is easy to make the mistake of writing $\frac{1}{2}k(x - x_0)^2$, but this is not the same thing. The latter involves the square of the difference, whereas the correct expression contains the *difference of the squares*. For example, if we stretch the spring from $x = 2$ cm to $x = 4$ cm, the increase in potential energy is $\frac{1}{2}k[(4 \text{ cm})^2 - (2 \text{ cm})^2] = \frac{1}{2}k(12 \text{ cm}^2)$ *not* $\frac{1}{2}k(4 \text{ cm} - 2 \text{ cm})^2 = \frac{1}{2}k(2 \text{ cm})^2$. We see also that the potential energy of the spring, defined in this way, is always positive, because whether we stretch or compress the spring, the restoring force and the displacement are in opposite directions. We can also think of the spring as being able to do work on its surroundings whether it is stretched or compressed.

4. *General work-energy theorem*. We have noted that the work done by the net force on a system is equal to the change in kinetic energy; that is, $W_{\text{net}} = \Delta K$. Let's look at a particular example. Consider a block moving along an inclined plane under the action of several forces. We can separate these forces into three convenient categories: the conservative forces, $\mathbf{F}_c$; the kinetic friction forces, $\mathbf{f}$; and any externally applied forces, $\mathbf{F}_{\text{ext}}$ (*other than* the conservative forces and the friction forces). The work done by all of these forces, according to the work–energy theorem, is

$$\int \Sigma \mathbf{F}_c \cdot d\mathbf{r} + \int \Sigma \mathbf{f} \cdot d\mathbf{r} + \int \Sigma \mathbf{F}_{\text{ext}} \cdot d\mathbf{r} = \Delta K$$

and since the work done by the conservative forces is equal to the negative of the potential energy change, we have

$$-\Delta U(r) + W_f + W_{\text{ext}} = \Delta K$$

or

$$W_f + W_{\text{ext}} = \Delta U + \Delta K = \Delta E$$

Now, the work done by the kinetic friction force is always negative, so it results in a *decrease* in the mechanical energy of the body. W_{ext} represents the work done by the *contact* forces acting externally to the body (including the normal force). Examples of such contact forces are the tension in a cord attached to the block, a hand pushing on the block, and so forth. W_{ext} can be either positive or negative.

Work done by the normal force in general is zero, but not always. If the displacement is *along* a surface, the normal force, being perpendicular to the surface, will do no work, because the force and the displacement are mutually perpendicular. In contrast, consider the situation shown in Figure 8.1, where the normal force is not perpendicular to the displacement, because the larger block is accelerating to the right. In this case $\mathbf{N} \cdot d\mathbf{r} \neq 0$, so $\int \mathbf{N} \cdot d\mathbf{r} \neq 0$.

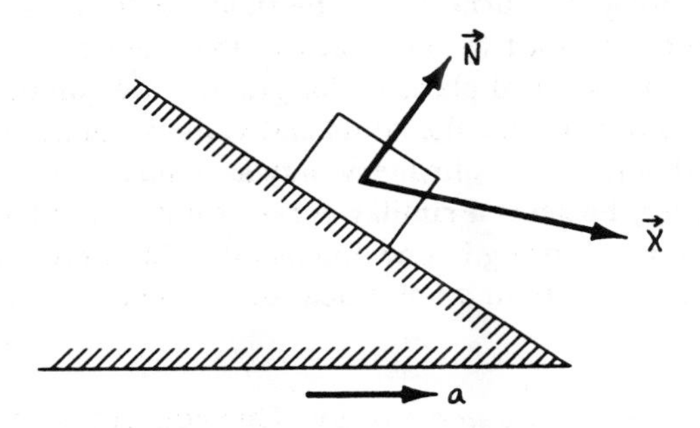

Figure 8.1 The normal force **N** does work, because the force and the displacement are not mutually perpendicular when the system is accelerating as shown.

In general, then, we can write a more complete work–energy theorem in the following form:

$$W_f + W_{\text{ext}} = \Delta E = \Delta U + \Delta K$$

or

$$U_0 + K_0 + W_f + W_{\text{ext}} = U + K$$

The latter form of the theorem is quite useful in solving problems. Let's examine this form a little more closely. The body starts at some point (arbitrary) with potential and kinetic energy. How this total mechanical energy changes depends on the work done by the nonconservative forces **f** and $\mathbf{F}_{\text{ext}}$. Let's consider them separately. Since W_f is always negative, it will "rob" the body of some of its mechanical energy; that is, the sum $U + K$ on the right will be less than $U_0 + K_0$. But if W_{ext} is positive, the sum $U + K$ will be larger than $U_0 + K_0$. If W_{ext} is negative, the sum will decrease. Often the two terms W_f *and* W_{ext} are combined and written W_{nc}, where *nc* means *nonconservative*. Then the general work–energy equation is written

$$U_0 + K_0 + W_{\text{nc}} = U + K$$

Comment: Note that U_0, K_0, U, and K are values of these variables at specific locations or times. In contrast, W_{nc} represents work done during (or over) the interval by nonconservative forces. Thus, to use this expression, we evaluate the potential and the kinetic energies at, for example, some initial position, and then again at some later position. If anywhere in this interval work has been done by nonconservative forces, this work must be included in the work–energy equation. (See sample problems 7 and 8.)

5. *The curve of potential energy*. When analyzing the curve of potential energy to see whether the object is in a "bound orbit," we must take care not to mistake the graph for the actual two-dimensional path of the particle. The graph merely shows the *energy* as a function of position. Because the *gravitational* potential energy is proportional to the height z, however, we can imagine a frictionless surface shaped like the potential-energy curve with a particle moving along this surface. The particle would indeed come to rest at the turning points and start to slide back in the other direction. Thus, in the case of motion in a vertical plane under gravity, we can pretend that the potential-energy curve is in fact the "hills and valleys" along which the particle moves. One should use this imagery only as a means of conceptualizing the motion, however, because a similar curve results from the motion of a particle bound to an ideal spring in one dimension. The curve may also represent the motion of a particle in an elliptical orbit about a center of gravitational attraction.

6. *The equivalence of mass and energy*. This concept is developed more fully in Chapter 41, but it is worth noting here that mass is not converted *to* energy, or vice versa. Rather, a transformation from one kind of energy into another takes place. Although mass and energy are essentially one and the same, mass usually is expressed in *kilograms*, whereas energy is expressed in *joules*. The relationship $\Delta E = \Delta mc^2$ gives us the *joule equivalent* of so many kilograms of mass. (The preceding argument holds for any consistent set of mass and energy units.)

7. *Power*. The *rate* at which work is performed or energy is transferred or expended is called the *power*. The term should not be used synonymously with energy, as in "a lot of power was used up in the process." The power indicates how quickly or how slowly the work is being carried out; it is not a description of a fundamentally different process. For example, 600 W of power is equivalent to work being done or energy being transformed at the *rate* of 600 J/s.

The equivalent expression for power, $P = \mathbf{F} \cdot \mathbf{v}$, describes the *instantaneous* power (the dot product of the instantaneous force and the instantaneous velocity). If the variables change with time, the power will change also. This expression for power should be used for finite periods of time *only* if the product $\mathbf{F} \cdot \mathbf{v}$ remains constant in time.

Sample Problems

1. *Worked Problem*. Suppose you have a potential-energy function given by

$$U(x) = \frac{a}{x^4} - \frac{b}{x^2}$$

a. Write the expression for the force $F(x)$.
b. Plot graphs of $U(x)$ and $F(x)$.

Solution:

a. Because U is a function of x, we can obtain the force by differentiation of the potential-energy function:

$$F(x) = -\frac{dU(x)}{dx} = \frac{4a}{x^5} - \frac{2b}{x^3}$$

The potential energy has an extremum when $\frac{dU(x)}{dx} = 0$; that is, when $F = 0$. This occurs at

$$\frac{4a}{x^5} - \frac{2b}{x^3} = 0$$

$$x^2 = \frac{2a}{b}$$

$$x = \sqrt{\frac{2a}{b}}$$

For example, if $a = b$, this occurs at $x = \sqrt{2} = 1.4$.

b. The graphs of $U(x)$ and $F(x)$ are shown in Figures 8.2 and 8.3, respectively. From these curves we see that $U(x)$ has a minimum value when $F = 0$.

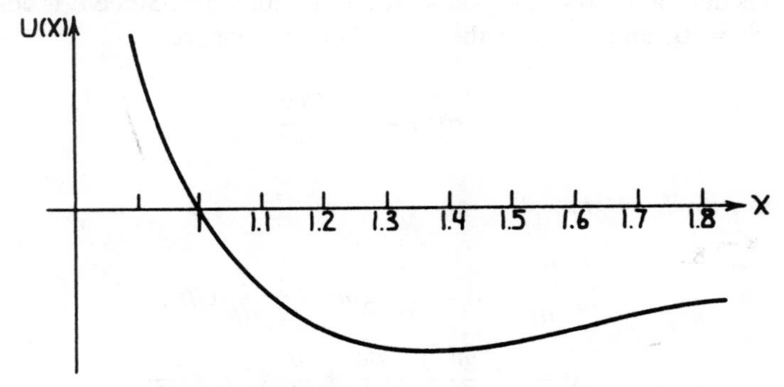

Figure 8.2 Problem 1. Plot of $U(x)$ versus x.

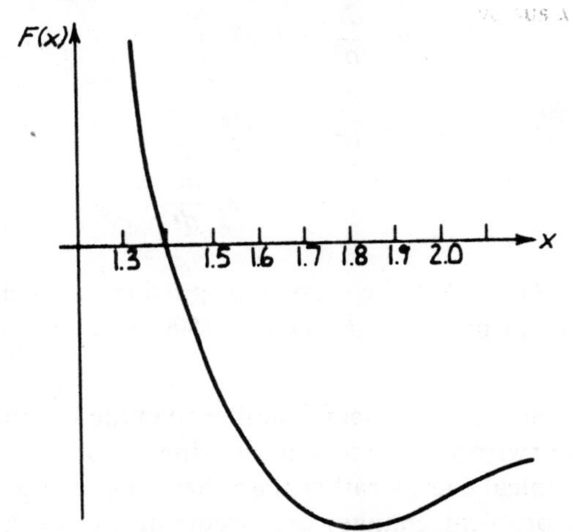

Figure 8.3 Problem 1. Plot of $F(x)$ versus x.

2. *Guided Problem.* Consider a particle of mass m moving under the influence of a force F_x such that the position of the particle at time t is given by $x(t) = A \cos \omega t$.

a. Calculate the force F_x.
b. What is the kinetic energy as a function of position?
c. If the potential energy is zero at $x = 0$, determine the potential energy $U(x)$.

Solution scheme

a. What is F_x in relation to the acceleration of the particle? How do you get $a(t)$ from $x(t)$?
b. How does the work–energy theorem relate the work done by the force to the change in kinetic energy?
c. Show that the force is conservative. Then calculate the potential energy.

3. *Worked Problem.* A particle is moving in one dimension under the action of a conservative force. The total energy of the particle is given by $E = K + U(x)$. Starting with this equation, derive Newton's Second Law.

Solution: Because the particle is under the influence of a conservative force, we know that the total mechanical energy is conserved and that the force is derivable from the potential-energy function. Since E is conserved, $dE{=}dt = 0$, and because the force is conservative,

$$F(x) = -\frac{dU(x)}{dx}$$

Then, with $K = \frac{1}{2}m\upsilon^2$,

$$\frac{dE}{dt} = 0 = \frac{d}{dt}(\tfrac{1}{2}m\upsilon^2) + \frac{d}{dt}U(x)$$

$$0 = \frac{d(\frac{1}{2}m\upsilon^2)}{d\upsilon} \cdot \frac{d\upsilon}{dt} + \frac{d}{dt}U(x) \cdot \frac{dx}{dt}$$

$$0 = m\upsilon\frac{d\upsilon}{dt} - F(x)\frac{dx}{dt}$$

$$0 = m\not{\upsilon}\frac{d\upsilon}{dt} - F(x)\not{\upsilon}$$

$$F(x) = m\frac{d\upsilon}{dt}$$

4. *Guided Problem.* A 30 kg mass is dropped from a height of 5.1 m onto a massless ideal spring whose k value is 1500 N/m. How far is the spring compressed?

Solution scheme: This problem is similar to guided problem 10 in Chapter 7, but this time we want you to solve it using the principle of the conservation of mechanical energy rather than the work–energy theorem.

a. In this problem changes are occurring in two kinds of potential energy. What kinds?

b. Establish a zero level for potential energy.
c. What is the form of the total mechanical energy at the beginning and at the end of the motion?

5. *Worked Problem.* A father wants to give his young son a thrill by swinging him all the way around the support bar of a swing without the chains going slack (Figure 8.4). The chains (two) are 3 m in length, and the boy weighs 245 N. The father knows that some energy will be dissipated through friction, so he plans for the swing to start with an initial speed faster (by 2 m/s) than that required to just make it around in the absence of friction. If he releases the boy 1 m above the lowest point of the swing, how fast must the boy be going at release?

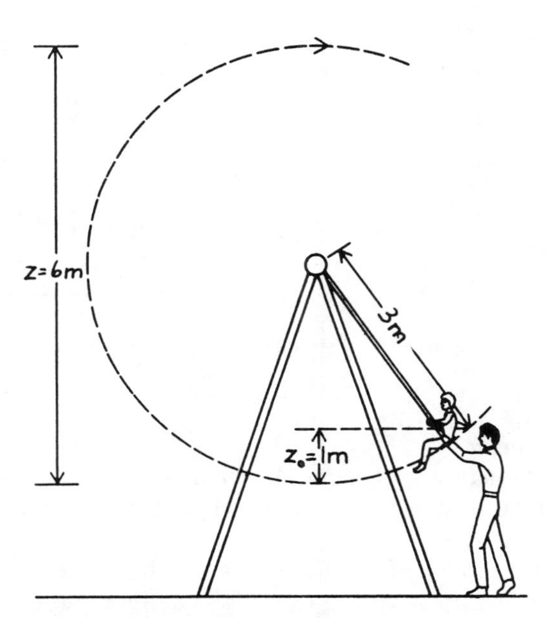

Figure 8.4 Problem 5. If the father pushes the boy fast enough, he can make the boy go around in a complete circle without falling out.

Solution: The general work–energy theorem states that $W_{nc} = (U + K) - (U_0 + K_0)$. If the father is starting the boy 2 m/s faster to compensate for the friction loss, then this loss must be equal to the extra kinetic energy he adds to the system. Formally, we have $W_{nc} = -\frac{1}{2}mv'^2$ where $v' = 2$ m/s. From Figure 8.4 we see that $z = 6$ m and $z_0 = 1$ m. If the chains are just about to go slack, $mg = mv^2/R$, or $v^2 = gR$. If v_0 is the required initial speed, then

$$-\tfrac{1}{2}mv'^2 = (mgz + \tfrac{1}{2}mgR) - (mgz_0 + \tfrac{1}{2}mv_0^2)$$
$$v_0^2 = 2g(z - z_0) + gR - v'^2$$
$$v_0 = 11.5 \text{ m/s (26 mph)}$$

6. *Guided Problem.* A roller-coaster car, having a mass of 950 kg when loaded, starts from rest at the top of a long ramp (Figure 8.5).

a. If the friction force is negligible, how fast is the car moving at its lowest point?
b. If the speed at this point is actually 25 m/s, how much energy has been dissipated through frictional losses?
c. The car enters a vertical loop-the-loop of radius 10 m. Assuming that the frictional losses from the bottom to the top of the loop equal 20% of the kinetic energy at the bottom, how fast will the car be traveling when it reaches the top of the loop? Will the car need an auxiliary rail to prevent its leaving the track?

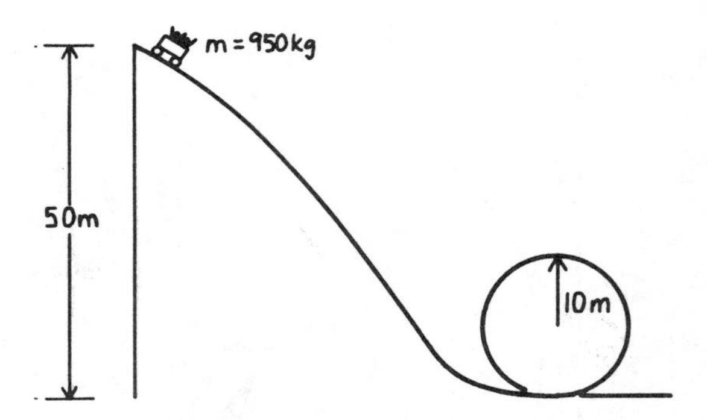

Figure 8.5 Problem 6. Roller coaster and loop.

Solution scheme

a. When there is no friction, mechanical energy is conserved. What is the relationship between the energy at the beginning and the energy at the lowest point?
b. In part b, how much of the total mechanical energy is missing?
c. The general work–energy theorem defines the relationship between the work done by the friction force as the car heads for the top and the initial kinetic energy at the bottom.
d. Solve for v at the top of the path. If the car is just to stay on the track, the speed must be $v^2 = gR$ (see problem 5). Is the speed you obtained at least this great?

7. *Worked Problem.* A horizontally directed force of 40 N pushes a 2.5 kg block from rest up a 37° incline as shown in Figure 8.6. The coefficient of kinetic friction is 0.25. When the block is 2 m higher than its starting point, how fast is it moving?

Solution: Because there is both a friction force and an externally applied nonconservative force, we apply the general work–energy theorem with $W_{nc} = W_f + W_{ext}$.

$$U_0 + K_0 + W_f + W_{ext} = U + K$$

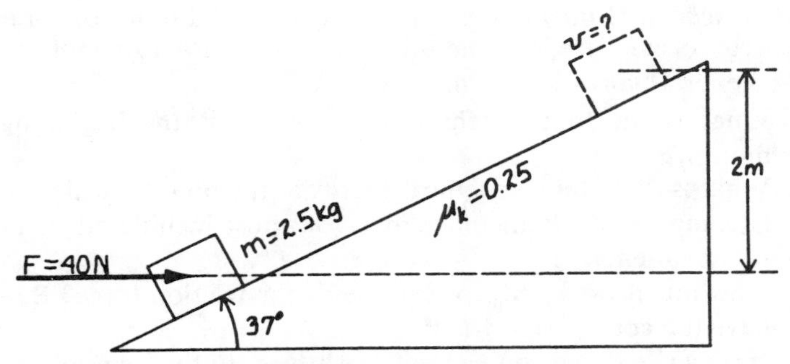

Figure 8.6 Problem 7. The block is pushed up the ramp by a horizontally directed constant force F.

If we take $U = 0$ at the initial position of the block, then

$$F \cos \theta \cdot s - \mu_k N s = mgz + \tfrac{1}{2} m v^2$$

But $N = mg \cos \theta + F \sin \theta$ and $s = \dfrac{z}{\sin \theta}$ so that

$$F \cos \theta \frac{z}{\sin \theta} - \mu_k (mg \cos \theta + F \sin \theta) \frac{z}{\sin \theta} = mgz + \tfrac{1}{2} m v^2$$

A rearranging of terms gives

$$v^2 = \frac{2Fz(\cot \theta - \mu_k)}{m} - 2(\mu_k \cot \theta + 1)gz$$

$$v = 4.1 \text{ m/s}$$

8. *Guided Problem.* A small cart having a mass of 65 kg starts from rest at a height of 5.1 m above the lowest point of a frictionless ramp (Figure 8.7). It goes down the ramp and back up to a height of 2.3 m, at which point the surface becomes horizontal. The cart then encounters a rough surface 6 m long, where the coefficient of kinetic friction is 0.30. Finally, after leaving the rough surface, the block encounters an ideal massless spring whose k value is 5×10^4 N/m. How far is the spring compressed?

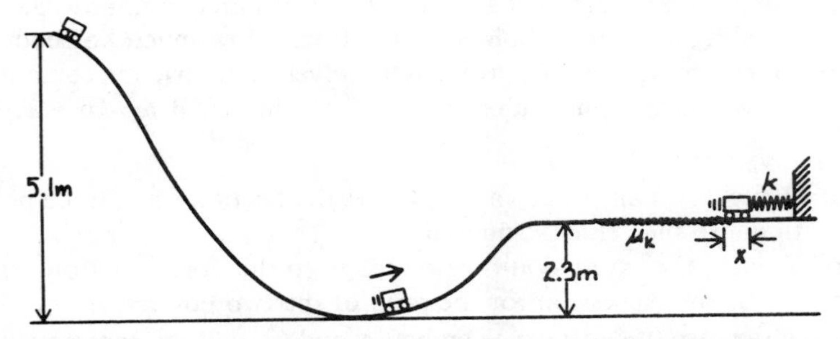

Figure 8.7 Problem 8. The cart slides along a frictionless ramp and over a rough surface, and compresses a spring in coming to rest.

Solution scheme: Many interchanges of energy and some loss of mechanical energy occur here, but the problem is solved very simply by means of the general work–energy theorem.

a. What is the total mechanical energy E at the beginning of the motion?
b. What is the total mechanical energy at the point at which the spring is compressed? Remember that you must include all forms of the potential energy.
c. How much work, W_{nc}, was done by the friction force? Be sure you have the correct sign for W_{nc}.
d. Now write down the general work–energy theorem and substitute the answers to parts a, b, and c.

9. *Worked Problem.* A car having a weight w can drive at a maximum speed of 30 km/h up a uniform slope of 1.5°. If the friction force is 0.05 times the weight of the car and is the same for all speeds, how fast can the car go down the slope?

Solution: The maximum speed is dictated by the horsepower of the car. The power for constant force and speed is $F\upsilon$. As the car moves up the slope, both gravity and friction act *opposite* to the car's motion. Thus, the car, to travel at constant speed, must supply a force up the plane equal to $bmg + mg \sin\theta$, where $b = 0.05$. While the car is moving down the slope, gravity *aids* the motion but the friction force retards it, so the car must supply a force equal to $bmg - mg \sin\theta$. (For the conditions of the problem to hold, bmg must be greater than $mg \sin\theta$; otherwise the car would, without the aid of the engine, accelerate down the hill.) The condition that the power be the same in each case gives us

$$F_{up}\upsilon_{up} = F_d\upsilon_d$$

$$\upsilon_d = \frac{F_{up}}{F_d}\upsilon_{up} = \frac{bmg + mg \sin\theta}{bmg - mg \sin\theta}\upsilon_{up}$$

$$\upsilon_d = \frac{(0.05 + 0.026)}{(0.05 - 0.026)} 30 \text{ km/h}$$

$$\upsilon_d = 95 \text{ km/h}$$

10. *Guided Problem.* The resistance force for objects moving through air is approximately proportional to the velocity squared. Suppose an automobile traveling 40 km/h requires about 10 hp. How much horsepower is required at 60 km/h? Can this problem be solved if the friction force is proportional to a linear combination of υ and υ^2, that is, if $f = k\upsilon + k'\upsilon^2$?

Solution scheme

a. Write down an equation that shows the friction force to be proportional to the velocity squared.
b. What is the expression for power, given this force relationship?
c. Set up an expression for the ratio of the two powers.
d. In the case of dependence on both υ and υ^2, will the constants divide out in the ratio?

Answers to Guided Problems

2. a. $F_x = -m\omega^2 x$

   b. $\Delta K = -m\omega^2 \dfrac{x^2 - x_0^2}{2}$

   $K = K_0 - m\omega^2 \dfrac{x^2 - x_0^2}{2}$

   c. $\Delta K + \Delta U = 0 \rightarrow \Delta U = m\omega^2 \dfrac{x^2 - x_0^2}{2}$ or $U = m\omega^2 \dfrac{x}{2}$

4. $x = 0.47$ m
6. a. 31 m/s

   b. 1.7×10^5 J

   c. 10.4 m/s, $v = \sqrt{gR} = 9.9$ m/s for just making it around the loop. Thus, the speed is sufficient (but just barely!) to eliminate the need for auxiliary rails.
8. $x = 0.22$ m
10. 34 hp; no

Chapter 9

Gravitation

Overview

In the preceding chapters we often made use of the gravitational force exerted on an object by the Earth, a force we call the *weight*. But what is the nature of this force? In this chapter we examine the more general situation in which there is a mutual gravitational attraction between two arbitrary bodies. We shall see that our expression for weight derived earlier, as well as the expression for the gravitational potential energy, has a more general form but is still valid for the situations in which it was first derived. Applications to the motions of large astronomical bodies are explored in this chapter also.

Essential Terms

law of universal gravitation
gravitation constant
Kepler's laws
apogee
perigee
aphelion
perihelion
gravitational potential energy
orbital energy
escape velocity
principle of equivalence

Key Concepts

1. *Newton's law of gravitation*. The most important thing to remember about this law is that it describes the forces between *particles*. It is not, in general, valid for extended bodies. Bodies possessing spherical symmetry do behave as if they are point particles gravitationally, but this behavior extends only to regions outside these bodies. For other shapes, one must resort to the integral calculus.

Occasionally the distance r in the gravitation law is taken to be the distance between the centers of mass. But consider this concept in relation to Figure 9.1, which shows the cross section of a coffee cup with its center of mass indicated. A coin whose center of mass is located at its geometric center is slowly lowered into the cup. As the center of gravity of the coin approaches that of the cup, the force of attraction becomes infinite! The error is in treating these objects as point particles, when in fact they are not. The infinite-force situation is not possible for real objects, because the particles cannot be in the same location simultaneously. If the coin and the cup are very far apart compared with their dimensions, the particle approximation is fairly good. For the same reason, the force exerted by one asteroid on another thousands of miles away obeys the gravitation law very well, even though the bodies are neither particles nor necessarily spherical. Even spherically symmetric bodies do not have to be homogeneous for us to treat them as mass points gravitationally. They must be spherically isotropic, however; that is, any variations in density must be the same along any radius.

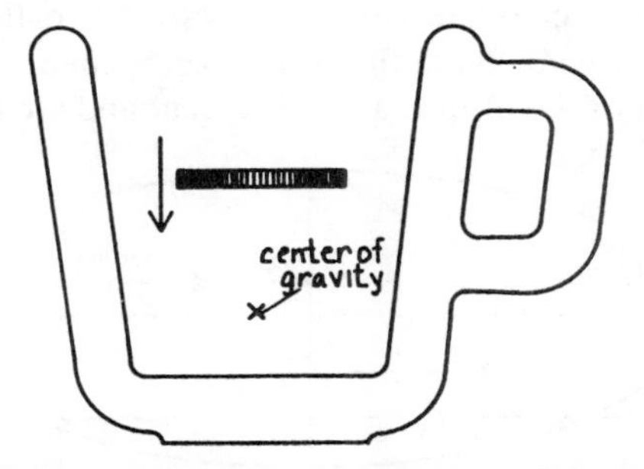

Figure 9.1 When the centers of gravity of the coin and the cup approach each other, does the force become infinite?

The term g, which we have called the acceleration of gravity, has another meaning, one that parallels closely a development described in Chapter 23, on electricity. The gravitational force can be written

$$\mathbf{F}_g = m\mathbf{g} = G\frac{M_E m}{R_E^2}\,\hat{\mathbf{r}}$$

Now, if we express this as

$$\mathbf{g} = \frac{\mathbf{F}_g}{m} = \frac{GM_E}{R_E^2}\,\hat{\mathbf{r}}$$

we have the force per unit mass. In this regard, $\mathbf{g}$ is called the *gravitational field intensity* and is expressed in newtons per kilogram. Thus, we have a *force field* in the vicinity of the mass that we can map by measuring the force on a unit mass placed in the field at the various points. It is important to recognize the distinction between $\mathbf{g}$ and G. The former is a variable that depends both on the attracting mass and on the position of the attracted body relative to this mass. In contrast, G is a universal *constant*, independent of any variables. It is the proportionality constant connecting the force to the other variables. In the SI system, the unit for G is $N \cdot m^2/kg^2$, so that the masses and the separation distance must be expressed in kilograms and meters, respectively.

2. *Circular orbits*. The expression derived for the period of a satellite in orbit about the Earth, for example, has no relationship to the rotation of the Earth. The derivation of the equation assumed a force between *particles*, and it is only because the Earth can be treated as a particle gravitationally that the equation is valid. So the periods, speeds, and the like of satellites are calculated with respect to an inertial reference frame *not* rotating with the Earth. For example, the 27,000 km/h speed required for a satellite to be in close orbit about the Earth is independent of the rotation speed of the Earth. In fact, if the Earth had the same angular speed as the satellite, bodies on the equator would be in orbit (weightless) also.

3. *Elliptical orbits*. The geometry of the ellipse is illustrated in Figure 9.2. The ellipse is characterized by a variable called the *eccentricity, e,* defined as the ratio c/a. Note that if c is zero, then e is zero as well. In this case the two foci of the ellipse are coincident and we have a circle. If $c = a$,

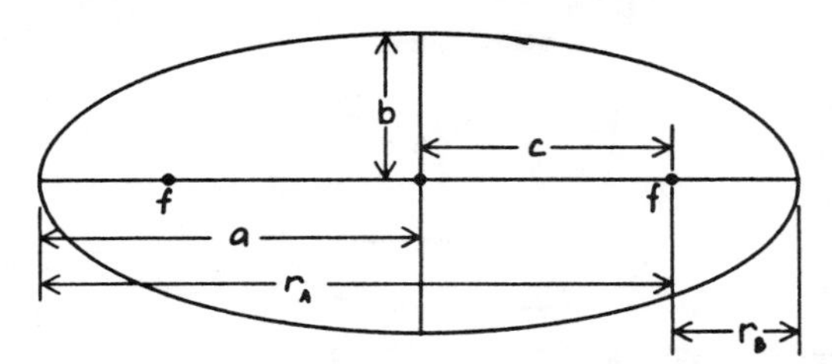

Figure 9.2 The geometry of the ellipse.

then $e = 1$ and the ellipse degenerates to a straight line. For elliptical orbits the distances r_A and r_B are called the *aphelion* and *perihelion*, respectively, if the body at the focus is the Sun (Gr. *helios*, "sun"). If it is the Earth, then these distances are called the *apogee* and *perigee* (Gr. *geos*, "earth"). The eccentricity of the ellipse can be expressed in terms of the orbital parameters r_A and r_B:

$$e = \frac{r_A - r_B}{r_A + r_B}$$

The eccentricities for all the planets are not very large, with Mercury and Pluto having the largest at $e = 0.21$ and 0.25, respectively. All others have e values less than 0.1. Other useful relationships include the *semimajor axis*, $a = \frac{1}{2}(r_A + r_B)$, and the *semiminor axis*, $b = \sqrt{r_A r_B}$.

The expression for the period of an elliptical orbit looks the same as that for a circular orbit (which is an elliptical orbit with a equal to zero) except that the radius of the orbit is replaced with the length of the semimajor axis. Because for most planets $e \ll 1$, the semimajor axis is almost the same as the radius of the approximate circle. For satellites, however, including comets and asteroids, these orbits can be highly elliptical. The eccentricity of Halley's comet is 0.97, close to a straight line.

In the expression $T^2 = \frac{4\pi^2}{GM} a^3$ it is assumed that the central attracting body M is fixed and the planet or satellite moves in an elliptical orbit around it. If the two masses are comparable, however, then $T^2 = \frac{4\pi^2}{G(M + m)} a^3$, where m is the second mass—the satellite or planet. This result comes from the mechanics of two-body interaction and uses a concept called the *reduced mass*. In this conception each body sees the other moving about it in an elliptical orbit with a period derived as we have described. In the case of circular orbits, it is easy to show (as we do in problem 9) that if we view the situation as two masses revolving around a common center of gravity, we get the same result.

4. *Gravitational potential energy*. In Chapter 7 we introduced the concept of gravitational potential energy (GPE), expressed in the form $U = mgh$. In the present chapter we define the GPE as $U = -\frac{GMm}{r}$; in this expression not only is the form of the definition different, but the signs are reversed! This latter form is more general and is based on a "natural" zero of potential energy, namely, that point at which $F \to 0$. This basis is somewhat analogous to that used to define the potential energy for a spring. It is still *differences* in potential energy that are important, however. So as we separate the two bodies, ΔU is still positive. In fact, we could (and often do) define the potential energy of a body relative to the Earth as

$$U(r) = GM_E m\left(\frac{1}{R_E} - \frac{1}{r}\right)$$

so that $U(R_E) = 0$ and $U(\infty) = GM_E/R_E$. This definition merely reflects a shift in the location of the origin. For example, the work required to move a body of mass m from the surface of the Earth to very far away is

$$U(r) - U(R_E) = \left(-\frac{GM_E m}{\infty}\right) - \left(-\frac{GM_E m}{R_E}\right) = \frac{GM_E m}{R_E}$$

or

$$U(r) - U(R_E) = GM_E m\left(\frac{1}{R_E} - \frac{1}{\infty}\right) - GM_E m\left(\frac{1}{R_E} - \frac{1}{R_E}\right)$$
$$= GM_E m/R_E$$

The same result is obtained either way.

Why did the expression mgh work in our earlier development? Equations 9.34 through 9.37 in the text show how, for $h \ll R_E$, the two expressions are equivalent.

The total mechanical energy, $U(r) + K$, for elliptical orbits is negative. Because the orbit is closed, this suggests that negative total energy is characteristic of bound orbits. That is indeed true. All orbits resulting from the gravitational attraction are *conic sections—ellipses, hyperbolas,* or *parabolas*. Only the ellipse is a bound orbit. The parabolic "orbit" ($E = 0$) is the dividing line between bound and unbound; theoretically, any infinitesimal change in the energy from $E = 0$ will cause the orbit to be bound or unbound, depending on whether dE is negative or positive, respectively.

5. *Spherical mass distributions*. We have discussed the significance of spherical mass distributions, but one point needs to be emphasized further. The text shows that a spherical shell exerts no force on a body placed *inside* the shell. This fact does not mean that such a shell acts as a gravitational screen, however. The field from any other body, inside or outside the shell, will exert a force on the body. The body feels no force only *from the shell*.

Another way of explaining the zero force inside the shell is to consider any point inside the shell and the elements of mass subtended by the same solid angle in opposite directions from the point, as illustrated in Figure 9.3. The gravitational field vector resulting from the mass element dm_1 is proportional to dA_1/r_1^2, and that resulting from dm_2 is proportional to dA_2/r_2^2. But the areas are proportional to r^2, so that $dA_1 \sim r_1^2$ and $dA_2 \sim r_2^2$. Therefore, $dg_1 = k\, r_1^2/r_1^2$ and $dg_2 = k\, r_2^2/r_2^2$. Thus, the contributions to g from the two opposite mass elements cancel. This is true for any point, so the g field inside the shell is everywhere equal to zero.

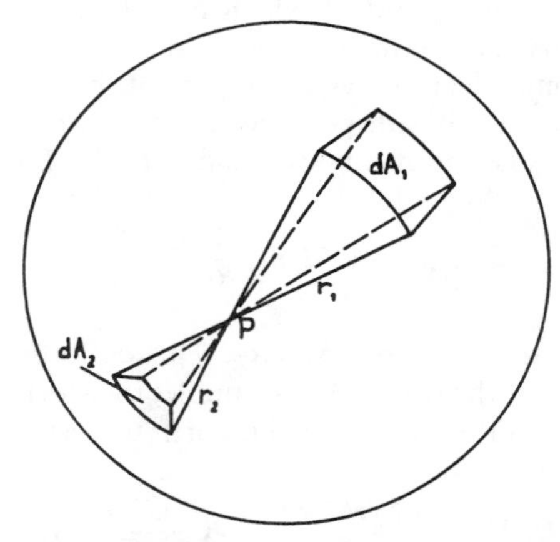

Figure 9.3 Geometry illustrating how diametrically opposite area elements from a spherical shell produce zero gravitational force at P.

6. *Principle of equivalence*. This principle states that no experiment can be performed whose outcome allows a distinction between the effects of acceleration and of local gravity. By local gravity we mean a uniform gravitational field existing over the region in which the experiment is performed. If the field varies significantly over the object in question, the force on a unit mass will not be the same in different parts of the field, a result different from that for uniform acceleration.

The nonuniformity of the gravitational field of the Moon at the Earth is responsible for the tides. Although the force exerted on the Earth by the Sun is, on the average, stronger than that of the Moon, the *difference* in force from one point to another results in tidal effects (basically a stretching effect caused by the difference in force). To see how small the tidal force is, consider the force exerted on a body of mass m by the Earth at two different altitudes:

$$F_g = \frac{GMm}{r^2}$$

Then

$$\frac{dF}{dr} = -\frac{2GMm}{r^3} = -\frac{2}{r}F$$

$$\frac{dF}{F} = -\frac{2}{r}\,dr$$

Integrating, we get

$$\ln F/F_0 = -\ln (R/R_0)^2 = \ln (R_0/R)^2$$

or

$$F = F_0\left(\frac{R_0}{R}\right)^2$$

where $R = R_0 + h$ and h is the small change in r. Then

$$F - F_0 = F_0\left[\left(\frac{R_0}{R}\right)^2 - 1\right]$$

$$\frac{\Delta F}{F} = \left[\frac{R_0^2}{(R_0 + h)^2} - 1\right] = \frac{-2R_0h - h^2}{(R_0 + h)^2}$$

Since $h \ll R_0$, $(R_0 + h)^2 \approx R_0^2$ and $h^2/R_0^2 \approx 0$, so that

$$\frac{\Delta F}{F_0} = -\frac{2h}{R_0}$$

For example, a 75 kg astronaut, 2 m tall, in low orbit about the Earth and with his long axis pointed radially away from the Earth would feel a difference in force from his feet to his head:

$$\Delta F = -F_0\frac{h}{R_0} = -735\ \text{N} \times \frac{2\ \text{m}}{6.4 \times 10^6\ \text{m}}$$

$$\Delta F = -2.3 \times 10^{-4}\ \text{N} = -0.000052\ \text{lb}$$

This is not much of a tidal force! However, the variation in the force across the diameter of the Earth resulting from the Moon's gravitational field is about 3%, enough to cause a noticeable effect. Because the Sun is so far away, the change in force across the Earth's diameter is only 0.009%, so the tidal effects are much weaker.

When astronauts are in Earth orbit, they are often described as "floating" in the capsule. This is a poor term to use, because it implies that something is supporting the astronauts, as when a balloon floats in air or a buoy floats in water. Even in this environment, however, there are small gravitational attractions between bodies in the capsule. The term *microgravity* has been coined by space scientists to indicate that, even in the noninertial frame of the space capsule, the effects of gravity are not entirely zero.

The so-called artificial gravity—induced by a rotating space station, for example—is like real gravity in its effects, according to the principle of equivalence. If a space station is in Earth orbit, we have a noninertial frame (the rest frame of the space station) in which the station is rotating about its axis. In the rest frame of the station (the frame orbiting the Earth but not rotating), there is no gravity, so the rotation must induce the artificial gravity. It is a general result that the directions "up" and "down" are related to the vector resultant of the gravity vector and the acceleration vector. Because in the rest frame of the station g is zero, "up" is in the direction of the acceleration vector associated with the rotation, which is toward the center of rotation.

Sample Problems

1. *Worked Problem.* The force that a person can exert against a surface by pushing with the legs (for example, by jumping) should be independent of local gravity as long as the attracting body (Earth or Moon) is extremely massive compared with the jumper. Thus, a person should be able to push away from the surface of the Moon or the Earth with the same force. Show that the ratio of heights attainable by jumping on the Moon and on the Earth is greater than 6:1, assuming that $g_M = (1/6)g_E$.

Solution: The acceleration during the jumping on each surface is

$$a_E = F - Mg_E$$

and

$$a_M = F - Mg_M$$

The speed as the jumper leaves the surface is

$$v_E^2 = 2a_E s$$
$$v_M^2 = 2a_M s$$
$$\frac{v_M^2}{v_E^2} = \frac{a_M^2}{a_E^2}$$

The height reached in each case is

$$h_M = v_M^2 / 2g_M$$

and

$$h_E = v_E^2 / 2g_E$$

or

$$\frac{h_M}{h_E} = \frac{v_M^2}{v_E^2} \cdot \frac{g_E}{g_M}$$

which gives

$$\frac{h_M}{h_E} = \frac{(F/M - g_M)}{(F/M - g_E)} \cdot \frac{g_E}{g_M}$$

Now $g_E/g_M \approx 6$, so

$$\frac{h_M}{h_E} = \frac{(6F/Mg_E - 1)}{(F/Mg_E - 1)}$$

which has the form

$$\frac{h_M}{h_E} = \frac{6X - 1}{X - 1}$$

where $X = F/Mg_E$. Note that $h_M/h_E > 6$. The limit 6 is obtained only when $X \gg 1$. But this means that $F \gg Mg_E$, the "Earth-weight" of the person, an unrealistic case. In fact, if $F = 2Mg_E$, we have $h_M/h_E = 12$.

2. *Guided Problem.* Two uniform spheres have densities ρ and ρ' and radii R and $3R$, respectively. If the two spheres are just touching and if no other body producing a gravitational field is in the vicinity, what is the force exerted on each sphere at the point of contact?

Solution scheme: Because the spheres are uniform and each is outside the other, they can be treated as particles.

a. Where is the mass effectively concentrated in each sphere with respect to the gravitational force?
b. If the spheres are just touching, what is the separation of the spheres?
c. Given the density and radius of each sphere, what is the mass?
d. How does the contact force compare with the gravitational force if the spheres are in equilibrium?

3. *Worked Problem.* A rod 20 m long, having a linear density λ of 500 g/m, lies along the y axis with its center at the origin, as shown in Figure 9.4a. A 2 kg uniform sphere lies on the x axis with its center of mass 3 m away from the rod. What force does the rod exert on the sphere?

Solution: Because the sphere is uniform, it can be treated as a particle at the center of mass. Figure 9.4b shows the relevant geometry. The gravitational force $d\mathbf{F}$ from the element of mass dm has both x and y components. But by symmetry each dm located symmetrically on opposite sides of the x axis will produce y components of $d\mathbf{F}$ that are equal and opposite, along with two equal x components. So the net force from these two symmetric elements is $(dF)_x = 2dF \cos\theta$. But

$$dF = GM\int \frac{\lambda dm}{r^2}$$

so that

$$dF_x = 2GM\lambda \int \frac{dm}{r^2} \cos\theta$$

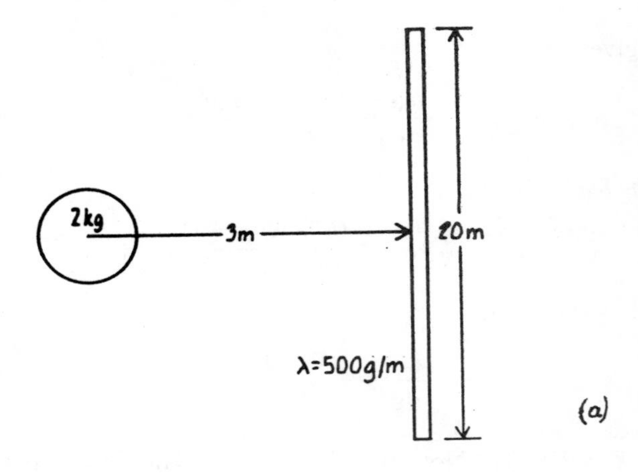

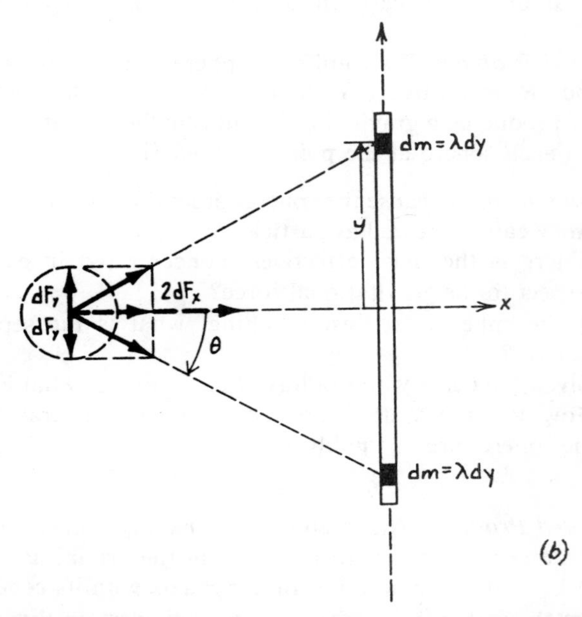

Figure 9.4 Problem 3. (a) A 2 kg spherical mass lies 3 m from a long, thin rod. (b) Each element symmetric about the x axis contributes the same x component of force. The y components cancel.

Now

$$y = x \tan \theta$$
$$dy = x\, d(\tan \theta) = x \sec^2 \theta d\theta$$
$$r^2 = x^2 + y^2 = x^2 + x^2 \tan^2 \theta = x^2 (1 + \tan^2 \theta)$$
$$r^2 = x^2/\cos^2 \theta$$

Then

$$dF_x = \frac{2GM\lambda}{x} \cos \theta d\theta$$

$$F_x = \frac{2GM\lambda}{x} \int_0^{\theta_m} \cos \theta d\theta$$

Note that because symmetric elements are included in dF_x, we need to integrate only from 0 to θ_m, *not* from $-\theta_m$ to θ_m, where

$$\tan \theta_m = \frac{L/2}{x} = \frac{10 \text{ m}}{3 \text{ m}} = 3.33$$

$$\theta_m = 73.3°$$

Then

$$F_x = \frac{2GM\lambda}{x} [\sin \theta]_0^{73.3°}$$

$$= \frac{2 \times 6.67 \times 10^{-11} \text{ N} \cdot \text{m}^2/\text{kg}^2 \times 2 \text{ kg} \times 0.5 \text{ kg/m } (0.96)}{3 \text{ m}}$$

$$F_x = 4.26 \times 10^{-11} \text{ N}$$

If the bar had been very long (in principle, infinitely long), the answer would be nearly the same (4.4×10^{-11} N), because then $\theta_m \to \pi/2$ and $\sin \pi/2 = 1$, not much greater than $\sin 73° = 0.96$. The problem here is identical to that of computing the force on a point charge located some distance from a line of charge.

4. *Guided Problem.* Show that the acceleration due to gravity at a height h above the surface of the Earth can be expressed as

$$g_h = \left(1 - \frac{2h}{R_0}\right) g_0$$

if $h \ll R_0$, where R_0 and g_0 are the values at the surface.

Solution scheme

a. Write separate expressions for g_0 and g_h in terms of the variables G, M_0, and $R_0(R_h)$.
b. Using $R_h = R_0 + h$, rewrite the expressions as $R_h = R_0(1 + h/R_0)$.
c. Take the ratio g_h/g_0.
d. If $h \ll R_0$, $h^2/R^2 \approx 0$. Also, use the approximation for small x that $\left(\frac{1}{1+x}\right)^n \approx 1 - nx$.

5. *Worked Problem*

a. Using the result from the preceding problem, show that the maximum range R of a projectile (ignoring air friction) increases with elevation above sea level according to the expression

$$\frac{dR}{R} = -\frac{dg}{g}$$

b. New Orleans is at sea level, and Denver is 1 mi above sea level. If a projectile has a range of 20 m in New Orleans, what is its *increase* in range in Denver, assuming that υ_0 is the same?

Solution

a. The maximum range of a projectile is given by

$$R = \upsilon_0^2/g$$

Then

$$\frac{dR}{dg} = -\frac{\upsilon_0^2}{g} = -\frac{1}{g}\frac{\upsilon_0^2}{g} = -\frac{R}{g}$$

which gives

$$\frac{dR}{R} = -\frac{dg}{g}$$

b. Integrating this equation, noting that

$$\int \frac{dx}{x} = \ln x$$

we have

$$\ln R/R_0 = -\ln g/g_0$$

But $e^{\ln x} = x$, so that

$$\frac{R}{R_0} = -\frac{g}{g_0}$$

$$R - R_0 = R_0(1 - g/g_0)$$

From problem 4,

$$g/g_0 = \left(1 - \frac{2h}{R_0}\right)$$

$$1 - g/g_0 = \frac{2h}{R_0} = \frac{2 \times 1.6 \text{ km}}{6.4 \times 10^3 \text{ km}} = 5 \times 10^{-4}$$

so that

$$R - R_0 = 20 \text{ m} \times 5 \times 10^{-4} = 0.010 \text{ m} = 1.0 \text{ cm}$$

6. *Guided Problem.* Consider a 1.0 kg pendulum bob suspended on a 2.0 m cord at the surface of the Earth. Assume the Earth to be spherical and isotropic. A large body 2.0 km away causes the pendulum to deviate from the vertical by 0.5°. What is the mass of the large body?

Solution scheme: The pendulum is in equilibrium under the influence of three forces: the cord, the gravitational pull of the Earth, and the gravitational pull of the large body.

a. Draw a "free-body" diagram showing the three forces.
b. The large body is not known to be spherically symmetric, but assume that the distance is such that the body may be approximated by a particle.

c. What is the expression for the gravitational pull from the Earth and from the other body?
d. Sum forces in the horizontal and vertical directions. Eliminate T from the two equations.

7. *Worked Problem.* A 100 kg rocket is placed in circular orbit at a distance of 1 Earth radius (R_E) from the *surface* of the Earth. The engines are then fired so that the rocket acquires a circular orbit 1 R_E higher.
a. How much work is required to achieve the new orbit?
b. What is the source of this work?

Solution

a. Solving this problem involves use of the work–energy theorem. Because the orbits are specified with respect to the surface of the Earth, the orbital radii are $2R_E$ and $3R_E$, respectively.

$$U_0(r) + K_0 + W_{NC} = U(r) + K$$
$$W_{NC} = \Delta U(r) + \Delta K$$

We have

$$U_0 = -\frac{GM_Em}{2R_E} \quad \text{and} \quad U = -\frac{GM_Em}{3R_E}$$

$$K_0 = \frac{GM_Em}{2(2R_E)} \quad \text{and} \quad K = \frac{GM_Em}{2(3R_E)}$$

$$W_{NC} = -\frac{GM_Em}{R_E}\left(\frac{1}{3} - \frac{1}{2}\right) + \frac{GM_Em}{2R_E}\left(\frac{1}{3} - \frac{1}{2}\right)$$

$$= \frac{GM_Em}{R_E}\left(\frac{1}{6} - \frac{1}{12}\right) = \frac{1}{12}\frac{GM_Em}{R_E}$$

$$W_{NC} = \frac{1}{12}\,\frac{6.67 \times 10^{-11}\ \text{N} \cdot \text{m}^2/\text{kg}^2 \times 5.98 \times 10^{24}\ \text{kg} \times 100\ \text{kg}}{6.4 \times 10^6\ \text{m}}$$

$$W_{NC} = 5.2 \times 10^8\ \text{J}$$

b. The work can come only from the internal energy of the rocket, which in this case is in the form of the chemical energy stored in the fuel.

8. *Guided Problem.* A 450 kg satellite is to be placed in a circular orbit 2.0 km above the Earth's surface.
a. What will be the satellite's period?
b. Compute the linear speed of the satellite.
c. How much work is required to place it into orbit? How much of this work goes into the potential-energy change, and how much goes into the kinetic-energy change?

Solution scheme: This problem deals with Kepler's Third Law, angular motion, and the work–energy theorem.

a. This is a near-Earth orbit, with $\mathbf{a} = R_E + h$. The period can be found from Kepler's Third Law.
b. What is the relationship between the period T and the angular velocity, ω? Relate ω to the linear velocity, υ.

c. Use the procedures in problem 7 to compute the work, W_{NC}. Compute the ΔU and ΔK terms separately. Note that the work term must be positive. Why?

9. *Worked Problem.* Two stars, one having mass 5 M_s and the other having mass 8 M_s, orbit about their common center of gravity in circular orbits. If the stars are separated by a distance of 5 AU, what is the period of rotation? M_s is the mass of the sun, and 1 AU is called the *astronomical unit* (the mean Earth–Sun distance), 1 AU = 1.5×10^8 km.

Solution: As we discussed in the Key Concepts section of this chapter, the situation can be viewed as a rotation about the center of mass, or as a rotation about one mass fixed, the other being represented by a reduced mass. We stated that, for the latter case, Kepler's Third Law is given by

$$T^2 = \frac{4\pi^2}{G(M+m)} a^3$$

We will show that the former view gives the same result. We can locate the center of mass relative to one of the stars by the following procedure. Let $M = 8\ M_s$ and $m = 5\ M_s$. Then, if the origin is located at the center of mass, we have

$$mr_1 = Mr_2 = M\ (D - r_1)$$

where D is the distance of separation. Then

$$mr_1 = MD - Mr_1$$

$$r_1 = \frac{M}{M+m} D$$

But r_1 is the orbital radius for m. The gravitational force on m is just

$$F = \frac{GMm}{D^2}$$

But this must be equal to the radial (centripetal) force holding m in its circular orbit:

$$\frac{GMm}{D^2} = \frac{m\upsilon^2}{r_1} = \frac{m}{r_1}\left(\frac{2\pi r_1}{T}\right)^2$$

where

$$\upsilon_1 = \frac{2\pi r_1}{T}$$

Then

$$\frac{GM}{D^2} = \frac{4\pi^2}{T^2} r_1 = 4\pi^2 \frac{M}{M+m} \frac{D}{T^2}$$

which gives

$$T^2 = \frac{4\pi^2\left(\dfrac{M}{M+m}\right)}{GM} D^3$$

Note that M divides out of the numerator and denominator, so we get

$$T^2 = \frac{4\pi^2}{G(M+m)} D^3$$

With the data given

$$T^2 = \frac{4\pi^2}{G(5\ \mathrm{M_s} + 8\ \mathrm{M_s})} (5\ \mathrm{AU})^3$$

$$T^2 = \frac{4\pi^2\ (5 \times 1.5 \times 10^{11}\ \mathrm{m})^3}{6.67 \times 10^{-11}\ \mathrm{N \cdot m^2/kg^2} \times (13 \times 2.0 \times 10^{30}\ \mathrm{kg})}$$

$$T = 2.1 \times 10^5\ \mathrm{s} = 59\ \mathrm{h} = 2.5\ \mathrm{days}$$

10. *Guided Problem.* The Cavendish experiment is very difficult to perform, but there is a way of estimating the value of G by using a method employed by Newton. The radius of the Earth is known, and the average density of rocks is about 5.5 g/cm^3 = 5500 kg/m^3. From these data, what value of G is obtained?

Solution scheme

a. What is the expression for the mass of the Earth (a sphere) in terms of R_E and the density ρ?

b. The gravitational force on a body of mass m at the surface of the Earth is just the weight mg_E. Equate the latter term to the general expression for the law of gravitation using the relationship for the mass developed in step a.

11. *Worked Problem.* A synchronous satellite is in orbit 3.59×10^4 km (22,200 mi) above the Earth's surface. At a point directly beneath the satellite at the Earth's surface, a rocket is fired vertically upward, aimed directly at the satellite.

a. If the rocket is just to make the same altitude as the satellite, with what speed must the rocket be traveling at engine shutdown, assuming it has not gained any appreciable height at this time?

b. Will the rocket intercept the satellite?

Solution

a. For the rocket just to reach the satellite's altitude, the kinetic energy at engine shutdown must be enough to give the rocket a final potential energy equal to that of the satellite. Because the only force presumed to be acting on the rocket is gravity (we ignore air friction in this problem), we have conservation of mechanical energy:

$$U_0(r) + K_0 = U(r) + K$$

$$-\frac{GM_Em}{R_E} + \tfrac{1}{2}mv_0^2 = -\frac{GM_Em}{r} + 0$$

$$\frac{1}{2}mv_0^2 = GM_Em\left(\frac{1}{R_E} - \frac{1}{r}\right)$$

$$v_0^2 = 2GM_E\left(\frac{r - R_E}{rR_E}\right)$$

$$v_0^2 = 2 \times 6.67 \times 10^{-11}\ \text{N} \cdot \text{m}^2/\text{kg}^2 \times 5.98 \times 10^{24}\ \text{kg}\left(\frac{3.59 \times 10^7\ \text{m}}{4.22 \times 10^7\ \text{m} \times 6.4 \times 10^6\ \text{m}}\right)$$

$$v_0^2 = 1.06 \times 10^8\ \text{m}^2/\text{s}^2$$

$$v_0 = 1.03 \times 10^4\ \text{m/s}$$

b. At launch the rocket and the satellite have the same *angular* velocity, because the satellite movement is synchronous with the rotation of the Earth. If the rocket is fired vertically upward, all the forces on the rocket are central, including those after shutdown, so angular momentum is conserved. But as the rocket moves away from the center of the Earth, its moment of inertia relative to an axis through the center of the Earth increases. Therefore, the angular velocity decreases and the rocket falls behind the satellite. An observer in the (noninertial) reference frame of the rotating Earth ascribes a lateral force to the sideways movement, a fictitious force in the inertial reference frame. In the frame of the Earth this supposed force is sometimes called the Coriolis force.

12. *Guided Problem.* A satellite orbits a planet such that its closest approach is 6.77×10^3 km and its farthest distance from the planet is 8.37×10^3 km. When it crosses the minor axis, its speed is 8.3 km/s.

a. At that time what is dA/dt?
b. What is the satellite's period?
c. What is the eccentricity of the orbit?
d. Calculate the mass of the planet from these data.

Solution scheme: The solution is based on various properties of the ellipse and on Kepler's laws.

a. How is the rate dA/dt related to the angular momentum? What is $r \sin \theta$ when the satellite crosses the minor axis?
b. The area of an ellipse is πab, where a and b are the semimajor and semiminor axes, respectively. How long does it take to sweep out the full area of the ellipse?
c. The definition of the eccentricity in terms of the closest and farthest points in the orbit is given in the Key Concepts section of this chapter under the heading *Elliptical orbits*.
d. If we know the period, we can calculate the mass from Kepler's Third Law. Assume that the mass of the planet is very large compared with the mass of the satellite.

Answers to Guided Problems

2. $3\pi^2\rho\rho' R^4$
6. $M = 5.1 \times 10^{15}$ kg!

Comment: Our treating the body as a particle is not realistic. Even if the body were spherically isotropic with a radius of 2.0 km (just touching the pendulum bob), the density would have to be 1.5×10^5 kg/m^3, or 13 times the density of lead! This problem illustrates the relative weakness of the gravitational force.

8. a. $T = 1.4$ h
   b. $v = 7.9 \times 10^3$ m/s = 17,800 mph
   c. $W_{NC} = 7.02 \times 10^9$ J; $\Delta U = 1.9 \times 10^4$ J; $\Delta K = 7.02 \times 10^9$ J
   Most of the energy is used in accelerating the satellite to orbit speed.

10. $G = \dfrac{g_E R_E^2}{M_E} = \dfrac{3\, g_E}{4\pi\rho R_E} = 6.7 \times 10^{-11}\ \text{N} \cdot \text{m}^2/\text{kg}^2$

12. a. 3.12×10^{10} m^2/s = 3.12×10^4 km^2/s
    b. 5.74×10^3 s = 1.6 h
    c. $e = 0.10$
    d. $M_p = \dfrac{4\pi^2}{GT^2} a^3 = 7.8 \times 10^{24}\ \text{kg} = 1.3\ M_E$

Chapter 10

Systems of Particles

Overview

The laws of motion and the general conservation laws are quite useful in the analysis of various types of motion. Further simplification results if we treat the objects under consideration as particles. Most of the analyses presented thus far have focused on a single particle. This chapter extends the arguments to several particles, giving us a way of looking at more extended systems and their internal motions.

Essential Terms

system
momentum
conservation of momentum
center of mass
centroid
mass density
center-of-mass velocity
system energy
internal kinetic energy
angular momentum
conservation of angular momentum

Key Concepts

1. *Conservation of momentum for a system of particles.* The law of conservation of momentum for a single particle is of little interest or use because essentially it repeats information we already know from the simple statements of Newton's laws: that the particle moves with constant velocity or remains at rest in the absence of a net force. If there is more than one particle, however, and if we treat the collection of particles as a single (closed) system, the absence of a net external force still means that the system momentum is conserved. Now, however, we have the possibility of the

transfer of momentum from one part of the system to another—and we are able to predict certain aspects of the motion, knowing that the motion is constrained by the requirement that the *total* momentum be conserved. In analyzing this more complicated system, we must distinguish between forces *internal* to the system and forces *external* to the system. Internal forces are those that are action–reaction pairs, that is, *interaction* forces among the constituents of the system. To say that momentum is *conserved* (or is a constant of the motion) means either that the only forces acting on the system of particles are the internal forces or, if external forces are acting, that their resultant is zero. We will discuss in the next chapter how we can have a nonzero resultant force on the system and, under certain conditions, still invoke the conservation of momentum.

2. *Center of mass*. This concept has considerable value in treating systems of particles, whether the system is continuous or consists of discrete particles. A closely allied concept is that of the *center of gravity*. Although these terms often are used interchangeably, the concepts are different; if the gravitational field is nonuniform over the system, these two points in general will not coincide. We defer a discussion of the center of gravity until we are in a position to examine the concept of gravitational torques on extended systems (see Chapter 13).

The center-of-mass determination involves computing a *weighted average*. Just as the average score in class is influenced (or *weighted*) by the number of persons having a particular score, the position of the center of mass is influenced by the distribution of the mass points. *There need be no mass at all at the location of the center of mass*. The significance of the concept lies in its role in describing the overall motion of the system.

3. *Motion of the center of mass*. Here we come to the heart of the usefulness of the center-of-mass concept. In a many-particle system, the complicated interactions among particles, as well as the variety of possible external forces acting on the particles, make it virtually impossible to predict behavior of the system on the basis of an analysis of individual particles. The center of mass of the system, however, regardless of the system's complexity, obeys two very simple relationships, namely, $\Sigma\mathbf{F}_{ext} = M\mathbf{a}_{CM}$ and $\mathbf{P} = M\mathbf{v}_{CM}$, where $M = \Sigma m_i$ is the total mass of the system. In other words, the center-of-mass motion is that of a *particle* of mass M located at the center of mass. In fact, to identify a center of mass it is not even necessary that the particles of the system interact or that the external forces act on all parts of the system, as we shall see in problem 9.

4. *Energy of a system of particles*. In developing the concept of the kinetic energy for a system of particles, we find that we cannot write the kinetic energy in the form $K = \frac{1}{2}Mv_{CM}^2$ even though the momentum can be written in the form $\mathbf{P}=M\mathbf{v}_{CM}$. The discussion in the text makes it clear why the term $\mathbf{u}_i$ (velocity relative to the center-of-mass frame) must be included, but it may not be entirely clear to you why this term is not included in the momentum expression. Let the velocity of the particle m_i relative to some stationary frame be $\mathbf{v}_i$. By the Galilean transformation, $\mathbf{v}_i = \mathbf{v}_{CM} + \mathbf{u}_i$. The momentum of the ith particle is just $\mathbf{p}_i = m_i(\mathbf{v}_{CM} + \mathbf{u}_i)$. The *total* momentum is then

$$\mathbf{P} = \sum \mathbf{p}_i = \sum_i m_i \upsilon_{\text{CM}} + \sum_i m_i \mathbf{u}_i$$
$$= M\mathbf{v}_{\text{CM}} + \sum_i m_i \mathbf{u}_i$$

The last term is the total momentum of the system relative to the center-of-mass frame, but in this frame the center of mass is at rest, so $\sum_i m_i \mathbf{u}_i = 0$. Therefore, $\mathbf{P} = M\upsilon_{\text{CM}}$. In the energy calculation we have the terms $\upsilon_i^2 = \mathbf{v}_i \cdot \mathbf{v}_i = (\mathbf{v}_{\text{CM}} + \mathbf{u}_i) \cdot (\mathbf{v}_{\text{CM}} + \mathbf{u}_i)$; as a result of this dot product, the term u_i^2 appears in the final expression. Thus, the kinetic energy of the system has two parts, one involving the kinetic energy associated with the center-of-mass motion and the other being the energy of motion relative to the center of mass. We will find this result to be extremely useful in analyzing the motion of rigid bodies under rotation (see Chapter 12).

Sample Problems

1. *Worked Problem.* A *ballistocardiograph* is an instrument that measures the recoil motion of the body resulting from blood being ejected into the ascending aorta from the left ventricle of the heart. A person lies horizontally on a table that is supported by a cushion of air, as shown in Figure 10.1. As the heart pumps blood into the ascending aorta (Figure 10.2), sensitive instrumentation records the motion of the table. Suppose 65 g of blood is ejected from the heart at a speed of 35 cm/s. If the body and table have a combined mass of 85 kg, what is the recoil speed of the body and table?

Solution: Because the body is lying horizontally, only those forces that have horizontal components can affect the motion of the body. In this case the only forces acting are the action–reaction pairs consisting of the heart–blood interaction. Therefore, momentum is conserved. Because

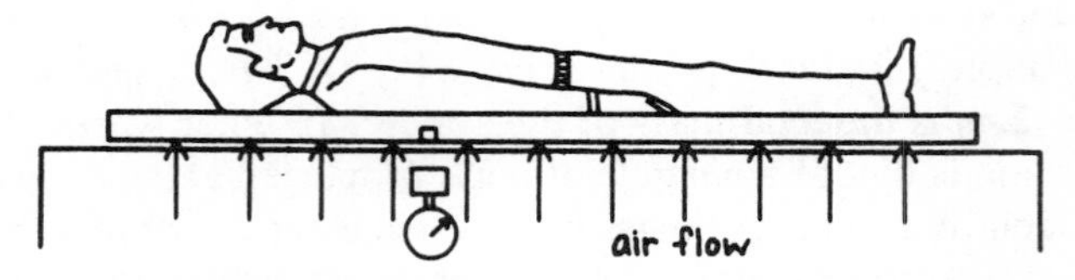

Figure 10.1 Problem 1. A ballistocardiograph.

there is no momentum initially, the total momentum after the heart stroke must still be zero. Therefore,

$$0 = m_\text{b}\mathbf{v}_\text{b} + m_\text{T}\mathbf{v}_\text{T}$$

where m_b is the mass of the ejected blood and m_T is the combined mass of the body and table. Using only the magnitudes, we have

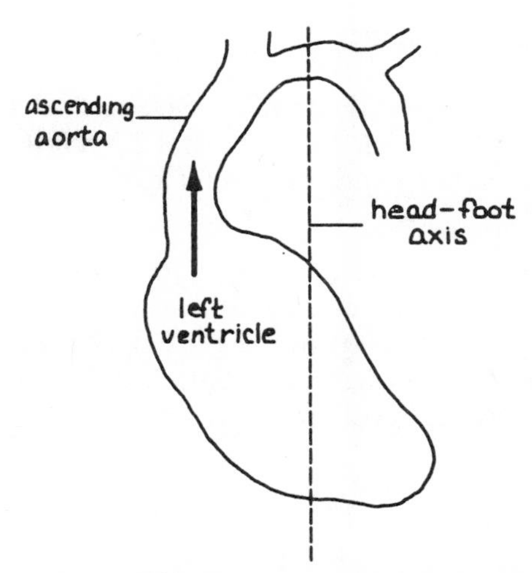

Figure 10.2 Problem 1. As the heart pumps blood into the ascending aorta, there is a reaction force that imparts momentum to the body.

$$0 = m_b v_b - m_T v_T$$

$$v_T = \frac{m_b}{m_T} v_b$$

$$v_T = \frac{0.065 \text{ kg}}{85 \text{ kg}} \times 35 \text{ cm/s} = 0.027 \text{ cm/s}$$

This is an extremely small velocity, but it can be detected by sensitive instruments.

2. *Guided Problem.* In sports that involve the "launching" of a projectile, such as baseball, golf, tennis, and shot put, the body of the person launching the projectile becomes a part of the interaction. For example, treating the collision between a bat and a ball must include a consideration of the mass of the person swinging the bat, because the mass of the bat is effectively made larger than its true mass. This greater mass is sometimes called the *effective mass of the striker*. Consider the following data obtained by measurement of a 47 g golf ball struck by a golf club. The ball is initially at rest, and the club head is traveling 50 m/s just prior to contact and 35 m/s just after the ball leaves the club. The initial velocity of the ball is 70 m/s. Treating this interaction as a collision between the club and the ball, find the effective mass of the club.

Solution scheme

a. If we assume that the only forces affecting the motion of the system are the internal action–reaction forces, what can we say about the total momentum of the system?

b. What is the momentum of the system (ball + club) just prior to impact?

c. What is the total momentum of the ball and club just after impact?

3. *Worked Problem.* A naval gun fires a 900 kg shell at a muzzle velocity of 920 m/s. The barrel, which has a mass of 5×10^4 kg, recoils into four springs, each of which has a spring constant k of 1×10^6 N/m. How far do the springs recoil?

Solution: The firing of the shell imparts a recoil velocity to the gun barrel. This velocity can be calculated using the principle of the conservation of momentum. The kinetic energy of the gun barrel is then transferred to the potential energy of the springs as the barrel comes momentarily to rest.

To find the speed of recoil, we have, using the conservation of momentum,

$$m_s v_s = M_b v_b$$

$$v_b = \frac{M_s}{m_b} \times v_s = \frac{900 \text{ kg}}{5 \times 10^4 \text{ kg}} \times 920 \text{ m/s} = 16.6 \text{ m/s}$$

The kinetic energy of the barrel must be equal to the potential energy of the springs at maximum compression. (*Note:* This is true *only* if we assume *massless* springs; otherwise, we have an inelastic collision between two masses, a situation we will deal with in Chapter 11.) The total k value for the four springs is 4×10^6 N/m. Then the conservation of energy gives us

$$\tfrac{1}{2} m_b v_b^2 = \tfrac{1}{2} k x^2$$

$$x^2 = \frac{m_b}{k} v_b^2$$

$$x^2 = \frac{5 \times 10^4 \text{ kg}}{4 \times 10^6 \text{ N/m}} \times (16.6 \text{ m/s})^2$$

$$x = 1.9 \text{ m}$$

4. *Guided Problem.* A machine gun fires 50 g bullets at a muzzle velocity of 900 m/s. Suppose the gunner can exert an average force of 150 N against the gun while firing. How many bullets per second can be fired?

Solution scheme

a. How is the average force related to the change in momentum?
b. If you know the mass and speed of each bullet, how does the number of bullets fired each second relate to the momentum change?

5. *Worked Problem.* Locate the center of mass of the disk shown in Figure 10.3.

Solution: The difficulty in this problem lies in how we are to treat the hole. The strategy here is to think of the hole as *negative* mass (meaning that mass has been removed). Let M be the mass of the disk and m' the (negative) mass of the hole. Then, since these are symmetric shapes, their

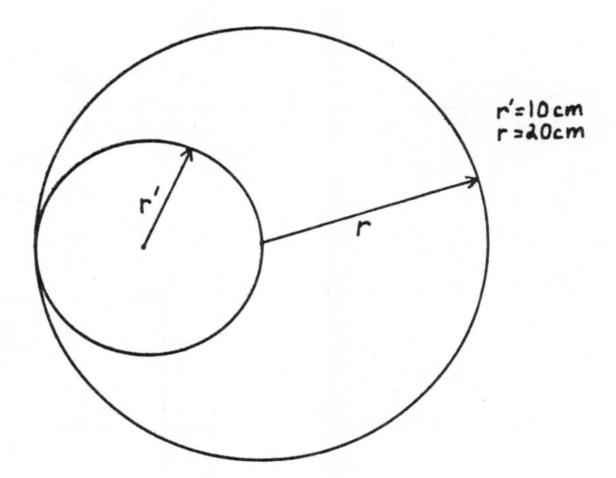

Figure 10.3 Problem 5. A uniform disk has a cut-out hole that shifts the center of mass away from the center of the disk.

center of mass lies at the centers of symmetry, which is at the center of the circle. If the origin of coordinates is taken to be the center of the larger disk and positive is to the right, we have

$$x_{\mathrm{CM}} = \frac{(-m)(-r')}{M - m}$$

The disk has a uniform density and thickness, so $V_{\mathrm{d}} = \pi r^2 t$, where t is the thickness, and $V_{\mathrm{h}} = \pi r'^2 t$. If ρ is the mass density, then

$$M = \rho \pi r^2 t$$
$$m = \rho \pi r'^2 t$$

Therefore

$$x_{\mathrm{CM}} = \frac{\rho \pi r'^2 t}{\rho \pi t (r^2 - r'^2)} \cdot r'$$

$$x_{\mathrm{CM}} = \frac{r'^3}{r^2 - r'^2} = \frac{(10\ \mathrm{cm})^3}{[(20\ \mathrm{cm})^2 - (10\ \mathrm{cm})^2]} = \frac{1000\ \mathrm{cm}^3}{300\ \mathrm{cm}^2}$$

$$x_{\mathrm{CM}} = 3.3\ \mathrm{cm}$$

or 3.3 cm to the right of 0.

6. *Guided Problem.* Locate the center of mass of the uniform plate shown in Figure 10.4.

Solution scheme

a. Divide the plate into sections having symmetry (rectangles).
b. Where is the center of mass of each of these sections?

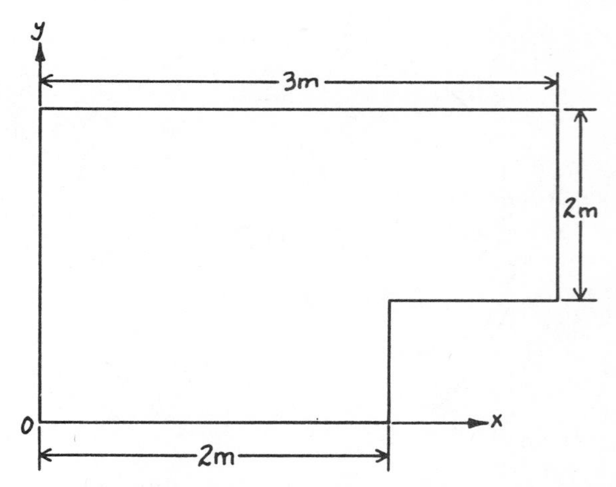

Figure 10.4 Problem 6. The center of mass of the plate may be found easily by segmenting it into rectangular plates.

c. What is the mass of each section?
d. Treating each section as a particle located at the center of mass, find the position of the overall center of mass.

7. *Worked Problem.* Suppose a rod of uniform cross-sectional area A and length L has a density given by $\rho(x) = kx$. Locate the position of the center of mass relative to the lighter end.

Solution: In the general expression for the center of mass

$$x_{CM} = \frac{1}{M}\int \rho x dV$$

the density ρ is now a function of x, so it cannot be taken outside the integral. With $dV = Adx$, we have

$$x_{CM} = \frac{1}{M}\int_0^L x(kx)Adx = \frac{kA}{M}\int_0^L x^2 dx$$

$$x_{CM} = \frac{kA}{M}\left[\frac{x^3}{3}\right]_0^L = \frac{kA}{M}\frac{L^3}{3}$$

Now

$$M = \int \rho(x)dV = kA\int_0^L xdx = kA\left[\frac{x^2}{2}\right]_0^L = \tfrac{1}{2}kAL^2$$

so that

$$x_{CM} = \frac{kAL^3/3}{\frac{1}{2}kAL^2} = \tfrac{2}{3}L$$

Note that if ρ were a constant we should expect $x_{CM} = L/2$. You should try to verify this equivalence.

8. *Guided Problem.* A solid cylinder of uniform density, radius r, and height $2r$ is capped by a uniform hemisphere (solid) of radius r, as shown in Figure 10.5. Locate the center of mass of this object.

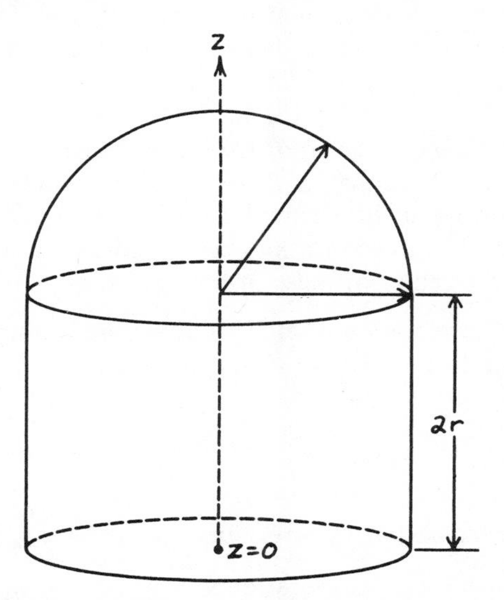

Figure 10.5 Problem 8. A solid cylinder is capped by a hemisphere of the same radius.

Solution scheme: The object consists of two basic shapes, a cylinder and a hemisphere. The cylinder has simple symmetry, so the center of mass is easily located. The center of mass of the hemisphere is more difficult to find.

a. The volume element to be integrated is shown in Figure 10.6. What is the volume dV of this element?
b. How is z related to a and r?
c. Locate the center of mass of the hemisphere relative to the center of its base, using

$$z_{\mathrm{CM}} = \frac{1}{M}\int \rho z dV$$

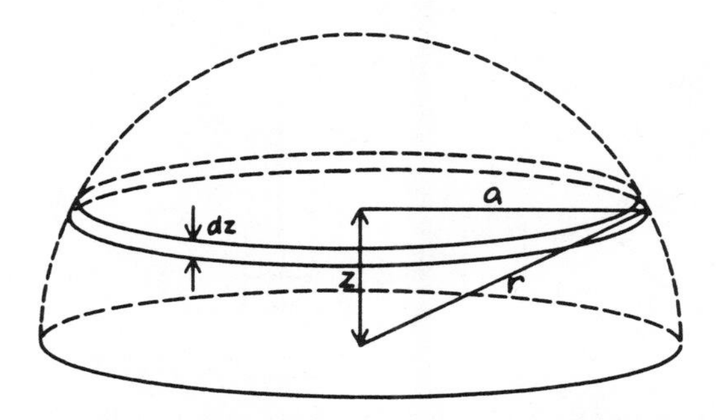

Figure 10.6 Problem 8. The volume element of the hemisphere in Figure 10.5.

d. What is the total mass of the hemisphere?
e. What is the total mass of the cylinder? Where is its center of mass?
f. Use the techniques for finding the center of mass of two particles of mass m_1 (hemisphere) and m_2 (cylinder) separated by a distance z, where z is the distance between the centers of mass of the two objects.

9. ***Worked Problem.*** A series of three 5 kg blocks connected by massless strings are pulled with a constant force of 25 N over a horizontal frictionless surface. At some point the string between the first and the second block breaks. If the 25 N force continues to act on the first block, what is the acceleration of the center of mass of the system after the string breaks?

Solution: The acceleration of the system before the string breaks is found by a simple application of Newton's Second Law:

$$a_{\mathrm{CM}} = \frac{F_{\mathrm{net}}}{M} = \frac{25\ \mathrm{N}}{15\ \mathrm{kg}} = 1.7\ \mathrm{m/s^2}$$

But what happens after the string breaks? The mass still attached to the string will now have an acceleration given by $a = \dfrac{25\ \mathrm{N}}{5\ \mathrm{kg}} = 5\ \mathrm{m/s^2}$, whereas the other two masses will have zero acceleration. There is still a net force of 25 N *on the system*, however, even though it acts on only one part. And since $F_{\mathrm{net}} = Ma_{\mathrm{CM}}$, the *center-of-mass* acceleration remains constant at 1.7 m/s^2. As we stated earlier, the forces on the system do not have to act on all parts of the system for the center-of-mass equation to be valid.

10. *Guided Problem.* Let's look again at the Atwood's machine problem discussed in Chapter 6, shown in Figure 10.7. We used Newton's Second Law applied separately to each mass to find the acceleration and the force acting upward on the pulley. Now find the force upward on the pulley using the center-of-mass concept.

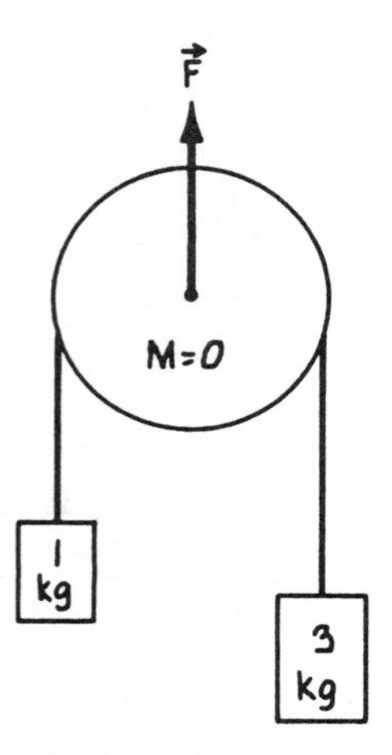

Figure 10.7 Problem 10. Atwood's machine.

Solution scheme

a. What is the expression for the acceleration of each of the masses?
b. How is the center-of-mass acceleration related to the individual accelerations? (*Note:* Relative to the center of mass, the two masses are accelerating in opposite directions. Be careful with signs.)
c. Relate the net force on the system to the center-of-mass acceleration.
d. As a check, solve the problem by the methods used earlier.

11. *Worked Problem.* A 20 g bullet traveling 300 m/s strikes a stationary 500 g block and remains embedded in the block. Assuming that the surface upon which the block rests is frictionless, calculate the energy associated with the center-of-mass motion and the internal energy both before and after the collision.

Solution: The center-of-mass velocity is found by means of the expression $M\mathbf{v}_{CM} = m_1\mathbf{v}_1 + m_2\mathbf{v}_2$, where $\mathbf{v}_1$ and $\mathbf{v}_2$ can be the speeds of the block and the bullet either before or after the collision, because momentum is conserved (the interaction forces at the collision are internal forces). Because only the bullet is moving before the impact, we can find V_{CM} quite easily:

$$V_{CM} = \frac{m_1 \upsilon_1 + m_2(0)}{m_1 + m_2}$$

$$V_{CM} = \frac{20 \text{ g} \times 300 \text{ m/s}}{520 \text{ g}} = 11.5 \text{ m/s}$$

Then

$$K_{CM} = \tfrac{1}{2}M\upsilon_{CM}^2 = \tfrac{1}{2}(0.52 \text{ kg})(11.5 \text{ m/s})^2$$

$$K_{CM} = 34 \text{ J}$$

The velocities of the block and bullet relative to the center of mass before impact are

$$u_1 = \upsilon_1 - \upsilon_{CM} = 300 \text{ m/s} - 10 \text{ m/s} = 290 \text{ m/s}$$

$$u_2 = \upsilon_2 - \upsilon_{CM} = 0 - 10 \text{ m/s} = -10 \text{ m/s}$$

so that

$$K_{int} = \tfrac{1}{2}m_1u_1^2 + \tfrac{1}{2}m_2u_2^2$$

$$= \tfrac{1}{2}(0.020 \text{ kg})(290 \text{ m/s})^2 + \tfrac{1}{2}(0.50 \text{ kg})(-10 \text{ m/s})^2$$

$$K_{int} = 841 \text{ J} + 25 \text{ J} = 866 \text{ J}$$

The total kinetic energy prior to the collision is thus

$$K = K_{CM} + K_{int} = 34 \text{ J} + 866 \text{ J} = 900 \text{ J}$$

As a check, we can work the problem in the laboratory frame and get simply

$$K = \tfrac{1}{2}m_1\upsilon_1^2 = \tfrac{1}{2}(0.02 \text{ kg})(300 \text{ m/s})^2$$

$$K = 900 \text{ J}$$

For the energies after the collision, we note that since the bullet and block are traveling at the same speed, there is no relative motion, so that

$$K = K_{CM} = 34 \text{ J}$$

Thus, the internal energy, calculated in the center-of-mass frame, is the amount of energy dissipated in the collision.

12. *Guided Problem.* A 100 kg missile is fired at a 5000 kg plane. The missile and plane are moving in the same direction at speeds of 1800 km/h and 1440 km/h, respectively. Calculate the total kinetic energy of the plane and missile prior to impact. What percentage of this energy is associated with the center-of-mass motion, and what percentage is internal energy?

Solution scheme

a. How is the center-of-mass velocity related to the individual velocities?
b. What is the total mass of the system?
c. Using this information, calculate K_{CM}.
d. What is the velocity of each object relative to the center of mass?
e. Calculate the energy of the plane and of the missile relative to the center of mass (the internal energy).
f. Form the ratio K_{int}/K_{tot}.

13. *Worked Problem.* A rocket with an initial total mass, including fuel, of 10^5 kg burns fuel at the rate of 500 kg/s at an exhaust velocity of 2500 m/s as it climbs vertically upward.

a. If g is assumed constant, calculate the weight, $w(t)$, of the rocket as a function of time.
b. What is the magnitude of the thrust?
c. Determine the acceleration of the rocket as a function of time, assuming air friction to be negligible.

Solution

a. The weight of the rocket is Mg, but M is a function of time and is given by

$$M(t) = M_0 - \frac{dM}{dt}\, t$$

where M_0 is the initial mass and dM/dt is the constant rate at which mass is being ejected from the rocket. The weight is then

$$\begin{aligned} w(t) &= M_0 g - \frac{dM}{dt}\, gt \\ &= 1 \times 10^5 \text{ kg} \times 9.8 \text{ m/s}^2 - 500 \text{ kg/s} \times 9.8 \text{ m/s}^2\, t \\ w(t) &= (9.8 \times 10^5 - 4.9 \times 10^3 t) \text{ N} \end{aligned}$$

b. Let F_t be the thrust, so that

$$F_t = u\frac{dM}{dt} = 2500 \text{ m/s} \times 500 \text{ kg/s} = 1.25 \times 10^6 \text{ N}$$

c. The acceleration is

$$a(t) = \frac{F_{net}(t)}{M(t)} = \frac{F_t - w(t)}{M(t)} = \frac{F_t - M(t)g}{M(t)}$$

$$a(t) = \frac{F_t}{M(t)} - g$$

$$a(t) = \frac{1.25 \times 10^6 \text{ N}}{1 \times 10^5 \text{ kg} - (500 \text{ kg/s})t} - g$$

Note that as t increases, the first term gets larger, so $a(t)$ increases. This relationship is consistent with the fact that, as fuel is burned, M decreases and the *net* force (thrust minus weight) increases.

14. *Guided Problem.* Consider the firing of a machine gun from the standpoint of the rocket equation. The gun has an initial mass (including bullets) of 5 kg, and it fires 50 g bullets at the rate of 10 per second at a muzzle velocity of 500 m/s.

a. How much thrust is developed in firing the gun?

b. If the gun is attached to a 10 kg cart on a horizontal frictionless surface, what acceleration will the gun and cart have after 5 s of firing? The bullets are fired horizontally.

Solution scheme

a. What is the relative velocity of the bullets and the gun?

b. At what rate is the mass of the gun decreasing?

c. What is the mass of the gun as a function of time?

d. What is the net force on the gun?

Answers to Guided Problems

2. $M_c = \dfrac{m_b v_b}{(v_c - v_c')} = 0.22$ kg

4. 3.3 bullets per second

6. With the coordinate system shown in Figure 10.4, the coordinates of the center of mass are $x_{CM} = 1.375$ m; $y_{CM} = 1.625$ m.

8. Hemisphere: $dV = \pi a^2 dz = \pi(r^2 - z^2)dz$

$$z_{CM} = \frac{\pi\rho}{M}\int_0^r (r^2 - z^2)z\,dz = \frac{3}{8}r$$

Composite body: $z_{CM} = \dfrac{43}{32}r$ above base of cylinder

10. $F = 29.4$ N

12. $v_{CM} = 402$ m/s

$K_{tot} = 4.125 \times 10^8$ J[1]

$K_{CM} = 4.12 \times 10^8$ J

$K_{int} = 4.9 \times 10^5$ J

$K_{int}/K_{tot} = 1.2 \times 10^{-3}$ or $K_{int} = 0.12\%$ of the total energy

[1] We carry four significant figures here to show that the total energy is almost completely center-of-mass energy.

14. a. $F_t = 250$ N (about 56 lb)

    b. At $t = 5$ s, $M(t) = 12.5$ kg

$$a(t)_{5s} = \frac{250 \text{ N}}{12.5 \text{ kg}} = 20 \text{ m/s}^2 \text{ (about 2 gee)}$$

Chapter 11

Collisions

Overview

The conservation laws are useful in solving problems for which the interaction forces are complicated functions of time. This usefulness is evident in the sudden and often violent interactions that take place during collisions or explosive processes. Building on the material in Chapter 10, we will now look more closely at the interactions between bodies undergoing collisions, whether these are "contact" collisions or the relatively slower and less violent action-at-a-distance interactions evident in such phenomena as the deflection of charged particles and the Sun's influence on the orbit of a comet.

Essential Terms

impulse
impulse approximation
time-averaged force
elastic collision
inelastic collision

Key Concepts

1. *Impulse and the impulsive force.* Strictly speaking, the impulse of the force $F(t)$ is defined as $I = \int_0^t F(t)dt$, and there is no restriction on the length of the time interval over which the force acts. But impulsive forces are most often thought of as those of very short duration, a few hundredths or thousandths of a second. The impulse concept is most useful in the study of these short-duration forces as they act during collision processes.

In this regard, we must consider the assumptions made about the relative sizes of these forces and of others that may be acting on the system. Let's look at a concept called the *impulse approximation*. Basically, this is an assumption either that any *external* forces acting on the system during the impulsive interaction are negligibly small compared with the internal interaction forces or that these external forces, acting over the short duration of the interaction, will produce negligibly small changes in the motion of the system. For example, suppose two carts are moving along a horizontal but *not* frictionless surface and undergo a collision. Friction is an external force that will change the overall momentum of the system, but if the collision time is very short, the momentum changes that occur *during the collision* are essentially those resulting from the interaction; that is, overall momentum is assumed to be unchanged, but the internal transfer of momentum does occur. If the collision is very slow, however, as in the slow compression of a spring attached to one of the carts, the external friction force will affect appreciably the motion of the carts during this longer time interval; conservation of momentum is not applicable in this situation. Therefore, it is important to recognize that when the principle of the conservation of momentum is applied to a collision process for which the impulse approximation is valid, the momentum changes that occur within the system must be regarded as those occurring *during the collision time only*. If, for example, υ_1 is the velocity of the cart before the collision and υ_1' is its velocity after the collision, these velocities are valid only *immediately before* impact and *immediately following* impact. For times earlier or later than the impact interval, these velocities may be quite different as a result of the action of the external forces.

2. *Time-averaged force*. In Chapter 7 we developed the concept of an average force based on the definition of work. This was a *displacement-averaged* force because of the definition $F_{avg}\Delta x$ = work. By a similar line of reasoning, we can consider an average force based on the impulse concept. Because $\int F(t)dt$ is the area under the $F(t)$ versus t curve, the constant force F that will give the same impulse in the interval Δt is $F_{avg} \cdot \Delta t = I$; that is, F_{avg} is the *time*-averaged force. If this force F_{avg} were to act over the entire interval Δt, the impulse would be the same as that resulting from the action of the variable force over the same interval of time. Note that a force can act over time without acting over a distance. Thus, it is possible for a force to produce a momentum change without doing work.

The average force may or may not be a useful concept, depending on the situation considered. Suppose a carpenter drives a nail into a board, striking the nail once per second. The force-versus-time curve is similar to that shown in Figure 11.1. The average force in this case is a fairly small number, because so much dead time occurs between blows of the hammer. Indeed, the average force might even be less than the weight of the hammer. Does this mean that the carpenter can just lay the hammer on the nail and drive it into the wood? Of course this does not happen. The important consideration here is the *maximum force* directed onto the nail by the hammer. A more meaningful average force is one that is computed over the interval during which the hammer is in contact with the nail. If the force buildup and decay during this interval is even approximately linear, reasonably good estimates of the maximum force can be made from knowledge of this average force. The same is true for the machine gun firing discussed in Chapter

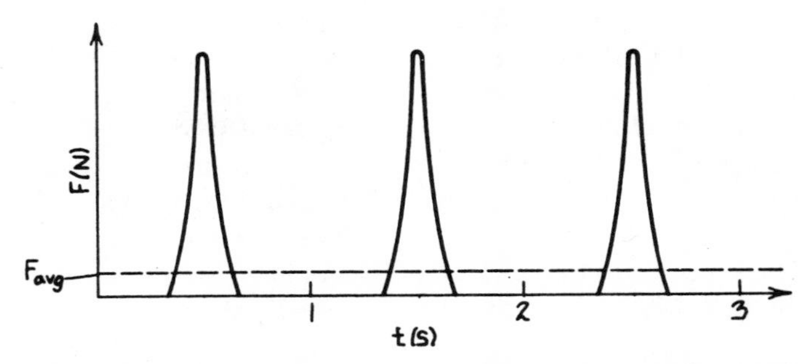

Figure 11.1 The force-versus-time curve for the blows of a hammer. The dashed line shows the average force.

10. If there is very little dead time between firings, the average force is a useful number, but if a bullet is fired every 5 s, the extremely small average force over the whole time interval between firings is practically worthless.

Also remember that the average force indicates nothing about how the force varies during the interval. Suppose you are traveling at a high rate of speed on a sled and you are to be brought to rest by a large spring. The average force exerted on you is $\frac{1}{2}kx_{max}$, giving an average acceleration equal to $\frac{1}{2}a_{max}$. If this average acceleration is 5 gee, that value may seem harmless enough, but the maximum acceleration is 10 gee, which could be destructive.

3. *Elastic and inelastic collisions.* These two terms refer specifically to the cases in which *kinetic energy* is conserved and not conserved, respectively. In these collision processes, we are assuming that the impulse approximation is valid, so momentum is conserved regardless. But it is only in *elastic* collisions that kinetic energy is conserved. As we noted in the last chapter, if there is no relative velocity between the colliding objects after the collision, the final kinetic energy is just the center-of-mass kinetic energy. The internal energy has been removed from the system, and this is the so-called *perfectly* or *totally inelastic collision*. Note that the requirement is that there be no relative velocity between the colliding objects after the collision. Whether they "stick together" is not crucial.

4. *Contact forces versus action-at-a-distance forces.* We have been speaking of collisions, but what do we mean by that term? We generally mean that two objects actually come in contact with each other. But what is "contact"? At the macroscopic (large-scale) level, we can make a distinction between two magnets actually hitting each other and two interacting through their respective fields—or can we? At the microscopic (molecular) level, all interactions are electrical, magnetic, or gravitational (except for those involving the short-range nuclear forces); that is, the forces are *action-at-a-distance forces*. Fortunately, we need not make this distinction, because the relationship $\int \mathbf{F}(t)dt = \Delta \mathbf{p}$ says nothing about the nature of the force, only that there is some kind of force acting on the particle. Thus, electrons being deflected by other charged particles and space probes being "captured" by a planet both constitute interactions and can be considered collision processes. For this reason a two-dimensional collision of billiard balls can be treated in the same way as the scattering of electrons.

5. *Equations for collision processes*. Although elastic collisions are analyzed by means of the equations of both conservation of momentum and conservation of kinetic energy, the latter involves squared terms and adds undesirable computational burdens. By combining the two conservative equations, we get a simpler expression to replace the equation for the conservation of kinetic energy and at the same time gain further insight into the collision process. For an *elastic* collision, we find that $\mathbf{v}_1 - \mathbf{v}_2 = -(\mathbf{v}_1' - \mathbf{v}_2')$, that is, that the *relative* velocities before and after the collision are the same except for sign. If the collision is totally inelastic, $\mathbf{v}_1' - \mathbf{v}_2' = 0$. Although this equivalence seems to imply that $\mathbf{v}_1 - \mathbf{v}_2 = 0$, which clearly is not true, remember that the relative-velocity equation is valid *only* for the *elastic* collision process, in which kinetic energy is conserved. Problem 25 in the text (p. 292), introduces the concept of *coefficient of restitution*, which enables you to use a modified form of the relative-velocity equation in solving problems involving inelastic collisions. We recommend that you learn to use this modified equation, because it is quite useful in solving a variety of collision problems.

6. *Collisions in two dimensions*. In one sense, handling problems involving two-dimensional collisions is no more difficult than working with one-dimensional collisions, because the conservation laws are the same. Indeed, because momentum is a *vector*, we can resolve it into x and y components, giving us two independent equations. We cannot do that with the kinetic-energy term, however, because the latter is a *scalar*. Thus, by moving into two dimensions, we add two new variables (the unknown angles) but only one independent equation. For this reason, one of the four unknowns ($\mathbf{v}_1'$, $\mathbf{v}_2'$, θ_1', or θ_2') must be specified in addition to the initial values of these quantities.

In what we have called action-at-a-distance collisions, we assume that the interaction forces act over a very limited region, called the *interaction region*. This region is not sharply defined, but in most cases it is small, either in extent or because the force drops off rapidly, as does the force $1/r^2$ in gravitation. In the latter situation the region may be considered relatively localized as far as measurable effects are concerned.

Sample Problems

1. *Worked Problem*. In guided problem 10.2 you calculated the effective mass of a golf club striking a golf ball. Now let's look at the interaction between the club and the ball. The contact time is generally estimated to be about 10^{-3} s.

a. What is the average force exerted on the ball by the club?
b. Make a reasonable estimate of the *maximum* force.

Solution:

a. From problem 10.2 we have the mass of the golf ball as 47 g. The ball goes from rest to 70 m/s in 0.001 s. The change in momentum of the ball is

$$\Delta p = 0.047 \text{ kg} \times 70 \text{ m/s} - 0 = 3.3 \text{ kg} \cdot \text{m/s}$$

But

$$F_{avg} \cdot \Delta t = \Delta p$$

so that

$$F_{avg} = \frac{\Delta p}{\Delta t} = \frac{3.3 \text{ kg} \cdot \text{m/s}}{0.001 \text{ s}} = 3300 \text{ N}$$

b. If we assume that force is approximately linear in its rise and decay, then $F_{max} \approx 2F_{avg}$ or

$$F_{max} \approx 6600 \text{ N}$$

This force is quite large and results in considerable temporary deformation of the ball.

2. *Guided Problem.* A 70 g baseball traveling 50 m/s is caught by the shortstop. In catching the ball, the fielder's glove moves backward, bringing the ball to rest in 0.002 s. What is the average force exerted on the glove by the ball?

Solution scheme

a. How does the impulse relate to the change in momentum?
b. What is the expression for impulse in terms of the average force?

3. *Worked Problem.* A 200 g glider and a 600 g glider are coupled together and move along a frictionless air track at 50 cm/s. An explosive charge on one of the gliders separates them, causing the 200 g glider to stop.

a. What is the subsequent velocity of the 600 g glider?
b. How much energy was added to the system in the separation process? Where did the energy come from?

Solution: The surfaces are frictionless and the explosive interaction is internal to the system, so momentum is conserved.

$$(m_1 + m_2)\upsilon = \cancelto{0}{m_1 \upsilon_1'} + m_2\upsilon_2'$$

a. $\upsilon_2' = \dfrac{m_1 + m_2}{m_2}\upsilon = \dfrac{200 \text{ g} + 600 \text{ g}}{600 \text{ g}} \cdot 50 \text{ cm/s} = 67 \text{ cm/s}$

b. $(K_{CM})_i = \frac{1}{2}(m_1 + m_2)\upsilon^2 = \frac{1}{2}(0.80 \text{ kg})(0.50 \text{ m/s})^2$

$(K_{CM})_i = 0.1 \text{ J}$

After the collision, $K_{CM} = 0.1$ J because $V_{CM} =$ constant.

$$\begin{aligned} K_{int} &= \tfrac{1}{2}m_1u_1^2 + \tfrac{1}{2}m_2u_2^2 \\ &= \tfrac{1}{2}(0.2 \text{ kg})(0 - 0.50 \text{ m/s})^2 + \tfrac{1}{2}(0.6 \text{ kg})(0.67 \text{ m/s} - 0.50 \text{ m/s})^2 \\ &= 0.025 \text{ J} + 0.0087 \text{ J} \\ K_{int} &= 0.034 \text{ J} = 3.4 \times 10^{-2} \text{ J} \end{aligned}$$

The added energy is internal energy from the chemical reaction taking place during the explosion.

Comment: The situation here is the reverse of that of a perfectly inelastic collision. If the 600 g glider traveling 67 cm/s collides with the 200 g stationary cart, the amount of energy *dissipated* will be 0.034 J. Once again we see the value of using the center-of-mass concept in collision processes. For totally inelastic collisions or explosive interactions, the energy dissipated or added is equal to the internal energy of the system before or after the interaction, depending on which of these two processes is taking place.

4. *Guided Problem.* A 1.5 kg block resting on a horizontal frictionless surface has a hole bored into it in which a 0.5 kg ideal spring is placed, as shown in Figure 11.2. A 500 g ball is fired into the hole at a speed of 50 m/s. The spring subsequently is compressed a distance of 6.0 cm, at which point a ratchet mechanism locks the spring in place. The block and the ball move off together at a speed v'.

a. Calculate v'.

b. Calculate the kinetic energy decrease of the system as a result of the collision. How much of that energy is stored in the spring as potential energy if $k = 1.0 \times 10^5$ N/m?

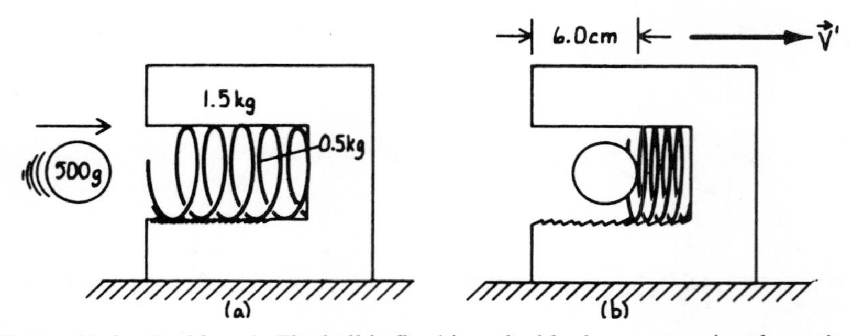

Figure 11.2 Problem 4. The ball is fired into the block, compressing the spring.

Solution scheme

a. What kind of collision is this?

b. Which conservation law(s) apply to this kind of collision?

c. How much kinetic energy does the system have before the collision?

d. What is the expression for the potential energy of a spring compressed a distance x?

e. As a check, compute $K_{CM} + K_{int}$ before the collision. For this type collision, $\Delta K = K_{int}$.

5. *Worked Problem.* Two men are standing on carts that move along a horizontal frictionless surface, as shown in Figure 11.3. The mass of each cart–man combination is 100 kg. Cart A is moving at 10 cm/s, and cart B follows in the same direction at 7 cm/s. The man on cart B throws a 2 kg ball to the man on A (assume the ball travels horizontally). The man on A is capable of throwing the ball at a speed of 15 cm/s. What is the maximum speed at which man B can throw the ball to man A so that A can bring himself completely to rest by throwing the ball in the forward direction?

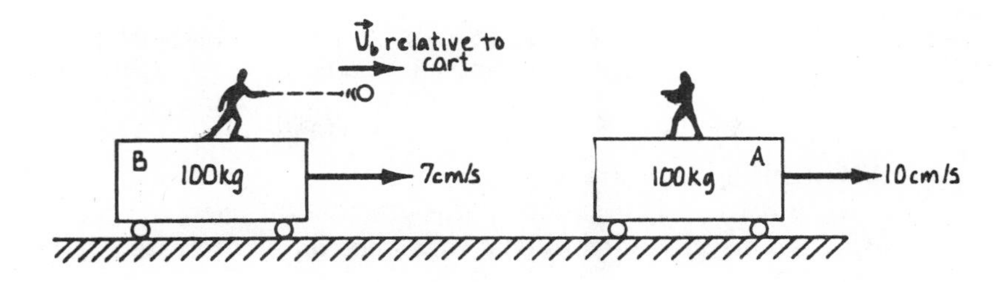

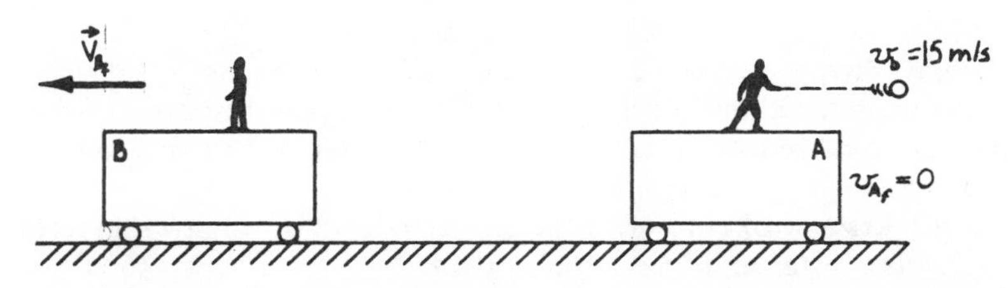

Figure 11.3 Problem 5. The man on A catches the ball, then turns around and throws it forward in such a way that he comes to rest.

Solution: We have three parts to the problem. First there is an "explosive" interaction in which B throws the ball (b):

$$(m_B + m_b)\upsilon_{B_i} = m_b(u_b + \upsilon_{B_i}) + m_B\upsilon_{B_f}$$

Then the second man, A, catches the ball, which constitutes a perfectly inelastic collision, resulting in a common velocity U of the ball and man:

$$m_b(u_b + \upsilon_{B_i}) + m_B\upsilon_{B_i} = (m_A + m_b)U$$

Finally man A throws the ball forward so that he stops:

$$(m_A + m_b)U = m_b\upsilon_{b_f} + 0$$

Combining the last two equations, we get

$$m_b(u_b + \upsilon_{B_i}) + m_B\upsilon_{B_i} = m_b\upsilon_{b_f}$$
$$2\text{ kg }(u_b + 0.07\text{ m/s}) + (100\text{ kg})(0.10\text{ m/s}) = (2\text{ kg})(15\text{ m/s})$$
$$2\text{ kg}\cdot u_b + 0.14\text{ kg}\cdot\text{m/s} + 10\text{ kg}\cdot\text{m/s} = 30\text{ kg}\cdot\text{m/s}$$
$$u_b = 9.9\text{ m/s}$$

Thus, man B can tnrow the ball no faster than 9.9 m/s. The recoil speed of B is found from the first equation:

$$(100\text{ kg} + 2\text{ kg})(0.07\text{ m/s}) = 2\text{ kg }(9.9\text{ m/s} + 0.07\text{ m/s}) + 100\text{ kg}\cdot\upsilon_{B_f}$$
$$\upsilon_{B_f} = \frac{7.14\text{ kg}\cdot\text{m/s} - 19.94\text{ kg}\cdot\text{m/s}}{100\text{ kg}}$$
$$\upsilon_{B_f} = -0.13\text{ m/s} = -13\text{ cm/s}$$

So, after throwing the ball to A, man B moves backward at 13 cm/s.

6. *Guided Problem.* A pile driver consists of a driver of mass M that is raised to a height h during every stroke and allowed to fall freely against the pile, which has mass m. A constant resistive force, f, acts on the pile.

a. If $M = 100$ kg, $m = 500$ kg, $h = 10$ m, and $f = 3300$ N, how far is the pile driven during each stroke?
b. If the pile is driven to a total depth d, show that the energy expended in raising the driver is independent of h and can be decreased by increasing M.

Solution scheme: Treat the collision as perfectly inelastic.

a. With what speed v does the driver strike the pile?
b. What is the speed of the pile immediately after impact?
c. Calculate the kinetic energy of the pile plus driver just after impact.
d. If the resistive force is constant, how much work does this force do in bringing the pile and driver to rest?
e. If it takes n strokes to drive the pile to a depth d, what is d in terms of x, the distance driven in one stroke?
f. How much energy is required to lift the driver n times? Express this in terms of f, d, M, and m.
g. Show that E gets smaller as M gets larger.

7. *Worked Problem.* A glider with mass m_1 of 500 g is moving to the right on a frictionless air track at 10 cm/s. It collides elastically with another glider of mass m_2, initially at rest (Figure 11.4). After the collision, m_2 rebounds elastically against a constraint at the end of the track and travels with the same velocity as m_1. Determine m_2.

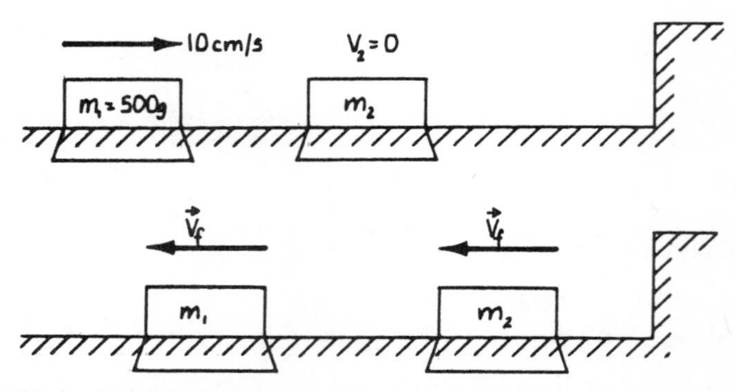

Figure 11.4 Problem 7. After the collision, m_2 rebounds from the end wall and travels at the same speed as m_1.

Solution: Because the track and its constraint can be considered an infinite mass (it doesn't move during the collision), the mass m_2 will rebound from the constraint at the same speed the mass had when it collided with the wall. Thus, our problem is to find a collision between m_1 and m_2 that causes the final velocities to be equal and opposite. The conservation-of-momentum equation gives

$$m_1 v_1 + 0 = m_1 v_1' + m_2 v_2'$$

Because the collision is elastic, kinetic energy is conserved, so the relative velocities before and after the collision are equal, except for sign:

$$v_1 - 0 = v_2' - v_1'$$

We specify that $v_2' = -v_1'$, so

$$m_1v_1 = m_1v_1' - m_2v_1'$$

and

$$v_1 = -v_1' - v_1' = -2v_1'$$

Then

$$-m_1(2v_1') = m_1v_1' - m_2v_1'$$
$$-2m_1 = m_1 - m_2$$
$$m_2 = 3m_1$$
$$m_2 = 1500 \text{ g}$$

To find the final velocity each mass has after the collision, we use the condition described:

$$v_1' = -v_1/2$$
$$v_1' = -5 \text{ cm/s}$$
$$v_2' = 5 \text{ cm/s}$$

Note that the block m_1 rebounds and moves to the left at 5 cm/s, whereas the mass m_2 moves to the right at 5 cm/s. After the latter rebounds elastically from the end constraint, however, it will move to the left at a speed of 5 cm/s, the same as that of the mass m_1.

8. *Guided Problem.* Tarzan, who has a mass of 90 kg, swings on a vine 30 m long to rescue Boy, who is standing on the ground surrounded by lions. If Boy has a mass of 35 kg and the two must swing up to the branch of a tree 5 m off the ground, from what height above the ground must Tarzan start his swing, assuming he starts from rest and grabs Boy at the lowest point in his path?

Solution scheme

a. What is Tarzan's kinetic energy just before he grabs Boy? What potential must he have had initially to acquire this kinetic energy?
b. When Tarzan grabs Boy, the collision is totally inelastic. How is their combined speed just after the collision related to Tarzan's speed just before the collision?
c. What is the combined energy of Tarzan and Boy when they come to rest in the tree? How does this relate to their kinetic energy just after the collision?

Note: This is a ballistic-pendulum problem.

9. *Worked Problem:* A 4000 kg open railroad car is moving along a frictionless track at 3.0 m/s when a hard rain begins to fall vertically at the rate of 10 cm/h. If the interior of the car measures 12 m × 6 m, determine the speed of the car after 15 min. The density of the rainwater is 10^3 kg/m^3.

Solution: If the rain falls vertically, its initial velocity in the horizontal direction is zero. Thus, the collision is a perfectly inelastic collision between a moving car and rain at rest. The rate of rainfall in kilograms per second is

$$\frac{dM}{dt} = \rho \frac{dV}{dt} = \rho A \frac{dy}{dt}$$

in which

$$\frac{dy}{dt} = 10 \text{ cm/h} = \frac{0.010 \text{ m}}{3600 \text{ s}} = 2.8 \times 10^{-6} \text{ m/s}$$

Thus

$$\frac{dM}{dt} = 10^3 \text{ kg/m}^3 \times (12 \text{ m} \times 6 \text{ m}) \times 2.8 \times 10^{-6} \text{ m/s}$$

$$\frac{dM}{dt} = 0.2 \text{ kg/s}$$

Because there are no external forces affecting the motion, we have conservation of momentum for the car and the accumulated rain:

$$M_c \upsilon_c = [M_c + M_r(t)]\upsilon(t)$$

where M_c is the initial mass of the car, υ_c is the initial speed of the car, $M_r(t)$ is the accumulated rain at time t, and $\upsilon(t)$ is the final speed of the combination at time t. Then

$$M_c \upsilon_c = \left(M_c + \frac{dM}{dt} \cdot t\right) \upsilon(t)$$

$$\upsilon(t) = \frac{M_c \upsilon_c}{M_c + \frac{dM}{dt} \cdot t} = \frac{6000 \text{ kg} \times 3 \text{m/s}}{6000 \text{ kg} + 0.2 \text{ kg/s} \times 900 \text{ s}}$$

$$\upsilon(t) = 2.91 \text{ m/s}$$

As we can see, this amount of rain will not affect the speed significantly.

Comment: We know that the addition of the rain will slow the car, but what happens if the car develops a leak and rainwater drains out the bottom (vertically)? It should be clear that the car does not speed up, but how do we account for momentum conservation? Think about the *horizontal* velocity component of the leaking rainwater. Has that changed?

10. *Guided Problem.* A 5 kg block and a 10 kg block are at rest, slightly separated, on a horizontal table. A 20 g bullet is fired into the 5 kg block and lodges there. This block then collides inelastically with the 10 kg block.

The coefficient of restitution for the two-block collision is 0.6, and the coefficient of friction between the blocks and table is 0.4. If the 10 kg block slides 40 cm before coming to rest, what was the initial speed of the bullet?

Solution scheme: Assume that the impulse approximation is valid for the collisions.

a. Write down an expression for the totally inelastic collision between the bullet and the 5 kg block.
b. Use the velocity of the bullet–block combination as the initial velocity for the two-block collision.
c. Write the momentum-conservation equation for the latter collision.
d. Referring to problem 25 in the text, write down the expression for the relative velocities before and after the collision when the coefficient of restitution is known.
e. Use the work–energy theorem to calculate the velocity of the 10 kg block just after the collision, given the coefficient of friction and the stopping distance.

11. *Worked Problem.* A 100 g puck glides along a frictionless surface and strikes a second puck m_2 initially at rest. After the collision the 100 g puck moves off at right angles to its original path and the other puck moves along a line making an angle of 20° with the original path, as shown in Figure 11.5. If the collision is perfectly elastic, what is the mass m_2 of the second puck?

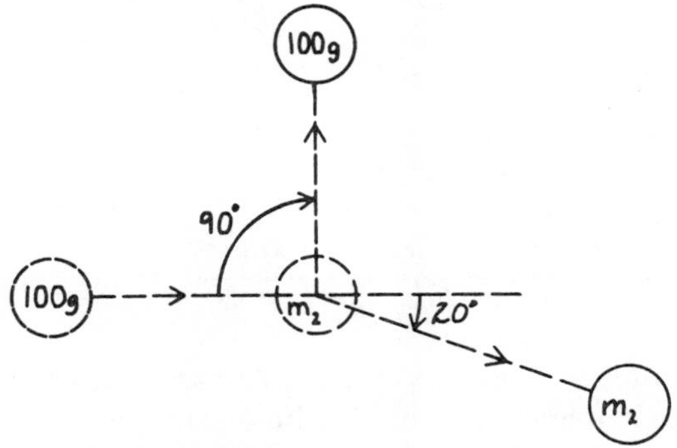

Figure 11.5 Problem 11. The 100 g puck collides elastically with the puck m_2.

Solution: Using the principle of the conservation of momentum in two dimensions, we have

$$x: \quad m_1 \upsilon_1 = 0 + m_2 \upsilon_2' \cos\theta \qquad (1)$$

$$y: \quad 0 = m_1 \upsilon_1' - m_2 \upsilon_2' \sin\theta \qquad (2)$$

$$\text{Energy:} \quad \tfrac{1}{2} m_1 \upsilon_1^2 = \tfrac{1}{2} m_1 \upsilon_1'^2 + \tfrac{1}{2} m_2 \upsilon_2'^2 \qquad (3)$$

$$m_1^2 \upsilon_1^2 = m_2 \upsilon_2'^2 \cos^2\theta \qquad (1)$$

$$m_1^2 \upsilon_1'^2 = m_2 \upsilon_2'^2 \sin^2\theta \qquad (2)$$

$$m_1 = m_1 \frac{\upsilon_1'^2}{\upsilon_1^2} + m_2 \frac{\upsilon_2'^2}{\upsilon_1^2} \qquad (3)$$

Divide Eq. (2) by Eq. (1) to get

$$\frac{v_1'^2}{v_1^2} = \tan^2\theta$$

Substituting this result into Eq. (3), we get

$$m_1 = m_1 \tan^2\theta + m_2 \frac{v_2'^2}{v_1^2}$$

But from Eq. (1)

$$\frac{v_2'^2}{v_1^2} = \frac{m_1^2}{m_2^2}\frac{1}{\cos^2\theta}$$

Then Eq. (3) becomes

$$m_1 = m_1 \tan^2\theta + \frac{1}{\cos^2\theta}\frac{m_1^2}{m_2^2}m_2$$

$$1 - \tan^2\theta = \frac{m_1}{m_2}\frac{1}{\cos^2\theta}$$

$$\frac{\cos^2\theta - \sin^2\theta}{\cancel{\cos^2\theta}} = \frac{m_1}{m_2}\frac{1}{\cancel{\cos^2\theta}}$$

But $\cos^2\theta - \sin^2\theta = \cos 2\theta$, so

$$\frac{m_1}{m_2} = \cos 2\theta$$

$$m_2 = \frac{m_1}{\cos 2\theta} = \frac{100\text{ g}}{\cos(2\times 20°)} = 131\text{ g}$$

12. *Guided Problem.* A proton of mass M has an initial energy K_M coming into a bubble chamber. It collides with a stationary electron of mass m and scatters through an angle ϕ. The electron moves off at an angle θ, as shown in Figure 11.6. The following data are known:

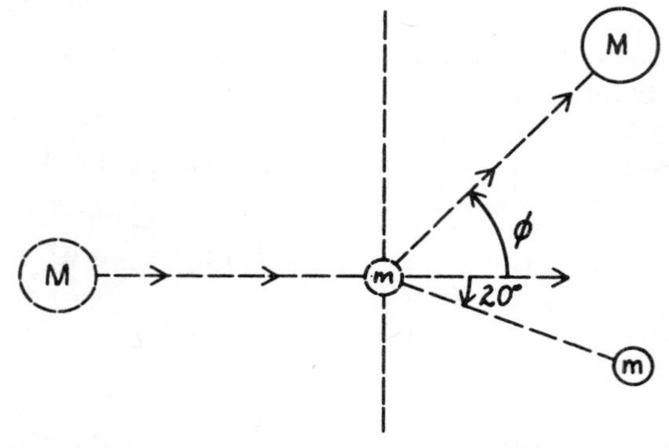

Figure 11.6 Problem 12. The proton scatters elastically after striking the electron.

$$M = 1.67 \times 10^{-27} \text{ kg}$$
$$m = 9.1 \times 10^{-31} \text{ kg}$$
$$\theta = 20°$$
$$K_M = 10 \text{ Gev} = 1.6 \times 10^{-9} \text{ J}$$

If the scattering is elastic, what is the scattering energy of the electron?

Solution scheme: The best way to handle a problem of this type is to solve it generally before inserting data. Use the symbols K for kinetic energy and p for momentum and the *primed* notation for after-collision data.

a. Write down the conservation-of-energy equation.
b. Write down the conservation-of-momentum equation in vector form.
c. Take the dot product of this last equation with itself.
d. Write the latter expression in terms of K.
e. Substitute the conservation-of-energy equation.
f. Solve for the momentum p'_m and set this equal to the p'_m obtained from $p'^2_m = 2mK'_m$.
g. Solve for K'_m, noting that $p_M^2 = 2MK_M$.

Answers to Guided Problems

2. 1750 N, or about 390 lb
4. a. $v' = 10$ m/s
   b. $K = 500$ J; $U_{\text{spring}} = 180$ J
6. a. $x = \dfrac{ghM^2}{f(M+m)} = 0.49$ m
   b. Let $d = nx$ and $E = n\,Mgh$

$$E = fd\left(1 + \frac{m}{M}\right)$$

8. 9.6 m
10. 553 m/s
12. $K'_m = \dfrac{4Mm}{(M+m)^2} K_M \cos^2\theta = 2.05 \times 10^{-3}\, K_M$

$$K'_m = 2.05 \text{ Mev}$$

Chapter 12

Kinematics of a Rigid Body

Overview

The notion of a particle has been useful in analyzing the motions of systems, because it has allowed us to focus on the salient features of the dynamical interactions taking place both within and outside the system. Real systems have extended dimensions, however, and often we need to know how the different parts of the system behave. The concepts of the center-of-mass motion and the relative velocities of particles were a step in that direction. We now look at the special case in which the *relative* positions of the particles that make up the system remain unchanged. We call this configuration a *rigid body*, and we begin by studying the *kinematics* of such bodies. The dynamics of these bodies is the subject of Chapter 13.

Essential Terms

rigid body
average angular velocity
instantaneous angular velocity
frequency
period
average angular acceleration
instantaneous angular acceleration
rotational kinetic energy
moment of inertia
parallel-axis theorem
perpendicular-axis theorem
angular momentum of a rigid body

Key Concepts

1. *General motion of a rigid body.* If we examine the motion of a rigid body carefully, we conclude that the most general displacement is a *translation* of a fixed point in the body, combined with a *rotation* about an axis passing through that point. Although this is true for any point in the body,

there is considerable advantage to selecting the center of mass as this point, primarily because the kinetic energy can be divided conveniently into two parts: the translational kinetic energy of the center of mass, and the rotational kinetic energy about an axis through the center of mass. If any point other than the center of mass is chosen, the kinetic energy term will have, in addition, a term involving the linear velocity of that point and the angular velocity of the body relative to that point. Therefore, choosing the center of mass as the reference point, we can state that the most general displacement of a rigid body is the translation of the center of mass plus a rotation about an axis through the center of mass.

2. *Rotation of a rigid body*. In the rotation of a rigid body about a fixed axis, we encounter such terms as *angular displacement, angular velocity,* and *angular acceleration*. In linear motion we found the displacement, velocity, and acceleration to be vectors, that is, to have magnitude and direction and to obey certain laws of addition, such as the commutative law. But if every particle in a rotating rigid body has its own unique position, velocity, and acceleration, is it possible to assign vector properties to their angular counterparts, which deal with aggregate motion? It is possible to do so, but we have to look at the nature of rotations of rigid bodies to see how these assignments are made. Figure 12.1 shows a point on a rigid body undergoing an angular displacement $d\boldsymbol{\theta}$ about an axis. Because of the nature of a rigid body, every particle in the body undergoes the same angular displacement about that axis. More important, an angular displacement about a *different* axis will result in a different *linear* displacement $d\mathbf{r}$ of the particle. It therefore seems reasonable to choose for the direction of the vector $d\boldsymbol{\theta}$ the direction of the rotation axis, because rotation about that axis produces a unique $d\mathbf{r}$.

It is not clear that arbitrarily choosing vectors in this way satisfies the critical requirements for a vector, however. For example, do two such angular displacement vectors commute under addition? According to Example 1 in the text (p. 298), *finite* rotations about different axes gives different results. If the rotations are *infinitesimal*, however, the rotations do commute, so it is reasonable to associate a vector property with $d\boldsymbol{\theta}$ and to assign its direction to be along the axis of rotation. We are still left with the decision as to *which* direction along the axis to assign. The convention that we follow is called the *right-hand rule* or the *right-hand-screw rule*. If a right-hand screw is turned in such a way that a point on the thread turns through an

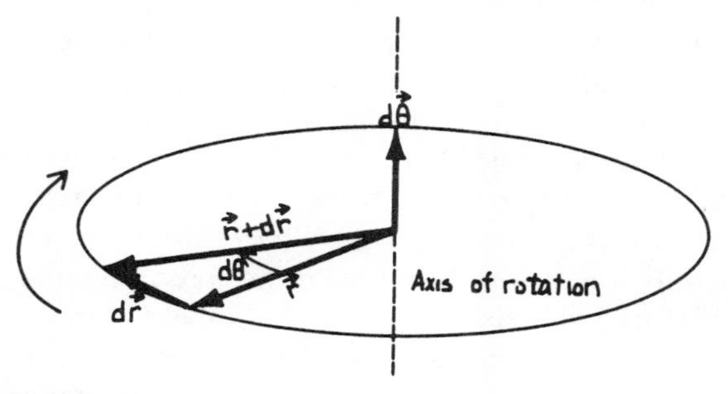

Figure 12.1 Development of the notion of a vector direction for $d\boldsymbol{\theta}$.

angle $d\theta$, the screw will advance in the direction $d\theta$. In an analogous fashion, the fingers of the right hand can substitute for the screw threads, and the thumb will point in the direction of $d\theta$ (Figure 12.2).

Once the vector nature of $d\theta$ has been established, it is easy to establish a direction for $\omega = d\theta/dt$. Thus, we see that ω has the same direction as $d\theta$. But what about $d\omega$? If the axis of rotation is fixed, the vectors ω and $d\omega$ are *parallel* or *antiparallel*, depending on whether ω is increasing or decreasing. We found this same type of connection between **v** and $d\mathbf{v}$ in straight-line motion. If, however, the axis of rotation changes direction in space, we are faced with a problem of dynamics, and we leave the discussion of that situation to Chapter 13.

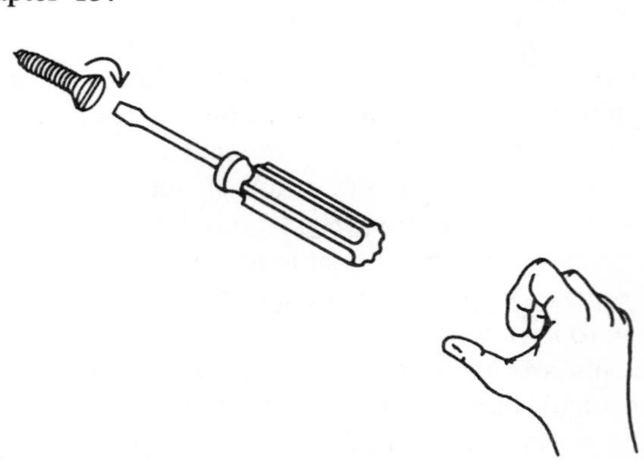

Figure 12.2 The right-hand-screw or right-hand rule for determining the direction of $d\theta$.

3. *Kinematics of rotation of a rigid body*. The relationships between the linear (tangential) motions of a particle in a rigid body rotating about a fixed axis and the angular motions about that same axis show a simple correspondence. Each linear variable—s, v, and a—associated with a point on the body is related to its rotational counterpart—θ, ω, and α, respectively—by a single multiplicative constant, R, the distance from the axis to the point. Therefore, all the kinematic equations presented in Chapter 2 may be used for rotational motion after a one-to-one substitution ($s \rightarrow \theta$, $v \rightarrow \omega$, and $a \rightarrow \alpha$) is made. No new concepts are needed; we need only to think in terms of *rotation* rather than *translation*.

4. *The radian*. One difficulty in thinking in terms of rotation is the need to use a new unit called the *radian*. The radian in one sense is really not a unit—it is the *ratio* of two lengths and therefore is dimensionless. For example, the so-called unit rad/s^2 is *dimensionally* the same as 1/s^2. But you should remember that any angular displacements involved in these equations are measured in *radians*, not in degrees. We note further that a common unit for angular velocity is *rev/s* or, more commonly, rev/min (*rpm*). To use the standard kinematic equations, it is necessary to have consistency in the units. Generally it is better to stay with the standard units of radians and seconds. Therefore, we need to be able to convert rpm to rad/s, for example. Because 2π rad corresponds to one revolution in angular displacement and 60 s is equal to one minute, we have 1 rpm = 2π rad/60 s = $\pi/30$ rad/s, or 1 rpm $\approx$ 0.1 rad/s.

5. *Moment of inertia.* As the term suggests, *moment of inertia* concerns resistance to acceleration; it is a measure of the resistance to *angular* acceleration. Just as a mass need not be accelerating to have inertia, a body does not have to be rotating to possess a moment of inertia. Indeed, moment of inertia is, in one sense, a *geometric* property of a body, because it is related both to the mass and to the mass *distribution* about a particular axis. And unlike ordinary inertia, which depends only on the mass, the moments of inertia for two bodies having identical mass may be quite different. Recall that the center of mass of a body depends heavily on the mass distribution because of the terms $\sum_i m_i\mathbf{r}_i$. The moment of inertia takes this concept one step further by its dependence on the square of the distance from a particular *axis*, that is, $I = \sum_i m_i r_i^2$. Note that the center of mass is reckoned relative to a *point* and involves the *position vectors* of the various mass elements, whereas the moment of inertia is always relative to an *axis* and, because of the r_i^2 term, depends on the *distance* of each mass element from the axis. Thus, the center of mass is a position *vector*, whereas the moment of inertia is a *scalar*. For example, two equal masses located on opposite sides of an axis will have their center of mass located on the axis. Moving the masses equal distances, but in opposite directions, will not change the location of the center of mass but will change the moment of inertia. And because of the r_i^2 dependence, moving the mass away from the axis has a dramatic effect on the moment of inertia. This sensitivity of the moment of inertia to r_i is illustrated in sample problem 7 in this chapter. For continuous mass distributions, we must use the integral calculus to compute the moment of inertia. The most difficult part of this procedure is deciding on the proper differential mass element *dm*. You should study the examples in the text carefully for insights into how these elements are chosen.

The additivity of moments of inertia is a feature that often allows us to break a complex shape into two or more simpler shapes for which the moments of inertia either are known or can be computed easily. The additivity of moments of inertia is evident from the definition $I = m_1r_1^2 + m_2r_2^2 + \ldots + m_ir_i^2 + \ldots$, where m_i is one of the mass elements of the body. These terms can be grouped in any order, so it is clear that the overall moment of inertia of the body is the sum of the moments of inertia of its parts. We must remember, however, that this is true only if all moments of inertia are referred to the *same axis*.

Although we relate the moment of inertia to the *torque* explicitly in Chapter 13, as a check you should note that the larger the moment of inertia, the more difficult it is to give the body an angular acceleration about the designated axis. We know from experience that the nearer the mass is concentrated to the axis of rotation, the easier it is to set the body into rotation. For example, it is easier to twirl a baton rapidly about its center than about one end (even if we could put an easy-grip handle at one end). This fact is reflected in the moments of inertia about axes through the end and the middle of the baton, which are $\frac{1}{3}ML^2$ and $\frac{1}{12}ML^2$, respectively.

6. *The parallel-axis and perpendicular-axis theorems.* The *parallel-axis theorem*, which states that $I = I_{CM} + Md^2$, is useful in computing moments of inertia about axes through points other than the center of mass. As the

name implies, the two axes must be *parallel*, but the new axis does not have to pass through the body. The *perpendicular-axis theorem,* $I_z = I_x + I_y$, has limited usefulness, because it involves the moments of inertia about mutually perpendicular axes of a thin, flat plate. There are situations, however, when this theorem greatly simplifies the determination of the moment of inertia, especially for symmetric bodies.

7. *Rotational kinetic energy*. You may recall that the rotational variables were introduced because of the difficulty of dealing with a multitude of individual particles in the rigid body, each having its own position, velocity, and acceleration. Because kinetic energy involves the square of the speed, the same difficulty occurs in attempting to define the kinetic energy of the rotating rigid body. Observing that each part of the body has the same angular speed, ω, however, enables us to write the kinetic energy as $K_{\text{rot}} = \frac{1}{2}I\omega^2$. This is not a different kind of kinetic energy, however, but simply a different way of expressing it. In Chapter 13 we will see how this form of expressing the kinetic energy has considerable utility when we consider both translation and rotation.

8. *Angular momentum of a system of particles*. We can define the angular momentum of a system of particles in much the same way that we define the angular momentum of a single particle. Both the center-of-mass motion and the motion of the system relative to the center of mass are involved here also. In fact, it can be shown that the angular momentum about *any* point O in space may be viewed as having two parts: the angular momentum about the center of mass plus the angular momentum associated with the motion of the center of mass about the point O. This equivalence can be written $\mathbf{L} = \mathbf{L}_{\text{CM}} + \mathbf{v}_{\text{CM}} \times M\mathbf{v}_{\text{CM}}$, where $\mathbf{v}_{\text{CM}}$ is the position of the center of mass relative to O.

9. *Angular momentum of a rigid body under rotation*. As is discussed in the text, the angular momentum vector $\mathbf{L}$ does not always point along the same direction as the angular velocity vector $\boldsymbol{\omega}$. Figure 12.25 from Example 12 in the text (p. 314) is one such instance. The relationship between the change in the angular momentum vector and the torque is discussed fully in Chapter 13. We will note here that the situation in Example 12 is called *dynamic imbalance* and causes a "shimmy" in the system as the rotation continues. This happens, for example, when a tire has mass asymmetry such that more mass is on one side of the tire. *Static* balance is possible, but the axis of rotation is not an axis of symmetry.

Perhaps the most useful outcome of the development in Section 12.6 is the fact that the z component of the angular momentum can be written as $\mathbf{L}_z = I\boldsymbol{\omega}$, when the axis of rotation is the z axis. If the axis of rotation happens also to be an axis of symmetry, then $\mathbf{L} = I\boldsymbol{\omega}$. For example, if masses are placed diametrically opposite to those in text Figure 12.25 and if these masses are equal to those already there, the z axis becomes an axis of symmetry and $\mathbf{L}$ points along the z axis. There is no longer any imbalance. This is the reason for placing weights on both sides when balancing a tire to avoid creating a dynamic imbalance.

Sample Problems

1. *Worked Problem.* A twirler throws a baton into the air so that it reaches a height of 10 m. What angular velocity must the baton be given initially if it is to make exactly 12 revolutions while in the air?

Solution: The baton will rotate at constant angular velocity while in the air, so our task is find out how long the baton is in the air. The angular velocity will be 12 rev/time, converted to radians per second. The time is found from the fact that the only energy *change* involved is the conversion of center-of-mass kinetic energy to potential energy.

$$\tfrac{1}{2}I\omega^2 + \tfrac{1}{2}M\upsilon_{CM_0}^2 = \tfrac{1}{2}I\omega^2 + Mgh$$

$$\upsilon_{CM_0}^2 = 2gh$$

$$\upsilon_{CM_0} = \upsilon_{CM_0} + at$$

$$0 = \sqrt{2gh} - gt$$

$$t = \sqrt{2gh}/g = \sqrt{\frac{2h}{g}} = \sqrt{\frac{2 \times 10 \text{ m}}{9.8 \text{ m/s}^2}} = 1.43 \text{ s}$$

$$2t = 2.86 \text{ s}$$

$$\frac{12 \text{ rev}}{2.86 \text{ s}} = 4.2 \text{ rev/s} = 26 \text{ rad/s}$$

2. *Guided Problem.* The speedometer on a car displays the linear speed of the car by measuring the rotation rate of the wheels. But the speedometer must be calibrated for a wheel with a particular diameter, since $\upsilon = \omega R$. Suppose your car comes equipped with 33 cm radius wheels and you decide to change to 38 cm radius wheels. If your speedometer reads 80 km/h, how fast are you really traveling?

Solution scheme

a. Consider the speedometer reading to be 80 km/h for both sets of tires. What does this say about ω in each case?
b. Compare the linear speed of the axle for the 33 cm tire with that for the 38 cm tire.

3. *Worked Problem.* Consider the belt-drive system shown in Figure 12.3. The smaller pulley has an angular acceleration α equal to 4.5 rad/s². After how long a time will the second pulley reach a rotation speed of 600 rpm if the system starts from rest and there is no slipping?

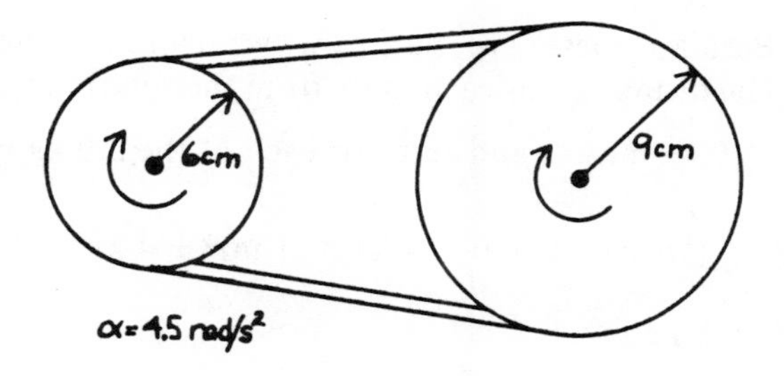

Figure 12.3 Problem 3. The smaller pulley connects to the larger by a belt that does not slip.

Solution: Assuming the belt does not stretch, each point on the belt has the same *tangential* speed, so the tangential speeds of the rims of the pulleys must be the same. Since $\upsilon = \omega R$, we have

$$\upsilon_1 = \omega_1 R_1$$
$$\upsilon_2 = \omega_2 R_2$$

So

$$\omega_1 R_1 = \omega_2 R_2$$
$$\omega_1 = \frac{R_2}{R_1}\omega_2$$
$$\omega_2 = 10 \text{ rps} \times 2\pi = 62.8 \text{ rad/s}$$
$$\omega_1 = \frac{9 \text{ cm}}{6 \text{ cm}} \times 62.8 \text{ rad/s} = 94.2 \text{ rad/s}$$
$$t = \frac{\omega_1}{\alpha} = \frac{94.2 \text{ rad/s}}{4.5 \text{ rad/s}^2} = 20.9 \text{ s}$$

4. *Guided Problem.* A uniform solid disk has an angular acceleration of 3.5 rad/s^2 about an axis through its center and perpendicular to the plane of the disk. The initial angular velocity is 2.0 rad/s.

a. Determine the *total* acceleration, **a**, as a function of time for a point on the disk 20 cm from the axis.
b. When $\theta = 6\pi$ rad, what is a?

Solution scheme: To find **a**, we need the vector combination of the tangential acceleration, $\mathbf{a}_t$, and the radial acceleration, $\mathbf{a}_r$.

a. Given the angular acceleration and the initial angular velocity, what is ω as a function of time?
b. What is the relationship between the radial acceleration, a_r, the angular velocity, ω, and the distance, r, from the rotation axis?
c. How is the tangential acceleration, a_t, related to the angular acceleration, α?
d. What are the respective directions of a_t and a_r?
e. Determine the angle between a_t and a_r to specify the direction for a.
f. Finally, what is the relationship between the angular displacement, θ, and the angular speed, ω? Use this to compute a_r.

5. *Worked Problem.* A configuration of four point masses is shown in Figure 12.4. Compute the moment of inertia about the two axes indicated.

Solution: Because these are particles, we can use the relationship $I = \sum_i m_i r_i^2$. The distance r_i of each mass from the rotation axis is 1.0 m for each of the 4.0 kg masses and zero for each of the 2.0 kg masses. Thus

$$I_1 = \sum_{i=1}^{4} m_i r_i^2 = 0 + 0 + 4 \text{ kg} \times (1 \text{ m})^2 + 4 \text{ kg} \times (1 \text{ m})^2$$
$$I_1 = 8 \text{ kg-m}^2$$

For the second axis, we have for each mass:

$$r_i^2 = (1\ \text{m})^2 + (0.5\ \text{m})^2 = 1.25\ \text{m}^2$$

$$I_2 = \sum_{i=1}^{4} m_i r_i^2 = r^2 \sum_{i=1}^{4} m_i = 1.25\ \text{m}^2 \times 12\ \text{kg}$$

$$I_2 = 15\ \text{kg-m}^2$$

6. *Guided Problem.* Table 12.1 in the text (p. 309) gives the moment of inertia of a cylinder about a diameter through the center. Derive this formula.

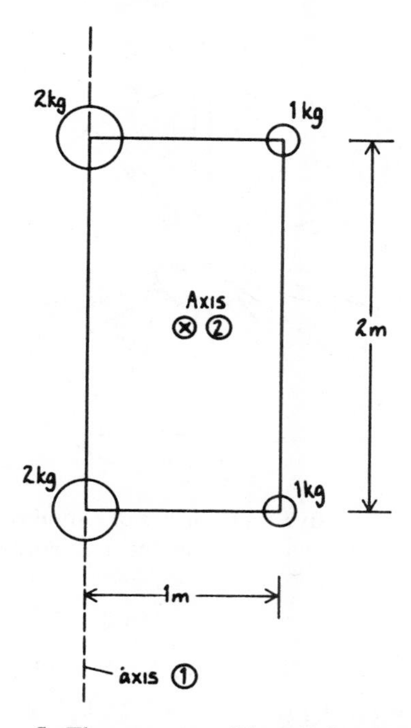

Figure 12.4 Problem 5. The moment of inertia is to be computed about each of two axes.

Solution scheme

a. The cylinder can be thought of as a series of thin disks. Consider a disk of thickness dz located a distance z from the axis. What is the moment of inertia of this disk about its own diameter?

b. Using the parallel-axis theorem, find the moment of inertia of this disk about the central axis.

c. What is the mass dm of this disk?

d. Use the continuous-mass form of the moment-of-inertia definition and integrate from $-L/2$ to $L/2$. What is the total mass of the cylinder?

e. As a check, note that if $R \rightarrow 0$, the expression is for the moment of inertia of a long, thin rod about a central axis perpendicular to the rod, obtained by using the parallel-axis theorem. If $L \rightarrow 0$, we have the moment of inertia of a thin disk about a diameter.

7. *Worked Problem.* A flywheel (Figure 12.5) consists of an outer rim of mass 30 kg, having inner and outer radii of 1.8 m and 2.0 m, respectively; 10 spokes, which may be thought of as long, thin rods 1.3 m long and each having a mass of 300 g; and an inner disk of radius 0.5 m and mass 10 kg. Calculate the moment of inertia about an axis through the center perpendicular to the wheel.

Solution: The wheel consists of three basic parts: a rim, a set of spokes, and an inner hub. The additive nature of the moment of inertia allows us to calculate these separately and then add them for the total amount of inertia.

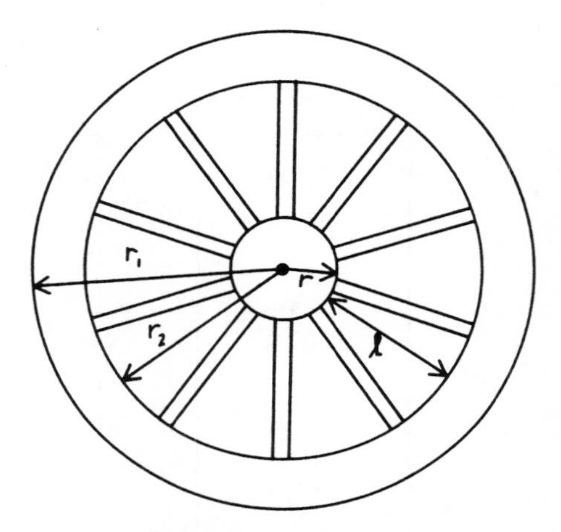

Figure 12.5 Problem 7. A flywheel consists of a solid rim, spokes, and an inner hub. Where is most of the moment of inertia about the central axis?

Rim: $I_r = \frac{1}{2}M(R_1^2 + R_2^2) = \frac{1}{2}30\text{ kg}\,[(1.8\text{ m})^2 + (2.0\text{ m})^2] = 108.6\text{ kg}\cdot\text{m}^2$

Hub: $I_h = \frac{1}{2}Mr^2 = \frac{1}{2}10\text{ kg}\,(0.5\text{ m})^2 = 1.25\text{ kg}\cdot\text{m}^2$

Spoke:
$$\begin{aligned} I_s' &= \tfrac{1}{12}M_s l^2 + M_s r^2 \\ &= \tfrac{1}{12}(0.30\text{ kg})(1.3\text{ m})^2 + (0.30\text{ kg})\,(1.15\text{ m})^2 \\ &= (0.042 + 0.40)\text{ kg}\cdot\text{m}^2 \\ &= 0.44\text{ kg}\cdot\text{m}^2 \end{aligned}$$

For 10 spokes,

$$\begin{aligned} I_s &= 10I_s' = 4.4\text{ kg}\cdot\text{m}^2 \\ I_{tot} &= 114\text{ kg}\cdot\text{m}^2 \end{aligned}$$

Thus, 96% of the total moment of inertia is in the rim. We are thus justified in approximating the flywheel as a thin hoop. In fact, we get roughly the same answer by considering only the rim and using the average radius, 1.9 m.

8. *Guided Problem.* In comparing moments of inertia of various bodies, we often find it convenient to reduce the body under consideration to an equiva-

lent thin hoop having the same moment of inertia. Because the hoop has a moment of inertia $I = MR^2$, we can write the moment of inertia of *any* body about a given axis as $I = Mk^2$, where k is the ***radius of gyration***. Essentially, k is the radius of an equivalent thin hoop centered on the rotation axis. For example, for a solid sphere, the moment of inertia about a diameter is 2/5 MR^2, so $k^2 = \frac{2}{5}R^2$. Suppose it is determined experimentally that a certain 5.6 kg body has a moment of inertia about a particular axis equal to 3.7 times the moment of inertia of a 5.0 kg, 10 cm-radius solid disk about its axis of symmetry.

a. Determine the radius of gyration of the object.
b. If the object is spinning at the rate of 20 rad/s, what is its kinetic energy?

Solution scheme

a. What is the moment of inertia of the disk?
b. What is the numerical value of the moment of inertia of the object under investigation?
c. What is the relationship among the moment of inertia, the mass, and the radius of gyration?
d. What is the expression for K in terms of I and ω?

Answers to Guided Problems

2. 92 km/h
4. b. $a = 3.8 \text{ m/s}^2$ at 5.3°
   a. $\mathbf{a} = \sqrt{(0.35)^2 + (0.4 + 2.8t + 1.2t^2)^2} \text{ m/s}^2$
6. See Table 12.1 in the text.
8. a. $k = 12.8$ cm
   b. $K = 18.5$ J

Chapter 13

Dynamics of a Rigid Body

Overview

This chapter synthesizes virtually all the concepts developed thus far. Furthermore, it shows that the laws governing the translational motions of particles have their counterparts in rigid-body rotation and that these laws are closely related in form. Thus, we bring together the kinematics and the dynamics, including energy relationships, of translational and rotational motions of a rigid body.

Essential Terms

torque
moment arm
equation of rotational motion
work in rotational motion
power in rotational motion
conservation of energy
conservation of angular momentum
rolling without slipping
precession

Key Concepts

1. *Torque*. The word *torque* comes from the Latin *torquere*, meaning "to twist," and it is from this mental picture that we can get a feel for the concept of torque. Consider a situation in which you are attempting to tighten (or "twist") a nut with a wrench. Your intuition probably tells you that the job is easier if you have a longer wrench, that is, if your effort is applied at a greater distance from the nut. Given that distance, you would not push on

the wrench in a direction toward the nut. In fact, you get the best results when you push perpendicular to the handle. All of this suggests that in computing the torque, the essential factors are the force, the distance at which the force is applied, and the angle that the force makes with the arm. In fact, the actual relationship is $\tau = rF \sin \theta$. But from vector algebra we know that this is the expression for the cross (vector) product of **r** and **F**. Thus, $\boldsymbol{\tau} = \mathbf{r} \times \mathbf{F}$.

Although we have described a familiar situation in which a twisting effect on a real body is expected, the torque is defined without reference to an actual twisting on real objects. Indeed, the origin about which the torque is specified need not be on the body at all. For most applications of this concept, however, we shall be assuming an origin at the center of mass, at least in the dynamics applications. The location of the origin is relatively unimportant in the statics analysis and is largely a matter of convenience and simplicity. The important thing to remember is that the direction of the torque vector is never the same as the direction of the force. The torque vector obeys the right-hand rule governing cross products and is in a direction perpendicular to both **r** and **F**. Thus, the direction of the torque is defined in a way similar to that of the angular velocity. Recall that in the latter case the particles making up the body have tangential velocity vectors that lie in a plane perpendicular to the rotation axis, but the $\boldsymbol{\omega}$ vector points along the axis of rotation. Just as the direction of $\boldsymbol{\omega}$ is not the same as that of the particle velocities, the direction of the torque vector $\boldsymbol{\tau}$ is not the same as the directions of the forces being applied to the particles. By defining these two quantities in an analogous way, we find very useful relationships between $\boldsymbol{\tau}$ and $\boldsymbol{\omega}$.

2. *The moment arm*. The concept of the *moment arm* (sometimes called the *lever arm*) is useful in dealing with torques. Technically, the moment arm is the $r \sin \theta$ term in the torque definition. It is computed by taking the shortest distance from the origin to the *line of action* of the force. To understand this conceptually, note that any force whose line of action passes through the origin will have zero torque (recall the wrench analogy earlier). Thus, if a force makes an arbitrary angle with the position vector, we can consider the force to have two components, one perpendicular to the position vector and one parallel to it. The parallel component exerts no torque; only the perpendicular component can do so. Therefore, if **r** is the position vector of the point of application of the force, then $F \sin \theta$ is the component perpendicular to it and so $\tau = r(F \sin \theta)$.

Thus, the torque may be viewed as either the position vector (magnitude) r multiplied by the perpendicular component of the force or the full force F multiplied by the moment arm, the component of the position vector perpendicular to the force. We encountered a similar idea in the concept of *work*, except that in that case we were dealing with the dot product. Nonetheless, the work, $\mathbf{F} \cdot \mathbf{d} = Fd \cos \theta$, is considered to be the actual displacement multiplied by the parallel component of the force, or the full force multiplied by the parallel component of the displacement.

3. *Equations for rotational motion.* As is discussed in the text, the equations for rotational motion are similar in form to those for translational motion of a particle, with the following correspondences:

$$\begin{aligned} \mathbf{F} &\rightarrow \tau \\ m &\rightarrow I \\ \mathbf{p} &\rightarrow \mathbf{L} \\ \mathbf{v} &\rightarrow \omega \\ \mathbf{a} &\rightarrow \alpha \\ d\mathbf{r} &\rightarrow d\theta \end{aligned}$$

Although this similarity in form makes the transition from translation to rotation appear fairly simple, it is often difficult to keep up with the vector directions of the rotational quantities. Doing so becomes important when we consider the rolling-without-slipping motion, because we must relate the translational variables to the rotational variables, and the selection of the proper signs is crucial. We will deal with that issue in a later section.

4. *Work–energy theorem in rigid-body rotation.* In the case of pure rotation about a fixed axis, the work–energy theorem is quite similar to that dealing with the purely translational motion of a particle. In the rotational case, we have to express the kinetic energy in rotational variables; that is, we must write $K = \frac{1}{2}I\omega^2$ rather than $K = \frac{1}{2}mv^2$. The work–energy theorem then states that $W_{NC} = \Delta E = \Delta U + \Delta K = \Delta U + \Delta(\frac{1}{2}I\omega^2)$. The work term, W_{NC}, for pure rotational motion, is given by $W_{NC} = \int \tau_z d\phi$, where τ_z is the sum of the torques of the nonconservative forces about an origin on the z axis. The term ΔU is the potential-energy change, and if the center of mass has any vertical change in position, then $\Delta U = Mgz_{CM}$. Another potential-energy term is possible, analogous to the $U = \frac{1}{2}kx^2$ equivalence for a Hooke's Law spring. If a rod is twisted about its axis so that the restoring torque is proportional to the angular displacement, that is, $\tau = K\phi$, then $W = \int \tau d\phi = k\int_{\phi_0}^{\phi} \phi d\phi = \frac{1}{2}K(\phi^2 - \phi_0)^2$. We will not encounter this situation, however, and have included it only to show the extensiveness of the analogies between linear and rotational variables.

5. *Conservation of angular momentum.* In the study of linear momentum, we found that applying a general principle such as the conservation of momentum allowed us to examine the motion of bodies without having detailed information about the interaction forces. The same is true for rotation. The changes in rotational motion of the various parts of the system can be predicted without knowledge of the details of the internal torques. As long as the net external torque on the body is zero, the angular momentum remains constant. Even if parts of the body individually change angular momentum, the *net* change must be zero.

A special case of interest in which the torque is zero is the *central force*, that is, a force directed toward the origin. Because the force is antiparallel to the position vector **r**, the torque is zero and angular momentum is conserved. The increasing spin rates of divers and skaters is a direct consequence of such forces, along with the angular-momentum conservation in gravitational systems. Kinetic energy, however, is not conserved in these examples of the conservation of momentum. For instance, the ice skater increasing her spin as she pulls in her arms conserves momentum but not kinetic energy. In fact, her kinetic energy increases as a result of the work done by the muscles in pulling in her arms.

6. *Rolling motion*. In deriving the basic expressions $\Sigma \mathbf{F}_{net} = M\mathbf{a}_{CM}$ and $\Sigma \boldsymbol{\tau}_{ext} = d\mathbf{L}/dt$, we note that in the latter case the torques and the angular momentum must be taken about the same origin. In fact, because Newton's Second Law was used in the derivation of this relationship, one might expect the expression to be valid only for inertial reference frames. In actuality, if the center of mass of the body is selected as the origin for these quantities, it doesn't matter what the center of mass is doing. It can even be accelerating. Again we emphasize the importance of the center of mass as a reference point in these analyses.

An important combination of translation and rotation is that found in a rigid body rolling along a surface. In the most general case, the object will also be slipping at the point of contact with the surface. If we assume that the center-of-mass acceleration is along the x axis, we can write the equation of motion for the center of mass in its x and y components, giving us two equations for three unknowns. The torque equation plus the condition $f = \mu N$ allows us to solve the problem.

The more interesting (and more readily solvable) problems are those in which no slippage occurs as the body rolls along the surface. Because such a condition implies the application of static friction, we no longer have a relationship between **f** and **N**, since in general the motion is not such that the body is on the verge of slipping. Because the point of contact is an *instantaneous axis of rotation*, however, the velocity of the center of mass is related to ω, the angular velocity about the center of mass, by the expression $v_{CM} = \omega R$. Therefore, $a v_{CM}/dt = R\, d\omega/dt$ or $a_{CM} = R\, d\omega/dt$, and once again the solution is possible.

One example of motion that is equivalent to rolling without slipping is the case of a string wrapped around a cylinder that can turn about its symmetry axis. Because the string doesn't slip, the velocity, v, of any point on the string obeys the relationship $v = \omega R$, where ω is the angular velocity of the cylinder about its central axis. In a sense this situation is equivalent to that involving a cylinder whose center of mass is stationary while the surface (string) moves by.

As was noted earlier, the kinetic energies can be separated neatly into the energy of the center-of-mass motion and the energy of rotation about the center of mass. From our discussion of this concept in the preceding chapter, we know the kinetic energy of rotation to be the internal energy K_{int}. Indeed, for problems involving rolling without slipping, it is quite easy (and often advantageous) to approach the solution from an energy, rather than a dynamic, standpoint. Mechanical energy is conserved because, even though there is friction, it is static friction and does no work. If there is any slipping, then kinetic energy is no longer conserved.

7. *Precession of a gyroscope*. Not only is gyroscopic precession interesting to watch because it seems to run counter to our expectations, but the

analysis is virtually impossible without use of the vector properties of torque and angular momentum. Because of the relationship $\boldsymbol{\tau}_{ext} = d\mathbf{L}/dt$, we note that the external torque $\boldsymbol{\tau}_{ext}$ must be in the direction of $d\mathbf{L}$, not $\mathbf{L}$. Because $\mathbf{L}$ points along the axis of rotation (for rotation about an axis of symmetry), any torque about the center of mass that tends to move this axis in the vertical plane is a torque whose vector direction is in the horizontal plane. Therefore, $d\mathbf{L}$ is horizontal and the gyroscope axis moves in the horizontal plane. There is virtually no way to visualize this situation from a particle-dynamics point of view, yet from the torque–angular momentum relationship it is easy to predict (if not visualize) the motion that occurs.

8. *Center of gravity.* A concept used extensively in the study of torques on rigid bodies is that of the *center of gravity*. If the gravitational field over the body is uniform (which is true in most cases except those involving large astronomical bodies), then the center of mass and the center of gravity are coincident. The center of mass depends entirely on the mass distribution, however, and has no connection with gravity. The concept of center of gravity, in contrast, is derived from an analysis of *gravitational torques* and has no meaning in the absence of gravity. For our purposes, however, it is almost always safe to use the two terms interchangeably in a practical sense. But remember that *for purposes of computing gravitational torques* the center of gravity is that point at which all the mass of the body appears to be concentrated. For other considerations, such as the determining of moments of inertia, that assumption is not valid.

If the center of mass is accelerating but there is no rotation about the center of mass, the static conditions can still be applied *rotationally*; that is, the sum of the torques is still zero *if the torques are taken about the center of mass*. To apply the torques about any other origin will very likely lead to erroneous results.

Sample Problems

1. *Worked Problem.* An 80 kg man stands at the end of a 25 kg diving board whose dimensions are 6.0 m × 30 cm (Figure 13.1). The board is supported at the opposite end and at a distance x from that end. If the end support suddenly gives way, what is the value of x that would result in both the man's and the end of the board's having a downward acceleration of g?

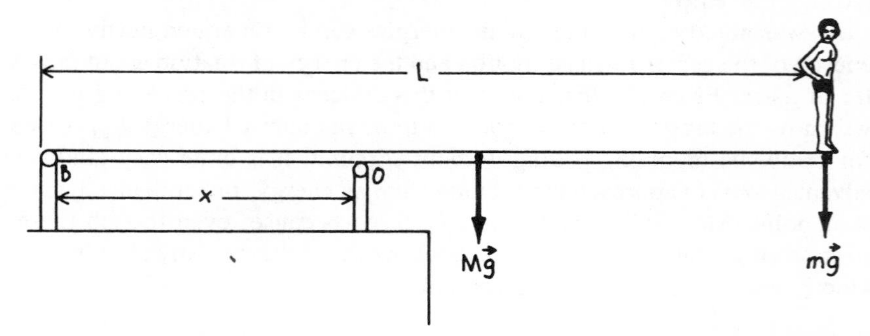

Figure 13.1 Problem 1. The support at B breaks, and the diver and the end of the board have acceleration g.

Solution: The board pivots freely about the inner support when the other support gives way. The man's weight and the weight of the board both exert torques about this pivot point. These torques result in an angular acceleration about the pivot; the *linear* (tangential) acceleration of the end of the board (and man) is related to the angular acceleration by the expression $a = R\alpha$, where R is the distance from the pivot to the end of the board; that is, $R = L - x$. Because the board is uniform, the mass may be considered concentrated at the center for purpose of computing gravitational torque. If the origin is taken at point O, then the torques about O are

$$Mg \cdot \left(\frac{L}{2} - x\right) + mg \cdot (L - x) = I\alpha$$

To find the moment of inertia, I_b, of the board about O, we can determine I_b about an axis through the center of mass, parallel to the pivot axis, and then use the parallel-axis theorem to find I_b about the pivot axis. Figure 13.1 shows the relevant geometry. The mass of the board, M, is given by $M = pLWt$, where p is the mass density and W is the width and t is the thickness. The volume element $dm = ptWdx$. Thus

$$I_{bCM} = \int x^2 dm = \rho t W \int_{-L/2}^{L/2} x^2 dm = \rho t W \left[\frac{x^3}{3}\right]_{-L/2}^{L/2}$$

$$I_{bCM} = \tfrac{1}{12} ML^2$$

Then I_b about the pivot is

$$I_b = \tfrac{1}{12} ML^2 + M\left(\frac{L}{2} - x\right)^2$$

The man can be considered a point mass for these calculations, so his moment of inertia about the pivot is $I_m = m(L - x)^2$. The total amount of inertia about the pivot, then, is

$$I_p = \tfrac{1}{12} ML^2 + M\left(\frac{L}{2} - x\right)^2 + m(L - x)^2$$

Since $\tau = I\alpha$, we have

$$Mg\left(\frac{L}{2} - x\right) + mg(L - x) = \left[\tfrac{1}{12} ML^2 + M\left(\frac{L}{2} - x\right)^2 + m(L - x)^2\right] \frac{a}{L - x}$$

Multiplying through by $L - x$ and setting a equal to g, we get

$$Mg\left(\frac{L}{2} - x\right)(L - x) + mg(L - x)^2 = \left[\tfrac{1}{12} ML^2 + M\left(\frac{L}{2} - x\right)^2 + m(L - x)^2\right] g$$

Note that the $mg(L - x)^2$ term is common to both sides, so it cancels. Then M divides out, and we have

$$\left(\frac{L}{2} - x\right)(L - x) = \tfrac{1}{12}L^2 + \left(\frac{L}{2} - x\right)^2$$

After the collecting of terms, this becomes

$$\tfrac{1}{3}L - x = 0$$
$$x = \tfrac{1}{3}L = 2 \text{ m}$$

Note the lack of dependence on m or M. Had the acceleration not been equal to g, these terms would not have divided out.

2. *Guided Problem.* A long, thin 1.0 kg rod of length 2.0 m is pivoted about its end on the floor. The rod is initially vertical. Given a slight push, it falls to the floor.

a. What is the angular acceleration of the rod as a function of ϕ, where ϕ is the angle made with the vertical?
b. At what angle does the end of the rod have a tangential acceleration that exceeds g?
c. What is the velocity of the center of mass as a function of ϕ?

Solution scheme: This situation involves a pure rotation about a fixed axis.

a. What is the torque exerted on the rod by gravity about the pivot when ϕ is some arbitrary angle?
b. What is the moment of inertia of the rod about the pivot?
c. What is the relationship between τ and α?
d. To find the angle for which $a > g$, relate the tangential acceleration to the angular acceleration and substitute $a = g$ for the limiting condition.
e. To find v_{CM} it is easier to use the conservation-of-energy approach, because a_{CM} is not constant. What is the initial potential energy of the center of mass relative to the floor?
f. What are the kinetic and potential energies at the arbitrary angle ϕ? Use rotational kinetic energy.
g. How is v_{CM} related to ω?

3. *Worked Problem.* A long, thin 1.5 kg rod of length 1.0 m hangs vertically from a pin passing through the rod at a distance h of 40 cm above the center of mass of the rod, as shown in Figure 13.2. At what point on the rod should a horizontal impulsive force be delivered so that there is no horizontal reaction force at the pin?

Solution: Consider an impulsive force $F(t)$ delivered at a point a distance z below the center of gravity. In general, there will be a horizontal reaction force $F'(t)$ at the pin. By the impulse–momentum relationship, we have

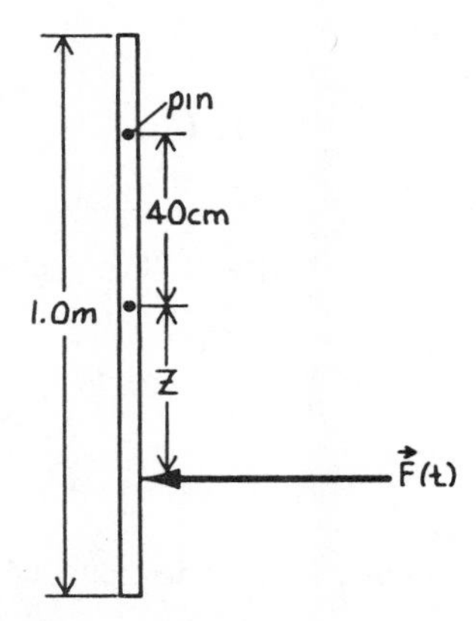

Figure 13.2 Problem 3. An impulsive force $F(t)$ is struck so that no reaction force is felt at the pin.

$$\int F(t)dt - \int F'(t)dt = \Delta p_{\mathrm{CM}} = M\upsilon_{\mathrm{CM}}$$

$$\upsilon_{\mathrm{CM}} = \frac{\int F(t)dt - \int F'(t)dt}{M}$$

But because the rod rotates about a fixed point O, the angular velocity, ω, about O is given by $\omega = \upsilon_{\mathrm{CM}}/h$. The impulsive torque is

$$(z + h)\int F(t)dt = \Delta L = I_0\omega$$

where I_0 is the amount of inertia about O. From the parallel-axis theorem, we have $I_0 = I_{\mathrm{CM}} + Mh^2 = \frac{1}{12}ML^2 + Mh^2$. Then, from the impulsive-torque equation, we have

$$\omega = \frac{(z + h)\int F(t)dt}{I_0}$$

or

$$\int F(t)dt = I_0\omega/(z + h)$$

From the center-of-mass expression, we have

$$\int F'(t)dt = \int F(t)dt - M\upsilon_{\mathrm{CM}} = \int F(t)dt - M\omega h$$

$$\int F'(t)dt = \int F(t)dt - \frac{Mh(z + h)\int F(t)dt}{I_0}$$

$$\int F'(t)dt = \int F(t)dt \left[1 - \frac{Mh(z + h)}{\frac{1}{12}ML^2 + Mh^2}\right]$$

If $\int F'(t)dt$ is to be zero—that is, if there is no horizontal reaction at the pin—then

$$\left[1 - \frac{Mh(z + h)}{\frac{1}{12}ML^2 + Mh^2}\right] = 0$$

Rearranging terms, we get

$$z = \frac{1}{12}\frac{L^2}{h}$$

Since the center of mass of the rod is 0.40 m below the support,

$$z = \frac{1}{12}\frac{(1\text{ m})^2}{0.40\text{ m}} = 0.21\text{ m}$$

or 61 cm below the support.

Comment: The expression just given for z can be written

$$z = \frac{k^2}{h}$$

lem 8, Chapter 12). When z is such that the condition $hz = k^2$ is satisfied, there is no reaction at the pin. The point at which the impulsive force is applied in this case is called the *center of percussion* for point O. In sports such as tennis and baseball, this is called the "sweet spot." The bat or racquet is pivoted at the hand (point O), and if the ball strikes at the center of percussion there is no reaction (sometimes called "sting") at the hand.

In another example, suppose you have a solid sphere whose radius of gyration relative to the center of mass is $k^2 = 2/5R^2$. If, as the sphere rolls on a table, it is struck at the center of percussion, there will be no sliding at the point of contact between the sphere and the table. Because $h = R$ in this case,

$$z = \tfrac{2}{5}R^2/R = \tfrac{2}{5}R$$

so that the center of percussion is $R + \frac{3}{5}R = \frac{7}{5}R$ above the table. This is the height of the side cushion on a pool table, so that when the ball rebounds there is no slipping and the angle of incidence is equal to the angle of reflection.

4. *Guided Problem.* A pendulum consists of a thin rod of negligible mass that is 1.0 m in length and is attached to the center of a 2 kg disk having a radius of 10 cm, as shown in Figure 13.3. The rod is pulled back through an angle of 53° and is released from rest.

a. What is the speed of the center of mass of the disk at its lowest point?
b. Now treat the disk as a 2 kg *particle* located at the center of mass of the disk. What percentage of error is made in the speed calculation making this assumption?

Solution scheme: This is another conservation-of-energy problem, with the kinetic energy being in rotational form (see problem 2).

a. Calculate the initial potential energy of the disk relative to its lowest point.

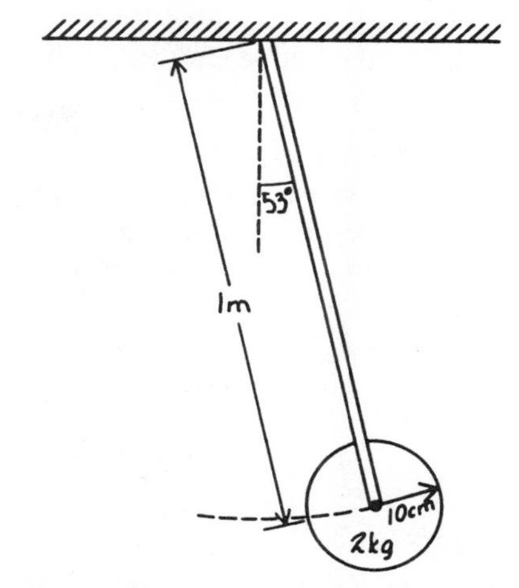

Figure 13.3 Problem 4. The disk is pulled back to a 53° angle and released.

b. What is the expression for the rotational kinetic energy when the disk is at the lowest point? Leave the expression in terms of I.
c. How is ω related to v_{CM}?
d. You should have an expression for v_{CM} as a function of I. For part b the expression has the same form except that the moment of inertia is different. Call it I' and the associated speed v'_{CM}.
e. Calculate the ratio v'/v.
f. To get the fractional difference, compute

$$\frac{v' - v}{v} = \frac{v'}{v} - 1$$

5. *Worked Problem*. A 2 kg disk having a radius of 20 cm is pivoted at a point on its circumference by a thin nail (Figure 13.4). The disk is pulled

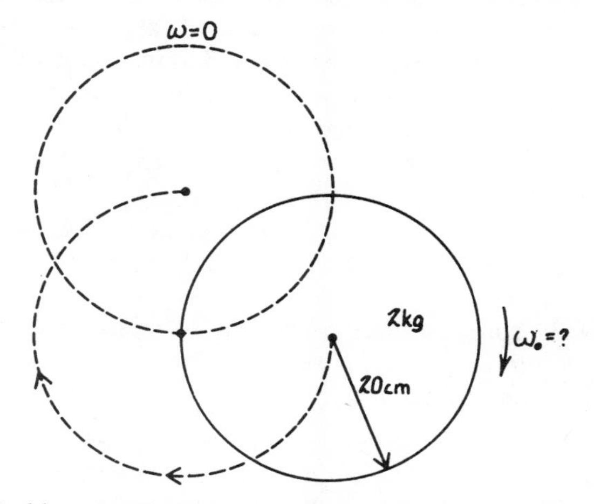

Figure 13.4 Problem 5. The disk is released at an initial angular velocity ω_0 such that its center of gravity just makes it to the top.

back so that its center of mass is at the same level as the nail. With what angular velocity must the disk be launched at this point so that the center of mass of the disk will just reach a point vertically above the nail?

Solution: This problem is perfectly suited for use of the conservation-of-energy principle. For convenience we set the zero of potential energy at the level of the nail. Then the initial mechanical energy is totally rotational kinetic energy. At the final position, all the energy is in the form of gravitational potential energy, or

$$\tfrac{1}{2}I\omega^2 = mgR$$

$$\omega^2 = \frac{2mgR}{I}$$

The moment of inertia of the disk about the nail is found by using the parallel-axis theorem:

$$I = I_{CM} + MR^2 = \tfrac{1}{2}MR^2 + MR^2 = \tfrac{3}{2}MR^2$$

Then

$$\omega^2 = \frac{2MgR}{\tfrac{3}{2}MR^2}$$

$$\omega = \sqrt{\frac{4}{3}\frac{g}{R}} = \sqrt{\frac{4}{3}\frac{9.8 \text{ m/s}^2}{0.20 \text{ m}}}$$

$$\omega = 8.1 \text{ rad/s}$$

6. *Guided Problem.* A solid disk and a hollow sphere having equal radii are allowed to roll without slipping down a 37° incline, starting from rest. If they start rolling simultaneously and their centers of mass are at the same elevation initially, how far apart will the centers of mass be after 3.0 s?

Solution scheme: Problems involving rolling objects generally can be solved by means of either the dynamic approach or the conservation-of-energy approach. In this case the dynamic approach appears preferable, because it gives us a_{CM} almost directly, and from a_{CM} we can get the displacement. The energy approach gives v_{CM} as a function of position, and we would still need to find v_{CM} as a function of time. If we take torques about the instantaneous axis of rotation, we avoid having to deal with the friction force.

a. What is the torque exerted by gravity about the axis of rotation?
b. Use the parallel-axis theorem to find the moment of inertia about this same axis.
c. Write the torque equation and use the relationship $\alpha_{CM} = a_{CM}/R$.
d. How is the distance traveled by the center of mass related to the acceleration and the time?

Note: The acceleration of the center of mass of either object is independent of both R and M. The stipulation in this problem that the two radii be equal served simply to enable each of the objects to start with their centers of mass at the same elevation.

7. *Worked Problem.* A 120 kg merry-go-round, a uniform disk with a radius of 1.5 m, is initially at rest. A 50 kg boy runs at a speed of 5.0 m/s along a path tangent to the rim and jumps on.

a. What is the angular speed of the merry-go-round after the boy jumps on?
b. How much energy is lost in the process of his jumping on?
c. If the merry-go-round comes to a stop in 40 s, what is the frictional torque exerted by the bearings?

Solution: This problem is the rotational analog of the perfectly inelastic collisions we studied in Chapter 11. The friction between the boy and the merry-go-round, as he is brought to rest relative to the merry-go-round, exerts equal and opposite torques about the central axis, so angular momentum is conserved.

a. If m is the mass of the boy and M is the mass of the merry-go-round,

$$m\upsilon R = I\omega$$

where I is the moment of inertia of the merry-go-round *and* of the boy.

$$m\upsilon R = (\tfrac{1}{2}MR^2 + mR^2)\omega$$

$$m\upsilon R = R_2(\tfrac{1}{2}M + m)\omega$$

$$\omega = \frac{m}{\tfrac{1}{2}M + m} \cdot \frac{\upsilon}{R} = \frac{50 \text{ kg}}{110 \text{ kg}} \times \frac{5 \text{ m/s}}{1.5 \text{ m}}$$

$$\omega = 1.5 \text{ rad/s}$$

b. The change in kinetic energy is

$$\Delta K = \tfrac{1}{2}I\omega^2 - \tfrac{1}{2}m\upsilon^2$$
$$\Delta K = \tfrac{1}{2}R^2(\tfrac{1}{2}M + m)\omega^2 - \tfrac{1}{2}m\upsilon^2$$
$$\Delta K = \tfrac{1}{2}(1.5 \text{ m})^2(110 \text{ kg})(1.5 \text{ rad/s})^2 - \tfrac{1}{2}(50 \text{ kg})(5 \text{ m/s})^2$$
$$\Delta K = -347 \text{ J}$$

Comment: Recall that the explosive interaction in linear-momentum conservation is the reverse of a perfectly inelastic collision. The present situation is the same; that is, it represents an explosive interaction in angular-momentum conservation. If the boy is initially on the merry-go-round traveling with it at an angular speed of 1.5 rad/s and if he jumps off tangentially in such a way that he is traveling 5 m/s relative to the ground and opposite to the rotation, the merry-go-round will come to rest and he will have used up 347 J of internal energy in the process.

c. The net torque about the center of mass is equal to the rate of change in angular momentum about the same axis, so

$$\tau_f = \frac{dL}{dt} = \frac{\Delta L}{\Delta t}$$

if τ_f is constant. Then

$$\tau_f = \frac{I\omega - I\omega_0}{\Delta t} = \frac{I(0 - \omega_0)}{\Delta t} = \frac{(\frac{1}{2}MR^2 + mR^2)(-\omega_0)}{\Delta t}$$

$$\tau_f = -\frac{R^2(\frac{1}{2}M + m)\omega_0}{\Delta t} = -\frac{(1.5 \text{ m})^2(110 \text{ kg})(1.5 \text{ rad/s})}{45 \text{ s}}$$

$$\tau_f = -8.25 \text{ N} \cdot \text{m}$$

8. *Guided Problem.* A roulette wheel consists of a disk-shaped solid of radius R and moment of inertia I_0 about a central axis. When it is set spinning at a constant angular velocity ω_0, a small ball of mass m is projected onto the wheel tangent to the rim and at a speed υ in a direction opposite to that of the rotation of the wheel. The ball eventually comes to rest in a slot located a distance r from the center. What is the final angular velocity of the wheel?

Solution scheme: Because the final velocity of the wheel is to be calculated and there are no net torques on the system (this is another problem involving "angular collision," similar to the last one), we can infer conservation of angular momentum.

a. What is the total angular momentum of the wheel and ball at the instant the ball is projected? Note that the ball's angular momentum has a sign opposite to that of the wheel's. Why?
b. After the ball comes to rest, what is the total angular momentum?

9. *Worked Problem.* The wheels on a motorcycle act as gyroscopes while the motorcycle is in motion. Show that to execute a right turn, the driver must first turn the handlebars to the left.

Solution: Using the right-hand rule, we find that the angular-momentum vector of the wheel points horizontally to the left (from the driver's point of view). To turn right and remain in rotational equilibrium, the motorcycle must lean to the right. This means that the *change* in the angular-momentum vector, $d\mathbf{L}$, must be directed vertically upward; therefore, the torque producing this change is directed vertically upward also. To produce a torque in this direction, the axis of the wheel must be turned counterclockwise (as seen from above), that is, to the left. This leftward motion is very brief and almost imperceptible, however, because as the motorcycle leans to the right as a result of this action, there is a torque on the wheel in the forward direction, making the wheel turn toward the right.

10. *Guided Problem.* A gyroscope is mounted on a rotating turntable, as illustrated in Figure 13.5. If the gyroscope's rotation is as indicated, which support will be subject to a greater reaction force?

Solution scheme

a. What is the direction of the angular-momentum vector of the gyroscope at the instant shown?
b. Because the gyroscope tends to maintain the direction of its $\mathbf{L}$ vec-

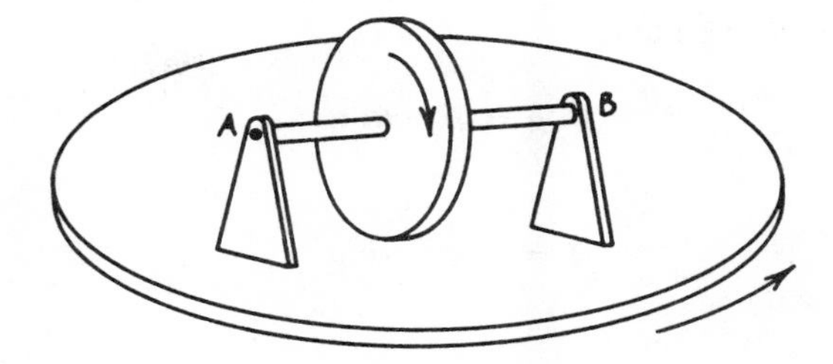

Figure 13.5 Problem 10. As the turntable rotates, the bearings A and B exert torques on the gyroscope.

tor, the supports will exert sideways forces on the axis of the gyroscope. What is the direction of the net torque on the gyroscope?

c. What is the $d\mathbf{L}$ resulting from this torque?

d. To determine which support must supply the greater reaction force, think about which direction the gyroscope axis will move in the absence of such a support.

11. *Worked Problem.* A disk with a mass of 15 kg and a radius of 25 cm is free to rotate about an axis through its center, as illustrated in Figure 13.6. A smaller disk whose mass is 0.5 kg and radius 5 cm is affixed coaxially to serve as a drum, about which a massless cord is wrapped. A 2 kg mass is attached to the end of the cord. If the system is released from rest and the cord comes off the drum after 5 revolutions, what torque must be applied to the disk to bring it to rest in 10 revolutions?

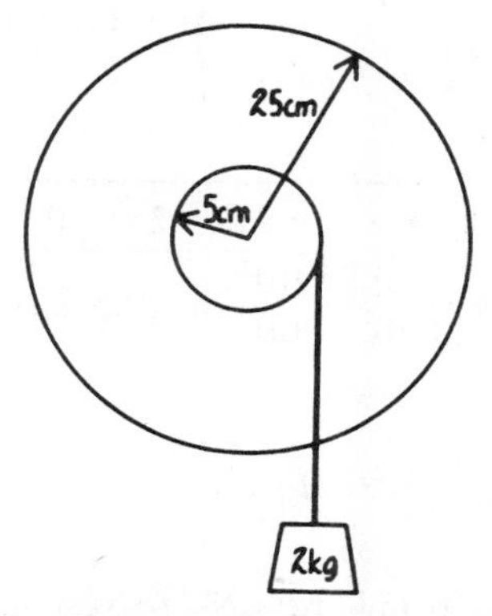

Figure 13.6 Problem 11. The mass on the cord rotates the pulley through 5 revolutions before dropping freely.

Solution: The problem can be approached in three different ways—dynamics ($\tau = I\alpha$), conservation of energy, and the torque–angular momentum relationship.

a. *The dynamic approach.* The torque on the disk is $\tau = Tr$, so that

$$Tr = I\alpha = (\tfrac{1}{2}MR^2 + \tfrac{1}{2}mr^2)\,\frac{a}{r}$$

$$T = (\tfrac{1}{2}MR^2 + \tfrac{1}{2}mr^2)\,\frac{a}{r^2}$$

For the mass m' we have

$$m'g - T = m'a$$

or

$$T = m'g - m'a$$

Then

$$m'g - m'a = \tfrac{1}{2}\left(M\frac{R^2}{r^2} + m\right)a$$

Finally

$$a = \frac{m'}{\tfrac{1}{2}M\dfrac{R^2}{r^2} + \tfrac{1}{2}m + m'}\,g$$

To find υ after 5 revolutions, we want

$$\upsilon^2 = \upsilon_0^2 + 2az$$

where

$$z = \frac{2\pi r}{\text{rev}} \times 5 \text{ rev}$$

But $\upsilon = r\omega$, where ω is the final angular velocity of the disk.

$$\upsilon = \sqrt{2\left(\frac{m'}{\tfrac{1}{2}M\dfrac{R^2}{r^2} + \tfrac{1}{2}m + m'}\right)g \cdot 2\pi r \cdot 5}$$

$$= \sqrt{\frac{2 \times 2\text{ kg} \times 9.8\text{ m/s}^2 \times 2\pi \times 0.05\text{ m} \times 5}{\tfrac{1}{2}(15\text{ kg})\dfrac{(0.25\text{ m})^2}{(0.05\text{ m})^2} + \tfrac{1}{2}(0.5\text{ kg}) + 2\text{ kg}}}$$

$$\upsilon = 2\text{ m/s}$$

$$\omega = \frac{\upsilon}{r} = \frac{2\text{ m/s}}{0.05\text{ m}} = 40\text{ rad/s}$$

The torque required to bring the disk to rest in 10 revolutions is found by again using the dynamical equation, $\tau = I\alpha$. But

$$\omega^2 = \omega_0^2 + 2\alpha\Delta\theta$$

so that

$$\alpha = \frac{0 - \omega_0^2}{2\Delta\theta}$$

and

$$\tau = (\tfrac{1}{2}MR^2 + \tfrac{1}{2}mr^2)\frac{-\omega_0^2}{2\Delta\theta}$$

$$\tau = -6\text{ N} \cdot \text{m}$$

b. *The conservation-of-energy approach*. After 5 revolutions the mass m' falls a distance z_0, so

$$m'gz_0 = \tfrac{1}{2}m'\upsilon^2 + \tfrac{1}{2}I\omega^2 = \tfrac{1}{2}m'\upsilon^2 + \tfrac{1}{2}I\,\frac{\upsilon^2}{r^2}$$

$$m'gz_0 = \tfrac{1}{2}\left(m' + \frac{I}{r^2}\right)\upsilon^2$$

$$m'gz_0 = \tfrac{1}{2}\left(m' + \tfrac{1}{2}M\,\frac{R^2}{r^2} + \tfrac{1}{2}m\right)\upsilon^2$$

$$\upsilon^2 = \frac{2\text{m}'gz_0}{m' + \tfrac{1}{2}M\,\dfrac{R^2}{r^2} + \tfrac{1}{2}m}$$

which is the same result we achieved before. The torque required to bring the disk to rest can also be found by means of the work–energy theorem.

$$\tfrac{1}{2}I\omega_0^2 + \int \tau\, d\theta = 0$$

If τ is assumed constant, then

$$\tau\Delta\theta = -\tfrac{1}{2}I\omega_0^2$$

$$\tau = -\tfrac{1}{4}(MR^2 + mr^2)\,\frac{\omega_0^2}{\Delta\theta}$$

$$\tau = -6\ \text{N}\cdot\text{m}$$

c. *The angular-momentum approach*. We start with the basic relationship

$$\tau = \frac{dL}{dt} = \frac{L - 0}{\Delta t}$$

This time we want the torque on the *system* (disk *plus* falling weight). When we consider the whole system, the tension force drops out as an internal force. Therefore, the *net* torque is just $m'gr$. Then

$$m'gr = \frac{m'\upsilon r + I\omega}{\Delta t}$$

But

$$\frac{\upsilon}{\Delta t} = a \quad \text{and} \quad \frac{\omega}{\Delta t} = \frac{a}{R}$$

$$m'gr = m'ra + \frac{Ia}{r}$$

$$m'gr = \left(m'r + \frac{I}{r}\right)a$$

$$a = \frac{m'}{\dfrac{I}{r^2} + m'}\,g = \frac{m'}{\tfrac{1}{2}M\,\dfrac{R^2}{r^2} + \tfrac{1}{2}m + m'}\,g$$

Which of the three methods is best is largely a matter of opinion and depends to some extent on the amount of calculation involved.

Comment: In methods a and c we calculated a torque while the string was still in contact with the drum. In the first approach we had $\tau = Tr$, whereas in the third method we had $\tau = m'gr$, which is not equal to rT. The difference is that in the first case we were computing the torque *only on the disk*. In the second we were computing the torque on the *entire system*.

12. *Guided Problem.* Figure 13.7 shows a 250 g mass suspended from a light string, which passes over a massless, frictionless pulley. The other end of the string is wound around a solid cylinder of mass 750 g and radius 20 cm. Calculate the linear acceleration of the center of mass of the cylinder if it rolls without slipping.

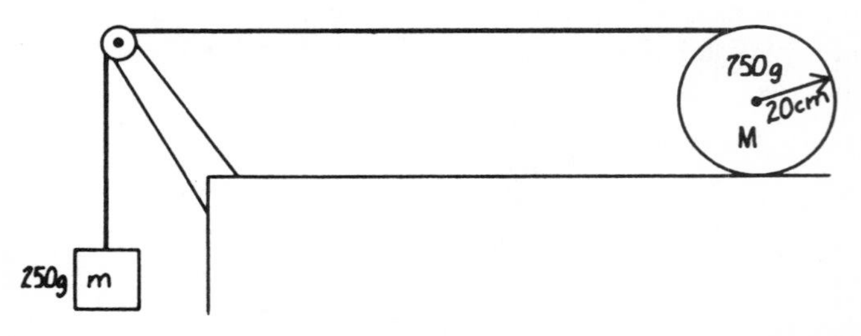

Figure 13.7 Problem 12. The cylinder rolls without slipping on the horizontal surface.

Solution scheme: To avoid having to deal with the static-friction force, it is best to take the torque about the instantaneous axis of rotation, which is the point of contact between the cylinder and the table.

a. What is the relationship among the tension, T, the mass, m, and the linear acceleration, a, of the mass m?
b. What is the torque about the rotation axis of M by the tension T?
c. What is the moment of inertia of the cylinder about the instantaneous axis?
d. How does the angular acceleration of the cylinder about this axis relate to a? (*Note:* The rim of the cylinder does not have the same linear acceleration as the center of mass.)
e. Eliminate T from the two equations, and solve for a. How is a related to a_{CM}?

Comment: If you go beyond the information asked for in the problem and calculate the friction force, you will find that $f = \frac{1}{4}Ma_{CM}$ and is in the same direction as the tension force. In fact, we can show as a general result that, if the rolling body has a *radius of gyration, k,* less than or equal to R, where R is the "rolling radius" (the distance from the point of contact on the surface to the center of mass), and the friction force is assumed to be in the direction of T, then

$$f = \tfrac{1}{2}Ma_{CM}\left(1 - \frac{k^2}{R^2}\right)$$

This relationship shows that under these conditions the friction force must be either zero or in the direction of T, because f can be negative only if $k > R$. If $k = R$, then $f = 0$, which would be the case, for example, if the rolling body were a thin hoop of radius R.

Answers to Guided Problems

2. a. $\alpha = \dfrac{3}{2}\dfrac{g}{L}\sin\phi$

   b. $a = \frac{3}{2}g\sin\phi$, so $a = g$ when $\sin\phi = \frac{2}{3}$. $\phi = 42°$.

   c. $\upsilon_{CM} = \sqrt{\frac{3}{4}gL(1 - \cos\phi)}$

4. a. $\dfrac{\upsilon'}{\upsilon} = \dfrac{I}{I'} = \dfrac{\frac{1}{2}MR^2 + ML^2}{ML^2} = 1 + \dfrac{\frac{1}{2}R^2}{L^2}$

   b. $\dfrac{\upsilon' - \upsilon}{\upsilon} = \dfrac{\frac{1}{2}R^2}{L^2} = 0.005$ or about 0.5%

6. Disk: $a_{CM} = \frac{10}{25}g$

   Sphere: $a'_{CM} = \frac{9}{25}g$

   $\Delta s = \frac{1}{2}(a_{CM} - a'_{CM})t^2$

   $\Delta s = 1.8$ m

8. $\omega = \dfrac{I_0\omega_0 - m\upsilon R}{I_0 + mr^2}$

   Comment: If $m\upsilon R = I_0\omega_0$, the wheel will stop. Note that this relationship is independent of r.

10. Support B, because $d\mathbf{L}$ points downward at that end.

12. $a_{CM} = \dfrac{1}{\left(2 + 3\dfrac{M}{m}\right)} g = \dfrac{1}{11} g = 0.90 \text{ m/s}^2$

Chapter 14

Statics and Elasticity

Overview

In our study of Newton's laws, we looked at the conditions necessary for a system to be in *translational* equilibrium. A body may be in translational equilibrium, yet not in *rotational* equilibrium. In this chapter we focus on the conditions necessary to achieve both types of equilibrium. In analyzing rotational equilibrium, we examine the general principles of operation of pulleys and levers and show how these devices behave as machines. Finally, we explore the ramifications of relaxing the assumption of perfect rigidity in "rigid bodies." This leads us to a study of stress–strain relationships and elastic moduli.

Essential Terms

static equilibrium
torque
force
moment arm
ideal mechanical advantage (IMA)
actual mechanical advantage (AMA)
lever
pulley
elastic
Young's modulus
bulk modulus
shear modulus

Key Concepts

1. *Statics of rigid bodies*. The most general condition for equilibrium of an extended body must take into account the fact that because forces may be applied at different points on the body, the condition $\Sigma \mathbf{F}_{ext} = 0$ is not sufficient; this relationship is appropriate only for the center-of-mass motion. For complete equilibrium there must be no angular acceleration of the body

either. In addition to the force requirement, therefore, we stipulate that the net torque on the body be zero.

In the case of *statics*, the body is not only in equilibrium but also at rest. Because it must be at rest relative to any point within that reference frame, it does not matter which point is taken to be the origin about which torques are computed. In fact, the origin may lie entirely outside the body. Although by convention counterclockwise torques are taken to be positive, the choice of sign in the static case is irrelevant, because the net torque is set equal to zero. The important consideration is that clockwise and counterclockwise torques be given opposite signs. The arbitrary choice of origin makes it possible to eliminate certain torques from the torque equation, because any force whose line of action passes through the origin has a zero torque. It is thus important to choose the origin carefully such that the number of unknowns is brought to a minimum. At some point, however, you may need to solve for a force that was eliminated earlier. If so, you can shift the origin to another point and include this force in the torque equation.

Included in the study of static equilibrium is the analysis of simple machines that are essentially pulleys and/or levers. The purpose of such devices is to multiply the effects of the input *force* (or, sometimes of the input *velocity*). Because these are machines, we must understand that the conservation of energy dictates an upper limit: *work in* = *work out*. For systems having energy losses (all *real* systems), we know that *work in* > *work out*, where we mean *mechanical* work. However, it is useful to assume the ideal case in order to get at the underlying concepts, much as we do in ignoring the effect of air friction on falling bodies. There are two basic, but closely related, ideas in the study of levers and pulleys as machines. One is the equilibrium-under-rotation concept, which allows one to decrease the required input force by increasing the moment arm, as in simple lever systems. The other is the concept of work as *force times distance*, which permits one to decrease the required input force by extending the distance over which the force acts, as in inclined planes or pulley systems. These ideas are linked, as can be seen in the lever

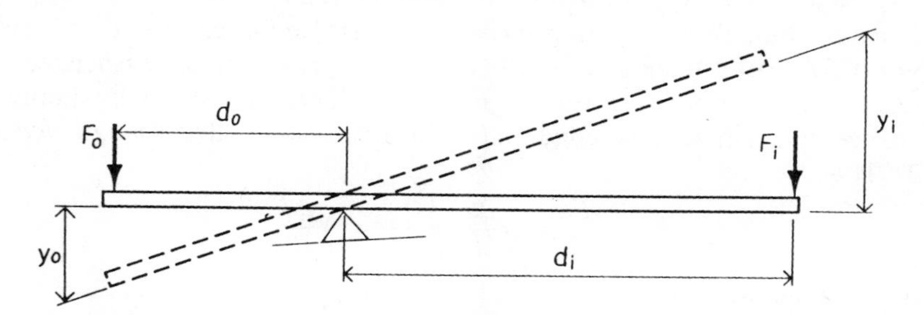

Figure 14.1 The lever as a machine involves the concepts of *torque* and *work*. For example, $F_o d_o$ is the torque of the force F_o, whereas $F_o y_o$ is the work done by that force.

example shown in Figure 14.1. The moment-arm concept tells us that the input force F_i is related to the output force F_o as follows:

$$\frac{F_o}{F_i} = \frac{d_i}{d_o}$$

The work concept shows that the relationship between the two forces and the distance they move is the following:

$$\frac{F_o}{F_i} = \frac{y_i}{y_o}$$

In this example, F_i is less than F_o, so F_i must move through a greater distance y_i to achieve the same work as the output force F_o moving through a distance y_o.

The *mechanical advantage* of a machine is simply the ratio F_o/F_i. This is the *actual mechanical advantage* (AMA), since it is true regardless of the efficiency of the system. If the system has no losses, the *ideal mechanical advantage* (IMA) is also equal to the ratio of the distances traveled (or ratio of the moment arms):

$$\text{IMA} = \frac{F_o}{F_i} = \frac{y_i}{y_o} = \frac{d_i}{d_o}$$

Note that the first ratio is also the AMA but that the other two pertain *only* to the IMA. It is easy to see why these are not the same in less than ideal systems. For example, suppose there is considerable friction in the fulcrum (pivot) of the lever. The ratios y_i/y_o and d_i/d_o are unaffected, since these are *geometrical* relationships. On the other hand, more input force is required, so the ratio F_o/F_i decreases, making AMA $<$ IMA.

2. *Elasticity of materials*. Many substances, on deformation by a force, return to their initial configuration. Such substances are said to be *elastic*. In studying the deformations resulting from external forces, we find it useful to introduce the concepts *stress* and *strain*, the former being the force per unit area and the latter being the change in dimension per unit dimension (change in length per unit length, change in volume per unit volume). The constant connecting stress and strain is the *elastic modulus*, defined in a manner consistent with the type of stress–strain. For example, if the substance is subject to a stretching or a squeezing force along a single dimension (as in a pulled wire), we call the stress a *tensile* or a *compressive* stress, associated with tensile and compressive forces, F_t and F_c. Then the strain is simply the change in length per unit length, $\Delta l/l$. The constant is called *Young's modulus, Y*. Thus,

$$\frac{F_t}{A} = Y\frac{\Delta l}{l} \text{ or } \frac{F_c}{A} = Y\frac{\Delta l}{l}$$

or, in general,

$$Y = \frac{F/A}{\Delta l/l}$$

If the compressive stress is associated with a *volume* strain, $\Delta V/V$, the stress is called the *pressure, P*. The associated elastic modulus is called the *bulk modulus, B*, and we have

$$B = -\frac{P}{\Delta V/V}$$

where the negative sign is to make B a positive number, since ΔV is a negative number. Often one uses the term *compressibility, K,* which is simply the inverse of the bulk modulus:

$$K = \frac{1}{B}$$

Finally, if the stress force acts tangentially to the surface, there is a tendency for the body to *shear*, and we have an associated *shear modulus, S,* defined by

$$S = \frac{F_s/A}{\Delta x/l}$$

where Δx is perpendicular to l, so that $\Delta x/l = \tan\theta \approx \theta$ (see Figure 14.20 in the text, p. 365).

Sample Problems

1. *Worked Problem.* Figure 14.2 shows a loose-fitting drawer 40 cm wide and 20 cm deep. If the drawer jams when being pulled out, the coeffi-

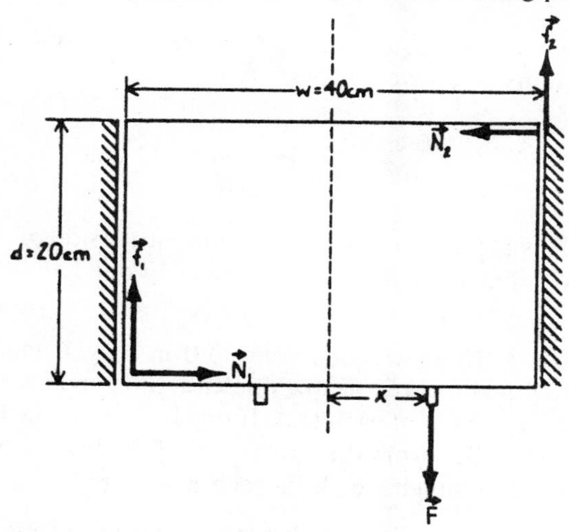

Figure 14.2 Problem 1. If the drawer is pulled by one handle, it sticks if x is too large.

cient of friction is 0.60. Suppose there are two handles on the drawer, located symmetrically on either side of the center line, as shown. If the drawer is pulled open by only one of the handles, how far from the center line can this handle be located to prevent jamming of the drawer?

Solution: If the drawer is pulled from a point to one side of the center line, there will be a tendency for the drawer to rotate in addition to its translation. This motion will bring the corners into contact with the sides. But if the drawer is not too loosely fit, the rotation angle will be negligible. The forces acting on the drawer are shown in Figure 14.2. When the drawer starts to jam, translational equilibrium requires that

$$F = f_1 + f_2 = \mu N_1 + \mu N_2$$

and

$$N_1 = N_2 = N$$

Therefore

$$F = 2\mu N$$

To have rotational equilibrium, we take torques first about the back corner, then about the front corner.

$$Nd - fw + F\left(\frac{w}{2} - x\right) = 0$$

and

$$Nd + fw - F\left(\frac{w}{2} + x\right) = 0$$

Combining the two torque equations, and using the results from the translational equations, we have

$$x = \frac{d/2}{\mu} = \frac{10\text{ cm}/2}{0.60} = 16.7\text{ cm} \approx 17\text{ cm}$$

Thus, if the handle is placed more than 17 cm to either side of the center line, the drawer will jam when it is pulled with only one handle.

2. *Guided Problem.* A 20 kg wooden beam 3.0 m long is tapered such that the center of gravity is located nearer one end. Two men are carrying the beam horizontally, one man at each end. If one is supporting 50% more weight than the other, how far from the lighter end of the beam is the center of gravity located and how much weight is each man supporting?

Solution scheme: The beam is in equilibrium, so both the forces and the torques about any point must be zero.

a. At what point does the weight of the beam appear to be concentrated?
b. What forces act on the beam vertically?
c. Let p be the fractional amount *more* that the man on the heavier end supports. How is the weight supported by this person related to that supported by the man on the lighter end?
d. Take torques about any point. (Taking those about one end will simplify the problem.)

3. *Worked Problem.* Show that in order for the cylinder to remain in the position illustrated in Figure 14.3, the coefficient of static friction between the wall and the cylinder must be such that $\mu \geqslant 1$.

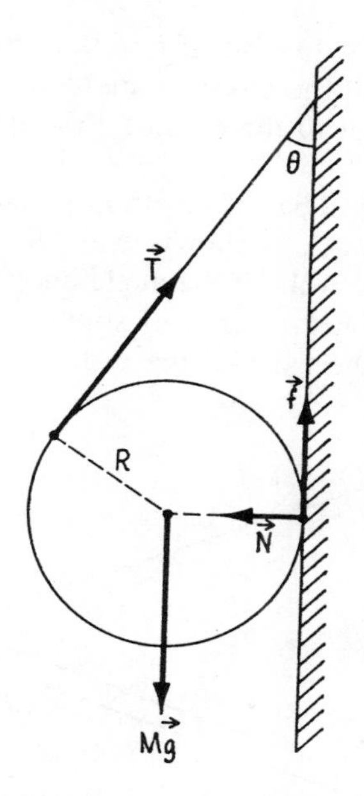

Figure 14.3 Problem 3. If the sphere is not to slip at the wall, $\mu \geqslant 1$.

Solution: Figure 14.3 shows the force diagram and the relevant geometry. The requirement for translational equilibrium gives

$$\Sigma F_x = 0 = T \sin \theta - N$$
$$\Sigma F_y = 0 = f + T \cos \theta - mg \qquad (1)$$

For rotational equilibrium we will take torques about the center of gravity of the cylinder. Then

$$\Sigma \tau_c = 0 = fR - TR$$

or

$$f = T \qquad (2)$$

If the cylinder is not to slip, the limiting condition is

$$f = \mu_s N \qquad (3)$$

Combining Eqs. (1), (2), and (3), we get

$$\frac{mg}{\cos \theta + 1} (\cos \theta + \mu \sin \theta) = mg$$

or

$$\mu = \frac{1}{\sin \theta}$$

Because $0 \leqslant \sin\theta \leqslant 1$, $\mu \geqslant 1$. Also, if $\theta \to 0$, $\mu \to \infty$ because the normal force $N \to 0$. But note that these results are based on the assumption that the cord is pulling tangent to the cylinder, so the torque is TR.

4. *Guided Problem.* A rigid bar of length L is suspended by a cord of the same length attached to a wall, as shown in Figure 14.4. The other end of the bar is placed against the wall. If the angle between the bar and the wall is 70° and the bar is just on the verge of slipping, what is the coefficient of static friction between the bar and the wall?

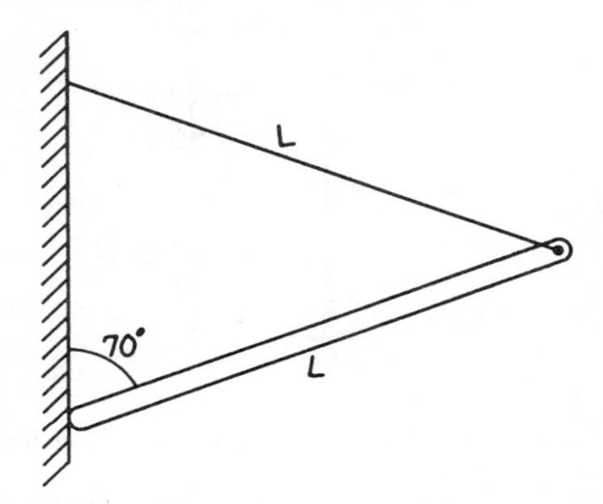

Figure 14.4 Problem 4. The bar is just on the verge of slipping.

Solution scheme: The problem is very similar to the preceding one. In the last problem the geometry made it simpler to take torques about the center of gravity. In this case it is simpler to take torques about the point of contact between the bar and the wall.

a. Given that the bar and the cord are the same length, what is the relationship among the angles of this triangle?
b. What is the sum of the forces in both the vertical and the horizontal directions?
c. Compute the torques about the point of contact. Be careful that you use the proper angles in computing the moment arms for the torques.
d. At the point of slipping, what is the relationship between f and N?
e. You now should have enough independent relationships among the variables to solve the problem.

5. *Worked Problem.* A lever 3.0 m long is used to lift the edge of a 204 kg crate (see Figure 14.5). Suppose the end of the lever is supporting half the weight of the crate. If the largest force that can be applied to the opposite end is 400 N,

a. Where should the fulcrum be placed?
b. What is the IMA of this system?

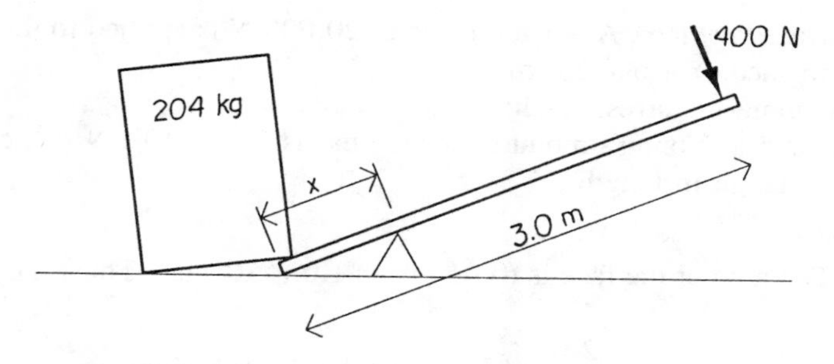

Figure 14.5 Problem 5. A lever lifts a large block.

Solution

a. As shown in the figure, let the distance from the crate to the fulcrum be x so that the applied force is a distance $3 - x$ from the fulcrum. The rotational equilibrium concept tells us that

$$F_i(3 - x) = mgx$$
$$400 \text{ N } (3.0 \text{ m} - x) = 2000 \text{ N}x$$
$$x = 0.5 \text{ m}$$

b. The IMA in this case is the ratio d_i/d_o, so

$$\text{IMA} = \frac{3 - x}{x} = \frac{3.0 - 0.5}{0.5} = 5$$

6. *Guided Problem*. Consider the wheel and axle shown in Figure 14.6. How much force must be exerted on the rope to just lift the block?

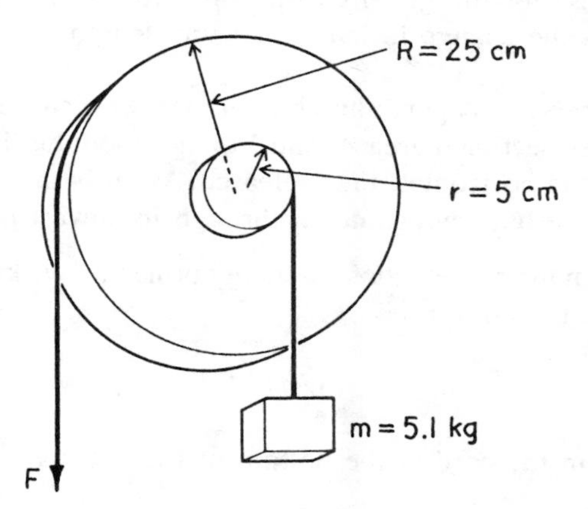

Figure 14.6 Problem 6. A wheel and axle used to lift a block.

Solution scheme

a. Treat the wheel and axle as a lever, with the two radii being the distances on either side of the fulcrum.
b. Let the rope force be the input force and the weight of the block the output force.

7. *Worked Problem.* A tension force of 20,000 N is applied to the 1.0 cm × 1.0 cm face of a bar 2.0 m long.

a. Calculate the stress in the bar.
b. Given that Young's modulus for the bar is 1.5×10^{11} N/m^2, calculate the change in length.

Solution

a. The area of the face is $(0.01\ \text{m})^2 = 1.0 \times 10^{-4}\ \text{m}^2$. Thus, the stress is

$$\frac{2.0 \times 10^4\ \text{N}}{1 \times 10^{-4}\ \text{m}^2} = 2.0 \times 10^8\ \text{N/m}^2$$

b. The relationship between the stress and the strain is the following:

$$\text{stress} = Y \times \text{strain} = Y\frac{\Delta l}{l}$$

so

$$\Delta l = \frac{l \times \text{stress}}{Y} = \frac{2.0\ \text{m} \times 2.0 \times 10^8\ \text{N/m}^2}{1.5 \times 10^{11}\ \text{N/m}^2}$$

$$\Delta l = 2.7 \times 10^{-3}\ \text{m} = 2.7\ \text{mm}$$

8. *Guided Problem.* A metal wire 3.0 m in length and having a diameter of 3 mm is stretched 1 mm under a tension of 500 N. Determine Young's modulus for the wire.

Solution scheme

a. What is the cross-sectional area of the wire in m^2? What is the tensile stress for the given tensile force and area?
b. What is the change in length per unit length?

9. *Worked Problem.* A pendulum bob of mass m is suspended from a cord of length l, cross-sectional area A, and Young's modulus Y. The bob is held so that the cord is horizontal, then released. What is the increase in length of the cord as the bob swings down through its lowest position?

Solution: From the conservation-of-energy principle, we know that the speed of the bob at the bottom is

$$v^2 = 2gl$$

The tension in the cord at the bottom of the path is

$$T - mg = mv^2/l$$
$$T = mg + mv^2/l = mg + m(2gl)/l = 3mg$$

Then the tensile stress is

$$T/A = 3mg/A$$

The change in length then is

$$\Delta l = \frac{1}{Y} l \frac{3mg}{A}$$

10. *Guided Problem.* Suppose 600 cm^3 of water expands to 625 cm^3 upon heating. What pressure is required to squeeze the water back to its initial volume? The compressibility of water is 50×10^{-6} atmospheres.

Solution scheme

a. What is the change in volume per unit volume?
b. How is the compressibility related to the bulk modulus?
c. Since the compressibility is in atm, in what units is the pressure (compressive stress) to be given?

Answers to Guided Problems

2. If x is the distance from the lighter end to the center of mass,

$$x = \frac{(1 + p)}{(2 + p)} L$$

so if $p = 0.50$, $x = 0.6L = 1.8$ m. Note that as $p \to 0$, $x \to \frac{1}{2}L$, as expected.

4. $\mu = \left(\frac{1}{\sin 70°} - \sin 20°\right) / \cos 20° = 0.77$

6. 10 N
8. 1.7×10^{11} N/m^2
10. 800 atm

Chapter 15

Oscillations

Overview

In the study of gravitation, we examined the various properties of orbits, one of which was *periodicity*: the orbit is characterized by motion that repeats itself at regular intervals. Periodic motion is quite common in a variety of natural phenomena, including wave motion. In this chapter we begin our study of periodic phenomena by analyzing a special case, that in which the particle undergoing the periodic motion vibrates back and forth about an equilibrium position. We call this *oscillatory* motion; it is an essential ingredient in the upcoming analysis of wave properties.

Essential Terms

oscillation
harmonic functions
amplitude
angular frequency
period
frequency
phase
phase constant
simple harmonic oscillator
isochronism
kinetic and potential energy
simple pendulum
stable and unstable equilibrium
physical pendulum
torsional pendulum
torsion constant
damped harmonic motion
Q factor
driving force
resonance

Key Concepts

1. *Simple harmonic motion* (*SHM*). This particular motion has extensive applications in a variety of phenomena, so it is essential that we clearly understand the principles governing the motion of a particle executing SHM.

Because this motion is periodic, it has the two properties characteristic of all periodic motion, namely, a *period* and a *frequency*. The equation of SHM has a particular functional form. First of all, the position of the particle is a function of time, the functional relationship being either $x(t) = A \cos(\omega t + \delta)$ or $x(t) = A \sin(\omega t + \delta)$. Not only is the *harmonic* (cosine or sine functions) nature of the relationship evident, but it is further required that these harmonic functions be of the form $(\omega t + \delta)$. The coefficient of t is called the *angular frequency* and is expressed in radians per second, just like the angular velocity. But there is a distinct conceptual difference in these two. The latter is simply the rate at which the angular displacement is changing, whereas the former is a measure of the frequency or period of the oscillatory motion. This identity of units and symbols for two apparently different variables arises from the connection between SHM and uniform circular motion. Even so, one should not confuse the angular frequency with the frequency. They differ by a factor of 2π, the relationship being $\omega = 2\pi\nu$. The frequency, ν, is expressed in the SI unit *hertz* (Hz), which is equivalent to 1 vibration per second or 1 cycle per second. Just as in the case of ω, the *dimensional* equivalent of this unit is 1/s.

The *phase* of the oscillations is the argument of the sine or cosine function in the expressions $x(t) = A \cos(\omega t + \delta)$ or $x(t) = A \sin(\omega t + \delta)$; that is, the phase is the term $\omega t + \delta$. Because the argument of the harmonic function is an angle, we can think of the phase as the location of the particle at a particular time, along the unit (360° or 2π rad) circle—that is, $0 \leqslant (\omega t + \delta) \leqslant 2\pi$ (or a multiple thereof). The constant δ, called the *phase constant*, fixes the time at which the particle has its maximum and minimum values of the amplitude A. If $\delta = 0$ and we use the sine function, then A will have its extrema at $t = n(\frac{1}{2}T)$, where $n = 0, 1, 2, \ldots$; if the cosine function is used, the extrema occur at $t = n(\frac{1}{4}T)$, where $n = 1, 3, 5, \ldots$. An alternative to using the phase constant is to express the displacement as a superposition of sine and cosine functions, the amplitude being a combination of harmonic functions of δ (see the discussion at the end of Section 15.1 in the text).

The position, velocity, and acceleration of the particle in SHM are each harmonic functions. But the velocity and acceleration are time derivatives of the position and velocity, respectively, and therefore the graphs of these functions show $\pi/2$ phase shifts relative to each other, because the derivatives of the sine and cosine are the cosine and negative sine, respectively. This relationship gives rise to a set of curves like those shown in Figure 15.1. Note the points at which maxima and minima occur for the different functions relative to one another.

2. *The simple harmonic oscillator*. In the preceding section we mentioned that uniform circular motion is related to SHM. If the motion of a particle undergoing uniform circular motion is projected onto the x axis, say, then this projection will exhibit SHM. A mass attached to a spring and set into vibration executes SHM also. In a standard demonstration, the projected shadow of a particle undergoing uniform circular motion is made to match exactly the motion of the mass on the spring, as illustrated in Figure 15.2. The SHM of such a mass results from a restoring force, $\mathbf{F}_r = -k\mathbf{x}$, which is simply Hooke's Law for an ideal spring. In both cases (the mass on the

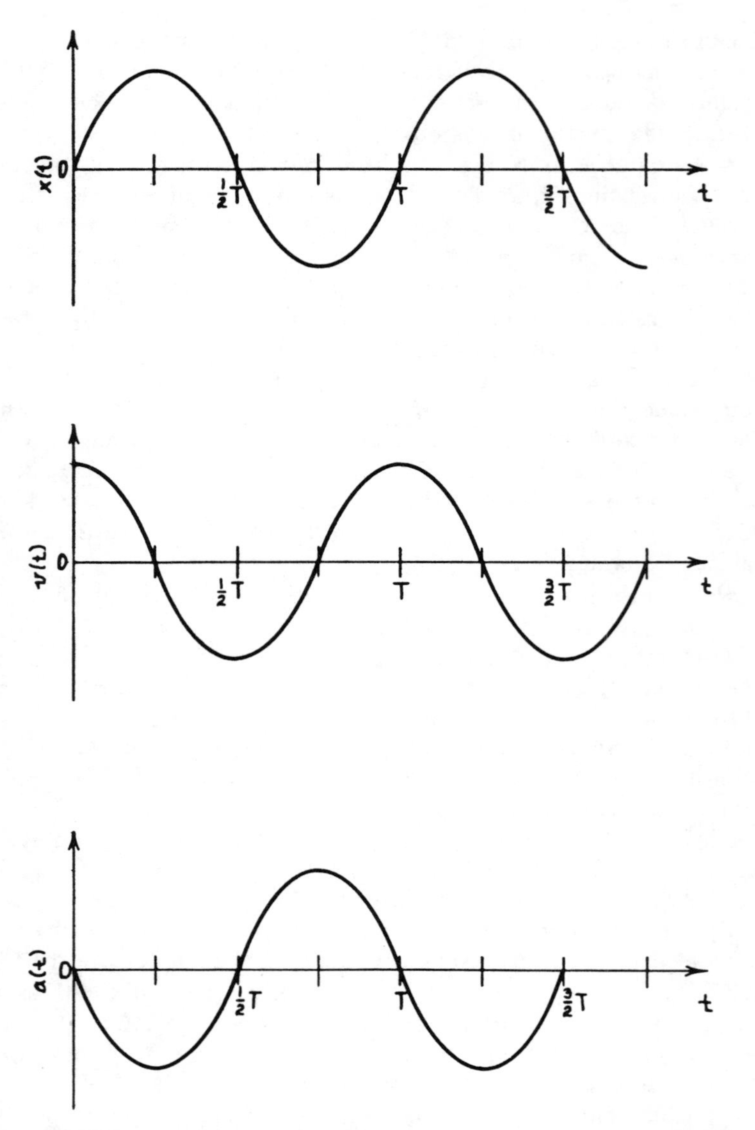

Figure 15.1 The phase relationship among $x(t)$, $\upsilon(t)$, and $a(t)$ for simple harmonic motion.

spring and the projection of uniform circular motion), we get the same *form* of the equation of motion. By comparing these forms, however, we see that in the circular-motion projection the coefficient of x in the differential equation is $-\omega^2$, whereas in the oscillating-mass situation it is $-k/m$. Thus, we can make the connection $\omega^2 = k/m$ and express the angular frequency, period, and so forth in terms of the mass and the spring constant. We should not, however, infer that SHM occurs only for masses attached to springs. The only requirement for SHM is that the restoring force be proportional to the displacement and in the opposite direction, that is, that $\mathbf{F}_r \propto -\mathbf{x}$.

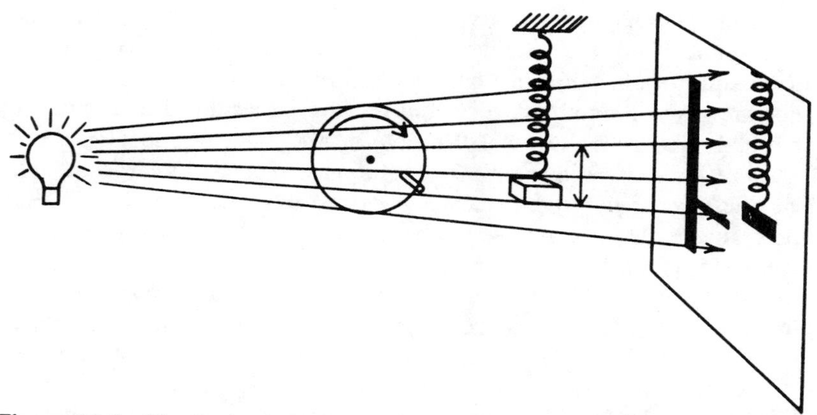

Figure 15.2 The projection of uniform circular motion is simple harmonic motion identical to the vertical oscillation of the mass on the spring.

In the vertically oscillating spring–mass system, the spring obeys the same relationship between the force and the displacement regardless of the value of g. Therefore, the *frequency* (or the period) of the oscillation is not changed with variations in g, but only with the *equilibrium point* about which the oscillation takes place. Basically, the Hooke's Law relationship causes the variation, whereas the uniform gravitational field sets the fixed equilibrium point. In the next section we will discuss this further, in the context of energy.

3. *Kinetic energy and potential energy*. Let's continue the previous discussion by looking at the potential energy terms once again. The potential energy of the spring is given by $U = \frac{1}{2}kz^2$ and is represented by the solid parabolic curve shown in Figure 15.3. Therefore, the total potential energy of the system is $U' = mgz + \frac{1}{2}kz^2$, which is the sum of the solid parabolic

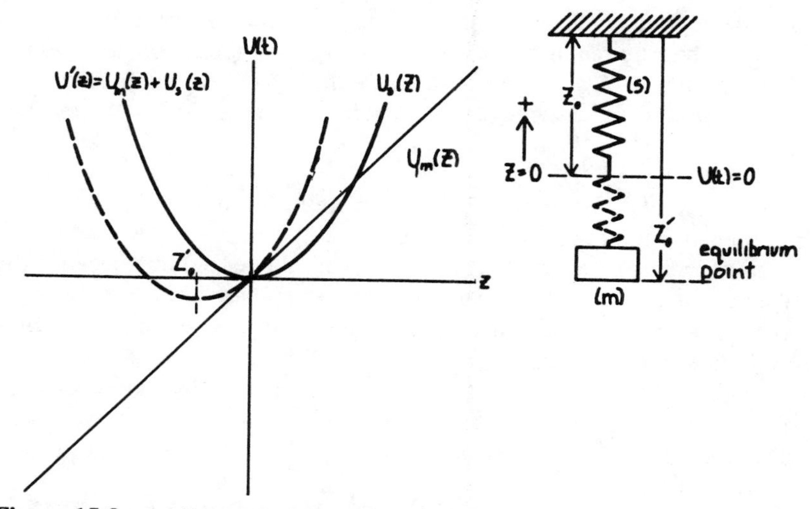

Figure 15.3 Adding the mass to the spring does not change the shape of the potential-energy curve; it merely shifts the equilibrium position.

curve and the straight-line curve representing mgz. The resulting curve is still parabolic. Thus, the system has the same general form of potential energy, but the equilibrium point has been shifted to a point $z = z_0'$.

For the sake of simplicity let's consider a mass attached to a spring on a horizontal frictionless surface. We note that $K = \frac{1}{2}mv^2$ and $U = \frac{1}{2}kx^2$, so $E = \frac{1}{2}mv^2 + \frac{1}{2}kx^2$ = constant. When the mass reaches its largest displacement (the *turning point*), we have $x = A$ and $v = 0$. Then $E = \frac{1}{2}kA^2$. The same result is obtained by use of the properties of the harmonic functions (see Eq. 15.45 in the text).

A real spring suspended from a support and displaced will oscillate with SHM having a well-defined frequency. But there is no mass attached, so how can it oscillate? The spring itself has mass (which we have ignored in our analyses), so part of the spring can be considered a mass suspended from the rest of the spring, so to speak. More detailed analysis will show that if the spring has mass, we can still treat the system as a massless spring plus an attached point mass if we add to the point mass one-third the mass of the spring.

4. *The simple pendulum*. There are three defining properties of a simple pendulum: the mass must be a *point* mass, that is, a particle; the string must be massless; and the amplitude must be sufficiently small that the angular displacement satisfies the condition $\sin\theta = \theta$ in radians. In principle, no pendulum can be truly simple, because for $\theta > 0$, $\sin\theta$ is not *exactly* equal to θ. If the bob on a pendulum 2 m long swings back and forth through an arc of about 30 cm, the *angular amplitude* is a little under 5°. The period for such a pendulum is $T = 2\pi\sqrt{\frac{l}{g}}(1 + 0.00048)$. The difference between this value and $T = 2\pi\sqrt{\frac{l}{g}}$ is only about 0.048%. Because the period of this pendulum is about 3 s (actually 2.8 s), however, we would find that in 24 h a clock reading predicted by use of the simple pendulum approximation would differ from the true reading by approximately 14 s. This does not mean the clock is losing time. The *length* of the pendulum is chosen carefully so that, with the expected amplitude, the clock will keep time fairly accurately.

5. *The physical pendulum*. Theoretically, any pendulum is a physical pendulum, because the mass has to occupy a finite volume. As we saw in the last section, we can treat many of these as simple pendulums because they satisfy very closely the criteria for such systems. There are cases, however, for which these approximations are no longer valid. For the physical pendulum the mass may be considered concentrated at the center of gravity *for computations of the torque*, but because the moment of inertia is involved, the actual mass distribution must be taken into account. It is for these reasons that the mass and the moment of inertia appear separately in the formula for the period.

The period of a physical pendulum is given by $T = 2\pi\sqrt{\frac{I}{mgl}}$, where l is the distance from the pivot to the center of gravity. The period for a simple pendulum is $T = 2\pi\sqrt{\frac{l'}{g}}$, in which we have used l' to represent the length of the cord in order to prevent confusion with the l in the physical-pendulum expression. Comparing these two expressions, we find that the periods are identical if $l' = I/ml$. Thus, I/ml is the length of an *equivalent simple pendulum*. The point that lies on a line connecting the pivot and the center of gravity and that is a distance l' from the pivot is called the *center of oscillation* of the physical pendulum. This point is coincident with the *center of percussion* (see problem 3, Chapter 13 of this guide). The proof of this assertion, along with a discussion of another property of the center of oscillation, is presented in problem 7.

6. *The torsional pendulum*. The torsional pendulum obeys an equation similar to that of the physical pendulum in all ways but one: the coefficient of θ in the expression for the torque on the physical pendulum contains the mass explicitly, whereas in the torsional pendulum this coefficient is related to the "rigidity" or *elasticity* of the system (for example, the fiber or rod being twisted) and does not include the mass explicitly. Thus, the expression for the period of the torsional pendulum includes I but not m.

7. *Damped and forced oscillations*. Clearly, oscillations of real systems do not go on forever unaided. Some force other than the restoring force must be applied to the system to overcome the dissipating mechanisms, giving rise to so-called *damped* oscillations. If the oscillatory system is left to itself, friction will cause the amplitude to decay according to the relationship $A = A_0e^{-\gamma t/2}$. A characteristic of this exponential function is that $A \rightarrow 0$ as $t \rightarrow \infty$; that is, an indefinitely long time should be required for the oscillations to die out completely. As the amplitude nears zero, however, the assumption of a steady friction force breaks down, so the system actually does come to rest. Thus, it makes more sense to speak of a *characteristic* time at which the amplitude decays to $1/e$ of its initial value, where e is the base of the natural logarithm, equal to approximately 2.718. (This concept is similar in some respects to other characteristic times associated with asymptotically decreasing quantities, such as the half-life for radioactive decay.) Because the longer the decay time, the longer the period over which the oscillations continue, the *ringing time* is proportional to $1/\gamma$, where γ is the decay constant. Thus, the term $Q = \omega/\gamma$ is a measure of the ringing time. For example, an aluminum rod, stroked with rosin so that standing resonant vibrations are set up, will typically continue to "ring" for several seconds, meaning that it has a fairly high Q.

To maintain the oscillations at a given amplitude, a *driving force* must be available to make up for the damping force, resulting in what are called *forced* oscillations. If the driving force is periodic and equal to the natural period of the oscillating system, large amplitudes can often be built up, in a phenomenon called *resonance*. One of two things can happen if the driving force is considerably larger than the damping force: the damping force can grow to match the driving force, or the system can rupture as a result of the increased energy input.

Sample Problems

1. *Worked Problem*. Figure 15.4 shows a 2 kg block at rest on a frictionless horizontal table, attached to a massless ideal spring for which $k = 200$ N/m. A 500 g block rests on top of the larger block. The coefficient of static friction between the two blocks is 0.45. What is the maximum amplitude SHM the system can undergo without having the smaller block slip on the larger one?

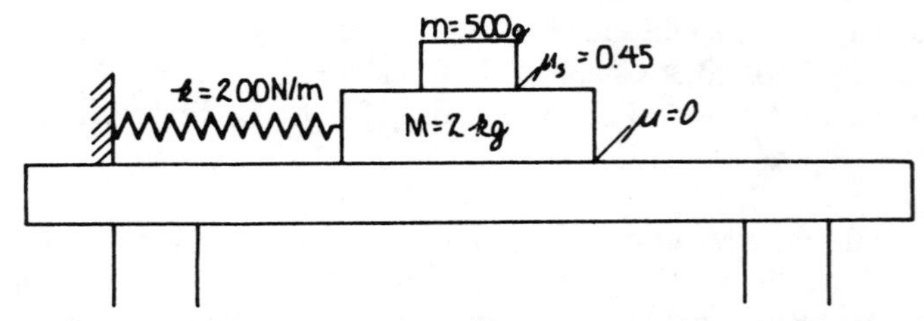

Figure 15.4 Problem 1. The 2 kg block undergoes simple harmonic motion. If the amplitude is too large, the upper block will slip.

Solution: The upper block (m) will start to slip when its acceleration is greater than $a_{max} = f_{max}/m$, where f_{max} is the maximum value of the static-friction force. But $f_{max} = \mu_s N = \mu_s mg$. Therefore, $a_{max} = \mu_s g$. If slipping has not yet occurred, then the two blocks are locked together and the acceleration of the SHM is given by $a = -\frac{k}{M'}x$. Therefore, the maximum amplitude is $x_{max} = -\frac{M'}{k}a_{max}$, where M' is the combined mass of the two blocks. Thus

$$x_{max} = \frac{M}{k}\mu_s g = \frac{2.5 \text{ kg}}{200 \text{ N/m}} \times 0.40 \times 9.8 \text{ m/s}^2$$

$$x_{max} = 0.049 \text{ m} = 4.9 \text{ cm}$$

2. *Guided Problem*. A 1 kg platform is attached to a massless spring so that the system can oscillate vertically, as shown in Figure 15.5. A 200 g

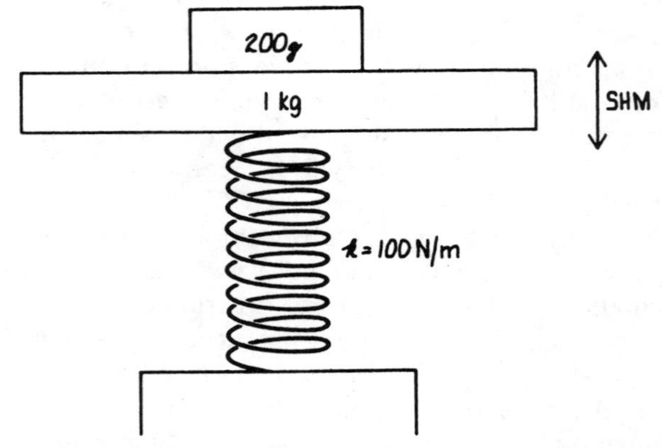

Figure 15.5 Problem 2. If the platform undergoes simple harmonic motion (SHM) and the amplitude is too large, the block will separate from the platform.

block sits atop the platform. If $k = 100$ N/m, what is the largest-amplitude SHM possible without the smaller block's becoming separated from the platform?

Solution scheme: This problem is similar to the preceding one. The condition for the block's leaving the platform is that the platform have a downward acceleration greater than g.

a. Write down the relationship between the displacement and the acceleration for SHM.
b. What is the total mass of the oscillating system?
c. Set the acceleration equal to the gravitational acceleration.

3. *Worked Problem.* A car accelerates uniformly in the horizontal direction at 6 m/s². A trinket attached to the end of a 15 cm light chain is suspended from the rearview mirror. Determine the period of small oscillations about the equilibrium point while the car is accelerating.

Solution: The principle of equivalence, discussed in Chapter 9, states that no distinction can be made between the effects of acceleration and of local gravity. Thus, in the frame of reference of the car in this problem, the effective gravitational field is the vector combination of $-\mathbf{a}$ and $\mathbf{g}$; in the car's noninertial reference frame, the actual acceleration is seen as being equivalent to a gravitational-field vector in the direction opposite to the acceleration. Thus, the *magnitude* of the effective field (and consequently the gravitational acceleration) $\mathbf{g}'$ is

$$g' = \sqrt{a^2 + g^2} = \sqrt{(6\ \mathrm{m/s^2})^2 + (9.8\ \mathrm{m/s^2})^2} = 11.5\ \mathrm{m/s^2}$$

Therefore, the period of small oscillations is

$$T = 2\pi\sqrt{\frac{l}{g'}} = 2\pi\sqrt{\frac{0.15\ \mathrm{m}}{11.5\ \mathrm{m/s^2}}} = 0.72\ \mathrm{s}$$

Comment: The equilibrium position of the pendulum in this case is that position in which the chain makes an angle θ with the vertical, where θ satisfies the relationship $a = g \tan \theta$.

4. *Guided Problem.* The equation for the period of a simple pendulum is given by the expression

$$T = 2\pi\sqrt{\frac{l}{g}}\left(1 + \frac{1}{2^2}\sin^2\frac{\theta_m}{2} + \ldots\right)$$

in which θ_m is the angular amplitude. Suppose a pendulum of length equal to 1.000 m is used to determine g at a certain location. If the amplitude of the oscillation is 8° (total swing equal to 16°), determine the error in g that would result if the $\sin^2 \theta_m/2$ term in the expression just given were not included.

Solution scheme

a. Rearrange the given expression to solve for g. For the square of the term in parentheses, use the approximation $(1 + x)^2 \approx 1 + 2x$ for $x \ll 1$.

b. Let g' be the acceleration of gravity obtained by use of the given expression, and let g be the value obtained by use of the approximate formula (without the $\sin^2 \theta_m/2$ term). Take the ratio g'/g.

5. *Worked Problem.* A spring–mass combination hanging vertically has a period for small oscillations equal to 3.0 s. If this system is allowed to come to equilibrium, how much of a displacement change would this mass undergo if the gravitational-field strength changed by 10^{-4} N/m (m/s²)?

Solution: The period of oscillation is given, instead of the spring constant k. But we know that

$$T = 2\pi\sqrt{\frac{m}{k}}$$

or

$$\frac{m}{k} = \frac{T^2}{4\pi^2}$$

In equilibrium, we have $mg = kx$ or

$$x = \frac{m}{k}g$$

Differentiating, we get

$$dx = \frac{m}{k}dg = \frac{T^2}{4\pi^2}dg$$

Then the change in displacement is

$$dx = \frac{(3.0\ \text{s})^2}{4\pi^2} \times 10^{-4}\ \text{m/s}^2$$

$$dx = 2.28 \times 10^{-5}\ \text{m} = 2.3 \times 10^{-2}\ \text{mm}$$

6. *Guided Problem.* A conical pendulum consists of a particle of mass m suspended from a cord of length l. The particle swings in a *horizontal* circle of radius r. Show that if the angle θ made by the string with the vertical is small, the period of the pendulum is given by

$$T = 2\pi\sqrt{\frac{l}{g}\cos\theta}$$

Solution scheme

a. What is the relationship between the centripetal acceleration and the tangential speed of the particle?
b. Use Newton's Second Law in both the vertical and the horizontal directions to eliminate T. You should have an expression for a in terms of r and θ. Set this expression equal to the expression for a in step a.
c. Substitute for v^2 the expression $(2\pi r)^2/T^2$. Express r as a function of l and θ.

7. *Worked Problem.* Show that the center of oscillation for a rigid body about a point O is also the center of percussion for this point.

Solution: In Chapter 13, problem 3, of this guide, we developed the idea of the center of percussion, that point at which an impulsive force causes no reaction at the pivot. The condition is that $k_{CM}^2 = ab$, where a is the distance from the center of percussion to the pivot point and b is the distance from the center of mass to the pivot. The term k_{CM} is the radius of gyration relative to an axis through the center of mass. In the notation used in the present chapter, we replace b with l. Now if we let l' be the distance from the pivot to the center of oscillation, we are to show that $l' = a + l$. We know that

$$l' = \frac{I_0}{Ml}$$

from our discussion of the center of oscillation. But $I_0 = I_{CM} + Ml^2$, so that

$$l' = \frac{Mk_{CM}^2 + Ml^2}{Ml}$$

Then

$$l' = \frac{k_{CM}^2 + l^2}{l}$$

But $k_{CM}^2 = al$, according to the definition of the center of percussion, so

$$l' = \frac{al + l^2}{l} = a + l$$

Comment: Another interesting property of the center of oscillation is that if the body is pivoted at the center of oscillation, the former pivot point becomes the new center of oscillation. Thus, if the physical pendulum is suspended at either of these two points, the period of oscillation will be the same.

8. *Guided Problem.* A physical pendulum has a moment of inertia $I_{CM} = Mk_{CM}^2$ and is suspended at a point that is a distance l from the center of mass. What value of l will result in the minimum value for the period?

Solution scheme

a. What is the expression for the period of a physical pendulum?
b. Write I in terms of I_{CM}, using the parallel-axis theorem.
c. Square both sides, and take the differential of both sides with T and l as variables.
d. Compute dT/dl, and set it equal to zero.
e. Solve for l.

Comment: In general, the period is given by

$$T = 2\pi\sqrt{\frac{Mk_{CM}^2 + Ml^2}{Mgl}} = 2\pi\sqrt{\frac{k_{CM}^2 + l^2}{gl}}$$

In solving for l, we get a quadratic equation, so there are two solutions for l, one on either side of the l for minimum T. If $l = k_{CM}$, however, there is only one solution, namely, $T = 2\pi\sqrt{k_{CM}/g}$. So if the physical pendulum is suspended at a point k_{CM} from the center of mass, the period is that of a simple pendulum of length k_{CM} and is the minimum period for that pendulum. This is the only point at which the pendulum can be suspended and be considered to have all its mass concentrated at the center of mass, and thus be treated as a simple pendulum.

9. *Worked Problem.* A large demonstration Foucault pendulum at a science museum consists of a heavy ball attached to a long, light wire approximately 20 m in length. If the pendulum is left to itself with no driving force, the amplitude will decay to one-half its original value in 45 min. What is the decay constant for this pendulum?

Solution: We know that the amplitude varies as

$$A = A_0 e^{-\gamma t/2}$$

Then the amplitude at $t = 0$ is

$$A_0 = Ae^0 = A$$

and at $t = 45$ min (2700 s) is

$$A' = Ae^{-\frac{2700\text{ s}}{2}\gamma}$$

The problem states that $A_0/A' = 2$, so that

$$\frac{A_0}{A'} = 2 = e^{\frac{2700\text{ s}}{2}\gamma} = e^{1350\gamma\text{ s}}$$

Taking the natural logarithm of each side, we get

$$\ln 2 = 1350\gamma\text{ s}$$

$$\gamma = \frac{\ln 2}{1350\text{ s}} = 5.1 \times 10^{-4}\text{ s}^{-1}$$

Comment: The damping is quite small (underdamped harmonic motion), so that $T \approx 2\pi\sqrt{l/g}$, or $\omega = 2\pi/T = \sqrt{g/l}$. Then $\omega = \sqrt{9.8\text{ m/s}^2/20\text{ m}} = 0.7$ rad/s and

$$Q = \omega/\gamma = \frac{0.7\text{ rad/s}}{5.1 \times 10^{-4}\text{ s}^{-1}} = 1.37 \times 10^3$$

which is a fairly high Q value. The "ringing time" is

$$t_r = 2/\gamma = 3.9 \times 10^3\text{ s} = 65\text{ min}$$

Thus, after 65 min the pendulum will have decayed such that its amplitude is $1/e\, A_0$.

10. *Guided Problem*. Suppose that in the preceding problem the ball is allowed to pass through a viscous medium that has a resistance force given by

$$F_r = -\gamma m v$$

Assume that $\gamma = 0.4\ s^{-1}$. The solution to the equation for this damping force is

$$x(t) = Ae^{-\gamma t/2} \cos\left[\sqrt{\omega^2 - \gamma^2/4} \cdot t + \delta\right]$$

What is the frequency of oscillation for this damped system?

Solution scheme

a. Find the angular frequency ω in the absence of damping ($\gamma = 0$).
b. Identify the new angular frequency ω' in the expression given in the problem.

Answers to Guided Problems

2. $x_{max} = -\frac{M}{k} g = \frac{1.2\ kg}{100\ N/m} 9.8\ m/s^2 = 0.118\ m$

4. $g'/g = 1.0024$, so that $\frac{g' - g}{g} = 0.0024 \approx 0.24\%$

8. $l = k_{CM}$ = the radius of gyration

10. $\omega' = \sqrt{\omega^2 - \gamma^2/4} = \sqrt{\frac{4\pi^2}{T^2} - \gamma^2/4} = \sqrt{(0.7)^2 - 0.01}$

$\omega' = 0.67\ rad/s$

Chapter 16

Waves

Overview

The study of simple harmonic motion has set the stage for the analysis of another kind of motion that is closely related to it. Using a model in which we start by analyzing a chain of coupled harmonic oscillators, we can generate a description of what is known as *wave motion*. We begin with the fairly simple description of a wave on a string and generalize our results to more complicated wave phenomena.

Essential Terms

wave motion
pulse
transverse waves
longitudinal waves
general wave function
amplitude of a wave
wave number
crest
trough
wavelength
harmonic wave equation
period of a wave
frequency of a wave
angular frequency
dispersive medium
group velocity
phase velocity
energy density
power
constructive interference
destructive interference
modulated wave
beats
beat frequency
Fourier's theorem
standing wave
node
antinode
normal mode
eigenfrequency

Key Concepts

1. *Wave pulse*. Although the wave pulse ultimately leads to a wave, the two are not synonymous. A pulse is a superposition of several harmonic waves. It is instructive, however, to begin our discussion with the concept of a pulse, in particular a traveling one, because it is easier to see the spatial movement of the pulse and to decide whether it maintains its shape as it propagates through a particular medium. Also, the time dependence of a nondispersing pulse is developed fairly easily, and this concept provides for a smooth transfer to the study of the functional relationships in wave motion.

There are three distinct types of pulses—*longitudinal* (compressional), *transverse*, and *torsional*. Certain waves and pulses are combinations of these, but the distinctions among the three types are fairly well defined.

2. *Traveling waves*. The term *traveling* used in this sense is synonymous with *propagating*. As the pulse propagates through a particular medium, it may change its shape, be attenuated, or continue unchanged. We will first direct our attention to the type of pulse that propagates unchanged in form and amplitude. This step is somewhat akin to studying dynamics without friction—it allows us to get at the underlying principles of the motion. Once we have the basic description in hand, we can go on to deal with the mechanisms in which the pulse dissipates.

A wave (pulse) propagating in the positive direction has a functional form $f(x - vt)$, whereas if it is propagating in the negative direction, the form is $f(x + vt)$. Although this may seem backward, it follows quite naturally from the consideration of two reference frames moving relative to each other with speed v. The functional form $(x \mp vt)$ is the only one that preserves the shape of the pulse as it propagates through the medium.

Harmonic waves (sine and cosine waves) are important not only because they are relatively simple but also because they form the basis for more complex waves. In addition, these waves are periodic, repeating over intervals of 2π in both space and time. In the preceding paragraph we stated that only the functional form $f(x \mp vt)$ preserves the shape of the wave or pulse. That is true, but $x \mp vt$ may be cast into different forms, all of which are equivalent. In other words, $x \pm vt$ is the same as $kx - \omega t$ or $2\pi\left(\frac{x}{\lambda} - \frac{t}{T}\right)$.

3. *Wave speed on a string*. The dependence of wave speed on mass density applies only to *transverse* waves. For example, the speed of compressional waves in solids depends on the elasticity of the medium. Consequently, it is often true that the wave travels faster in a denser medium. For a string precisely the opposite is true: as the mass density increases, the speed decreases. For now we will confine our attention to waves on a string, for which the inverse relationship between speed and linear mass density holds.

If the disturbance propagating along the string is a *pulse*, it means that two or more harmonic waves have been superimposed, as we shall see in the next section. If the medium is dispersive, the pulse shape changes with time. Furthermore, the speed with which this pulse propagates is not the same as the speed of the harmonic waves of which the pulse is composed. The speed of the pulse is called the *group velocity*, whereas the speed of the individual harmonic waves is called the *phase velocity*. The phase velocity is wavelength dependent in a dispersive medium, further complicating the analysis. We shall not explore this matter further; rather, we wish simply to point out the implications of the assumptions made in deriving the wave equation.

4. *Energy and power in a wave*. The energy in a wave arises from the interaction of adjacent particles; each segment of the string, for example, exerts a force on the succeeding segment, displacing it and thereby doing work on that segment. This is the idea of the coupled harmonic oscillator. The *spatial* distribution of energy density, dE/dx, for a given instant of time varies as $\sin^2 kx$, and the maximum value of the energy density is proportional to A^2 for a given μ and ω. Thus, a doubling of the amplitude results in a quadrupling of the energy density. Put another way, four times the energy is required to double the amplitude. The *rate* at which energy arrives at a given point in space, dE/dt, is a function of the same variables as the energy density and of the wave velocity.

5. *Superposition of waves*. The principle of superposition is a powerful tool in the analysis of wave motion. It allows us to analyze the more complex wave forms in terms of a summation, or *superposition*, of simpler harmonic waves. Although the superposition principle implies that two waves arriving at a point do not interact, they still *interfere*. This statement sounds like a paradox, but basically it means that the medium responds independently to each wave without changing the future progress of *that* wave. A particular particle in the medium has a displacement that is the sum of the individual displacements, meaning that if one wave by itself causes a particular displacement, and the other wave by itself causes a different displacement, the net displacement is simply the sum of these two when the waves arrive simultaneously. When we say that two waves interfere, we do not mean that one impedes the progress of the other; rather, we mean that the superposition principle holds. If the resulting amplitude is enhanced, we call the interference *constructive*. If the amplitude is diminished, the interference is *destructive*.

6. *Standing waves*. The term *standing wave* is somewhat a misnomer, because the effect being described is produced by two waves propagating in opposite directions through the medium. The fact that the locations of zero amplitude do not move with time makes it appear that there is a stationary wave. The only requirement for a standing wave is that there be two waves of the same frequency and amplitude traveling in opposite directions. If the medium is a restricted one, however, such as a string tied down at its two ends, then there are certain *boundary conditions* that must be met, namely, that the ends be points of *no displacement* (*nodes*). This fact restricts the frequencies of waves possible and gives rise to a set of *normal frequencies*, sometimes called *eigenfrequencies*.

Sample Problems

1. *Worked Problem.* A traveling wave on a string has the mathematical form

$$y(x,t) = 0.10 \sin\left(\frac{3\pi}{7}x + 36\pi t\right)$$

a. Determine the amplitude, wavelength, period, and frequency of this wave.
b. Determine its velocity of propagation.
c. What is the acceleration of a particle as a function of time at a point x on the string?

Solution

a. We first must decide which of the standard forms of the traveling-wave equation matches the given equation. The one to use is

$$y(x,t) = A \cos 2\pi\left(\frac{x}{\lambda} \mp \frac{t}{T} + \delta\right)$$

If $\delta = \pi/2$, then

$$y(x,t) = A \sin 2\pi\left(\frac{x}{\lambda} \mp \frac{t}{T}\right)$$

The wave equation given in the problem can be rewritten in the form

$$y(x,t) = 0.10 \sin 2\pi(\tfrac{3}{14}x + 18t)$$

or

$$y(x,t) = 0.10 \cos 2\pi(\tfrac{3}{14}x + 18t + \pi/2)$$

By comparing terms with the standard form, we get

$$A = 0.10 \text{ m}$$
$$\lambda = \tfrac{14}{3} \text{ m} = 4.7 \text{ m}$$
$$T = \tfrac{1}{18} \text{ s} = 0.055 \text{ s}$$
$$\nu = \frac{1}{T} = 18 \text{ Hz}$$

b. The velocity of the wave is found by means of the expression

$$v = \frac{\lambda}{T}$$

which gives

$$v = \frac{4.7 \text{ m}}{0.055 \text{ s}} = 85 \text{ m/s}$$

Because of the functional form $f(x + vt)$, the wave is traveling in the $-x$ direction.

c. The variable v represents the *transverse* displacement of a point on the string, so for a given position x, we get the acceleration by differentiating v with respect to time, holding x constant:

$$\frac{\partial^2 y(x,t)}{\partial t^2} = -0.10(36)^2\pi^2 \sin\left(\frac{3\pi}{7}x + 36\pi t\right)$$

$$\downarrow$$

$$a(x,t) = -36^2\pi^2 y(x,t) \text{ m/s}^2$$

Note that the particle at position x obeys the simple harmonic-motion equation

$$\frac{\partial^2 y(x,t)}{\partial t^2} + ky = 0$$

2. *Guided Problem.* Consider the equation for a traveling wave

$$y(x,t) = 0.01 \cos(0.5x - 20t)$$

where y and x are in meters and t is in seconds.

a. Determine the amplitude, wavelength, frequency, and period of this wave.
b. Determine the wave velocity (speed *and* direction).
c. For a point x on the string, determine the *transverse* velocity as a function of time.

Solution scheme

a. Which standard form of the traveling-wave equation most closely matches that given in the problem?
b. How is k related to the wavelength?
c. How is ω related to the period?
d. What is the relationship between the velocity of the wave, the wavelength, and the period?
e. How does the sign of the ωt term relate to the direction of travel?
f. What is the partial derivative of y with respect to t?

3. *Worked Problem.* A uniform rope of length l, hanging vertically, is given a sharp impulse at its lower end so that a pulse travels up the rope. Determine the speed of the pulse as a function of position in the rope.

Solution: The rope, having a uniform density μ, will have a tension that varies linearly from bottom to top because each part of the rope must support that which is below it. The tension at any given point y is given by

$$F(y) = m'g$$

where m' is the mass of the rope below the point y. Thus, $m' = \mu y$. Then

$$F(y) = \mu g y$$

The velocity $v(y)$ is

$$v(y) = \sqrt{\frac{F}{\mu}} = \sqrt{\frac{\mu g y}{\mu}} = \sqrt{gy}$$

Note that the velocity is independent of μ. Can you explain this?

4. *Guided Problem.* A 3.0 g string, 2.0 m long, is under a tension of 5 N.
a. Calculate the speed of a wave pulse traveling along this string.
b. Determine the change in speed with respect to a change in tension; that is, determine dv/dF.

Solution scheme

a. What is the relationship between the speed v and the linear mass density μ?
b. Differentiate this expression with respect to F.
c. What is the value of the linear mass density?

5. *Worked Problem.* A traveling wave on a string having linear mass density μ is represented by the equation

$$y(x,t) = 0.05 \cos (3\pi x - 40\pi t)$$

Determine the power at a given position as a function of time if $\mu = 1.5 \times 10^{-3}$ kg/m.

Solution: The power is given by

$$P = \frac{dE}{dt} = \frac{dE}{dx} \cdot \frac{dx}{dt} = v\frac{dE}{dx} = v\mu\omega^2 A^2 \sin^2 (kx - \omega t)$$

We have

$$\lambda = \frac{2\pi}{k} = \frac{2\pi}{3\pi}\text{ m} = \frac{2}{3}\text{ m}$$

$$T = \frac{2\pi}{\omega} = \frac{2\pi}{40\pi} = \frac{1}{20}\text{ s}$$

$$v = \frac{\lambda}{T} = \frac{\frac{2}{3}\text{ m}}{\frac{1}{20}\text{ s}} = 13.3\text{ m/s}$$

Then

$$P = 13.3\text{ m/s} \times 1.5 \times 10^{-3}\text{ kg/m} \times (40\pi)^2\text{ rad}^2/\text{s}^2 \times 0.05\text{ m} \times \sin^2 (3\pi x - 40\pi t)$$

$$P = 0.79 \sin^2 (3\pi x - 40\pi t)\text{ J/s}$$

6. *Guided Problem.* For the preceding problem, calculate the *average* power $\langle P \rangle$ transported along the string.

Solution scheme: The average power is calculated by averaging the power expression over one period, that is,

$$\langle P \rangle = \frac{1}{T}\int_0^T P(x,t)dt$$

a. What is the period T in terms of ω?
b. What is the average value of $\sin^2 \theta(x,t)$ over one period T? $\left(\textit{Note:} \text{ The integral } \int \sin^2 x\, dx = \frac{x}{2} - \frac{\sin x \cos x}{2}.\right)$

7. *Worked Problem.* Two strings having the same linear density are placed side by side to provide extra loudness for certain notes on a piano. Suppose the strings are tuned initially to middle C (262 Hz). After a time, one of the strings loosens a bit and there is a beat frequency of 3.5 Hz when the two strings are struck. By what percentage must the tension be increased to bring the string back in tune?

Solution: If the strings are identical and are the same length, then they will have the same tension F when tuned to the same pitch. If the length is the same, the *wavelength of the fundamental* is the same regardless of the pitch, because we are dealing with standing waves on a string. Thus

$$\lambda_1 = \frac{v_1}{\nu_1} \qquad \text{(representing 262 Hz)}$$

$$\lambda_2 = \frac{v_2}{\nu_2} \qquad \text{(representing the other string)}$$

But $\lambda_1 = \lambda_2$, so that

$$\frac{v_1}{\nu_1} = \frac{v_2}{\nu_2}$$

Substituting $v = \sqrt{F/\mu}$ we have

$$\frac{1}{\nu_1}\sqrt{\frac{F_1}{\mu}} = \frac{1}{\nu_2}\sqrt{\frac{F_2}{\mu}}$$

or

$$F_1\nu_2^2 = F_2\nu_1^2$$

Now $\left|\frac{F_1 - F_2}{F_2}\right|$ is the fractional change in tension required.

$$F_1 - F_2 = F_2\left(\frac{\nu_1^2}{\nu_2^2} - 1\right)$$

or

$$\frac{F_1 - F_2}{F_2} = \frac{\nu_1^2}{\nu_2^2} - 1$$

The beat frequency is $\nu_b = \nu_1 - \nu_2$, so that

$$\frac{F_1 - F_2}{F_2} = \frac{\nu_1^2}{(\nu_1 - \nu_b)^2} - 1$$

$$\frac{F_1 - F_2}{F_2} = \frac{262 \text{ Hz}}{[(262 - 3.5) \text{ Hz}]^2} - 1 = 0.027$$

Thus, the tension must be increased 2.7%.

8. *Guided Problem.* In problem 7 we saw how the frequency in a constant-length string varies with the tension. Now let's explore how the length of a string must change when the mass density is changed, given that the tension and the fundamental frequency are to stay the same. In other words, if we have a string with a different μ and it is to vibrate at the same frequency under the same tension, by what factor must its length change?

Solution scheme

a. If the tension stays the same and the frequency is not to change, which of the variables v, ν, and λ changes?
b. How is v expressed in terms of the tension F and the mass density μ?
c. What is the rate of change of L with respect to μ?
d. Finally, integrate from L_0 to L and μ_0 to μ.

9. *Worked Problem.* A steel wire having a mass of 4.0 g and a length of 1.25 m is tightened to a tension of 1020 N.

a. Find the wavelength and the frequency of the fundamental and the third harmonic of the standing waves on this string.
b. What is the wave equation for the third harmonic?

Solution

a. We first must find the velocity of the wave in the string, because the length information gives us the wavelength. From these two data we can find the frequency. The velocity is given by $v = \sqrt{F/\mu}$, where m = m/L = 0.0040 g/1.25 m = 3.2×10^{-3} kg/m. Then

$$v = \sqrt{\frac{1020 \text{ N}}{3.2 \times 10^{-3} \text{ kg/m}}} = 565 \text{ m/s}$$

For the fundamental, $\lambda = 2L$, so that

$$\nu = \frac{v}{\lambda} = \frac{v}{2L} = \frac{565 \text{ m/s}}{2.50 \text{ m}} = 226 \text{ Hz}$$

For the third harmonic,

$$\lambda = \frac{2L}{n} = \frac{2L}{3}$$

and

$$\nu_3 = \frac{565 \text{ m/s}}{\frac{2}{3}1.25 \text{ m}} = 678 \text{ Hz}$$

b. For the standing wave of the third harmonic, we have

$$y(x,t) = 2A \cos kx \cos \omega t$$
$$= 2A \cos \frac{2\pi}{0.83} x \cos \frac{2\pi}{678} t$$
$$y(x,t) = 2A \cos 7.5x \cos (0.0093)t$$

10. *Guided Problem.* A string is fixed only at one end.
a. Show that the eigenfrequencies are given by

$$\nu_n = n \frac{v}{4L} \qquad n = ?$$

b. If three successive harmonics on this string are 360 Hz, 600 Hz, and 840 Hz, what is the fundamental frequency?
c. Which harmonics are those given?
d. If the wave speed is 672 m/s, what is the length of the string?

Solution scheme

a. What is the wavelength to be associated with a wave pattern having a node at one end and the *adjacent antinode* at the other end?
b. Draw the wave patterns for the cases in which the other end is the second antinode, third antinode, and so forth.
c. Use the frequency, velocity, and wavelength relationship to determine the frequencies for the various standing waves possible on this string.
d. What restriction is placed on n for the three given frequencies (which are multiples of the fundamental)?
e. After calculating the fundamental, use the basic harmonic relationship to determine the length of the string.

Answers to Guided Problems

2. a. $A = 0.01$ m

$$\lambda = \frac{2\pi}{k} = 12.6 \text{ m}$$

$$T = \frac{2\pi}{\omega} = 0.31 \text{ s}$$

$$\nu = \frac{1}{T} = 3.2 \text{ Hz}$$

b. $v = \dfrac{\lambda}{T} = \dfrac{12.6 \text{ m}}{0.31 \text{ s}} = 40$ m/s in the +x direction

c. $\dfrac{\partial y}{\partial t} = -0.2 \sin (0.5x - 20t)$

4. a. $v = \sqrt{\dfrac{5\ \text{N}}{0.003\ \text{kg}/2\ \text{m}}} = 58\ \text{m/s}$

   b. $\dfrac{dv}{dF} = \dfrac{1}{2\sqrt{\mu F}} = 5.8\ \dfrac{\text{m/s}}{\text{N}}$

6. $\langle P \rangle = \frac{1}{2}\mu\omega^2 A^2 v = 0.39\ \text{J/s} = 0.39\ \text{W}$

8. $\dfrac{L - L_0}{L} = \sqrt{1 + \dfrac{F}{L_0^2 4\nu^2}\left[\dfrac{\mu_0 - \mu}{\mu_0\mu}\right]} - 1$

10. a. $L = \frac{1}{4}\lambda$ or $\lambda = 4L$
    Succeeding patterns show that

$$\nu_n = n\,\frac{v}{4L} \qquad n = 1, 3, 5, 7 \ldots$$

   b. The ratios are

$$600/360 = \tfrac{5}{3}$$
$$840/360 = \tfrac{7}{3}$$

   If the patterns of odd harmonics are 1/3, 3/3, 5/3, 7/3, then $\nu_0 = 120$ Hz.

   c. The harmonics given are the third, fifth, and seventh.

   d. $\nu_0 = \dfrac{v}{4L} \rightarrow L = \dfrac{v}{4\nu_0} = \dfrac{672\ \text{m/s}}{4 \times 120\ \text{Hz}} = 1.4\ \text{m}$

Chapter 17

Sound and Other Wave Phenomena

Overview

The analysis of transverse waves propagating along a string sets the stage for a discussion of more complicated wave phenomena, including wave motion in two and three dimensions. The properties of sound waves permit an analysis of a variety of common phenomena, such as the sound produced by musical instruments and the apparent frequency shifts resulting from relative motion between the sound source and the observer.

Essential Terms

wave front
plane waves
ultrasound
sound intensity
decibel
white noise
octave
speed of sound
eigenfrequencies
Doppler effect
Mach cone
sonic boom
speed of water waves
S and P waves
Richter magnitude scale
diffraction

Key Concepts

1. *Wave fronts*. In our study of waves propagating along a string, we did not consider the wave to have "breadth"; it was transmitted along a single dimension. Many waves are developed as a disturbance moving out in more than one direction, however, such as that produced by a pebble dropped onto

the surface of a pond. The *crests* and *troughs* can now be defined in terms of a spatial distribution in which a *circular* or *spherical wave front* moves outward from the source. A line drawn from the source intersecting the wave front and running perpendicular to it defines the *direction* of movement of the front. Thus, we can indicate the outward movement of the fronts by drawing these lines rather than by drawing the complicated fronts themselves. In optics these lines are called *rays*, a name that technically can be used in other wave propagation as well but usually is not. Because we so often analyze such waves at rather large distances from the source, these wave fronts, for all practical purposes, are two dimensional (plane) rather than three dimensional (spherical). We call these *plane waves*. The propagation of plane waves is indicated by *parallel* lines or rays, because the rays are perpendicular to the direction of the wave fronts.

2. *Sound waves in air*. To be more complete, we should title this section "Sound waves in a gas," because much of our discussion is applicable to other (ideal) gases in addition to air. The fundamental difference among them is in the speed of sound, which varies considerably from gas to gas. Sound waves are nothing more than disturbances traveling through a medium as longitudinal pulses or waves. These disturbances result in momentary, localized density changes that, as a result of coupling, propagate through the medium. In a gas the molecules already are in rapid thermal vibration, or motion. The amount of motion depends on the temperature, as we shall see in Chapter 19. A sound wave, or pulse, is a localized increase or decrease in this thermal motion; that is, the wave disturbance is *superimposed* on the random thermal motions.

Sound is defined as the disturbance having a frequency spectrum that the normal human ear can perceive. Whether the ear in fact perceives it is another matter. The old conundrum about whether a tree falling in a forest makes a sound if no one is there to hear it has no definitive answer, except that normally when a tree falls, the spectrum of frequencies produced in the air includes those frequencies perceptible by the human ear. So with our definition, the tree does indeed make a sound.

The *intensity* of sound is defined as the power transported per unit area of the wave front. The SI unit of intensity is W/m^2. It is instructive to calculate the maximum amplitude of vibration of the human ear drum for sounds ranging in intensity from about 10^{-12} W/m^2 (threshold of hearing) to about 1 W/m^2 (threshold of pain). It can be shown that the *pressure amplitude* P is given by $P = k\rho v^2 A$, where k is a constant, ρ is the density, v is the velocity, and A is the amplitude. The maximum pressure variation that corresponds to an intensity of 1 W/m^2 is about 30 Pa (1 Pa = 1 N/m^2), which is approximately 0.03% of normal atmospheric pressure. For a sound wave this corresponds to an amplitude A of around 10^{-5} m. For the threshold intensity, A is only about 10^{-11} m, a figure that testifies to the amazing sensitivity of the human auditory system.

The range of intensities over which the auditory system is sensitive makes the W/m^2 an inconvenient unit, so the *intensity level*, known as the *decibel* (*dB*), is more frequently used. It is based on a logarithmic scale such that the intensity in *bels* (10 dB) above a standard level (0.468×10^{-12} W/m^2) is equal to the power of 10 representing the multiplicative factor of the increase in intensity. For example, if the sound has an intensity 100 times that of the standard, then its intensity level is 2 bels ($100 = 10^2$) or 20 decibels (deci $= 10^{-1}$). The decibel is a more convenient unit because it allows gradations in intensity level that largely avoid decimals and it is closer to the differences in intensity level perceptible by the auditory system. Moreover, the logarithmic scale makes it possible to express ratios of intensities as differences in intensity levels. For example, street noise has an intensity level of 70 dB, whereas a whisper is 20 dB. The difference is 50 dB, although the ratio of intensities is 10^5, or 100,000!

The relationship between intensity and distance from the source, as described by the inverse-square law, is based on the distribution of a *fixed* amount of energy as it spreads over larger and larger areas. The source must be a *point* source, in principle, for the inverse-square law to hold. The relationship is valid even for nonlocalized sources, however, if the nonlocalization does not depart too greatly from a point source or if the intensity is measured at a considerable distance from the source. The relationship being discussed here is *not* the result of attenuation that occurs with a loss of energy; rather, it is simply the spreading of a constant amount of energy over increasing areas.

The standing-wave patterns set up in a resonant cavity—a hollow tube, for example—depend on whether the tube is open or closed at the end opposite the source. (We assume here that the sound source is at an open end.) If the end opposite the source is open also, then each end of the tube must be a displacement *antinode* (pressure node). By drawing the wave patterns consistent with these restrictions, one can easily show the eigenfrequencies:

$$\nu_n = \frac{nv}{2l} \quad n = 1, 2, 3, 4, \ldots$$

If the tube is closed at the opposite end, however, the possible standing-wave patterns give rise to eigenfrequencies as follows:

$$\nu_n = \frac{nv}{4l} \quad n = 1, 3, 5, 7, \ldots$$

Thus, only the *odd* harmonics are present in the tube with the closed end. Many pipe organs have pipes of both kinds, each with its own characteristic eigenfrequency spectrum.

3. *The Doppler effect.* This effect is the familiar rise and fall in pitch one hears when a train passes while blowing its whistle. One important consideration in analyzing the Doppler effect is that the expressions for the apparent frequency are derived with the assumption that the observer and the source are moving directly toward or away from each other. If the two are moving transverse to each other, no Doppler effect is present. If you have ever experienced the Doppler effect, you may have noticed that the perceived, or apparent, frequency does not change abruptly from one value to another but, rather, that there is a kind of uniform "slide" that takes place very quickly as the source or observer moves past. This effect has to do with the

fact that the source moves *by* rather than *through* you, so that as it comes near, there is an appreciable angle between you and it, reaching 90° at the moment of passing and then growing smaller again. The equations developed in the text assume that the angle is 0°, meaning that the source is moving toward you on a close approach and is fairly distant. But up close, $\theta \neq 0°$, so the frequency difference, $\Delta\nu$, diminishes in proportion to cosine θ. Thus, there is a smooth change from the approach condition to the recession condition. This effect is masked to some extent by the increase and decrease in sound intensity that occurs as the source passes by.

4. *Shock waves*. The shock wave (or bow wave) results when the wave source moves faster than the speed of the wave in the particular medium. We associate these waves most often with supersonic aircraft, but speed boats produce similar, but not exactly the same, waves in water. In the case of the aircraft, the engine noise is not necessary to produce the shock wave (often called a *sonic boom*). Just as the boat plowing through the water sets up a bow wave, so the plane can produce the same sort of wave in air merely by plowing through the air. Indeed, Figure 17.18 in the text (p. 450) shows the shock wave set up by a bullet traveling faster than the speed of sound in air.

5. *Water waves*. Water waves (the gravity type) basically are a combination of transverse and longitudinal waves. Each point in the water moves essentially in a circular path, giving water waves their characteristic "rolling" effect. In deep water the speed is a function of wavelength, an effect called *dispersion*. In shallow water the speed depends on the depth, resulting in *refraction* of the waves as they encounter regions of differing depths. An analogous phenomenon occurs in light.

6. *Diffraction*. Because the two terms sound so similar, the phenomena of *diffraction* and *refraction* often are confused. The latter is a change of direction as a wave changes speed, usually caused by a depth change in the case of water waves. Diffraction, in contrast, is a spreading effect as a wave passes an obstacle. The wave may in fact continue in the same medium (in refraction, the wave changes medium, or conditions), but the portion of the front encountering the obstacle experiences a kind of "drag" effect, whereas the other parts of the wave continue unimpeded. These effects are more pronounced as the wavelength approaches the size of the gap. This description is highly simplified and does not account for the interference effects associated with diffraction. A more detailed analysis employing light waves will be given in Chapter 40.

Sample Problems

1. *Worked Problem*. The terms p and p_0 represent pressure *amplitudes* associated with sound intensities I and I_0.

a. Show that

$$\frac{I}{I_0} = \frac{p^2}{p_0^2}$$

b. Using the results from part a, show that the intensity level in decibels is given by

$$I_{db} = 20 \log \left(\frac{p}{p_0}\right)$$

c. Suppose the pressure amplitude of a wave is raised from 1.5×10^{-5} Pa to 5×10^{-5} Pa. What is the corresponding change (in decibels) in the intensity level?

d. Are equal changes in pressure amplitude accompanied by equal changes in intensity level?

Solution

a. Recall from Chapter 16 that the power transmitted in a wave is proportional to the *square* of the amplitude. The intensity is proportional to the power, so $I \propto p^2$, where p is the pressure amplitude. The result follows.

b. The definition of I_{dB} is

$$I_{dB} = 10 \log \frac{I}{I_0}$$

but from part a

$$\frac{I}{I_0} = \frac{p^2}{p_0^2} = \left(\frac{p}{p_0}\right)^2$$

From the property of logarithms $\log a^n = n \log a$, so that

$$I_{dB} = 20 \log \frac{p}{p_0}$$

c. $\Delta I_{dB} = 20\left(\log \frac{p_2}{p_0} - \log \frac{p_1}{p_0}\right)$

But $\log x - \log y = \log \frac{x}{y}$, so that

$$\Delta I_{dB} = 20 \log \frac{p_2}{p_1}$$

$$\Delta I_{dB} = 20 \log \frac{5 \times 10^{-5}\ \text{Pa}}{1.5 \times 10^{-5}\ \text{Pa}} = 20 \times 5.23 \times 10^{-1}\ \text{dB}$$

$$\Delta I_{dB} = 10\ \text{dB}$$

d. Note that $\Delta I_{dB} \propto \frac{p_2}{p_1}$, *not* $p_2 - p_1$. In general $p_2/p_1 \neq p_2 - p_1$, so equal increments in the pressure amplitude will not produce equal increments in the intensity level.

2. *Guided Problem.* A roaring crowd has an intensity level of 100 dB. A single voice in the crowd has an intensity level of 70 dB. What is the ratio of the sound intensity of the crowd to that of the single voice?

Solution scheme

a. Write down the expression for *intensity level differences* in terms of intensities.

b. You should have a logarithmic expression relating I_2 to I_1.

3. *Worked Problem.* A pipe containing air is open at one end and has a piston that can be moved back and forth in the pipe. A sound source of 500 Hz is placed at the open end, and the piston is moved away from this end until, at a distance of 17.5 cm, a resonance is observed.

a. Determine the speed of sound in the tube.

b. At what distance should the next resonance occur?

Solution

a. The eigenfrequencies for a pipe closed at one end are given by

$$\nu_n = \frac{nv}{4l} \quad n = 1, 3, 5, \ldots$$

where l is the length of the pipe. Because the distance 17.5 cm represents the first resonance, it must be the fundamental. Therefore, $n = 1$ and

$$v = 4l\nu_1 = 4 \times 0.175 \text{ m} \times 500 \text{ Hz}$$
$$v = 350 \text{ m/s}$$

b. For the next resonance, $n = 3$ (only the odd harmonics are present). We have the same frequency as before, however, so the length of the pipe must change to achieve this second resonance. Thus

$$\nu_3 = \frac{3v}{4l}$$

or

$$l = \frac{3v}{4\nu_3} = \frac{3 \times 350 \text{ m/s}}{4 \times 500 \text{ Hz}} = 0.525 \text{ m} = 52.5 \text{ cm}$$

4. *Guided Problem.* A long tube, open at one end, has a loudspeaker placed near the open end, as illustrated in Figure 17.1. Standing waves are set up, indicated by small piles of powder at the displacement nodes, which are 7.1 cm apart. If the velocity of sound in the pipe is 340 m/s, what is the frequency of the loudspeaker?

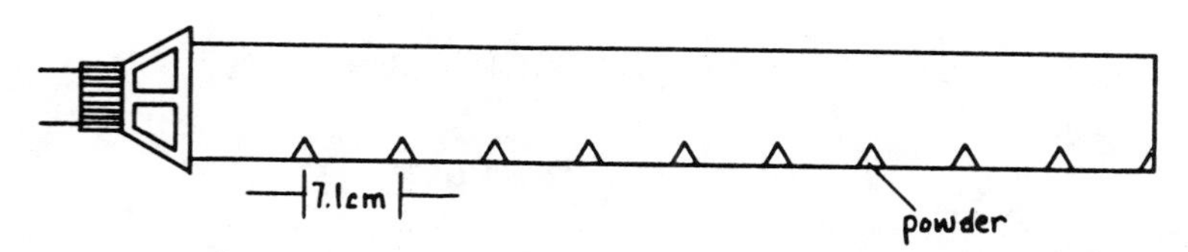

Figure 17.1 Problem 4. The standing wave set up in the tube causes powder to pile up at the displacement nodes.

Solution scheme

a. What is the relationship between the spacing of the displacement nodes and the wavelength?
b. What is the relationship among the frequency, the wavelength, and the velocity?

5. *Worked Problem.* A train traveling at a constant speed along a straight track enters a canyon. The engineer blows the whistle, which has a fundamental frequency of 440 Hz. The sound of the whistle reflects back from the canyon wall to the engineer, who notes that the combined sounds (the reflected sound plus that coming directly from the whistle) produce a musical *minor second*, that is, an A and an A♯ (see Table 17.2 of the text, p. 441). How fast is the train moving?

Solution: The engineer hears the whistle from the train directly with no Doppler shift. The echo from the cliff, however, is a stationary source emitting a frequency higher than that of the whistle itself, because the train is moving toward the cliff and therefore the wave fronts striking the cliff are more closely spaced. To compute this latter frequency, we use the Doppler equation for a moving source (the train) and a stationary "observer" (the cliff):

$$\nu' = \nu\left(\frac{1}{1 - V_E/v}\right)$$

If we assume that the train travels at a speed lower than the speed of sound, then

$$\nu' \approx \nu(1 + V_E/v)$$

Next, the cliff is considered a stationary source with the receiver moving toward it at a speed $V_E = V_R$. Thus, the frequency heard by the receiver (the engineer) is

$$\nu'' = \nu'\left(1 + \frac{V_R}{v}\right) = \nu\left(1 + \frac{V_E}{v}\right)\left(1 + \frac{V_R}{v}\right)$$

But since $V_R = V_E$, we have

$$\nu'' = 440 \text{ Hz}\left(1 + \frac{V_E}{v}\right)^2$$

Now A♯ is 466 Hz, so $\Delta\nu = \nu'' - \nu = 26$ Hz. Therefore,

$$\Delta\nu = 440 \text{ Hz}\left[\left(1 + \frac{V_E}{v}\right)^2 - 1\right] = 26 \text{ Hz}$$

$$\left(1 + \frac{V_E}{v}\right)^2 = 1.059$$

$$\left(1 + \frac{V_E}{v}\right) = 1.029$$

$$V_E = 0.029\ V$$

$$V_E = 0.029 \times 332\ \text{m/s} = 9.7\ \text{m/s} = 35\ \text{km/h}$$

6. *Guided Problem.* You are standing on a platform at a train station when two trains approach from opposite directions on parallel tracks. Each sounds its whistle, which has a fundamental frequency of 400 Hz. If one train is traveling at 50 km/h, how fast is the other traveling if you hear a beat frequency of 4 Hz?

Solution scheme: There are two solutions to this problem.

a. What is the apparent frequency of the whistle on the train traveling at 50 km/h?
b. What are the two different ways to get a beat frequency of 4 Hz?
c. Calculate the two different speeds the second train could have such that the 4 Hz beat frequency would result.

7. *Worked Problem.* A motorboat is moving through shallow water 2 m deep at a speed of 27 km/h. What is the half angle of the Mach cone formed by the bow wave?

Solution: The velocity of waves in shallow water is given by

$$v = \sqrt{gh}$$

$$v = \sqrt{9.8\ \text{m/s}^2 \times 2\ \text{m}} = 4.4\ \text{m/s} = 16\ \text{km/h}$$

The half angle of the Mach cone is

$$\sin\theta = v/V_E$$

so that

$$\sin\theta = \frac{16\ \text{km/h}}{27\ \text{km/h}} = 0.6$$

$$\theta = 36^\circ$$

8. *Guided Problem.* A supersonic aircraft traveling at 1500 km/h is climbing at a certain angle with respect to the horizontal. If the sonic boom reaches the observer just as the aircraft passes directly overhead, at what angle is the plane climbing?

Solution scheme

a. What is the half angle of the Mach cone?
b. How is the reference line for this angle related to the angle of climb?
c. When the plane is directly overhead, what is the angle formed between the plane, the observer, and the horizontal?
d. What is the sum of the interior angles of a triangle?
e. Use this information to compute the angle of climb.

Answers to Guided Problems

2. $\Delta I_{\text{dB}} = 30 \text{ dB} = 10 \log \frac{I}{I_0}$, so that $\log \frac{I_2}{I_1} = 3$ or $\frac{I_2}{I_1} = 10^3 = 1000$
4. 428 Hz
6. 38 km/h *or* 62 km/h
8. 37°

Chapter 18

Fluid Mechanics

Overview

The study of materials at the macroscopic level divides naturally into the study of solids and of fluids because the two show distinct differences in behavior. In this chapter we focus on fluid behavior, which encompasses both *liquids* and *gases*. Following our common pattern, we will first examine behaviors for *ideal* fluids to get at the underlying principles and then point out the differences between the real and the ideal.

Essential Terms

ideal (perfect) fluid
velocity field
streamline flow
continuity equation
pressure
pascal
torr
millibar
atmosphere of pressure
Pascal's Principle
hydrostatic pressure
overpressure (gauge pressure)
buoyant force
Archimedes' Principle
Bernoulli's equation

Key Concepts

1. *Ideal (perfect) fluid*. For the same reasons that we introduced mechanics by exploring the motions of bodies under ideal conditions (point masses, frictionless surfaces, massless strings, and so forth), we begin the study of fluid motion by looking at the so-called *ideal* (perfect) fluid. In the most general case, such a fluid is characterized by being *incompressible* (no density change) and *nonviscous* (no fluid resistance) and by having *irrotational*

steady-state flow. Irrotational means that there is no internal rotational motion (there are no vortices) in the flow. These restrictions on the fluid may seem stringent, but it is surprising how well we can approximate real fluid behavior with them.

2. *Fluid flow*. The description of fluid flow includes two important parameters—the *density* and the *flow velocity*. The former is rather straightforward; the latter, dealing with *volume* flow, is nonetheless expressed in the same units as ordinary velocity, such as meters per second. The reason for this is that the flow velocity is defined as the *volume per unit area per unit time*, which dimensionally is

$$V = \frac{\Delta V}{\Delta A \Delta t} = \frac{L^3}{L^2 T} = \frac{L}{T}$$

where L and T represent length and time, respectively. Should there be a situation in which the flow velocity varies within the stream, the volume element, ΔV, and the surface element, ΔA, must be considered in the limit as $\Delta V \to 0$ and $\Delta A \to 0$.

3. *Streamline flow*. The concept of streamlines is very useful in the study of fluid flow, but the use of such a model can easily lead to misunderstandings. The streamlines trace out a *velocity field* and exist as *constructs* to aid in the mapping of the flow. Although we can simulate the streamline flow by placing dyes in a flowing fluid, the "streamers" that show up are not in fact streamlines but, rather, bundles of streamlines—in principle, an infinite number of them.

4. *Density of streamlines*. If the density of the streamlines is proportional to the flow velocity, then to measure the density requires counting the (parallel) streamlines in a given finite region. Should the flow velocity vary over a small region, we may have to consider infinitesimally small volumes, so that in practice a large number of streamlines would have to be drawn, all very closely spaced. This isn't practical, so a few are carefully selected to give as precise a rendition of the flow field as possible.

5. *Flow tubes*. A bundle of streamlines is called a *flow tube* and is useful in keeping track of a particular volume of fluid as it flows. Streamlines in such a tube cannot cross. If they were to do so, then at the point of intersection there would be two values for the velocity, an impossibility.

6. *Continuity equation*. The continuity equation, which is based on the conservation of fluid mass and fluid density, derives from the assumption of the incompressibility of the fluid. Even so, if compressible fluids are allowed to reach a steady-state condition, they behave like incompressible fluids, and the continuity equation is still valid, at least to a good approximation.

7. *Pressure*. We often incorrectly speak of pressure as if it were a vector. Pressure is a *scalar*, so it is not correct to say that a fluid exerts a pressure in a given direction. Pressures arise from *forces*, which *are* vectors.

Pressure is defined as the ratio of the *magnitudes* of the force and the area. Otherwise, there would be a ratio of two vectors, which is not defined. A variety of units have been developed for pressure; their usefulness depends on the particular situation for which they were developed. The official SI unit for pressure is the *pascal* (Pa), equal to 1 N/m^2. A unit developed primarily for meteorology is the *bar*, equal to 10^5 Pa. This value is approximately equal to atmospheric pressure at standard temperature, 1.013×10^5 Pa. To avoid decimals, a smaller unit called the *millibar* (mbar) is used more often in weather studies. Thus, one standard atmosphere (atm) is equal to 1013 mbar. We often hear weather forecasters speak of the barometric pressure as being so many "inches of mercury," or low pressures in laboratories reported as "millimeters of mercury." These are not valid pressure units; they are not expressible as force per unit area. Mercury barometers are common pressure-measuring devices, however, and it is convenient to express the pressure in terms of how high the mercury column stands in the tube. Because the static pressure is given by $p = \rho gh$, the pressure is proportional to h, so the unit is quite useful.

8. *Overpressure*. Often it is convenient to express the pressure in terms of its deviation from the atmospheric or ambient pressure. This deviation is called the *overpressure* or, commonly, the *gauge pressure*. For example, an automobile tire that is said to be "out of air" actually has air inside that is at the same pressure as the air outside. A tire gauge would read zero pressure, because it is designed to measure the gauge pressure, or overpressure, rather than the actual pressure. Note, however, that gauges can, and often do, read absolute pressures. A closed-tube manometer does this, for example.

9. *Pressure in a static fluid*. One of the more important principles in fluid statics is *Pascal's Principle*. The principle has to do with *changes* in pressure in an otherwise static fluid: *Pressure applied to a fluid at rest is transmitted undiminished to all parts of the fluid and to the walls of the container enclosing it*. Note that the pressure does not have to be the same throughout the fluid; it is the *change* in pressure that is transmitted through the fluid. If the fluid is incompressible, this change gets transmitted without diminishing.

Because the pressure is propagated as longitudinal pressure waves in the fluid, how fast this change in pressure is transmitted depends on the compressibility of the fluid. For a perfectly incompressible fluid, the bulk modulus is infinite, so the wave velocity is infinite also. Therefore, the closer the fluid is to being incompressible, the faster the pressure variation will be transmitted to the rest of the fluid.

The expression for the variation in fluid pressure with z, the vertical dimension, assumes that the fluid is in equilibrium in an *inertial* reference frame. If the fluid is accelerating, especially in a direction other than vertical, the pressure gradient will be along a direction other than that of $\mathbf{g}$. In fact, the pressure gradient in an accelerated fluid will be along the direction defined by $\mathbf{g} - \mathbf{a}$. As is shown in Figure 18.1, this term also defines the "downward" direction for the noninertial rest frame of the fluid. Thus, the pressure gradient is always in the direction defined as "downward" in the particular reference frame. Consider the centrifuge spinning at such a high speed

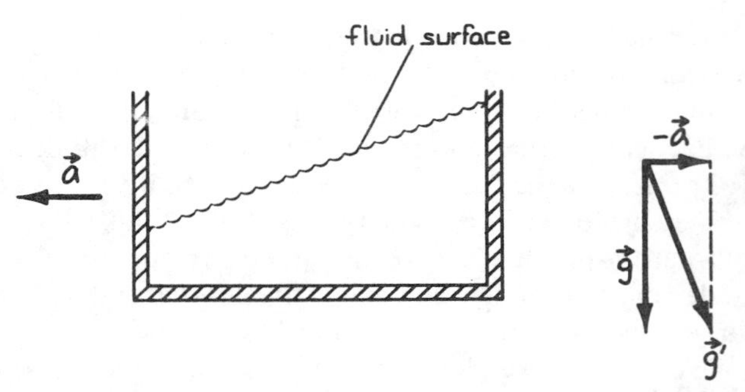

Figure 18.1 A container of fluid with an open surface is accelerated to the left. In the rest frame of the fluid, **g**′ defines the "downward" direction and is the direction of the pressure gradient.

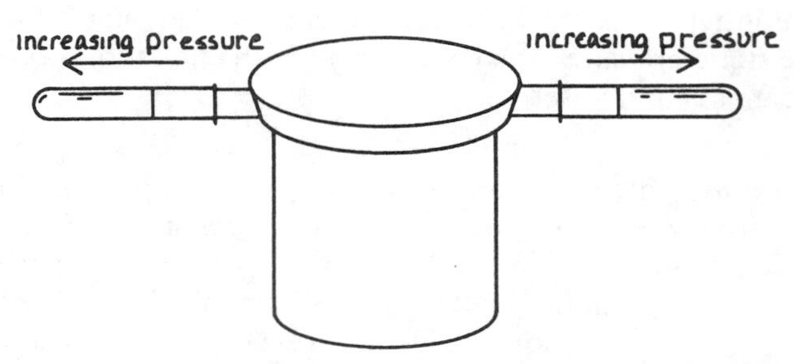

Figure 18.2 In the centrifuge the pressure gradient is radially outward.

that the test tube is virtually horizontal, as in Figure 18.2. Then $a \gg g$ and the pressure gradient is radically outward; that is, the pressure becomes progressively larger toward the "bottom" of the tube.

10. *Archimedes' Principle.* This principle states that a body immersed in a fluid medium is subject to a buoyant force equal to the weight of the fluid displaced. Note very carefully that the buoyant force does *not* depend on the weight of the object immersed. For example, a 5 g balsa wood ball and a 39 g iron ball will each displace 5 cm^3 of water, so the buoyant force on each is 0.049 N. This force is enough to support the weight of the balsa wood (0.049 N). However, the iron ball will sink ($w_{\text{iron}} = 0.3$ N). Basically, an object immersed in a fluid has two forces acting on it: the gravity force (its weight) and the buoyant force. The vector sum of these two forces is the *net* force, which, according to Newton's Second Law, determines the state of the object's motion. If the balsa ball were to be forced under water and then released, the upward buoyant force would be greater than the weight, so the ball would accelerate upward. The iron ball, in contrast, has a weight greater than the buoyant force, so it would accelerate downward.

Archimedes' Principle, as stated, assumes that the fluid forms a continuous layer *beneath* the object. To see why this is so, consider the situation shown in Figure 18.3. If the block makes a tight seal at the bottom of the container so that no fluid can get underneath, there will be no upward force on the block, even though it has displaced an amount of fluid equal to its

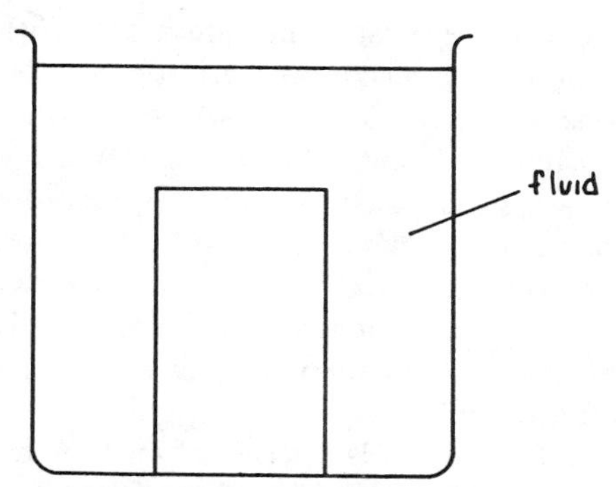

Figure 18.3 The block makes a tight seal at the bottom of the container. The "buoyant force" in this situation is downward.

own volume. In Figure 18.4 there is a buoyant force, but it is not equal to the weight of the fluid displaced.

If a gas-filled container, such as a balloon, is immersed in another gas, the buoyant force is equal once again to the weight of the *displaced* gas. In deciding whether the balloon floats or rises, we must consider the weight of the balloon fabric as well as the weight of the gas inside. If the balloon is to be able to float at all, the density of the gas inside must be such that its weight is less than the weight of the displaced gas by an amount equal to the weight of the material from which the balloon is made. The largest buoyant force would be realized if the interior of the balloon were a vacuum. The structural requirements for keeping the walls from collapsing would make it virtually impossible to construct a "vacuum" balloon, however. The simpler solution is to use a gas of lower density, such as helium, or air of a higher temperature than the ambient temperature.

11. *Fluid dynamics; Bernoulli's equation*. So much controversy has arisen over the correct interpretation of Bernoulli's principle that it is difficult to separate the plausible from the implausible. Strictly speaking, Bernoulli's equation applies only to the stream tubes we discussed earlier. The fluid

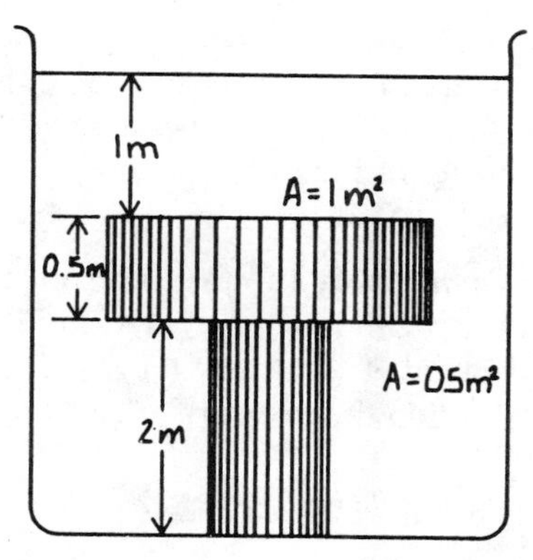

Figure 18.4 If the cylinder makes a tight seal at the bottom, the buoyant force is not equal to the weight of the displaced fluid. (The numerical data are for problem 5.)

should be ideal, and there must be no interaction with the outside environment except through the external pressure forces that drive the fluid along. Considerable debate has swirled around the question of what makes airplanes fly.[1] The usual argument is that because of the wing shape (called the *camber*), the airflow is increased over the top, resulting in a lowering of the pressure there and thereby generating an upward lift force on the wing. Others argue that because planes can and do fly upside down, the major effect is the transfer of momentum to the wing by the air as it is deflected downward by the wing. This principle explains why airplane wings are canted slightly upward even when the plane is in level flight. In the next chapter we will show that pressure from a gas is a result of molecular bombardment, so the two views are in one sense really not that different.

Sample Problems

1. *Worked Problem.* A standard door (80 cm × 200 cm) makes a tight fit in its frame. If the atmospheric pressure drops by 0.5% on one side of the door, what is the net horizontal force exerted on the door?

Solution: The net pressure force is just the pressure *difference* multiplied by the surface area. Thus

$$F_{\text{net}} = (p_1 - p_2)A = \Delta p \times A$$
$$F_{\text{net}} = (0.005)(1.015 \times 10^5 \text{ N/m}^2) \times (0.80 \text{ m} \times 2.00 \text{ m})$$
$$F_{\text{net}} = 812 \text{ N } (182 \text{ lb})$$

2. *Guided Problem.* Find the total force exerted on the 10 m-wide side of a swimming pool filled with water to a depth of 4 m.

Solution scheme

a. Because the pressure varies only with depth, what is the force dF on a narrow horizontal strip of dimensions 10 m × dz at a depth z below the surface of the water?

b. Integrate the force over the depth of the pool.

3. *Worked Problem.* A U tube is filled with water so that in equilibrium the water level is that shown in Figure 18.5a. A substance of unknown density is then poured into one side of the tube. The substance does not mix with water. The final configuration is that shown in Figure 18.5b. Determine the density of the added fluid.

Solution: Within the same fluid, the pressure must be the same at the same depth. Consider the depth labeled by AA' in Figure 18.5b. The pressure on the left-hand side at A is just $p_0 + \rho gh$, where ρ is the unknown density. On the opposite side the pressure at A' is $p_0 + \rho_w(h + h')$. Then

$$p_0 + \rho gh = p_0 + \rho_w g(h + h')$$

[1]For a good account of this controversy, see Norman F. Smith, "Bernoulli and Newton in Fluid Mechanics," *The Physics Teacher*, November 1972.

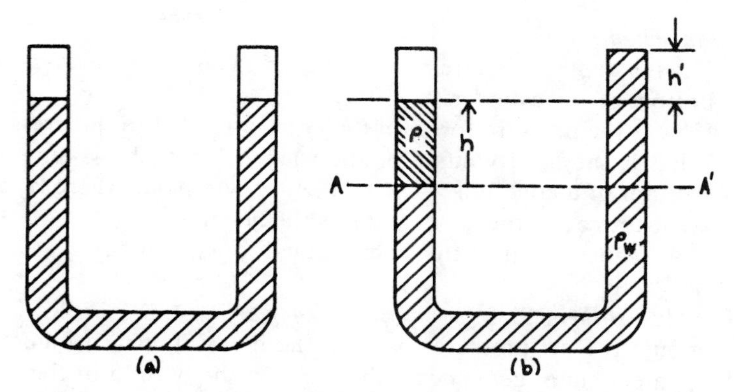

Figure 18.5 Problem 3. (a) Water in a U tube at equilibrium. (b) A fluid of density ρ is poured into the left-hand tube.

$$\rho = \rho_w \frac{h + h'}{h} = \frac{8 \text{ cm}}{3 \text{ cm}} \rho_w$$

$$\rho = 2.67 \times 10^3 \text{ kg/m}^3$$

Note that ρ / ρ_w is equal to 2.67. This ratio, the density of the fluid divided by the density of water, is called the *specific gravity* of the fluid.

4. *Guided Problem.* A conical funnel having a height of 9 cm and radii of 10 cm and 1 cm at the upper and lower ends, respectively, has its lower end pressed tightly into a cylindrical container having a radius of 10 cm and a height of 1 cm (Figure 18.6).

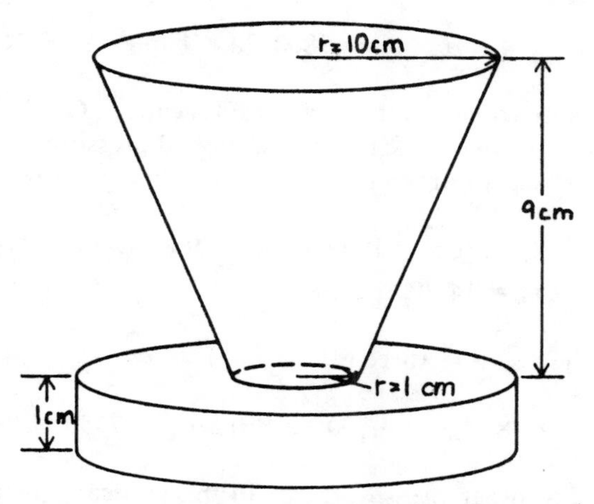

Figure 18.6 Problem 4. The cone and the cylinder are filled with water. What is the pressure on the bottom of the cylinder? What is the force?

a. Calculate the total pressure force on the bottom of the cylinder (ignore atmospheric pressure).
b. Calculate the total weight of the water.
c. Do the answers for parts a and b agree? Should they?

Solution scheme

a. Use the depth–pressure relationship to calculate the pressure at the bottom of the cylinder.
b. How is the pressure force on the bottom related to the pressure?
c. What is the total volume of the water? *Hint:* The volume of the frustum of a cone is $V = \frac{1}{3}(A_1 + A_2 + \sqrt{A_1A_2})h$, where A_1 and A_2 are the areas of the bases and h is the height.
d. How is the weight of the water related to its volume?

5. *Worked Problem.*

a. Compute the net pressure force on the block shown in Figure 18.4. The water cannot get beneath the block at the bottom of the vessel.
b. If the block is raised slightly so that water is allowed to flow underneath, what is the resulting pressure force?

Solution

a. Direct application of Archimedes' Principle, as stated, is not possible while the object sits on the bottom. First we will work the problem in the most general way, calculating the pressure forces on the various parts of the block. Then we will directly apply Archimedes' Principle and show that different results are obtained. At the top of the block the pressure is

$$p_t = \rho g z_t = 1000 \text{ kg/m}^3 \times 9.8 \text{ m/s}^2 \times 1 \text{ m}$$
$$p_t = 9800 \text{ Pa}$$

so that the pressure force is

$$F_t = p_t \times A_t = 9800 \text{ Pa} \times 1 \text{ m}^2 = 9800 \text{ N}$$

The pressure forces on the vertical sides cancel. On the underside of the *upper* portion of the block, an upward pressure force is possible only on 0.5 m^2 of area. The pressure at this depth is

$$p_b = \rho g z_b = 1000 \text{ kg/m}^3 \times 9.8 \text{ m/s}^2 \times 1.5 \text{ m}$$
$$p_b = 14{,}700 \text{ Pa}$$

so that the pressure force is

$$F_b = p_b \times A_b = 14{,}700 \text{ Pa} \times 0.5 \text{ m}^2 = 7350 \text{ N (upward)}$$

There is no other pressure force in the vertical direction, because water cannot get beneath the block at the bottom. The net pressure force is then

$$F_t - F_b = 9800 \text{ N} - 7350 \text{ N} = 2450 \text{ N (downward)}$$

Thus, the block is held to the bottom by the pressure force. This is true regardless of the density of the block, because its own weight will always act downward. Now let's see what a direct application of

Archimedes' Principle gives us; in other words, let's assume that there is a buoyant force upward equal to the weight of the displaced fluid.

$$V_{fl} = 1\ m^2 \times 0.5\ m + 0.5\ m^2 \times 2\ m = 1.5\ m^3$$
$$W_{fl} = \rho g V_{fl} = 1000\ kg/m^3 \times 9.8\ m/s^2 \times 1.5\ m^3$$
$$W_{fl} = 14{,}700\ N\ (upward)$$

Both the magnitude and the direction of the net pressure force are incorrect.

b. Now suppose we allow water to flow beneath the block at the bottom of the vessel. Then we have an additional upward pressure force of

$$F_l = p_l A_l = \rho g z_l A_l = 1000\ kg/m^3 \times 9.8\ m/s^2 \times 3.5\ m \times 0.5\ m^2$$
$$F_l = 17{,}150\ N$$

Therefore, the net pressure force on the block is

$$F_b + F_l - F_t = 73{,}500\ N + 17{,}150\ N - 9800\ N$$
$$F_{net} = 14{,}700\ N\ (upward)$$

which is precisely what we obtained using Archimedes' Principle. Once again we see that the principle works *if* the fluid is allowed to be continuous across the bottom of the immersed object.

6. *Guided Problem.* A research structure in the shape of a rectangular box 6 m × 4 m × 3 m high is anchored to the floor of the ocean. The bottom of the structure is 50 m below the surface, and the water is allowed to pass freely beneath. The entire structure and its contents have a mass of 50,000 kg. What is the force required to keep the structure anchored to the seafloor? Suppose the mass given does not include the mass of the air in the structure. How much difference will this make in the results? Assume the density of the air to be 1.029 kg/m³, with the volume equal to that of the structure less 10% to allow for the space occupied by the thickness of the walls and the equipment.

Solution scheme

a. What is the volume of sea water displaced?
b. Given the density of sea water to be $1.02 \times 10^3\ kg/m^3$, what is the weight of the displaced sea water?
c. What is the buoyant force on the structure?
d. What is the weight of the structure?
e. What is the difference between the weight and the buoyant force?

7. *Worked Problem.* A ball having a radius of 5 cm and a density of $0.4 \times 10^3\ kg/m^3$ is dropped from a height of 5.1 m into a tank of fresh water. Ignore any dissipative effects, and calculate the depth to which the ball will sink.

Solution: Because the ball is less dense than the water, the upward buoyant force will be greater than the ball's weight. Thus, as the ball enters the water, it will have an upward acceleration equal to $a = F_{net}/m$. Taking the downward direction as positive, we have

$$F_{net} = mg - F_B = \rho_b V_b g - \rho_w V_w g$$

After the ball is immersed, $V_b = V_w = V$, so that

$$F_{net} = gV(\rho_b - \rho_w)$$

and

$$a = \frac{F_{net}}{m} = \frac{F_{net}}{\rho_b V} = \left(1 - \frac{\rho_w}{\rho_b}\right)g = -\left(\frac{\rho_w}{\rho_b} - 1\right)$$

The distance the ball travels in coming to rest is found by use of the kinematic equation $\upsilon^2 = \upsilon_0^2 + 2az$:

$$z = \frac{\upsilon^2 - \upsilon_0^2}{2a} = \frac{0 - \upsilon_0^2}{-2\left(\frac{\rho_w}{\rho_b} - 1\right)} = \frac{0 - 2gh}{-2\left(\frac{\rho_w}{\rho_b} - 1\right)}$$

$$z = \frac{gh}{\left(\frac{\rho_w}{\rho_b} - 1\right)}$$

$$z = \frac{9.8\ \text{m/s}^2 \times 5.1\ \text{m}}{\left(\frac{1000}{400} - 1\right)} = 33\ \text{m}$$

8. *Guided Problem.* A spherical instrument housing, attached by a cable to the seafloor, has a radius of 1.5 m and a mass of 8.5×10^3 kg. The cable snaps, and the instrument accelerates toward the surface 15 m above. How fast will it be moving when it reaches the surface? Neglect any friction effects.

Solution scheme: This problem is highly similar to the preceding one.

a. What is the volume of the sphere?
b. What is the buoyant force resulting from the displacing of the sea water? $\rho_s = 1.025 \times 10^3\ \text{kg/m}^3$.
c. What is the *net* force on the sphere?
d. What is the acceleration of the sphere?

9. *Worked Problem.* A large cylindrical storage tank, 10 m in diameter and 10 m high, is to have a catch basin whose diameter must be such that the liquid from any leak that develops anywhere on the side of the tank will be contained within the basin. Calculate the minimum diameter of the basin if the tank is situated centrally within it.

Solution: First we must find the greatest horizontal distance the liquid will travel if a leak develops. We will make the assumption that the flow is nonviscous and that Torricelli's theorem holds. Two factors affect the range: the initial horizontal velocity and the time of flight. The distance above the ground determines the time of flight, whereas the distance of the leak from the *top* of the liquid determines the fluid discharge velocity. Torricelli's theorem states that the velocity of efflux of a liquid element flowing from a small opening at a depth h in a tank is the same as that which the element would acquire by falling freely through the same height h. This is the result from Example 9 in the text (p. 481). Thus, the speed at which the fluid leaves the hole is $\upsilon = \sqrt{2gh}$. The horizontal distance is given by $x = \upsilon_x t$, where t is the time of flight. We can find t by using the equation

$$z = \upsilon_{0_z} t + \tfrac{1}{2} a_z t^2$$

or since $a_z = g$ and $\upsilon_{0_z} = 0$, we have

$$t = \sqrt{\frac{2z}{g}}$$

But the distance fallen, z, is the total height of the tank, H, minus the distance from the top to the hole, h; that is

$$z = H - h$$

Then

$$t = \sqrt{\frac{2(H-h)}{g}}$$

and

$$x = \upsilon_x t = \sqrt{2gh} \times \sqrt{\frac{2(H-h)}{g}}$$

$$x = 2\sqrt{h(H-h)}$$

To find the maximum horizontal distance, we take dx/dh and set it equal to zero:

$$2 \frac{d}{dh}(hH - h^2)^{1/2} = 0 = \tfrac{1}{2}(hH - h^2)^{-1/2}(H - 2h)$$

From this we get

$$h = H/2$$

so that the maximum range is

$$x_{\max} = 2\sqrt{\frac{H}{2}\left(H - \frac{H}{2}\right)} = 2\sqrt{\frac{H}{2} \cdot \frac{H}{2}} = H$$

Thus, the maximum range is equal to the height of the water in the tank, assuming that the tank rests on the ground. Moreover, the leak would have to occur midway up the (full) tank to achieve this range. If the tank is 10 m in diameter, then the catch basin must have a radius equal to that of the tank (5 m) plus an extra 10 m. So the diameter of the catch basin must be 30 m.

10. *Guided Problem*. Torricelli's theorem, stated in the preceding problem, is based on the assumption that the liquid level in the tank is falling very slowly compared with the fluid velocity at the hole. Calculate the velocity of the fluid leaving the hole if we take the velocity of the upper surface into account.

Solution scheme

a. With reference to Example 9 in the text (p. 481), what is the resulting Bernoulli's equation when $\upsilon_1 \neq 0$?
b. How is υ_1 related to υ_2 in the continuity equation?

Answers to Guided Problems

2. $F = 7.8 \times 10^5$ Pa
4. a. 31 N
   b. 10 N
   c. No; no
6. 2.3×10^4 N downward. The weight of the air contributes about 3%.
8. 14 m/s
10. $\upsilon_2 = \left[\dfrac{2gh}{1 - \left(\dfrac{A_2}{A_1}\right)^2}\right]^{1/2}$. Clearly if $A_1 = A_2$, $\upsilon_2 \to \infty$. What is wrong?

Chapter 19

The Ideal Gas and Kinetic Theory

Overview

Having developed fairly complete and quite successful models for particle dynamics, we might ask whether such models are applicable to all systems. The answer is a qualified yes; that is, in principle, Newton's laws govern the behavior of all particles at ordinary speeds, no matter how large the collection. But as a practical matter, the application of these laws becomes virtually impossible when we are dealing with more than just a few particles. For large populations, whether of molecules or of people, the statistical behaviors not only are more useful but also are likely to be the only information we can acquire with any degree of precision. In this chapter we explore the behavior of gases, which consist literally of billions of billions of individual molecules even in small regions of space. We will develop a statistical model that describes the average motion of these molecules, while ignoring the details of the motion of individual molecules.

Essential Terms

ideal gas
ideal-gas law
universal gas constant
Boyle's Law
Gay-Lussac's (Charles') Law
Avogadro's number
Boltzmann's constant
standard temperature and pressure (STP)
triple point
absolute (Kelvin) temperature scale
Celsius scale
Fahrenheit scale
root-mean-square (rms) velocity
equipartition theorem
mean free path

Key Concepts

1. *Ideal-gas law*. This law for the behavior of gases (expressed mathematically as $pV = nrT$) works very well for almost all gases at reasonable temperatures and pressures. When we use the mathematical expression for this law, we must express the temperature in *kelvins*, not Fahrenheit or Celsius degrees. It is possible to express the law using other temperature scales, but if we do so we lose the ability to take ratios of temperatures in comparing the parameters under different conditions. According to the law, the absolute temperature will be zero when either the pressure or the volume is zero. These conditions are not actually attainable for real substances. One reason is that gases will liquefy, or solidify, prior to reaching that temperature.

2. *The temperature scale*. The temperature of any substance is difficult to define precisely without appealing to the energies of molecules. The ideal-gas law provides a natural way to define temperatures, because it defines a uniform behavior for all (ideal) gases and approximately uniform behavior for real gases. For example, if we place a gas in a container and maintain the gas at constant volume, we can monitor the pressure for various amounts of internal energy (temperature) of the gas. The Celsius scale is based on 0° being defined as the triple point of water and 100° as the boiling point (when the pressure is 1 standard atmosphere). The bulb of gas is immersed in the ice-water bath (approximating the triple point), and the pressure is noted. The bulb is then placed in the boiling water, and the new pressure is recorded. The two values are plotted on a pressure-versus-temperature graph, as shown in Figure 19.1. A straight line is drawn between these two points,

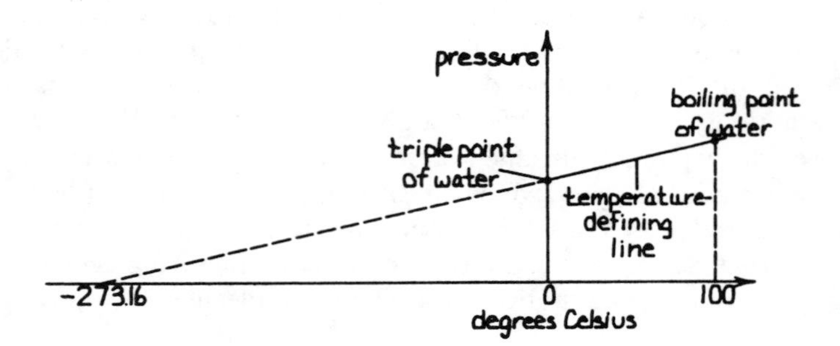

Figure 19.1 Temperature-defining line based on ideal gas at constant volume.

this line being the *temperature-defining line*. In other words, when the gas bulb is placed in another environment (and its volume is kept constant), there will be a certain pressure reading, which when located on the p vs T graph will define the temperature. Any other type of thermometer (mercury, for example) must be calibrated against this line.

If we extend the temperature line downward, eventually it will cross the T axis. This point corresponds to $p = 0$ and represents an obvious lower limit to the temperature. This temperature is the natural, or *absolute*, zero, and careful measurements show this intersection with the T axis to occur at

−273.16°C, or approximately −273°C. By doing nothing more than shifting the origin of the temperature scale to this intersection point, we can make the associated temperature the *zero* on the absolute scale. This scale is called the *Kelvin* scale, and the temperatures read from this scale are expressed in *kelvins* (K). (Formerly this unit was called degrees Kelvin and written °K, but the official SI unit is the *kelvin* and is abbreviated *K* with no degree symbol.) Because the change from Celsius to kelvins involves only a shift in origin, the conversion is quite simple:

$$T_C = T_K - 273$$

The Fahrenheit scale, however, does not have the same size *interval* as the *C* and *K* scales. The simplest way to see this difference is to examine the graph of T_F versus T_C. The freezing and boiling points of water are (0, 32) and (100, 212), where the first number is the Celsius temperature. Figure 19.2 shows the graph to have a T_F-axis intercept of 32 and a slope of (212 − 32)/100 = 9/5. The slope-intercept equation for a straight line gives us

$$T_F = \tfrac{9}{5}T_C + 32$$

or

$$T_C = \tfrac{5}{9}(T_F - 32)$$

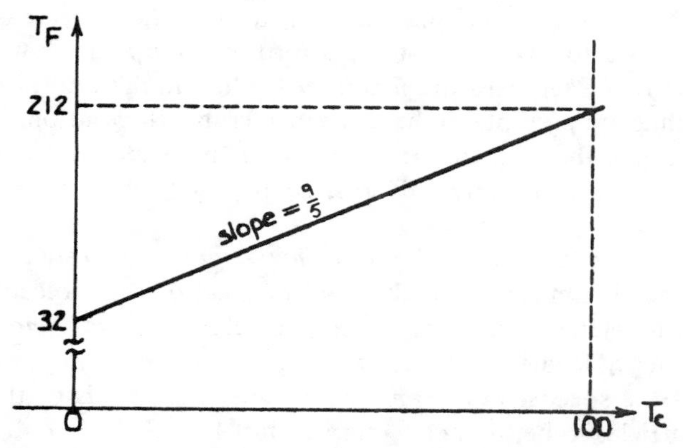

Figure 19.2 Graph of Fahrenheit versus Celsius temperature.

3. *Kinetic pressure*. The derivation of a macroscopic expression for pressure based on the collective behavior of microscopic particles is a classic example of how the averaging process over a large collection of diverse particle motions can yield macroscopically measurable results. These results can be simulated by various means, one of which is illustrated in Figure 19.3. When the lower platform moves rapidly up and down, it causes the steel balls to bounce randomly off the walls of the container as well as on the upper piston, which is free to move up and down in the cylinder. The piston settles into a position such that the *average* force exerted by the steel balls just balances the weight of the piston. As the volume gets larger, fewer

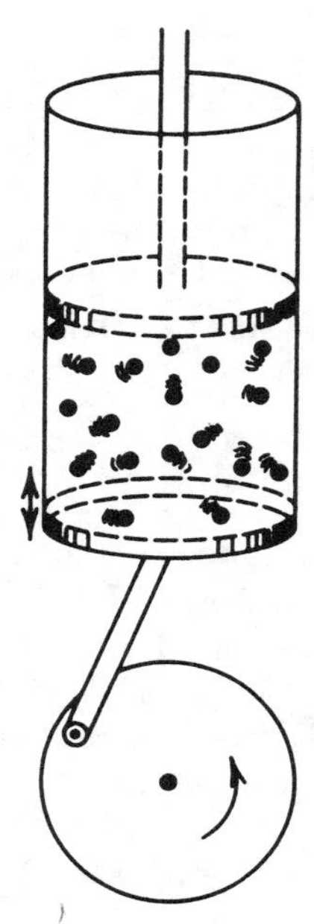

Figure 19.3 The steel balls bounce off the walls of the cylinder and the piston, simulating gas pressure.

collisions per unit time take place and the average force decreases. This can be compensated for by having the platform move up and down at a higher rate, simulating a temperature increase. In the simulation device, relatively few steel balls are present, so there is considerable fluctuation in the average force. In a gas the tremendous number of molecules ($\sim 10^{23}$) causes this force to be macroscopically uniform, giving rise to uniform pressures.

4. *Energy of an ideal gas*. The *root-mean-square* (*rms*) speed of the molecules in a gas is a measure of the *typical* speed of the molecules. It is this speed that is related to the Kelvin temperature of the gas—*not* the average speed of the molecules, but the square root of the average (mean) of the *squares* of the speeds. For example, consider the speeds of thirteen hypothetical particles to be (in meters per second) 1, 1, 2, 3, 4, 4, 4, 5, 5, 6, 7, 7, 8. The *average* of these numbers is their simple sum divided by the number of particles: $57/13 = 4.38$ m/s. The *rms* speed is found by summing the squares of these numbers, dividing by 13, and then taking the square root: $\sqrt{311/13} = 4.89$. For large numbers of particles in a gas, the difference in these two numbers usually is not very large.

The temperature of a gas is defined by

$$T = \frac{1}{3}\frac{mv_{\text{rms}}^2}{k} = \frac{2}{3}\frac{1}{k}\left(\frac{1}{2}mv_{\text{rms}}^2\right)$$

In other words, we can specify the temperature of a gas in terms of an *average* translational kinetic energy of the gas molecules. Any individual molecule may have at some instant a low speed, a high speed, or some intermediate speed. But *on the average*, for a particular temperature the molecule will have a kinetic energy given by the expression just presented. As the temperature changes, the *average* kinetic energy changes accordingly. Indeed, we can say that the temperature is a *measure* of the average kinetic energy.

Sample Problems

1. *Worked Problem.* Calculate the pressure at 2 km above the Earth's surface if the temperature is a uniform 290 K from the surface to that altitude.

Solution: We start with the expression for the variation in pressure with depth in a fluid. We use the differential form, because ρ is not a constant in this case.

$$dp = -\rho g dz$$

or

$$\frac{dp}{dz} = -\rho g$$

But since $pV = nRT$ and $V = m/\rho$, we can write

$$p = \frac{n}{m} RT\rho$$

The quantity n/m is just the molecular mass, that is, the number of grams per mole, which we will designate as m_0. Then

$$p = \frac{RT}{m_0} \rho$$

or

$$\rho = \frac{m_0}{RT} p$$

Therefore

$$\frac{dp}{dz} = -\frac{m_0 g}{RT} p$$

Separating the variables gives us

$$\frac{dp}{p} = -\frac{m_0 g}{RT} dz$$

Then

$$\int_{p_0}^{p} \frac{dp}{p} = -\frac{m_0 g}{RT} \int_0^z dz$$

where z is the height above the surface. Integrating, we get

$$\ln \frac{p}{p_0} = -\frac{m_0 g}{RT} z$$

Noting that $e^{\ln x} = x$, we have

$$p = p_0 e^{-\frac{m_0 g}{RT} z}$$

We see that the pressure diminishes exponentially with height, assuming a uniform temperature. The mean molecular mass of air is 28.97 g/mol. The constant R has a value that depends on the system of units. If the kilogram-mole is being used, then $R = 8314$ J/K, whereas if the gram-mole is the unit, then $R = 8.314$ J/K. To ensure that the exponent just presented is dimensionless, we use the *kilogram-mole* so that

$$p = p_0 e^{-\frac{28.97 \times 9.8 \text{ m/s}^2}{8314 \text{ J/K} \times 290 \text{ K}} \times 2000 \text{ m}}$$

$$p = p_0 e^{-0.235} = 0.79\, p_0$$

2. *Guided Problem.*

a. Show that the mean molecular mass of air is 28.97 g/mol.

b. Use this result to calculate the approximate percentages of N_2 and O_2, assuming that these make up virtually 100% of air.

Solution scheme

a. What is the density of air at STP?

b. What mass of air occupies 1 liter?

c. How much volume does 1 mole of gas occupy at STP?

d. What are the molecular masses of N_2 and O_2?

e. If x is the fraction of N_2 and y is the fraction of O_2, how are x and y related?

3. *Worked Problem.* A gas having a molecular mass of 44 g is confined in a 150 l box at a temperature of 300 K and at a pressure of 1 atm. A partition is placed in the middle of the box, and the temperature on the left side is raised to 390 K. If the partition has the dimensions 50 cm × 20 cm, what is the net force exerted on the partition by the gas?

Solution: First we look at the gas in the left-hand container at the initial conditions. The pressure is standard, but the temperature is not. So let's see what the density of the gas would be at STP. Since 1 mole of a gas occupies 22.4 l under these conditions, the density of the gas would be

$$\rho = \frac{44 \text{ g}}{22.4\ l} = 1.96 \text{ g}/l$$

and there would be 1.96 g/l × 75 l = 147 g in the left side of the box. If we use the result from problem 1, $\rho = m_0 p/RT$, so at the same pressure we have $\rho = \rho_0 \dfrac{T_0}{T}$; therefore, the density at 300 K is

$$\rho = 1.96 \text{ g}/l \times \frac{273 \text{ K}}{300 \text{ K}} = 1.78 \text{ g}/l$$

Thus, the mass of gas contained in 75 l at 300 K is

$$m = 1.78 \text{ g}/l \times 75\ l = 134 \text{ g}$$

Then

$$n = 134 \text{ g}/44 \text{ g/mol} = 3 \text{ mol}$$

Now if we raise the temperature of the gas, at constant volume, to 390 K, the pressure becomes

$$p_2 = p_1 \frac{T_2}{T_1} = 1 \text{ atm} \times \frac{390 \text{ K}}{300 \text{ K}} = 1.3 \text{ atm}$$

so that the force on an area 50 cm × 20 cm is

$$F = \Delta p \times A = 0.3 \text{ atm} \times 0.1 \text{ m}^2$$
$$F = 0.3 \times 1 \times 10^5 \text{ Pa} \times 0.1 \text{ m}^2$$
$$F = 3 \times 10^3 \text{ N} = 670 \text{ lb}$$

4. *Guided Problem.* A cylinder 1.5 m high with a radius of 20 cm is filled with air at STP. A tight-fitting but frictionless 100 kg piston is set in the cylinder (Figure 19.4).

a. At what height above the bottom of the cylinder will the piston come to rest?

b. What is the minimum temperature that will just cause the piston to be pushed out of the cylinder?

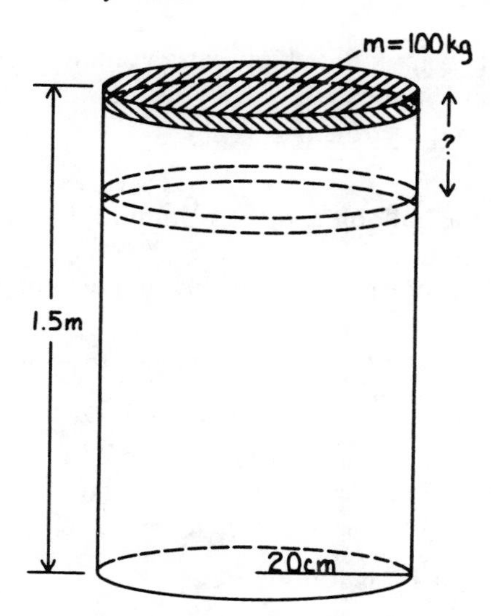

Figure 19.4 Problem 4. The piston is allowed to sink, compressing the gas. How far will it sink?

Solution scheme

a. If the temperature is held constant, what is the relationship between the pressure and the volume?
b. What is the pressure above and below the piston when it settles?
c. How is this pressure difference related to the upward force on the piston?
d. What must this pressure force balance?
e. What temperature difference will give rise to the same pressure difference with a volume equal to that of the entire cylinder?

5. *Worked Problem.* A bicycle pump consists of a cylinder fitted with a piston that can move a total of 50 cm. The pump is attached to a tire whose internal gauge pressure is 2.3 atm. Assume that the pumping action takes place at constant temperature. How far does the piston move before air begins to enter the tire?

Solution: The pressure in the cylinder must be greater than the gauge pressure of the tire before air can be forced from the pump to the tire. At constant temperature, $p_1V_1 = p_2V_2$, with $p_1 = 1$ atm, $p_2 = 2.3$ atm, and $V_2 = 0.5 \text{ m} \times A$, where A is the cross-sectional area of the pump cylinder. Then

$$V_2 = lA = \frac{p_1}{p_2} V_1 = \frac{1 \text{ atm}}{2.3 \text{ atm}} \times 0.5 \text{ m} \times A$$

$$l = \frac{0.5 \text{ m}}{2.3} = 0.22 \text{ m}$$

Thus, the piston must move 50 cm − 22 cm = 28 cm before the air begins to enter the tire.

6. *Guided Problem.* A tank of helium has a volume of 30 l, and the internal pressure is 30 atm. Suppose we wish to fill a 400 m^3 balloon at STP. How many tanks are required?

Solution scheme

a. Assume that the balloon and the helium tanks are at the same temperature.
b. What is the relationship between the volume in liters and the volume in cubic meters?
c. How is the change in volume related to the change in pressure at constant temperature?

7. *Worked Problem.* Find the *rms* velocity of H_2 at 30°C. At what temperature does O_2 have the same *rms* velocity?

Solution: The mass of H_2 is found by

$$m_{H_2} = \frac{2 \text{ g}}{6 \times 10^{23}} = 3.33 \times 10^{-24} \text{ g} = 3.33 \times 10^{-27} \text{ kg}$$

The *rms* velocity is then

$$v_{rms} = \sqrt{\frac{3kT}{m}} = \sqrt{\frac{3 \times 1.38 \times 10^{-23}\ \text{J/K} \times 303\ \text{K}}{3.33 \times 10^{-27}\ \text{kg}}}$$

$$v_{rms} = 1.9 \times 10^{3}\ \text{m/s}$$

To find the temperature at which O_2 has the same *rms* velocity, we note that

$$\sqrt{\frac{3kT_{O_2}}{m_{O_2}}} = \sqrt{\frac{3kT_{H_2}}{m_{H_2}}}$$

$$T_{O_2} = \frac{m_{O_2}}{m_{H_2}} T_{H_2}$$

The mass of O_2 is

$$m_{O_2} = \frac{32\ \text{g}}{6.02 \times 10^{23}} = 5.31 \times 10^{-23}\ \text{g} = 5.31 \times 10^{-26}\ \text{kg}$$

Therefore

$$T = \frac{5.31 \times 10^{-26}\ \text{kg}}{3.33 \times 10^{-27}\ \text{kg}} \times 303\ \text{K} = 4837\ \text{K} = 4564°\text{C}$$

8. *Guided Problem.*

a. What is the minimum radius for a planet having the average density of the Earth to retain an oxygen atmosphere? Express the result in terms of the temperature T.

b. What is the minimum radius for a temperature of 300 K?

Solution scheme

a. Determine the escape velocity for a particle at the surface of a planet of mass M and radius R (see Chapter 9).

b. How are the mass, radius, and density related?

c. Assume the minimum condition to be that for which the *rms* velocity equals the escape velocity.

d. Use 5500 kg/m^3 for the average density of the Earth.

Answers to Guided Problems

2. 21% O_2, 79% N_2
4. a. 1.39 m
   b. 295 K
6. 89 tanks
8. a. $1.59 \times 10^4\ T^{1/2}$ meters
   b. 275 km

Chapter 20

Heat

Overview

The kinetic theory developed in Chapter 19 lays the foundation for a theory of energy transfer from one part of a body to another or from one part of a system to another. This energy transfer basically involves the random thermal motion of molecules, and it results from differences in temperatures. We call this type of energy transfer *heat*. In addition to energy transfer from molecular interactions, we will explore the mechanism of energy transfer by electromagnetic radiation.

Essential Terms

calorie
kilocalorie
specific heat capacity
heat
temperature
mechanical equivalent of heat
linear expansion
volume expansion
coefficient of linear expansion
coefficient of cubical expansion
conduction
heat flow
thermal conductivity
heat conduction
heat of fusion
heat of vaporization
specific heat at constant volume
specific heat at constant pressure
adiabatic processes

Key Concepts

1. *Heat as energy transfer*. Heat is not a substance that a body possesses or contains. It is, rather, a form of energy *transferred* from one body to another as a result of *temperature differences*. In other words, heat is energy

in transit. Any body having a temperature above absolute zero possesses energy, called the *internal* energy. This energy is not the same as heat, because heat is the transferred energy from one body to another. To say that a body possesses a quantity of heat is basically to adopt a view popular in earlier times, namely, that every body contained a fluid called the *caloric* that could be absorbed or given off by the body.

If two substances in contact are at different temperatures, the heat energy will flow from the hotter to the colder, regardless of the internal energy possessed by each body. For example, a tub of water at 20°C has considerably more total thermal energy than a glass of water at 20°C, but if the two are placed in contact, no net energy will flow between them. Thus, the critical requirement for heat energy transfer is that there be a *temperature difference*.

The official SI unit for energy is the *joule*, and strictly speaking it is the only unit needed for all forms of energy, including heat energy. Historically, however, the heat energy unit was based on the temperature changes resulting from heat transfer, and today we still find it convenient to use the *calorie* (or *kilocalorie*) as a measure of heat energy. But because heat is a form of energy, it must be expressible in *joules*. The conversion is called the *mechanical equivalent of heat*, and its numerical value is determined by doing work on a system and measuring the temperature change. But we should not presume that the joule is strictly a unit of mechanical energy and the calorie strictly a unit of heat energy. One could just as well express the work done in raising a block in calories, but to do so would quite likely create confusion.

To raise or lower the temperature of a body, heat energy must flow into or out of the body to change its internal energy. The amount of heat energy required to change the temperature of a body of unit mass by 1 degree is called the *specific heat capacity* or, more commonly, the *specific heat*. The unit, called the *kilocalorie*, is the heat energy required to change the temperature of 1 kg of water 1C°. The *calorie* is one-thousandth of a kilocalorie.

2. *Thermal expansion*. It is common knowledge that most bodies expand or contract when the temperature changes. This *thermal expansion* or contraction is rooted in the thermal vibrations of the molecules in the body. In some cases the molecular arrangement plays a role, as in the unusual density change in water between 4°C and 0°C. But how much an object expands depends on more than the temperature difference. If the body is heated uniformly, each element of length expands by the same amount, so the longer the body, the greater the expansion. Some materials expand differently than others, so the nature (composition) of the material is important. Finally, the temperature difference determines the overall expansion, a greater difference in temperature giving rise to a greater expansion. If we are to characterize a material according to its expansion properties, we need to have a measure that is characteristic of the material itself—not dependent on its length or the temperature difference. Thus, we need to look at the *change in length per unit length per unit temperature change*. We call this quantity the *coefficient of linear expansion*, α. Thus, $\alpha = \frac{\Delta L}{L\Delta T}$. Its dimension is 1/C°, because $\Delta L/L$ is dimensionless. For expansion in three dimensions we have a similar expression, except that the coefficient of linear expansion is replaced by the *coefficient of cubical expansion*, β. Because the cubical

expansion is made up of linear expansions along three dimensions, we have the simple relationship $\beta = 3\alpha$.

3. *Conduction of heat.* The propagation of a mechanial wave through a medium depends on the transfer of energy from one molecule to the next. The transfer of heat energy by conduction is also accomplished by molecular transfer of energy. So what is the essential difference? In the transfer of heat energy, the molecular motion remains *random*—the molecules simply increase their random motion as the energy is passed along. In wave propagation an *ordered* motion of the molecules is superimposed on the random thermal motion. In conduction the amount of heat energy transferred depends on the type of material; on the area of contact, A, between the two temperature sources; on the temperature difference, ΔT; on the time interval, Δt; and on the thickness of the material, Δx. The characteristic measure of conduction, called the *thermal conductivity, k,* is defined as

$$k = \frac{\Delta Q}{\Delta t \cdot \Delta T \cdot A \cdot \dfrac{1}{\Delta x}}$$

where the $1/\Delta x$ term reflects the fact that the heat flow diminishes with increasing thickness. If the temperature gradient is nonuniform, we can write the *rate* of heat flow in differential form:

$$\frac{dQ}{dt} = kA\,\frac{dT}{dx}$$

where dT/dx is the thermal gradient.

4. *Changes of state (phase).* Although the transfer of heat energy requires a temperature difference, a substance that absorbs this energy does not necessarily undergo a temperature change. Boiling water has a constant temperature, yet copious amounts of heat energy are being transferred from the burner to the water. Similarly, a glass containing ice and water will continue to absorb heat energy from the surrounding warmer air, yet the temperature stays fixed at 0°C until all the ice is melted. These two examples illustrate changes of state, or phase, of water. The energy absorbed by the water is used to break the bonds responsible for the solid or liquid phase (plus a small amount to do work in the expansion or contraction, as we shall see in Chapter 21) rather than to increase the average thermal energy of the molecules (the temperature). The process can occur in either direction. As water condenses from steam or solidifies from liquid water, heat energy is transferred out of the water to the surroundings. For example, 1 g of steam, upon condensing, gives off approximately 540 calories of heat energy, whereas when water freezes, each gram of ice formed results in the liberation of 80 calories.

The amount of heat energy transferred in changes of phase for a given substance depends only on the amount (mass) of material. Thus, $\Delta Q \sim m$. The amount of heat transferred *per unit mass* (thus making it a *characteristic property*) is called the *latent heat of fusion* (or *vaporization*), given by L_f

and L_v, respectively. Therefore, $\Delta Q = mL_f$ and $\Delta Q = mL_v$. These expressions are valid only for changes of phase. Within one phase, the heat transfer is described by our earlier expression $\Delta Q = cm\Delta T$.

5. *The specific heat of a gas.* In the preceding section we indicated that the heat energy transferred into or out of a system in constant phase is $\Delta Q = cm\Delta T$. For a gas this expression remains valid, but because of some unusual properties of gases, it is important to note whether the gas is being heated (or cooled) at constant volume or at constant pressure. Moreover, when dealing with gases, we find it more convenient to work with *moles*, rather than mass, as the basic unit. So the expression for heat transfer into or out of gases is given by one of the two expessions

$$\Delta Q = nc_v\Delta T$$

and

$$\Delta Q = nc_p\Delta T$$

The quantities c_v and c_p are related by

$$c_p - c_v = R$$

where R is the universal gas constant. For a monatomic gas, $c_v = \frac{3}{2}R$ and $c_p = \frac{5}{2}R$, whereas for a diatomic gas, $c_v = \frac{5}{2}R$ and $c_p = \frac{7}{2}R$.

6. *The adiabatic equation.* An *adiabatic* process is one that takes place without heat energy's being transferred into or out of the system. Although a true adiabatic process is an idealization, processes that are approximately adiabatic can occur, especially if the system is well insulated or if the changes take place quickly, before any appreciable amount of energy enters or leaves the system. For example, the rapid compression of a gas by the piston in an automobile engine and the compression of air in a tire pump both approximate adiabatic processes, because they occur so quickly that very little heat enters or leaves the system. The relationship between pressure and volume for an adiabatic process is pV^γ = constant, where γ is the ratio c_p/c_v. If, for example, the initial pressure and volume and the final volume are known, the final pressure can be determined. Then the ideal-gas equation can be used to determine the temperature change. It can be shown that the adiabatic equation is expressible in terms of the other two pairs of variables, namely

$$TV^{\gamma-1} = \text{constant}$$

and

$$PT^{(\gamma-1)/\gamma} = \text{constant}$$

Sample Problems

1. *Worked Problem.* A 150 l hot-water heater has two 1300 W heating elements. If no heat is lost through the walls of the tank, how much time is required to heat the water in the tank from 10°C to 50°C?

Solution: The problem involves the conversion of electrical energy into heat energy. For heat energy we will use the more common unit, the *kilocalorie*. The number of kilocalories supplied by the heating coil is

$$\frac{\Delta Q}{\Delta t} = \frac{2 \times 1300 \text{ J/s}}{4185 \text{ J/kcal}} = 0.62 \text{ kcal/s}$$

To heat 150 l of water over the indicated temperature range requires

$$\Delta Q = \frac{1.0 \text{ kcal}}{\text{kg} \cdot \text{C}^\circ} \times 150 \text{ l} \times (50^\circ\text{C} - 10^\circ\text{C}) = 6000 \text{ kcal}$$

The time required is found by equating the two expressions for ΔQ:

$$0.62 \text{ kcal/s} \times \Delta t = 6000 \text{ kcal}$$
$$\Delta t = 9677 \text{ s}$$
$$\Delta t = 2.7 \text{ h}$$

2. *Guided Problem.* Suppose the electricity goes off and you wish to heat water in your 1 gal (3.8 l) ice cream freezer by carefully insulating the inner container and cranking the paddles to stir the water (Figure 20.1). You can turn the crank at 2 rev/s, and each revolution requires you to do 8 J of work. Assuming no losses through the walls, by how much will the temperature of the water change if you crank for 1 h?

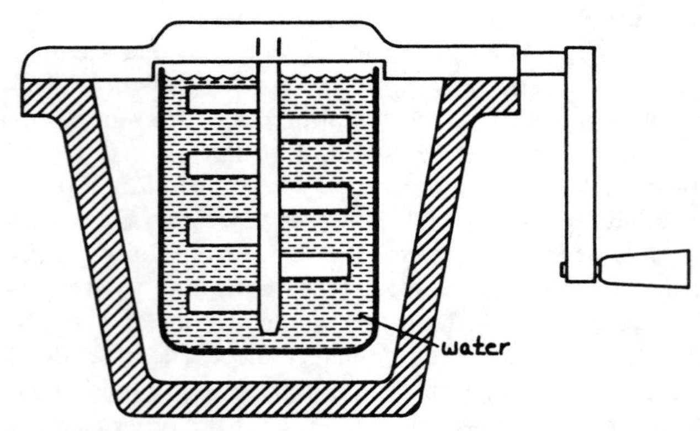

Figure 20.1 Problem 2. The crank, when turned, rotates the paddles, stirring the water in the insulated container.

Solution scheme

a. How much work in joules is done in 1 h?
b. How many kilocalories does this represent?
c. What is the mass in kilograms of 1 gal of water?
d. Use the specific-heat equation to get the temperature change.

3. *Worked Problem.* The *method of mixtures* is a technique used to determine the specific heat of a substance. In this method two substances at different temperatures are mixed in an insulated vessel so that no heat enters or

leaves the system. The heat "lost" by the hotter substance is then equal to the heat "gained" by the cooler one. Suppose a 500 g sample of lead is heated in boiling water and then *quickly* transferred to a Styrofoam cup containing 200 g of water initially at 20°C. If the lead and water reach an equilibrium temperature of 26°C, what is the specific heat of the lead?

Solution: Because the system is isolated thermally from its surroundings, the heat transferred out of the lead is absorbed by the water. For the lead

$$\Delta Q_L = c_L m_L \Delta T_L$$

For the water

$$\Delta Q_W = c_W m_W \Delta T_W$$

Because there is no net heat flow, $\Delta Q_{tot} = 0$. Therefore

$$\Delta Q_{tot} = 0 = \Delta Q_L + \Delta Q_W$$

or

$$\Delta Q_L = -\Delta Q_W$$

Let T_e be the equilibrium temperature of the water and lead. Then

$$c_L m_L (T_e - T_L) = -c_W m_W (T_e - T_W)$$

where T_L and T_W are the initial temperatures of the lead and water, respectively. We have, then,

$$c_L = -\frac{c_W m_W (T_e - T_W)}{m_L (T_e - T_L)}$$

$$c_L = -\frac{(1.0 \text{ kcal/kg} \cdot \text{C}°) \times (0.200 \text{ kg}) \times (26°\text{C} - 20°\text{C})}{0.500 \text{ kg} \times (26°\text{C} - 100°\text{C})}$$

$$c_L = 0.032 \text{ kcal/kg} \cdot \text{C}°$$

4. ***Guided Problem.*** **Suppose that in the preceding problem the water is in** a copper cup rather than a Styrofoam cup but that the cup is still insulated from its surroundings. If the cup has a mass of 250 g and the specific heat of copper is 0.22 kcal/kg-C°, what is the final temperature of the system?

Solution scheme

a. When the lead is placed in the water, how much energy is transferred to the water?
b. How much energy is transferred to the copper cup?
c. How is the sum of these two energies related to the energy transferred out of the lead?

5. ***Worked Problem.*** A copper transmission line suspended between two towers is 300 m long at −10°C. On a day when the temperature is 40°C, by how much does the length of the wire increase?

Solution: The *change* in length of the wire is given by

$$\Delta L = L - L_0 = \alpha L_0 \Delta T$$

and

$$\Delta L = 1.7 \times 10^{-5}/\text{C}° \times 300 \text{ m} \times 50\text{C}°$$
$$\Delta L = 0.26 \text{ m}$$

Therefore, the change in length is 26 cm, or about 0.09% of the initial length. Although the *percentage* is quite small, the actual change in length of 26 cm causes a noticeable sag in the wire.

6. *Guided Problem.* A demonstration illustrating thermal expansion uses a brass ball and an iron ring. At room temperature (~20°C), the ball (d = 6.000 cm) will not pass through the ring (d = 5.990 cm). If the ring is heated to a sufficiently high temperature, the ball will pass through.

a. At what temperature of the ring will this happen?
b. At what *common* temperature can the ball pass through the ring?

Solution scheme: Because the diameter is the relevant parameter in this situation, we need consider only the *linear* expansion.

a. What temperature change in the iron ring will cause its diameter to change by 0.010 cm?
b. Write the linear-expansion equation so that d is expressed as a function of ΔT. Do this for both the brass and the iron. Set the d's equal to each other and solve for the *common* ΔT, with $T_i = 20°\text{C}$.

7. *Worked Problem.* A picnic cooler is used to store a block of ice at 0°C. It is found that the ice melts at a rate of 4 g/min. How much heat energy per second is passing through the walls of the cooler?

Solution: The heat of fusion of ice is 80 kcal/kg, so if the ice is melting at the rate of 4 g/min, we have

$$0.004 \frac{\text{kg}}{\text{min}} \times 80 \frac{\text{kcal}}{\text{kg}} = 0.32 \text{ kcal/min}$$

The power in watts is

$$P = 0.32 \frac{\text{kcal}}{\text{min}} \times \frac{1 \text{ min}}{60 \text{ s}} \times \frac{4185 \text{ J}}{\text{kcal}} = 22 \text{ W}$$

8. *Guided Problem.* A 1.5 kg iron ball heated to a temperature of 500°C is dropped into a thermally insulated vessel. Then 200 g of water at 20°C is poured into the vessel, which is then immediately closed to prevent heat losses. How much liquid water remains at equilibrium?

Solution scheme

a. Assume that some of the water will boil, but not all. What will the equilibrium temperature be?
b. How much heat energy is transferred from the ball to the water?

c. How much energy is required to heat the water to the boiling point?
d. What is the expression for the heat energy required to boil a mass m of water?
e. Equate the heat lost by the iron to the heat gained by the water. If any liquid water is left, what is the possible range of values for m?

9. *Worked Problem.* A box whose dimensions are 1.0 m × 2.0 m × 0.5 m has walls 10 cm thick made of an insulating material. A small electric heater is used to maintain the interior temperature at a uniform 20°C; the outside temperature is 0°C. If the heater supplies 20 W of power, what is the k value for the walls?

Solution: Each wall of the box has the same thickness, so if the temperature is uniform throughout the box, we need know only the surface area. But clearly the outside surface area is greater than the inside surface area, because of the wall thickness. To a good approximation, however, we can take the inside area to be the effective area in the thermal conduction, ignoring the corner effects. The inside area is found by

$$A = 2(0.9\text{ m} \times 1.9\text{ m} + 1.9\text{ m} \times 0.4\text{ m} + 0.9\text{ m} \times 0.4\text{ m})$$
$$A = 5.7\text{ m}^2$$

Then

$$\frac{\Delta Q}{\Delta t} = P = k\frac{A\Delta T}{x}$$

$$k = \frac{P \cdot x}{A\Delta T} = \frac{20\text{ J/s} \times 0.10\text{ m}}{5.7\text{ m}^2 \times 20\text{ C}^\circ} = 0.018\text{ J/s} \cdot \text{m} \cdot \text{C}^\circ$$
$$k = 0.0042\text{ cal/s} \cdot \text{m} \cdot \text{C}^\circ$$

which is approximately the same order of magnitude as k for Styrofoam ($k_{\text{Sty}} = 0.0024\text{ cal/s} \cdot \text{m} \cdot \text{C}^\circ$).

10. *Guided Problem.* A picture window designed for extreme temperature conditions consists of two panes of glass with an air layer between. Each pane is 3 mm thick, and the air spacing is 5 mm. The room temperature is 20°C, and the outside temperature is 0°C.

a. What is the heat flow per unit time per unit area for this window?
b. How thick would a single pane of glass have to be to give the same heat flow?

Solution scheme

a. For part a use the results given in Problem 34 of the text (p. 539).
b. For part b use k for glass to calculate the thickness given the heat flow rate.

Answers to Guided Problems

2. 3.6 C°
4. 24.5°C
6. a. 176°C
   b. −199°C
8. 107 g
10. a. 93 W/m^2
    b. 0.17 m = 17 cm

Chapter 21

Thermodynamics

Overview

In this chapter we deal with the properties of matter in bulk as they relate to heat and temperature, an area of study called *thermodynamics*. The laws of thermodynamics are quite general and independent of the details of the particular system. We will derive relationships among the thermodynamic variables of a system, relationships called *equations of state.*

Essential Terms

perpetual motion machine
First Law of Thermodynamics
thermal reservoir
thermodynamic efficiency
reversible process
Carnot cycle
Carnot engine
thermodynamic temperature scale
theorem of Clausius
entropy
Second Law of Thermodynamics
Carnot's theorem
Third Law of Thermodynamics

Key Concepts

1. *Perpetual motion machines.* Humankind has forever sought a "free lunch," that is, has looked for ways to get work out of devices with no expenditure of energy. Often the devices have been so clever (at least on paper) that it is difficult to see just where the system breaks down. Most of these devices, however, violate either the First or the Second Law of Thermodynamics. These laws are so firmly established that physicists feel extremely confident that no device violating these laws will ever be built. Basically, there are two kinds of perpetual motion machines. The first kind violates

the First Law of Thermodynamics in that it creates its own internal energy source. The second kind violates the Second Law because zero or negative entropy changes are produced.

2. *First Law of Thermodynamics*. This law is simply a statement of the conservation of energy. It also implies that energy can be stored internally in materials. This *internal energy* can be a function of several variables, but for an ideal gas it is a function only of the temperature. The *change* in internal energy, ΔU, is a function only of the *difference* $\Delta Q - \Delta W$ and does not depend on how the system got from one internal energy state to another. Only the *change* in internal energy has physical significance, so the zero point is arbitrary. We encountered the same situation in dealing with the mechanical potential energy. The First Law applies only to *equilibrium states*—states in which the system has "settled down."

3. *The Carnot engine*. In mechanics we speak of systems in which the work input equals the work output as being 100% efficient. Such machines must be friction-free. In thermodynamics we inquire whether a system exists that takes heat from a source and converts it entirely into work. The problem is that to make any thermodynamic system useful, we must be able to repeat the cycle. For example, consider the system consisting of a cylinder fitted with a frictionless piston and connected thermally to a heat reservoir as illustrated in Figure 21.1. As the gas expands, the temperature remains constant, as a result of being in contact with the reservoir). Thus, the internal energy change is zero and $\Delta Q = \Delta W$. For this to be true, the expansion must be a *reversible process*; that is, the expansion must be so slow that the whole system is at the same temperature at any given time. To get the piston back to its original position, we could put the work back into the system, but then $\Delta Q = 0$ and $\Delta W = 0$ and nothing useful has been accomplished. Suppose instead we let the piston continue to expand while the cylinder is *insulated* from the reservoir. Then the work comes from the internal energy of the gas, so the temperature drops. We now have extracted the most work we can from the gas.

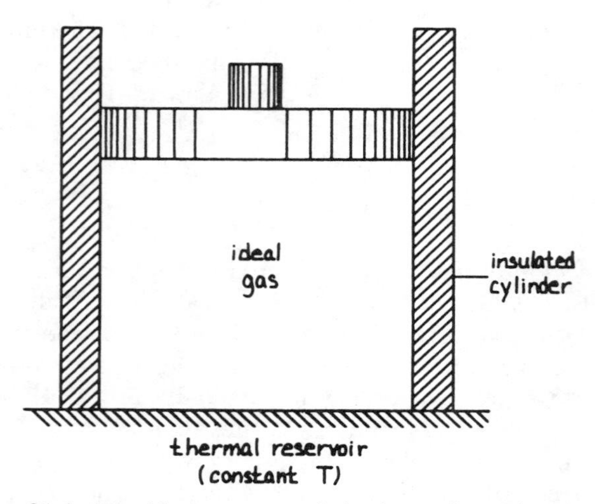

Figure 21.1 The ideal gas expands isothermally against the piston.

To get the system back to its initial state, we first do work to compress the gas *isothermally* by connecting the cylinder to the reservoir at the lower temperature. Heat energy then flows out of the cylinder into the reservoir. By insulating the cylinder once again, we can do work on the gas and bring the internal energy back up to its original value. No energy has been lost, but because $\Delta U = 0$, $\Delta Q = \Delta W$, or $Q_1 - Q_2 = \Delta W$. Because $Q_2 \neq 0$, the amount of heat energy, Q_1, extracted from the high-temperature reservoir was converted partially into work, the remainder being expelled at a lower temperature. The cycle just described is called the *Carnot cycle* and is the most efficient cycle possible. The efficiency of a Carnot engine (one that operates by the Carnot cycle) is a function only of the temperatures of the two reservoirs and is given by

$$e = 1 - \frac{T_2}{T_1}$$

The temperatures in this expression are expressed in *kelvins*.

4. *Entropy*. In the First Law of Thermodynamics, which in the differential form is $dQ = dU + dW$, we know that dU can be expressed in terms of thermodynamic variables and changes in these variables. Also, dW can, for reversible processes, be expressed in terms of other thermodynamic variables and their changes. One example is the change in volume of a fluid system, for which $dW = pdV$. A similar relationship for dQ would be useful. We define a quantity called the *entropy, S,* such that $dQ = TdS$ for reversible processes. By integrating between the initial and final states, we get

$$\Delta S = S_f - S_i = \int_i^f \frac{dQ}{T}$$

For example, in an isothermal process, we have

$$\Delta S = \frac{1}{T}\int_i^f dQ = \frac{\Delta Q}{T}$$

Because entropy is a function of the state of the system, if the process is cyclic, the initial and the final states are the same and $\Delta S = 0$.

5. *The Second Law of Thermodynamics*. The Second Law can be expressed in many forms, but the most common version, which appears most closely related to the ban on perpetual motion, is that it is impossible to devise a process whereby the *only* result is to extract heat from a reservoir and convert it entirely into work. Note the word *only*. It is not impossible to convert heat entirely into work. An isothermal, reversible expansion does that. The thermodynamic state of the system has changed, however—the volume increases and the pressure drops. So converting heat into work is not the only result of this process.

Another statement of the Second Law concerns heat transfer. No process is possible whereby the only result is to extract heat from a low-temperature reservoir and eject it to a higher-temperature reservoir. Heat flow is an irreversible process—heat flows spontaneously from a hot reservoir to a cold one, but never the reverse without work being performed.

The Second Law relates also to entropy. By applying the Second Law, we can show that the entropy of an isolated system never decreases; the change in entropy must be either zero or positive. If we look at entropy as a measure of *disorder*, we can see why this statement is true. Statistically, any state in which all the particles have *specified* parameters is as likely as any other. For example, any specified order in a deck of cards is as likely as any other in a random shuffle. Most of these arrangements are considered to be disordered, however. Shuffling the deck will very likely not produce all diamonds arranged in sequence, followed by clubs in sequence, and so forth, but this sequence is as likely as any other *specified* sequence. It is because almost all the other arrangements are considered disordered that shuffling tends to "mix up" the cards. For the same reason, a drop of ink in water is at first a fairly ordered arrangement. But as time goes on, the ink moves into the almost infinite number of other arrangements that leave it uniformly scattered in the medium—a disordered arrangement. The chance exists that the ink will reassemble itself into a drop, but that chance is infinitesimally small. So it is with energy. Useful energy is ordered, but less useful energy is more disordered. The natural progression of a system left to itself is toward a state of more disorder.

Sample Problems

1. *Worked Problem.* One kilogram of ice at 0°C is turned completely into water at 0°C by the addition of 80 kcal of heat energy. The density of ice is 920 kg/m^3, and the density of water is 1000 kg/m^3. How much of this heat energy is used to increase the internal energy of the system?

Solution: We appeal to the First Law of Thermodynamics, which states that

$$\Delta U = \Delta Q - \Delta W$$

We know that the amount of heat energy transferred is 80 kcal. The work done in the contraction process is $\Delta W = p \cdot \Delta V$, where

$$\Delta V = m\left(\frac{1}{\rho_w} - \frac{1}{\rho_i}\right)$$

Then

$$\Delta U = \Delta Q - m\left(\frac{1}{\rho_w} - \frac{1}{\rho_i}\right)p$$

$$\Delta U = 80 \text{ kcal} - 1 \text{ kg} \times \left(\frac{1}{1000 \text{ kg/m}^3} - \frac{1}{920 \text{ kg/m}^3}\right) \times 1 \times 10^5 \text{ Pa}$$

$$\Delta U = 80 \text{ kcal} - (-8.7 \times 10^{-5} \text{ m}^3) \times 1 \times 10^5 \text{ Pa}$$

$$\Delta U = 80 \text{ kcal} + 8.7 \text{ J}$$

Since 1 kcal = 4185 J, we have

$$8.7 \text{ J} = 2.08 \times 10^{-3} \text{ kcal}$$

Finally,

$$\Delta U = 80 \text{ kcal} + 2.1 \times 10^{-3} \text{ kcal}$$

Only a very small fraction (~0.003%) of the internal energy change comes from the work done by the atmospheric pressure. Most of the change comes from the heat energy absorbed by the ice.

2. *Guided Problem.* An ideal gas expands according to the curve shown in Figure 21.2.

a. How much work is done by the gas as it expands from A to B?
b. How much work is done by the gas if it is compressed from B to A following the same path?

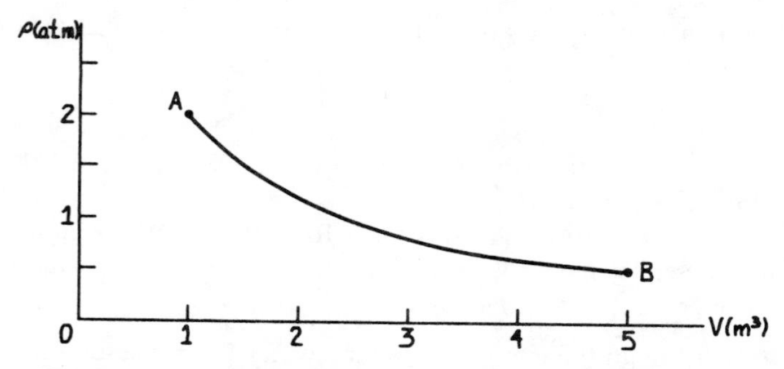

Figure 21.2 Problem 2. An ideal gas expands from A to B along the path shown.

Solution scheme: We know that the work done by the gas is the integral $\Delta W = \int p dV$.

a. How is this integral related to the area under the $p-V$ curve?
b. In a reversing of the path, in what way, if any, have the pertinent parameters changed?

3. *Worked Problem.* Suppose a certain gas has the property that the pressure in *pascals* is related to the volume, in cubic meters, by the expression $p = 2.4 \times 10^4\ V^{2/3}$. Calculate the work done by the gas in expanding from 1.5 m³ to 3.0 m³.

Solution: The work done by the gas is given by the integral $\Delta W = \int p dV$. Then

$$\Delta W = \int_{1.5}^{3.0} 2.4 \times 10^4\ V^{2/3} dV = 2.4 \times 10^4 \int_{1.5}^{3.0} V^{2/3} dV$$

$$\Delta W = 2.4 \times 10^4 \left[\frac{V^{5/3}}{5/3}\right]_{1.5}^{3.0}$$

$$\Delta W = 2.4 \times 10^4 \times \tfrac{3}{5}[6.24 - 1.97]$$

$$\Delta W = 6.1 \times 10^4 \text{ J}$$

Comment: We did not carry the units through the calculation, but you should check that the units are consistent. In particular, if $k = 2.4 \times 10^4$, then the units for k must be the units associated with $p/V^{2/3}$, which are J/m^5. Verify this.

4. *Guided Problem.* Suppose one liter of an ideal gas is expanded isothermally from a volume of 1 m^3 at atmospheric pressure to a volume of 3 m^3.
   a. How much work is done by the gas?
   b. How much heat energy must enter the system?
   c. What is the final pressure of the gas?

Solution scheme
   a. How is work related to pressure and volume?
   b. If the gas is ideal, how are the pressure, volume, and temperature related?
   c. If the process is isothermal, how has the internal energy of the system changed?
   d. For constant temperature, how are the pressure and the volume related?

5. *Worked Problem.* A heat engine draws water from a reservoir at 4°C and passes it through the engine, extracting heat from it until the water reaches a temperature of 2°C. At that point the water is ejected to the outside air, whose temperature is a constant −10°C. If 20 kW is required in the heating and the efficiency is 30% of that of a Carnot engine, what must be the water flow rate through the engine?

Solution: Although the engine is drawing water from a reservoir, it is not extracting heat at constant temperature. The water flowing through the engine acts like a reservoir at constant temperature 3°C (the average of 4°C and 2°C). The efficiency of the engine is

$$e = 1 - \frac{Q_1}{Q_2} = 0.30\left(1 - \frac{T_2}{T_1}\right)$$

Now the heat extracted per unit time from the high-temperature reservoir is

$$\frac{Q_1}{t} = \frac{m}{t} c\Delta T = \frac{m}{t} \times (1 \text{ kcal/kg} \cdot \text{C}°) \times 2 \text{ C}°$$

Then

$$1 - \frac{Q_2}{Q_1} = \frac{Q_1 - Q_2}{Q_1} = \frac{(Q_1 - Q_2)/t}{Q_1/t}$$

$$Q_1/t = \frac{(Q_1 - Q_2)/t}{0.30\left(1 - \frac{T_2}{T_1}\right)}$$

$$\frac{m}{t} c\Delta T = \frac{(Q_1 - Q_2)/t}{0.3\left(1 - \frac{T_2}{T_1}\right)}$$

$$\frac{m}{t} = \frac{(Q_1 - Q_2)/t}{0.30\left(1 - \frac{T_2}{T_1}\right)c\Delta T} = \frac{20 \text{ kW}}{0.30\left(1 - \frac{263 \text{ K}}{275 \text{ K}}\right) \times 1 \frac{\text{kcal}}{\text{kg} \cdot \text{C}^\circ} \times 2 \text{ C}^\circ}$$

$$\frac{m}{t} = 1.82 \times 10^2 \text{ kg/s} \Rightarrow 182 \ l/\text{s} = 48 \text{ gal/s}$$

6. *Guided Problem.* In a refrigerator, heat is extracted from a colder reservoir and ejected to a warmer reservoir, with work being done on the refrigerant. How efficiently this process takes place is given by a quantity called the *coefficient of performance*, $\varkappa$, defined as the amount of heat extracted from the low-temperature reservoir divided by the amount of work performed in ejecting it to the higher-temperature reservoir.

a. Show that

$$\varkappa = \frac{1}{e} - 1$$

where e is the thermodynamic efficiency.

b. If 300 J of energy is extracted from the low-temperature reservoir at 250 K, how many joules are ejected to the high-temperature reservoir at 375 K if the refrigerator has a coefficient of performance equal to 1/5 that of a Carnot refrigerator?

Solution scheme

a. For a heat engine operating over a full cycle, how is the work done related to the net heat transfer?
b. Using the symbols Q_1 and Q_2 for the heat transferred from the low-temperature and to the high-temperature reservoirs, respectively, write down the definition of $\varkappa$.
c. Rewrite this expression in terms of T_1 and T_2.
d. How does this expression compare with that for the efficiency e?
e. To solve part b, equate the expression for $\varkappa$ for the real refrigerator (using heat flows) to that of $\varkappa$ for the ideal Carnot refrigerator (using Ts). Solve for Q_2.

7. *Worked Problem.* Suppose 100 g of water at 30°C is mixed with 50 g of water at 5°C. What is the change in entropy upon mixing?

Solution: We will assume that the process takes place by a succession of near equilibrium states, so that

$$\Delta S = \int \frac{dQ}{T}$$

where $dQ = mcdT$. Then for the hotter water

$$\Delta S_{\text{H}} = m_{\text{H}}c \int_{T_{\text{H}}}^{T_{\text{f}}} \frac{dT}{T} = m_{\text{H}}c \ln\left(\frac{T_{\text{f}}}{T_{\text{H}}}\right)$$

and for the colder water

$$\Delta S_C = m_C c \int_{T_C}^{T_f} \frac{dT}{T} = m_C c \ln\left(\frac{T_f}{T_C}\right)$$

The method of mixtures gives us T_f, the equilibrium temperature:

$$-m_H c(T_f - T_H) = m_C c(T_f - T_C)$$

so that

$$T_f = \frac{m_H T_H + m_C T_C}{m_C + m_H} = \frac{100 \text{ g} \times 30°\text{C} + 50 \text{ g} \times 5°\text{C}}{150 \text{ g}}$$

$$T_f = 21.7°\text{C} = 295 \text{ K}$$

Then

$$\Delta S_{tot} = \Delta S_H + \Delta S_C = c\left(m_H \ln \frac{T_f}{T_H} + m_C \ln \frac{T_f}{T_C}\right)$$

$$\Delta S_{tot} = 1.0 \text{ cal/g} \cdot \text{K}\left[100 \text{ g} \ln \frac{295 \text{ K}}{303 \text{ K}} + 50 \text{ g} \ln \frac{295 \text{ K}}{278 \text{ K}}\right]$$

$$\Delta S_{tot} = 1.0 \text{ cal/g} \cdot \text{K}[100 \text{ g} \times (-0.0279) + 50 \text{ g} \times (0.0583)]$$

$$\Delta S_{tot} = 1.0 \text{ cal/g} \cdot \text{K}[-2.79 + 2.92]\text{g}$$

$$\Delta S_{tot} = 0.13 \text{ cal/K} = 0.54 \text{ J/K}$$

8. *Guided Problem.* A thermal conductor is placed so that one end is in contact with a reservoir at 600 K; the other end is attached to a reservoir at 250 K. A total heat energy transfer of 1500 kcal is made from one end of the conductor to the other. If no change in temperature occurs along the conductor (for example, the conductor is infinitesimally thin), calculate the entropy change of

a. each reservoir
b. the conductor
c. the universe

Solution scheme

a. What is the amount of heat transferred (in joules)?
b. If the temperature remains constant in the reservoir, what is the expression for the entropy change?
c. What is the sign of the entropy change for heat flow into or out of a system?
d. For the conductor, what is the *net* heat transfer? What is the entropy change for the conductor if it is at constant temperature?
e. What is the *net* change in entropy for the whole system? What sign do you expect this change to have?

Answers to Guided Problems

2. The work is found by computing the area under the curve.
   a. 3.8×10^5 J
   b. -3.8×10^5 J

4. a. 1.1×10^5 J
   b. 1.1×10^5 J = 26.6 kcal
   c. 1/3 atm
6. b. 1050 J
8. a. High-temperature reservoir: $\Delta S = -10.5$ J/K
   Low-temperature reservoir: $\Delta S = 25.1$ J/K
   b. Conductor: $\Delta S = 0$
   c. Universe: $\Delta S = 14.6$ J/K

Chapter 22

Electric Force and Electric Charge

Overview

We now move into a second major area of physics, that of electromagnetic forces. Experiments on static electricity were performed by the Greeks, but just as in the case of mechanics, the application of experimental methods and the development of a quantitative theory had to wait until modern times. Electric forces are caused by charge, a property that some bodies possess and that uncharged bodies do not. Benjamin Franklin chose to take the charge left on glass rubbed with silk as positive. Much later it was found that according to this convention the electron has negative charge, and this convention has remained.

To understand the nature of electric forces, we need to realize that the basic building block of substances, the atom, consists of a massive, tiny, positively charged nucleus surrounded by a cloud of light, negative electrons, much as the sun is surrounded by a planetary system. Electrons are held to the nucleus by the electric force of attraction between the negative electron and the positive nucleus, and it is the same electric force that holds the atoms together in molecules or crystals. Although when we drop a rock, the rock falls to Earth because of the gravitational force of attraction between the rock and the Earth, nearly all the mechanical forces of everyday experience, such as the push of a hand against a door, are caused by the combined electric forces of many atoms. Electric forces, therefore, are the forces of everyday life.

Electric forces can usually be described rather simply. When particles exerting electric forces are at rest or are moving very slowly, the electric forces are described in terms of a precise empirical law called Coulomb's Law of force. Most materials can be classified according to their electric behavior as either insulators or conductors. If a charged body is suspended by a silk thread, an insulator, the body retains its charge. If it is suspended by a metal wire, a conductor, it loses its charge, because the charge is free to flow away from the body along the wire. The charges on an insulator are usually at rest, so that the electrical forces between charged insulating bodies can be described by Coulomb's Law unless the charged bodies are moving very rapidly with respect to one another. The charges on a conductor are free to move, and we will find that sometimes this factor must be considered when dealing with the electric forces between conductors.

Essential Terms

electric force
electrostatic force
electric charge
Coulomb's Law
permittivity constant
charge quantization
charge conservation
conductor
insulator

Key Concepts

1. *Electric force.* The electric force between the protons in the nucleus of an atom and the electrons is the force of attraction that holds atoms together. This force also holds atoms together in molecules or crystals, and combines molecules or crystals to form everyday matter. Thus, our immediate environment is dominated by electric forces. Most mechanical forces are electric in nature; for example, the push of the hand against a door tends to compress the molecules of the surface of the door. The molecular forces, which resist any deformation of the molecule, are due to attractive forces between the nucleus of the atom and surrounding electrons, and repulsive forces between electrons; thus, they are electric in nature. So, although we experience the resistance to deformation as a mechanical force pushing back against our hand, the origin of the force is actually the sum of the electric forces at the molecular level.

2. *Electrostatic force.* If the particles exerting electric force are at rest or are moving very slowly, we can neglect what we will later call magnetic effects. The electric forces remaining are what are called *electrostatic forces*. Because the electrostatic force between an electron and a proton is about 2.2×10^{39} times as strong as the gravitational attraction, we can usually neglect gravitational effects when dealing with electric forces. In addition, whereas gravitational forces are always attractive and the force between an electron and a proton is always attractive, the electric force between two electrons or two protons (or between *any* two identical particles) is repulsive. It is useful to describe the electrostatic force between two particles as being caused by *electric charge*, and because the force can be either attractive or repulsive, it is necessary to specify two kinds of charge. The charge of the proton is defined to be plus (+), and the charge of the electron, negative (−). By experiment, the magnitudes of these charges are found to be *exactly* equal, and they are designated as $+e$ and $-e$, respectively. Like charges repel one another; unlike charges attract.

3. *Coulomb's Law*. The fundamental law of electrostatic forces is called, after its discoverer, Coulomb's Law. It says that the electrical force between two point charges is proportional to the product of the magnitudes of the two charges and inversely proportional to the square of the distance between the two charges. The direction of the force is along the line joining the two charges. In order to write Coulomb's Law mathematically, we must define the unit of charge, and we do this by defining the charge of the electron to be 1.60×10^{-19} coulombs, which we designate as e. The force between the two charges q and q' can then be expressed as

$$\mathbf{F}_{q'} = [\text{constant}] \times \frac{qq'}{r^2}\,\hat{\mathbf{r}}$$

where $F_{q'}$ is the force on charge q' caused by charge q, as indicated in Figure 22.1. $F_{q'} = -F_q$, so that Newton's Law of action reaction is satisfied. $\hat{\mathbf{r}}$ is the unit vector in the direction from q to q'. Note that if either q or q' is negative (unlike charges), the force is attractive; otherwise (like charges), the force is repulsive.

Figure 22.1 The force on charge q' caused by charge q is in the direction indicated by the unit vector $\hat{\mathbf{r}}$.

4. *Permittivity constant*. Because force (F) is measured in newtons, distance (r) in meters, and charge (q) in coulombs, the constant in Coulomb's Law must have the dimensions of $\text{N} \cdot \text{m}^2/\text{C}^2$. The constant is written as $1/(4\pi\varepsilon_0)$ and has the value of $8.99 \times 10^9\ \text{N} \cdot \text{m}^2/\text{C}^2$, where $\varepsilon_0 = 8.85 \times 10^{-12}\ \text{C}^2/\text{N} \cdot \text{m}^2$, and is called the *permittivity constant*. As we will see later on, the choice of $1/(4\pi\varepsilon_0)$ for the constant allows us to express many of the fundamental equations of electric forces more simply than we could otherwise.

5. *Charge quantization*. We find by measurement that charge is granular, and that every grain, whether positive or negative, is of the same magnitude as any other. Measurements by Millikan showed that the electron possesses a unique charge, designated $-e$, and that the charges existing on all objects, whether positive or negative, were always integral multiples of e. Since then, the measured charges of all particles found in nature have always been found to be 0, $\pm e$, $\pm 2e$, and so forth. We therefore say that charge is *quantized* and that the fundamental unit of charge, e, is the quantum of charge.

6. *Conductors and insulators*. Metals belong to a class of materials called *conductors*, which permit electric charge to move easily through their volume. *Insulators*, such as glass, do not. In a conductor one or more electrons from each atom are free to move through the material; in an insulator all the electrons are bound to the individual atoms. We can, however, transfer electrons to or from the surfaces of insulators if we loosen them, using friction, from the atoms to which they are bound. If we rub glass with silk, the silk becomes negatively charged, and the glass positively charged, through the transfer of electrons from the surface of the glass to the silk. When we walk across a nylon carpet and then touch a metallic object, we may see a

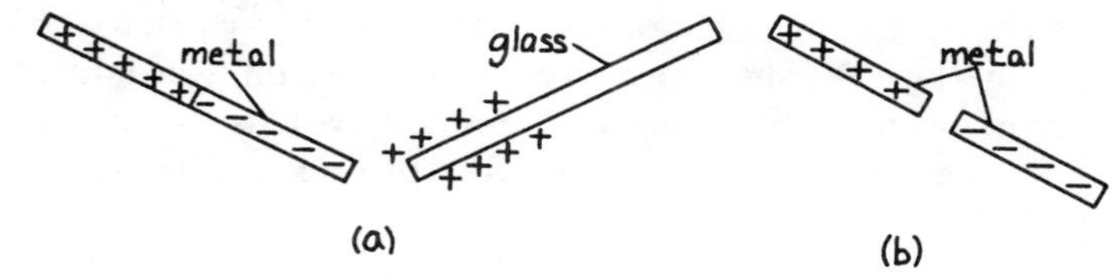

Figure 22.2 The separation of charge by electrostatic induction. In (a) the charged glass rod is brought near the two metal rods; in (b) the two metal rods are separated, and the glass is removed.

spark and feel a sudden electric shock. This happens because friction between our feet and the carpet has caused a transfer of charge between our body and the carpet. Our bodies are sufficiently good insulators to permit a significant charge to build up until we allow it to "leak off" when we touch a metallic object.

But we don't even need to touch a conductor in order to move charges around on it. Suppose we have two metal cylinders on insulating stands. We place them in contact to form one long conductor, and then we bring a positively charged glass rod near one end of the cylinder. The positive charge on the glass rod will attract electrons on the conductor, so that the near end of the cylinder becomes negatively charged and the far end positively charged. If we now separate the two cylinders, the near end is negatively charged and the far end positively charged. This form of separation of charge is called *electrostatic induction* and is indicated schematically in Figure 22.2.

Sample Problems

1. *Worked Problem*

a. Calculate the ratio of the electric force to the gravitational force between an electron ($m = 9.1 \times 10^{-31}$ kg) and a proton ($m = 1.67 \times 10^{-27}$ kg) at a separation of 1.0 m.

b. Did you really need to know how far apart the two particles are?

Solution

a. We begin by drawing a "free-body" diagram (FBD) of an isolated electron–proton pair (Figure 22.3). The only forces acting on the two particles are the electric and gravitational forces between them. The gravitational force (F_G) is always attractive and is given by

$$F_G = Gm_1m_2/r^2$$

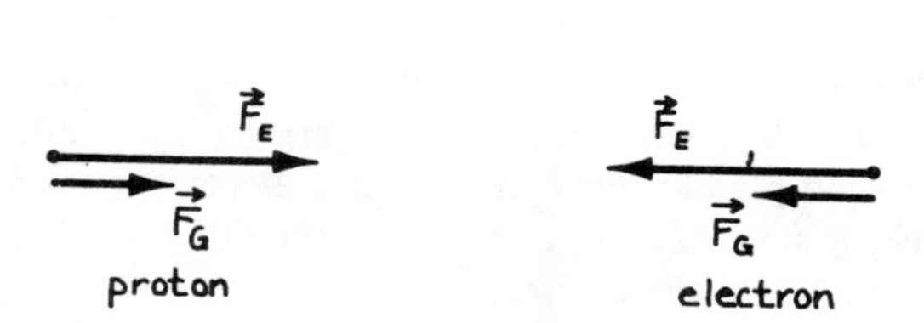

Figure 22.3 Problem 1. A "free-body" diagram of an isolated electron–proton pair, showing only the electric and the gravitational forces between them.

where G, the universal gravitational constant, is equal to 6.67×10^{-11} $N \cdot m^2/kg^2$. Because the charge of the electron is negative and that of the proton is positive, Coulomb's Law tells us that the electric force between the electron and the proton is also attractive; it is given by

$$F_E = \frac{1}{4\pi\varepsilon_0}\frac{q_1 q_2}{r^2} = \frac{1}{4\pi\varepsilon_0}\frac{e^2}{r^2}$$

If we use vector notation for the electric force,

$$\mathbf{F}_E = \frac{1}{4\pi\varepsilon_0}\frac{q_1 q_2}{r^2}\hat{\mathbf{r}} = \frac{1}{4\pi\varepsilon_0}\frac{(e)(-e)}{r^2}\hat{\mathbf{r}}$$
$$= \frac{1}{4\pi\varepsilon_0}\frac{e^2}{r^2}(-\hat{\mathbf{r}})$$

which tells us, as we already know, that the electric force between the electron and the proton is attractive.

If we now take the ratio of the electric force to the gravitational force, we obtain the following:

$$\frac{F_E}{F_G} = \frac{\dfrac{1}{4\pi\varepsilon_0}\dfrac{e^2}{r^2}}{G\dfrac{m_1 m_2}{r^2}} = \left(\frac{1}{4\pi\varepsilon_0}\frac{e^2}{r^2}\right)\left(\frac{r^2}{Gm_1 m_2}\right)$$
$$= \frac{e^2}{4\pi\varepsilon_0 G m_1 m_2} = 1.35\ (10^{20})\left(\frac{kg}{C^2}\right)\frac{e^2}{m_1 m_2}$$

If we now substitute in the values for e, m_1, and m_2, we obtain

$$\frac{F_E}{F_G} = 1.35\ (10^{20})\ \frac{kg}{C^2}\ \frac{(1.6 \times 10^{-19})^2\ C^2}{(1.67 \times 10^{-27})(9.1 \times 10^{-31})\ kg^2}$$
$$= 2.3 \times 10^{39}$$

The electrical force is immensely stronger than the gravitational force!

b. Note that the ratio of the two forces is independent of their separation, r, because both forces are proportional to $(1/r^2)$.

2. *Guided Problem.* A plastic rod rubbed with fur acquires a negative charge of 0.5×10^{-9} C. Assuming that the charging involves the transfer of electrons only to the plastic, how many extra electrons has the plastic acquired? What is the increase in mass of the plastic? (See Table A.1 of the textbook for the value of the electron mass.)

Solution scheme

a. The total charge transferred is given by [number of electrons] × [charge of each electron], where the charge of each is 1.6×10^{-19} C.

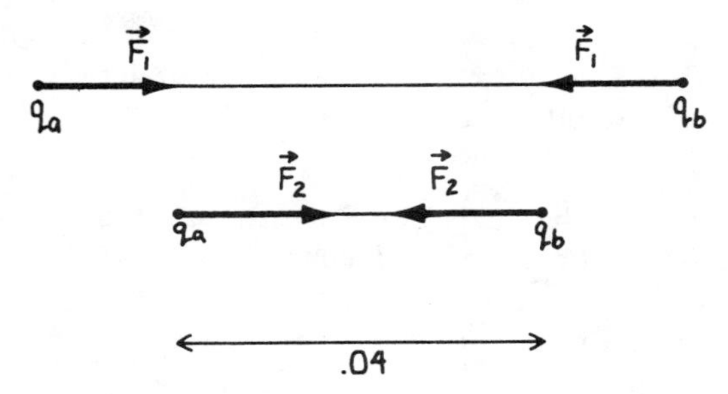

Figure 22.4 Problem 3. The electric forces of attraction on two pith balls. In the upper diagram the balls are 0.08 m apart, and in the lower diagram the separation is 0.04 m. Force $F_1 = 2 \times 10^{-5}$ N.

b. You should now know how many extra electrons the plastic has acquired. The increase in mass is given by [number of electrons transferred] × [mass of each].

3. *Worked Problem.* Two small electrically charged pith balls are 0.08 m apart and attract each other with a force of 2×10^{-5} N. If we bring them closer together so that they are now 0.04 m apart, what is the new force of attraction?

Solution: Again it is worthwhile to draw FBDs showing the forces on the pith balls. You should remember that from Newton's Law of action–reaction, the force on the second ball is equal and opposite to the force on the first, so that it doesn't matter which ball we specify in the problem. The FBD is shown in Figure 22.4. From Coulomb's Law the force of attraction between the two balls can always be written as

$$F = \frac{1}{4\pi\varepsilon_0}\frac{q_a q_b}{r^2}$$

where r is the separation between the balls. If the original separation of the balls, r_1, is 0.08 m, and the final separation of the balls, r_2, is 0.04 m, we can find the ratio of F_1, the force when $r = r_1$, to F_2, the force when $r = r_2$, as follows:

$$\frac{F_1}{F_2} = \frac{\dfrac{1}{4\pi\varepsilon_0}\dfrac{q_a q_b}{r_1^2}}{\dfrac{1}{4\pi\varepsilon_0}\dfrac{q_a q_b}{r_2^2}} = \left(\frac{r_2}{r_1}\right)^2$$

or

$$F_2 = \left(\frac{r_1}{r_2}\right)^2 F_1 = \left(\frac{0.08}{0.04}\right)^2 F_1 = 4F_1$$

$$= 4\,(2 \times 10^{-5}) = 8 \times 10^{-5}\ \text{N}$$

By halving the separation of the two pith balls, we have increased the electric force by a factor of 4. Note that we did not need to solve for the charges q_a and q_b on the two pith balls to solve the problem.

4. *Guided Problem.* A charge of 5 μC (5×10^{-6} C) is at the origin, and charge of −3 μC is at $x = 20$ cm. Find a position on the x axis at which the electric force on a third charge of 1 μC is exactly zero.

Solution scheme

a. Let the point on the x axis at which the electric force is zero be called P. P can be to the left of both charges (on the negative x axis), between the two charges, or to the right of both charges. Draw an FBD showing the forces on the 1 μC charge at point P for each of these cases in turn.

b. First draw the FBD with P on the negative x axis. For all points along the negative x axis, the repulsive force caused by the 5 μC charge is always greater than the attractive force caused by the −3 μC charge, because the 5 μC charge is closer to point P. Can you conclude that there is *no position* along the negative x axis at which the force on the 1 μC charge is zero?

c. Now draw an FBD for point P between the two charges. Show that the force on the 1 μC charge will always be directed toward the right, along the positive x axis. Is there a solution for point P between the two charges?

d. Now draw a diagram showing point P to the right of both charges. We show this in Figure 22.5. The two forces on the 1 μC charge, F_{-3} and F_5, are in opposite directions. Because the smaller charge (−3 μC) is closer to P than is the larger charge, the magnitude of F_{-3} may be as large or larger than that of F_5, so that the two forces could cancel one another. Have you found where there may be a solution to the problem?

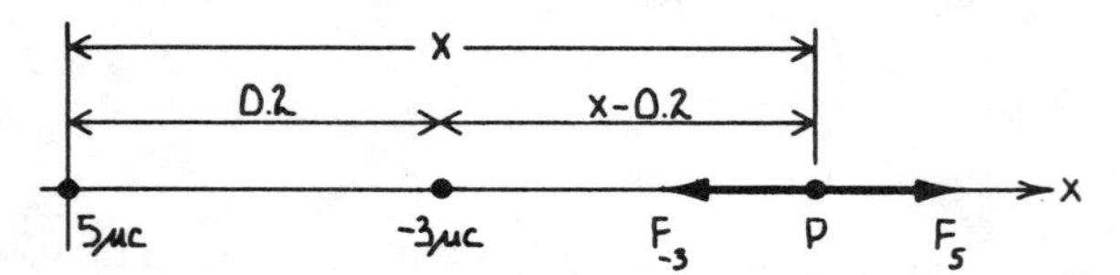

Figure 22.5 Problem 4. A "free-body" diagram showing the forces on a 1 μC charge at point P caused by charges of 5 μC and −3 μC.

e. Now you are ready to solve the problem mathematically. At point P, a distance x from the origin, the two forces F_{-3} and F_5 cancel, so that the net force on the 1 μC charge is zero. Write down, using Coulomb's Law, the expression for F_{-3}. Similarly, write down F_5. $F_{-3} = F_5$, or $F_5 - F_{-3} = 0$; can you solve the equation for the distance x?

5. *Worked Problem.* Two small, spherical Styrofoam balls, each of mass 4 g, are hung from a common point in the ceiling by silk threads 1 m long. After being given identical charges, the balls repel each other and hang so that each thread makes an angle of 15° with the vertical. Find the charge given to each Styrofoam ball. The acceleration of gravity, g, equals 9.80 m/s².

Solution: In a problem like this, carefully drawn diagrams, especially FBDs, are most important. Each ball has a charge q, so that it is clear that the repulsive force between the balls can be expressed using Coulomb's

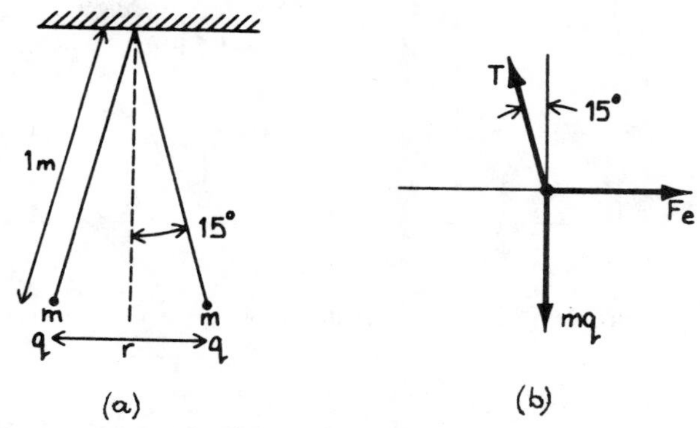

Figure 22.6 Problem 5. (a) Two small Styrofoam balls hung from a common point in the ceiling by threads 1 m long. (b) A "free-body" diagram of the forces on one of the balls.

Law, where the distance r between the two balls is given by $r = 2\times$ [length of thread] $\times$ sin 15° = 2 sin 15° = 0.52 m. The two balls, and an FBD showing the forces on one of the balls, are shown in Figure 22.6.

The electric force of repulsion, F_e, is given by

$$F_e = \frac{1}{4\pi\varepsilon_0}\frac{q^2}{r^2}\ \text{N}$$

The FBD in Figure 22.6(b) shows all the external forces acting on one of the balls. Because the ball is in equilibrium, the horizontal forces should sum to zero, and so must the vertical forces. The weight of the ball is mg, and the tension in the thread is T. Then $T \sin 15° = F_e$ for the horizontal forces, and $T \cos 15° = mg$ for the vertical forces. If we divide these two equations, we obtain

$$\frac{\sin 15°}{\cos 15°} = \tan 15° = \frac{F_e}{mg}$$

or

$$F_e = mg \tan 15° = 1.05 \times 10^{-2}\ \text{N}$$

But we already know that

$$F_e = \frac{1}{4\pi\varepsilon_0}\frac{q^2}{r^2}$$

where $r = 0.52$ m. Solving for q, we obtain $q = 5.5 \times 10^{-7}$ C.

6. *Guided Problem*. An electric dipole consists of two equal and opposite charges separated by a distance $2l$, as shown in Figure 22.7. Find the magnitude and direction of the force exerted on a test charge of 1μC placed at (a) point A, with coordinates (0,10 cm), and (b) point B, with coordinates (10,0) cm. The dipole is oriented as indicated in the figure. The length $l = 0.50$ cm.

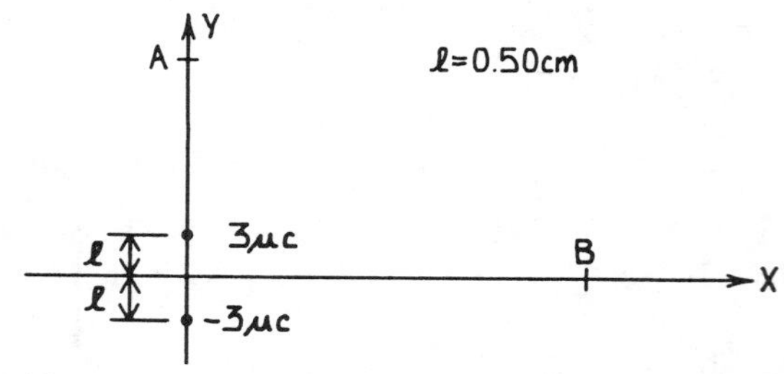

Figure 22.7 Problem 6. An electric dipole with the two charges lying on the y axis.

Solution scheme

a. Draw a force diagram showing the forces on the test charge at point A. Both forces are parallel to the y axis, with the force caused by the -3 μC charge being along the $-y$ direction.

b. Using Coulomb's Law, write down the expression for the total force on the test charge. Because we know all the distances (remember that $l = 0.005$ m) and charges, we can find the forces exactly, and so calculate the net force on the test charge.

c. Figure 22.8 shows the force diagram for the test charge at point B. Are the magnitudes of F_3 and F_{-3} equal?

d. Express F_3 and F_{-3} in terms of their x and y components. Should the x components just cancel each other? Should the y components cancel each other, or should they be additive?

e. Now find the magnitude of the total force on the test charge. You will need to use Coulomb's Law for this—again, you should know all the distances and charges.

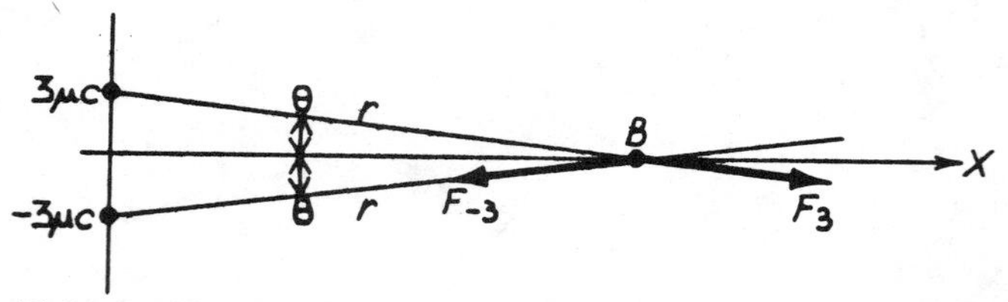

Figure 22.8 Problem 6. The electric forces acting on a test charge at point B caused by the electric dipole shown.

7. *Worked Problem.* Suppose that the protons and the electrons in 1 kg of monatomic hydrogen are separated. All the electrons are transported to the Moon, while the protons remain to distribute themselves uniformly on the Earth. What would be the resultant electric force of attraction between the Earth and the Moon? (The Earth-to-Moon distance is 3.84×10^8 m.)

Solution: In order to solve this problem, we must be able to calculate the amount of charge in a kilogram of monatomic hydrogen. Avogadro's number, N_A equals 6.02×10^{23} atoms per mole; because for monatomic hydrogen the mass of 1 mole is 1 g, we have 6.02×10^{26} atoms per kilogram, or 6.022×10^{26} protons, and the same number of electrons. The total charge of the Moon is then $(6.022 \times 10^{26})(-e)$ C, and that of the Earth is $(6.022 \times 10^{26})(+e)$ C.

We next need to consider whether we can call the Earth and Moon point charges, even though rather large ones, and use Coulomb's Law to obtain the total force of attraction. The Earth's radius is 6×10^6 m, small in comparison to the Earth-to-Moon distance of 4×10^8 m. Thus, the approximation of considering the Earth and Moon as point charges should not be a bad one, especially because the protons, at least, are distributed *uniformly* on the Earth. (We will find later that our approximation, in fact, is exactly correct.)

Using Coulomb's Law, we can then give the total force of attraction by

$$F = \frac{1}{4\pi\varepsilon_0}\frac{q_1 q_2}{r^2} = 9\ (\mathrm{N}^9)\ \frac{[6.022\ (10^{26})]^2[1.6(10^{-19})]^2}{[3.84\ (10^8)]^2}$$
$$= 5.67 \times 10^8\ \mathrm{N}$$

8. *Guided Problem.* In the Bohr model of atomic hydrogen, an electron of mass 9.1×10^{-31} kg revolves about a proton of mass 1.67×10^{-27} kg in a circular orbit of radius 5.3×10^{-11} m. If the centripetal force is provided by the Coulomb force between the proton and the electron, what is the velocity of the electron for a stable orbit?

Solution scheme

a. The centripetal force required to keep a particle in a stable circular orbit is given by $F_c = m\upsilon^2/r$, where r is the radius of the circular orbit. If this centripetal force is provided by the Coulomb force, then should the Coulomb force of attraction between the electron and the proton equal F_c?
b. You should now have enough information to solve for υ, the velocity of the electron for a stable orbit.

9. *Worked Problem.* Suppose that the Earth is held in orbit around the Sun not by gravitational attraction but by electrostatic attraction. How many extra electrons would have to be placed on the Sun, and how many extra positive charges (say, protons) on the Earth, to create the right attractive force? Assume that the numbers of extra charges on each body are in the same proportion as the radial dimensions of the body ($R_{\mathrm{Sun}}:R_{\mathrm{Earth}} = 109:1$).

Solution: Since the required electrical force = the gravitational force, then $F = GM_sM_e/R^2 = kQ_sQ_e/R^2$, where R is the Earth–Sun distance, G is the universal gravitational constant, $k = 1/4\pi\varepsilon_0 = 8.99 \times 10^9$ MKS units, and M and Q refer to the Sun–Earth masses and charges. $M_s = 1.99 \times 10^{30}$ kg, and $M_e = 5.98 \times 10^{24}$ kg. It is clear that the Earth–Sun distance cancels out, so that we can write $GM_sM_e = kQ_sQ_e$, and since the number of charges are in proportion to the radial dimensions, $Q_s = 109Q_e$, so that $kQ_sQ_e = k(109Q_e^2)$. Then,

$$Q_e^2 = GM_sM_e/[109k]$$
$$= (6.67 \times 10^{-11})(1.99 \times 10^{30})(5.98 \times 10^{24})/[109 \times 8.99 \times 10^{11}]$$
$$= 4.0 \times 10^{32},\ \text{or}\ Q_e = 2 \times 10^{16}\mathrm{C}$$

But since $Q_e = N_{\mathrm{protons}}(1.6 \times 10^{-19}\mathrm{C})$, then $N_{\mathrm{protons}} = 1.25 \times 10^{35}$, for the number of positive charges on the Earth, and since $Q_s = 109Q_e$,

$N_s = 109 \times 1.25 \times 10^{35} = 1.36 \times 10^{37}$ for the excess electrons on the Sun.

10. *Guided Problem.* We believe that the neutron (a neutral particle) is made up of two d quarks, each with charge $-e/3$, and one u quark, with charge $2e/3$, where $e = 1.6 \times 10^{-19}$C. Supposing that the quarks can be treated like classical particles and that the three quarks inside the neutron are positioned at the corners of an equilateral triangle, and are each 1.0×10^{-15} m apart, find the electrical force of attraction that the two d quarks exert on the u quark. (See Figure 22.9.)

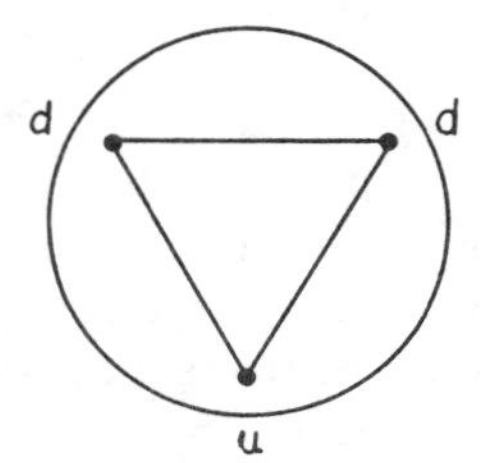

Figure 22.9 Problem 10. Two d quarks and one u quark inside a neutron.

Solution scheme

a. Draw a force diagram showing the two forces acting on the u quark. Draw x and y axes on your diagram. You should note that each force is 30° away from the vertical, or y axis.
b. Using Coulomb's Law, write down the expression for each force. Because we know all the distances and charges, we can calculate each force exactly.
c. Are the two forces equal? Look at the x and y components of each force. You should be able to see that the x components cancel exactly, leaving only the y components, which add.
d. The net force on the u quark is therefore just twice the y component of one of the forces. Evaluate this force.

Answers to Guided Problems

2. 3.12×10^9 electrons; 2.84×10^{-21} kg. You should realize that the number of extra electrons is negligible in comparison with the total number of electrons in the rod. A rod of 1 cm^3 volume (a small rod) would have about 10^{23} electrons; similarly, the mass increase is negligible compared with the total mass.
4. The distance $x = 0.89$ meters.
6. a. 0.54 ($\hat{y}$) N
   b. 0.27 ($-\hat{y}$) N
8. $\upsilon = 2.18 \times 10^6$ m/s
10. $\mathbf{F} = (\hat{y})88.5$ N

Chapter 23

The Electric Field

Overview

In the last chapter we learned that the force between two charges at rest or moving slowly with respect to each other can be described in terms of Coulomb's Law. Usually we have more than two charges, and often there are very many charges. In this chapter we find that the principle of linear superposition applies to electrostatic forces—that the force contributed by each charge is independent of the presence of other charges. This is a very important empirical law, for it means that Coulomb's Law can always be applied to systems of static charges, no matter how many charges are present. In this chapter we are also introduced to the concept of an electric field, which allows us to deal with the idea of action at a distance. An electric field can be represented schematically by electric field lines, which are lines drawn parallel to the direction of the field. We also discuss the behavior in an electric field of a type of charge distribution often found in nature, the electric dipole.

Essential Terms

linear superposition
electric field
field lines
electric dipole
dipole moment

Key Concepts

1. *Linear superposition*. Experiments show that electrostatic forces have the same linear character as gravitational forces, which we have already studied; that is, electrostatic forces obey the principle of *linear superposi-*

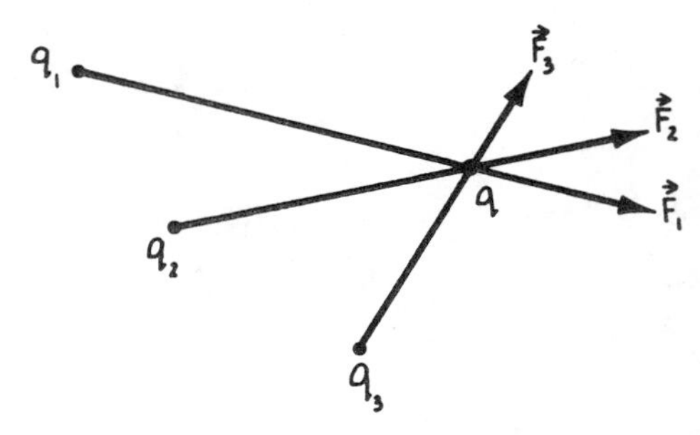

Figure 23.1 Based on the principles of superposition, the net force on charge q can be given by $\mathbf{F} = \mathbf{F}_1 + \mathbf{F}_2 + \mathbf{F}_3$.

tion. If several charges $q_1, q_2, q_3, \ldots$ act on a charge q, the collective force on q is the linear vector sum of the individual forces. The force contributed by each charge is independent of the presence of the other charges, so that Coulomb's Law is valid no matter how many point charges are present, or even if we have a continuous distribution of charge. For example, in Figure 23.1, the force on q, which we call $\mathbf{F}$, is given by

$$\mathbf{F} = \mathbf{F}_1 + \mathbf{F}_2 + \mathbf{F}_3 = \frac{1}{4\pi\varepsilon_0}\left[\frac{qq_1\hat{\mathbf{r}}_1}{r_1^2} + \frac{qq_2\hat{\mathbf{r}}_2}{r_2^2} + \frac{qq_3\hat{\mathbf{r}}_3}{r_3^2}\right]$$

For the charge distribution shown in Figure 23.2, the force on q is obtained by integrating over all the charge enclosed within the shaded volume, so that

$$\mathbf{F} = \int d\mathbf{F} = \frac{q}{4\pi\varepsilon_0}\int \frac{dq\hat{\mathbf{r}}}{r^2}$$

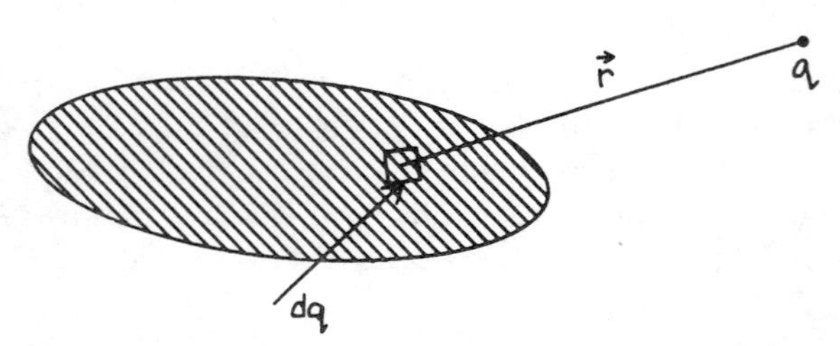

Figure 23.2 The force on q is obtained by integrating over the charge enclosed within the shaded area.

2. *Electric field*. Most of the forces in everyday life are *contact* forces, such as the push of a hand against a wall. The electrostatic force, like the gravitational force, acts *at a distance*, that is, even when the bodies are not touching. To explain the concept of action at a distance, we introduce the idea of an *electric field*. When an electric charge is placed at some point, the charge establishes an electric field everywhere. Any other charge, placed anywhere, will interact with the field existing at that point. An electric field in a coordinate frame at rest with respect to the charge producing it (as long as the charge is not accelerating!) is called an *electrostatic field*. For an electrostatic field we do not need to consider questions relating to the time required for the field to propagate through all space.

3. *The electric field of a single charge.* Rather than thinking of two charges as interacting directly, we think in terms of an electric charge setting up an electric field, which in turn exerts a force on a charge placed in the field. This concept creates the problems of calculating (1) the fields that are set up by given distributions of charge and (2) the forces that given fields exert on charges placed in them. Mathematically, we define *electric field strength* as the force per unit charge exerted on a unit positive test charge, that is, $\mathbf{E} = \mathbf{F}/q$ newtons/coulomb. The test charge, really an imaginary test charge, is assumed not to disturb the arrangement of charges creating the field. The electric field at a point P a distance r from a point charge q' is given by

$$\mathbf{E} = \frac{1}{4\pi\mathcal{E}_0}\frac{q'}{r^2}\,\hat{\mathbf{r}}$$

and is indicated schematically in Figure 23.3.

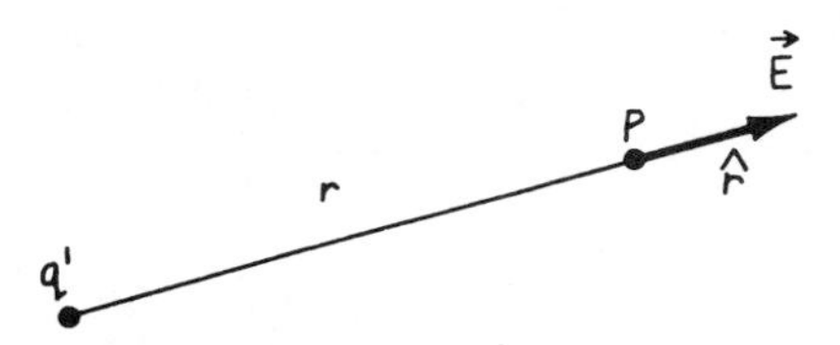

Figure 23.3 The electric field $\mathbf{E}$ a distance r from a point charge q'.

4. *The electric field caused by several charges.* The electric field caused by several charges is evaluated by linear superposition of the individual fields, in the same way that the force on a charge q caused by several other charges is obtained by summing the individual forces vectorially. For example, the field $\mathbf{E}$ at some point P caused by three charges q_1, q_2, and q_3 is given by $\mathbf{E} = \mathbf{E}_1 + \mathbf{E}_2 + \mathbf{E}_3$, as shown in Figure 23.4. The field $\mathbf{E}$ in this case does not vary as the inverse square of the distance from any one of the three charges q_1, q_2, and q_3.

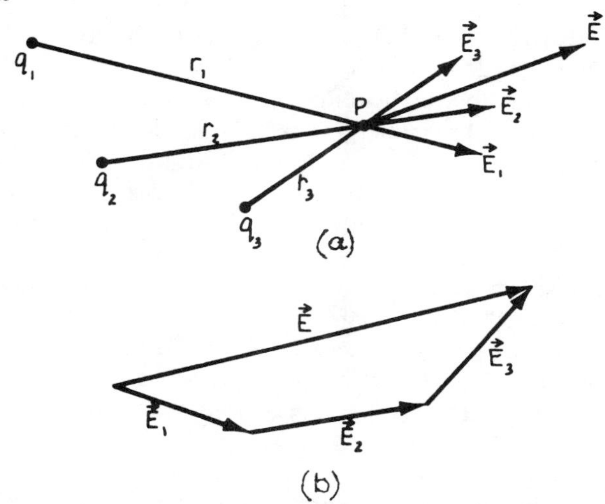

Figure 23.4 (a) The electric field $\mathbf{E}$ caused by several point charges. The field $\mathbf{E}$ is the vector sum $\mathbf{E}_1 + \mathbf{E}_2 + \mathbf{E}_3$, as indicated in (b).

5. *Field lines.* The electric field near a point charge q' always points directly away from the charge q' (assumed to be positive) and varies inversely as the square of the distance away from q'. We can represent the electric field by drawing force vectors, showing the magnitude and direction of the field at a large number of points, as shown in Figure 23.5(a). We can also represent the electric field by drawing a number of electric *field lines.* Field lines are always drawn parallel to the direction of the electric field. The field lines caused by a point charge q' are shown in Figure 23.5(b).

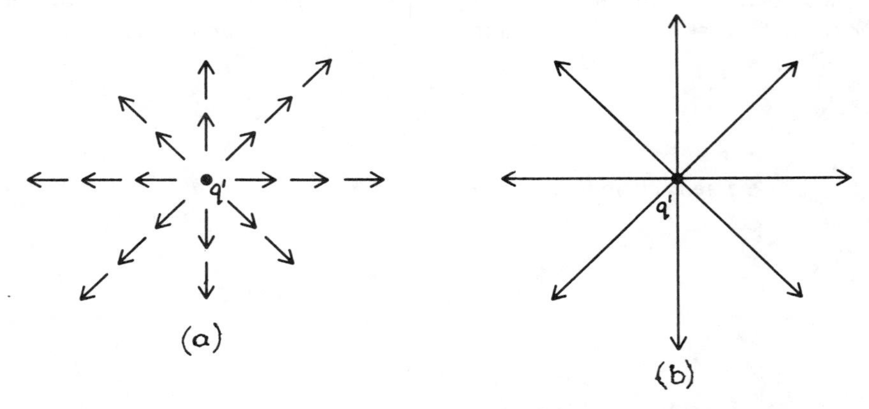

Figure 23.5 (a) Force vectors used to represent an electric field. In (b) field lines are used to represent the same field.

We usually use field lines to give a graphical representation of the electric field. The density of lines, that is, the number of lines per unit area perpendicular to the lines, is proportional to the magnitude of the electric field. Field lines start on positive charges and end on negative charges. In Figure 23.5 the field lines start on the positive charge q' and are directed radially outward. We usually start with the supposition that the universe is uncharged; therefore, for an isolated positive charge q', there must exist somewhere an equal negative charge $-q'$. We implicitly assume that this negative charge is uniformly distributed in space a long distance from the positive charge, so that the field lines do eventually end on this distributed negative charge. By definition, the number of field lines emerging from a charge q' is $q'/\mathcal{E}_0$, so that there are $q'/\mathcal{E}_0$ lines emerging from a point charge q' and they are distributed evenly over all radial directions. At a distance r from the charge, these lines are distributed uniformly over the area $4\pi r^2$, giving $q'/4\pi\mathcal{E}_0 r^2$ lines per square meter. In other words, the *line density* is equal to the magnitude of the electric field.

When drawing field lines, we can often use symmetry arguments to tell us the direction of the electric field. For example, because an infinitely long, straight, uniformly charged line defines a single direction, or axis, in space, the field direction must be perpendicular to this axis—that is, the field must radiate outward uniformly in all directions from this axis, as indicated in Figure 23.6.

6. *Electric dipole.* An important type of charge distribution, the *electric dipole*, consists of two charges, $+Q$ and $-Q$, with equal magnitudes and

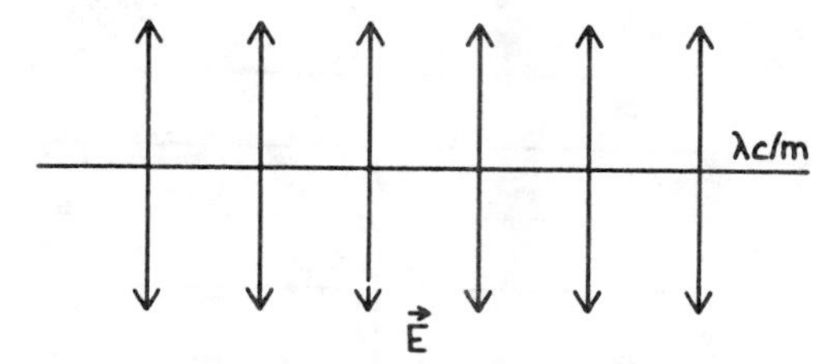

Figure 23.6 The electric field caused by a long, straight, uniformly charged line.

opposite signs, separated by a fixed distance l. There is no net force on a dipole in a uniform electric field, because the forces on the positive and negative charges are equal and opposite, but there is a torque that will tend to align the axis of the dipole parallel to the electric field, as shown in Figure 23.7.

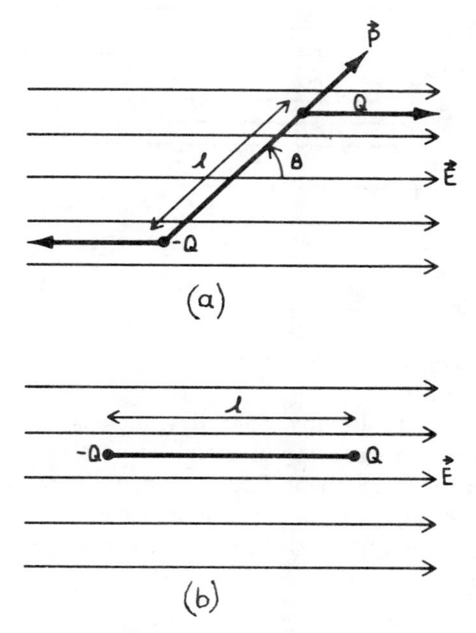

Figure 23.7 The torque on a dipole in an electric field tends to align the axis of the dipole parallel to the field, as in (a). There is no torque on the aligned dipole shown in (b).

The magnitude of the force on each end of the dipole is QE, so that the torque on the dipole is given by $QE(l \sin \theta) = QEl \sin \theta$, which is clearly zero in Figure 23.7(b). If we define the quantity $Q\mathbf{l}$ as $\mathbf{p}$, and call $\mathbf{p}$ the *dipole moment*, then in vector notation the torque τ, is $\tau = Q\mathbf{l} \times \mathbf{E} = \mathbf{p} \times \mathbf{E}$. The potential energy of a dipole in an electric field is given by $U = \mathbf{p} \cdot \mathbf{E} = -pE \cos \theta$, so that the potential energy is arbitrarily chosen to be zero when the axis of the dipole is perpendicular to the field. U is then a maximum (pE) when the axis of the dipole is antiparallel to the field, because the dipole will tend to rotate in the field and can do external work in doing so. The potential energy of a dipole in an electric field is indicated in Figure 23.8.

The main reason that the electric dipole is such an important type of

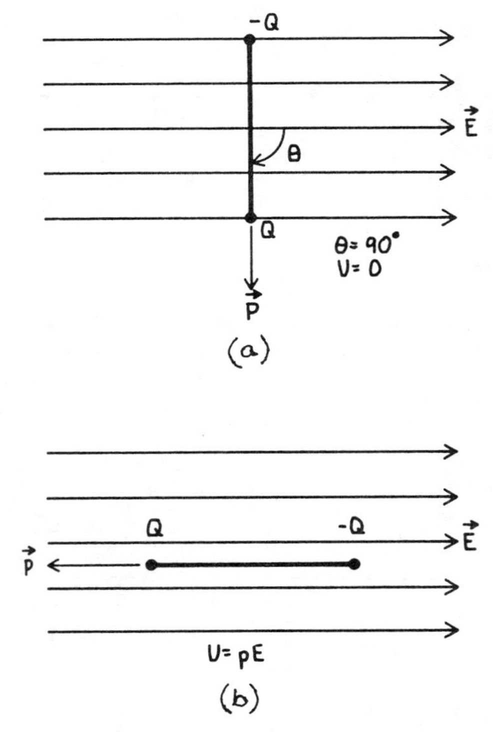

Figure 23.8 The potential energy, **p** • **E**, of a dipole in an electric field. In (a) $U = 0$. In (b) the dipole axis is antiparallel to **E** and the potential energy U is a maximum, pE.

charge distribution is that many molecules, such as water (H_2O), have a permanent separation between the effective centers of positive and negative charge and so have a permanent dipole moment. Those that do not may still acquire an induced dipole moment when placed in an external electric field; that is, the electric field forces a net separation of the centers of positive and negative charge.

Sample Problems

1. *Worked Problem.* Three charges, *a, b,* and *c,* each of 3 μC, are located at $x = 0$, $x = 0.5$ m, and $x = 1.0$ m, respectively. What is the force exerted by charges *b* and *c* on charge *a*? The charges are indicated schematically in Figure 23.9.

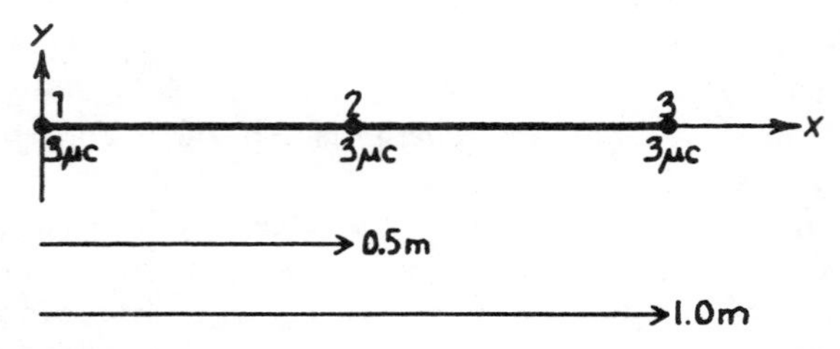

Figure 23.9 Problem 1. Three charges at $x = 0$, $x = 0.5$ m, and $x = 1.0$ m.

Solution: The principle of superposition of forces tells us that the force on charge a is the vector sum of the force caused by charge b and that caused by charge c—that is, $\mathbf{F}_a = \mathbf{F}_b + \mathbf{F}_c$, where $\mathbf{F}_b$ and $\mathbf{F}_c$ are, respectively, the forces on charge a caused by charges b and c. Because all the charges are positive, $\mathbf{F}_b$ and $\mathbf{F}_c$ are both in the $(-x)$ direction. From Coulomb's Law

$$\mathbf{F}_a = \frac{1}{4\pi\varepsilon_0}\left[\frac{q_a q_b}{r_b^2} + \frac{q_a q_c}{r_c^2}\right](-\hat{\mathbf{x}})$$

$$q_a = q_b = q_c = 3 \times 10^{-6}\ \mathrm{C}$$

$$r_b = 0.5\ \mathrm{m};\ r_c = 1.0\ \mathrm{m}$$

$$\mathbf{F}_a = -0.41(\hat{\mathbf{x}})\ \mathrm{N}$$

2. *Guided Problem.* A small charged pith ball of mass 0.090 g is "suspended" in an electric field so that the gravitational force on the pith ball is exactly balanced by the electric force.

a. If the electric field strength is 3.0×10^4 N/C pointing upward, what is the charge on the pith ball?
b. Is the pith ball positively or negatively charged, and to how many excess charges does this correspond? The acceleration of gravity, g, equals 9.8 m/s^2.

Solution scheme

a. Construct a "free-body" diagram showing the forces on the pith ball. Because the pith ball is in equilibrium, the resultant force on the ball is zero, and you can solve directly for the charge q on the pith ball.
b. The direction of the electric field $\mathbf{E}$ is upward, as is the direction of the electric force on the pith ball. $\mathbf{F} = q\mathbf{E}$; is q positive or negative?
c. By "excess charges" we mean the number of excess electrons, if q is negative, or the number of excess positive ions (charge $= +e$), if q is positive. The magnitude of q is known; what is the number of charges needed to make up this total charge q?

3. *Worked Problem.* Find the force on a point charge of 1 μC placed, as shown in Figure 23.10, 20 cm from the midpoint of a circle of radius 10 cm

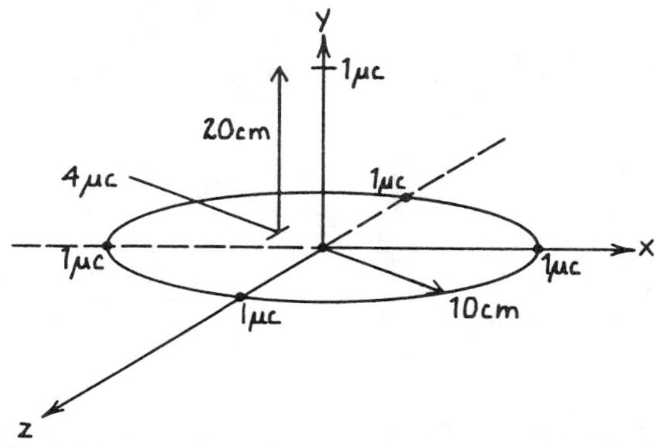

Figure 23.10 Problem 3. A 1 μC charge 20 cm from the midpoint of a circle of radius 10 cm lying in the x–z plane.

lying in the x–z plane. A charge of 4 μC is positioned at the center of the circle (the origin), and four charges of 1 μC each are at +10 cm and −10 cm from the origin, lying along the x and z axes, respectively.

Solution: In this problem also we use the principle of linear superposition to find the resultant force on the 1 μC charge. The force caused by the 4 μC charge at the origin is directly parallel to the y axis and can be calculated directly from Coulomb's Law, giving $\mathbf{F} = 0.9\hat{\mathbf{y}}$ N. The force caused by each of the remaining 1 μC charges placed on the circumference of the 10 cm circle is equal in magnitude, because each charge is the same distance from the 1 μC charge on the y axis. If we call the forces caused by the two charges on the x axis F_2 and F_3, we see that the horizontal components of $\mathbf{F}_2$ and $\mathbf{F}_3$ are equal and opposite and so cancel each other, leaving only the vertical components, which are equal and parallel and so can be added. Forces $\mathbf{F}_2$ and $\mathbf{F}_3$ are shown schematically in Figure 23.11.

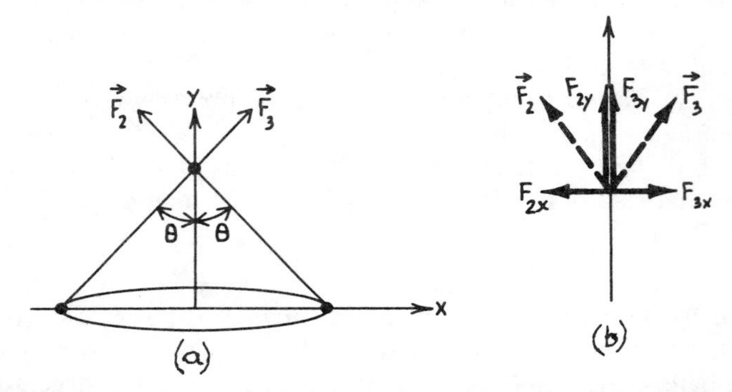

Figure 23.11 Problem 3. The forces on the point charge caused by the two 1 μC charges lying along the x axis. In (b) the forces are shown resolved into their horizontal and vertical components.

Similarly, if we call the forces caused by the two charges on the y axis F_4 and F_5, the horizontal components cancel each other, leaving only the vertical components, which can be added. The net effect of the four charges on the circumference of the 10 cm circle is therefore a single force.

$$4F_2 \cos\theta\hat{\mathbf{y}} = 4F_2(0.894)\hat{\mathbf{y}}$$

F_2 can be obtained directly from Coulomb's Law:

$$F_2 = \frac{1}{4\pi\varepsilon_0}\frac{(1\times 10^{-6})^2}{[0.2 + 0.1]^2} = 0.18 \text{ N}$$

$$4F_2(0.894) = 0.64 \text{ N}$$

The resultant force on the 1 μC charge at $y = 20$ cm is given by the sum of individual forces and is $(0.9 + 0.64)\hat{\mathbf{y}} = 1.54\hat{\mathbf{y}}$ N.

4. *Guided Problem.* A small, charged pith ball is suspended by a thread in a uniform electric field, as shown in Figure 23.12. The electric field deflects

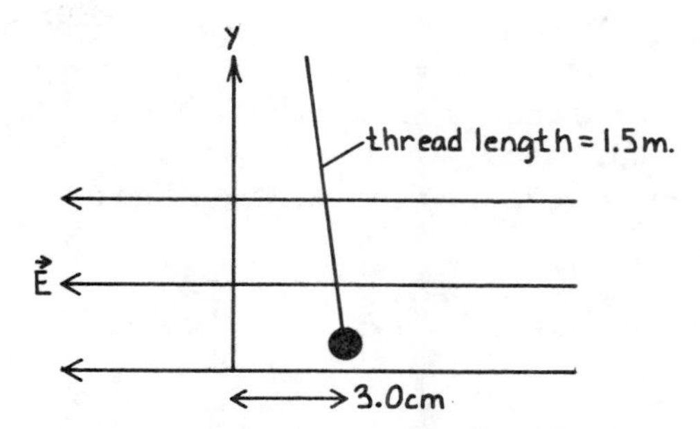

Figure 23.12 Problem 4. A charged pith ball suspended by a thread in a uniform electric field.

the ball 3.0 cm from its equilibrium position with the field off. Find the electric force on the ball. If the ball carries a net excess of 2×10^{10} electrons, find the magnitude of the electric field. The pith ball has a mass of 1.0 g.

Solution scheme

a. Construct a "free-body" diagram showing the forces on the pith ball. Remember that because the pith ball is in equilibrium, the resultant force equals zero. You know the weight of the ball and the angle that the thread makes with the vertical; find the thread tension.
b. If you know the thread tension, T, you can solve directly for the electric force F_e on the ball.
c. From the number of excess electrons on the pith ball, find the net charge, q.
d. What is the relationship between electric field strength **E**, electric force $\mathbf{F}_e$, and net charge q? Knowing $\mathbf{F}_e$ and q, can you find **E**?

5. *Worked Problem.* Calculate the electric field **E** at a distance d above the midpoint of a line of length L of charge of uniform charge density λ C/m. What would the field strength be if the length L were infinite, the charge density remaining at λ C/m? (The first part of this problem is identical to problem 23.30[b] in the text [p. 606]; it also appears in many textbooks on electromagnetic theory. The solution of this problem is an interesting and instructive mathematical exercise, which is why we give it.)

Solution: We begin by drawing a diagram, Figure 23.13, in which we show the contribution to the field strength made by a small element dx of

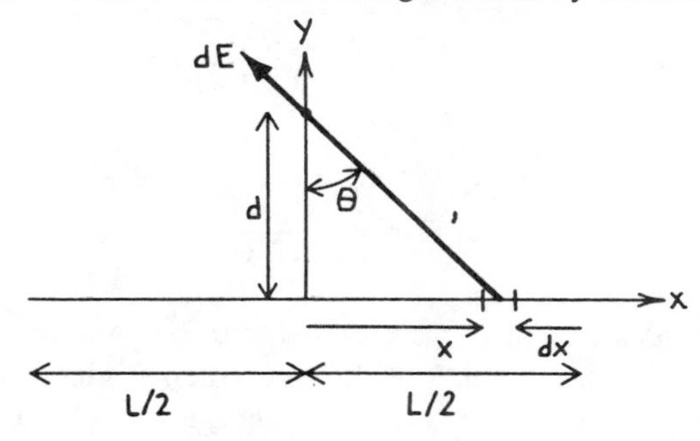

Figure 23.13 Problem 5. The field contribution, $d\mathbf{E}$, resulting from a small element dx of the line of charge.

the line of charge. We call this contribution $d\mathbf{E}$. For every such element of length between $x = 0$ and $x = L/2$, there is a corresponding element of length between $x = 0$ and $x = -L/2$ such that the horizontal, or x, components of dE exactly cancel, whereas the vertical, or y components of $d\mathbf{E}$ add. Then the field resulting from the whole line of charge is given by

$$\mathbf{E} = 2(\hat{\mathbf{y}}) \int_0^{L/2} dE_y = 2(\hat{\mathbf{y}}) \int_0^{L/2} dE \cos\theta$$

We can determine dE by Coulomb's Law:

$$dE = \frac{1}{4\pi\varepsilon_0} \frac{\lambda dx}{(d^2 + x^2)}$$

From the figure we see that

$$\cos\theta = \frac{d}{\sqrt{d^2 + x^2}}$$

$$\mathbf{E} = (\hat{\mathbf{y}}) \frac{\lambda d}{2\pi\varepsilon_0} \int_0^{L/2} \frac{dx}{(d^2 + x^2)^{3/2}}$$

The integral can be solved either by substitution or through use of a table of integrals, from which we find that

$$\int \frac{dx}{(x^2 + a^2)^{3/2}} = \frac{x}{a^2\sqrt{x^2 + a^2}}$$

$$\mathbf{E} = \frac{\hat{\mathbf{y}}\lambda d}{2\pi\varepsilon_0} \left[\frac{L/2}{d^2\sqrt{d^2 + L^2/4}} \right] \mathrm{N/C}\cdot$$

To find the value of $\mathbf{E}$ if we allow the length L to become infinite, divide both numerator and denominator by L to obtain

$$\mathbf{E} = \frac{\hat{\mathbf{y}}\lambda}{4\pi\varepsilon_0 d} \frac{1}{\sqrt{\left(\frac{d}{L}\right)^2 + \frac{1}{4}}}$$

which reduces to

$$\mathbf{E} = \frac{\hat{\mathbf{y}}\lambda}{2\pi\varepsilon_0 d} \frac{\mathrm{N}}{\mathrm{C}}$$

6. *Guided Problem*. Neutrons are believed to be composed of an "up," or "u," quark, with a charge of 2/3 of the fundamental electronic charge, and two "down," or "d," quarks, each with a charge of $-1/3$ of the electronic charge, so that the total net charge is zero. The three quarks are confined within a ''bag.'' Given the arrangement of quarks shown in Figure 23.14,

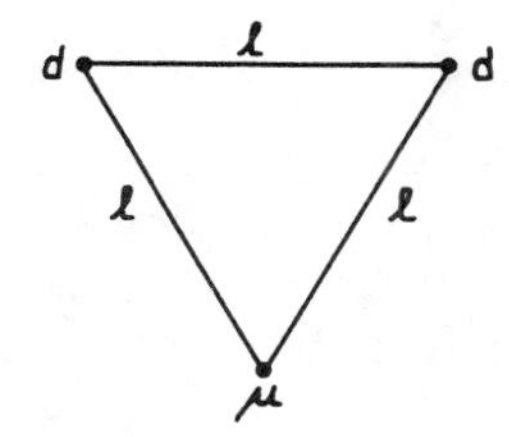

Figure 23.14 Problem 6. A neutron, composed of two d quarks and one u quark inside a bag.

what is the electric field at the u quark resulting from the two d quarks? What is the net electric force on the u quark? The distance l in the figure is 1 fermi, or 1×10^{-15} m.

Solution scheme

a. We can solve this problem using the principle of linear superposition. The field at the u quark, $\mathbf{E}$, equals $\mathbf{E}_1 + \mathbf{E}_2$, where $\mathbf{E}_1$ and $\mathbf{E}_2$ are, respectively, the field contributions from each of the d quarks. Draw a schematic diagram of $\mathbf{E}_1$ and $\mathbf{E}_2$.
b. Are the magnitudes of $\mathbf{E}_1$ and $\mathbf{E}_2$ equal? Resolve them into their horizontal and vertical components. Which components, if any, cancel, and which components add?
c. Now express the total field $\mathbf{E}$ at the u quark in terms of E_1, the magnitude of the field resulting from either of the d quarks.
d. From the definition of field resulting from a point charge, calculate E_1 and so determine $\mathbf{E}$.
e. All that remains is to find the net electric force on the u quark. You know its charge and the electric field that it sees. Can you find the electric force?

7. *Worked Problem.* A charge of 3 μC is located at the center of a hollow aluminum sphere, with inner radius 20 cm and outer radius 30 cm, as shown in Figure 23.15. Calculate the *net* number of electric lines of flux that enter the metal of the hollow aluminum sphere.

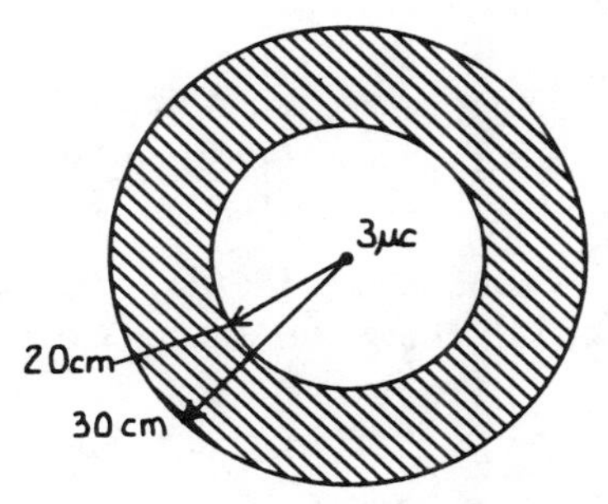

Figure 23.15 Problem 7. A hollow aluminum sphere with a 3 μC charge at its center.

Solution: First, let us find the density of lines of electric flux at respective distances of 20 cm and 30 cm from a 3 μC charge. There are q'/ε_0 field lines emerging from a point charge q'; at a distance r from the charge, these lines are distributed uniformly over the area $4\pi r^2$, giving $\frac{q'}{4\pi\varepsilon_0} r^2$ lines per square meter. Since $q' = 3 \times 10^{-6}$ C, at $r = 0.2$ m the line density is 6.75×10^5 lines/m^2 and at $r = 0.30$ m the line density is 3.00×10^5 lines/m^2. The number of lines of flux that enter the metal is just the product of the line density and the area of the metal, $4\pi r^2$; that is,

$$\left(\frac{q'}{4\pi\varepsilon_0 r^2}\right)(4\pi r^2) = \frac{q'}{\varepsilon_0}$$

independent of r. The number of lines entering the metal is equal to the number of lines leaving the metal is equal to 3.39×10^5, so that the net number of flux lines that enter the metal is equal to zero.

8. *Guided Problem.* A proton traveling with a speed of 5×10^5 m/s along the x axis enters a uniform electric field $\mathbf{E}$ equal to $100\hat{\mathbf{y}}$ N/C. The field extends for a distance of 0.5 m along the x axis. By how much is the proton deflected from its original trajectory at the point at which it leaves the electric field?

Solution scheme

a. Draw a schematic diagram showing the electric field and the path of the proton in the field. Let y equal the deflection of the proton from its original trajectory where it leaves the field.
b. Because the field exerts a force $q\mathbf{E}$ on the proton, you should be able to write an expression for the acceleration produced by the field on the proton. (The mass of the proton is 1.67×10^{-27} kg.)
c. From what you remember of mechanics, write down the equation of motion of a particle, giving its displacement in terms of its acceleration, initial velocity, and the time during which it experiences the acceleration.
d. How long does the proton experience the acceleration? Remember that the electric field does not change the motion of the proton very much. What is the initial velocity of the proton in the y direction?
e. Now solve the equation of motion for the required displacement, y.

9. ***Worked Problem.*** Calculate the electric field strength 1 m above an infinite thin sheet of charge with charge density σ C/m^2. Use the result of problem 5 in your solution.

Solution: In problem 5 we found that the electric field strength at a distance d above an infinite line charge of λ C/m is given by

$$\mathbf{E} = \frac{\hat{\mathbf{y}}\lambda}{2\pi\varepsilon_0 d} \text{ N/C}$$

Figure 23.16 shows how we can use this result to solve the present problem. Divide the infinite sheet into thin parallel strips, each of width dx and of infinite length, so that each strip runs parallel to the x axis. Let dE be the field contribution resulting from any one of these strips. Then, because the charge per meter of length on any strip is σdx,

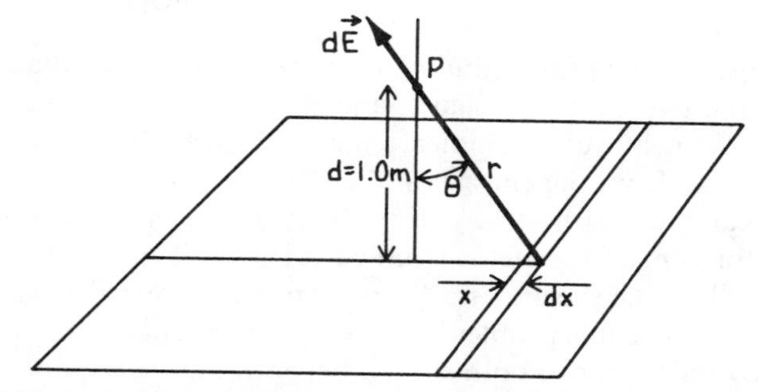

Figure 23.16 Problem 9. Calculation of the electric field strength above an infinite thin sheet of charge.

$$d\mathbf{E} = \frac{\sigma dx}{2\pi\varepsilon_0 r}\hat{\mathbf{r}}$$

Now because the sheet of charge is infinite in size, we can choose point P to be above the center of the sheet, that is, directly above the origin. For every strip positioned at x and contributing $d\mathbf{E}$ to the total field, there is another strip at $-x$ contributing $d\mathbf{E}$ of the same magnitude to the total field at point P. From the symmetry it is clear that the horizontal components from these two strips will exactly cancel, so that we need consider only the vertical components, which are given by $dE \cos\theta$. That is, $E = \int dE_y$, where $dE_y = \frac{\sigma dx}{2\pi\varepsilon_0 r}\cos\theta$. For convenience we make the substitution $x = d\tan\theta$, so that $dx = d\sec^2\theta d\theta$, and since $d/r = \cos\theta$, $r = d/\cos\theta$.

$$dE_y = \frac{\sigma(d\sec^2\theta d\theta)\cos^2\theta}{2\pi\varepsilon_0 d} = \frac{\sigma d\theta}{2\pi\varepsilon_0}$$

Then

$$E = \int\frac{\sigma d\theta}{2\pi\varepsilon_0} = \frac{\sigma}{2\pi\varepsilon_0}\int d\theta$$

As x varies between −infinity and +infinity, θ varies from $-\pi/2$ to $+\pi/2$, so that

$$E = \frac{\sigma}{2\varepsilon_0} \quad \text{or} \quad \mathbf{E} = \hat{\mathbf{y}}\frac{\sigma}{2\varepsilon_0}$$

Example 23.4 of your textbook finds the electric field given by an infinite sheet of charge by subdividing the sheet into an infinite number of concentric charged rings of radius R and width dR. We have here shown that the same result can be obtained by dividing the sheet into an infinite number of narrow, adjacent, charged strips of infinite length and width dx.

10. *Guided Problem.* Find the electric field strength at all points in space caused by two parallel, charged, infinite sheets as described in problem 9. One sheet has a charge density of 1×10^{-9} C/m^2; the other has a charge density of -1×10^{-9} C/m^2. The two sheets are 2 m apart.

Solution scheme

a. As usual, a helpful beginning is to draw a schematic diagram, showing the two sheets of charge and the field strengths resulting from each one. Be sure to indicate the field on both sides of each sheet resulting from the charge on the sheet.

b. Linear superposition says that the field at any point in space is the vector sum of the fields from each sheet taken one at a time, as if only that one sheet existed. You can therefore use linear superposition to solve this problem. There are three separate regions of space to consider: above both sheets, between the sheets, and below both sheets. Note that for the sheet with positive charge, the field is directly away from the sheet. For the sheet with negative charge, the field is directly toward the sheet. In neither case does the field vary with distance away from the sheet.

c. Can you now solve this problem? Remember to draw a schematic diagram.

11. *Worked Problem.* Two very large, flat sheets intersect at right angles. One carries a uniform positive charge distribution of 2.0×10^{-5} C/m^2, and the other carries a uniform negative charge distribution of -2.0×10^{-5} C/m^2 (Figure 23.17). Find the magnitude and the direction of the electric field in each of the four quadrants.

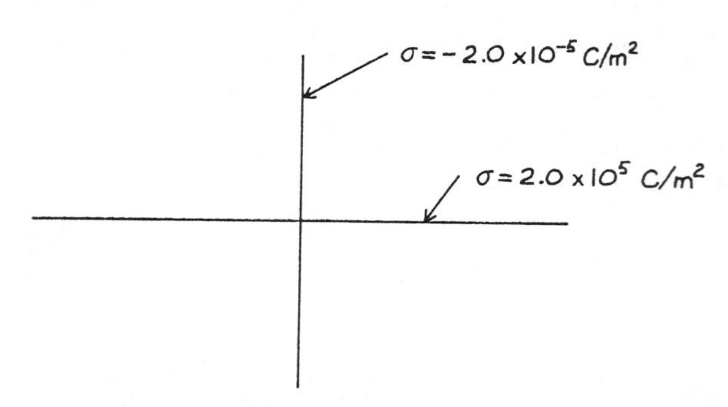

Figure 23.17 Problem 11. Two infinite, charged sheets intersecting at right angles.

Solution: In Figure 23.18(a) we have sketched the two sheets of charge and the electric fields resulting from each sheet. The magnitude of the field from each sheet is $E = \sigma/2\varepsilon_0 = 1.13 \times 10^6$ N/C. We have labeled the four quadrants I, II, III, and IV. From linear superposition, the field anywhere is the vector sum of the fields from each sheet taken one at a time, as if only that sheet existed. The fields in the four quadrants are sketched in Figure 23.18(b). Since the field from one sheet always intersects the field from the second sheet at 90°, the magnitude of the resultant field in each quadrant is given by E (cos 45°) + E (cos 45°) = $1.42E$ = 1.60×10^6 N/C. In each quadrant the direction of the resultant field is at an angle of 45° to the x axis, with the fields in quadrants I and IV pointing to the right and the fields in quadrants II and III pointing to the left.

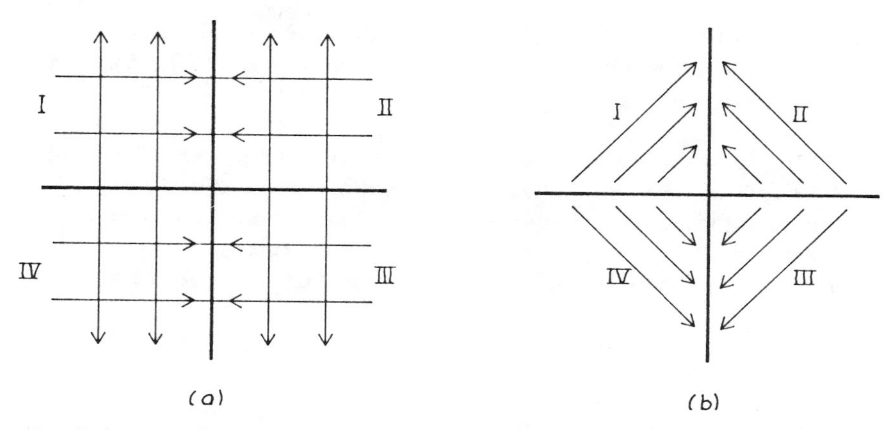

Figure 23.18 Problem 11. (a) The electric fields due to the two charged sheets, taken one at a time. In (b) the resultant field in each of the four quadrants is sketched.

12. *Guided Problem.* Calculate the electric field at point P, a distance d above two semi-infinite lines of charge of uniform charge density λ as shown in Figure 23.19. The lines of charge, which lie on the x axis, start at $x = -L/2$ and $x = L/2$, respectively.

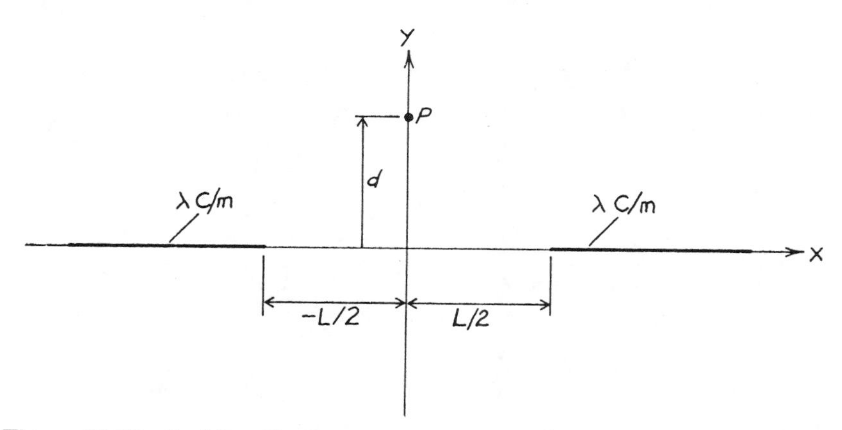

Figure 23.19 Problem 12. Two semi-infinite lines of charge of uniform charge density.

Solution scheme

a. This problem can be solved rather easily through superposition. Consider the two semi-infinite lines of charge as a single infinite line of charge of charge density λ extending along the whole x axis plus a length L of charge density $-\lambda$ extending from $-L/2$ to $L/2$.
b. We know that at point P the field due to the infinite line of charge is given by $\mathbf{E} = (\hat{\mathbf{y}})\lambda/2\pi\varepsilon_0 d$, and, from worked problem 5, we know the field due to a finite line of charge of length L. In this case the charge density is negative.

c. Add the two fields to obtain the resultant field at point P. Check that the field increases as L decreases and becomes equal to that for a single infinite line of charge if $L = 0$.

13. *Worked Problem.* The centers of two electric dipoles are placed 10 cm apart, as shown in Figure 23.20. For each dipole the distance between the positive and the negative charges is 0.01 cm. Find the potential energy of the dipole at $x = 10$ cm. What is the torque exerted on this dipole by the dipole at the origin? $l/2 = 0.005$ cm. The magnitude of each charge is 1×10^{-9} C.

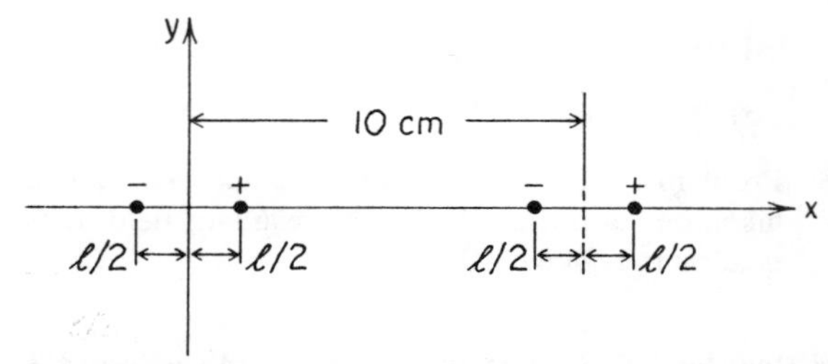

Figure 23.20 Problem 13. Two dipoles on the x axis, 10 cm apart.

Solution: At a distance x along the x axis, the electric field resulting from the dipole at the origin can be found, using Coulomb's Law, as follows:

$$\mathbf{E} = \frac{\hat{\mathbf{x}}}{4\pi\varepsilon_0}\left[\frac{q}{\left(x - \frac{l}{2}\right)^2} - \frac{q}{\left(x + \frac{l}{2}\right)^2}\right]$$

$$= \frac{\hat{\mathbf{x}}q}{4\pi\varepsilon_0 x^2}\left[\frac{1}{\left(1 - \frac{l}{2x}\right)^2} - \frac{1}{\left(1 + \frac{l}{2x}\right)^2}\right]$$

Since $x \gg l$ we use the approximation (for $x \ll 1$) that $(1 + x^n) \cong 1 + nx$ to obtain

$$\mathbf{E} = \frac{\hat{\mathbf{x}}q}{4\pi\varepsilon_0 x^2}\left[\left(1 + \frac{l}{x}\right) - \left(1 - \frac{l}{x}\right)\right] = \frac{\hat{\mathbf{x}}2ql}{4\pi\varepsilon_0 x^3} = \frac{2\mathbf{p}}{4\pi\varepsilon_0 x^3}$$

The potential energy of a dipole in an electric field is given by $U = -\mathbf{p} \cdot \mathbf{E}$, where $\mathbf{E}$ is the electric field strength seen by the dipole. Here the electric field $\mathbf{E}$ is not exactly constant over the length of the dipole at $x = 10$ cm, but it is nearly so, and it is a good approximation to take the field strength $\mathbf{E}$ at $x = 10$ cm, that is, at the center of the dipole. We therefore evaluate $\mathbf{E}$ at $x = 10$ cm. The dipole moment $\mathbf{p} = \hat{\mathbf{x}}lq = \hat{\mathbf{x}}(1 \times 10^{-13})$ m $\cdot$ C, and we obtain, for $x = 10$ cm, $\mathbf{E} = \hat{\mathbf{x}}1.8$ N/C.

The value of $U = -p \cdot E = -1.8 \times 10^{-13}$ J. Note that the value of U is negative because the dipole is aligned parallel to E. The torque on the dipole at $x = 10$ cm is given by $\boldsymbol{\tau} = \mathbf{p} \times \mathbf{E}$ or $\boldsymbol{\tau} = \hat{\mathbf{x}}(1 \times 10^{-11}) \times \hat{\mathbf{x}}1.8 = 0$.

14. *Guided Problem.* For the dipole shown in Figure 23.21, find the electric field strength at points along the x axis for values of x much greater than l $(x \gg l)$.

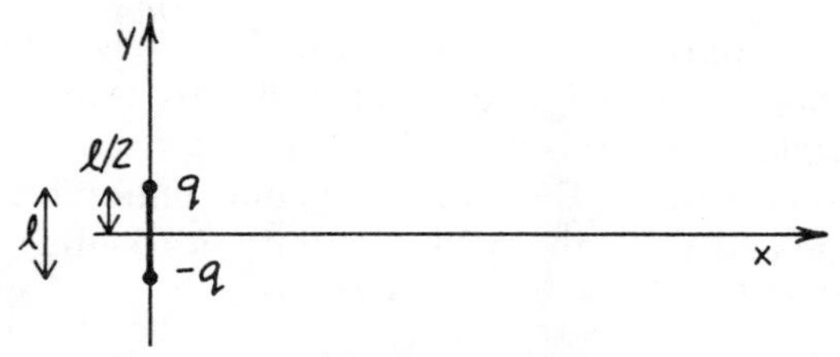

Figure 23.21 Problem 14. An electric dipole at the origin whose axis is parallel to the y axis.

Solution scheme

a. Choose some point P, a distance x along the x axis. Sketch the contributions to the field at P resulting from the two charges of the dipole.

b. The x components of the fields resulting from the positive and the negative charges of the dipole exactly cancel, whereas the vertical, or y, components add, so that the total field at point P is given by $\mathbf{E} = -\hat{\mathbf{y}}2E_y = -\hat{\mathbf{y}}2E \sin\theta$. E is the magnitude of the field caused by either one of the dipole charges. What is the angle θ?

c. Using Coulomb's Law, now solve for $\mathbf{E}$. Your answer can be expressed more simply if you again make use of the approximation that for $x \ll 1$, $(1 + x)^n \cong 1 + nx$.

15. *Worked Problem.* One of the largest electric dipoles that we observe in nature is that caused by a thundercloud. Suppose that for a certain cloud the charge distribution is approximately 50 C at a height of 10 km and −50 C at a height of 4 km directly below the positive charge. Where will an airplane approaching directly toward the cloud at a height of 7 km experience the greatest electric field? What is the value of this field?

Solution: The charge distribution is shown in Figure 23.22. Problem 14 demonstrated that for points along the x axis a great distance from the dipole, the field increases as $1/x^3$ as you approach the dipole and is always directed downward, along the $-y$ axis. The $1/x^3$ dependence is no longer valid as you approach the dipole axis. The field still increases as you approach the dipole, however, and remains pointing along the negative y axis. These points will be clear if you sketch the electric lines of force for a dipole.

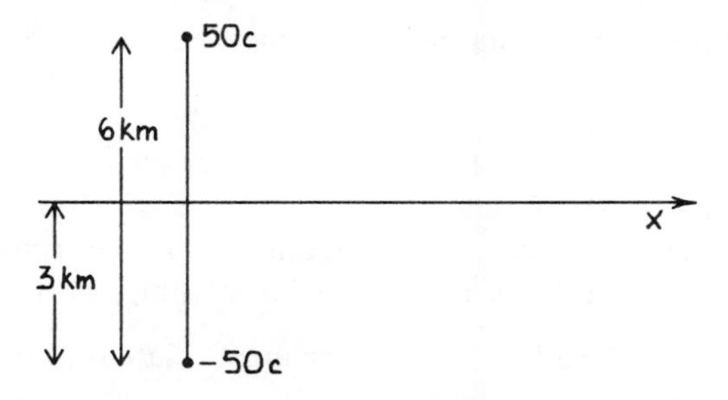

Figure 23.22 Problem 15. The charge distribution.

The greatest electric field will therefore be *directly between the two charges*. Its magnitude can be calculated directly using Coulomb's Law and the principle of superposition. Let **E** be the electric field directly between the two charges. If **E**+ is the field contribution resulting from the positive charge and **E**− is the field contribution resulting from the negative charge, then $\mathbf{E} = \mathbf{E}+ + \mathbf{E}- = (-\hat{\mathbf{y}})(E+ + E-)$.

$$\mathbf{E} = (-\hat{\mathbf{y}})\left(\frac{1}{4\pi\varepsilon_0}\right)\left[\frac{Q}{d^2} + \frac{Q}{d^2}\right]$$

where $Q = 50$ C and $d = 3 \times 10^3$ m. Solving, we obtain the result

$$\mathbf{E} = (-\hat{\mathbf{y}})(1 \times 10^5) \text{ N/C}$$

The field lines near an electric dipole are shown in Figure 23.23.

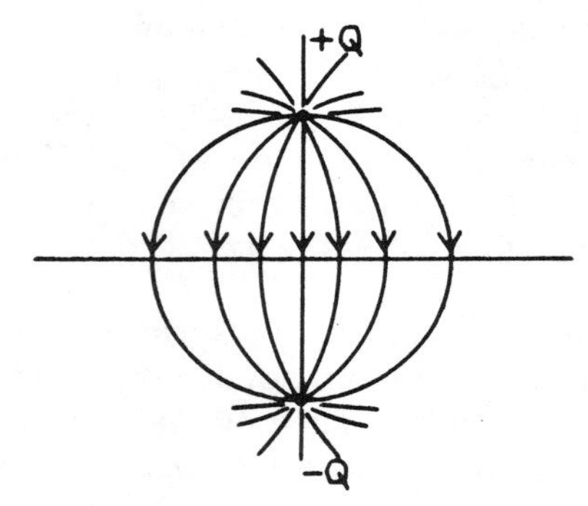

Figure 23.23 Problem 15. The electric field lines near a dipole.

Answers to Guided Problems

2. a. $Q = 2.94 \times 10^{-8}$ C
   b. The ball is positively charged. This corresponds to 1.84×10^{11} excess charges.
4. $E = 6.13 \times 10^4$ N/C
6. $\mathbf{E} = (\hat{\mathbf{y}})8.27 \times 10^{20}$ N/C. The net force $\mathbf{F} = (\hat{\mathbf{y}})88.7$ N.
8. $y = 4.8 \times 10^{-3}$ m or 4.8 mm
10. $\mathbf{E} = 0$ in the region above both plates and also in the region below both plates (assuming that the plates are horizontal). *Between* the plates $\mathbf{E} = (-\hat{\mathbf{y}})\dfrac{\sigma}{\varepsilon_0} = (-\hat{\mathbf{y}})113$ N/C, where the upper plate is positively charged and the lower plate is negatively charged.
12. $\mathbf{E} = \hat{\mathbf{y}}\left(\dfrac{\lambda}{2\pi\varepsilon_0}\right)\left[1 - \dfrac{1}{2\sqrt{(d/L)^2 + 1/4}}\right]$
14. $\mathbf{E} = -\hat{\mathbf{y}}\dfrac{p}{4\pi\varepsilon_0 x^3}$

Chapter 24

Gauss' Law

Overview

This chapter extends and quantifies the concept of electric field lines encountered in Chapter 23 by calculating the electric flux, or number of field lines, through a surface. The flux calculations then allow us to prove an important theorem called Gauss' Law. This theorem, which is based on Coulomb's Law, allows easy and rapid calculation of electric field strength when the charge distribution causing the field has a high degree of symmetry. For example, it is very easy to calculate the field near an infinite sheet of charge using Gauss' Law. Without Gauss' Law, as we found in the last chapter, the calculation is tedious and slow.

Essential Terms

electric flux
closed surface
Gauss' Law
Gaussian surface
static equilibrium

Key Concepts

1. *Electric flux.* We can best understand electric flux by considering a small element of area ΔS located in an electric field $\mathbf{E}$, as shown in Figure 24.1. We represent the element of area by the vector $\hat{\mathbf{n}}\Delta S$, where $\hat{\mathbf{n}}$ is the unit vector perpendicular to the surface ΔS. We define the electric flux, $\Delta\Phi$, through ΔS as given by

$$\Delta\Phi = \mathbf{E} \cdot \hat{\mathbf{n}}\Delta S = E\Delta S \cos\theta$$

From the definition, $\Delta\Phi$ will be positive if θ is less than 90° and negative if θ is greater than 90°. In the last chapter we found that the electric field line

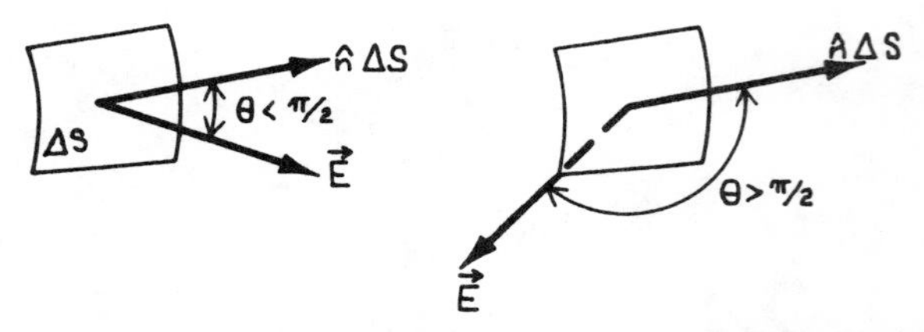

Figure 24.1 A small element of area ΔS in an electric field **E**. In each diagram $\hat{\mathbf{n}}$ is the unit vector perpendicular to the surface ΔS.

density is equal in magnitude to the electric field, **E**. The flux, $\Delta\Phi$, through the element of area ΔS is therefore *exactly* equal to the number of electric field lines through ΔS.

For an isolated small element of area ΔS, you may wonder what it means for $\Delta\Phi$ to be negative. Suppose we consider an arbitrary *closed surface*, a surface that encloses a well-defined volume, as shown in Figure 24.2. For every surface element ΔS on the closed surface, the unit normal vector $\hat{\mathbf{n}}$ is directed outward from the surface; thus, a negative value of $\Delta\Phi$ implies that lines of flux are entering the volume enclosed by the surface, and a positive value of $\Delta\Phi$ means that flux lines are leaving the volume.

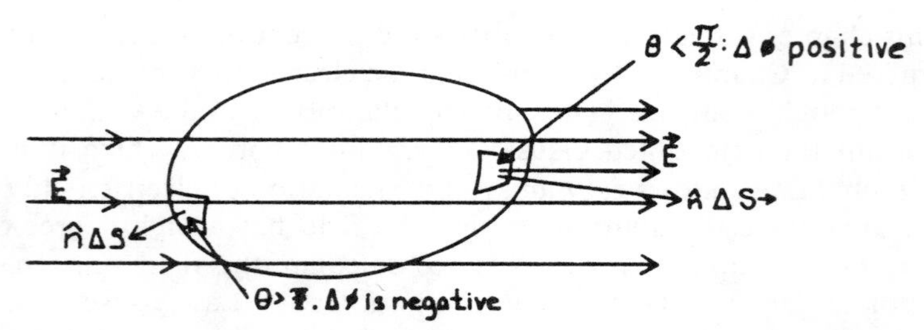

Figure 24.2 The evaluation of the flux through a surface element ΔS on a closed surface.

The total net electric flux, Φ, leaving the arbitrary closed surface is given by

$$\Phi = \oint \Delta\Phi = \oint \mathbf{E} \cdot \hat{\mathbf{n}} ds$$

where the integration is carried out over the area of the closed surface. If we evaluate the electric flux Φ for a closed surface and obtain a positive result, our answer tells us that more electric field lines are leaving the surface than are entering it; because field lines originate on positive charge, we can infer that the closed surface must contain some positive charge. Similarly, a negative answer implies that our closed surface must contain some negative charge. If $\Phi = 0$, the volume enclosed by the surface must contain no net electric charge. Furthermore, because field lines are directed radially away from an isolated point charge, field lines from charges *outside* the closed surface enter the closed surface at some point and pass right through it, making no net contribution to the electric flux enclosed by the surface.

2. *Gauss' Law*. The statements we have just made are quantified in Gauss' Law, which says that the net electric flux, Φ, through a closed surface, defined by $\Phi = \oint \mathbf{E} \cdot \hat{\mathbf{n}} ds$, is numerically equal to $Q/\mathcal{E}_0$, where Q is the *total* charge enclosed by the surface. This law follows directly from our definition that the number of field lines emerging from a charge q' is $q'/\mathcal{E}_0$. It does not matter how many small charges q' are contained within our volume; each one of them contributes $q'/\mathcal{E}_0$ field lines that pass out through the closed surface. For convenience, we usually write Gauss' Law as

$$\oint \mathbf{E} \cdot d\mathbf{s} = \frac{Q}{\mathcal{E}_0}$$

rather than as

$$\oint \mathbf{E} \cdot \hat{\mathbf{n}} ds = \frac{Q}{\mathcal{E}_0}$$

We use the small circle in the middle of the integral sign to indicate that the integration is carried out over a closed surface; when we are applying Gauss' Law, the closed surface is usually called a *Gaussian surface*. Gauss' Law is usually used in the calculation of electric field strength; it is most useful when the charge distribution causing the field has a high degree of symmetry. For example, a point charge has spherical symmetry, so we expect the electric field resulting from a point charge to have spherical symmetry. A line of charge defines a single direction in space; we expect any electric field resulting from this line of charge to have symmetry with respect to this one direction in space. A sheet of charge defines a plane in space; from symmetry, we expect that the electric field resulting from a sheet of charge will have symmetry with respect to the plane in space and therefore must be oriented perpendicular to the plane. Gauss' Law allows easy and rapid calculation of electric field strength for any of these charge distributions.

3. *Static equilibrium*. An electrical conductor has rather special electrostatic properties. At all points *inside* an electrical conductor, the electric field must be zero, because if it were not, the conduction electrons would move. When we first place a conductor in an electric field, the conduction electrons *do* move and redistribute themselves on the surface of the conductor such that the field set up by the surface charges on the conductor exactly cancels the external electric field inside the conductor. When this state occurs, the charge distribution on the conductor is said to be in *static equilibrium*. We can use Gauss' Law to prove that for a conductor in static equilibrium, the following are true: (a) any net electric charge resides on the surface of the conductor, and (b) the electric field at the surface of the conductor is normal to the surface.

These two results can have some interesting consequences. Figure 24.3 shows a needle stuck into a charged metal slab. Note that when we force the field lines to be perpendicular to the surface of the conductor, even to the surface of the needle, the field lines tend to be crowded together near the needle. Consequently, the electric field strength is much greater at the needle than on the smooth surface of the slab.

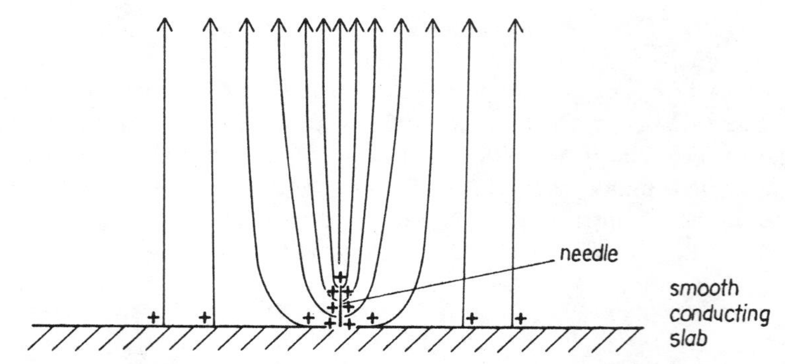

Figure 24.3 The electric field lines near a metal needle stuck into a smooth, conducting, charged slab.

Sample Problems

1. *Worked Problem.* A cube is positioned with one corner at the origin. The length of one side is a meters. If the electric field present is given by $(x + y)\hat{\mathbf{x}}$ N/C, what is the net electric flux leaving the cube? How much charge is present inside the cube? The cube is shown in Figure 24.4.

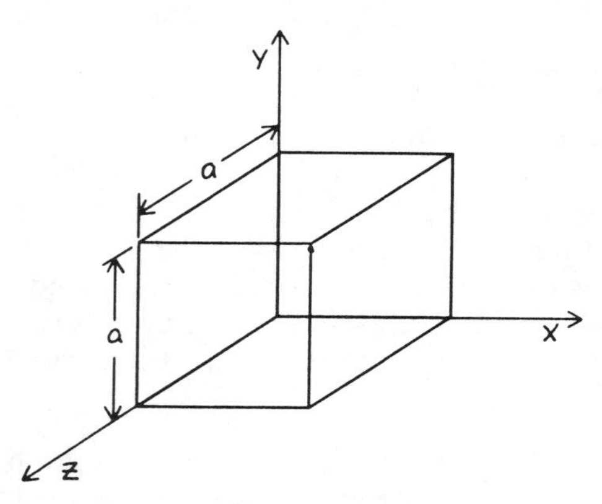

Figure 24.4 Problem 1. A cube positioned with one corner at the origin.

Solution: The total flux Φ leaving the cube is given by $\Phi = \oint \mathbf{E} \cdot \hat{\mathbf{n}} ds$, where the integral must be carried out over all six faces of the cube. Since $\mathbf{E} = E\hat{\mathbf{x}}$ only, we need worry only about the flux through the two faces in the y–z plane at $x = 0$ and $x = a$, because for the remaining four faces, $\mathbf{E} \cdot \hat{\mathbf{n}} = 0$.

Let Φ_1 be the flux leaving the cube through the face at $x = 0$ and Φ_2 be the flux leaving the cube through the face at $x = a$. Then

$$\begin{aligned}
\Phi = \Phi_1 + \Phi_2 &= \int y\hat{\mathbf{x}} \cdot (-\hat{\mathbf{x}}) ds_1 + \int (a + y)\hat{\mathbf{x}} \cdot \hat{\mathbf{x}} ds_2 \\
&= -\int y ds_1 + \int (a + y) ds_2 \\
&= -\int y ds_1 + \int y ds_2 + \int a ds_2
\end{aligned}$$

ds_1 and ds_2 are small increments of area on the faces at $x = 0$ and $x = a$, respectively. ds_1 and ds_2 can be chosen to be equal, and since the areas of the two faces are equal, $-\int y ds_1 + \int y ds_2 = 0$. Then

$$\Phi = a\int ds_2 = a^3$$

Alternately, we could have said that because the field component $y\hat{\mathbf{x}}$ is independent of x, the same number of lines that enter the cube on the left (resulting from this field component) must leave it on the right. The net electric flux leaving the cube is then a^3, and since $\Phi = Q/\varepsilon_0$, the charge inside the cube is $a^3\varepsilon_0$ coulombs.

2. *Guided Problem.* The atmospheric electric field of the Earth is about 150 N/C and points downward. Calculate the number of electric lines of flux passing through a Frisbee 25 cm in diameter when the plane of the Frisbee is (a) horizontal and (b) at an angle of 30° to the horizontal. Does the existence of the electric field imply that the Earth's surface is normally charged? If so, about how many excess charges are present in every square meter of surface area? Are they positive or negative?

Solution scheme

a. Use the definition of flux through a small element of area, $\Delta\Phi = \mathbf{E} \cdot \hat{\mathbf{n}}\Delta S$, to calculate the number of electric lines of flux through the Frisbee in cases a and b.
b. To answer the questions of charge on the Earth's surface, draw a Gaussian cylinder with one end of the cylinder on the Earth's surface and the other end above the Earth's surface, and calculate the net flux into or out of the cylinder. Since from Gauss' Law the net flux = [charge enclosed within the cylinder]/ε_0, you should be able to find the surface charge density on the Earth's surface.
c. From the charge density, and from the magnitude of the electronic charge (1.6×10^{-19} C), find the number of charges per square meter.

3. *Worked Problem.* Find the electric field at a point 5 cm above the center of a 12-inch-diameter stereo record that has accumulated an evenly distributed static charge of 6×10^{-10} C. Then, using Gauss' Law, find the electric field at a point 5 cm above a stereo record of *infinite* diameter with the same surface charge *density* on it. By how much would your answer have been in error if you had assumed an infinite record (instead of a 12-inch record) and just used Gauss' Law to answer the first part of the question?

Solution: The easiest way to answer this problem is to start with the results of Example 3, Chapter 23 of the text (p. 593): for a ring of charge, indicated in Figure 24.5(a), the electric field on the axis of the ring is given by

$$E_z = \frac{1}{4\pi\varepsilon_0}\frac{Qz}{(R^2 + z^2)^{3/2}}$$

Divide the record, a circular disk shown in Figure 24.5(b), into concentric rings. The radius of a ring is r and its area is $2\pi r dr$. The charge on any ring, dq, equals $2\pi r dr\sigma$. Then the electric field at the point P in Figure 24.5(b) resulting from the charged ring is given by

$$dE_z = \frac{dqz}{4\pi\varepsilon_0(r^2 + z^2)^{3/2}} = \frac{2\pi\sigma z r dr}{4\pi\varepsilon_0(r^2 + z^2)^{3/2}} = \frac{\sigma z}{2\varepsilon_0}\frac{rdr}{(r^2 + z^2)^{3/2}}$$

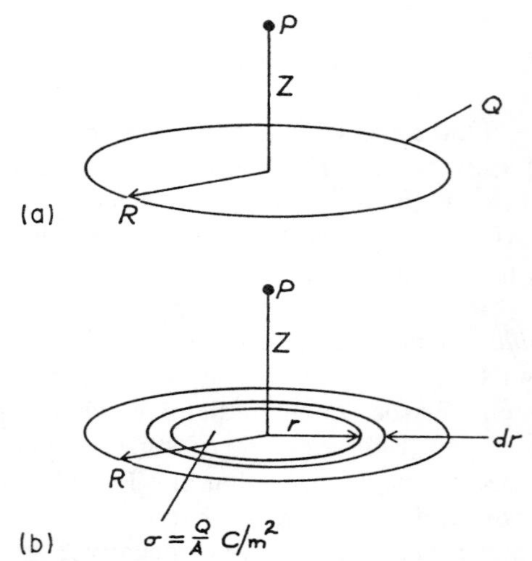

Figure 24.5 Problem 3. (a) A uniform ring of charge of radius R. (b) A uniformly charged disk that has been subdivided into rings of radius r and thickness dr.

The field E_z resulting from the whole record is then just

$$E_z = \frac{\sigma z}{2\varepsilon_0}\int_0^R \frac{rdr}{(r^2 + z^2)^{3/2}}$$

From integral tables,

$$\int \frac{rdr}{(r^2 + z)^{3/2}} = -\frac{1}{\sqrt{r^2 + z^2}}$$

so that

$$E_z = \frac{\sigma}{2\varepsilon_0}\left[1 - \frac{z}{\sqrt{R^2 + z^2}}\right]$$

$R = 6 \times 2.54 = 15.2$ cm, and $z = 5$ cm, so that

$$E_z = \frac{\sigma}{2\varepsilon_0}[1 - 0.31] = \frac{0.69\sigma}{2\varepsilon_0}$$

$$\sigma = \frac{Q}{A} = \frac{6 \times 10^{-10}}{\pi(.152^2)} = 8.26 \times 10^{-9}\ \text{C/m}^2$$

Then

$$E_z = 322\ \text{N/C}$$

We now apply Gauss' Law to a record of infinite diameter. We must choose the record to be of infinite diameter if we are to use Gauss' Law, because only for an infinite surface can we be sure, from symmetry, that the field will be everywhere normal to the surface and everywhere the same on the two faces of the cylinder. The Gaussian surface is shown in Figure 24.6.

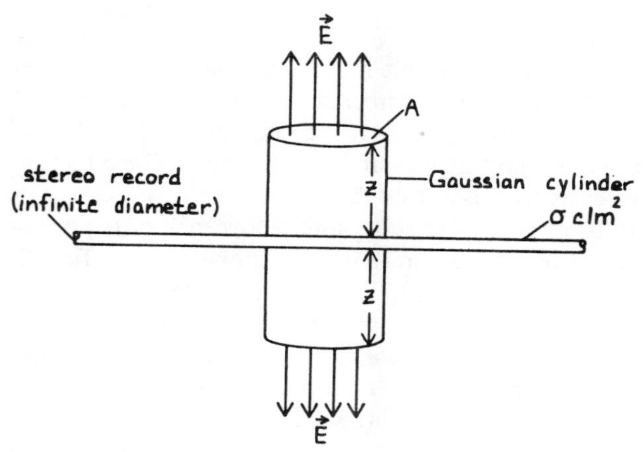

Figure 24.6 Problem 3. The use of Gauss' Law to calculate the field near an infinite, uniformly charged disk.

The flux through the Gaussian surface is

$$\oint \mathbf{E} \cdot d\mathbf{s} = EA + EA = 2EA$$

and the total enclosed charge is σA. (Here we are really assuming that the record is thin and that only the top surface is charged. From the wording of the problem, we should probably assume that the total accumulated static charge of 6×10^{-10} C is distributed evenly over *both* top and bottom of the record—the result will be the same.) Then from Gauss' Law, $2EA = \sigma A / \varepsilon_0$ or $E_z = \sigma / 2\varepsilon_0$. When we substitute in our value of 8.26×10^{-9} C/m² for σ, we obtain the value of $E_z = 467$ N/C, so that our answer, in assuming an infinite record rather than a 12-inch-diameter record, is in error by 45%. This error depends on the ratio of z to the record diameter, R, so that for distances closer to the record, the error becomes less.

4. *Guided Problem.* Calculate the electric field everywhere in space caused by two infinite sheets of charge, one lying in the x–y plane and having a charge density $\sigma_1 = 2 \times 10^{-9}$ C/m². The second plane is at an angle of 30° with respect to the x–y plane and carries a charge density of -1×10^{-9} C/m². The two sheets of charge are indicated schematically in Figure 24.7.

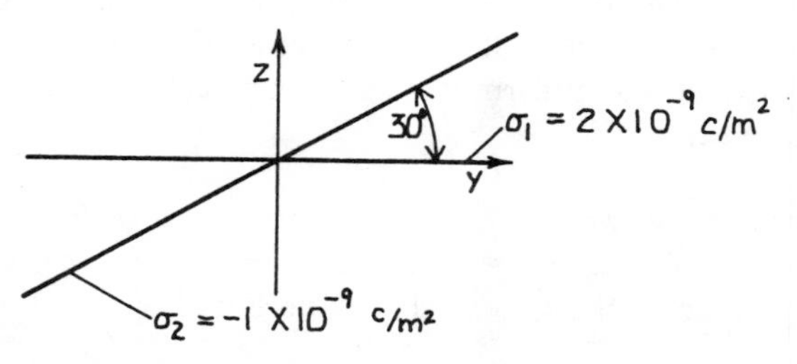

Figure 24.7 Problem 4. Two infinite sheets of charge at an angle of 30° relative to each other.

Solution scheme

a. Referring back to problem 3, we see that the field caused by an infinite charged sheet is directed out perpendicular to the sheet and is given by $E = \sigma/2\varepsilon_0$, so that the magnitude of the field does not decrease as we go farther out from the sheet. The direction of the field reverses as we pass through the sheet.

b. Find the electric field above and below each of the two sheets resulting from the charge on that sheet alone; that is, find the field contribution resulting from each of the sheets.

c. Use the principle of superposition to find the resultant field in each of the four distinct regions of space defined by the intersection of the two sheets. (In each region of space add the field components resulting from the separate sheets.)

5. *Worked Problem.* A point charge q is positioned at the center of an uncharged, hollow aluminum sphere of inner radius a and outer radius b. Sketch the lines of electric flux everywhere, and graph the electric field E versus r, the distance from the charge q. How much charge, if any, is on the inner surface of the sphere at $r = a$? on the outer surface at $r = b$? Explain your answers.

Solution: We can assume that the aluminum sphere, a conductor, is in static equilibrium, so that for all points inside the conductor the electric field is equal to zero. The field lines that start on the charge q must therefore end on the inner surface of the sphere, at $r = a$, so that there must be a negative charge of $-q$ distributed uniformly on the inner surface of the sphere. But because the sphere is uncharged, the outer surface must now carry a net charge q, so that lines of flux must now start again on the outer surface of the sphere and extend outward to infinity. Figure 24.8(a) shows the electric field lines and the induced charges.

To deduce the field lines shown in the figure, we apply Gauss' Law, in turn, to spheres of radii $r < a$, $a < r < b$, and $r > b$. For $r < a$, Gauss' Law gives the result that $E(4\pi r^2) = q/\varepsilon_0$, so that $E = q/(4\pi\varepsilon_0 r^2)$. For $a < r < b$, we know that $E = 0$, and therefore $Q = 0$. This implies that the charge induced on the inner surface of the sphere must be $-q$, giving a total charge enclosed by the Gaussian surface of $(q - q) = 0$. For $r > b$, we know that the total charge enclosed by the Gaussian surface is q, and therefore that $E = q/(4\pi\varepsilon_0 r^2)$. The field E is shown plotted versus r in Figure 24.8(b). The induced charge on the inner surface of the sphere is $-q$, leaving a net charge of $+q$ on the outer surface. This must be the amount of charge required to cancel exactly the external field set up inside the conductor by the charge q at $r = 0$. It does not matter whether the charge q is at $r = 0$ or at $r = b$ (and distributed uniformly over the surface); the electric field outside $r = b$ is the same.

6. *Guided Problem.* Rework problem 5 for the situation in which, in addition to the charge q placed at $r = 0$, a charge $-Q$ is uniformly distributed over the outer surface of the sphere. The sphere is shown in Figure 24.9.

Solution scheme

a. Again assume that the aluminum sphere is in static equilibrium. Do you know the electric field inside the conductor?

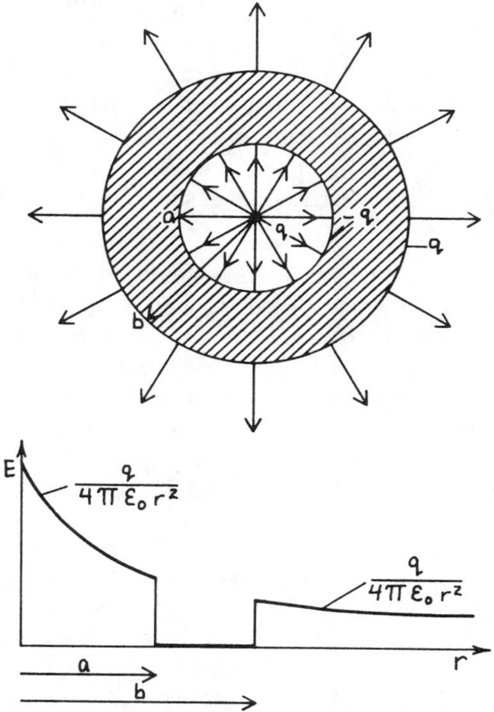

Figure 24.8 Problem 5. (a) Electric field lines and induced charges resulting from a point charge q at the center of an uncharged, hollow aluminum sphere. (b) The electric field E versus r, the distance from the charge q, for the charge in (a).

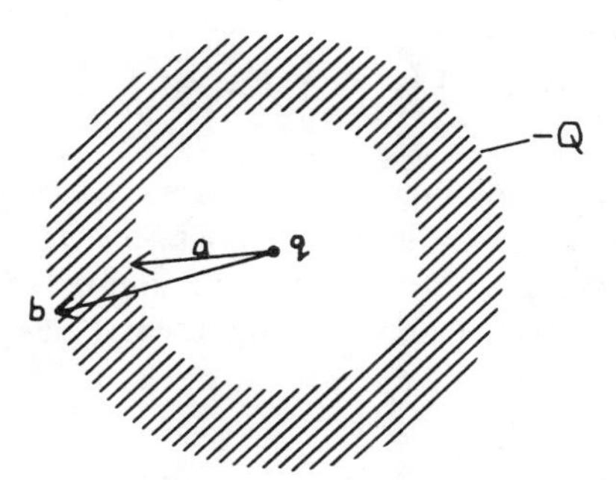

Figure 24.9 Problem 6. The charge q is inside the hollow aluminum sphere. The charge $-Q$ is distributed uniformly over the surface of the sphere.

b. Apply Gauss' Law, in turn, to spheres of radii $r < a$, $a < r < b$, and $r > b$. This will determine the electric field for $r < a$ and for $r > b$. Because you already know the electric field for $a < r < b$, use Gauss' Law in this region to tell you what the charge on the inner surface of the sphere at $r = a$ must be.

c. Knowing what the induced charge on the inner surface of the sphere is, you can now calculate the total charge on the *outer* surface of the sphere at $r = b$.

d. You should now be able to graph the electric field E versus r.

7. *Worked Problem.* (a) A small rubber balloon of diameter 8 cm has a charge of 1.2×10^{-8} C distributed uniformly over its surface. It is placed with its center at $x = y = z = 0$ in a region of space where the electric field is $80\hat{\mathbf{x}}$ N/C. What is the net electric field inside the balloon? At $x = 50$ cm? At $x = -50$ cm? (b) The rubber balloon is replaced with a thin aluminum sphere, also of diameter 8 cm, and also with a charge of 1.2×10^{-6} C on its surface. What is the net electric field inside the aluminum sphere?

Solution

a. This problem is solved by using Gauss' Law and the principle of superposition. The total electric field is given by the sum of the existing field, $80\hat{\mathbf{x}}$, and the field due to the charge on the balloon. *Inside* the balloon, the field due to the charge on the *surface* of the balloon is zero, so that the total field is just $80\hat{\mathbf{x}}$ N/C. *Outside* the balloon, the field due to the balloon is given by $\mathbf{E} = Q\hat{\mathbf{r}}/(4\pi\varepsilon_0 r^2) = 108\hat{\mathbf{r}}/r^2$ N/C, so that at $x = 50$ cm (0.5 m), $\mathbf{E}_{\text{balloon}} = 432\hat{\mathbf{x}}$ N/C and at $x = -50$ cm, $\mathbf{E}_{\text{balloon}} = -432\hat{\mathbf{x}}$ N/C. The total field at $x = 50$ cm is then $512\hat{\mathbf{x}}$ N/C, and at $x = -50$ cm it is $-352\hat{\mathbf{x}}$ N/C.

b. Inside the aluminum sphere, the total electric field must be *zero*; if it were not, there would be an electric field over the surface of the sphere, and this would cause the surface electrons to move and redistribute themselves. When the sphere is first placed in the electric field, the charges on the surface redistribute themselves so as to cancel out the external field inside the sphere; that is, there are more positive charges on the $+x$ side of the sphere than on the $-x$ side. Note that this is true whether or not the sphere has any initial charge on it.

8. *Guided Problem.* Two long, concentric metal cylinders have radii R_1 and R_2 m, with $R_1 < R_2$. The inner cylinder has a uniform charge density of ρ C/m^2, and the outer cylinder $-\rho$ C/m^2. Find the electric field at (a) $r < R_1$, (b) $R_1 < r < R_2$, and (c) $r > R_2$.

Solution scheme

a. We draw end views of the two cylinders, as shown in Figure 24.10. We will need three Gaussian surfaces, one for $r < R_1$, one for $R_1 < r < R_2$, and one for $r > R_2$. Remember that each surface is a cylinder of length L. Can you write down the expression for the area of a cylinder?

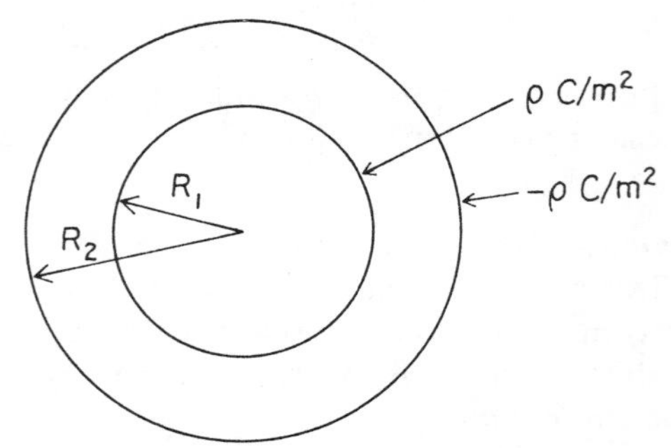

Figure 24.10 Problem 7. An end view of two concentric, charged cylinders of radii R_1 and R_2. Cylinder R_1 has charge density ρ C/m^2, and cylinder R_2 charge density $-\rho$ C/m^2.

b. Apply Gauss' Law to each of the three surfaces in turn, remembering that since the charge density is given in terms of C/m^2, you must calculate the surface area of the cylindrical surfaces inside your respective Gaussian surfaces. What is the symmetry of the problem?
c. Can you now evaluate the field for all three regions? Be careful in the case $r > R_2$. Cylinder R_2 has a larger area than cylinder R_1. What does this say about the net charge inside your third Gaussian surface?

9. *Worked Problem.* Positive charge of λ C/m is uniformly distributed over the volume of a long cylinder of radius R. A cylindrical cavity of radius $R/2$ is cut out of the solid cylinder, the center of gravity of the cavity being at a distance $R/2$ from the center of the original solid cylinder; the cut-out material and its charge are discarded. Find the electric field at the point P at a distance r from the center of the original cylinder. What new electric field does the cylinder with the cavity produce at the point P? The cylinder with the cavity is shown in Figure 24.11.

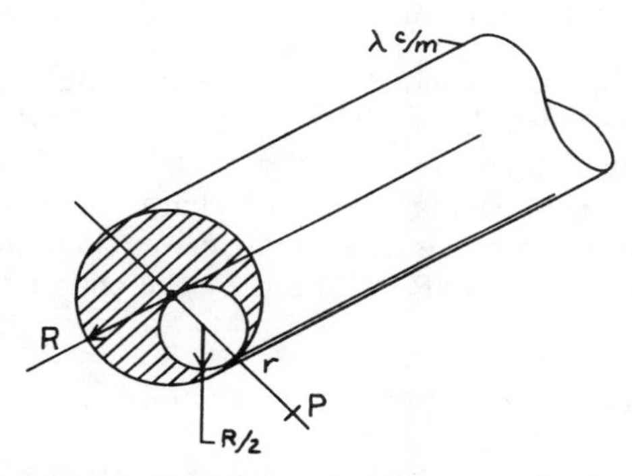

Figure 24.11 Problem 9. A long, uniformly charged cylinder of radius R. A cylindrical cavity has been cut out of the solid cylinder.

Solution: This apparently complex problem can be handled easily if we use Gauss' Law and the principle of superposition. To find the field at a distance r from the original cylinder, we draw a Gaussian cylinder of length L and radius r centered on the original cylinder, as shown in Figure 24.12. Point P lies on the Gaussian cylinder. From symmetry we expect that the electric field is directed outward from the cylinder and so is everywhere normal to the side of the Gaussian cylinder. The magnitude of the field should be the same at all points a distance r from the center of the solid cylinder. Then, applying Gauss' Law,

$$\oint \mathbf{E} \cdot d\mathbf{s} = E \oint ds = E(2\pi rL)$$

For the ends of the cylinder, $\mathbf{E} \cdot d\mathbf{s} = 0$, so that

$$2\pi rLE = \lambda L / \varepsilon_0$$
$$\mathbf{E} = \lambda \hat{\mathbf{r}} / 2\pi \varepsilon_0 r$$

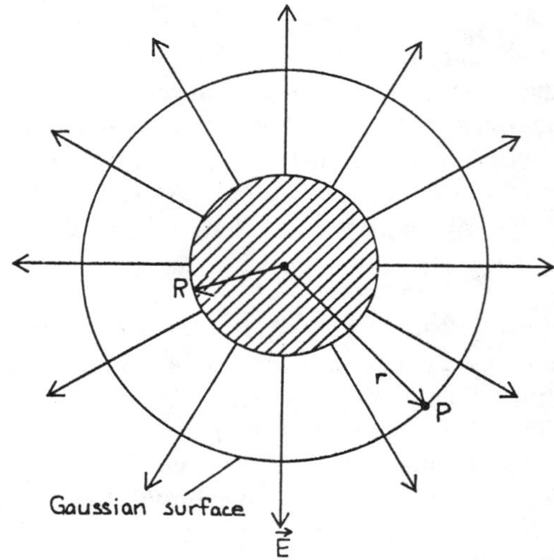

Figure 24.12 Problem 9. An end view of a long, solid, charged cylinder of radius R. P is a point on a Gaussian surface of radius r.

To find the field that the cylinder with the cavity produces, we simply *add* the field at point P resulting from the original solid cylinder to the field produced at point P by a solid cylinder of radius $R/2$ positioned at the cavity. The small cylinder has the same charge density as the original cylinder had, but the charge is now negative. We therefore *cancel out* the effect on the field at point P produced by the charge contained in the cavity.

$$[\text{Volume of cavity/original volume of cylinder}] = (R/2)^2/(R^2) = 1/4$$

The small cylinder then must have a charge of $-\lambda/4$ C/m. To find the field contribution at P caused by our small, negatively charged cylinder, we once again apply Gauss' Law using a Gaussian cylinder of radius r', as indicated in Figure 24.13. The field on the Gaussian surface is

$$\mathbf{E}' = \frac{\lambda/4\hat{\mathbf{r}}'}{2\pi\varepsilon_0(r - R/2)}$$

Because point P is on a straight line passing through the centers of both cylinders, at point P the field $\mathbf{E}$ and the field $\mathbf{E}'$ resulting from the small cylinder are antiparallel.

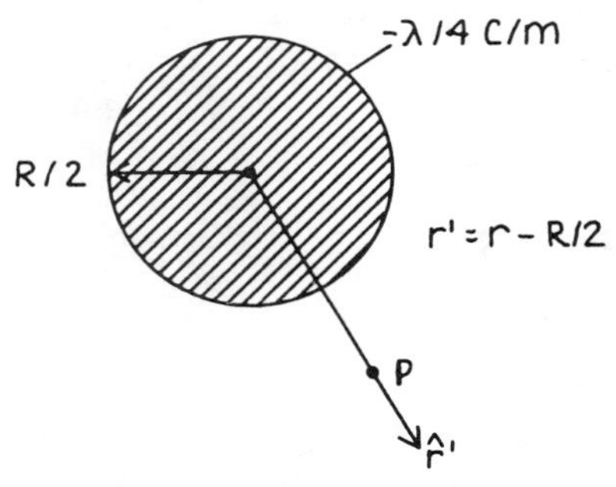

Figure 24.13 Problem 9. P is the same point as in Figure 24.12, but here the Gaussian surface is centered on the small cylinder of radius $R/2$.

The field at point P, obtained from $\mathbf{E} + \mathbf{E}'$, is then

$$\mathbf{E} + \mathbf{E}' = \frac{\lambda \hat{\mathbf{r}}}{2\pi\varepsilon_0}\left[\frac{1}{r} - \frac{1/4}{(r - R/2)}\right]$$
$$= \frac{\lambda \hat{\mathbf{r}}}{2\pi\varepsilon_0}\left[\frac{1}{r} - \frac{1}{(4r - 2R)}\right]$$

10. *Guided Problem*

a. A charge Q is placed on a large, flat, metal plate of area A. Find the charge density on the plate and the electric field both above and below the plate.

b. A second, identical large plate is placed close by and parallel to the first plate. Charge $-Q$ is placed on the second plate. Find the charge densities on the upper and the lower faces of both plates and the electric fields in the regions above, below, and between the plates.

Solution scheme

a. We know that the charge will distribute itself on the surface only. For the single isolated plate, do we expect a uniform distribution on both the top and the bottom of the plate, that is, over the whole surface? Perhaps the easiest way to answer this question is to assume that the charge does *not* distribute itself uniformly. Won't this result in electric fields in the metal, which will immediately cause a redistribution of charge? Can you now calculate the charge density on the plate surface?

b. To find the electric field above and below the plate, draw as a Gaussian surface a small cylinder with one end just inside the surface of the plate. Only one cylinder face can contribute to $\oint \mathbf{E} \cdot d\mathbf{s}$. Which one is it? The Gaussian surface is shown in Figure 24.14. On the basis of this information, can you now evaluate the electric field E both above and below the plate?

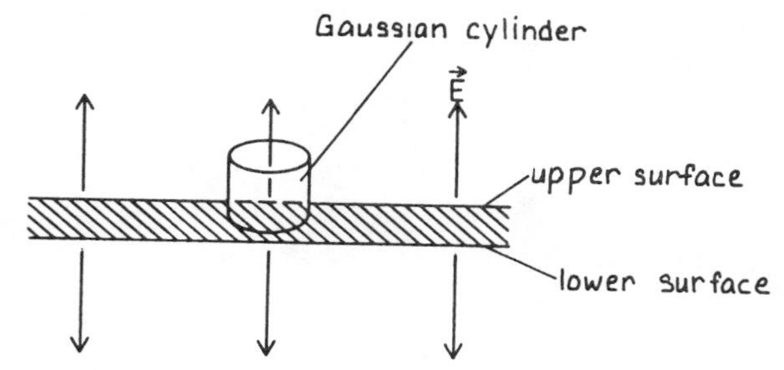

Figure 24.14 Problem 10. A small Gaussian cylinder with one end just inside the surface of a charged metal plate.

c. To answer part b of this problem, note that each plate is in the electric field caused by the charge on the other plate; the charges on each plate will move so as to cancel the external electric field set up inside the plate by the charges on the other plate. The free charges will all be on the lower face of the upper plate and on the upper face of the lower plate when both plates are in static equilibrium. Again, draw small cylinders with one end just inside the surface of the plate; do this for the region above both plates, below both plates, and between the plates. Is there any field at all above and below the plates? This configuration of two overlapping plates is called a *parallel plate capacitor.*

Answers to Guided Problems

2. a. 7.36 lines
   b. 6.37 lines
   Yes, the Earth's surface is normally negatively charged; this charge corresponds to about 8.3×10^9 electrons per square meter of area.
4. **Label the regions of space 1, 2, 3, and 4, as shown in Figure 24.15. The fields in the respective regions are as follows:**

$$\begin{aligned} 1:&\quad 162\hat{\mathbf{z}} - 28.2\hat{\mathbf{y}}\ \text{N/C} \\ 2:&\quad 64.1\hat{\mathbf{z}} + 28.2\hat{\mathbf{y}}\ \text{N/C} \\ 3:&\quad -162\hat{\mathbf{z}} + 28.2\hat{\mathbf{y}}\ \text{N/C} \\ 4:&\quad -64.1\hat{\mathbf{z}} - 28.2\hat{\mathbf{y}}\ \text{N/C} \end{aligned}$$

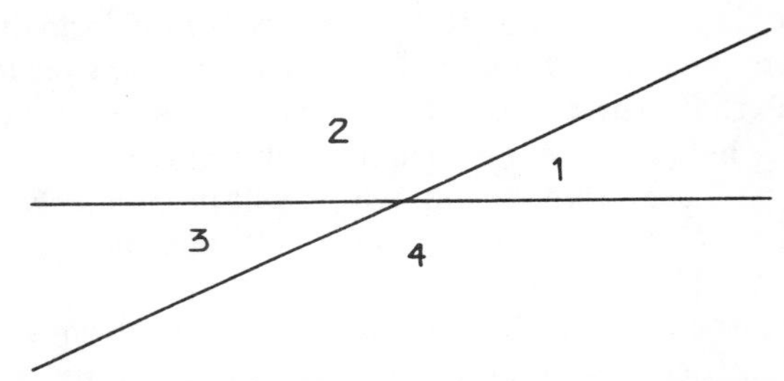

Figure 24.15 Problem 4. The four regions of space defined by two infinite, intersecting charged planes.

6. For $r < a$: $\quad \mathbf{E} = \dfrac{q\hat{\mathbf{r}}}{4\pi\varepsilon_0 r^2}$

   For $a < r < b$: $\quad \mathbf{E} = 0$

   For $r > b$: $\quad \mathbf{E} = \dfrac{(q - Q)\hat{\mathbf{r}}}{4\pi\varepsilon_0 r^2}$

8. For $r < R_1$: $\quad \mathbf{E} = 0$

   For $R_1 < r < R_2$: $\quad \mathbf{E} = \dfrac{\rho R_1 \hat{\mathbf{r}}}{\varepsilon_0 r}$

   For $r > R_2$: $\quad \mathbf{E} = \dfrac{-\rho[R_2 - R_1]\hat{\mathbf{r}}}{\varepsilon_0 r}$

10. a. The surface charge density is $Q/2A$, where A is the area of either the top or bottom surface of the plate. The field is $E = \sigma/\varepsilon_0$ and is directed away from the plate. The field is indicated schematically in Figure 24.16(a).
    b. The field is zero above and below the plates; in the region between the plates the field is given by $E = \sigma/\varepsilon_0$. The field is shown in Figure 24.16(b).

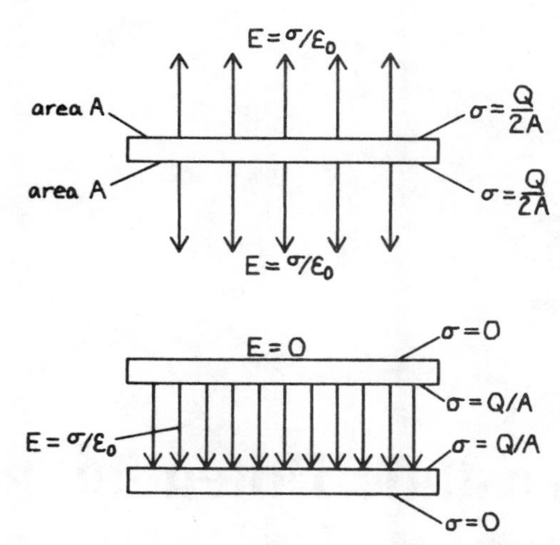

Figure 24.16 Problem 10. Surface charge densities and electric fields.

Chapter 25

The Electrostatic Potential

Overview

In this chapter we show that the electrostatic force, like gravitational force, is conservative. and we define the concept of electric potential energy of a charge in an electric field. The potential-energy concept is useful in the calculation of the motion of charged particles in an electric field. We define electrostatic potential as the potential energy per unit charge, and we evaluate it for several different charge distributions. We then show how the electric field can be calculated by differentiation of the potential function. In an optimal section we prove the mean-value theorem, an elegant theorem used for the calculation of electrostatic potential in a region free of electric charge.

Essential Terms

conservative force
electrostatic potential
electron volt
gradient of the potential
equipotential surface

Key Concepts

1. *Conservative force*. Let a point charge q move from A to B in the field of a charge Q along the path shown in Figure 25.1(a). The work done on q by the electrostatic force exerted by charge Q is $W_{AB} = \int_A^B \mathbf{F} \cdot d\mathbf{l}$. We might expect that W_{AB}, the value of the line integral, would depend on the path between A and B, but we can show that W_{AB} depends *only* on r_A and r_B, *not* on the shape of the path that connects A and B. As we have defined it (in Chapter 8), a conservative force is one in which the work done by the force

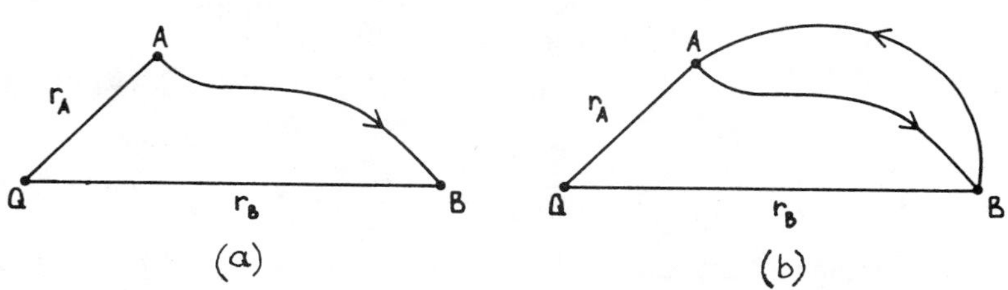

Figure 25.1 (a) A point charge moving from A to B in the field of the charge Q. (b) The point charge moves back from B to A along a different path.

depends only on the initial and the final positions of the particle. Thus, the electrostatic force is clearly a conservative force. Suppose now that the particle travels back to point A along a different path, as shown in Figure 25.1(b). The whole path, then, starts at A, follows some curve, and ends at A, so that $W_{AA} = \oint \mathbf{F} \cdot d\mathbf{l} = 0$, because the two end points are one and the same point. This is an alternative way of defining a conservative force: as a force for which the closed-line integral vanishes, so that the work must be independent of the path.

2. *Electrostatic potential.* The idea of potential energy is useful when we deal with conservative forces. In Figure 25.1(a) we define the work done by the electric force in moving the charge q from A to B, W_{AB}, as being equal to the decrease in the potential energy of q. Thus, $W_{AB} = U(A) - U(B)$, or, as it is usually written, $U(B) = U(A) - W_{AB}$. Here $U(A)$ and $U(B)$ are defined to be the potential energy of q at A and B, respectively. This is exactly how we define the change in potential energy of a particle of mass m falling through a height h in a gravitational field; in this case the gravitational force does work W_{AB}, equal to mgh on the particle. For the electrostatic case, we have $U(B) = U(A) = \int_A^B \mathbf{F} \cdot d\mathbf{l}$. If we divide each term by the charge q, the equation becomes $V(B) = V(A) - \int_A^B \mathbf{E} \cdot d\mathbf{l}$, where $V(B) = U(B)/q$ and is the potential energy of a unit charge at point B. $V(B)$ is called the *electrostatic potential* of point B. The units of electrostatic potential, joules/coulombs, are called *volts*, and it is important to remember that a volt is a unit of work per unit charge.

The single most important example of electrostatic potential is the potential resulting from a point charge Q. For simplicity we assume that A, B, and Q all lie on a straight line, and we choose a straight path from A to B, as shown in Figure 25.2. (Remember that because the electrostatic force is conservative, any path from A to B will give the same result.)

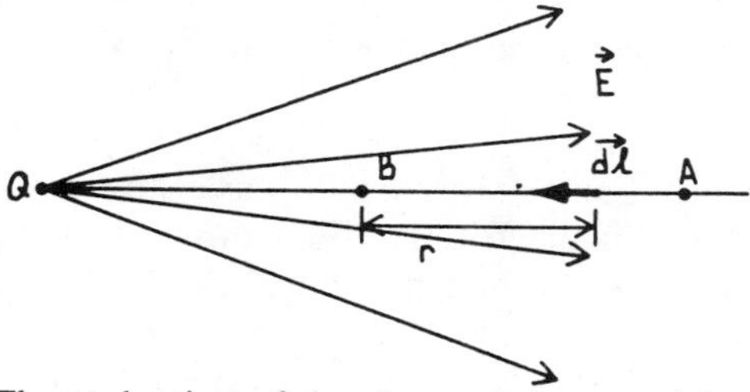

Figure 25.2 The evaluation of the electrostatic potential resulting from a point charge Q.

$$V(B) = V(A) - \int_A^B \mathbf{E} \cdot d\mathbf{l}$$

$$= V(A) + \int_A^B E dl$$

Because $\mathbf{E}$ and $d\mathbf{l}$ are antiparallel, $\mathbf{E} \cdot d\mathbf{l} = -Edl$. But $dl = -dr$, so that

$$V(B) = V(A) - \int_A^B E dr$$

Since

$$E = \frac{Q}{4\pi\varepsilon_0} \frac{1}{r^2}$$

$$V(B) = V(A) - \frac{Q}{4\pi\varepsilon_0} \int_A^B \frac{dr}{r^2}$$

or

$$V(B) = V(A) + \frac{Q}{4\pi\varepsilon_0} \left[\frac{1}{r_B} - \frac{1}{r_A} \right]$$

If we choose A to be an infinite distance from A and *arbitrarily* choose $V(A) = V(\infty) = 0$ (we can do this because we are really interested only in potential differences), then $V(B) = Q/4\pi\varepsilon_0 r_B$. For a distance r from a point charge Q, the potential is usually written as

$$V(r) = \frac{Q}{4\pi\varepsilon_0 r}$$

Here we have really defined the potential at point B a distance r from a charge Q to be equal to the work needed to bring a unit test charge from infinity to the point B, or $V(B) = -\int_\infty^B \mathbf{E} \cdot d\mathbf{l}$.

3. *Electron volt*. If a charge q is accelerated through a potential difference of 1 volt, the increase in energy of the charge is given by q (1 V), because 1 J/C of work will have been done on the charge q. If the charge q happens to have the fundamental electronic charge e, the increase of energy of the charge will be $e \times 1\ \text{V} = (1.6 \times 10^{-19}\ \text{C}) \times (1\ \text{V}) = 1.6 \times 10^{-19}$ J. This unit of energy is given the name *electron volt*, or eV, and 1 eV $= 1.6 \times 10^{-19}$ J.

4. *Gradient of the potential*. The electrostatic potential V is a scalar, or one-component, quantity, and so it is usually easier to calculate (for a given charge distribution) than the electric field $\mathbf{E}$, a three-component quantity. If we know the potential, we can obtain the electric field by differentiation. The potential difference between two points A and B in an electric field is given by $V(B) - V(A) = -\int_A^B \mathbf{E} \cdot d\mathbf{l}$. In Figure 25.3, points P_0 and P_1 are

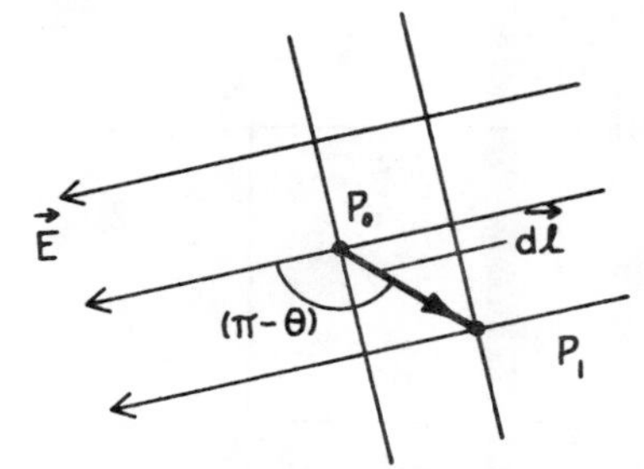

Figure 25.3 The relationship between electric field strength and change of potential.

separated by a *very small* displacement $d\mathbf{l}$. The change in potential between these points is then

$$dV = -\mathbf{E}\cdot d\mathbf{l} = -Edl\cos(\pi - \theta) = Edl\cos\theta$$

This can be rewritten as $E\cos\theta = dV/dl$. But $-E\cos\theta$ is the component of $\mathbf{E}$ in the direction of $d\mathbf{l}$, so that the component of $\mathbf{E}$ in the direction of $d\mathbf{l}$, E_l, can be written as the *partial derivative* of V with respect to l, or $E_l = -\partial V/\partial l$.

In general, the components of an electric field $\mathbf{E}$ can be written as

$$E_x = -\frac{\partial V}{\partial x} \quad E_y = -\frac{\partial V}{\partial y} \quad E_z = -\frac{\partial V}{\partial z}$$

In vector notation this is expressed as

$$\mathbf{E} = -\hat{\mathbf{x}}\frac{\partial V}{\partial x} - \hat{\mathbf{y}}\frac{\partial V}{\partial y} - \hat{\mathbf{z}}\frac{\partial V}{\partial z}$$

The quantity of the right side of the equation is called the *gradient of the potential*. If we know the potential, we can always obtain $\mathbf{E}$ by differentiation.

Sample Problems

1. *Worked Problem.* Find the potential difference between the origin and the point with coordinates (4,4) in the electric field $\mathbf{E} = 3x\hat{\mathbf{x}} + 4y\hat{\mathbf{y}}$. Is this electric field a conservative field? How much work is required to move a proton from the origin to the point (4,4) in the electric field? How much work is required to move an electron between the same two points?

Solution: The potential difference between points A and B is given by the expression $V(B) - V(A) = -\int_A^B \mathbf{E}\cdot d\mathbf{l}$, where A is the origin and B is the point with coordinates (4,4), as indicated in Figure 25.4. Since $d\mathbf{l} = \hat{\mathbf{x}}dx + \mathbf{y}dy$,

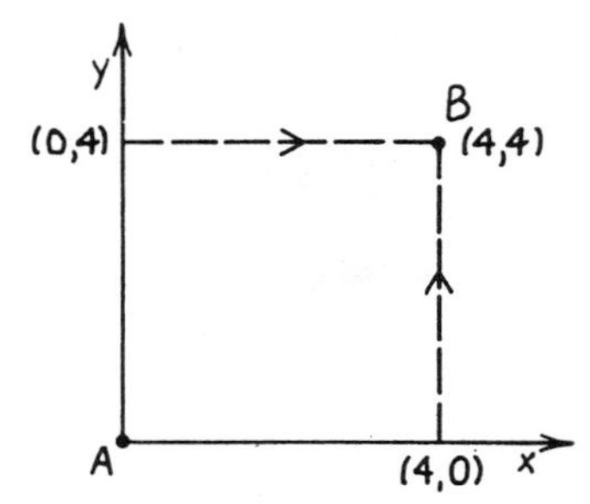

Figure 25.4 Problem 1. Two paths of integration between A and B.

$$-\int_A^B \mathbf{E} \cdot d\mathbf{l} = -\int (E_x dx + E_y dy)$$
$$= -\int (3x dx + 4y dy)$$

If we integrate from the origin to the point (4,0) and then from (4,0) to the point (4,4), we obtain the result that

$$-\int_A^B \mathbf{E} \cdot d\mathbf{l} = -\left[\int_{(0,0)}^{(4,0)} 3x dx + \int_{(4,0)}^{(4,4)} 4y dy\right]$$
$$= -[24 + 32] = -56 \text{ volts}$$

If we choose as an alternate path to integrate from the origin to the point (0,4) and then from (0.4) to (4.4), we obtain the result that

$$-\int_A^B \mathbf{E} \cdot d\mathbf{l} = -\left[\int_{(0,0)}^{(0,4)} 4y dy + \int_{(0,4)}^{(4,4)} 3x dx\right] = -[32 + 24] = -56 \text{ volts}$$

so that the result is independent of the path we follow. (If we choose some other path, such as from the origin to (2,0) and then from (2,0) to (2,4) and from (2,4) to (4,4), we obtain a potential difference of −56 volts.) Because the result is independent of the path followed, the electric field is conservative.

The work required to move a charge q from A to B is given by $q(V_B - V_A)$, where $(V_B - V_A) = -56$ V. Because for a proton $q = 1.6 \times 10^{-19}$ C, the work required is -9.0×10^{-18} J, which means that no external work is necessary to move the proton and that in fact the proton increases its kinetic energy by 9.0×10^{-18} J in moving from A to B. The work required to move an electron from A to B is $+9.0 \times 10^{-18}$ J, which means that external work *is* necessary and that the potential energy of the electron is increased in the process by 9×10^{-18} J.

2. *Guided Problem.* In the last chapter we found that the field a distance r away from a long line of charge of uniform charge density λ C/m is given by $\mathbf{E} = \lambda\hat{\mathbf{r}}/2\pi\varepsilon_0 r$ N/C. If $\lambda = 1 \times 10^{-7}$ C/m, what is the potential difference between a point A, 80 cm from the line of charge, and point B, 40 cm from the line of charge? The two points are shown in Figure 25.5. Which point is at the higher potential, and why?

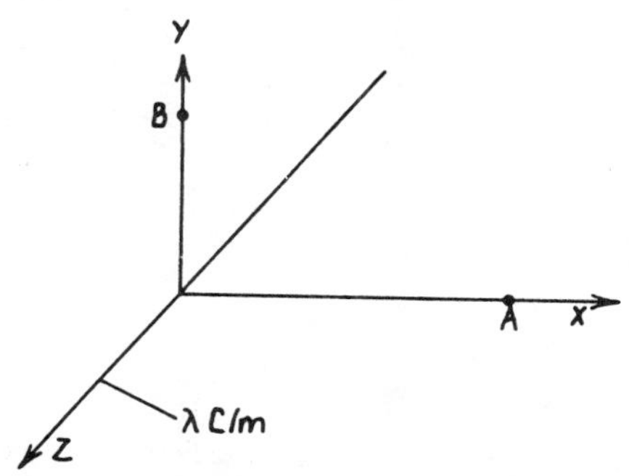

Figure 25.5 Problem 2. Two points, A and B, at respective distances of 80 cm and 40 cm from a line of charge at the origin and parallel to the z axis.

Solution scheme

a. The potential between any two points A and B is given by $V(B) - V(A) = -\int_A^B \mathbf{E} \cdot d\mathbf{l}$, and the result is independent of the path followed from A to B. Choose the simplest path, namely, the one along the x axis from A (80,0) to A' (40,0) and from there along a circular arc to B (0,40).

b. Along the x axis, because dl is antiparallel to $\hat{\mathbf{r}}$, $\hat{\mathbf{r}} \cdot d\mathbf{l} = -rdl$. But $dl = -dr$, so that $\hat{\mathbf{r}} \cdot d\mathbf{l} = dr$, and the integral $-\int \mathbf{E} \cdot d\mathbf{l}$ then can be easily evaluated.

c. Along the circular arc, because $d\mathbf{l}$ is perpendicular to $\hat{\mathbf{r}}$, $\hat{\mathbf{r}} \cdot d\mathbf{l} = 0$. Then

$$V(B) - V(A) = -\left[\int_A^{A'} \mathbf{E} \cdot d\mathbf{l} + \int_{A'}^{B} \mathbf{E} \cdot d\mathbf{l}\right] = -\int_A^{A'} \mathbf{E} \cdot d\mathbf{l}$$

Finally, $V(B) - V(A) = V(A') - V(A)$ can be found.

d. If your answer is positive, B is at a higher potential than A, which means that external work was needed to move a test charge from A to B. Your answer shows that all points at a constant distance r from the line of charge are at the same potential—no work was needed to move a test charge along the circular arc from A' to B.

3. *Worked Problem.* In Chapter 23 we found that the electric field resulting from a large, uniform sheet of charge with a uniform charge density of σ C/m^2 was given by $E = \sigma/2\varepsilon_0$. Each of two very large sheets of paper carries a uniform charge distribution, one of 2×10^{-6} C/m^2 and one of -2×10^{-6} C/m^2 (see problem 23.25 in the text, p. 606). The sheets intersect at an angle of 90° and are oriented so that they are each at an angle of 45° relative to the x and y axes, as shown in Figure 25.6.

a. Find the electric field at all points close to the sheets of paper.

b. Find the potential difference between points A, with coordinates (1,1) m, and B, with coordinates (−1,1) m.

c. If we choose the origin to have zero potential as a reference, what are the potentials of A and B? Why do we not choose infinity as our reference point of zero potential?

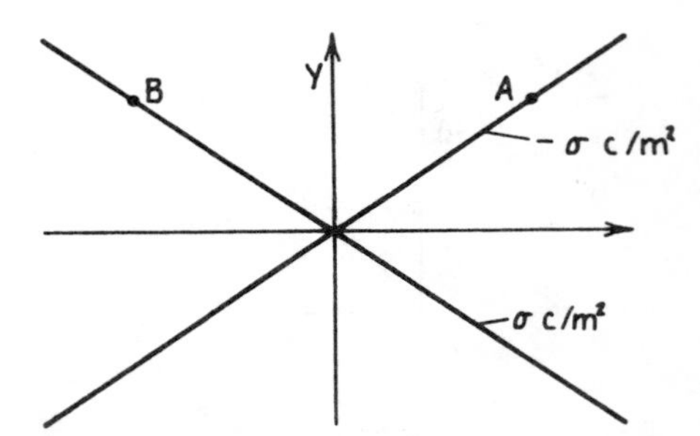

Figure 25.6 Problem 3. Two large, charged sheets of paper intersect at an angle of 90°. The sheets are at 45° with respect to the x and y axes.

Solution

a. The field resulting from the positively charged sheet is given by $E = \sigma/2\mathcal{E}_0$ and points directly away from the sheet; the field resulting from the negatively charged sheet is $E = -\sigma/2\mathcal{E}_0$ and points toward the sheet. From superposition we know that the field at any point is given by the vector sum of the two fields; the fields in the four regions of space defined by the intersection of the two sheets are shown in Figure 25.7. The magnitude of the field in each of the regions 1 through 4 is the same and is given by

$$\frac{\sqrt{2}\sigma}{2\mathcal{E}_0} = 1.6 \times 10^5 \text{ N/C}$$

The direction of the field is different in each of the four regions.

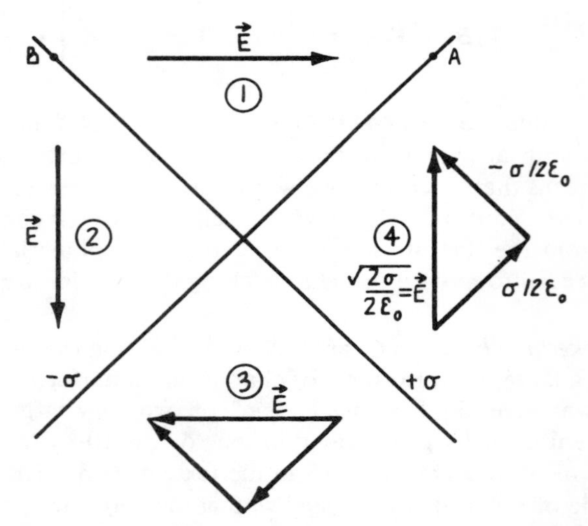

Figure 25.7 Problem 3. The electric fields resulting from the two intersecting, charged sheets of paper.

b. The potential difference $V(B) - V(A) = -\int_A^B \mathbf{E} \cdot d\mathbf{l}$ can be most easily evaluated by choosing the path directly from A to B, parallel to the x axis. Then

$$\int_A^B \mathbf{E} \cdot d\mathbf{l} = = -\int_A^B \frac{\sqrt{2}\sigma}{2\varepsilon_0}\, \hat{\mathbf{x}} \cdot d\mathbf{l}$$

$$\hat{\mathbf{x}} \cdot d\mathbf{l} = -dl = dx$$

$$-\int_A^B \mathbf{E} \cdot d\mathbf{l} = -1.6 \times 10^5 \int_{(1,1)}^{(-1,1)} dx$$

$$= 3.2\ (10^5)\ \text{V}$$

c. If we choose the origin as a reference point, then the values of the potential at points B and A, respectively, are given by

$$V(B) = -\int_{(0,0)}^{(-1,1)} \mathbf{E} \cdot d\mathbf{l} \quad V(A) = -\int_{(0,0)}^{(-1,1)} \mathbf{E} \cdot d\mathbf{l}$$

It is most convenient to choose a path from the origin upward along the y axis to the point (0,1), and then to follow the x axis (all in region 1). Along the y axis E is perpendicular to dl, so that $E \cdot dl = 0$. Then

$$V(B) = -\int_{(0,1)}^{(-1,1)} \mathbf{E} \cdot d\mathbf{l} = -1.6 \times 10^5 \int_{(0,1)}^{(-1,1)} dx$$

$$= 1.6 \times 10^5\ \text{V}$$

and

$$V(A) = -\int_{(0,1)}^{(1,1)} \mathbf{E} \cdot d\mathbf{l} = -1.6 \times 10^5 \int_{(0,1)}^{(1,1)} dx$$

$$= -1.6 \times 10^5\ \text{V}$$

The reason for not choosing infinity as a reference point of zero potential is that the field of an infinite sheet of charge does not decrease in magnitude as we move away from the sheet. Therefore, because our paper sheets are imagined to have infinite size, there would be an infinite potential difference between any finite point near the origin and our reference point at infinity.

4. *Guided Problem.* Consider two large, flat copper plates; the plates are parallel and 5 cm apart. Each plate has an area of 2 m^2; the upper plate has a charge of -1×10^{-6} C on it; the lower plate has an equal and opposite charge.

a. What is the potential difference between the plates?

b. A proton (mass = 1.67×10^{-27} kg) is shot along the negative y axis through a small hole in the upper plate toward the lower plate. The velocity of the proton as it passes through the hole is $-\hat{\mathbf{y}}\ (1 \times 10^6\ \text{m/s})$. With what speed will the proton hit the lower plate? The two plates are shown in Figure 25.8.

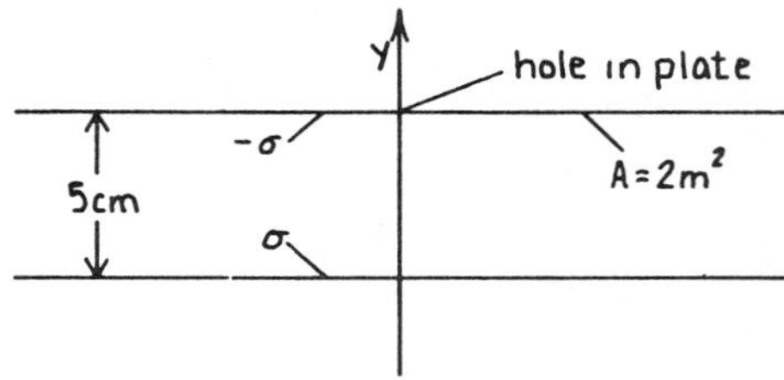

Figure 25.8 Problem 4. Two large, flat copper plates. The upper plate has a small hole in it.

Solution scheme

a. The electric field between the two large, parallel conducting plates is given by $\sigma/\varepsilon_0 = Q/A\varepsilon_0$. (See problem 24.10 of this guide.) The electric field is along the $+\hat{\mathbf{y}}$ direction, from the lower toward the upper plate. Can you calculate the potential difference between the plates by evaluating $V(B) - V(A) = -\int_A^B \mathbf{E} \cdot d\mathbf{l}$ where B is a point on one plate and A is a point on the other plate directly opposite to B? Note that here because $\mathbf{E}$ is constant, the magnitude of the potential difference is just $E(d)$, where d is the plate separation.

b. To calculate the speed of the proton when it hits the lower plate, we need to find the work done on the proton in passing through the region between the plates. This work equals the change in kinetic energy of the proton. Because the proton is moving antiparallel to the electric field, it will slow down. The decrease in kinetic energy is then given by qV, where q is the charge of the proton and V is the potential difference between the plates. Because the speed of the proton is much less than the speed of light, c, we can use $1/2\ mv^2$ to represent the kinetic energy of the proton.

c. Once we have calculated the kinetic energy of the proton when it hits the lower plate, we can calculate the speed from $KE = 1/2\ mv^2$.

5. *Worked Problem.* A thin aluminum sphere of radius 20 cm has a charge of 1.0×10^{-7} C uniformly distributed over its surface.

a. Assuming that the center of the sphere is at $r = 0$, calculate and sketch the electric field for all values of r.

b. Calculate and plot the electric potential, V, for all r, choosing as a reference point $V = 0$ for r = infinity.

c. How much work is needed to bring a proton from infinity to the surface of the sphere? Express your answer both in joules and in electron volts.

d. If a small hole is made in the sphere, how much extra work (from the surface of the sphere) is required to bring the proton in through the hole to the center of the sphere at $r = 0$?

Solution

a. From Gauss' Law we know that the electric field outside the sphere is given by $\mathbf{E} = \dfrac{Q}{4\pi\varepsilon_0}\dfrac{1}{r^2}\hat{\mathbf{r}}$. Because there is no charge inside the sphere, $\oint \mathbf{E} \cdot d\mathbf{s} = Q/\varepsilon_0 = 0$ for all Gaussian surfaces with $r \leqslant R$,

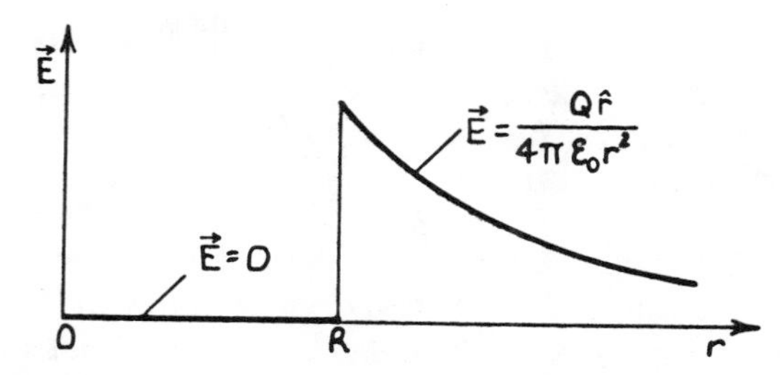

Figure 25.9 Problem 5. A plot of E versus r, where $r = 0$ is the center of a charged aluminum sphere.

where $R = 0.20$ m. Then $\mathbf{E} = 0$ for $r \leqslant R$. The plot of $\mathbf{E}$ versus r is shown in Figure 25.9.

b. For a point charge Q, we know that the potential V at a distance r from the point charge is given by $V = Q/4\pi\varepsilon_0 r$. Even though the sphere is not a point charge, the field for $r \geqslant R$ is *exactly* the same as the field for a point charge, namely, $\mathbf{E} = Q\hat{\mathbf{r}}/4\pi\varepsilon_0 r^2$. Since we obtained the potential for a point charge by the integration $V = \int_\infty^r \mathbf{E} \cdot d\mathbf{l}$ the potential for the uniformly charged sphere must be exactly the same, for $r \leqslant R$, as that of a point charge Q; that is, for points outside the sphere, it does not matter whether the charge is uniformly distributed over the sphere or all concentrated in a point charge at the origin (provided that the total charge Q is the same). For $r < R$ we need to extend the integration inward; that is,

$$V = -\left[\int_\infty^R \mathbf{E} \cdot d\mathbf{l} + \int_R^r \mathbf{E} \cdot d\mathbf{l}\right] = \frac{Q}{4\pi\varepsilon_0 R} - \int_R^r \mathbf{E} \cdot d\mathbf{l}$$

Since $\mathbf{E} = 0$ for $r \leqslant R$, however, the contribution of the second integral is zero, so that $V = Q/4\pi\varepsilon_0 R$ for all $r \leqslant R$. In other words, V is a constant for the region inside the sphere. (Because the field is zero, no further work is needed to move a unit test charge from the surface of the sphere to anywhere inside it.) The plot of V versus r is shown in Figure 25.10. For $r = 0.2$, $Q = 1.0 \times 10^{-7}$ C and $V = 4500$ volts.

c. To bring a proton from infinity to the surface of the sphere, the work required is just $(q)V$, where $V = Q/4\pi\varepsilon_0 R$ (since $V = 0$ at infinity) and where q = the charge of the proton $= 1.6 \times 10^{-19}$ C. Then work $= 1.6 \times 10^{-19}$ (4500) $= 7.2 \times 10^{-16}$ J, or 4500 eV.

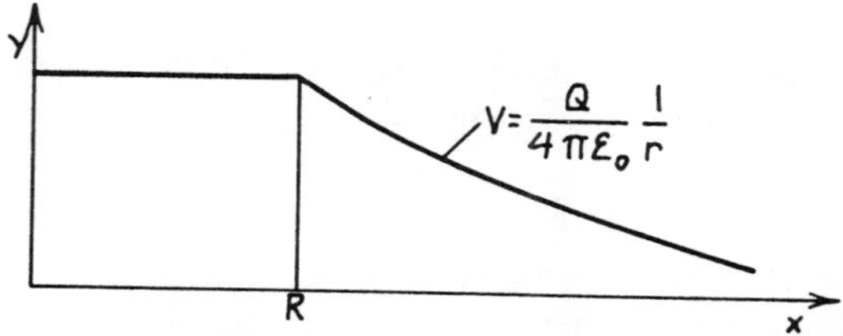

Figure 25.10 Problem 5. V versus r for the charged aluminum sphere.

d. Because the potential is constant inside the sphere (or, alternately, because the field is zero), no further work is needed to bring the proton in through the hole to the center of the sphere; that is, 4500 eV will bring a proton from infinity to the center of the sphere.

6. *Guided Problem.* Suppose that the nucleus of a uranium atom is a uniform charged sphere with a charge of $92e$ and a radius of 7.4×10^{-15} m. What are the values of the electric field and the electrical potential at (a) the surface of the nucleus and (b) the center of the nucleus?

Solution scheme

a. This problem is rather similar to worked problem 5, which involved a small, charged aluminum sphere. Again, for $r \geqslant R$, the field is exactly the same as for a point charge, namely $\mathbf{E} = Q\hat{\mathbf{r}}/4\pi\varepsilon_0 r^2$, and the potential is, similarly, that for a point charge, $V = Q/4\pi\varepsilon_0 r$. Can you now find the electric field and the potential for $r = R$?
b. For $r > R$, we need to calculate $\mathbf{E}$ using Gauss' Law. Here $\mathbf{E}$ is *not* 0 for $r < R$, since the charge extends all the way into $r = 0$. By drawing a Gaussian surface and applying Gauss' Law, you should easily find that for $r < R$, $\mathbf{E} = Qr\hat{\mathbf{r}}/4\pi\varepsilon_0 R^3$, so that $\mathbf{E}$ decreases linearly down to $\mathbf{E} = 0$ at $r = 0$.
c. To find the potential at $r = 0$, we need to extend the integration $V = \int \mathbf{E} \cdot d\mathbf{l}$ inward from $r = R$ to $r = 0$, just as we did in problem 5, part b. That is, $V_{r<R} = V_R - \int_R^r \mathbf{E} \cdot d\mathbf{l}$ where the expression for $\mathbf{E}$ is given in part b, above. Qualitatively, V must increase as we move a positive test charge inward from $r = R$, since we are doing additional work against an electric field, so you will find that $V_{r=0} > V_{r=R}$. Can you now deduce the expression for $V_{r=0}$? Now find the explicit value of $V_{r=0}$.

7. *Worked Problem.* An alpha particle (charge $2e$, mass 6.68×10^{-27} kg) is shot directly toward a stationary uranium nucleus (charge $92e$, $R = 7.4 \times 10^{-15}$ m), at a speed of 2.1×10^7 m/s. Will the alpha particle make contact with the nuclear surface? If not, how close will it get to the nucleus? What speed would the alpha particle need to just make contact with the nuclear surface?

Solution

In this problem, the total energy of the alpha particle, namely, the sum of the electrical potential energy and the kinetic energy, $U + K$, is conserved. The total energy is just the initial kinetic energy, given by

$$K_0 = (1/2mv^2 = (1/2)(6.68 \times 10^{-27})(2.1 \times 10^7)^2 = 1.473 \times 10^{-12} \text{ J}$$

The potential energy, U, is given by $qV = (2e)V$, where, if we assume that the alpha particle does not penetrate the nucleus (that is, $r \geqslant R$), then

$$V = Q/4\pi\varepsilon_0 r$$

The alpha particle will slow down as it approaches the nucleus because of its increasing potential energy until all the kinetic energy is in the form of potential energy. At this point, U = initial kinetic energy = 1.473 $\times 10^{-12}$ J = $(2e)(92e)/4\pi\varepsilon_0 r$, so that $r = 2.88 \times 10^{-14}$ m. Clearly, the alpha particle does not reach the nuclear surface. For it to reach the nuclear surface, the initial kinetic energy must equal the electrical potential energy at the nuclear surface, $U_R = (2e)(92e)/4\pi\varepsilon_0 R$, where $R = 7.4 \times 10^{-15}$ m. Therefore, $U_R = 5.73 \times 10^{-12}$ J, and since $U_R = (1/2)mv^2$, then $v = 4.14 \times 10^7$ m/s, or more than twice the speed that it had.

8. *Guided Problem.* The length of one edge of a cube is 10 cm. All the eight corners of the cube are occupied by charges of 1×10^{-8} C. Find the electric potential V and the field **E** at the exact center of the cube.

Solution scheme

a. To visualize the problem more easily, draw a large cube and locate the cube center, P, in your diagram. Calculate, in terms of the edge length, the distance from any edge to the center.
b. Use the principle of superposition (the potential at P resulting from several charges is the sum of the potentials resulting from each charge) to calculate the potential at P. The potential resulting from a single point charge Q is given by $V = Q/4\pi\varepsilon_0 r$.
c. You can also use the principle of superposition to find the field **E**. If your sketch is drawn fairly carefully, you will see that the field contribution from a given corner of the cube is exactly canceled by the field resulting from the charge on the body diagonal corner of the cube, so that the total electric field at P must be exactly zero.

9. *Worked Problem.* An electron is in a stable circular orbit of radius r about a proton. The external work needed to bring the electron in from infinity to a distance r from the proton is called the potential energy of the electron. Find, in terms of r and e (the charge of the electron), the potential energy of the electron, U, and also the kinetic energy of the electron in its orbit (KE). Find the ratio U/KE and the total energy $U + KE$ of the electron. Is the total energy positive or negative? If your answer is negative, explain the meaning of a "negative energy."

Solution: The electron and proton are shown schematically in Figure 25.11. The potential at a distance r away from the proton is given by $V = e/(4\pi\varepsilon_0 r)$. The work needed to bring the electron in from infinity is

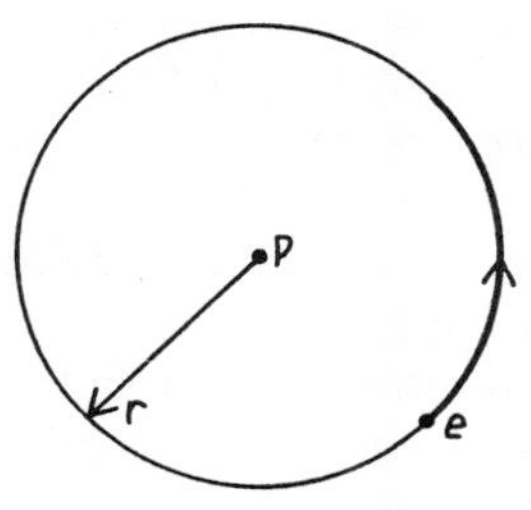

Figure 25.11 Problem 9. An electron in an orbit of radius r about a proton.

then $W = qV = -eV = -e^2/(4\pi\varepsilon_0 r)$. The potential energy of the electron is then

$$U = -e^2/4\pi\varepsilon_0 r$$

The fact that U is negative means that the electron is "bound" to the proton, so that external work is needed to remove it from the proton.

For any particle in a stable orbit in circular motion, the required centripetal force is mv^2/r. This centripetal force is supplied by the Coulomb force of attraction between the proton and the electron, $e^2/4\pi\varepsilon_0 r^2$. Then $mv^2/r = e^2/4\pi\varepsilon_0 r^2$; this can be rewritten as $mv^2 = e^2/4\pi\varepsilon_0 r = -U$. Then $\frac{1}{2}mv^2 = KE = \frac{1}{2}|U|$; that is, the magnitude of the kinetic energy is one-half that of the potential energy. The total energy of the electron is the following:

$$U + KE = \frac{-e^2}{4\pi\varepsilon_0 r} + \frac{1}{2}\frac{e^2}{4\pi\varepsilon_0 r} = -\frac{1}{2}\frac{e^2}{4\pi\varepsilon_0 r}$$

The meaning of the "negative energy" is that the electron is bound in a stable orbit to the proton and that external work must be applied to the electron if we wish to remove it from the proton.

10. *Guided Problem*. A large aluminum sphere ($R = 1.0$ m) is temporarily connected by a long, thin, insulated copper wire to a small aluminum sphere ($r = 1.0$ cm). A charge of $Q_T = 1 \times 10^{-7}$ C is placed onto the system and is shared between the two spheres. The wire is then removed. Find the charges Q and q on the large and small spheres, respectively, and their potentials in volts. What are the field strengths at the surfaces of the large and small spheres? If we reconnect the thin wire, increase the total charge Q_T on the system to 2×10^{-7} C, and then remove the wire, will electrical breakdown in the air adjacent to the spheres occur? Dry air can sustain a maximum electrical field strength of 3×10^6 V/m.

Solution scheme

a. Draw a sketch of the two spheres. The statement that the wire is *long* means that we can treat the spheres as separate bodies, that the field of one sphere does not significantly affect the field of the other. Will the spheres have a common potential, V?
b. Write down the expression for the potential V at the surface of each sphere in terms of the charge on the sphere. If V is the same for each sphere, you can equate the two expressions, giving you the ratio of the charges on each sphere in terms of their radii. Remember that you know the total charge Q_T. Can you now obtain exactly Q and q?
c. Knowing the charge on each sphere, can you find the potential at the surface of each sphere?
d. Now calculate the surface charge densities, σ_1 and σ_2, on the two spheres. You will notice that they are very different!
e. For any conductor the field strength at the surface is given by $E = \sigma/\varepsilon_0$. Calculate the field strengths at the surfaces of the two spheres.

f. Now suppose that the total charge $Q_T = 2 \times 10^{-7}$ C; that is, it is doubled. Can you reevaluate the field strengths at the surface of the two spheres and determine if the maximum allowable field strength is exceeded?

11. *Worked Problem.* Two large point charges of 6×10^{-3} C and -3×10^{-3} C are 4 m apart. Where, along the line that passes through both charges, is

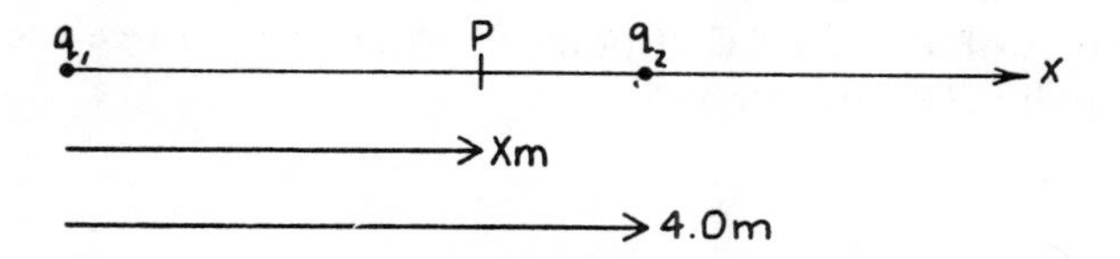

Figure 25.12 Problem 11. Two point charges 4.0 m apart. $q_1 = 6 \times 10^{-3}$ C and $q_2 = -3 \times 10^{-3}$ C.

the electric potential V zero? (Exclude infinity as a solution.) Calculate the electric field **E** at this point using two different methods.

Solution: The two charges are indicated in Figure 25.12. Let the potential at point P equal zero. But, using superposition, the potential at P is given by

$$V = \frac{1}{4\pi\varepsilon_0}\left[\frac{q_1}{x} + \frac{q_2}{4 - x}\right]$$

$$\frac{q_1}{x} + \frac{q_2}{4 - x} = 0$$

$$\frac{6}{x} - \frac{3}{4 - x} = 0$$

$$x = 2.67 \text{ m}$$

But $E_x = -\partial V/\partial x$, where $V = \frac{1}{4\pi\varepsilon_0}\left[\frac{q_1}{x} + \frac{q_2}{4 - x}\right]$. Clearly, since $V = V(x)$ only, $\partial V/\partial y = \partial V/\partial z = 0$, so that E does not have y or z components. Differentiating the expression for V, we obtain

$$E_x = \frac{1}{4\pi\varepsilon_0}\left[-\frac{q_1}{x^2} + \frac{q_2}{(4 - x)^2}\right]$$

Substituting $q_1 = 6(10^{-3})$ C, $q_2 = -3 \times 10^{-3}$ C, and $x = 2.67$ m, we obtain $E_x = 2.29(10^7)$ N/C, or $\mathbf{E} = 2.29(10^7)\hat{\mathbf{x}}$ N/C. We can also obtain **E** by calculating the field components at P resulting from the two charges q_1 and q_2 and then adding the components vectorially. Then

$$\mathbf{E} = \frac{1}{4\pi\varepsilon_0}\left[\frac{q_1}{x^2}\hat{\mathbf{x}} - \frac{q_2\hat{\mathbf{x}}}{(4 - x)^2}\right] = \frac{1}{4\pi\varepsilon_0}\left[\frac{6(10^{-3})\hat{\mathbf{x}}}{(2.67)^2} + \frac{3(10^{-3})\hat{\mathbf{x}}}{(1.35)^2}\right]$$

giving

$$\mathbf{E} = 2.29(10^7)\hat{\mathbf{x}} \text{ N/C}$$

12. *Guided Problem.* A dipole is aligned parallel to the y axis, as shown in Figure 25.13. Find the potential at a point P on the y axis, where $y \gg a$. Find by differentiation the electric field $\mathbf{E}$ at point P. If the dipole happens to be a water molecule with a dipole moment of $p = 6.2 \times 10^{-30}$ C • m and point P is 0.6 nanometers (1 nm $= 10^{-9}$ m) from the center of the dipole, find the values of V and of $\mathbf{E}$ at point P. What is the force exerted on an electron at point P by the dipole?

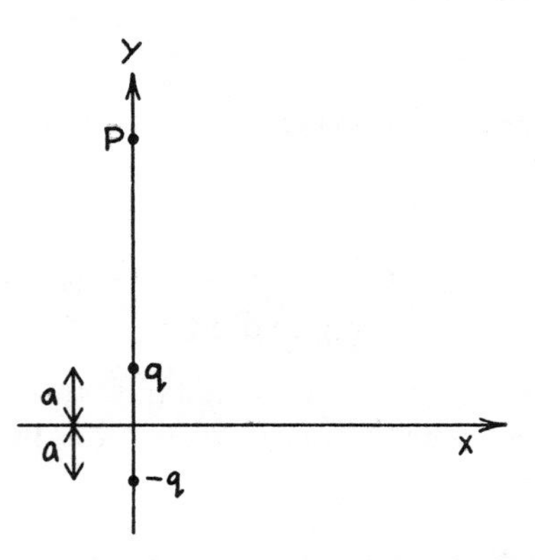

Figure 25.13 Problem 12. A dipole aligned parallel to the y axis. The distance y from the origin to point P is very much greater than a.

Solution scheme

a. Write down the expression for the potential at P in terms of the charges q and $-q$ and their respective distances $(y - a)$ and $(y + a)$. Remember that it is usual to write $2qa$ as p, the dipole moment, and that since $y \gg a$, $y^2 \cong y^2 - a^2$.
b. Evaluate $E_y = -\partial V/\partial y$ from your expression for the potential. What are the values of E_x and E_z?
c. Because we are given p for the water molecule (you have neglected a in your expressions for V and $\mathbf{E}$, since $y \gg a$) and the electron is at $y = 0.6 \times 10^{-9}$ m, you can obtain explicit values for V and for $\mathbf{E}$.
d. Since $\mathbf{F} = q\mathbf{E}$, calculate the force on the electron, remembering to note the direction of the force.

Answers to Guided Problems

2. $V(B) = V(A) = \dfrac{\lambda}{2\pi\varepsilon_0} \ln 2 = \dfrac{(1 \times 10^{-7})(.093)}{2\pi(8.85 \times 10^{-12})} = 1.25(10^3)$ volts

Point B is at the higher potential, because we had to do external work on the unit test charge to move it from A to B.

4. a. $V = 2820$ volts
   b. $\upsilon_f = 6.8 \times 10^5$ m/s
6. $V = 8300$ volts
   $\mathbf{E} = 0$
8. $E_R = 2.42 \times 10^{22}$ V/m $E_{r=0} = 0$

   $V_R = 1.78 \times 10^7$ V $V_{r=0} = \dfrac{Q}{4\pi\varepsilon_0}\left[\dfrac{1}{R} + \dfrac{1}{2R}\right] = 2.68 \times 10^7$ V
10. $Q = 9.9 \times 10^{-8}$ C
    $q = 9.9 \times 10^{-10}$ C
    $V = 890$ V at the surface of each sphere
    $E_R = 890$ V/m
    $E_r = 8.9 \times 10^5$ V/m
    If $Q_T = 2 \times 10^{-7}$ C, then
    $E_R = 1780$ V/m and $E_r = 1.76 \times 10^6$ V/m.
    There should *not* be electrical breakdown.
12. $$V = \frac{2qa}{4\pi\varepsilon_0}\left[\frac{1}{y^2a^2}\right] \simeq \frac{p}{4\pi\varepsilon_0}\frac{1}{y^2}$$

    $$E_y = \frac{p}{2\pi\varepsilon_0 y^3} \quad \text{or} \quad \mathbf{E} = \frac{p}{2\pi\varepsilon_0 y^3}\,\hat{\mathbf{y}}$$

    For the water molecule, and $y = 0.6 \times 10^{-9}$ m,

    $V = 0.16$ volts
    $\mathbf{E} = 5.17(10^8)\hat{\mathbf{y}}$ N/C
    $\mathbf{F} = 8.27(10^{-11})(-\hat{\mathbf{y}})$ N (downward)

Chapter 26

Electric Energy

Overview

In the last chapter we calculated the potential energy of a test charge in the electric field of a given charge distribution. We are now ready to evaluate the potential energy of the charge distribution itself, without the test charge. We do this by regarding the charge distribution as a collection of point charges and calculating the work needed to bring all these point charges close together to their final configuration. By applying our results to charged conductors, we then show that this potential energy can be considered to be stored in the electric field produced by the charges (rather than in the charges themselves), with the energy being most concentrated where the electric field is strongest.

Essential Terms

electric potential energy
self-energy
electric fringing field
energy density of an electric field

Key Concepts

1. The *electric potential energy* of a system of point charges is the work needed to assemble the point charges from a state in which every charge is at an infinite distance from every other charge and so has zero potential energy. For a pair of point charges q_1 and q_2 separated by a distance r_{12}, it requires zero work to bring q_1 to any position in the absence of q_2. It then takes an amount of work

$$U = \frac{1}{4\pi\varepsilon_0}\frac{q_1 q_2}{r_{12}}$$

to bring q_2 from infinity to a position r_{12} from q_1, where the expression for U is the same as that described in the preceding chapter. To bring in a third charge q_3 from infinity to a position r_{13} from q_1 and r_{23} from q_2 requires an *additional* amount of work

$$\frac{1}{4\pi\varepsilon_0}\frac{q_1q_3}{r_{13}} + \frac{1}{4\pi\varepsilon_0}\frac{q_2q_3}{r_{23}}$$

The total amount of work needed to assemble all three charges is, then,

$$U = \frac{1}{4\pi\varepsilon_0}\left[\frac{q_1q_2}{r_{12}} + \frac{q_1q_3}{r_{13}} + \frac{q_2q_3}{r_{23}}\right]$$

The three charges, and their separations r_{12}, r_{13}, and r_{23}, are shown in Figure 26.1.

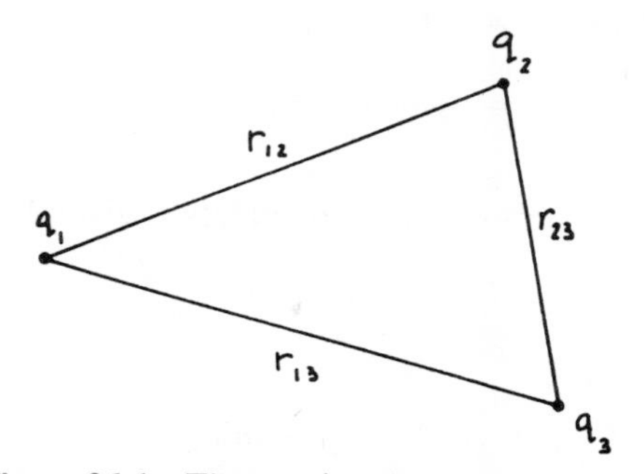

Figure 26.1 Three point charges, q_1, q_2, and q_3.

This expression for U can be rewritten as

$$U = \tfrac{1}{2}V_{\text{other}}(1)q_1 + \tfrac{1}{2}V_{\text{other}}(2)q_2 + \tfrac{1}{2}V_{\text{other}}(3)q_3$$

$$V_{\text{other}}(1) = \frac{1}{4\pi\varepsilon_0}\frac{q_2}{r_{12}} + \frac{1}{4\pi\varepsilon_0}\frac{q_3}{r_{13}}$$

which is just the potential at charge q_1 produced by charges q_2 and q_3. Similarly, $V_{\text{other}}(2)$ is the potential at charge q_2 produced by charges q_1 and q_3, and $V_{\text{other}}(3)$ is the potential at q_3 caused by q_1 and q_2. Since this expression for U can be applied to any number of charges, the electric potential energy for n point charges can be written as

$$U = \tfrac{1}{2}V_{\text{other}}(1)q_1 + \tfrac{1}{2}V_{\text{other}}(2)q_2 + \ldots + \tfrac{1}{2}V_{\text{other}}(n)q_n$$

To see how useful this expression can be, let us apply it to evaluate the potential energy of a single charged conductor with total charge Q whose surface is at a potential V. We can imagine the total charge Q to be made up of a large number of very small charges q, so that $V_{\text{other}} = V$ for each of the many small charges q contributing to the total charge Q. Then the total electric energy is just given by

$$U = \tfrac{1}{2}Vq_1 + \tfrac{1}{2}Vq_2 + \tfrac{1}{2}Vq_3 + \ldots$$
$$= \tfrac{1}{2}V(q_1 + q_2 + \ldots) = \tfrac{1}{2}VQ$$

That is, the electric energy of a charged conductor is one-half the product of the charge times the electric potential. If we have a large number of charged conductors with charges Q_1, Q_2, Q_3, . . . and respective potentials V_1, V_2, V_3, . . . , the total electric energy is given directly by

$$U = \tfrac{1}{2}Q_1V_1 + \tfrac{1}{2}Q_2V_2 + \tfrac{1}{2}Q_3V_3 + \ldots$$

2. *Energy density of an electric field.* We can easily calculate the electric energy U of a pair of large, parallel metal plates separated by a distance d and carrying respective charges of $+Q$ and $-Q$.

$$\mathrm{U} = \tfrac{1}{2}Q_1V_1 + \tfrac{1}{2}Q_2V_2$$
$$= \tfrac{1}{2}Q(V_1 - V_2)$$
$$= \frac{1}{2}\frac{Q^2d}{\varepsilon_0 A}$$

The electric field in the region between the plates is given by

$$E = Q/\varepsilon A$$

Thus

$$U = \tfrac{1}{2}\varepsilon_0 E^2 \times [\text{volume}]$$

giving an energy per unit volume of the field

$$u = \tfrac{1}{2}\varepsilon_0 E^2$$

It can be shown that this is a very general result, true for all electric fields, whether uniform or nonuniform.

Sample Problems

1. *Worked Problem.* What is the electric potential energy of three specks of dust, each carrying a charge of 1×10^{-9} C, arranged at the corners of an equilateral triangle, the sides of which are 2 mm long?

Solution: The electric potential energy is equal to the work needed to assemble the three charges and is given by the expression

$$U = \frac{1}{4\pi\varepsilon_0}\left[\frac{3q^2}{r}\right] = \frac{(9.0 \times 10^9)\ 3\ (1.0 \times 10^9)^2}{(2 \times 10^{-3})}$$
$$= 1.35 \times 10^{-5}\ \text{J}$$

We can also solve the problem using the expression

$$U = \tfrac{1}{2}V_{\text{other}}(1)q_1 + \tfrac{1}{2}V_{\text{other}}(2)q_2 + \tfrac{1}{2}V_{\text{other}}(3)q_3$$

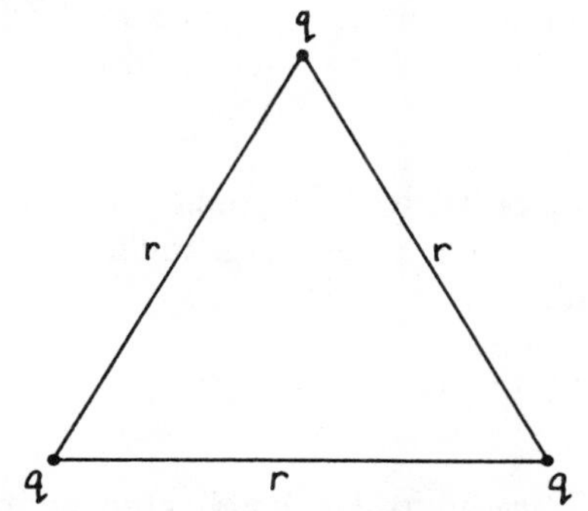

Figure 26.2 Problem 1. Three equal charges at the three corners of an equilateral triangle.

From Figure 26.2 we see that

$$V_{\text{other}}(1) = \frac{1}{4\pi\varepsilon_0}\left[\frac{q_2}{r_{12}} + \frac{q_3}{r_{13}}\right]$$
$$= \frac{1}{4\pi\varepsilon_0}\left[\frac{2q}{r}\right]$$
$$= V_{\text{other}}(2) = V_{\text{other}}(3)$$
$$U = \left(\frac{1}{2}\right)\frac{1}{4\pi\varepsilon_0}\left[\left(\frac{2q}{r}\right)q + \left(\frac{2q}{r}\right)q + \left(\frac{2q}{r}\right)q\right]$$
$$= \frac{1}{4\pi\varepsilon_0}\left[\frac{3q^2}{r}\right]$$

In agreement with the expression derived using the first method of solution.

2. *Guided Problem.* Four charges, each of magnitude q coulombs, are placed at the corners of a square. The length of each side of the square is r meters. What is the potential energy of the system when

a. all four charges are positive?
b. all four charges are negative?
c. two adjacent charges are positive and the other two charges are negative?

Solution scheme

a. Sketch a diagram showing the four charges, one at each corner of the square. What is the distance between opposite corners of the square?
b. What is the electrical energy of a pair of charges? Add the electrical energy of each pair of charges in the ensemble to obtain the total energy.
c. How many pairs of charges are there, and what are their separations?
d. Now evaluate U for the ensemble of four charges, each of magnitude $+q$.
e. Will the result be any different if all four charges are negative?

f. For part c you must consider the same six pairs of charges, but you must now be careful of the sign of the energy for each pair of charges.

3. *Worked Problem.* Solve the preceding problem, that of the electric energy of four charges at the corners of a square, by using the expression for the electric potential energy

$$U = \sum_i \tfrac{1}{2} V_{\text{other}}(1) q_i$$

Solution: Again, it is useful to sketch a diagram showing the four charges and the distances between them. This is done in Figure 26.3. The electric energy is given by

$$U = \tfrac{1}{2} V_{\text{other}}(1) q_1 + \tfrac{1}{2} V_{\text{other}}(2) q_2 + \tfrac{1}{2} V_{\text{other}}(3) q_3 + \tfrac{1}{2} V_{\text{other}}(4) q_4$$

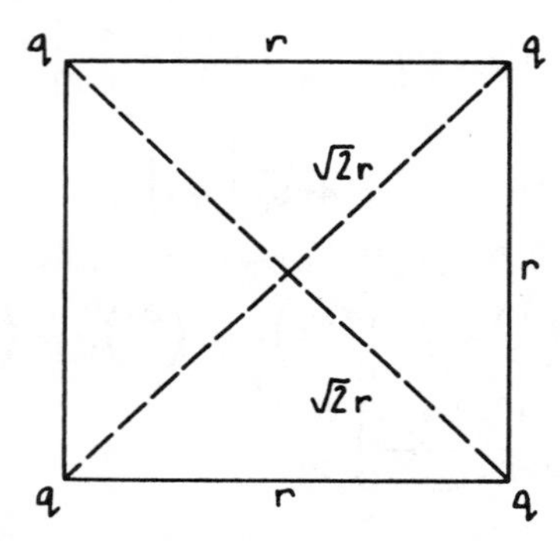

Figure 26.3 Problem 3. Four equal charges, each at one corner of a square.

For part a, $q_1 = q_2 = q_3 = q_4 = q$, so that

$$V_{\text{other}}(1) = \frac{q}{4\pi\varepsilon_0}\left[\frac{1}{r} + \frac{1}{r} + \frac{1}{\sqrt{2}r}\right] = \frac{q}{4\pi\varepsilon_0}\left[\frac{1 + 2\sqrt{2}}{\sqrt{2}r}\right]$$

$$= V_{\text{other}}(2) = V_{\text{other}}(3) = V_{\text{other}}(4)$$

$$U = \frac{1}{2}\left[\frac{q}{4\pi\varepsilon_0}\left(\frac{1 + 2\sqrt{2}}{\sqrt{2}r}\right)\right](4q) = \frac{2q^2}{4\pi\varepsilon_0}\left(\frac{1 + 2\sqrt{2}}{\sqrt{2}r}\right)$$

For part b, $q_1 = q_2 = q_3 = q_4 = -q$, so that

$$V_{\text{other}} = \frac{-q}{4\pi\varepsilon_0}\left(\frac{1 + 2\sqrt{2}}{\sqrt{2}r}\right)$$

and

$$U = \frac{1}{2}\left[\frac{-q}{4\pi\varepsilon_0}\left(\frac{1 + 2\sqrt{2}}{\sqrt{2}r}\right)\right](-4q) = \frac{2q^2}{4\pi\varepsilon_0}\left(\frac{1 + 2\sqrt{2}}{\sqrt{2}r}\right)$$

as in part a.

For c, let $q_1 = q_2 = q$, and $q_3 = q_4 = -q$. Then

$$V_{\text{other}}(1) = V_{\text{other}}(2) = \frac{1}{4\pi\varepsilon_0}\left[\frac{-q}{r} + \frac{q}{r} - \frac{q}{\sqrt{2}r}\right] = \frac{-q}{\sqrt{2}r}\left(\frac{1}{4\pi\varepsilon_0}\right)$$

$$V_{\text{other}}(3) = V_{\text{other}}(4) = \frac{1}{4\pi\varepsilon_0}\left[\frac{q}{r} - \frac{q}{r} + \frac{q}{\sqrt{2}r}\right] = \frac{q}{\sqrt{2}r}\left(\frac{1}{4\pi\varepsilon_0}\right)$$

and

$$U = \left(\frac{1}{2}\right)\frac{1}{4\pi\varepsilon_0}\left[\frac{-q}{\sqrt{2}r}(q) - \frac{q}{\sqrt{2}r}(q) + \frac{q}{\sqrt{2}r}(-q) + \frac{q}{\sqrt{2}r}(-q)\right]$$

$$= \frac{1}{8\pi\varepsilon_0}\left[\frac{-4q^2}{\sqrt{2}r}\right] = \frac{-q^2}{2\sqrt{2}\pi\varepsilon_0 r} \text{ J}$$

4. *Guided Problem.* Two concentric metal spheres of radii a and b meters, respectively, carry respective charges of $-Q$ and $+Q$, as shown in Figure 26.4. Calculate the electric energy of the two conductors, and compare your answer with that you obtain by evaluating the total electric energy stored in the electric field of the system.

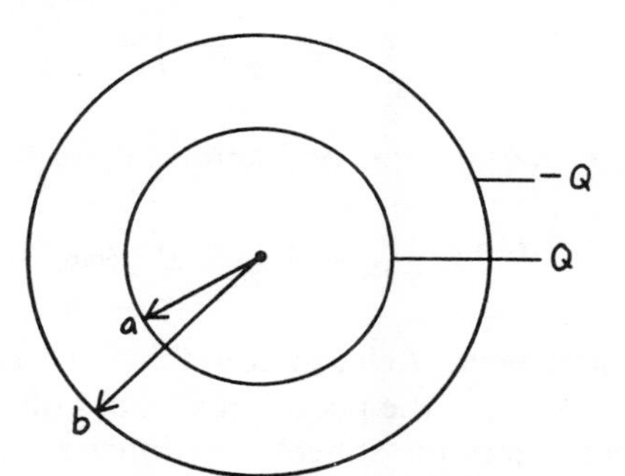

Figure 26.4 Problem 4. Two concentric metal spheres of radii a and b, respectively, and with respective charges of Q and $-Q$.

Solution scheme

a. The electric energy of the conductors is given by

$$U = \tfrac{1}{2}Q_1V_1 + \tfrac{1}{2}Q_2V_2 = \tfrac{1}{2}Q(V_a - V_b)$$

The potential difference $V_a - V_b$ can be found from the definition of electric potential,

$$V_a - V_b = -\int_a^b E\,dl$$

where

$$E = \frac{1}{4\pi\varepsilon_0}\frac{Q}{r^2}$$

b. The electric field is given by $\mathbf{E} = \frac{1}{4\pi\varepsilon_0}\frac{Q}{r^2}\hat{\mathbf{r}}$ for $a \leqslant r \leqslant b$ and is zero everywhere else. Since the energy per unit volume of field is given by $u = \frac{1}{2}\varepsilon_0 E^2$, the total energy is given by evaluating $\int u\,dV$ over the volume occupied by the electric field. Choose for the volume element $dV = 4\pi r^2 dr$.

c. You should find that your answers derived in steps a and b agree.

5. *Worked Problem.* Compare the electric potential energies of a conducting sphere of radius R with charge Q on its surface with those of a nonconducting sphere, also of radius R and with a charge Q uniformly distributed over the volume of the sphere.

Solution: For a conducting sphere the potential at the surface is given by

$$V(R) = \frac{Q}{4\pi\varepsilon_0}\frac{1}{R}$$

where R is the radius of the sphere. The electric potential energy is then just

$$U = \tfrac{1}{2}QV = \tfrac{1}{2}Q\left(\frac{Q}{4\pi\varepsilon_0 R}\right) = \frac{Q^2}{8\pi\varepsilon_0 R}$$

For the uniformly charged nonconducting sphere, we use the expression

$$U = \tfrac{1}{2}\sum_i V_{\text{other}}(1)q_1 = \tfrac{1}{2}\int V_{\text{other}}\,dq$$

where dq is the increment of charge contained in an infinitesimal volume of the sphere and V_{other} is the potential at that volume element resulting from all other charges in the sphere. The integration is to be taken over the volume of the sphere. Example 25.5 of the text shows that the potential at any point inside a uniformly charged sphere is given by

$$V(r) = V(R) - \int_R^r \mathbf{E} \cdot d\mathbf{l}$$

The electric field $\mathbf{E}$ inside the sphere is obtained from Gauss' Law. If ρ is the charge density in the sphere, so that $\rho = Q/\text{volume} = Q/(4/3)\pi R^3$, then inside the sphere the electric field

$$E = \frac{1}{4\pi\varepsilon_0}\frac{[(4/3)\pi r^3 \rho]}{r^2} = \frac{\rho r}{3\varepsilon_0}$$

and

$$\begin{aligned} V(r) &= V(R) - \int_R^r \frac{\rho r}{3\varepsilon_0}\hat{\mathbf{r}} \cdot d\mathbf{l} \\ &= \frac{Q}{4\pi\varepsilon_0} + \frac{\rho}{6\varepsilon_0}[R^2 - r^2] = \frac{3QR^2 - Qr^2}{8\pi\varepsilon_0 R^3} \end{aligned}$$

This is the value of V_{other} to be used in the expression for U. We let dq be the charge in a spherical shell of radius r and thickness dr, so that $dq = 4\pi r^2 dr\rho = \dfrac{3Qr^2}{R^3}\,dr$, and U is finally given by

$$U = \frac{1}{2}\int_0^R \left(\frac{3QR^2 - Qr^2}{8\pi\varepsilon_0 R^3}\right)\left(\frac{3Qr^2}{R^3}\,dr\right)$$

$$= \frac{3Q^2}{16\pi\varepsilon_0 R^6}\int_0^R (3R^2 - r^2)r^2 dr$$

$$= \frac{3Q^2}{16\pi\varepsilon_0 R^6}\left[R^5 - \frac{R^5}{5}\right]$$

$$= \frac{3Q^2}{20\pi\varepsilon_0 R}$$

Note that the ratio $U_{\text{conducting}}/U_{\text{nonconducting}} = 5/6$, so the total energy in the nonconducting sphere is slightly larger than that stored in the conducting sphere of equal radius.

6. *Guided Problem*. In worked problem 5, we showed that the total energy of a sphere of radius R and with charge Q uniformly distributed over its volume is given by $3Q^2/20\pi\varepsilon_0 R$. Suppose we have such a sphere and are able to break it up and reshape it into two smaller spheres, each of 1/2 the volume of the original sphere and each with charge $Q/2$. Both the new spheres have uniform charge distributions (Figure 26.5). Compare the total energy of the two new spheres with that of the original large sphere when the two new spheres are far apart. There is a process in nature called nuclear fission, in which a heavy nucleus splits into two smaller ones. Can you see why such a process releases energy?

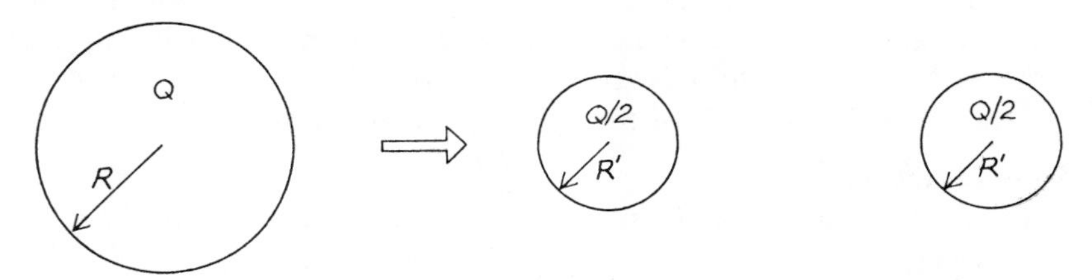

Figure 26.5 Problem 6. A single large sphere of radius R and charge Q breaks up into two smaller spheres, each with charge $Q/2$.

Solution scheme

a. The volume of a sphere is given by $V = (4/3)\pi R^3$. Can you use this to determine the radius R' of the new spheres in terms of the radius R of the original sphere?
b. Knowing the radius R' and the expression for the total energy of a charged sphere, you should be able to calculate the electrical energy U' of each of the new spheres.
c. If the new spheres are "far apart," what is the total electrical energy

of the new system? Express this in terms of U_0, the electrical energy of the original large, charged sphere. Has energy been released in the process? What do you think has happened to this energy?

7. *Worked Problem.* In Example 26.2 of your text, we find that the electric energy U of two large, parallel metal plates of area A, separated by a distance d, and with respective charges of Q and $-Q$ is given by $U = Q^2d/2\varepsilon_0A$. What is the electric energy if the plate separation is increased to $2d$? If the plate separation is decreased to $d/2$? Are your answers consistent with the principle of energy conservation? Explain.

Solution: For a plate separation d, the electrical energy is given by $U(d) = (1/2)Q[V_1 - V_2]$, where $V_1 - V_2 = Ed = Qd/\varepsilon_0A$. Then $U(d) = Q^2d/2\varepsilon_0A$. When the plate separation changes to $2d$, with no change in Q or $-Q$, the electric field E remains constant, so that

$$U(2d) = Q^2(2d)/2\varepsilon_0A = 2U(d)$$

When the plate separation changes to $(d/2)$, again the electric field remains constant, so

$$U(d/2) = Q^2(d/2)/2\varepsilon_0A = (1/2)U(d)$$

Yes, these answers are consistent with energy conservation, since it requires external work to force the plates farther apart [to $(2d)$ from d], and the plates can do external work if the separation is allowed to decrease. That is, the plates must be held apart, or they will come together.

8. *Guided Problem.* A spherical aluminum ball of 40 cm radius mounted on an insulating mount is charged to a potential of 10,000 volts.

a. How much electrical energy is stored on the ball?
b. Calculate the energy stored in the electric field surrounding the ball.
c. Although the electric field extends outward to infinity, what percentage of the total energy is stored in the first 1 m of electric field outward from the surface of the ball? In the first 3 m?

Solution scheme

a. Use the expression $U = \frac{1}{2}QV$ to evaluate the total energy of the system.
b. The energy density of an electric field is given by $u = \frac{1}{2}\varepsilon_0E^2$. The energy stored in a volume element dv of the electric field is therefore given by udv, and the total energy stored in the field is $\int udv$, where the integral extends from R to infinity. (Remember that the field is zero inside the surface of the ball.) The integral can be evaluated by making the small volume elements spherical shells of radius r and thickness dr.
c. What new limits on the integrand should you use in order to calculate part c of the problem?

9. *Worked Problem.* A long, cylindrical cable has a solid copper central cylinder of radius 4 mm. Outside of this and insulated from it is a thin, con-

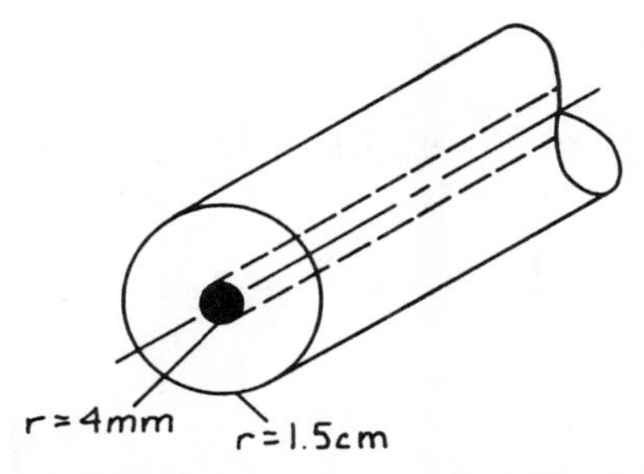

Figure 26.6 Problem 9. A long, cylindrical cable consisting of two concentric copper cylinders.

centric outer copper cylinder of radius 1.5 cm, as shown in Figure 26.6. The central conductor receives 2000 volts, while the outer cylinder is kept at ground potential.

a. Find the electric field strength in the space between the two conductors, and the charge per meter on the central conductor.
b. Find the electric energy per meter of length stored in the system.

Solution: Because the outer conductor is kept at ground potential, there is no electric field outside this conductor and no charge on the outer surface of the conductor. There is a charge of λ C per meter on the outer surface of the inner cylinder and an equal and opposite induced charge on the inner surface of the outer conductor. The electric field in the space between the two conductors can be easily obtained by applying Gauss' Law to an arbitrary intermediate cylindrical shell of radius r and length l. Then we obtain the result that

$$E(r) = \frac{\lambda}{2\pi\varepsilon_0 r} \text{ V/m}$$

The potential difference between the two conductors is

$$\begin{aligned}\Delta V &= -\int_b^a \mathbf{E}(r) \cdot d\mathbf{l} = \int_b^a E(r)dl \\ &= -\int_b^a E dr \text{ [since } dl = -dr] \\ &= -\frac{\lambda}{2\pi\varepsilon_0}\int_b^a \frac{dr}{r} = \frac{\lambda}{2\pi\varepsilon_0} \ln (b/a)\end{aligned}$$

where b and a are the outer and inner radii of the two conductors. Since the known quantity is the potential difference ΔV (2000 volts), we can now evaluate both E and λ. We know that $(b/a) = (1.5/0.4)$, so $\lambda = 8.42 \times 10^{-8}$ C/m and $E = 1515/r$ V/m for the region between the two conductors. We can then evaluate the electric energy per unit length either through

$$U = \tfrac{1}{2}Q(\Delta V) = \tfrac{1}{2}\lambda\Delta V = 8.42 \times 10^{-5} \text{ J/m}$$

or through

$$U = \int_a^b u\,dv = \int_a^b \tfrac{1}{2}\varepsilon_0 E^2 dv = \int_a^b (\tfrac{1}{2}\varepsilon_0 E^2)(2\pi r dr)$$

$$= \tfrac{1}{2}\varepsilon_0 (1515)^2 (2\pi) \int_a^b \frac{dr}{r} = 8.42 \times 10^{-5}\ \text{J/m}$$

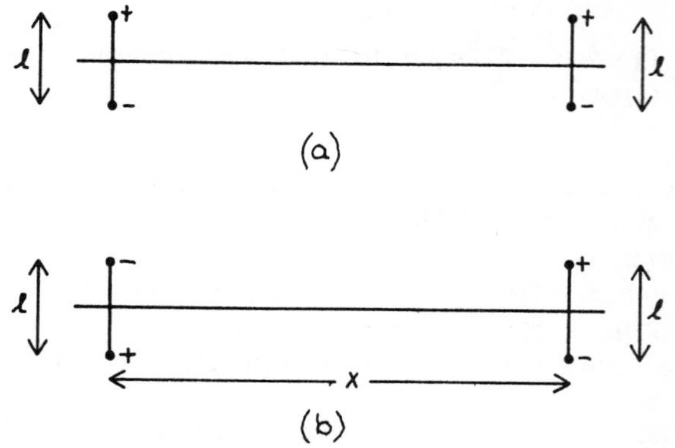

Figure 26.7 Problem 10. Two identical interacting dipoles. In (a) the dipole axes are both parallel, and in (b) the dipole axes are antiparallel. For both (a) and (b) the distance between the dipoles is x m.

10. *Guided Problem.* You are given two interacting electric dipoles, oriented as shown in Figure 26.7. The separation of the dipoles, x, is much greater than the dipole length l. Find the electric potential energy of the dipole system
   a. if the axes of the dipoles are parallel, as in Figure 26.7(a).
   b. if the axes are antiparallel, as in Figure 26.7(b).

Solution scheme

a. The electric field at a distance x away from the dipole can be found most easily by considering the dipole as two point charges ($+q$ and $-q$, respectively). You should be able to show that for $x \gg l$, the field is parallel to the dipole axis and is given by

$$E \simeq \frac{p}{4\pi\varepsilon_0 x^3}$$

where $p = ql$.

b. The potential energy U of a dipole in an electric field is given by $U = -\mathbf{p} \cdot \mathbf{E}$. In the first case the dipole axis is parallel to the electric field, and in the second case it is antiparallel. From your answer it should be clear which is the more stable configuration—that is, dipoles oriented as are those in this problem like to align themselves with their axes antiparallel!

Answers to Guided Problems

2. a. $\dfrac{q^2}{2\pi\varepsilon_0 r}\left[\dfrac{1+2\sqrt{2}}{\sqrt{2}}\right]$ J

   b. $\dfrac{q^2}{2\pi\varepsilon_0 r}\left[\dfrac{1+2\sqrt{2}}{\sqrt{2}}\right]$ J

   c. $\dfrac{-q^2}{2\sqrt{2}\pi\varepsilon_0 r}$ J

4. $U = \dfrac{Q^2}{8\pi\varepsilon_0}\left[\dfrac{1}{a}-\dfrac{1}{b}\right]$ J

6. $R' = 0.794R$

   U' = electrical energy of the two new spheres $= 2\left[3\left(\dfrac{Q}{2}\right)^2/20\varepsilon_0 R'\right]$

   $= 0.63\mu_0$ when $\mu_0 = 3Q^2/20\varepsilon_0 R$

   The total electrical energy has decreased and is available in the form of kinetic energy of the new spheres.

8. a. $U = 2.22 \times 10^{-3}$ J

   b. $U = 2.22 \times 10^{-3}$ J

   c. 71%; 88%

10. a. $U = \dfrac{p^2}{4\pi\varepsilon_0 x^3}$

    b. $U = \dfrac{-p^2}{4\pi\varepsilon_0 x^3}$

    Case b, in which the potential energy is a minimum, represents the more stable condition.

Chapter 27

Capacitors and Dielectrics

Overview

When we store charge on an arrangement of conductors, we also store energy. Any arrangement of conductors that is used to store charge is called a *capacitor*, and thus capacitors also store electrical energy. The conducting surfaces that store charges in a capacitor are usually in close proximity to insulating material. This insulating material, which is called a *dielectric*, may drastically alter the electric field from what it would be in a vacuum. Because the electric field is changed by the presence of a dielectric, both Gauss' Law and the expression for the energy density of an electric field must be modified when dielectrics are present.

Essential Terms

capacitor
capacitance
capacitors in series and in parallel
dielectric
dielectric constant
bound charge
electric displacement
energy in a capacitor

Key Concepts

1. A *capacitor* is any arrangement of conductors used to store electric charge; thus, from our very first discussions of electrostatics, we have talked about capacitors without explicitly defining them. An isolated conducting sphere, for example, is a capacitor. The most common type of capacitor, however, consists of two parallel conducting plates and is called the *parallel-plate capacitor*. We find that for any capacitor the relationship $Q = CV$, where C is a constant, holds. Q is the charge on the capacitor, and V is the

potential of the conductor if the capacitor is an isolated conductor, or the potential difference between the two conductors if the capacitor is a parallel-plate capacitor.

2. The constant C, the ratio of charge stored to the potential V, is called the *capacitance* of a capacitor. A large value of C implies that a large amount of charge can be stored at a low potential. We almost always find that the capacitance C can be expressed in terms of the geometry of the conductors, so that when the geometry is fixed, so is the capacitance of the system.

The capacitance can usually be obtained by finding the relationship between the charge Q stored on the conductor and the potential difference V between the conductors. In the case of an isolated conducting sphere of radius R, the potential $V = Q/4\pi\varepsilon_0 R$. Here we can imagine a second sphere of infinite radius at zero potential with an equal and opposite charge, so that even in this case V can be considered the potential difference between two spheres. Then $C = Q/V = 4\pi\varepsilon_0 R$ coulombs/volt.

Three of the more common types of capacitors are concentric spheres, parallel plates, and concentric cylinders, as shown in Figure 27.1. In each case we have equal and opposite charges on two conductors and have defined the potential difference between the two conductors to be ΔV, so that $C = Q/\Delta V$. In each case we can obtain the electric field in the region between the two conductors from Gauss' Law and then obtain ΔV from the definition of potential difference, $\Delta V = -\int \mathbf{E} \cdot d\mathbf{s}$. For the concentric spheres

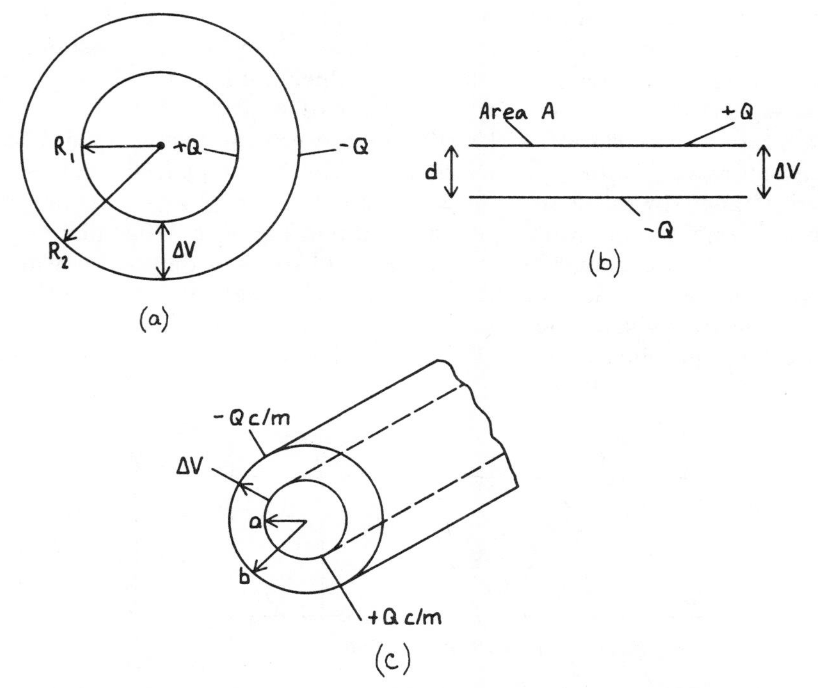

Figure 27.1 Three common types of capacitor: (a) concentric spheres, (b) parallel plates, (c) concentric cylinders.

$$\Delta V = \frac{Q}{4\pi\varepsilon_0}\left[\frac{1}{R_1} - \frac{1}{R_2}\right]$$

so that

$$C = \frac{Q}{\Delta V} = 4\pi\varepsilon_0\left[\frac{R_1R_2}{R_2 - R_1}\right]$$

Note that if we let R_2 approach infinity, C approaches $4\pi\varepsilon_0 R_1$, the value for an isolated sphere.

For the two parallel plates, $\Delta V = Ed = \sigma d/\varepsilon_0$, where σ is the surface charge density, Q/A, so that

$$C = \frac{Q}{\Delta V} = \frac{\sigma A}{\sigma d/\varepsilon_0} = \frac{\varepsilon_0 A}{d}$$

For the concentric cylinders we can show that

$$\Delta V = \frac{Q \ln (b/a)}{2\pi\varepsilon_0}$$

from which

$$C = Q/V = \frac{2\pi\varepsilon_0}{\ln (b/a)} \text{ coulombs per volt per meter length}$$

3. *Dielectric*. We have so far assumed that the space between the conductors of a capacitor is a vacuum. In practice, this is seldom true. For example, in a parallel-plate capacitor the two plates are almost always separated by a slab of insulating material, or *dielectric*, which not only provides mechanical support for the plates but also modifies the electrical behavior of the capacitor. If we place a dielectric slab between the plates of an isolated capacitor, we find that the electric field between the plates is reduced. To understand why this is so, consider Figure 27.2. Some molecules are *polar* molecules, molecules with a permanent dipole moment. When placed in an external electric field, polar molecules tend to align themselves with the field. The result is an induced surface *bound charge*, q', on the surface of the dielectric, which produces an electric field E' inside the dielectric in the direction opposite to that of the external field E.

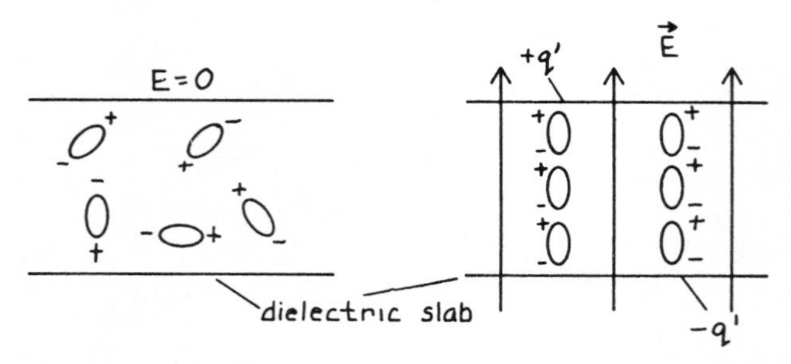

Figure 27.2 A polar dielectric material. In (a) there is no external electric field; in (b) the polar molecules are aligned parallel to the electric field E.

In a nonpolar dielectric material, the molecules do not have a permanent dipole moment, but even when nonpolar materials are subject to an external electric field, the positive and negative centers of charge are pulled slightly apart to form dipole moments. Again, the net result when the dielectric is placed in an external electric field is an induced bound surface charge and a decrease in the net electric field in the dielectric material.

4. *Dielectric constant*. The dielectric constant, κ, is defined as the factor by which the electric field is reduced by the induced surface charge of a dielectric. That is, $E = \frac{1}{\kappa}E_{\text{free}}$, where E_{free} is the electric field that the free charges on the capacitor plates produce by themselves, and where $\kappa \geqslant 1$. Suppose we consider the isolated charged parallel-plate capacitor shown in Figure 27.3(a). We then slide a slab of dielectric between the plates, as shown in Figure 27.3(b). With the dielectric the potential difference between the plates is $Ed = \frac{(E_{\text{free}})d}{\kappa} = \frac{\Delta V_0}{\kappa}$, so that the potential difference is reduced by the factor κ. The capacitance with the dielectric, $C = Q_{\text{free}}/\Delta V$, is given by $C = \frac{Q_{\text{free}}}{\Delta V_0/\kappa} = C_0\kappa$, where C_0 is the capacitance without the dielectric. Thus the capacitance is increased by the factor κ.

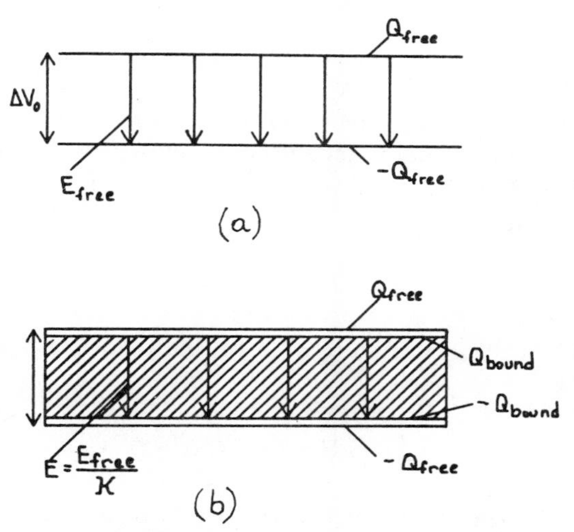

Figure 27.3 (a) An isolated charged parallel-plate capacitor. (b) The same capacitor with a slab of dielectric between the plates.

Now consider a Gaussian surface consisting of a cylinder extending from inside the upper plate of the capacitor shown in Figure 27.3(b) down into the dielectric. The Gaussian surface is shown in Figure 27.4. The free-charge density is $Q_{\text{free}}/A = \sigma_{\text{free}}$, and the bound-charge density is $Q_{\text{bound}}/A = \sigma_{\text{bound}}$. From Gauss' Law we find that the electric field E in the dielectric is given by

$$E = \frac{\sigma}{\varepsilon_0} = \frac{\sigma_{\text{bound}} + \sigma_{\text{free}}}{\varepsilon_0} = \frac{\sigma_{\text{bound}}}{\varepsilon_0} + \frac{\sigma_{\text{free}}}{\varepsilon_0}$$

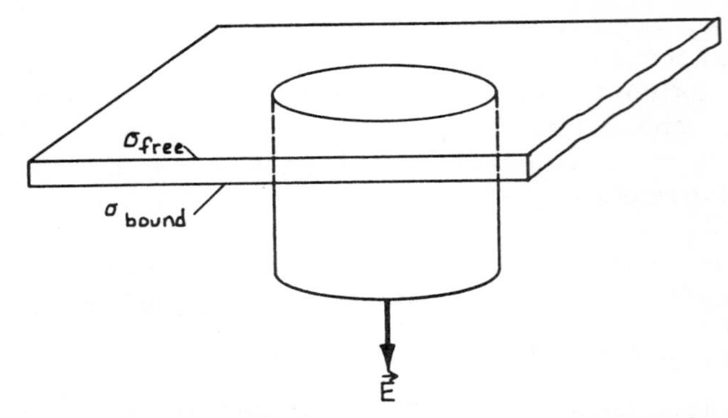

Figure 27.4 Gauss' Law applied to a parallel-plate capacitor with a dielectric.

But since

$$E = E_{\text{free}}/\kappa = \sigma_{\text{free}}/\varepsilon_0\kappa$$

$$\frac{\sigma_{\text{free}}}{\varepsilon_0\kappa} - \frac{\sigma_{\text{free}}}{\varepsilon_0} = \frac{\sigma_{\text{bound}}}{\varepsilon_0}$$

$$\sigma_{\text{bound}} = \sigma_{\text{free}}\left(\frac{1}{\kappa} - 1\right) = -\frac{\kappa - 1}{\kappa}\,\sigma_{\text{free}}$$

so that we have obtained the bound charge density in terms of the free charge density.

It is usually more convenient, however, to modify Gauss' Law when a dielectric is present so that it depends only on the free charge and not on the bound charge. For the parallel-plate capacitor of Figure 27.3(a), Gauss' Law can be written

$$\oint \mathbf{E}_{\text{free}} \cdot d\mathbf{s} = Q_{\text{free}}/\varepsilon_0$$

With the dielectric, as in Figure 27.3(b), the field is $E = E_{\text{free}}/\kappa$. Substituting, we obtain

$$\oint \kappa\mathbf{E} \cdot d\mathbf{s} = Q_{\text{free}}/\varepsilon_0$$

where $\mathbf{E}$ is the field in the dielectric expressed in terms of Q_{free} only. We must be a bit cautious here: In the last expression Q_{free} is the free charge contained inside a Gaussian surface—we have modified Gauss' Law. Q_{free} is *not* the total free charge stored on the capacitor plates unless we use the whole area of the plates in our Gaussian surface. Gauss' Law for a dielectric is usually written

$$\oint \mathbf{D} \cdot d\mathbf{s} = Q_{\text{free}}$$

where $\mathbf{D} = \varepsilon_0\kappa\mathbf{E}$ and is called the *electric displacement vector.*

5. *Energy in a capacitor.* The energy stored in a capacitor with charge $+Q$ and $-Q$ on two conductors is given directly from the equation of the last chapter:

$$U = \frac{1}{2}Q_1V_1 + \frac{1}{2}Q_2V_2 = \frac{1}{2}Q(V_2 - V_1) = \frac{1}{2}Q\Delta V = \frac{1}{2}\frac{Q^2}{C}$$

where ΔV is the potential difference between the two conductors. If there is a dielectric between the two conductors, the expression for U is still correct, where Q is the free charge Q_{free}. The energy required to create Q_{bound} is accounted for by the fact that the capacitance C is altered by the presence of the dielectric.

We obtained in the last chapter the expression

$$u = \tfrac{1}{2}\varepsilon_0 E^2$$

for the energy density of an electric field in a vacuum. Let us calculate the energy density inside a dielectric filling the region between the plates of a parallel-plate capacitor. As before, the plate area is A and the plate separation is d. The total energy stored, U is given by

$$U = \tfrac{1}{2}Q\Delta V = \tfrac{1}{2}Q(Ed)$$

But from Gauss' Law for a dielectric,

$$\kappa E = \frac{\sigma_{\text{free}}}{\varepsilon_0}$$

or

$$\varepsilon_0 \kappa EA = \sigma_{\text{free}} A = Q$$

so that

$$U = \tfrac{1}{2}\varepsilon_0 \kappa \text{EA}(Ed)$$

or

$$\frac{U}{\text{Vol}} = \tfrac{1}{2}\kappa\varepsilon_0 E^2 = u$$

Thus, our previous expression for energy density is multiplied by the dielectric constant κ.

Sample Problems

1. *Worked Problem.* A spherical capacitor consists of two concentric aluminum spheres with respective radii of 16 and 18 cm. Charges of $+Q$ and $-Q$ are applied to the inner and outer spheres, respectively. Find the capacitance. How long must a cylindrical capacitor consisting of metal cylinders with respective radii of 18 and 16 cm be in order to have the same capacitance as the spherical capacitor? Ignore edge effects. What plate area is required if a parallel-plate capacitor with plate separation of 2 cm is to have the same capacitance? Compare this plate area with the *average* area of the conducting surfaces for each of the first two capacitors.

Solution: The capacitance of a spherical capacitor is given by

$$C = 4\pi\varepsilon_0\left[\frac{R_1R_2}{R_2 - R_1}\right] \quad R_1 = 0.16\text{ m};\ R_2 = 0.18\text{ m}$$

Then

$$C = \frac{(.16)(.18)}{9.0(10^9)(0.02)} = 1.6 \times 10^{-10}\text{ F}$$

The capacitance of the cylindrical capacitor is given by

$$C = \frac{2\pi\varepsilon_0}{\ln(b/a)}\text{ F/m} \quad b = 0.18\text{ m};\ a = 0.16\text{ m}$$

Then

$$C = 4.72 \times 10^{-10}\text{ F/m}$$

Then

$$\text{Length } L = 1.6/4.72 = 0.34\text{ m}$$

For a parallel-plate capacitor, the capacitance is given by

$$C = \frac{\varepsilon_0 A}{d} = 1.6 \times 10^{-10}\text{ F}$$

Then

$$A = \frac{(1.6 \times 10^{-10})(.02)}{(8.85 \times 10^{-12})} = 0.36\text{ m}^2$$

The average plate area for the spherical capacitor is $(A_1 + A_2)/2$, where $A_1 = 4\pi R_1^2 = 0.32\ m^2\ and\ A_2 = 0.41\text{ m}^2$. The average plate area is then 0.36 m^2. For the cylindrical capacitor the plate area is given by $2\pi Rl$, where $l = 0.34$ m. Again, the average area is 0.36 m^2.

It is clear that for plate separations that are fairly small compared with the size of the conductors, the most important criteria in determining the capacitance of a capacitor are the plate area and the plate separation. It makes little difference whether the capacitor is a spherical capacitor, a parallel-plate capacitor, or a cylindrical capacitor.

2. *Guided Problem.* A capacitor is constructed of interlocking metal plates, as shown in Figure 27.5. The plate separation $d = 0.2$ cm for all plates, and the plate overlap area, A, is 10 cm^2. Find the capacitance of the system. The right half of the capacitor has a charge of $+Q$, and the left half $-Q$, yet there are five plates on the right and only four on the left. Does this mean that the surface charge densities are different? Explain.

Solution scheme

a. The system is clearly a parallel-plate capacitor, so the capacitance can easily be calculated if we know the plate separation and the effective plate area.

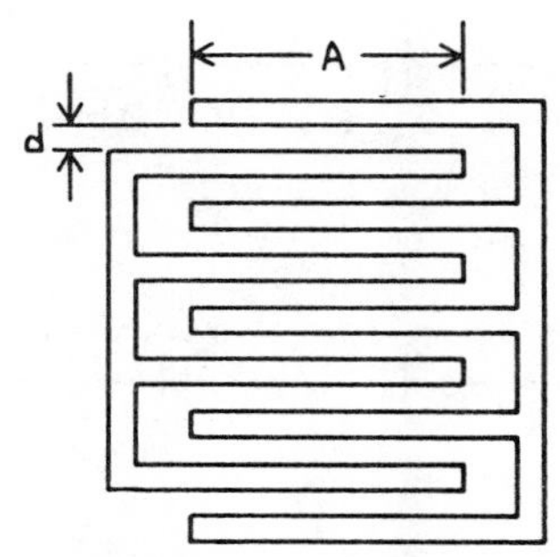

Figure 27.5 Problem 2. A metal parallel-plate capacitor with overlapping plates.

b. The plate separation d is clearly 0.2 m; the plate area for each overlapping plate is given. How many overlapping plates are there? An alternate way to view the problem is to consider the capacitor as a number of capacitors connected in parallel. In this case, how many parallel capacitors are there?
c. To answer the last part of the question, remember that in a parallel-plate capacitor, the charge is normally distributed on one face only of each plate. Do we expect there to be charge on the extreme upper and lower faces of the right half of the capacitor? What is the charge distribution on the remaining eight plate faces?

3. *Worked Problem.* You are given the capacitor network shown in Figure 27.6. A potential difference of 50 volts is applied between points A and B. If the value of $C = 1 \times 10^{-7}$ F, find the charge on each capacitor, the potential difference across each capacitor, and the equivalent capacitance of the system.

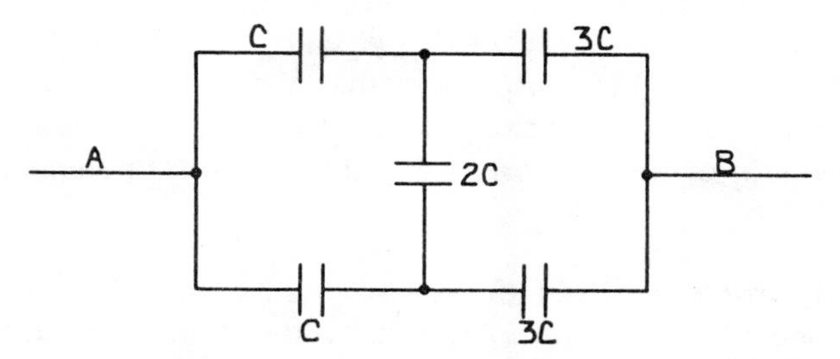

Figure 27.6 Problem 3. A capacitor network.

Solution: Because the upper and the lower loops of the network have identical capacitors, the potential at the upper and the lower ends of the $2C$ capacitor has the same value, so that there is zero potential difference across the $2C$ capacitor and zero charge on the capacitor. We can therefore neglect this capacitor; the network behaves as if the $2C$ capacitor did not exist. Then the equivalent network consists of the parallel network shown in Figure 27.7(a), which can be reduced to two parallel capacitors, as shown in Figure 27.7(b).

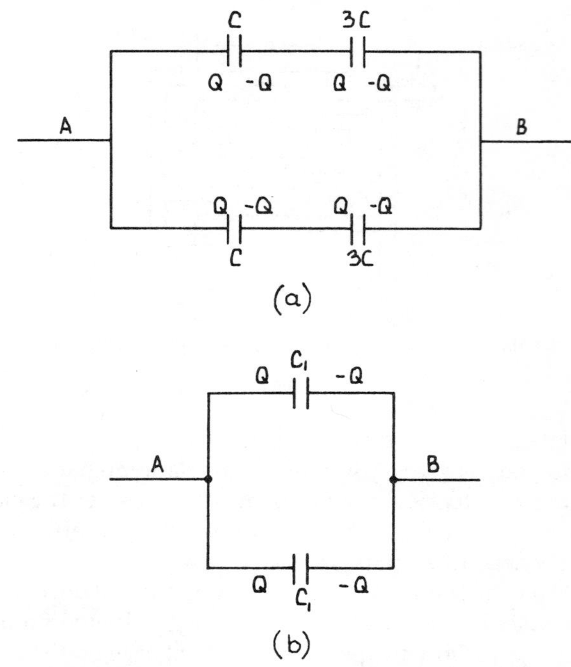

Figure 27.7 Problem 3. (a) A parallel capacitor network. (b) C_1 is the equivalent capacitance of C and $3C$ in series.

C_1 is the equivalent capacitance of C and $3C$ in series and is given by the relationship $1/C_1 = 1/C + 1/3C = 4/3C$. Thus, $C_1 = 3/4C$. The equivalent capacitance of the whole network is then just $C_1 + C_1$ or $C_{eq} = 2C_1 = 1.5C$ F.

The charge Q is obtained from $C_1 = Q/V = Q/50$, so that $Q = 3.75\,(10^{-6})$ C. This is the charge across each of the capacitors C and $3C$. The potential difference across each capacitor is given by the relationship $V = Q/C$, so that the potential difference across the capacitors C is $V = 37.5$ volts, and 12.5 volts across the capacitors $3C$. The total potential drop across each series arm of the network is 50 volts, the total potential drop between A and B.

4. *Guided Problem.* Given three capacitors, two with capacitance C farads and one with capacitance $2C$ farads, how many different values of capacitance can be obtained through different series and parallel combinations of the capacitors? (Use all three capacitors at all times.) What different values of equivalent capacitance are possible?

Solution scheme

a. In solving this problem you should draw out all possible arrangements of connecting the three capacitors and calculate the equivalent capacitance for each case,
b. With all three capacitors in series, calculate the equivalent capacitance. You will find only one value, the minimum possible capacitance value for the three capacitors together.

c. Go through the same exercise for all three in parallel. You will find the *maximum* value of equivalent capacitance.
d. Consider networks with two capacitors in parallel and the third in series with the parallel combination of the other two. How many different equivalent capacitance values can you obtain?
e. Finally, consider networks with two capacitors in series, the third being in parallel with the series combination of the other two.
f. How many different values of equivalent capacitance did you find?

5. *Worked Problem.* The parallel-plate capacitor shown in Figure 27.8(a) is connected to a 100-volt source. After the capacitor is charged, the voltage source is disconnected.

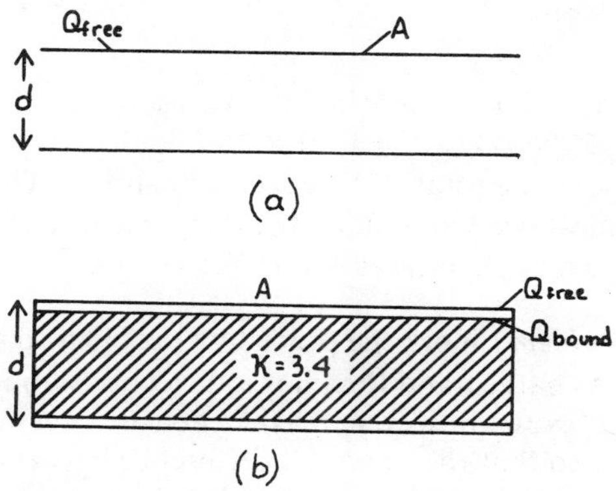

Figure 27.8 Problem 5. (a) A parallel-plate capacitor. (b) The same capacitor, with a dielectric filling the gap between the plates.

a. Find the charge on the capacitor, the electric field strength, the capacitance, and the energy stored on the capacitor. The plate area $A = 4.0$ m^2, and the plate separation $d = 4$ mm.
b. A slab of plastic with dielectric constant $\kappa = 3.4$ is slid between the plates of the isolated capacitor without discharging it in any way [Figure 27.8(b)], so that the gap is completely filled with the dielectric. Find the values of free charge and bound charge, the potential difference between the capacitor plates, the electric field strength, the new capacitance, and the new energy stored on the capacitor. If the energy has been changed by the insertion of the dielectric, has energy conservation been violated? Explain.

Solution

a. The capacitance C_0 is given by the expression

$$C_0 = \frac{\varepsilon_0 A}{d} = \frac{8.85(10^{-12})4}{4(10^{-3})} = 8.85(10^{-9})\text{ F}$$

Then the charge Q_{free} is just $C_0 = 8.85 \times 10^{-7}$ C. Since $V = E_0 d$, the electric field $E_0 = V/d = 25 \times 10^3$ V/m. The total energy stored, U_0, is equal to $\frac{1}{2}QV = 4.42 \times 10^{-5}$ J.

b. With the dielectric between the plates, Q_{free} remains unchanged, but a bound charge Q_{bound} is induced on the upper and the lower surfaces of the dielectric, reducing the electric field by the factor κ. Then $E = E_0/\kappa = 7.35 \times 10^3$ V/m. The capacitance is increased by the factor κ, so that $C = \kappa C_0 = 3.54 \times 10^{-8}$ F. The potential difference between the plates, V, is equal to $Ed = V_0/\kappa = 29.4$ volts. The bound-charge density is

$$\sigma_{\text{bound}} = -\frac{\kappa - 1}{\kappa}\sigma_{\text{free}}$$

or

$$Q_{\text{bound}} = -\frac{\kappa - 1}{\kappa}Q_{\text{free}} = -0.70Q_{\text{free}} = -6.2 \times 10^{\%}\ \text{C}$$

The new energy stored on the capacitor is given by $U = \frac{1}{2}Q_{\text{free}}V = \frac{1}{2}Q_{\text{free}}V_0/\kappa = U_0/\kappa = 1.30 \times 10^{-5}$ joules. There has been a large decrease in the total electrical energy stored. The capacitor must do mechanical work in pulling the dielectric into the space between the plates—energy conservation is not violated.

6. *Guided Problem.* Work the preceding problem, that of a parallel-plate capacitor that is charged and then has a dielectric slab inserted between the capacitor plates. Now, however, let the capacitor remain connected to the 100-volt source, so that the potential difference between the plates remains constant while the dielectric is inserted. Again, $d = 4$ mm and the plate area $A = 4.0$ m^2. The capacitor is shown in Figure 27.9.

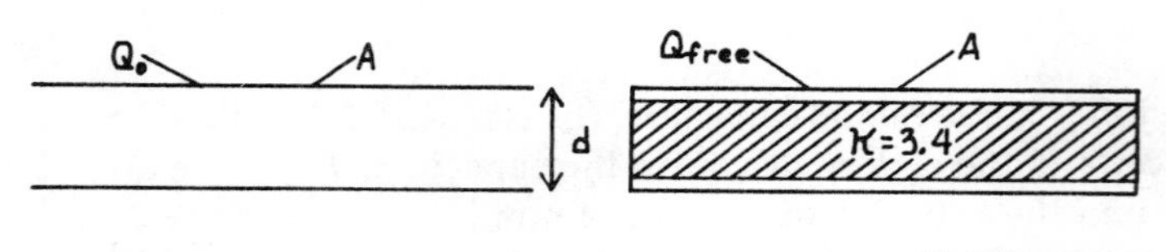

Figure 27.9 Problem 6. A parallel-plate capacitor, first without and then with the space between its plates filled with a dielectric. The capacitor remains connected to a 100-volt source at all times.

Solution scheme

a. The answers to part a of the problem are unchanged from those of the preceding problem. Call the capacitance without the dielectric C_0, the free charge Q_0, and the energy stored U_0.

b. When the dielectric is inserted, the potential difference between the capacitor plates is forced to remain unchanged. The electric field must also therefore remain unchanged—that is, $E = E_0$. The new capacitance C is increased by the factor κ. Can you now calculate the new free charge, Q_{free}, on the capacitor?

c. Knowing the free charge, you should be able to evaluate the bound charge Q_{bound}, using the relationship $\sigma_{\text{bound}} = -\frac{\kappa - 1}{\kappa}\sigma_{\text{free}}$.

d. Knowing the free charge, you should also be able to calculate the stored energy U. Whereas in the preceding problem the final energy was decreased compared with the initial energy, here you should find the final energy, U, to be greater than U_0—it is clear that energy had to be added to the system by the 100-volt source, and thus, again, energy conservation is not violated.

7. *Worked Problem.* In Figure 27.10 the outer metallic sphere of radius $2R$ has an inner concentric dielectric shell of inner radius $1.5R$ and dielectric constant $\kappa = 3.0$. The inner metal shell has radius R. There is a charge Q on the inner metal shell and a charge $-Q$ on the outer spherical shell.

a. Plot the electric field strength, **E**, between $r = 0$ (the center of the spheres) and $r = 3R$. What are the values of **E** at $r = R$, $r = 1.5R$ (outside the dielectric), $r = 1.5R$ (inside the dielectric), and at $r = 2R$?

b. Find the overall capacitance.

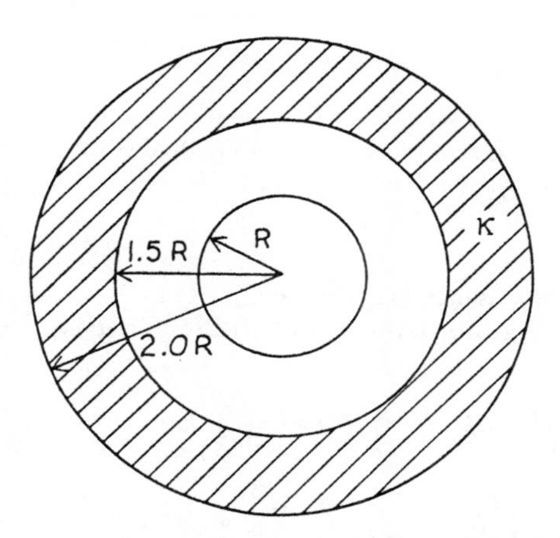

Figure 27.10 Problem 7. Two concentric metal spheres of radii R and $2R$. Inside the outer sphere is a dielectric shell of inner radius $1.5R$.

Solution

a. From Gauss' Law, the field is clearly zero for $r < R$. Between R and $1.5R$, the field, from Gauss' Law, is given by $Q/4\pi\varepsilon_0 r^2$. To find the field between $1.5R$ and $2R$, we apply Gauss' Law for a dielectric and find that the field strength is reduced by a factor κ from what it would be in the absence of the dielectric. Therefore, in this region, the magnitude of **E** is $Q/4\pi\kappa\varepsilon_0 r^2$. Outside $2R$, there is zero net charge, so that $\mathbf{E} = 0$. For $r = R$, the magnitude of **E** is $Q/4\pi\varepsilon_0 R^2$; for $1.5R$ (outside the dielectric),

$$E = Q/4\pi\varepsilon_0(1.5R)^2 = (1/2.25)Q/4\pi\varepsilon_0 R^2.$$

Just inside the dielectric, still at $r = 1.5R$,

$$E = (1/2.25)Q/4\pi\kappa\varepsilon_0 R^2 = (1/6.75)Q/4\pi\varepsilon_0 R^2.$$

At $r = 2R$,

$$E = (1/4)Q/4\pi\kappa\varepsilon_0R^2 = (1/12)Q/4\pi\varepsilon_0R^2.$$

In all cases **E** is directed radially outward. The field is shown plotted as a function of r in Figure 27.11.

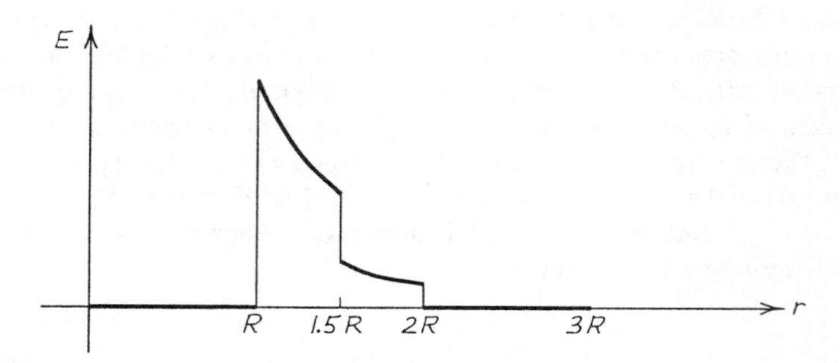

Figure 27.11 Problem 7. The electric field plotted as a function of r for two concentric spheres of radii R and $2R$. Inside the outer sphere is a dielectric shell of inner radius $1.5R$.

b. The overall capacitance is given by $C = Q/\Delta V$, where ΔV is the change in potential in going from R out to $2R$. That is,

$$\Delta V = -\int_R^{2R} \mathbf{E}\cdot d\mathbf{l} = -\int_R^{1.5R} \mathbf{E}\cdot d\mathbf{l} - \int_{1.5R}^{2R} \mathbf{E}\cdot d\mathbf{l}$$

$$= \frac{Q}{4\pi\varepsilon_0}\left[\int_R^{1.5R} \frac{dr}{r^2} + \frac{1}{3}\int_{1.5R}^{2R} \frac{dr}{r^2}\right]$$

$$= \frac{Q}{4\pi\varepsilon_0}\left\{\left[-\frac{1}{r}\right]_R^{1.5R} + \left(\frac{1}{3}\right)\left[-\frac{1}{r}\right]_{1.5R}^{2R}\right\}$$

$$= \frac{Q}{4\pi\varepsilon_0}\left[\left(\frac{1}{R} - \frac{1}{1.5R}\right) + \frac{1}{3}\left(\frac{1}{1.5R} - \frac{1}{2R}\right)\right] = \frac{Q}{4\pi\varepsilon_0}\left[\frac{1}{3R} + \frac{1}{3}\left(\frac{1}{6R}\right)\right]$$

$$\Delta V = \frac{Q}{4\pi\varepsilon_0}\left(\frac{7}{18R}\right) \qquad \text{But} \qquad c = \frac{Q}{\Delta V} = \frac{Q}{\frac{Q}{4\pi\varepsilon_0}\left(\frac{7}{18R}\right)}$$

$$= \left(\frac{18}{7}\right) 4\pi\varepsilon_0 R$$

8. *Guided Problem.* A long, cylindrical coaxial cable has an inner conductor of 0.36 cm diameter copper wire and an outer conductor of diameter 1.20 cm. The space between the inner and the outer conductors is filled with a plastic material. The capacitance of a 100 m length of the cable is measured and found to be 1.25×10^{-8} F. What is the dielectric constant of the plastic? A potential of 1000 volts is applied to the inner conductor, while the outer

conductor is kept at ground potential. How much energy is now stored in the cable? Where is the energy density of the electric field the highest?

Solution scheme

a. What is the expression for the capacitance of a cylindrical capacitor? Can you write it down, knowing that this is the capacitance per meter of length? What is the capacitance of a 1 m section of the cable?
b. Can you now solve for the dielectric constant κ?
c. What is the expression for the energy stored in a capacitor? From this expression, you should be able to find the energy stored in the cable.
d. What is the expression for the energy density of an electric field? Write this down. It should tell you that the electric energy density will be highest where the electric field is the highest. Where is the field a maximum?

9. *Worked Problem.* The region between the plates of a parallel-plate capacitor is filled half with polystyrene ($\kappa = 2.6$) and half with rubber ($\kappa = 6.7$), as indicated in Figure 27.12. The plate separation $d = 6.0$ cm, and the plate area $A = 1.0\ \text{m}^2$.

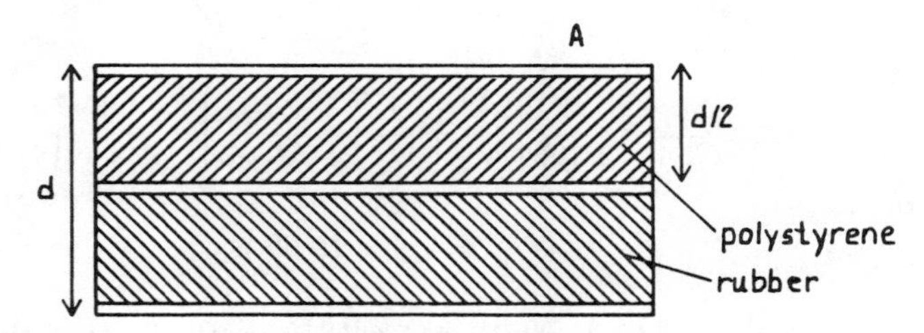

Figure 27.12 Problem 9. A parallel-plate capacitor, with two different dielectric materials.

a. Find the capacitance of the capacitor.
b. A charge of 1.0×10^{-6} C is stored on the capacitor. Find the electric field in each dielectric, the potential difference V between the plates, and the potential across each dielectric layer.

Solution

a. In solving this problem, consider the following: A thin metal plate placed midway between the plates of a parallel-plate capacitor, as shown in Figure 27.13, changes the single capacitor into two capacitors in series. The capacitance of each of the two capacitors is twice that of the single capacitor, so that the total capacitance is unchanged.

The easiest way of looking at this problem is to treat the capacitor as two capacitors in series, each with plate separation $d/2$ and plate area A, and with respective dielectric constants κ_1 and κ_2. Because of the different dielectrics, the capacitance of the two capacitors in series is *not* the same. The equivalent capacitance of the two capaci-

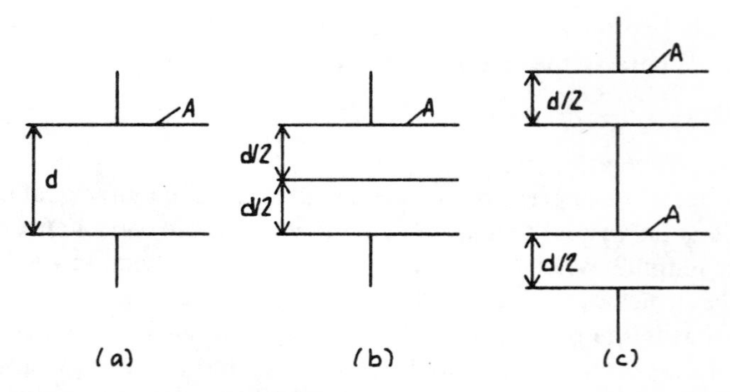

Figure 27.13 Problem 9. (a) A parallel-plate capacitor. In (b) a thin metal plate is placed midway between the capacitor plates, changing the single capacitor into two capacitors in series, as indicated in (c). The capacitance is the same in (a), (b), and (c).

tors in series is given by

$$\frac{1}{C_{eq}} = \frac{1}{C_1} + \frac{1}{C_2}$$

$$C_1 = \frac{\varepsilon_0 \kappa_1 A}{d/2} \quad C_2 = \frac{\varepsilon_0 \kappa_2 A}{d/2}$$

$$\frac{1}{C_{eq}} = \frac{d}{2\varepsilon_0 A}\left[\frac{1}{\kappa_1} + \frac{1}{\kappa_2}\right] = \frac{d}{2\varepsilon_0 A}\left[\frac{\kappa_1 + \kappa_2}{\kappa_1 \kappa_2}\right]$$

or

$$C_{eq} = 5.5 \times 10^{-10}\ \text{F}$$

b. Since $C = Q_{free}/V$, $V = Q_{free}/C = 1810$ volts. To find the electric field inside the dielectrics, apply Gauss' Law for a dielectric:

$$\oint \kappa\varepsilon_0 \mathbf{E} \cdot d\mathbf{s} = Q_{free}$$

$$E = \frac{\sigma_{free}}{\kappa\varepsilon_0} = \frac{Q_{free}}{\kappa\varepsilon_0 A}$$

Because we are told that $Q_{free} = 1 \times 10^{-6}$ C, we know that $\sigma_{free} = 1 \times 10^{-6}\ \text{C/m}^2$. The field in the polystyrene is therefore given by

$$E = \frac{1 \times 10^{-6}}{2.6(8.85 \times 10^{-12})} = 4.34 \times 10^4\ \text{V/m}$$

The field E in the rubber, with κ equal to 6.7, is given by

$$E = \frac{1 \times 10^{-6}}{6.7(8.85 \times 10^{-12})} = 1.69 \times 10^4\ \text{V/m}$$

The potential difference across the polystyrene layer is then given by $Ed = (4.34 \times 10^4)(.03) = 1302$ volts, and that across the rubber by

$(1.69 \times 10^4)(0.03) = 507$ volts. These two potential differences should sum to the total potential difference across the capacitor plates, which they do.

10. *Guided Problem.* For any dielectric material there is a maximum electric field strength that the material can withstand without breaking down and losing its insulating quality. For air this is about 3×10^6 V/m; for transformer oil this is 12×10^6 V/m. Compare the capacitance and the total charge that can be stored on a parallel-plate capacitor of plate area 6 m^2 and plate separation of 0.5 cm
   a. if the capacitor is in air.
   b. if the capacitor is immersed in transformer oil, for which the dielectric constant $\kappa = 2.2$.

Solution scheme

a. The capacitance in air can be easily evaluated using the equation for the capacitance of a parallel-plate capacitor. Assume that the field between the plates $E = 3 \times 10^6$ V/m, the maximum possible value for an air gap. From this assumption you should be able to obtain the maximum voltage difference between the plates, since $V = Ed$. Now, from the definition of capacitance, you can obtain the maximum charge that can be stored on the capacitor.

b. When the capacitor is immersed in oil, the capacitance is increased by the dielectric constant factor. Furthermore, the maximum electric field is now increased to 12×10^6 V/m, allowing a larger maximum potential difference between the plates. You should be able to obtain Q_{max} just as you did in part a. By what ratio has the amount of charge that can be stored increased?

Answers to Guided Problems

2. We have eight pairs of identical plates; the capacitance of each pair is 4.42×10^{-12} F. The total capacitance is then 3.54×10^{-11} F. There is no charge on the extreme upper and lower plate surfaces of the right half of the capacitor system. The charge distribution on each of the remaining sixteen plate faces is $(Q/8A)$ C/m^2, $(+Q/8A)$ on the right side and $(-Q/8A)$ on the left side.
4. The possible values of capacitance are 0.4 C, 0.75 C, 1.0 C, 1.67 C, 2.5 C, and 4 C (0.4 C when all are in series; 4 C when all in parallel).
6. a. Without the dielectric $C_0 = 8.85 \times 10^{-9}$ F; $E_0 = 25 \times 10^3$ V/m; the charge $Q_{free} = 8.85 \times 10^{-7}$ C; the energy $U_0 = 4.42 \times 10^{-5}$ joules.
   b. With the dielectric $C = \kappa C_0 = 3.54 \times 10^{-8}$ F; $E = E_0 = 25 \times 10^3$ V/m; $Q_{free} = CV = \kappa C_0 V = 3.54 \times 10^{-6}$ C; $U = \kappa U_0 = \frac{1}{2} Q_{free} V = 1.77 \times 10^{-}$ joules.
8. $\kappa = 2.7$
   $U = \frac{1}{2}CV^2 = 6.25 \times 10^{-3}$ J
   E is greatest at the surface of the inner conductor, and here the energy density of the field is greatest.

10. a. $C_0 = 1.06 \times 10^{-8}$ F
$Q_{max} = 1.59 \times 10^{-4}$ C
b. $C = 2.33 \times 10^{-8}$ F. $Q_{max} = 1.4 \times 10^{-3}$ C, which is nearly nine times larger than in part a.

Chapter 28

Currents and Ohm's Law

Overview

Until this chapter we have considered conductors only under static conditions, with no moving charge and with zero electric field inside the conductors. If we continuously supply positive and negative charge to opposite ends of a long conductor, however, the electrical charges will maintain an electric field across the conductor, causing charge to flow continuously along the electric field. This flow of electric charge is called an electric current. We examine the behavior of an electric current in a conductor such as a metal and in a class of materials called semiconductors.

The current in a conductor is usually described by an extremely useful relationship called Ohm's Law, which states that the ratio of the potential difference between the ends of a conductor to the magnitude of the current through the conductor is equal to a quantity called the resistance of the conductor. We discuss the properties and behavior of resistance and finally examine the effect of series and parallel combinations of conductors with different resistance values.

Essential Terms

electric current
current density
Ohm's Law
resistance and resistivity
semiconductor, donors, acceptors
resistor
resistors in series and in parallel

Key Concepts

1. *Electric current*. The instantaneous rate at which positive charge passes through a surface is called the electric current through the surface. Figure 28.1 shows a surface S that intersects a conductor through which charge is

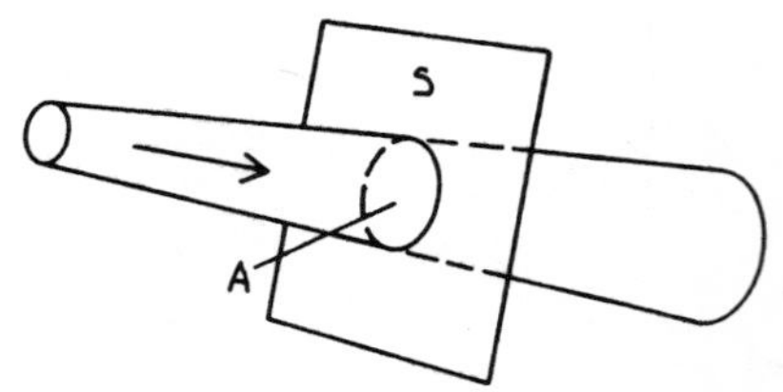

Figure 28.1 Current flowing through a surface S that intersects a conductor.

flowing. The arrow is used to indicate the direction of the flow of positive charge. Then the current $I = dq/dt$. For an isolated conductor, the principle of charge conservation tells us that the current is the same everywhere along the conductor. In a metal the charge carriers are electrons, so that although the current flow is defined to be in the direction of the flow of positive charge carriers, the real flow is that of electrons in the opposite direction.

Usually the current is uniformly distributed over the whole cross-sectional area of the conductor. In these cases the *current density* j is defined as the current per unit area of the conductor, that is, $j = I/A$. In general, the current density is defined as $j = dI/dA$, where dI is the current flowing through a very small element of area dA of the conductor, so that the total current $I = \int \mathbf{j} \cdot d\mathbf{A}$, where the integration is carried over the cross-sectional area of the conductor.

2. *Ohm's Law.* Let us examine the flow of current in a metal on a microscopic scale. The current is due to the flow of conduction electrons, which behave as an electron "gas" in the conductor. The electrons are in constant high-speed thermal motion and experience frequent collisions with the crystal lattice. Suppose that the average time between collisions is τ. An external electric field causes a slow average drift velocity, υ_d, of the electrons in the gas along the field. Let the average time between collisions with the lattice be τ, and suppose that each collision destroys the forward drift velocity. Then, during time τ, the electron loses the momentum to the lattice at the rate

$$(dp/dt)_{\text{loss}} = m\upsilon_d/\tau$$

and gains momentum at the rate

$$(dp/dt)_{\text{gain}} = \text{force} = -eE$$

For a steady-state current, $m\upsilon_d/\tau = -eE$, giving for the average drift velocity the result that

$$\upsilon_d = \frac{-eE\tau}{m}$$

Now let us count the number of electrons per second that flow past point P of the conductor shown in Figure 28.2. If the average drift velocity is $\mathbf{v}_d$, in 1 s all the conduction electrons in the conductor within a distance υ_d to the left of P will pass point P. Then the current $I = \Delta q$ per second $= dq/dt = -enA\upsilon_d$. If we substitute in our value of υ_d, we obtain

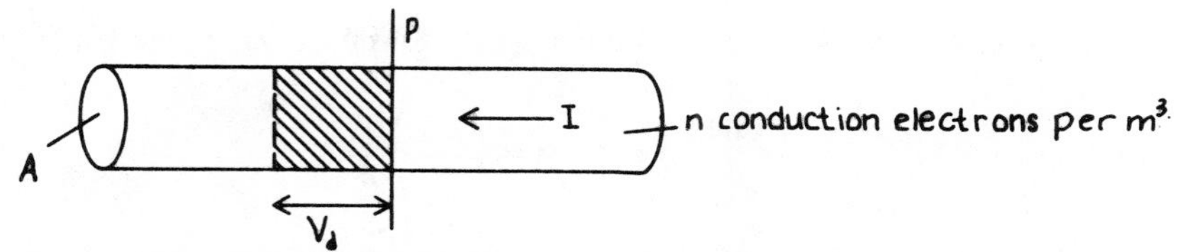

Figure 28.2 A metal conductor with n conduction electrons per cubic meter and with cross-sectional area A m^2.

$$I = \frac{e^2 n\tau}{m} AE$$

where the direction of the current, as indicated in Figure 28.2, is opposite the direction of v_d, the drift velocity of the free electrons. For a conductor of uniform cross-sectional area, the potential between the ends of the conductor $\Delta V = El$, where l is the length. Then the current I can be written as

$$I = \left(\frac{e^2 n\tau A}{ml}\right) El = \left(\frac{e^2 n\tau A}{ml}\right) \Delta V$$

or $I = \Delta V/R$, where R is called the *resistance* and is given here by

$$R = \frac{ml}{e^2 n\tau A}$$

The expression $I = \Delta V/R$ is called *Ohm's Law*. For a material that obeys Ohm's Law, the relationship $I = \Delta V/R$, where R is a constant independent of ΔV, must be true. Many materials, including plasmas, obey Ohm's Law, but only for a metal conductor do we expect the resistance R to be given by the simple expression $R = ml/e^2 n\tau A$. The resistance R is defined to be simply the ratio $\Delta V/I$. *Any* material has a resistance, but only in those cases in which the resistance is constant does the material obey Ohm's Law.

3. *Resistivity*. We usually write the resistance in the form

$$R = \rho l/A$$

where the quantity ρ is called the resistivity and is independent of the length and cross-sectional area of the conductor. We can therefore compute the resistance of a given sample from its physical dimensions if we know the resistivity.

For a metal, we obtained the relationship $R = ml/e^2 n\tau A$, so we expect the resistivity to be given by

$$\rho = m/e^2 n\tau$$

Again, this is only an approximate relationship and is useful only for metals. In fact, for metals the resistivity increases somewhat with increasing temperature, and tables of resistivity values usually specify a temperature (often 20°C) at which the value for ρ is correct.

Since $I = \Delta V/R$ and $R = \rho l/A$, we can write Ohm's Law in the form

$$I = \frac{\Delta V(A)}{\rho l} \quad \text{or} \quad \frac{I}{A} = \left(\frac{\Delta V}{l}\right)\frac{1}{\rho}$$

Since I/A is the current density,

$$j = \frac{E}{\rho}$$

This equation, which says that the current density is proportional to the electric field strength in the conductor, does not require a conductor of uniform cross-sectional area.

4. *Resistor; resistors in series and in parallel.* All conductors have resistance, and any resistive circuit element is called a *resistor*. A resistor is indicated by the symbol —\/\/\/\/\/—. A series connection of two resistors is shown in Figure 28.3 and is used to indicate that the two resistors are so

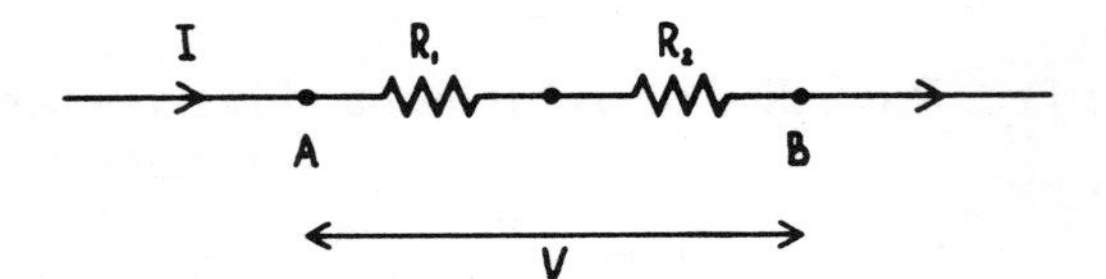

Figure 28.3 Two resistors in series.

connected that all current flowing through resistor R_1 also flows through R_2 —that is, there is a common current I in the two resistors. The potential difference across R_1 is given by $V_1 = IR_1$, and the potential difference across R_2 is given by $V_2 = IR_2$. The total potential difference V between A and B is

$$V = V_1 + V_2 = I(R_1 + R_2)$$

The two resistors are therefore equivalent to a single resistor with resistance

$$R_{eq} = R_1 + R_2$$

A parallel combination of two resistors is shown in Figure 28.4. The two resistors are so connected that although the same potential difference V be-

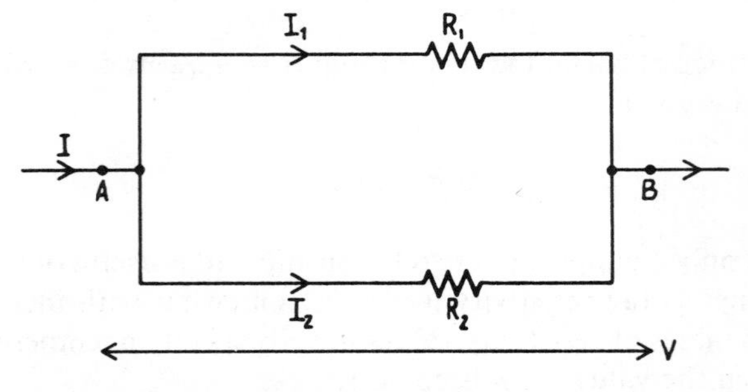

Figure 28.4 Two resistors in parallel.

tween A and B appears across each resistor, the current divides, some flowing through R_1 and some through R_2. The total current $I = I_1 + I_2 = V/R_1 + V/R_2 = V[1/R_1 + 1/R_2] = V/R_{eq}$. The two resistors in parallel are therefore equivalent to a single resistor with resistance given by

$$\frac{1}{R_{eq}} = \frac{1}{R_1} + \frac{1}{R_2}$$

Sample Problems

1. *Worked Problem.* The Los Alamos Meson Physics Facility accelerator has a maximum average proton current of 1 ma (1×10^{-3} ampere) at an energy of 800 MeV. How many protons per second strike a target exposed to this beam of protons? If the beam is of circular cross section and has a diameter of 5 mm, what is the current density? As with many accelerators, the beam current is not a steady current but, rather, occurs in pulses. There are 120 pulses per second, and between each pulse there is a relatively long time interval during which the beam current is zero. If the "duty factor" is 8%, which means that the beam is "on" and protons are striking the target 8% of the time, what is the length of each pulse, and what is the length of the time interval between pulses? What is the "average" beam current *during* a pulse? How many protons per pulse strike the target?

Solution: From the definition of current, $I = dq/dt = n(e) = 1 \times 10^{-3}$, we obtain $n = 1 \times 10^{-3}/1.6 \times 10^{-19} = 6.25 \times 10^{15}$ protons per second. The current density $j = I/A$, where $A = \pi d^2/4 = \pi(5 \times 10^{-3})^2/4 = 1.96(10^{-5})\ \mathrm{m}^2$.

$$j = 51\ \mathrm{A/m^2}$$

The beam "on" time is 0.08 s/s, and since there are 120 pulses per second, the length of each pulse is $0.08/120 = 6.67 \times 10^{-4}$ s, or 667 μs. The time interval between pulses is $0.92/120 = 7.67 \times 10^{-3}\ \mathrm{s} = 7670\ \mu\mathrm{s}$. The average current during a pulse is given by $I = dq/dt = 1 \times 10^{-3}/0.08 = 0.0125$ A, and in every pulse there are $6.25 \times 10^{15}/120 = 5.2 \times 10^{13}$ protons.

2. *Guided Problem.* Niobium metal has an atomic weight of 92.9 and a density of 8.57 gm/cm^3. At extremely low temperatures it is a "superconductor"; that is, it has zero electrical resistance. A wire of superconducting niobium 0.2 cm in diameter can carry a current of 1900 A. Assuming that every niobium atom has two free conduction electrons, estimate the average drift velocity of the conduction electrons in the wire. What is the current density in the wire? Assume that the expression that we derived for the resistance of a metal is still roughly applicable to this situation. What assumption must we make about the conduction process in order for a resistance value R equal to zero to be possible?

Solution scheme

a. Write down the expression for the current I in terms of the number of conduction electrons per cubic meter, the area of the conductor, and the average drift velocity. Clearly, to estimate v_d, we need to obtain n, the number of conduction electrons per cubic meter. Calculate n from the density, the atomic weight, and Avogadro's number, being careful about units. Remember to take two conduction electrons per atom. Calculate υ_d.

b. From the definition of current density, evaluate j.

c. Write down the expression for R in terms of the conductor length and area, the electron charge and mass, the number n, and the mean time between collisions of the conduction electrons with the lattice. It should be clear what assumption we *must* make if R is to be zero —do you think that this is a reasonable assumption in view of the very low temperatures at which a metal becomes superconducting?

3. *Worked Problem.* A copper bar 0.5×1.0 cm in cross-sectional area and 3.0 m long has a steady potential of 0.1 volt between its ends. Find the current flowing through the bar, the current density, and the electric field strength in the bar. How many electrons per second flow past any point of the bar? The atomic weight of copper is 63.5, and the density is 8.96 g/cm³. Assuming that there is one free conduction available per atom of copper, find the average drift velocity of the electrons, υ_d, and how long it takes, on the average, for an electron to flow from one end to the other end of the bar. The resistivity of copper is $1.7 \times 10^{-8}\ \Omega \cdot$ m.

Solution: Because the bar is a uniform conductor, $El = V = 0.1$ volt. Then the electric field $E = 0.033$ V/m along the bus bar. Since the current density $j = E/\rho = 0.033/1.7 \times 10^{-8}$

$$j = 1.96 \times 10^6\ \mathrm{A/m^2}$$

The total current $I = jA = (1.96 \times 10^6)(.5 \times 1 \times 10^{-4})$ or $I = 98$ A. The number of electrons is $N = I/e = 6.12 \times 10^{20}$ per second.

To obtain the drift velocity υ_d, we use the expression that $I = enA\upsilon_d$, so that we need to estimate n, the number of conduction electrons per cubic meter. One cubic centimeter of copper contains $(8.96/63.5)(6.02 \times 10^{23}) = 8.49 \times 10^{22}$ atoms and the same number of conduction electrons. Since 1 m³ contains 10^6 cm³, $n = 8.49 \times 10^{28}$ electrons per cubic meter. Substituting in our values for I, A, n, and e, we obtain $\upsilon_d = 1.44 \times 10^{-4}$ m/s for the average drift velocity. The time required for an electron for flow through the bar is just $t = \text{length}/\upsilon_d = 3.0/1.44 \times 10^{-4} = 2.1 \times 10^4$ s.

4. *Guided Problem.* An electrical cable 1 km long needs to carry a current of 100 A. The maximum allowable voltage drop along the cable is 10 volts. If the cable is to be made of copper, what diameter copper cable should be used? If the cable were aluminum, what is the minimum diameter that could be used? What are the masses of the cables, and which cable would be lighter, if weight is an important consideration? The resistivities of copper and aluminum are, respectively, $1.7 \times 10^{-8}\ \Omega \cdot$ m and $2.8 \times 10^{-8}\ \Omega \cdot$ m. The density of copper is 8.96 g/cm³, and that of aluminum is 2.70 g/cm³.

Solution scheme

a. Calculate the resistance of the cable from the current and the voltage difference between the ends of the cable.

b. What is the expression for resistance in terms of resistivity and the cable dimensions? Now calculate the area and from it the diameter for each of the two materials, copper and aluminum.
c. From the volume and density, calculate the mass of each cable. Which one is lighter?

5. *Worked Problem.* A strand of a superconducting cable consists of 2100 thin filaments of niobium alloy, each with diameter of 0.01 mm, embedded in a copper matrix of overall diameter 0.7 mm. The current density in the strand is 1×10^5 A/cm^2. For a single such strand, find the total cross-sectional area of the niobium alloy and of the copper matrix. What is the total current in the strand? How much of this is carried by the alloy, and how much by the copper matrix? Suppose that, because of a sudden increase in temperature, the superconductor suddenly fails. How much current is now carried by the copper, and how much by the alloy, assuming that the resistivity of the alloy (nonsuperconducting) is $20 \times 10^{-8}\ \Omega \cdot$m. The resistivity of copper is $1.7 \times 10^{-8}\ \Omega \cdot$m. What would be the voltage drop along a strand 1 m long (a) while still superconducting and (b) after failure of superconductivity, assuming that the current remains constant?

Solution: The total cross-sectional area of the strand is 3.848×10^{-3} cm^2, while that of the alloy filaments alone is $2100(\pi \times 0.001 \times 0.001)/4 = 1.649 \times 10^{-3}$ cm^2. The area of the copper is then 2.199×10^{-3} cm^2 per strand. The total current per strand is $(1 \times 10^5)(3.838 \times 10^{-3}) = 385$ A. Since the resistivity of the superconductor = 0, the resistance of the niobium alloy = 0, and therefore all of the current is carried by the alloy. If the superconductor fails, the resistance of the 2100 alloy strands is given by $\rho L/A = 20 \times 10^{-8}/(1.649 \times 10^{-3} \times 10^{-4}) = 1.213$ ohms (note that we multiplied by 10^{-4} to convert the area of m^2), while that of the copper matrix is $1.7 \times 10^{-8}/(2.199 \times 10^{-7}) = 0.077$ ohms. Since the voltage drop across the copper equals that across the alloy, that is, $I_{cu}R_{cu} = I_{al}R_{al}$, or $I_{cu}/I_{al} = R_{al}R_{cu}$, and $I_{cu} + I_{al} = 385$ A, the copper would carry $385 \times (1.21/1.29) = 362$ A, and the alloy filaments would carry only $385 \times (0.077/1.29) = 23$ A.

While superconducting, the voltage drop along a strand 1 m long = 0.0 V, since $R = 0$. After the failure of superconductivity, the resistance is given by $R = R_{al}R_{cu}/(R_{al} + R_{cu}) = 0.0724\ \Omega$. The voltage drop is thus given by $V = IR = 27.9$ V.

6. *Guided Problem.* Two coils of wires, each 80 m long, are connected in parallel to form a single coil with two strands of wire each 80 m long. The diameter of each wire is 0.10 cm. One wire is made of copper ($\rho = 1.7 \times 10^{-8}\ \Omega \cdot$m); the material in the other wire is unknown. When you connect the ends of the two wires across a 12.0 V car battery, the current through the battery is 8.11 A. Find the resistivity of the second wire, and from the table of resistivities in your text (p. 690), see if you can deduce what it is made of.

Solution scheme

a. Since you know the total voltage and current, you should be able to find the parallel resistance of the two coils of wire.
b. From the resistivity of copper, the total length, and the wire diameter, can you find the resistance of the copper wire?

c. Can you now find the resistance of the "unknown" wire coil?
d. Since you know the length and the diameter of the "unknown" wire, can you find its resistivity? Look at the table in your text—can you tell what material the wire is probably made of?

7. *Worked Problem.* According to a simple model of the atom, in a hydrogen atom the electron may be thought of as rotating around a fixed proton in a circular orbit of radius r. If the velocity of the electron is υ m/s, find an expression for the average current flow around the proton. Find the current in amperes. $r = 5.3 \times 10^{-11}$ m, and the kinetic energy of the electron is 13.6 eV.

Solution: It is worthwhile to make a sketch of the problem (Figure 28.5). The current I is defined as dq/dt; the current past any point of the circumference of the circle is the charge per unit time past that point. Suppose we take a time interval of 1 s; the current must be the charge multiplied by the number of times the electron passes that point every second.

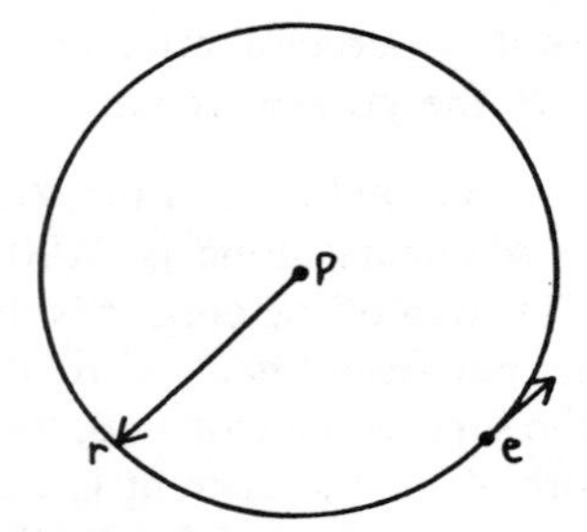

Figure 28.5 Problem 7. A simplified model of atomic hydrogen.

Then

$$I = (e)(\upsilon)/2\pi r = \frac{(1.6 \times 10^{-19})\ \upsilon}{2\pi(5.3 \times 10^{-11})}\ \text{A}$$

The kinetic energy $K = \frac{1}{2}m\upsilon^2 = 13.6$ eV. We know that 1 eV $= 1.6 \times 10^{-19}$ J, and since $m = 9.1 \times 10^{-31}$ kg,

$$\upsilon^2 = \frac{2(13.6)(1.6 \times 10^{-19})}{(9.1 \times 10^{-31})}$$

$$\upsilon = 2.19 \times 10^6 \text{ m/s}$$

By substitution, we find that

$$I = e\upsilon/2\pi r = 1.05 \times 10^{-3} \text{ A}$$

8. *Guided Problem.* Figure 28.6 shows the graph of potential difference across the conductor versus current for a certain sample. What is the resistançe of the sample (approximately) when the potential difference is 1.0 volt? when it is 10.0 volts? Does the sample obey Ohm's Law, and, if not, why not? Draw a graph of potential difference versus current for a material that does obey Ohm's Law.

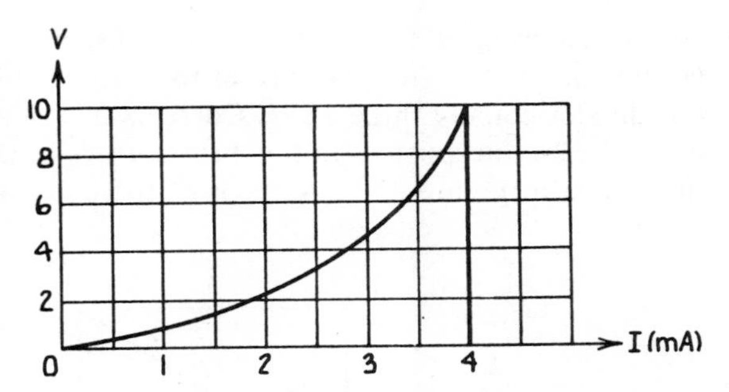

Figure 28.6 Problem 8. Potential difference versus current for a material. The current is in milliamperes (1 mA $= 10^{-3}$ A).

Solution scheme

a. Resistance is defined as the ratio of potential difference to current through a conductor. Can you read the respective values of current for the potential differences of 1.0 and 10.0 volts from the graph? From these current values you should be able to obtain the values of resistance and so determine whether or not the conductor obeys Ohm's Law.
b. How should the graph of voltage versus current look if the material obeys Ohm's Law?

9. *Worked Problem*. The cube shown in Figure 28.7 has a resistor along each edge. Those adjacent to corner A and those adjacent to corner B are 2-ohm resistors; the remaining six resistors are all 1-ohm resistors. What is the equivalent resistance of the resistor network between points A and B?

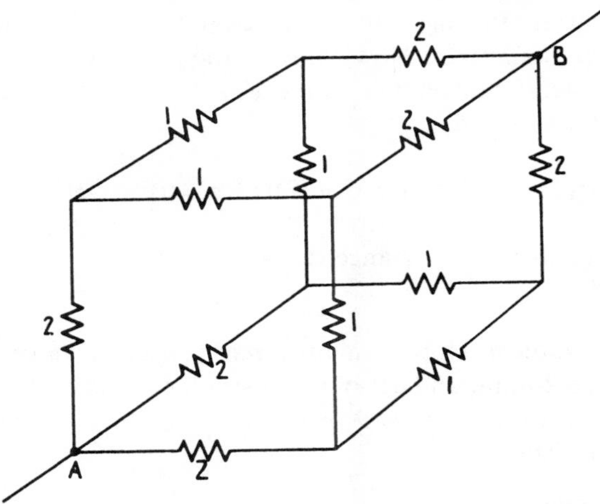

Figure 28.7 Problem 9. A cube with a resistor along each edge. Those marked "2" are 2-ohm resistors, and those marked "1" are 1-ohm resistors.

Solution: This is a slight variation of an old problem that appears in many books of electromagnetic theory. It is not really a problem on series and parallel combinations of resistors as much as it is a problem on observing symmetries.

Let the current entering at corner A be I amperes. From A there are exactly three equivalent paths for the current to follow, and therefore at A the current divides equally three ways. Currents of $I/3$ amperes flow through each of the 2-ohm resistors to the three corners of the cube adjacent to A, as shown in Figure 28.8. At each of these corners, there are

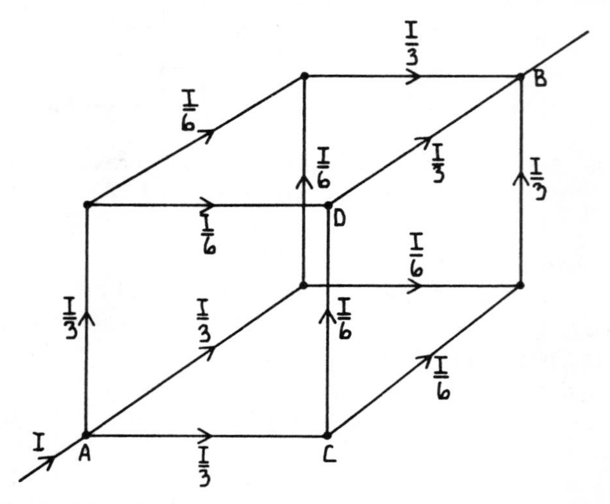

Figure 28.8 Problem 9. The respective currents along the edges of the cube.

exactly two equivalent paths for the current to follow. The current through *each* of the 1-ohm resistors is therefore $\frac{1}{2}(I/3) = I/6$ amperes. Since the current leaving each corner is the same as the total current entering it, the current through each of the 2-ohm resistors adjacent to B must be $2(I/6) = I/3$ amperes. Now consider the path $ACDB$ shown in Figure 28.8. The potential difference $V_{AB} = V_{AC} + V_{CD} + V_{DB}$, and by applying Ohm's Law we obtain

$$V_{AB} = 2I/3 + (1)I/6 + (2)I/3 = 9I/6 = 3I/2 = IR_{eq}$$

so that the equivalent resistance $R_{eq} = 1.5$ ohms.

10. *Guided Problem*. How many different resistance values can be made from a 1-ohm, a 2-ohm, and a 3-ohm resistor? You may take any combination of one, two, or all three resistors. Make a table of the resulting values of equivalent resistance.

Solution scheme

a. Taking one resistor at a time, find what resistance values are possible. (This is the easiest part of the problem!)
b. Taken two at a time, both resistors may be in series or in parallel. What resistance values do you find?
c. Taken three at a time, all three may be in series or in parallel. In addition, two may be in parallel, with the third in series with the parallel resistors, or two may be in series, with the third in parallel

with the two in series. Evaluate carefully all the different possibilities. (This is the hardest part of the problem!)

d. Tabulate your results, noting where you obtain the same value of equivalent resistance in different ways.

Answers to Guided Problems

2. $v_d = 0.068$ m/s, or 6.8 cm/s. Note that this is between two and three orders of magnitude greater than the drift velocity in problem 3.
   The current density $j = 6.05 \times 10^8$ A/m^2.
   We must assume that the electrons do not collide with the lattice, that is, that the time between collisions approaches infinity.
4. For copper, the diameter is 1.47 cm.
   For aluminum, the diameter is 1.89 cm.
   The cable mass, if copper, is 1520 kg and, if aluminum, is 756 kg. Even though the diameter of the aluminum cable is greater, the density of aluminum is less, so that the aluminum cable is much lighter than the copper one.
6. $R_{cu} = 1.73\ \Omega$
   $R_{unknown} = 10.24\ \Omega$
   $\rho_{unknown} = 10 \times 10^{-8}\ \Omega \cdot m$
   The material is iron.
8. At $V = 1.0$ volt, $I = 1.25$ mA, $R = 800$ ohms.
   At $V = 10.0$ volts, $I = 4$ mA, $R = 2500$ ohms.
   The resistance R is not a constant, independent of voltage, so the conductor does not obey Ohm's Law. For a material that obeys Ohm's Law, the graph of V versus I would be a straight line running through the origin, as shown in Figure 28.9.

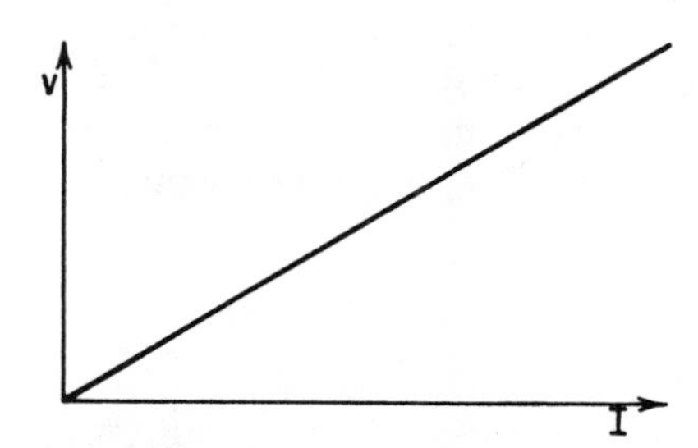

Figure 28.9 Problem 8. A graph of V versus I for a material that obeys Ohm's Law.

10. The values of equivalent resistance are as follows (units are ohms):
    1, 2, 3 (resistors taken one at a time)
    0.67, 0.75, 1.2, 3.0, 4.0, 5.0 (taken two at a time)
    2.2, 3.67, 2.75, 6, 0.55, 0.83, 1.33, 1.50 (taken three at a time)
    Sixteen different values of equivalent resistance are therefore possible, ranging from 0.67 ohms to 6.0 ohms.

Chapter 29

DC Circuits

Overview

In the last chapter we discussed steady current flow through an electrical conductor, flow caused by continuously supplying positive and negative charges onto opposite ends of the conductor, or by supplying positive or negative charge to one end of the conductor and removing it from the other end. We now discuss the mechanism, or pump, of electricity, by which charges are continuously supplied to the conductor. We then deduce the rules governing the flow of electric charge around the complete loop from the electric pump, through the conductor, and back to the pump. This loop is called an electric circuit. We show that an electric circuit, like any other system, must obey the laws of charge and energy conservation, and we develop an expression for the energy and power expended in an electric circuit.

Essential Terms

circuit
direct current
electromotive force
battery
dry cell
electric generator
fuel cell
solar cell
single-loop circuit
Kirchhoff's rules
multiloop circuit
Joule heating
electrical measurements
RC circuits

Key Concepts

1. In order to maintain a steady, continuous current through a conductor, usually called a *direct current*, it is necessary to have an external device, or "electric pump," that can continuously supply electric charges to one end

of the conductor and remove them from the other end. A charged capacitor is a device that can, and will, supply charges to the ends of a conductor if we hook the conductor across the capacitor plates. The resulting current flow, however, will not be continuous and steady, because as the capacitor becomes discharged, the supply of charges available on the capacitor decreases and the electric field in the conductor decreases. The current stops flowing as soon as the supply of charges on the capacitor becomes exhausted.

2. *Electromotive force*. One instrument that can maintain a direct current is a *battery*, a device with two terminals, or points with a potential difference between them where connections can be made to an external conductor. Charges fed into the terminal of lower potential are pumped by the battery to the other terminal and are available to flow from there into an external conductor. In short, a battery acts very much like a hydraulic pump. The electromotive force, or emf, of the battery is simply the potential difference between the two terminals—the energy per coulomb supplied by the battery as charge flows through it from the low-voltage terminal to the terminal at higher voltage. A battery is referred to as a source of emf. The essential constituents of a battery are two dissimilar metals immersed in an ionic solution. The ionic solution acts to deposit electrons on the negative electrode and to absorb electrons from the positive electrode. Other sources of emf are electric generators, fuel cells, and solar cells.

3. *Single-loop circuit*. The DC circuit shown in Figure 29.1 has a steady current I flowing from terminal B of the source of emf through the conductor and into the lower terminal A, which is at a lower potential than B by the

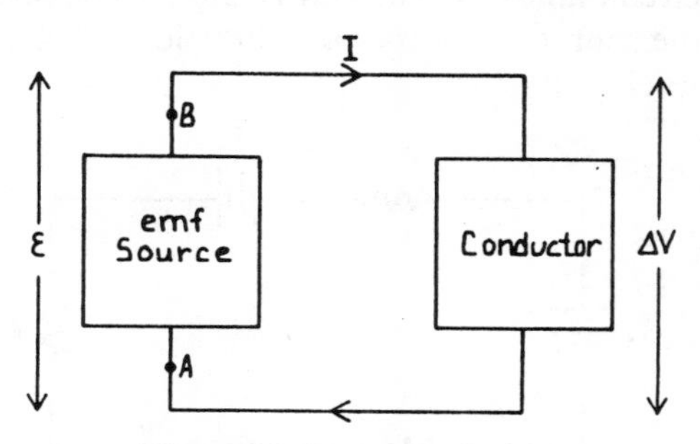

Figure 29.1 A DC circuit.

amount $\mathcal{E}$. The lines between the source of emf and the conductor represent paths along which the current can flow with no resistance. In the steady state, charge must always return to point B with the same energy it had originally, so that the energy loss by the charge in the conductor must exactly match the increase in potential energy given to the charge by the source of emf—that is, $\mathcal{E} + \Delta V = 0$. When we discussed current flow along a conductor, we pointed out that the conduction electrons lose all their drift momentum during each of their frequent collisions with the lattice of the conductor. The electrons emerge from the conductor with exactly the same kinetic energy with which they enter the conductor, having given up the electrical potential energy ΔV per unit charge to the lattice during the passage along the conductor.

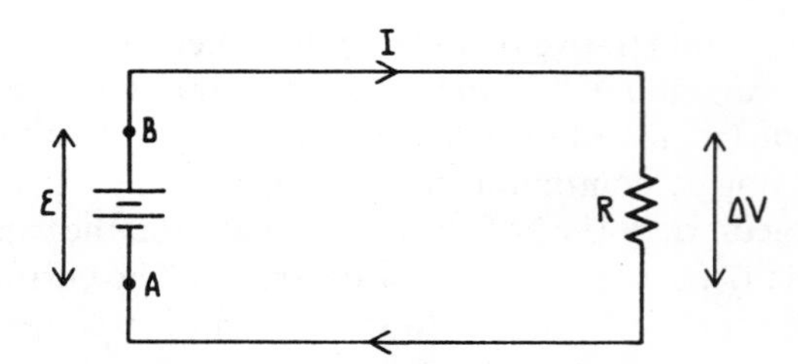

Figure 29.2 The same DC circuit as in Figure 29.1.

The circuit of Figure 29.1 is usually drawn as shown in Figure 29.2. The nonuniform horizontal lines on the left are used to indicate a source of emf. Since the potential drop between the two ends of the conductor $\Delta V = -IR$, the expression $\mathcal{E} + \Delta V = 0$ can be rewritten as $\mathcal{E} - IR = 0$, where R is the resistance of the conductor. This is the circuit equation for a *single-loop circuit*; the term *single-loop* means that there is only one path available for the current to follow. The equation can be stated more generally: around any closed loop the sum of the emfs and all the potential drops across circuit elements must equal zero. This last statement allows for the possibility of there being more than one source of emf and more than one conducting element in the circuit. Since conducting elements have resistance, R, usually they are just referred to as resistors.

4. ***Kirchhoff's rules.*** The statement just made, that around any closed loop the sum of all the emfs and all the potential drops across resistors and other circuit elements must equal zero, is often referred to as Kirchhoff's *second rule* of circuit analysis. The emf is taken to be positive if the current flows through the source in the forward direction. Figure 29.3 shows a possible closed-loop circuit.

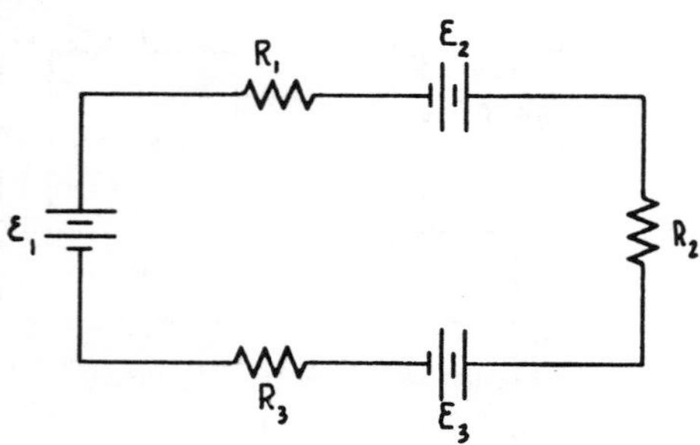

Figure 29.3 A closed loop circuit with several sources of emf.

Let us assume that the current flow in the circuit is clockwise. Then from Kirchhoff's second rule, since the current I is the same everywhere in the circuit,

$$\mathcal{E}_1 + \mathcal{E}_2 - \mathcal{E}_3 - IR_1 - IR_2 - IR_3 = 0$$

$$\mathcal{E}_1 + \mathcal{E}_2 - \mathcal{E}_3 = I(R_1 + R_2 + R_3)$$

$$I = \frac{\mathcal{E}_1 + \mathcal{E}_2 - \mathcal{E}_3}{R_1 + R_2 + R_3}$$

If our result for the current I is positive, our choice that the current flow is clockwise was correct. A negative value for I means that our choice was wrong; the direction of I is counterclockwise.

It is also possible to have *multiloop circuits*, in which currents can flow along several alternative paths. Such a circuit is shown in Figure 29.4. Unless we allow charge to build up in the circuit, the sum of currents entering and leaving any branch point in the circuit must equal zero. This is really a statement of charge conservation, equivalent to stating that the current is the same everywhere in a single-loop circuit, but it is usually referred to as *Kirchhoff's first rule* of circuit analysis.

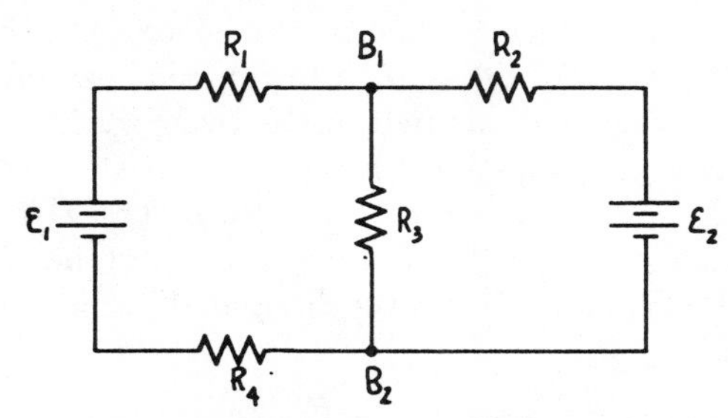

Figure 29.4 A multiloop circuit. B_1 and B_2 are branch points.

Kirchhoff's rules allow us to solve for the currents flowing in the various loops of multiloop circuits. For example, we can solve for the currents in the circuit of Figure 29.4 as follows: first, label the three possible values of current I_1, I_2, and I_3, as shown in Figure 29.5, guessing at the most likely direction of positive current flow. If we are wrong, the sign of the current will turn out negative but the magnitude will be correct. Next apply Kirchhoff's first rule to one of the branch points, giving

$$I_1 + I_2 = I_3$$

Now apply Kirchhoff's second rule to any two closed loops:

$$\mathcal{E}_1 - I_1R_1 - I_3R_3 - I_1R_4 = 0$$
$$\mathcal{E}_2 - I_2R_2 - I_3R_3 = 0$$

We have three equations and three unknowns, I_1, I_2, and I_3, so that we can determine the values of current, provided that we know the resistances R_1,

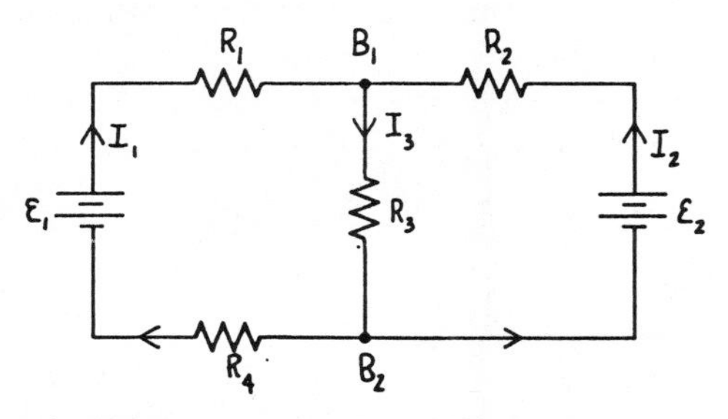

Figure 29.5 Solving for the currents I_1, I_2, and I_3 in the circuit of Figure 29.4.

R_2, R_3, and R_4 and the values of emf. No additional information would be obtained by applying the first rule to branch point B_2 in addition to B_1, or by applying the second rule to a different closed loop—for example, to the loop containing both sources of emf. The results would be the same. Kirchhoff's rules provide a powerful technique in the analysis of DC circuits.

5. *Joule heating*. A DC circuit is not an isolated system, because energy is both fed into it and extracted from it, but it is a steady-state system, and we have shown that in such a system the source emf must equal the potential drop across the resistor. For every unit of charge that passes through it, the source of emf does the work $\mathcal{E}$. The source of emf therefore supplies an amount of power $P = \mathcal{E}(dq/dt) = \mathcal{E}I$ to the current passing through it. An equal amount of power is dissipated in the resistor. The loss in potential energy per unit charge in passing through the resistor is ΔV, so that the power dissipated is given by $\Delta V(dq/dt) = IR(dq/dt) = I^2R$. The energy dissipated in the resistance R is called Joule heating, and the power supplied by the source of emf is equal to the power dissipated in the circuit resistance—that is,

$$P = I^2R$$

6. *Electrical measurements* are usually made with ammeters and voltmeters, both of which respond to an electric current passing through the instrument. Nearly all modern-day ammeters and voltmeters are electronic and are designed so that the same instrument can be used for either current or voltage readings. The essential difference is that an ammeter, which measures the current flowing in a circuit, is connected in series in the circuit and has a very low internal resistance so that it does not affect the total current flowing. A voltmeter, which is hooked up in parallel with the circuit, has a very high resistance and so draws very little current. Both instruments are designed to allow the measurements to be made while minimally disturbing the operation of the circuit.

7. An *RC circuit* is any circuit consisting of a resistor and a capacitor in series (with or without a battery), as shown in Figure 29.6. When the circuit shown in Figure 29.6(a) is first hooked up, current will flow from the battery through the resistor and will charge up the capacitor. When the charge on the capacitor is such that the voltage across the capacitor plates equals the emf of the battery, the current will stop flowing. We can apply Kirchhoff's circuit rule—namely, that the sum of all the emfs and voltage drops around the circuit is zero—to look at the time dependence of the current flow in the circuit.

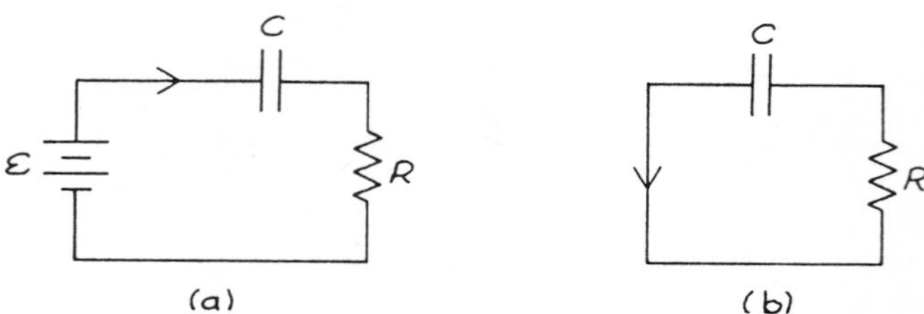

Figure 29.6 (a) An RC circuit with a battery of emf $\mathcal{E}$. (b) An RC circuit with the capacitor discharging through the resistor.

For a current I, the voltage across the resistor is IR, and if the charge on the capacitor plates is Q, the voltage across the capacitor plates is Q/C, so that the circuit equation is

$$\varepsilon - IR - Q/C = 0$$

Note that $I = dQ/dt$, so that we have a differential equation, which can be solved for the charge Q on the capacitor:

$$Q = C\varepsilon(1 - e^{-t/RC})$$

By differentiating this equation, we find that

$$I = dQ/dt = (\varepsilon/R)e^{-t/RC}$$

which tells us that the current has an initial value (at $t = 0$) of ε/R, and then decreases exponentially to zero.

If we now disconnect the battery and connect the free terminals of the capacitor and the resistor, as is shown in Figure 29.6(b), the capacitor will discharge through the resistor. By again applying Kirchhoff's rule, we find that the current is described by

$$I = dQ/dt = -(\varepsilon/R)e^{-t/RC}$$

where we get the negative sign since the current flows in a direction opposite to that of the original current in Figure 29.6(a).

Sample Problems

1. *Worked Problem.* A typical solar cell 1 inch in diameter has an output of 150 mA at 0.6 V when exposed to full sunlight. You want to design a solar panel capable of providing 1 A of current at an emf of 60 V to an external load. How many cells will you need? How big a panel will you need, and how should you connect the cells to one another? (Assume that the cells are circular.) What possible drawback does your system have, and can you think of a possible solution?

Solution: In order to provide 60 V, you will need $N = 60/0.6 = 100$ cells in series. To provide 1 A, you will need $1/0.15 = 7.0$ cells in parallel. You will therefore need a total of $7 \times 100 = 700$ cells, connected as indicated in Figure 29.7 with seven strings of 100 series cells in parallel. The mini-

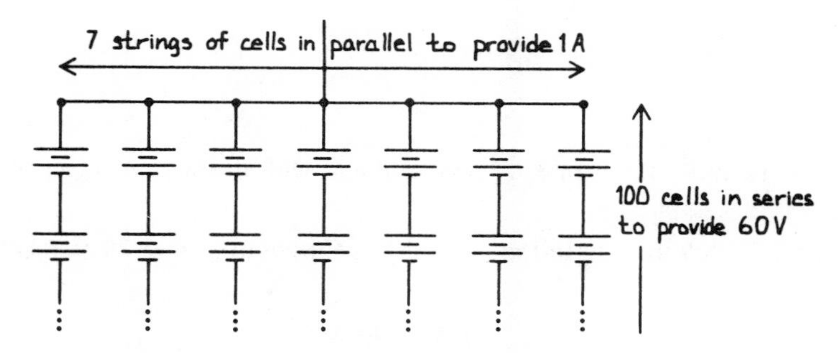

Figure 29.7 Problem 1. A bank of 700 solar cells providing 60 V at 1 A.

mum area needed is 100 inches × 7 inches, or 0.45 m^2. The drawback is that the system will work only during periods of full sunlight; a possible solution is to have the cells hooked to a bank of batteries, so that the batteries can be used when sunlight is not available. If the *average* load is 60 V at 1 A, then many more cells will be needed in order to keep the batteries charged.

2. *Guided Problem.* A typical 12 V battery is rated at 55 amp-hours, so that the battery will deliver 55 A for 1 h. How many such batteries are needed to drive an electric automobile for a period of 4 h, if an average power of 10 hp (1 hp = 746 watts) is required for the automobile? If each battery weighs 30 lbs, what weight in batteries must the automobile carry?

Solution scheme

a. Calculate the total energy needed, in joules, to drive the automobile for 4 h.
b. Calculate the total energy stored in a single battery, in joules.
c. You should now be able to evaluate the total number of batteries needed, and their weight.

3. *Worked Problem.* Nearly all sources of emf have an internal resistance; such a source can be regarded as an emf in series with a resistor, as shown in Figure 29.8. The external terminals of the source are labeled A and B. A 1020-ohm resistor is connected between A and B of a battery; the potential difference between A and B (with the resistor connected) is found to be 1.99 V. When a 9.87-ohm resistor is used instead of the 1020-ohm resistor, the potential difference between A and B is 1.95 V. What is the internal resistance and the emf of the battery? If a 2.0-ohm resistor were used, what value of potential difference would we expect?

Figure 29.8 Problem 3. A source of emf with internal resistance R_i.

Solution: Sketch the single-loop circuit, as shown in Figure 29.9. The circuit equation is

$$\mathcal{E} - I(R_i + R) = 0$$

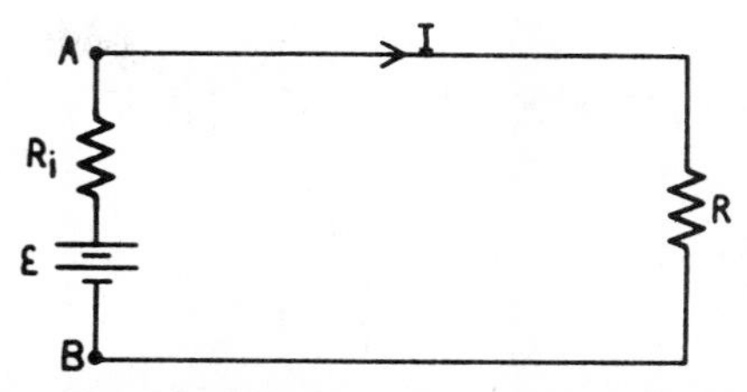

Figure 29.9 Problem 3. A source of emf with internal resistance R_i and load resistance R.

The measured potential difference between A and B is given by $V = IR$. I_1 = current with 1020-ohm resistor = V/R = 1.99/1020 = 1.95 (10^{-3}) A. I_2 = current with 9.87-ohm resistor = 1.95/9.87 = 0.198 A. For both cases

$$\mathcal{E} = I(R_i + R)$$

Then

$$I_1(R_i + R_1) = I_2(R_i + R_2)$$

$$R_i(I_2 - I_1) = I_1R_1 - I_2R_2$$

$$R_i(.196) = 1.990 - 1.950$$

$$R_i = 0.20\ \Omega$$

$$\mathcal{E} = I_1(R_i + R_1) = 1.95\ (10^{-3})(.20 + 1020)$$

$$\mathcal{E} = 1.99\ \text{V}$$

If a 2.0-ohm resistor is used

$$I = \frac{\mathcal{E}}{R_i + R} = \frac{1.99}{0.20 + 2.0} = 0.904\ \text{A}$$

$$V = IR = 0.904(2.0) = 1.81\ \text{V}$$

4. *Guided Problem.* Find the currents in all parts of the circuit shown in Figure 29.10. Resistance values are in ohms. In the figure, ε_1 = 4.0 V, ε_2 = 6.0 V, and $\mathcal{E}_3$ = 11.0 V.

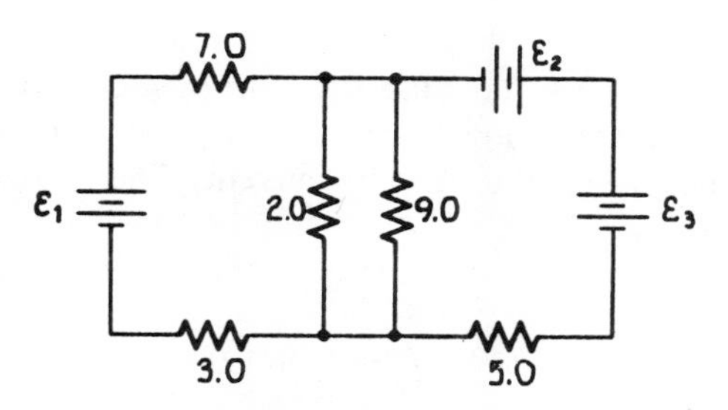

Figure 29.10 Problem 4. A multiloop circuit.

Solution scheme

a. Replace the parallel 2- and 9-ohm resistors by their equivalent resistance R_{eq}. Then redraw the circuit as shown in igure 29.11, indicating the unknown currents.

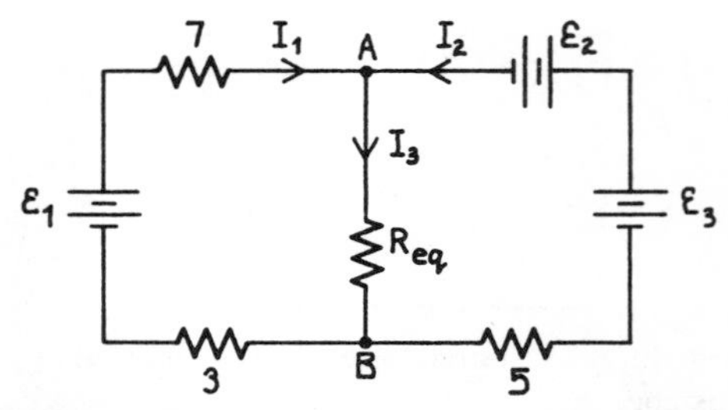

Figure 29.11 Problem 4. The same multiloop circuit, with the unknown currents labeled.

b. Apply Kirchhoff's first rule of circuit analysis to the branch point A.
c. Apply Kirchhoff's second rule of circuit analysis to any two closed loops. Suggested loops are the left side loop and the right side loop.
d. You should now have three equations and three unknowns, I_1, I_2, and I_3. Solve for the three currents.
e. To obtain the currents in the two central resistors (2 and 9 ohms), obtain the potential difference V_{AB} using $V_{AB} = I_3R_{eq}$ (you should now know both I_3 and R_{eq}). Having obtained V_{AB}, calculate the currents in the two resistors.

5. *Worked Problem.* Find the unknown emf, $\mathcal{E}$, and all unspecified currents in the multiloop network shown in Figure 29.12. All emf values are in volts, and all resistance values are in ohms.

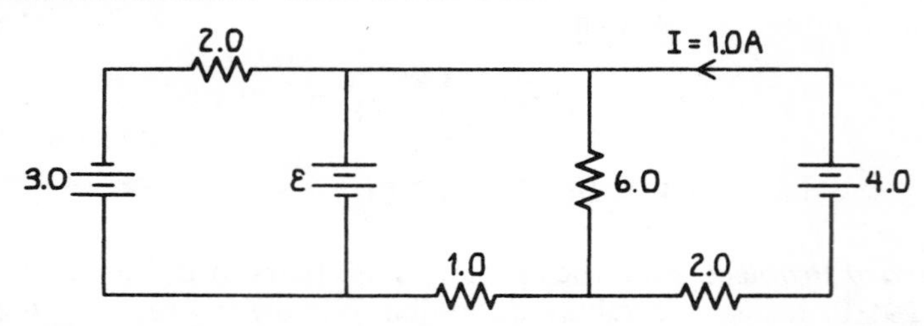

Figure 29.12 Problem 5. A multiloop network. The emf ε is unknown.

Solution

a. Redraw the network, labeling the unknown currents and assigning directions of current flow. Remember that a wrong guess will result only in a negative value for the current. The network is shown in Figure 29.13.
b. Applying Kirchhoff's first rule at branch points A and B gives the equations

$$I_2 + I_3 = 1.0$$
$$I_1 + I_4 = I_3$$

The same equations are obtained by applying the rule at points C and D.

c. We can now apply Kirchhoff's second rule to any of the closed loops. With the extreme right side loop we obtain $4 - 6I_2 - 2 = 0$,

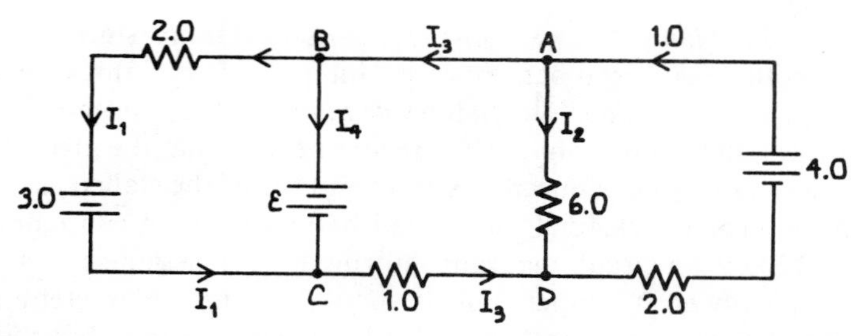

Figure 29.13 Problem 5. The same network as in Figure 29.12, with the unknown currents labeled.

which gives $I_2 = 0.33$ A. By substituting for I_2, $0.33 + I_3 = 1.0$, so that $I_3 = 0.67$ A and $I_1 + I_4 = 0.67$ A. Now again apply the second rule, this time to the loop starting with the 3.0 V emf source, through $C - D - A - B$ and back to the 3.0 V source of emf. We obtain $3 - 0.67 + 6\,(0.33) = 2I_1$, where we take 6 (0.33) to be positive because in the direction we are following the loop, we are going from a lower to a higher potential. Then we obtain $I_1 = 2.16$ A. Then $2.16 + I_4 = 0.67$, so that $I_4 = -1.49$ A. We have now obtained all the unknown currents and can now obtain the unknown emf by applying the second rule to the loop on the extreme left. $\mathcal{E} = 2I_1 + 3.0 = 0$, so that $\mathcal{E} = 1.32$ V. Our guess concerning the direction of I_4 was incorrect—the flow is opposite.

6. *Guided Problem.* Light bulbs usually have tungsten filaments and operate at high temperature. The resistance of a 60-watt light bulb, measured when cold, is about 30 ohms, yet the bulb dissipates only about 60 watts of power when connected across a 110-volt source. How can this be so? Compare the "hot" and "cold" resistances of the bulb, as well as the power consumptions when the bulb is hot and cold (assuming that the filament could be kept cold). The resistivity of tungsten at 20°C is $5.6 \times 10^{-8}\ \Omega \cdot$ m. The temperature coefficient of resistivity is 450×10^{-5} per degree centigrade. Estimate the approximate operating temperature of the light bulb filament.

Solution scheme

a. From the resistance measured when cold, calculate the expected "cold" power dissipation of the bulb.
b. The bulb dissipates 60 watts when hot; use this figure to calculate the resistance of the filament when hot. Compare this with the 30-ohm resistance when the bulb is cold.
c. Assuming that the value for the temperature coefficient of resistivity remains valid over a rather large temperature range, calculate the rise in temperature needed to produce the resistance when hot, and from this deduce the "hot" operating temperature of the filament.
d. You should now be able to explain why measuring the "cold" resistance of a light bulb will not allow you to estimate the normal power consumption of the bulb, even though you know the voltage across the bulb.

7. *Worked Problem.* A common practice is to "jump start" a stalled automobile with a run-down battery with "jumper cables," thick, insulated copper cables used to connect the battery of an operating automobile to the battery of the stalled automobile. (It is recommended that the ground cable from the good battery be connected onto the motor of the stalled car, rather than onto the ground terminal of the bad battery, but we can ignore this advice here.) Sketch a circuit diagram showing how the connections should be made. Typically each jumper cable is about 0.50 cm in diameter and 3 m long. A starter motor may require 80 A of current. Assume that the good battery has an emf of 12.4 V and an internal resistance of 0.02 ohms. Corresponding values are 10 V and 0.08 ohms for the run-down battery. (The resistivity of copper is $1.7 \times 10^{-8}\ \Omega \cdot \mathrm{m}$.) Calculate the currents through each battery, the resistance of the starter motor, the power output of each battery, and the power loss in the starter cables. Can you explain why the jumper cables sometimes become rather warm when used in this way?

Jump starting is dangerous if the batteries are connected incorrectly. Sketch a diagram of how *not* to hook up the cables. Under these circumstances the run-down battery can sometimes explode. Estimate the approximate current through the batteries and the power absorbed by the run-down battery; for your estimate you can ignore the presence of the starter motor.

Solution

a. A circuit diagram showing the proper connections is given in Figure 29.14(a); in Figure 29.14(b) we include the internal resistances of the batteries and the resistance of the jumper cables. The cable resistance is given by

$$R = \frac{\rho l}{A} = \frac{(1.7 \times 10^{-8})(6)}{(1.96 \times 10^{-5})} = 5.2 \times 10^{-3}\ \Omega$$

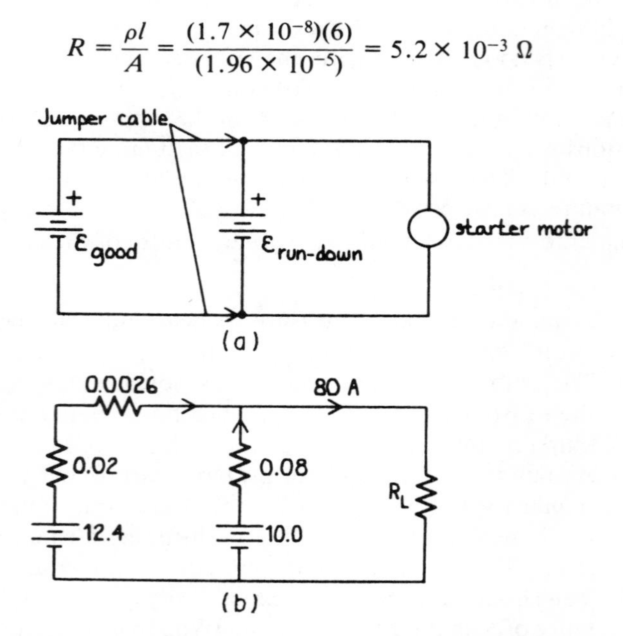

Figure 28.14 Problem 7. (a) The correct connections for jump-starting an automobile. (b) The equivalent circuit for (a). Resistances are in ohms; emfs are in volts.

b. The unknown currents are indicated in the equivalent circuit of Figure 29.14(b); in addition, the starter motor is represented by a load resistor R_L. The cable resistance is represented by a single resistor; it does not matter whether we place this resistor in the upper or lower part of the left side loop of the circuit. From Kirchhoff's first rule

$$I_1 + I_2 = 80 \text{ A}$$

From the second rule

$$12.4 - 0.02I_1 - 0.0026I_1 - 80.0R_L = 0$$
$$10 - 0.08I_2 - 80R_L = 0$$

c. By collecting terms and eliminating R_L,

$$12.4 - 0.0252\, I_1 = 10 - 0.08\, I_2$$

and since $I_2 = 80 - I_1$, $I_1 = 83.6$ A, so that $I_2 = 3.6$ A. The voltage across the starter motor is given by $V = 12.4 - 83.6(0.02 + 0.0052) = 10.3$ V, so that $R_L = V/I = 0.129$ ohms.

d. The power output of the good battery is equal to the product of the emf and the current, so that $P = (12.4 \times 83.6) = 1040$ watts. The power *input* to the run-down battery is $P = (3.6 \times 10) = 36$ watts (this battery is being charged by the good battery, because its emf is so low). In addition there is $P = I_2^2R_i = 1.0$ watt of Joule heating. The power dissipated in the booster cables is given by $I_2^2R = 36$ watts, so we expect the cables to become warm.

e. An incorrect hookup is shown in Figure 29.15, with the positive terminal of the good battery connected to the negative terminal of the run-down battery. For simplicity we have combined the internal resistance of the good battery and the cable resistance into a single resistor of resistance 0.025 ohm. For an estimate, we neglect the presence of R_L; that is, since R_L is much larger than the internal resistance of the battery, we use as our circuit equation

$$12.4 - 0.025\, I + 10 - 0.08\, I = 0 \quad \text{or} \quad I = 213 \text{ A}$$

There is now a very large current through both batteries. The Joule heating in the run-down battery is $I^2R_i = 3630$ watts. Such a sudden absorption of power can sometimes cause a battery to explode. (If you solve the circuit of Figure 29.15 correctly, you will find that our value for I, 213 A, is very close to the correct value.)

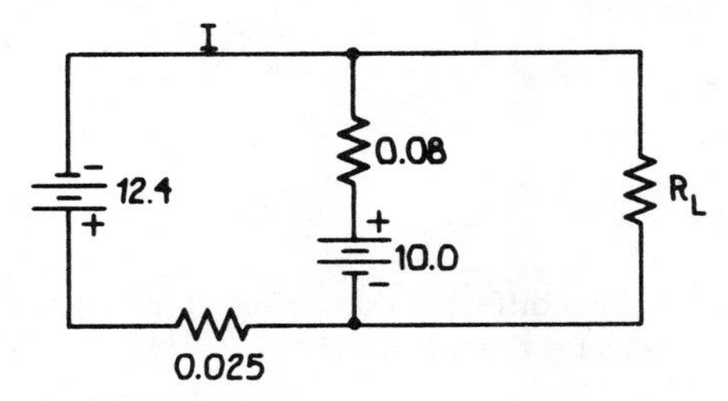

Figure 29.15 Problem 7. The wrong way to jump-start an automobile.

8. *Guided Problem.* Copper wire is manufactured according to a "gauge number"; the larger the gauge number, the smaller the wire diameter. No. 12 wire has a diameter of 0.2053 cm, no. 16, 0.1291 cm, and no. 20, 0.08118 cm. The ends of a piece of no. 12 wire 20 m long are attached to the terminals of a 12-volt car battery with internal resistance of 0.02 ohms. Find the electric field in the wire, the resistance of the wire, the current through the wire, and the power loss in the wire. The resistivity of copper is 1.7×10^{-8} $\Omega \cdot$ m. Answer the same question assuming that the 12-gauge wire is disconnected and replaced first by an equal length piece of no. 16 wire, and then by an equal length piece of no. 20 wire. Which wire is likely to overheat and burn out first?

Solution scheme

a. From the resistivity, the length, and the area, calculate the resistance of each wire.
b. Draw the equivalent single-loop circuit, showing the battery emf, the battery internal resistance, and the wire resistance. You should be able to apply the loop equation to the circuit and so solve for the current through each of the wires.
c. Since power is given by I^2R, you should now be able to find the power dissipated in each wire and to determine which wire is likely to burn out first. All three might well burn out, but the current is much the highest in the wire with the lowest resistance, so it would probably burn out first.
d. To obtain the electric field in each wire, you need to calculate the voltage drop across the ends of each wire. You should now be able to estimate the electric field in each wire.

9. *Worked Problem.* A student is given a 1.000 V standard cell, with internal resistance 1 Ω, a voltmeter of internal resistance 10 kΩ, and an ammeter of internal resistance 1 Ω. He is also given a nominal "100 Ω" resistor and asked to measure its resistance, R, as accurately as he can. Suppose that R is exactly 100.0 Ω. What value of R will the student measure if he uses the circuit shown in Figure 29.16(a)? What value will he get if he uses the circuit of Figure 29.16(b)?

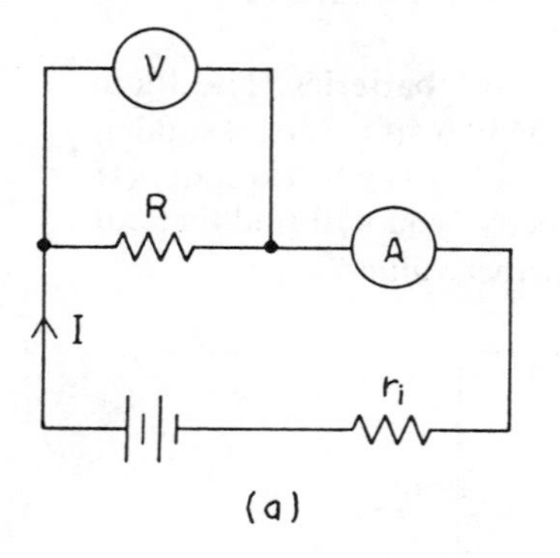

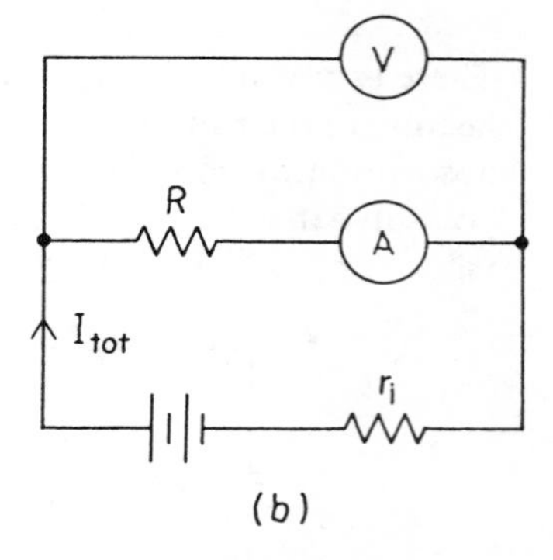

Figure 29.16 Problem 9. Two different ways of using an ammeter and a voltmeter to measure resistance.

Solution: In circuit a, the parallel resistance of the voltmeter and the 100 Ω resistor is given by R_p = 100(10,000)/10,100 = 99.0 Ω. R_{tot} = 99 + 1 + 1 = 101 Ω, and I = 1,000/101 = 0.0099 A. ΔV, the voltage across R, is given by $I(R_t)$ = 0.0099(99) = 0.98 V. The measured value of R is therefore $\Delta V/I$ = 0.98/0.0099 = 99 Ω.

In circuit b, the parallel resistance R_p = 10,000(101)/10,101 = 99.99 Ω. The total circuit resistance is $R_p + r_i$ = 100.99 Ω, so that I_{tot} = 1.000/100.99 = 0.0099 A. The voltage across R_p = $(I_{tot})R_p$ = 0.0099(99.99) = 0.990 V. The current through the ammeter, I_a = 0.990/101 = 0.009803 A. The measured value of R is therefore V/I_a = 0.990/0.009803 = 101 Ω.

10. *Guided Problem.* An unknown capacitor is charged to a potential of 10.0 V. The capacitor is then connected across an 8.00 kΩ resistor, and the charging voltage across the capacitor is measured with an oscilloscope. After 5.0 × 10^{-3} s (10 ms), the potential across the capacitor has dropped to 5.0 V, and after 10.0 × 10^{-3} s (10 ms), the potential is 2.60 V. What is the value of the capacitor?

Solution scheme

a. The charge on the capacitor is given by $Q = \varepsilon C e^{-t/RC}$, and since $C = Q/V$, $V = V_0 e^{-t/RC}$, where V = voltage across the capacitor at time t. We can rewrite this as $V_0/V = e^{t/RC}$. Taking logarithms, we get $\ln(V_0/V) = t/RC$. We know R, as well as the initial voltage, V_0, and are told the new voltage after a time t. Can you find the values of C for the given two values of t?

b. You have obtained two values of C. Do they agree exactly? Note that you should restrict your answer to no more than two significant figures. What is your best overall estimate of C?

Answers to Guided Problems

2. Forty-five batteries are needed, with a total weight of 1350 lbs.
4. I_2 = 0.69 A; I_1 = 0.29 A; I_3 = 0.98 A.
   The current in the 2-ohm resistor is 0.80 A; the current in the 9-ohm resistor is 0.18 A.
6. R_{cold} = 30 ohms; R_{hot} = 202 ohms. The approximate operating temperature is 1300°C. As the temperature of the filament rises, its resistance increases, so that the current and power dissipation decrease.
8. R_{12} = 0.103 ohms; I_{12} = 97.8 A; P_{12} = 982 W; E_{12} = 0.50 V/m
   R_{16} = 0.260 ohms; I_{16} = 42.9 A; P_{16} = 478 W; E_{16} = 0.56 V/m
   R_{20} = 0.386 ohms; I_{20} = 29.5 A; P_{20} = 337 W; E_{20} = 0.57 V/m
10. At t = 5 ms, C = 9.02 × 10^{-7} F
    At t = 10 ms, C = 9.28 × 10^{-7} F
    ∴ C = 9.15 × 10^{-7} F

Chapter 30

The Magnetic Force and Field

Overview

Having introduced the idea of the flow of charge, which we called an electric current, we are ready to talk about magnetism, a phenomenon associated with moving electric charge. Humankind has known about the effects of magnetism for at least two thousand years, but only recently have we come to understand its association with moving electric charge. We begin here by giving an equation for the magnetic force between two moving charges, just as, in our study of electrostatics, we began with Coulomb's Law for the force between two stationary charged particles. Having given the fundamental force law between two moving charged particles, we define the magnetic field vector and deduce how to calculate the magnetic field of an electric current. Finally, in an optional section, we show how electric and magnetic effects are related through special relativity; in fact, we show that magnetic effects follow inevitably from Coulomb's Law and special relativity—that electricity and magnetism are one and the same thing.

Essential Terms

magnetic force
permeability constant
magnetic field
tesla
Gauss' Law for magnetism
Biot–Savart law
magnetic dipole
magnetic dipole moment

Key Concepts

1. *Magnetic force.* According to Coulomb's Law, the electric force exerted by a point charge q' on a point charge q, where both charges are either at rest or are moving very slowly, is

$$\mathbf{F} = \frac{1}{4\pi\varepsilon_0}\frac{qq'}{r^2}\hat{\mathbf{r}}$$

When the charges are in motion, Coulomb's Law is still valid but there is an *additional* force, which we call a magnetic force, acting between the two charges. The expression for the magnetic force, which is more complicated than Coulomb's Law, is

$$\mathbf{F} = \frac{\mu_0}{4\pi}\frac{qq'}{r^2}\mathbf{v} \times (\mathbf{v}' \times \hat{\mathbf{r}})$$

where μ_0 is called the *permeability constant*. Figure 30.1 shows the force exerted on q by q' when the two charges are traveling in the $+x$ direction and are a distance r apart.

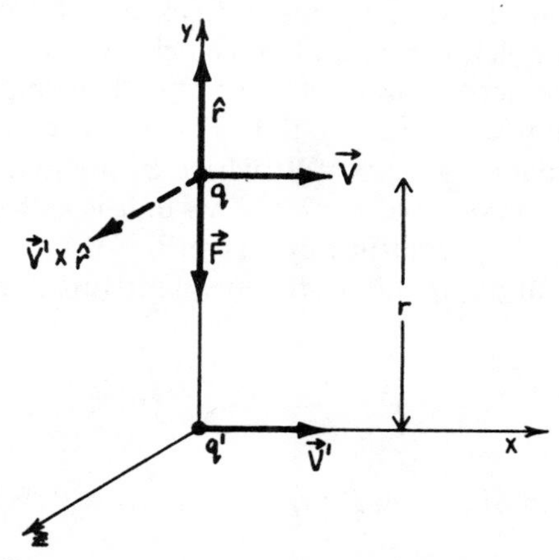

Figure 30.1 The magnetic force exerted by charge q' on charge q.

The magnetic force is attractive and so here is exactly opposite in direction to the Coulomb force, which would be repulsive. If either q or q' were negative (and the other positive), the direction of both the magnetic and the Coulomb force would reverse, so both forces would still be exactly opposite in direction. If we reverse the direction of either $\mathbf{v}$ or $\mathbf{v}'$, the direction of the magnetic force would reverse, whereas the Coulomb force would be unchanged.

To complicate matters further, in general we find that the magnetic force is not parallel to the line joining the two charges (unlike the Coulomb force) and that the magnetic force that q' exerts on q is *not* equal and opposite to the magnetic force that q exerts on q', although it is for the example shown in Figure 30.1. Newton's Law of action–reaction fails when we consider the magnetic forces between two isolated particles, and in order to show that momentum is conserved (one of our most cherished principles), we must consider also the momentum of the electric fields of the particles.

2. *The magnetic field.* The discussion of magnetic force is much simplified if we introduce the idea of a magnetic field, through which the magnetic

force is communicated from one charge to another. If we restrict ourselves to discussing point charges, we find that we always have electric fields to consider, and magnetic forces and fields if the charges are moving. We will find, however, that magnetic forces and fields are also associated with electric currents, for example, in the region around a wire carrying an electric current. We can, therefore, find magnetic forces, and so define a magnetic field, in regions in which we can neglect any electric field. An electric field can be defined by placing a charge at some point and measuring the force on the charge; the force on the charge is given by $\mathbf{F} = q\mathbf{E}$, where $\mathbf{E}$ is the electric field strength. It then follows that, for a point charge,

$$\mathbf{E} = \frac{q}{4\pi\varepsilon_0}\frac{\hat{\mathbf{r}}}{r^2}$$

Similarly, we can define a magnetic field at some point by placing a test charge q at that point and giving it a velocity $\mathbf{v}$. The charge will experience a magnetic force dependent on its velocity. (It is implied that if $\mathbf{v} = 0$, the force is zero, so that the electric field $\mathbf{E} = 0$.) The magnetic field, $\mathbf{B}$, is defined by the equation $\mathbf{F} = q\mathbf{v} \times \mathbf{B}$, where B, the magnetic field strength, is expressed in the units N/(C • m/s). This unit is called the *tesla*. If $\mathbf{v}$ is perpendicular to $\mathbf{B}$, the magnitude of F is qvB; if $\mathbf{v}$ is parallel to $\mathbf{B}$, $F = 0$. Since for two point charges, q and q', the magnetic force on q resulting from q' is given by

$$\mathbf{F} = \frac{\mu_0}{4\pi}\frac{qq'}{r^2}\mathbf{v} \times (\mathbf{v}' \times \hat{\mathbf{r}})$$

the magnetic field of a point charge q' moving with velocity $\mathbf{v}'$ must be given by

$$\mathbf{B} = \frac{\mu_0}{4\pi}\frac{q'}{r^2}(\mathbf{v}' \times \hat{\mathbf{r}})$$

Figure 30.2 shows a charge q' at the origin, with velocity $\mathbf{v}'$ parallel to the x axis. The magnetic field at point P resulting from q' is shown. The force

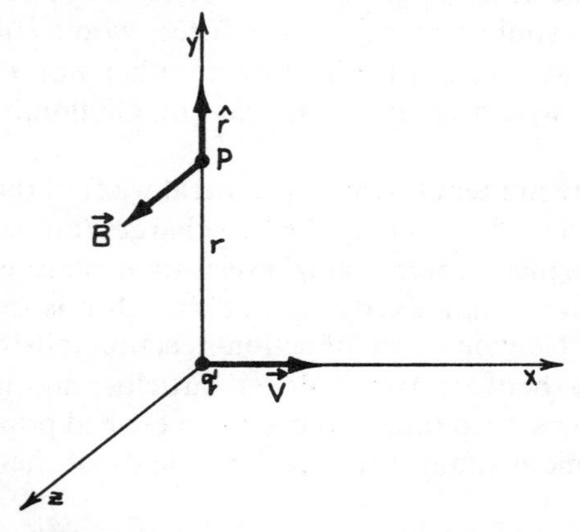

Figure 30.2 The magnetic field of a charge q' moving with velocity $\mathbf{v}'$ along the x axis.

on a charge q at point P is then $q\mathbf{v} \times \mathbf{B}$. In agreement with Figure 30.1, the force is directed toward charge q' if $\mathbf{v}$ is also parallel to the x axis.

The magnetic field can be represented graphically by field lines in exactly the same way that an electric field can be represented by electric field lines. The tangent to the field line indicates the direction of the field, and the line density is proportional to the strength of the field. The magnetic field lines for the charge q' of Figure 30.2 are closed circles. For all points on the y–z plane, the magnitude of the field depends only on r, the distance from the origin. The field for all points along the $+y$ axis is directed along the $+z$ direction; the field for all points along the $+z$ axis is directed along the $-y$ direction. Since the field lines are closed circles, they have neither an origin nor a terminal point, unlike electric field lines. Because the lines are continuous, as many enter as leave any closed surface. You can convince yourself that this is so by trying to draw a surface for which the net flux is not zero! Consequently, $\int \mathbf{B} \cdot d\mathbf{s} = 0$ for any closed surface. This equation holds for any arbitrary magnetic field and is called *Gauss' Law for magnetism*.

3. *The Biot–Savart law*. According to the principle of superposition, the electric field resulting from several charges is the vector sum of the fields resulting from the charges taken one at a time, or $\mathbf{E} = \mathbf{E}_1 + \mathbf{E}_2 + \mathbf{E}_3 + \ldots$. The principle of superposition also holds for the magnetic field; the net magnetic field generated by several charges is also given by $\mathbf{B} = \mathbf{B}_1 + \mathbf{B}_2 + \mathbf{B}_3 + \ldots$.

The magnetic field resulting from a current I, defined as the flow of many charges through a conductor, can be easily found, from superposition, from the expression for the magnetic field of a single moving charge. The result is called the *Biot–Savart law* and is expressed as

$$d\mathbf{B} = \frac{\mu_0}{4\pi} \frac{I d\mathbf{l} \times \hat{\mathbf{r}}}{r^2}$$

$d\mathbf{B}$ is the contribution to the magnetic field at point P from the length $d\mathbf{l}$ of the conductor, which carries a current I. The magnetic field for the whole conductor is obtained by integrating over the length of the conductor. $d\mathbf{B}$, shown in Figure 30.3, is perpendicular to the plane containing $d\mathbf{l}$ and $\hat{\mathbf{r}}$.

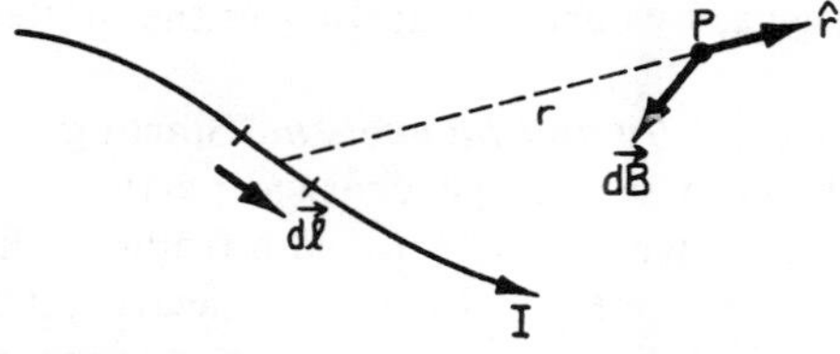

Figure 30.3 The magnetic field resulting from a length $d\mathbf{l}$ of a conductor.

4. *Magnetic dipole*. A small ring or loop of current is called a magnetic dipole; the magnetic field lines are depicted in Figure 30.4, together with the corresponding field lines caused by an electric dipole. At large distances the magnetic field along the axis of the dipole is given by

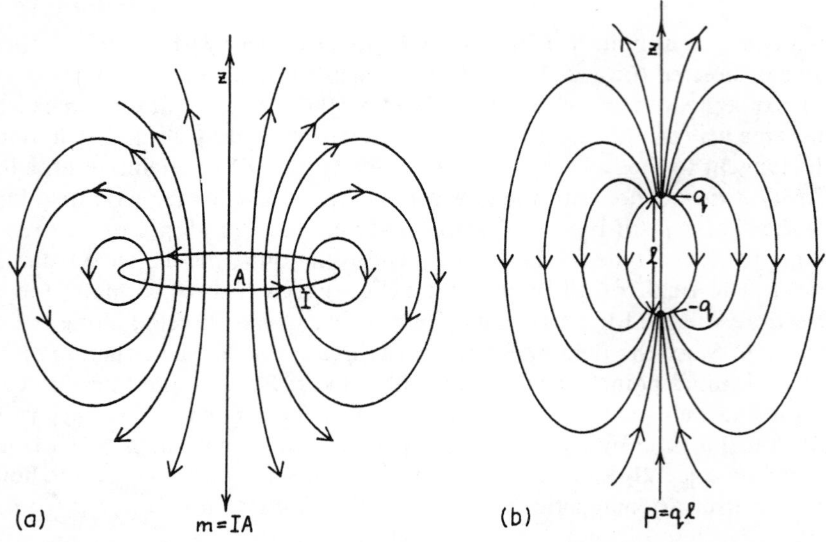

Figure 30.4 The field of (a) a magnetic dipole and (b) an electric dipole.

$$B_z = \frac{\mu_0}{2\pi}\frac{I\pi R^2}{z^3} = \frac{\mu_0}{2\pi}\frac{m}{z^3}$$

with $m = I$ (area). For the electric dipole the corresponding expression for the field along the dipole axis is

$$E_z = \frac{2ql}{4\pi\varepsilon_0 z^3} = \frac{2p}{4\pi\varepsilon_0 z^3}$$

In both cases the field is proportional to the dipole moment and decreases as the inverse cube of the distance.

An electron, rotating in an orbit around a nucleus, constitutes a small-current loop and so can be expected to have a magnetic-dipole moment. Electrons, protons, and most elementary particles, however, have what is called an *intrinsic* dipole moment—the particles behave at all times like rotating charge distributions. The Earth itself behaves like a large magnetic dipole, behavior caused, we believe, by currents flowing in the molten iron core.

5. *The link between electricity and magnetism.* Starting with an expression for the magnetic force between two charged particles comparable to Gauss' Law of electrostatics, we have developed a framework for magnetic forces very similar to the one we developed for electrostatic forces. It is worthwhile to explore the link between electricity and magnetism. Suppose we have a charged particle moving parallel to a wire carrying a current I, as shown in Figure 30.5. The charge q feels a magnetic field caused by the current in the wire. The field B, as given by the Biot–Savart law, is

$$B = \frac{\mu_0}{2\pi}\frac{I}{y}$$

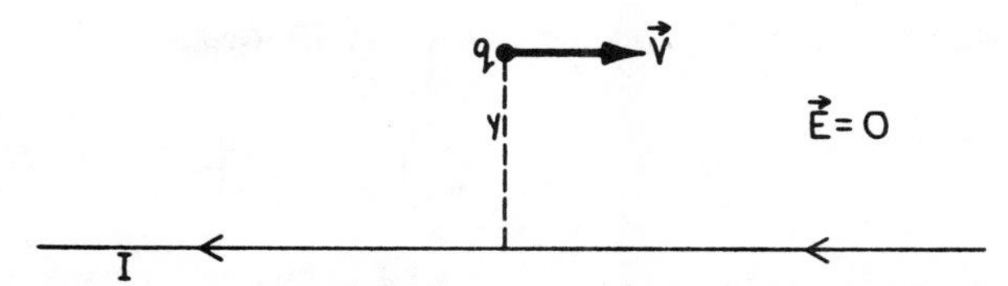

Figure 30.5 A charge moving parallel to a wire carrying a current I. $E = 0$ near the wire.

The charge therefore feels a magnetic force $F_m = q\upsilon B$, forcing it away from the wire. The electric field seen by the charge is zero. Now consider the reference frame in which the charge is at rest; that is, suppose that we are riding on the moving charge q. Since $\upsilon = 0$, the magnetic force $F_m = 0$ in this new reference frame, shown in Figure 30.6. We can show, however, that the wire now generates an electric field **E**. From special relativity, we know that moving objects appear shorter than stationary objects, so that the positive charges on the wire are compressed into a shorter length of the wire in Figure 30.6 than they were in Figure 30.5. The negative charge density is also changed and, depending on the magnitude of υ, may actually be decreased in the new reference frame. In any event, in the reference frame of Figure 30.6 the wire is no longer uncharged, and we can show that the charge density on the wire is given approximately by $\lambda\upsilon^2/c^2$, where $-\lambda$ is the charge density of the electrons carrying the current in the wire in Figure 30.5.

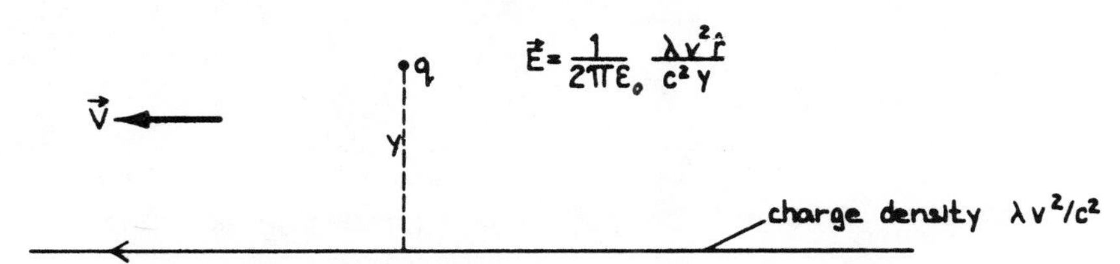

Figure 30.6 A new reference frame, in which the charge q is at rest. The wire is moving to the left with speed υ.

The electric field in Figure 30.6 is given by

$$\mathbf{E} = \frac{1}{2\pi\varepsilon_0}\frac{\lambda\upsilon^2\hat{\mathbf{r}}}{c^2 y}$$

so that the force on the charge q is

$$\mathbf{F}_e = q\mathbf{E} = \frac{q\lambda\upsilon^2\hat{\mathbf{r}}}{2\pi\varepsilon_0 c^2 y}$$

If we take as a special case (and our final result does *not* depend on this) the velocity υ to be equal to the electron drift velocity in the wire, then $-\lambda\upsilon = I$, which is the current in Figure 30.5, so that

$$\mathbf{F}_e = \frac{q\upsilon I\hat{\mathbf{r}}}{2\pi\varepsilon_0 c^2 y}$$

But the magnetic force, F_m, on the charge in Figure 30.5 is given by

$$\mathbf{F}_m = q\upsilon\mathbf{B} = q\upsilon\left(\frac{\mu_0 I}{2\pi y}\right)\hat{\mathbf{r}} = \frac{\mu_0 q\upsilon I}{2\pi y}\hat{\mathbf{r}}$$

Now accelerations must be independent of the frame of reference in which they are seen; that is, it must be true that $F_e = F_m$. But the ratio $F_m/F_e = \mu_0\varepsilon_0 c^2$. If we substitute in values for μ_0 and ε_0, we see that this is true, that $\mu_0\varepsilon_0 c^2 = 1$. In fact, we have shown not only that $\mu_0\varepsilon_0 = 1/c^2$ but also that the magnetic force is a direct consequence of the electric force.

Sample Problems

1. *Worked Problem.* One of the problems of producing an intense ion beam is that the beam tends to diverge because of electrical repulsion between the ions. Two protons in such a beam are moving parallel to each other at a separation of 1×10^{-12} m, as shown in Figure 30.7. What is the net repulsive force between the protons if the protons have speeds of (a) 0.1 c, (b) 0.4 c, (c) 0.7 c, and (d) 1.0 c, where c is the velocity of light, 3×10^8 m/s?

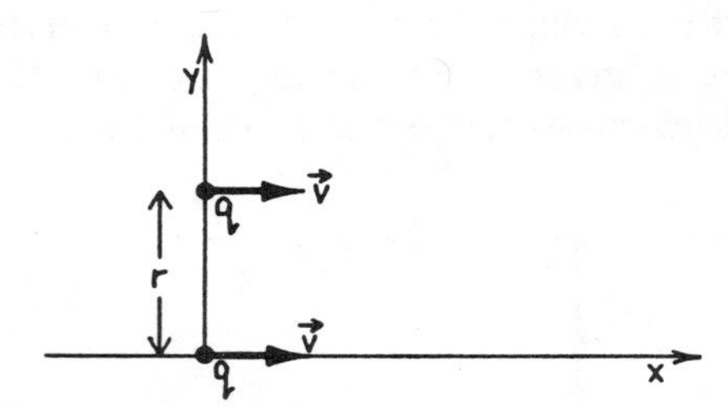

Figure 30.7 Problem 1. Two protons moving parallel to each other.

Solution: The electric, or Coulomb, force is given by

$$F_c = \frac{1}{4\pi\varepsilon_0}\frac{q^2}{r^2} = \frac{(9.0 \times 10^9)(1.6 \times 10^{-19})^2}{(1 \times 10^{-12})^2} = 2.3 \times 10^{-4}\ \text{N, repulsive}$$

The magnetic force, F_m, is given by

$$\mathbf{F}_m = \frac{\mu_0 q^2}{4\pi r^2}\mathbf{v} \times (\mathbf{v}' \times \hat{\mathbf{r}})$$

$$F_m = \frac{\mu_0 q^2}{4\pi r^2}\upsilon^2,\ \text{attractive}$$

The ratio of the two forces $F_m/F_c = \mu_0\varepsilon_0\upsilon^2$.

For (a), $\upsilon = 0.1\ c$, so that $F_m/F_e = \mu_0\varepsilon_0 c^2(.01)$ or $F_m = 0.01\ F_c$.
Therefore, $F_{net} = 2.3 \times 10^{-4}$ N, repulsive.
For (b), $\upsilon^2 = 0.16\ c^2$, so that $F_m = 0.16\ F_c$; $F_{net} = 1.93 \times 10^{-4}$ N.
For (c), $\upsilon^2 = 0.49\ c^2$, so that $F_m = 0.49\ F_c$; $F_{net} = 1.17 \times 10^{-4}$ N.
For (d), $\upsilon^2 = 1.0\ c^2$, so that $F_m = 1.0\ F_c$; $F_{net} = 0$.

Thus, for speeds far less than the velocity of light, the electric force is very strong, but as the particle speeds approach the speed of light, the electric force is nullified by the magnetic force of attraction and the beam no longer "blows up" because of electrical repulsion.

2. *Guided Problem.* Figure 30.8 shows two protons moving at constant speed. q_1 moves along the x axis; q_2 moves along the y axis. The relative positions of the two protons are shown at t = 0, 1, and 2 ns (1 ns = 1 $\times 10^{-9}$ s) in Figures 30.8(a), (b), and (c), respectively. Show the direction of the magnetic force each proton exerts on the other at each relative position.

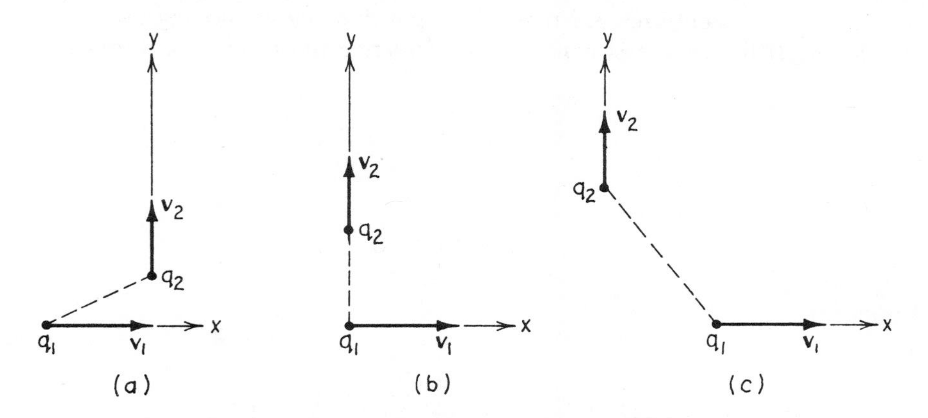

Figure 30.8 Problem 2. The relative positions of two protons at (a) t = 0, (b) t = 1, and (c) t = 2 ns.

Solution scheme

a. Look at the expression for the magnetic forces that moving charges exert on each other. Note that $\mathbf{F}_2 \sim \mathbf{v}_2 \times (\mathbf{v}_1 \times \hat{\mathbf{r}})$, where $\mathbf{F}_2$ is the force that q_1 exerts on q_2, and $\hat{\mathbf{r}}$ is the unit vector toward q_2 from q_1. Sketch the direction of $(\mathbf{v}_1 \times \hat{\mathbf{r}})$ for each of t = 0, 1, and 2 ns. In each case, does $(\mathbf{v}_1 \times \hat{\mathbf{r}})$ point along the $+z$ axis?

b. For each case, sketch the direction of $\mathbf{v}_2 \times (\mathbf{v}_1 \times \hat{\mathbf{r}})$. Does this vector point along the x axis in every case? At what time will the magnitude be the largest?

c. Now go through the same procedure for $\mathbf{F}_1$, where $\mathbf{F}_1 \sim \mathbf{v}_1 \times (\mathbf{v}_2 \times \hat{\mathbf{r}}')$, where $\hat{\mathbf{r}}'$ is now the unit vector toward q_1 from q_2. Note that in every case $\mathbf{v}_2 \times \hat{\mathbf{r}}'$ is parallel (or antiparallel) to the z axis. Is $\mathbf{v}_2 \times \hat{\mathbf{r}}'$ zero at one of the times? (Note that if two vectors are parallel, their cross product = 0.)

d. Can you now sketch the direction of $\mathbf{F}_1$ at all three times? Is $\mathbf{F}_1$ ever zero? Is $\mathbf{F}_1$ ever equal and opposite to $\mathbf{F}_2$ in this problem?

3. *Worked Problem.* Sketch the magnetic field lines in the y–z plane of a charge q' moving along the x axis with a speed v'. Calculate the magnitude of the magnetic field at a point P that is 1×10^{-10} m along the y axis from the charge if the charge has an energy of 100 eV and is (a) an electron, (b) a proton, (c) an alpha particle (a helium nucleus). The masses of the three

particles are as follows: $m_e = 9.1 \times 10^{-31}$ kg; $m_p = 1.67 \times 10^{-27}$ kg; $m_{alpha} = 4 \times 1.67 \times 10^{-27}$ kg.

Solution: The magnetic field of a charge q' moving with speed υ' is given by

$$\mathbf{B} = \frac{\mu_0}{4\pi}\frac{q'}{r^2}(\mathbf{v} \times \hat{\mathbf{r}})$$

For a particle moving along the x axis, the field lines are circles in the y–z plane, concentric about the particle; the lines are sketched in Figure 30.9. The density of lines increases closer to the charged particle. The directions of the field lines are those for a positive charge moving out of the paper. If the charge is negative, the direction of the lines is reversed.

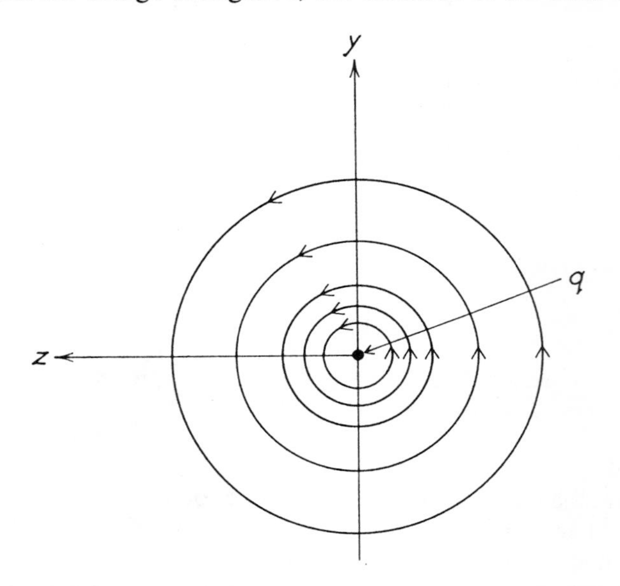

Figure 30.9 Problem 3. Magnetic field lines for a positively charged particle moving out of the paper along the positive x axis.

For a 100 eV particle, the kinetic energy is $\frac{1}{2}m\upsilon^2 = 100\ (1.6 \times 10^{-19}) = 1.6 \times 10^{-17}$ J. The speeds of the three particles are therefore $\upsilon_e = 5.93 \times 10^6$ m/s; $\upsilon_p = 1.38 \times 10^5$ m/s; $\upsilon_{alpha} = 6.92 \times 10^4$ m/s. The magnitude of the field is given by

$$\mathbf{B} = \frac{\mu_0}{4\pi}\frac{q'\upsilon'}{r^2}$$

where $r = 1 \times 10^{-10}$ m and $q' = 1.6 \times 10^{-19}$ C for an electron or a proton but 3.2×10^{-19} C for an alpha particle. Then $B_e = 9.5$ T; $B_p = 0.22$ T; $B_{alpha} = 0.22$ T.

4. *Guided Problem.* A proton with a velocity of $10\,\hat{\mathbf{i}} + 20\,\hat{\mathbf{j}}$ m/s enters a uniform magnetic field represented by $\mathbf{B} = 0.4\,\hat{\mathbf{i}} + 0.3\,\hat{\mathbf{j}} + 0.02\,\hat{\mathbf{k}}$ T. Deter-

mine the force on the proton and the magnitude of the acceleration experienced by the proton when it enters the magnetic field. The mass of the proton is 1.67×10^{-27} kg.

Solution scheme

a. The magnetic force on a charged particle in a magnetic field is given by $\mathbf{F} = q\mathbf{v} \times \mathbf{B}$. Evaluate the cross product of these two quantities in terms of their vector components. This will give the force on the proton.
b. Now calculate the magnitude of the force from the vector components.
c. You should now be able to determine the acceleration using Newton's Law of motion.

5. *Worked Problem.* Using the Biot–Savart law, show that the magnetic field a distance y from a wire of length L, as shown in Figure 30.10, is given by the expression

$$B = \frac{\mu_0 I}{4\pi y}[\sin\theta_1 + \sin\theta_2]$$

The wire carries a current I.

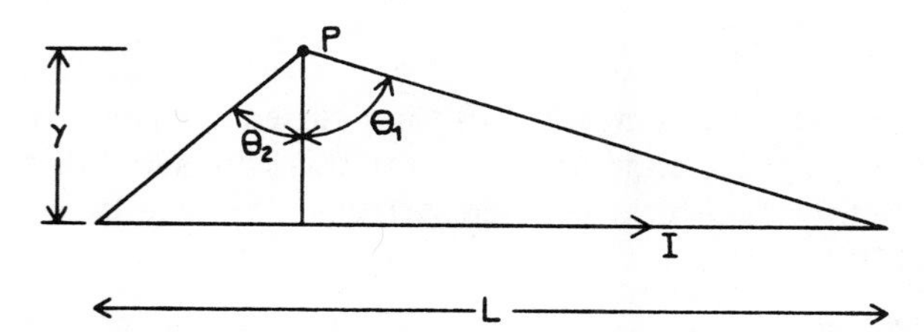

Figure 30.10 Problem 5. Point P is a distance y from a wire of length L carrying a current I.

Solution: Figure 30.11 shows how we can apply the Biot–Savart law to this problem. The field contribution from an element of wire of length dx is in the $+z$ direction and is given by

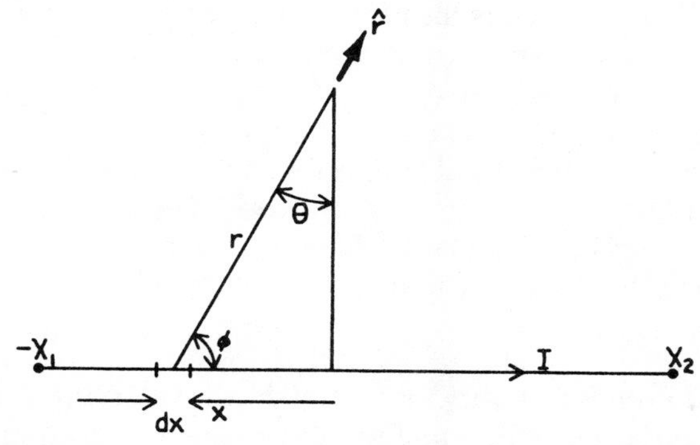

Figure 30.11 Problem 5. Use of the Biot–Savart law to find the field of a wire of length L with current I. The current enters the wire at $x = -x_1$ and leaves at $x = x_2$.

$$d\mathbf{B} = \frac{\mu_0}{4\pi}\frac{I(\mathbf{dx} \times \hat{\mathbf{r}})}{r^2} = \frac{\mu_0}{4\pi}\frac{Idx \sin\phi}{r^2}\,\hat{\mathbf{k}}$$

$$\sin\phi = y/r$$

$$db = \frac{\mu_0}{4\pi}\frac{Idxy}{(x^2+y^2)^{3/2}}$$

$$B = \frac{\mu_0}{4\pi} Iy \int_{-x_1}^{x_2} \frac{dx}{(x^2+y^2)^{3/2}}$$

We can easily evaluate the integral using integral tables:

$$\int \frac{dx}{(x^2+y^2)^{3/2}} = \frac{x}{y^2\sqrt{x^2+y^2}}$$

Then

$$B = \frac{\mu_0}{4\pi} Iy\left[\frac{x}{y^2\sqrt{x^2+y^2}}\right]_{-x_1}^{x_2} = \frac{\mu_0 I}{4\pi y}\left[\frac{x_2}{\sqrt{x_2^2+y^2}} + \frac{x_1}{\sqrt{x_1^2+y^2}}\right]$$

$$\mathbf{B} = \frac{\mu_0 I}{4\pi y}\left[\sin\theta_1 + \sin\theta_2\right]\hat{\mathbf{k}}$$

6. *Guided Problem.* Evaluate $\oint \mathbf{B} \cdot d\mathbf{S}$ for a closed surface surrounding a moving charged particle, and show that for this case $\oint \mathbf{B} \cdot d\mathbf{S} = 0$. Show qualitatively that for an arbitrary closed surface in the vicinity of a magnetic dipole, $\oint \mathbf{B} \cdot d\mathbf{S} = 0$.

Solution scheme

a. The magnetic field lines of a moving charge form concentric circles centered on the axis of motion of the particle. The simplest closed surface about a moving charge is a sphere of radius r centered on the particle. Show that for all points on the surface of the sphere, the field lines are tangent to the surface of the sphere and so are normal to $d\mathbf{S}$. This should allow you to show that $\oint \mathbf{B} \cdot d\mathbf{S} = 0$.

b. Alternatively, you can show that for all possible closed surfaces, every field line that enters the closed surface also leaves it, so the net flux through the surface is zero.

c. For the magnetic dipole, sketch the field lines in the vicinity of a magnetic dipole. All your field lines should be continuous lines; you should therefore be able to show that for any closed surface, every field line that enters must also leave.. Therefore, $\oint \mathbf{B} \cdot d\mathbf{S} = 0$. This statement is called Gauss' Law for magnetism and is equivalent to saying that magnetic monopoles do not exist.

7. *Worked Problem.* A very long, straight wire carrying a current of 1.0 A is bent at right angles in the middle of the wire as shown in Figure 30.12. What is the magnetic field at (a) $x = y = 0$, $z = \pm 0.2$ m, and at (b) $x = y = 0.2$ m, $z = 0$?

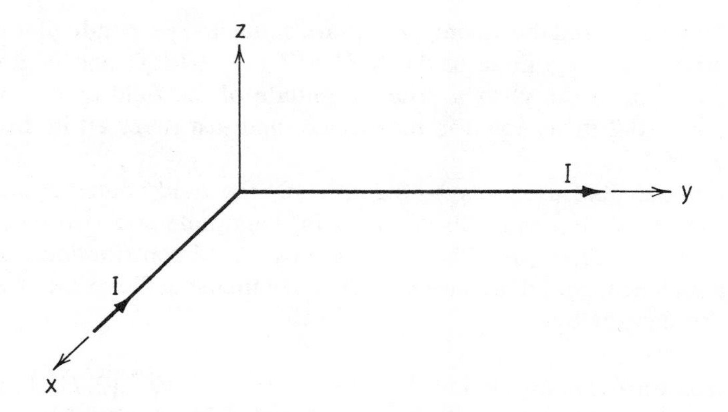

Figure 30.12 Problem 7. A very long, straight wire carrying a current of 1.0 A and bent at right angles in the middle as shown.

Solution

a. Qualitatively, for points on the z axis, the field of the wire segment on the x axis will be directed along the $+y$ axis for values of $z > 0$, and along the $-y$ axis for $z < 0$. The field of the wire segment on the y axis will point along the $+x$ axis for $z > 0$, and along the $-x$ axis for $z < 0$, as is shown in Figure 30.13. The magnitude of each field component, above, must be the same.

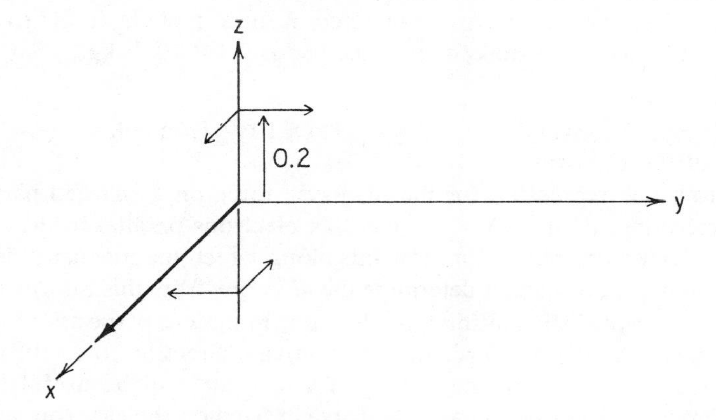

Figure 30.13 Problem 7. Magnetic field components on the $\pm z$ axis due to current-carrying wire segments parallel to the x and y axes.

We can use the expression found in worked problem 5 to evaluate these components. For the field at $x = y = 0$; $z = +0.2$ m,

$$B_y = [\mu_0 I/4\pi z](\sin\theta_1 + \sin\theta_2)$$

where $\theta_1 = 0$; $\theta_2 = \pi/2$. Then

$$B_y = [\mu_0 I/4\pi(0.2)](1) = 5 \times 10^{-7}\ \text{T}.$$

Since the x and the y components are equal and perpendicular to each other, the magnitude of the field is 7.1×10^{-7} T, and is at 45° to the x (or to the y) axis. The magnitude of the field at $x = y = 0$; $z = -0.2$ m, is equal in magnitude, and just reversed in direction, to this.

b. For the field at $x = y = 0.2$; $z = 0$, the field contributions from each wire segment will be equal in magnitude and will each point in the $-z$ direction. Therefore, the two field contributions will just add. The magnitude of each is given (again using the result of worked problem 5) by

$$[\mu_0 I/4\pi(0.2)][\sin(\pi/4) + \sin(\pi/2)] = 5 \times 10^{-7}\,[0.71 + 1.00]$$
$$= 8.55 \times 10^{-7}\ \mathrm{T}$$

The total field is just 1.71×10^{-6} T.

8. *Guided Problem.* Electrons carrying energy of 1 keV are injected from different directions into a uniform magnetic field. Electrons traveling parallel to the x axis experience no deflection. Electrons injected in the x–y plane at 45° to the x axis experience a force of 1×10^{-14} N in the negative z direction.

a. Find the magnitude and direction of the magnetic field.
b. Find the direction in which the electrons injected at 90° to the x axis (and on the x–y plane) are deflected.
c. What directions are electrons injected in the x–z plane at 45° to the x axis deflected? The mass of the electron is 9.1×10^{-31} kg.

Solution scheme

a. Using the conversion that 1 eV = 1.6×10^{-19} J, calculate the velocity v of the electron.
b. From the expression for the magnetic force on a charged particle, determine, by means of the fact that electrons parallel to the x axis experience no deflection, the axis along which the magnetic field is pointing. (You cannot determine the *direction* from this information!)
c. Now use the information that electrons in the x–y plane at 45° to the x axis experience a force in the negative z direction to establish the *direction* of the magnetic field. (A diagram will be useful here.) Knowing the direction and the force exerted on the electron, as well as the velocity of the electron, determine the magnitude of the magnetic field.
d. You now know the magnetic field vector **B**. Solve for the force exerted on the electrons to answer parts b and c of this problem.

9. *Worked Problem.* A charge of 0.05 μC is smeared uniformly over a thin plastic disk 12 cm in diameter, which is then spun about its axis at a speed of 60 revolutions per second. Find the magnetic field produced by the charge at the center of the disk. What current at the rim of the disk would produce the same magnetic field if the disk were held stationary?

Solution: From Example 5 in the text (p. 741), we know that the field at the center of a circular wire loop of radius R and current I in the x–y plane is

$$B_z = \frac{\mu_0}{2\pi} \frac{I(\pi R^2)}{R^3}$$

Divide the disk into concentric rings of radius r and thickness dr, as shown in Figure 30.14. The area of any ring is $2\pi r dr$, so that the current

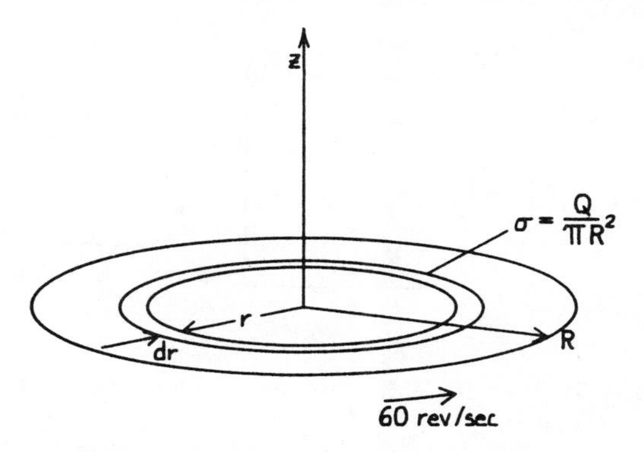

Figure 30.14 Problem 9. A charged spinning plastic disk.

resulting from any ring is given by i = [charge on ring] × [no. of revolutions/s] = $(dq) \times 60 = \sigma dA(60) = 60\ \sigma(2\pi r dr)$. The field contribution dB_z resulting from any ring is

$$dB_z = \frac{\mu_0}{2\pi} \frac{i(\pi r^2)}{r^3} = \frac{\mu_0(60\ \sigma 2\pi r dr)(\pi r^2)}{2\pi r^3} = 60\ \mu_0 \sigma \pi dr$$

$$B_z = \int dB_z = 60\ \mu_0 \sigma \pi \int_0^R dr = 60\ \mu_0 \sigma \pi R$$

since $\sigma = Q/\pi R^2$

$$B_z = \frac{60\ \mu_0 Q}{R}$$

Substituting in the values for μ_0, Q, and R, we obtain $B_z = 3.14 \times 10^{-11}$ T. The magnetic field resulting from a current I at the rim of a fixed disk is given by

$$B_z = \frac{\mu_0}{2\pi} \frac{I(\pi R^2)}{R^3} = \frac{\mu_0 I}{2R}$$

If we equate the two expressions, we obtain

$$\frac{\mu_0 I}{2R} = \frac{60\ \mu_0 Q}{R}$$

or

$$I = 120\ Q = 120(.05 \times 10^{-6})$$
$$I = 6 \times 10^{-6}\ \text{A} \quad \text{or} \quad 6\ \mu\text{A}$$

10. *Guided Problem.* With the help of the result of worked problem 5, compare the magnetic field at the center of each of the wire loops shown in Figure 30.15. Each loop has a current of I equal to 1.0 A. The length $L = 50$ cm. Compare the magnetic moment of each of these loops.

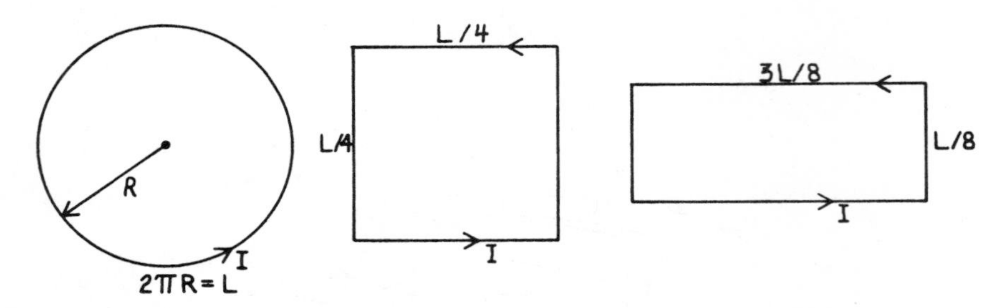

Figure 30.15 Problem 10. Three different current loops, each with current I and total length L.

Solution scheme

a. In each case, note that the direction of the magnetic field at the loop center is the same.
b. Problem 9 gives the expression for the field at the center of a circular wire loop. Use this expression to obtain the field at the center of the circuit loop of wire.
c. To obtain the field at the center of each of the rectangular loops, use the expression developed in problem 5. You will need to evaluate the appropriate distances from the wire to the center of the loop and the appropriate angles. Remember that you need to do the calculation only once for the square and then multiply your result by 4. The rectangle is a little more difficult.
d. You should now be able to compare the magnetic fields for the three cases.
e. Comparison of the magnetic moments is straightforward. Does the loop with the largest magnetic moment give you the largest magnetic field at the center of the loop? Do the results surprise you?

11. *Worked Problem.* An infinite, flat, thin sheet of copper carries a current density $\mathbf{J} = 100\ \hat{\mathbf{i}}$ A/m; that is, every strip 1 m in width along the z axis carries a current of 100 A in the positive x direction. Find the magnetic field at all points along the y axis, both above and below the sheet. The sheet lies in the x–z plane.

Solution: Divide the sheet into thin strips parallel to the x axis, as indicated in Figure 30.16. Each strip has a width of dz and carries a current

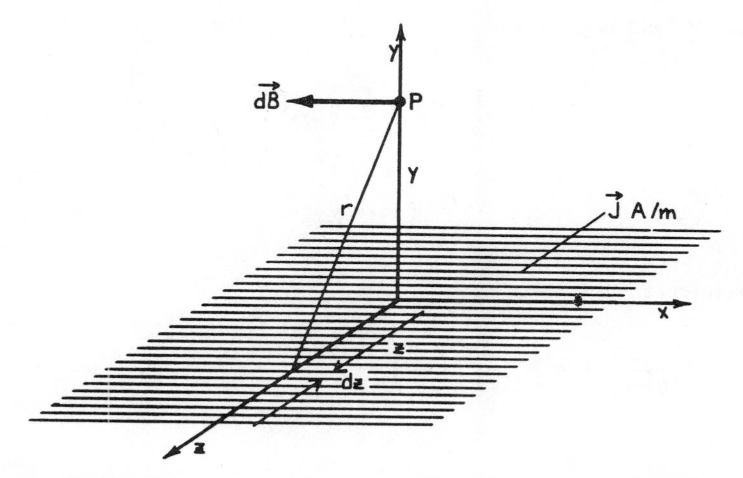

Figure 30.16 Problem 11. The magnetic field at P resulting from a strip of current of width dz.

$di = Jdz$ in the x direction. The field contribution dB resulting from a strip is given by

$$dB = \frac{\mu_0 di}{2\pi r} = \frac{\mu_0 Jdz}{2\pi\sqrt{z^2 + y^2}}$$

dB is in the y–z plane, at an angle θ to the horizontal, as indicated in Figure 30.17. Since the copper sheet has infinite dimensions, for every strip a distance z along the positive z axis, there is another strip a distance $-z$ along the negative z axis. The vertical contributions of each strip to dB exactly cancel, leaving only the horizontal contributions, which are additive.

$$dB_{\mathrm{H}} = dB \cos\theta = dB\left(\frac{y}{r}\right) = \frac{dBy}{\sqrt{z^2 + y^2}}$$

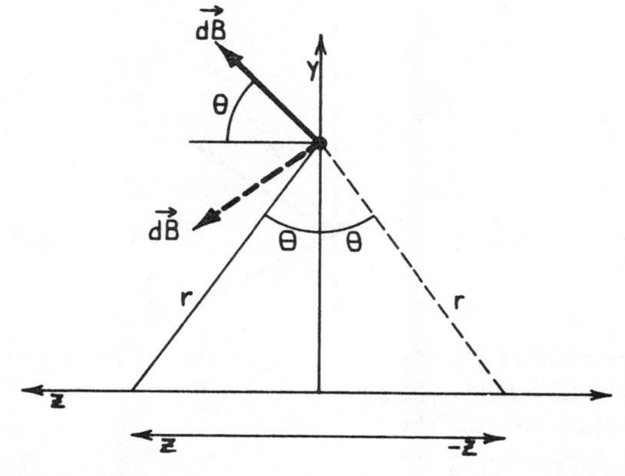

Figure 30.17 Problem 11. The copper sheet as viewed from along the positive x axis. dB is the magnetic field contribution from a strip dz.

The total field is given by

$$B = \int_{-\infty}^{\infty} dB_{\mathrm{H}} = \int_{-\infty}^{\infty} \frac{\mu_0 J dz y}{2\pi(z^2 + y^2)}$$

$$B = \frac{\mu_0 J y}{2\pi} \int_{-\infty}^{\infty} \frac{dz}{z^2 + y^2}$$

From integral tables

$$\int \frac{dz}{z^2 + y^2} = \frac{1}{y} \tan^{-1}\left(\frac{z}{y}\right)$$

or

$$\int_{-\infty}^{\infty} \frac{dz}{z^2 + y^2} = \frac{1}{y} [\tan^{-1}(\infty) - \tan^{-1}(-\infty)] = \frac{1}{y}\left[\frac{2\pi}{2}\right] = \frac{\pi}{y}$$

$$B = \frac{\mu_0 J y \pi}{2\pi y} = \frac{\mu_0 J}{2}$$

or

$$\mathbf{B} = \frac{\mu_0 J}{2}\hat{\mathbf{k}} = \frac{4\pi\,(10^{-7})(100)}{2}\hat{\mathbf{k}} = 6.28(10^{-5})\hat{\mathbf{k}}\ \mathrm{T}$$

The magnetic field below the sheet has the same magnitude as that above the strip, but the direction is now reversed: It is now $6.28 \times 10^{-5}\,(-\hat{\mathbf{k}})$ T, as shown in Figure 30.18.

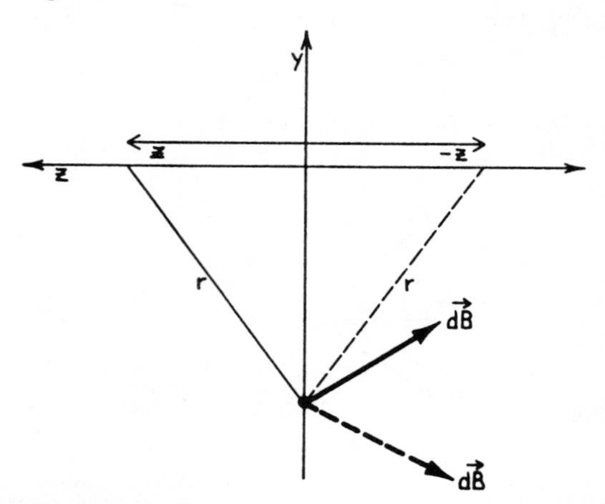

Figure 30.18 Problem 11. The same view of the copper sheet as in Figure 30.17. The field below the sheet is in the direction opposite to that of the field above the sheet.

12. *Guided Problem.* A star wars spaceship happens to be flying at a speed of 0.3 c (c = speed of light) parallel to and 20 m from an intense 100 A ion beam, as shown in Figure 30.19. Somehow the spaceship has acquired a

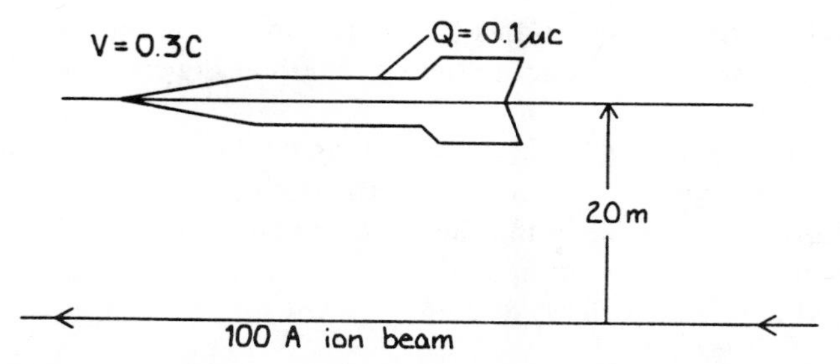

Figure 30.19 Problem 12. A spaceship flying parallel to an intense, 100 A ion beam.

charge of 0.1 μC. What is the magnetic force on the spaceship, and what is its direction? The spaceship navigator has devices that can measure magnetic or electric fields and forces. What magnetic force does he see acting on his ship? What electric force? If he sees an electric force caused by the ion beam, what is the value of the electric field causing it?

Solution scheme

a. Evaluate the magnetic field at the spaceship caused by the ion beam.
b. Knowing the magnetic field and the charge on the ship, calculate the force on the ship and its direction.
c. In the frame of reference of the ship, $\upsilon = 0$. What magnetic force does the navigator see? What electric force must he therefore see, and what is its direction?
d. Since we know the electric force $F_e = qE$, and we know q, the charge on the spaceship, can you calculate the electric field seen by the navigator?

Answers to Guided Problems

2. The direction of $\mathbf{F}_2$, the force on q_2, is always parallel to the x axis.
   At $t = 0$, the force $\mathbf{F}_1$ (on q_1) is parallel to the $-y$ axis.
   At $t = 1$, $\mathbf{F} = 0$.
   At $t = 2$, $\mathbf{F}_1$ is parallel to the $+y$ axis.
4. $\mathbf{F} = q(4\,\hat{\mathbf{i}} - 2\,\hat{\mathbf{j}} - 5\,\hat{\mathbf{k}})$ N, where $q = 1.6 \times 10^{-19}$ C
   $F = 6.71\ q = 1.07 \times 10^{-18}$ N
   $a = 6.43 \times 10^{8}$ m/s
6. For all cases we should be able to show that $\int \mathbf{B} \cdot d\mathbf{S} = 0$.
8. a. $\mathbf{B} = -\hat{\mathbf{i}}B_0$; that is, the direction of $\mathbf{B}$ is along the negative x axis. The magnitude of B, that is, B_0, equals 4.69×10^{-3} T.
   b. $\mathbf{F} = (-\hat{\mathbf{k}})evB$; that is, the deflection is in the negative z direction.
   c. $\mathbf{F} = (\hat{\mathbf{j}})evB$; that is, the deflection is in the positive y direction.
10.

Circular loop	*Square loop*	*Rectangular loop*
$B_z = \dfrac{\mu_0 I \pi}{L}$	$B_z = \dfrac{\mu_0 I}{L}\ (3.60)$	$B_z = \dfrac{\mu_0 I}{L}\ (2.60)$

where $\mu_0 I/L = 2.51(10^{-6})$ T.

$\mu = IA = IL^2/4\pi$ $IL^2 = .25\ \text{A} \cdot \text{m}^2$	$\mu = IL^2/16$	$\mu = 3IL^2/64$

Although the circular loop has the largest magnetic moment, the field strength is slightly less than the field strength at the center of the square loop.

12. The magnetic force is 9.02×10^{-6} N and is directed toward the ion beam. In the frame of reference of the navigator, the charge on the spaceship is stationary, so that he sees no magnetic force. He does see an electric force of attraction 9.02×10^{-6} N toward the ion beam, however, and an electric field, directed toward the ion beam, of 90 V/m.

Chapter 31

Ampère's Law

Overview

In the last chapter we established a mathematical framework for the study of magnetism and magnetic effects, based on the magnetic force between two moving point charges. Using this framework, we can deduce Ampère's Law, a law very similar to Gauss' Law of electrostatics, which allows us to calculate the magnetic field of a current when certain symmetry conditions are met. We also investigate in more detail the motion of charges in magnetic fields and then the force exerted by a magnetic field on a wire carrying a current. We find that a current loop experiences a torque when placed in a magnetic field, just as an electric dipole experiences a torque when placed in an electric field. A current loop is called a magnetic dipole and has magnetic properties very similar to the electrical properties of an electric dipole.

Essential Terms

Ampère's Law
right-hand rule
field of a solenoid
cyclotron frequency
Lorentz force
velocity selector
Hall effect
torque on a current loop
potential energy of a current loop

Key Concepts

1. *Ampère's Law*. In the last chapter we saw that if we apply the Biot–Savart law to evaluate the field of a long, straight wire with current I_0, we find that the magnetic field lines form concentric circles around the wire and that the magnitude of the field is given by $B = \frac{\mu_0 I_0}{2\pi r}$ at a distance r from

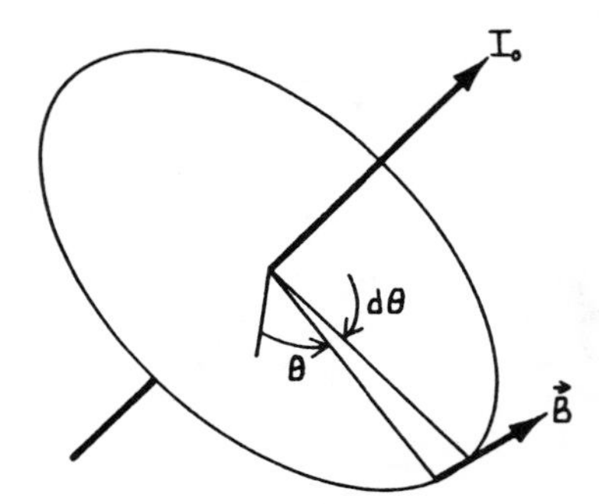

Figure 31.1 The field **B** a distance r from a long, straight wire with current I_0.

the wire. The field is indicated in Figure 31.1; the magnetic field, in vector notation, is given by

$$\mathbf{B} = \frac{\mu_0 I_0 \hat{\mathbf{u}}_\theta}{2\pi r}$$

$\hat{\mathbf{u}}_\theta$ is a unit vector tangent to a circle of radius r centered on the wire. If we evaluate $\oint \mathbf{B} \cdot d\mathbf{l}$ around *any* closed loop around the wire, only components of $d\mathbf{l}$ parallel to $\hat{\mathbf{u}}_\theta$ can contribute to the integral, so that

$$\oint \mathbf{B} \cdot d\mathbf{l} = \oint \frac{\mu_0 I_0}{2\pi r} \hat{\mathbf{u}}_\theta \cdot \hat{\mathbf{u}}_\theta r d\theta = \frac{\mu_0 I_0}{2\pi} \oint d\theta$$

$$\oint \mathbf{B} \cdot d\mathbf{l} = \mu_0 I_0$$

This result is a very powerful theorem used for the calculation of magnetic field strengths when there is a high degree of symmetry. It also plays a role in the evaluation of magnetic field strength similar to that of Gauss' Law in the calculation of electric field strength. In the case of Gauss' Law the integral was over a closed surface; the integral in Ampère's Law is a closed *line* integral.

As with Gauss' Law, a few basic symmetries allow us to calculate magnetic fields easily using Ampère's Law. Three such symmetries are cylindrical symmetry, planar symmetry, and spherical symmetry. The field of a long, straight wire with current I_0 is a case in which cylindrical symmetry can be involved. Ampère's Law tells us (just as the Biot–Savart law did) that the field of a long, straight wire with a current I_0 is given, at a distance r from the wire, by $B = \mu_0 I_0 / 2\pi r$. The field of an infinite sheet of current, illustrated in Figure 31.2, is a situation in which we can invoke planar symmetry. We rarely find a spherically symmetric current distribution.

In the last chapter we used the Biot–Savart law to show, after a rather tedious calculation, that the magnetic field at point P is given by $\mathbf{B} = \mu_0 J/2\, \hat{\mathbf{k}}$ and that the field below the sheet is given by $\mathbf{B} = \mu_0 J/2\, (-\hat{\mathbf{k}})$. The field can be evaluated easily using Ampère's Law. If we look in along the x axis and draw a closed loop running through point P, as shown in Figure 31.3, around which to evaluate $\oint \mathbf{B} \cdot d\mathbf{l}$, we see that each strip of current running

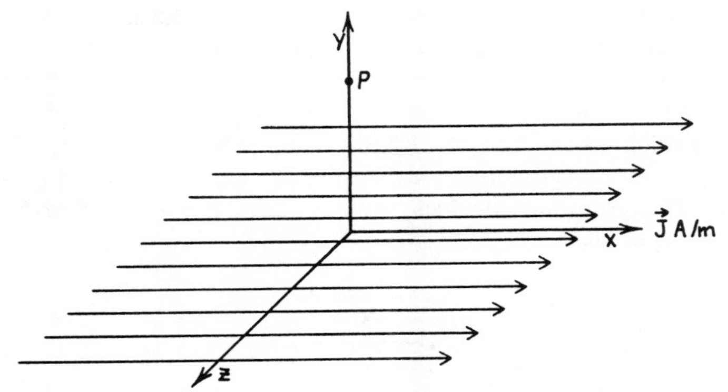

Figure 31.2 An infinite, flat sheet of current, with current density $\mathbf{J} = J\hat{\mathbf{i}}$ A/m.

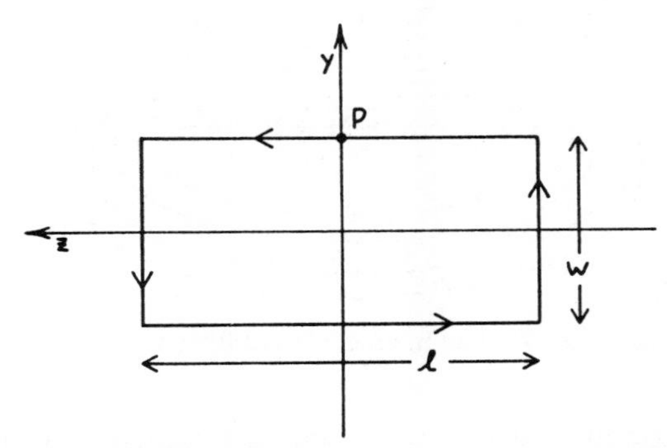

Figure 31.3 Use of Ampère's Law to find the magnetic field of an infinite sheet of current with current density $\mathbf{J} = J\hat{\mathbf{i}}$ A/m.

toward us and out of the page parallel to the x axis contributes a field parallel to the z axis below the sheet.

From the principle of symmetry we can also argue that the only available field directions are either perpendicular to the sheet (the direction of the field lines in the electrostatic case) or parallel (and antiparallel) to the sheet. We expect the field to be perpendicular to the direction of the current, and since the current is parallel to the x axis, we expect the field lines to be in the y-z plane. It is reasonable, then, to expect that B is parallel to the z axis above the sheet and antiparallel to the z axis below the sheet. Then

$$\oint \mathbf{B} \cdot d\mathbf{l} = Bl + Bl = 2Bl$$

(The contributions from the sides of height w are either 0 or cancel one another.)
Then $2Bl = \mu_0$ (total enclosed current)

$$= \mu_0 Jl \quad \text{or} \quad B = \frac{\mu_0 J}{2}$$

2. *Field of a solenoid.* A solenoid is a conducting wire wound in a tight helical coil. We can use symmetry arguments to show that the field lines

should be parallel to the axis of the solenoid; alternatively, we can view a solenoid as a superposition of many single loops of wire, and the resulting field as arising from the superposition of the fields of single loops. The field lines are indicated in Figure 31.4. From Ampère's Law the magnitude of the field inside the solenoid can be given by $B = \mu_0 n I_0$, where I_0 is the current in each loop of wire and n is the number of turns of wire per meter of length of the solenoid. Many common devices, such as relays, inductors, and electromagnets, are solenoids.

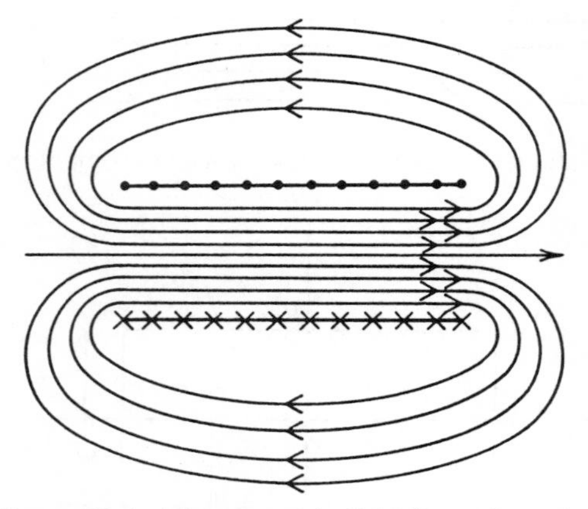

Figure 31.4 The magnetic field lines of a solenoid.

A toroid, a conducting wire wrapped in a tight coil in the shape of a doughnut, behaves like a very long solenoid—that is, like a solenoid with no ends. From Ampère's Law the field inside a toroid can be given by

$$B = \frac{\mu_0 N I_0}{2\pi r}$$

where N is the total number of turns of wire in the toroid. If a toroid is "thin," that is, as indicated in Figure 31.5, if the mean radius R is much greater than $(r_2 - r_1)$, then we can write $N = 2\pi R n$, where n is the number of turns per meter in length, just as in the expression for a solenoid. Since

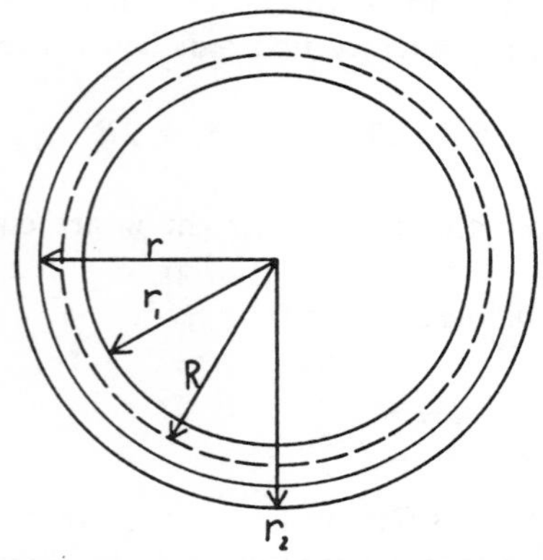

Figure 31.5 A toroid, with outer and inner radii r_2 and r_1. R is the mean radius, and r is any radius between r_2 and r_1.

$R \simeq r$, the expression for the field reduces to $B = \mu_0 n I_0$, the same expression that we obtained for a solenoid. Transformers, used for changing AC voltage levels, usually use toroids, as do many other electromagnetic devices.

3. *Cyclotron frequency.* Because the magnetic force $\mathbf{F} = q\mathbf{v} \times \mathbf{B}$, is always perpendicular to the velocity $\mathbf{v}$ of a particle, the force changes the direction of the velocity but not its magnitude. A particle moving perpendicular to a uniform magnetic field is deflected in a circular path; the magnetic force provides the centripetal force, so that $F_{\text{centripetal}} = qvB = mv^2/r$. Then $qB = mv/r$, or $r = mv/qB$. The time required for a complete circle, $T = 2\pi r/v = 2\pi m/qB$, is independent of the velocity of the particle. The frequency $v = 1/T = (qB/2\pi m)$ is called the cyclotron frequency. Figure 31.6 shows a charged particle moving perpendicular to a uniform magnetic field. The field direction is into the page.

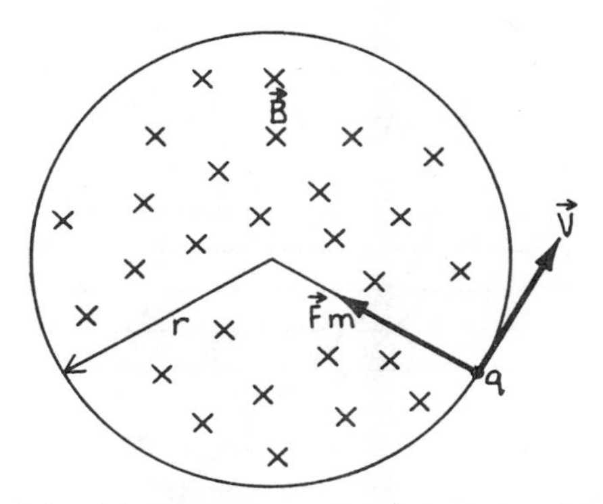

Figure 31.6 A particle with charge q moving in a circle of radius r in a uniform magnetic field. The direction of $\mathbf{B}$ is into the paper.

4. *Lorentz force.* In general, a moving charged particle is subject to both electric and magnetic forces; the *total* force on the particle is then given by

$$\mathbf{F} = q\mathbf{E} + q\mathbf{v} \times \mathbf{B}$$

This force is called the Lorentz force.

5. *Torque on a current loop.* It follows from the definition of magnetic force on a charged particle, $\mathbf{F} = q\mathbf{v} \times \mathbf{B}$, that a length of wire dl with a current I experiences a force $d\mathbf{F} = d\mathbf{l} \times \mathbf{B}$ when placed in an external magnetic field $\mathbf{B}$. If we place a rectangular loop of wire with current I in a uniform external magnetic field, there will be a magnetic force on each side of the loop, as shown in Figure 31.7. The net force on the loop is zero, because the forces on opposite sides of the loop are equal and opposite to each other. If the axis of the loop happens to be aligned at an angle θ with respect to the field, as shown in Figure 31.8, even though the forces on opposite sides of the loop are equal and opposite to each other, there is a torque, or twist, on the loop, making it want to align itself perpendicular to the field, as in Figure 31.7. The torque on the loop in Figure 31.8 is

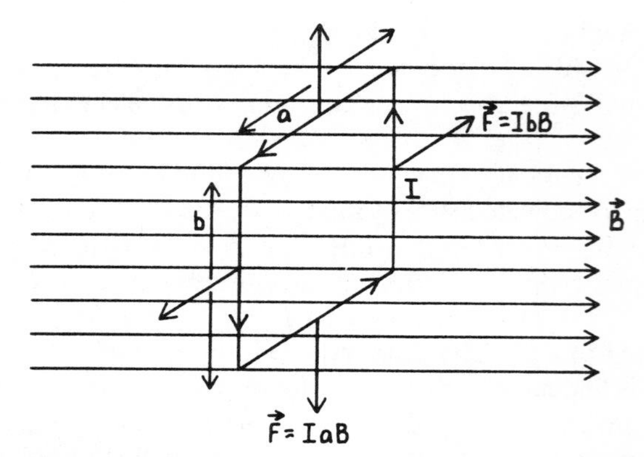

Figure 31.7 A rectangular current loop perpendicular to a uniform magnetic field **B**.

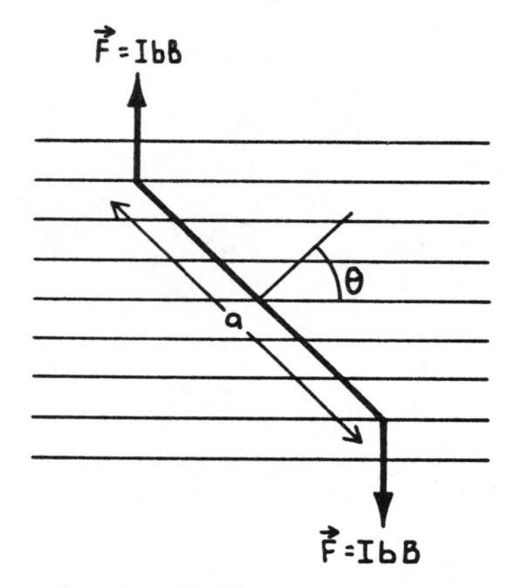

Figure 31.8 The same loop as in Figure 31.7 at an angle θ with respect to the magnetic field.

$$\begin{aligned} F(a \sin \theta) &= (IbB)(a \sin \theta) \\ &= IAB \sin \theta \\ &= mB \sin \theta \\ &= \mathbf{m} \times \mathbf{B} \end{aligned}$$

A is the area of the loop, and $m = IA$ is the magnetic moment of the loop. This expression for torque is almost identical to the expression for the torque on an electric dipole in an electric field, where we found the torque to be $\mathbf{p} \times \mathbf{E}$. Similarly, just as we found the potential energy of an electric dipole in an electric field to be given by $U = -\mathbf{p} \cdot \mathbf{E}$, the potential energy of a magnetic dipole in a magnetic field is $U = -\mathbf{m} \cdot \mathbf{B}$. The direction of $\mathbf{m}$ is perpendicular to the plane of the loop, so that the potential energy is taken to be zero with the loop aligned in its "preferred" orientation, with its axis perpendicular to the magnetic field.

Sample Problems

1. *Worked Problem.* The beam of the Los Alamos Meson Physics Facility is a 1 mA, 800 MeV proton beam. If the beam has a uniform charge distribution over its cross-sectional area and is 3 mm in diameter, find the magnetic field strength at the edge of the beam. Make a graph of the field strength from the center of the beam ($r = 0$) to a point 1 m from the beam center. What is the direction of the magnetic force on charged particles in the beam?

Solution: From Ampère's Law we know that the field at a distance B from a current i is given by $B = \mu_0 i / 2\pi R$. At the edge of the beam, $R = 1.5$ mm, so that

$$B = \frac{(1.256 \times 10^{-6})(1 \times 10^{-3})}{2\pi\ (1.5 \times 10^{-3})} = 1.33 \times 10^{-7}\ \text{T}$$

For $r < R$ the current enclosed within a circle of radius r is $(r^2/R^2)i$, so that from Ampère's Law, the field can be given by

$$B = \frac{\mu_0 (r^2/R^2) i}{2\pi r} = \frac{\mu_0 i r}{2\pi R^2}$$

The use of Ampère's Law is demonstrated in Figure 31.9. Then $B = 0$ at $r = 0$ and increases linearly with r until $r = R$. For $r > R$, B decreases as $B = \mu_0 i / 2\pi r$. At $r = 1$ m, $B = 2 \times 10^{-10}$ T. A graph of B versus r is shown in Figure 31.10. Since $\mathbf{F} = q\mathbf{v} \times \mathbf{B}$, the direction of the magnetic force on charged particles is inward.

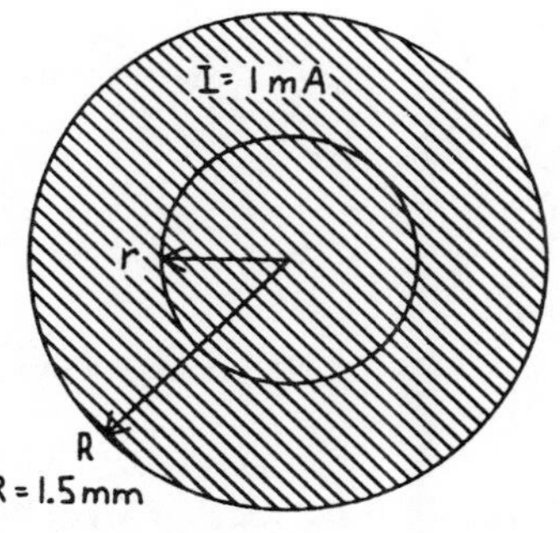

Figure 31.9 Problem 1. Using Ampère's Law to find the magnetic field for $r < R$. The total current is i.

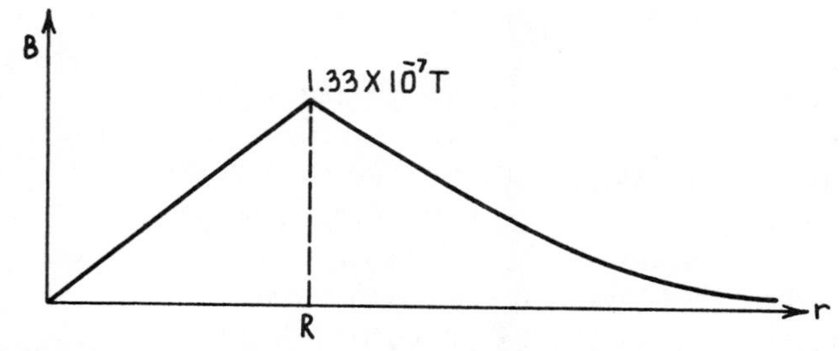

Figure 31.10 A graph of B versus r for the proton beam of Figure 31.9.

2. *Guided Problem.* A superconducting solenoid is to be used to generate a magnetic field of 3.0 T. (For superconductors, the resistance, R, is exactly zero.) If the solenoid winding has 200 turns per meter, what is the required current in the windings? What force per unit length is exerted on the coil windings by the magnetic field, and in what direction is the force?

Solution scheme

a. Write the expression for the magnetic field of a solenoid in terms of the current in the windings and the number of turns per unit length, and use this to find the required current.
b. Write the expression for the force exerted on a conductor of length $d\mathbf{l}$ with current I in a magnetic field $\mathbf{B}$. To determine the direction of the force, draw a diagram showing an end-view of the solenoid.

3. *Worked Problem.* A long, round copper cylinder 10 cm in diameter has a hole 3 cm in diameter bored along its length; the center of the hole is 2 cm from the center of the cylinder, as shown in Figure 31.11. A current of 100 A uniformly distributed over the cross-sectional area of the metal is flowing along the cylinder. Find the magnetic field

a. along the center line of the hole
b. along the outer periphery of the hole, indicated by point P in the figure

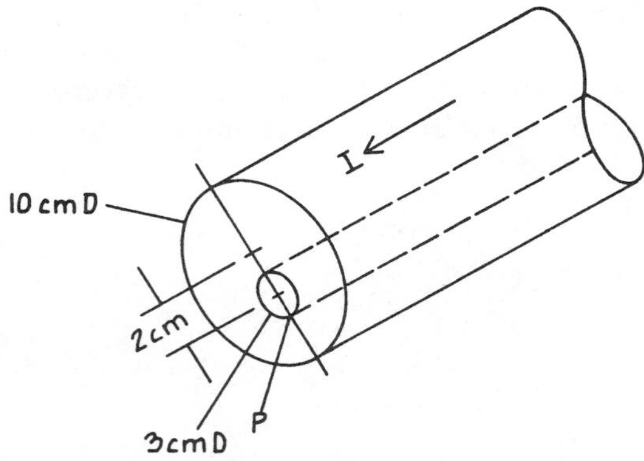

Figure 31.11 Problem 3. A long cylinder with a hole 3 cm in diameter bored along its length.

Solution: Figure 31.12 shows how we can treat the system as a round cylinder with no hole through it, plus a second cylinder with the *same* current density, except that the current in the second cylinder flows in the opposite direction. In other words, we can use superposition to simplify the problem.

Next, we evaluate Ampère's Law for a closed loop of radius r running through the position of the center line of the hole in the cylinder; that is, let $r = 0.02$ m. Then $\int \mathbf{B} \cdot d\mathbf{l} = 2\pi r B = \mu_0$ (total enclosed current), where the total enclosed current for the round cylinder with no hole is given by the product of the current density and the enclosed area. The current density $j = I/A$, where $I = 100$ A, and A = [area of round cylinder] − [area of hole] $= (\pi/4)(0.1^2 - 0.03^2)$ or $A = 7.62 \times 10^{-3}$ m^2, so that $j = 100/7.62\ (10^{-3})$ $= 1.31 \times 10^4$ A/m^2. Then $2\pi r B = \mu_0(j)(\pi r^2)$, or

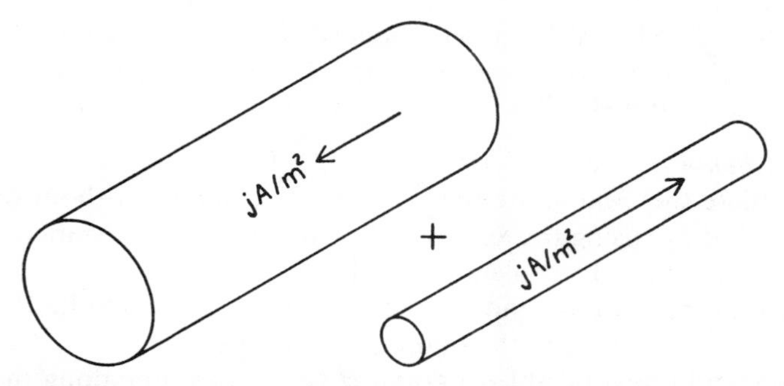

Figure 31.12 Problem 3. The long cylinder with the hole bored through it is equivalent to two cylinders with currents flowing in opposite directions.

$$B = \mu_0 jr/2 = 1.256 \times 10^{-6}\,(1.31 \times 10^4)(0.02)/2$$
$$B = 1.65 \times 10^{-4}\text{ T}$$

We need to add to this the field contribution resulting from the current flowing in the second cylinder in the opposite direction, but at the *center* of this cylinder, $B = 0$; if we apply Ampère's Law, the total enclosed current at the center of the cylinder is zero. Then the total magnetic field along the center line of the hole is $B = 1.65 \times 10^{-4}$ T.

To find the magnetic field along the outer periphery of the hole, we repeat the process; the current density $j = 1.31 \times 10^4$ A/m², and we now let r equal 0.035 m. Then Ampère's Law tells us that $B = \mu_0 jr/2 = 1.256 \times 10^{-6}\,(1.31 \times 10^4)(0.035)/2 = 2.88 \times 10^{-4}$ T. This is the field contribution from the large round cylinder. The contribution from the current flowing in the second cylinder in the opposite direction is given by $B = \mu_0 jr/2$ also, where $j = 1.31 \times 10^4$ A/m² and $r = 0.015$ m, because the diameter of this cylinder is 3 cm. We obtain, for this field contribution, $B = 1.23 \times 10^{-4}$ T. The second field contribution subtracts from the first, so that the net field is then $B = (2.88 \times 10^{-4}) - (1.23 \times 10^{-4}) = 1.65 \times 10^{-4}$ T, which is exactly the result we obtained for the center of the hole in the cylinder!

4. *Guided Problem.* A solenoid 50 cm long with 800 turns per meter has a second coil 10 cm long with 90 turns wound tightly over the first winding in the middle of the solenoid, as shown in Figure 31.13. Graph the magnetic field along the length of the solenoid (ignoring end effects) for the following currents in the two windings:

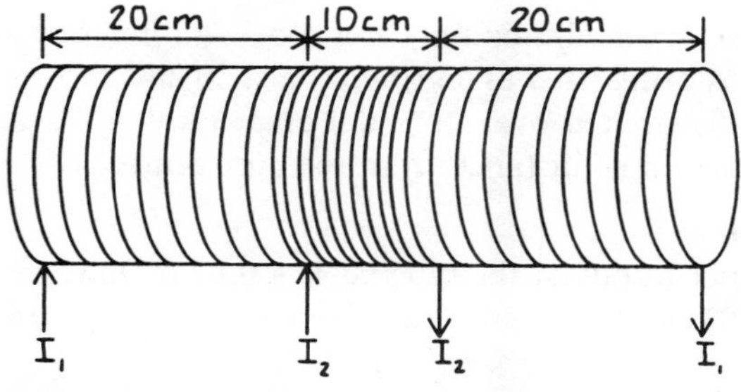

Figure 31.13 Problem 4. A solenoid with two windings.

a. $I_1 = 1.0$ A, $I_2 = 0$ A.
b. $I_1 = 1.0$ A, $I_2 = 1.0$ A, both in the same direction around the solenoid.
c. $I_1 = 1.0$ A, $I_2 = 1.0$ A, as before, but the direction of I_2 is reversed. $\mu_0 = 1.256 \times 10^{-6}$ H/m.

Solution scheme

a. Calculate the field along that part of the solenoid without the second winding, using the expression we obtained from Ampère's Law for the field inside a solenoid.
b. Use superposition to find the magnetic field in the middle 10 cm of the solenoid, where both windings contribute.
c. You should now be able to graph $\boldsymbol{B}$ versus position along the solenoid for each of the three cases.

5. *Worked Problem.* Two large, thin copper sheets intersect at an angle of 60°, as shown in Figure 31.14. Each sheet carries a uniformly distributed current J equal to 100 A/m of length, the direction of the current in each sheet being to the right in the figure, as indicated. Find the magnetic field everywhere resulting from the current in the two copper sheets.

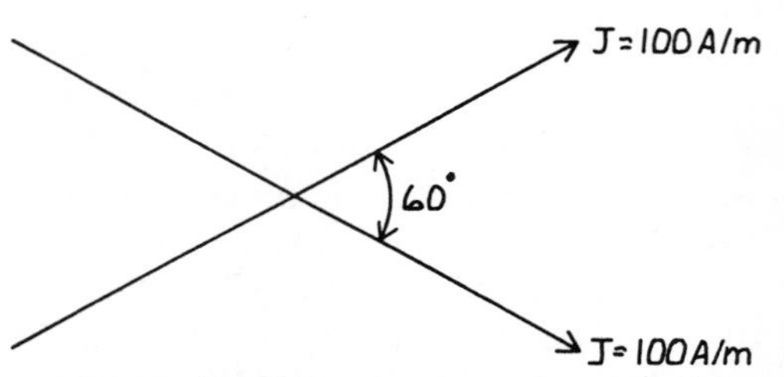

Figure 31.14 Problem 5. Two large copper sheets with current density 100 A/m. The two sheets intersect at an angle of 60°.

Solution: As we found in our discussion of Ampère's Law, the field of a large conducting sheet $B = \mu_0 J/2$, as indicated in Figure 31.15(a). The field does not decrease as we move farther away from the sheet. For the two sheets we can easily use superposition to find the total fields in the four regions 1, 2, 3, and 4, shown in Figure 31.15(b).

The figure shows that the fields from the two sheets cancel in regions 1 and 3, so that in these two regions $\boldsymbol{B} = 0$. In regions 2 and 4 the fields from the two sheets are additive, so that here $B = \mu_0 J = 1.256 \times 10^{-6}$ (100) $= 1.256 \times 10^{-4}$ T. The direction of $\boldsymbol{B}$ is out of the paper in region 2 and into the paper in region 4.

6. *Guided Problem.* An aluminum water pipe with inner radius 2 cm and outer radius 3 cm is being used as a grounding strip and carries a current of 150 A uniformly distributed over its metal area. Find and graph the magnetic field as a function of distance from the pipe center.

Solution scheme

a. Let the inner radius of the pipe be $r_1 = 0.02$ m and the outer radius be $r_2 = 0.03$ m. Now apply Ampère's Law, in turn, to each of the three regions; $r < r_1$; $r_1 < r < r_2$; $r > r_2$.

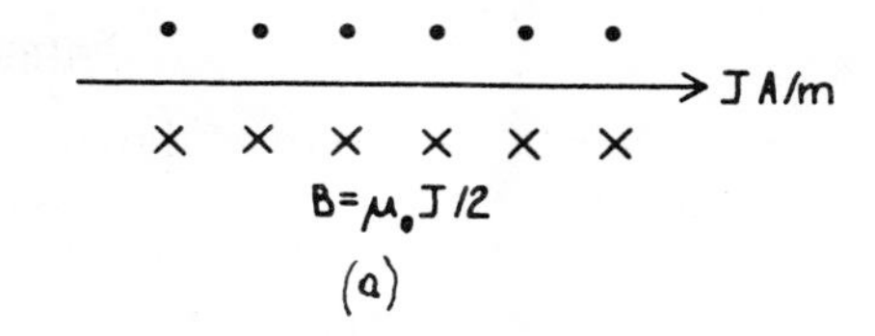

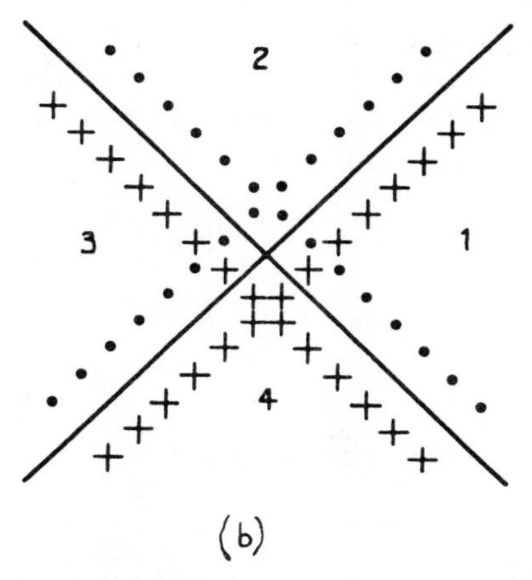

Figure 31.15 Problem 5. (a) The field resulting from a single large conducting sheet. B is out of the paper above the sheet and into the paper below the sheet. (b) The use of superposition to find the field resulting from two large conducting sheets. The fields cancel in regions 1 and 3 and are additive in regions 2 and 4.

b. For $r < r_1$ the total enclosed current is 0. What is B?
c. For $r_1 < r < r_2$ the total enclosed current is the product of the total current and the fraction of the area of the pipe inside r. Can you find B for this region?
d. For $r > r_2$ all the current is enclosed. What is B here?
e. You should now be able to graph B from $r = 0$ out to any distance.

7. *Worked Problem.* In a high-energy physics experiment, particles are often identified by measuring both their velocity and their momentum. In a certain magnetic field, protons moving perpendicular to the field are found to have a radius of curvature of 40 cm. What radius of curvature would (a) deuterons (mass = 2 × proton mass) and (b) positive pions (mass = 0.148 × proton mass) have, if their velocities were the same as that of the protons? (c) Compare the energies of both deuterons and pions with the proton energy if their radii of curvature are found to be the same as that of a proton.

Solution: Recall that for a charged particle moving perpendicular to a uniform magnetic field, $Bqv = mv^2/R$, or $BR = mv/q$—that is, the centripetal force is provided by the magnetic force. Then the ratios $(BR)_d/(BR)_p = (mv/q)_d/(mv/q)_p$, where d implies a deuteron and p implies that the charged particles is a proton. Since in this problem both particles have the same velocity,

$$R_d/R_p = m_d/m_p = 2$$

In other words, the radius of curvature of the deuteron would be *twice* that of the proton. Similarly,

$$(BR)_{pion}/(BR)_p = m_{pion}/m_p = 0.148$$

The radius of the pion would be only 0.148 times that of the proton. If the radii of curvature of the deuteron and pion were the same as that of the proton, then (because all three particles have the same charge) $(BRq)_d = (BRq)_{pion} = (BRq)_p$. But $BRq = mv = \sqrt{2\,mT}$ for any charged particle, where T is the kinetic energy. Then

$$(\sqrt{2\,mT})_d = (\sqrt{2\,mT})_p$$

$$T_d = \frac{m_p}{m_d}\,T_p = \tfrac{1}{2}T_p$$

Similarly,

$$(\sqrt{2\,mT})_\pi = (\sqrt{2\,mT})_p$$

$$T_\pi = \frac{m_p}{m_\pi}\,T_p = \frac{T_p}{0.148}$$

or

$$T_\pi = 6.76\ T_p$$

8. *Guided Problem.* Cosmic ray protons in a region of outer space are moving in large circles, indicating that they are moving perpendicular to a uniform magnetic field. The average time required for the protons to complete a circle is found to be 4.70 seconds. If the protons have an average energy of 500 keV, find the strength of the magnetic field causing the circular motion, and the diameter of the circular orbits. The proton mass is 1.67×10^{-27} kg. Compare the magnetic field strength that you find with the average magnetic field strength found at the Earth's surface (5×10^{-5} T).

Solution scheme

a. Find the velocity of the protons. The energy can be obtained by converting the 500 keV to joules (1 eV = 1.6×10^{-19} J), and since the proton mass is given, you should be able to find the velocity.
b. Knowing the velocity and the time required to complete a circle, calculate the radius of the circle (distance = velocity × time).
c. For a particle moving in a circular path in a uniform magnetic field, the centripetal force is provided by the magnetic force, qvB. You can now find the magnetic field strength, B.

9. *Worked Problem.* A velocity selector has parallel metal plates 2 cm apart, with a potential difference of 10 kV between the plates (10 kV = 10,000 V). What magnetic field strength is needed, and in what direction, in order for (a) 100 keV protons and (b) 100 keV electrons to pass between the plates undeflected? Assume that both particles are nonrelativistic. The plates are shown in Figure 31.16.

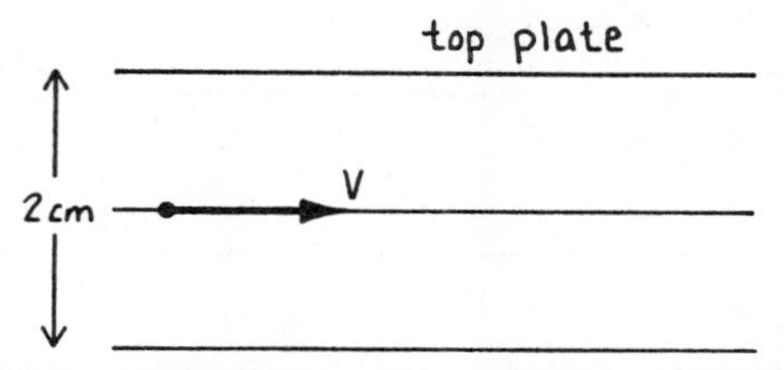

Figure 31.16 Problem 9. The metal plates of a velocity selector.

Solution: In order for the particles to be undeflected, the magnetic force on the charged particles must be equal and opposite to the electric force, **Fe** = $q\mathbf{E}$. The electric force is directed between the plates, and since the magnetic force $\mathbf{F}_m = q\mathbf{v} \times \mathbf{B}$, the direction of **B** must be either into or out of the paper, that is, perpendicular to **E**. If **E** is directed downward, then **B** must be directed into the paper for protons and out of the paper for electrons.

The magnitude of the magnetic field strength is given by $\upsilon B = E$, where $E = V/d = 1000/0.02 = 5 \times 10^5$ V/m. We can calculate the velocity of a particle from its energy, where the energy 100 keV = 100 $(10^3) \times 1.6 \times 10^{-19}$ J $= (1/2)m\upsilon^2$. The proton mass is 1.67×10^{-27} kg, giving a velocity υ of 4.38 (10^6) m/s. The electron mass is 9.1×10^{-31} kg, giving an electron velocity υ of 1.875×10^8 m/s. Then, since $B = E/\upsilon$, $B = 0.14$ T for protons, into the paper, and $B = 2.67 \times 10^{-3}$ T for electrons, out of the paper.

10. *Guided Problem.* You are given a long piece of wire and a battery, as well as a switch, and are asked to check the direction of the magnetic field of a large electromagnet. How might you do this?

Solution scheme

a. Make a large single-loop DC circuit (with a switch), with the ends of the wire hooked to the positive and negative terminals of the battery. Make sure that you know the direction of the flow of current in the loop.
b. Suppose that you pass one end of the loop through the field of the magnet, so that the current is flowing roughly perpendicular to the magnetic field. (You usually know from the geometry the direction of the field lines, but not the direction that the field line vectors are pointing.) If you switch on the current through the loop, the end of the loop in the magnetic field will experience a force. Write down the expression for this force.
c. From the direction that the wire moves, you can deduce the direction of the magnetic field. (This technique is very easy and simple, and it works!)

11. ***Worked Problem.*** A Hall probe is a device often used to measure the strength of a magnetic field. A conducting strip with a known current is placed in the unknown magnetic field, and the Hall effect voltage is measured between the top and bottom of the strip, as indicated in Figure 31.17.

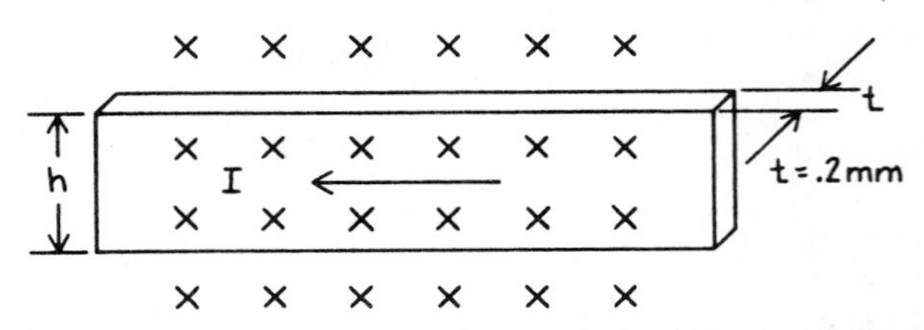

Figure 31.17 Problem 11. A Hall probe. The direction of the unknown magnetic field is into the paper, as indicated by the crosses.

A certain Hall probe has a thickness of 0.2 mm. The strip material has a charge carrier density of 4×10^{24} per cubic meter. A Hall voltage of 63 μV is measured between the top and the bottom of the strip when the current through the strip is 20 mA. Find the magnetic field strength. Is the top of the strip at a positive or a negative potential with respect to the bottom of the strip?

Solution: The magnetic field exerts a force qv_dB on the charge carriers, forcing them toward the bottom of the Hall probe. This is true whether the charge carriers are positive or negative, because if the current is actually caused by electrons, the electrons move from left to right, opposite the direction that the current is flowing. (Since the current in the probe results from electrons, the bottom of the Hall probe becomes negatively charged relative to the top.) The concentration of negative charge carriers at the bottom of the probe causes an electric field $E = Vh$, where V is the Hall voltage between the top and the bottom of the probe.

At equilibrium the force on the electrons resulting from this field is equal to the magnetic force, so that $v_dB = E = V/h$. But the current i in the probe can be expressed as nev_dA, where n is the density of charge carriers, e is the electron charge, v_d is the drift velocity, and $A = ht =$ the cross-sectional area of the strip. We can therefore eliminate v_d in the expression $v_dB = V/h$, so that since $v_d = i/(neA)$, $Bi/(neht) = V/h$ or

$$B = Vnet/i$$

$V = 63 \times 10^{-6}$ V; $n = 4 \times 10^{-24}$ m^{-3}; $e = 1.6 \times 10^{-19}$ C; $t = 0.2 \times 10^{-3}$ m; $i = 20 \times 10^{-3}$ A. Substituting in these values, we obtain $B = 0.4$ T.

12. *Guided Problem.* A small, circular, 100-turn coil of wire 6 cm in diameter carries a current of 0.1 A. When the coil is oriented at 45° with respect to a uniform magnetic field, the torque on the coil is 0.020 N • m.

a. What is the strength of the magnetic field?
b. How much energy is needed to flip the coil from a position parallel to the same magnetic field to a position antiparallel to the field, that is, a flip of 180°? Does it make any difference how quickly the coil is flipped?

Solution scheme

a. Calculate the magnetic moment of the coil.

b. Recall the expression for the torque exerted on a coil of magnetic moment m in a magnetic field. Because you know the torque, as well as the magnetic moment of the coil, you should now be able to calculate the field strength B.

c. The potential energy of a magnetic moment is given by $U = -\mathbf{m} \cdot \mathbf{B}$. Using this expression, you should be able to calculate U for the two positions of the coil and then to calculate the difference in energy between the two positions. Does the speed with which the coil is flipped enter into any of the calculations of potential energy?

13. *Worked Problem.* Figure 31.18 shows an idealized schematic diagram of an electromagnetic rail gun, designed to fire projectiles at speeds of up to 10 km/s. The projectile P sits between and in contact with two parallel rails along which it can slide freely. A generator G provides a current that flows up one rail, across the projectile, and back down the other rail. w = distance between the rails, r = radius of the rails (assumed to be circular), and l is the current supplied by the generator.

a. Find an expression for the magnetic field between the rails.

b. Show that the force on the projectile is to the right and is given by $F = (i^2\mu_0/2\pi) \ln [(w + r)/r]$.

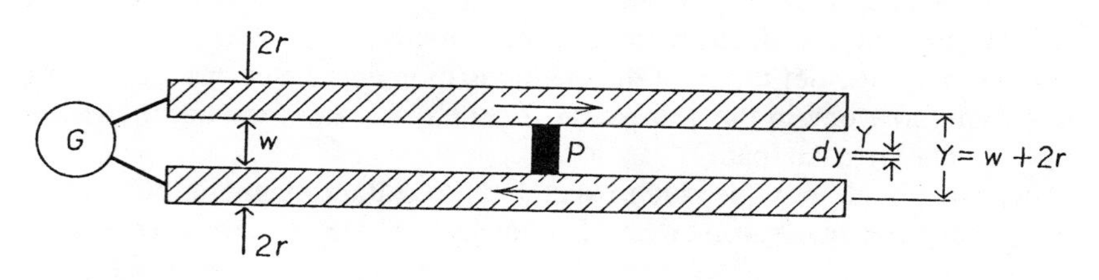

Figure 31.18 Problem 13. An electromagnetic rail gun. P is the projectile. The current flows up one rail, across the projectile, and back down the other rail.

Solution: Let Y = distance from the center of one rail to the center of the other rail. For a point a distance y from the center of the top rail, and in the space between the two rails, $B(y) = B_1(y) + B_2(y)$. The field contribution from each rail is pointing *into* the paper. From Ampère's Law,

$$B_1(y) = \mu_0 i/4\pi y$$

and

$$B_2(y) = \mu_0 i/4\pi(w + 2r - y)$$

and

$$B(y) = (\mu_0 i/4\pi)[1/y + 1/(w + 2r - y)]$$

where $r \leqslant y \leqslant r + w$.

To calculate the force on the projectile, it is easiest to calculate the force from $B_1(y)$ and $B_2(y)$ separately. Then $d\mathbf{F}_1 = i d\mathbf{y} \times \mathbf{B}_1(\mathbf{y})$, and from the right-hand rule, this is to the right. Then

$$dF_1 = iB_1(y)dy = (\mu_0 i^2/4\pi)[dy/y]$$

and

$$F_1 = \int_r^{r+w} dF_1 = (\mu_0 i^2/4\pi) \ln [(r + w)/r]$$

From symmetry, the force on P due to the field from the other rail is exactly the same, so that the total force on the projectile,

$$F = F_1 + F_2 = (\mu_0 i^2/2\pi) \ln [(r + w)/r]$$

14. *Guided Problem.* A classical electron moves in a circular path of radius 5.2×10^{-11} m about a proton. The charge of the electron is 1.6×10^{-19} C and its velocity is 2.23×10^6 m/s. Find the magnetic moment of the electron-proton system due to the orbital motion of the electron. Sketch the magnetic field of the magnetic dipole.

Solution scheme

a. The definition of electric current is $i = dq/dt$, so if you consider dq as the total charge past a point on the path of the electron during a time interval $dt = 1$ s, you should be able to calculate i from e, the charge of the electron, and v, its velocity.
b. What is the definition of the magnetic moment of a current loop? Can you now obtain the value of μ for the electron, knowing the radius of the circular path?
c. You should now be able to sketch the resultant magnetic field of the magnetic dipole; remember that the charge of the electron is *negative*.

Answers to Guided Problems

2. $i = 1.19 \times 10^4$ A. The force per unit length on the coils is 3.57×10^3 N/m and is directed outward, so that the solenoid will tend to explode.
4. $B_1 = 0.0010$ T (the field caused by the current I_1)
$B_2 = 0.00113$ T (the field caused by the current I_2)
For case a, with I_2 equal to zero, the field is uniform; $B = 0.0010$ T along the length of the solenoid. For cases b and c, the field along the solenoid length is plotted in Figure 31.19. In case b, B rises to 0.213 $\times 10^{-2}$ T over the 10 cm-long center part of the solenoid; in part c, B is reversed in direction and is only 0.013×10^{-2} T over the 10 cm-long center part.
6. For $r < r_1$ $B = 0$: for $r_1 < r < r_2$ $B = \dfrac{\mu_0 i}{2\pi r}\left[\dfrac{r^2 - r_1^2}{r_2^2 - r_1^2}\right]$ so that for $r = r_1$ $B = 0$; for $r = r_2$ $B = 10 \times 10^{-4}$ T; for $r = 0.025$ (midway between r_1 and r_2), $B = 5.4 \times 10^{-4}$ T. For $r > r_2$ $B = \mu_0 i/2\pi r$. A graph of B versus r is given in Figure 31.20.
8. $R = 7.32 \times 10^6$ m for the radius of the orbit. The magnetic field strength $B = 1.4 \times 10^{-8}$ T, which is 2.5×10^{-4} of the average strength of the Earth's magnetic field at the Earth's surface.

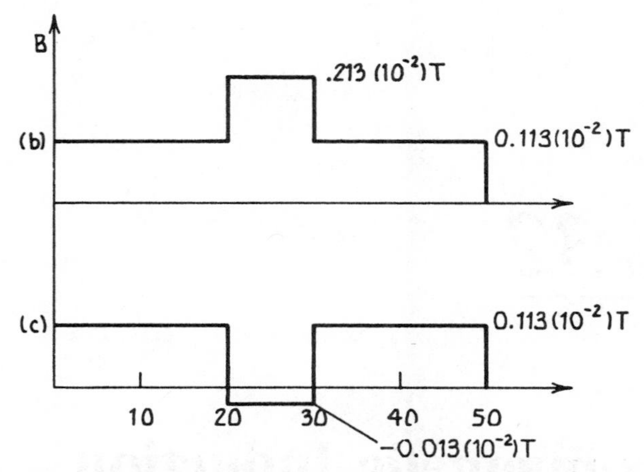

Figure 31.19 Problem 4. The magnetic field B along the length x of the solenoid with two windings. The field is plotted for cases b and c.

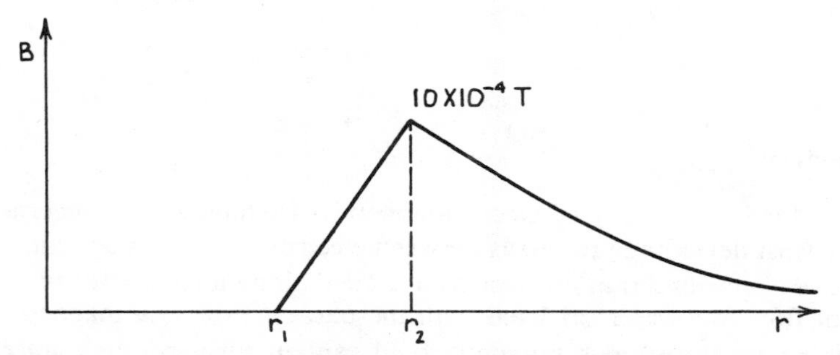

Figure 31.20 Problem 6. Graph of B versus r for a pipe with inner and outer radii r_1 and r_2 and with current i equal to 150 A.

10. Because the magnetic force is given by $d\mathbf{F} = i d\mathbf{l} \times \mathbf{B}$, the direction that the wire deflects allows us to deduce the field direction.
12. a. $B = 1.0$ T
    b. $U = 0.057$ J. No, it does not matter how quickly the coil is flipped.
14. $\mu = IA = \frac{e\upsilon r}{2} = 9.27 \times 10^{-24}$ A•m^2

    Figure 13.21 gives a sketch of the magnetic field.

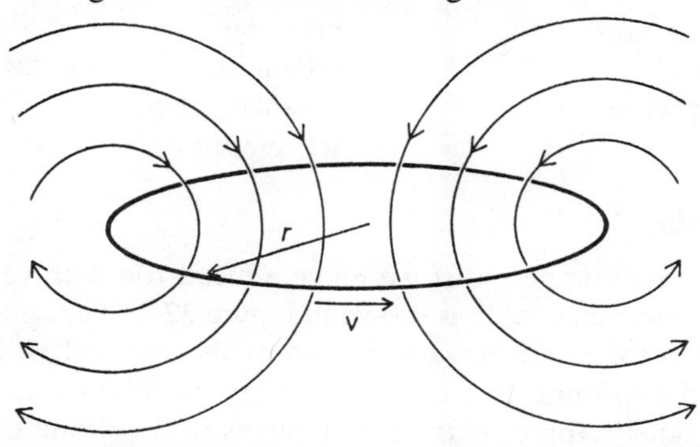

Figure 31.21 Problem 14. The magnetic field of a classical electron-proton system.

Chapter 32

Electromagnetic Induction

Overview

Thus far in our study of electromagnetism we have been concerned primarily with developing parallels between electrostatic and magnetic theory. We have also found that an electrostatic field along a conductor produces a current flow and that associated with the current flow is a magnetic field. We will now find that a magnetic field can, in turn, produce an electric field, but only if the magnetic field is changing. An electric field caused by a changing magnetic field is called an induced electric field, and the law describing the process is called the law of electromagnetic induction. In this chapter we will study the process of induction and the properties of electric fields caused by electromagnetic induction.

Essential Terms

motional emf
homopolar generator
Faraday's Law
magnetic flux; weber
Lenz' Law
betatron
self-inductance; mutual inductance
magnetic energy
RL circuit

Key Concepts

1. *Motional emf.* Suppose that we move a metal rod with velocity v perpendicular to a magnetic field, as shown in Figure 32.1. The conduction electrons will feel a force evB toward the top of the rod and will accumulate at the top of the rod until the electric field set up by the electrons exactly cancels the magnetic force. If the rod happens to be sliding over a pair of long wires, so that a complete circuit exists (a situation shown in Figure

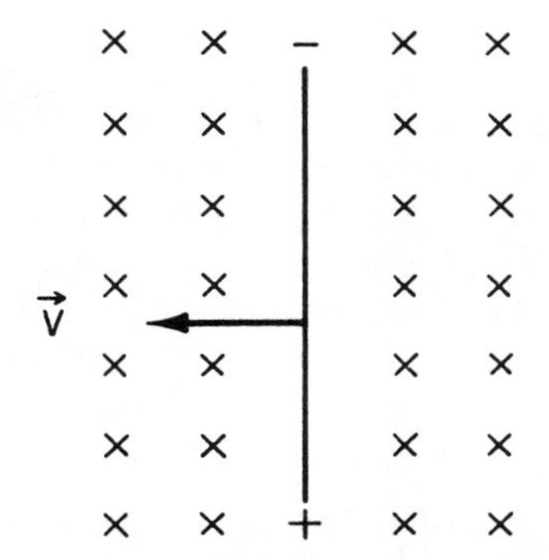

Figure 32.1 A metal rod moving with velocity υ perpendicular to a magnetic field.

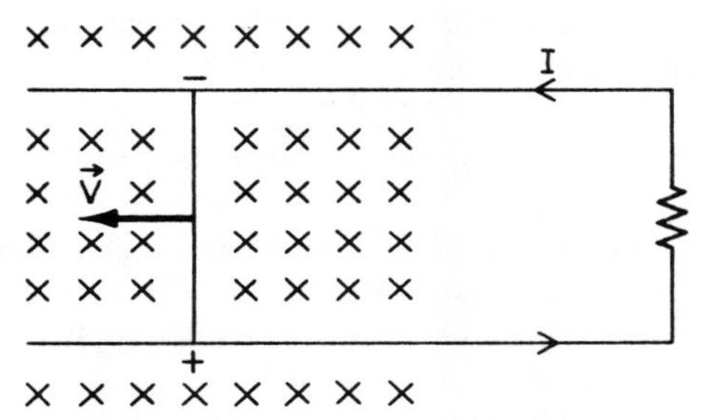

Figure 32.2 The rod is now moving over a pair of long wires, so that a complete circuit is formed. A current i flows in the circuit.

32.2), the electrons will flow around the circuit, so that the rod acts as source of emf. Current will not flow if we move a rectangular loop through the magnetic field, because then electrons will be forced toward the top of the loop at both ends.

Because emf is defined as the work done on a unit positive charge, the emf is given by

$$\begin{aligned} \mathcal{E} &= [\text{force}][\text{distance}] \\ &= (\upsilon B)(l) \end{aligned}$$

where l is the length of the rod. The direction of the emf, the direction that positive charge will move, is from the top to the bottom of the loop.

This kind of induced emf is called motional emf because it is due to the motion of the rod through the magnetic field. The area that the rod sweeps out per unit time is $l\upsilon$; the density of magnetic field lines is B. $Bl\upsilon$ is the number of magnetic field lines that the moving rod cuts per unit time, so that the emf can also be expressed as

$$\mathcal{E} = [\text{no. of magnetic field lines cut per second}]$$

This equation can be shown to hold for any conductor moving through any magnetic field.

2. *Faraday's Law*. A moving rod will cut magnetic field lines whenever there is relative motion between the rod and a magnetic field. On the basis of relativity theory we do not expect it to matter whether the rod moves past a fixed magnetic field or whether a magnetic field moves past a fixed rod, and we find by experiment that in the two cases the induced emf is the same. We can also cut field lines by increasing or decreasing the strength of the magnetic field, usually by changing the electric current in the magnet producing the magnetic field. Again, we find that cutting field lines by changing the strength of a magnetic field near a fixed conductor induces an emf given by

$$\mathcal{E} = [\text{no. of field lines cut per second}]$$

This statement is known as Faraday's Law and is usually written as follows: *The induced emf along any path equals the rate of cutting of magnetic field lines*. Because the induced emf is the path integral of the induced electric field, $\mathbf{E}'$, Faraday's Law becomes

$$\int \mathbf{E}' \cdot d\mathbf{l} = [\text{rate of cutting of magnetic field lines}]$$

For a closed path the rate of cutting of magnetic field lines can be expressed in terms of the magnetic flux, Φ_B, defined as the number of magnetic field lines inside the closed path. Then $\Phi_B = \int \mathbf{B} \cdot d\mathbf{S}$, and the rate of cutting of magnetic field lines is just equal to the rate of charge of magnetic flux in the area enclosed by the path.

For a closed path Faraday's Law becomes

$$\int \mathbf{E}' \cdot d\mathbf{l} = -\frac{d\Phi_B}{dt}$$

This is the form in which Faraday's Law is usually written. In contrast to the electrostatic field, which is conservative, so that $\int \mathbf{E} \cdot d\mathbf{l} = 0$ around any closed path, the induced electric field $\mathbf{E}'$ is not conservative. For a closed path

$$\int \mathbf{E}' \cdot d\mathbf{l} \neq 0$$

3. *Lenz' Law*. The minus sign in the equation

$$\mathcal{E} = \int \mathbf{E}' \cdot d\mathbf{l} = -\frac{d\Phi_B}{dt}$$

indicates that the sign of the induced emf is related to the sign of the flux. The sign of the induced emf can be most easily found by applying a rule called Lenz' Law, which says that the induced emf is in such a direction as to oppose the change that produced it. For example, in Figure 32.2, as the rod slides to the left, the magnetic flux through the loop increases. The induced emf causes current to flow from the top to the bottom of the rod, so that the magnetic field lines caused by the induced emf point out of the paper at the right side of the rod, opposing the increase in magnetic flux through the loop.

In Figure 32.3 we have redrawn Figure 32.2 except for the original magnetic field lines pointing into the paper; instead, we show the field lines resulting from the induced current through the rod. The field lines to the right of the rod act to cancel the increase in magnetic flux through the loop as the rod moves.

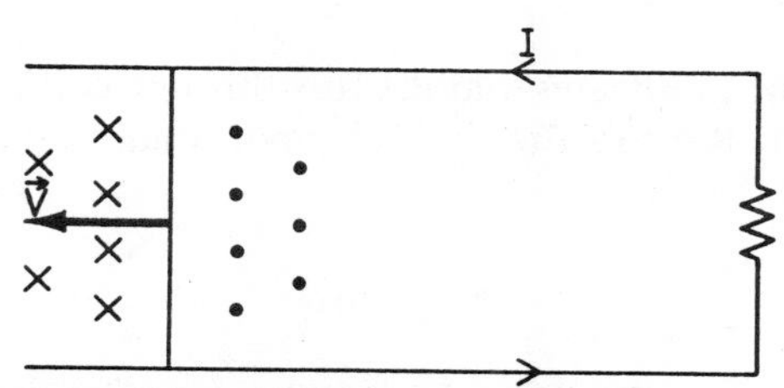

Figure 32.3 The induced magnetic field lines caused by the current in the rod shown in Figure 32.2.

As a second example, consider the two coils shown in Figure 32.4. When switch S is closed, current flows through coil 1 in the direction shown, producing the indicated flux lines through coil 2. The induced emf in coil 2 causes a current to flow in the direction shown (*opposite* to the direction of the current in coil 1) so that the induced flux lines from coil 2 oppose those from coil 1. When switch S is opened, the current in coil 1 stops, so that there is a sudden decrease in the flux lines through coil 2 resulting from coil 1. The induced emf in coil 2 now causes a current to flow in the opposite direction in coil 2, again opposing the change in the total flux through coil 2.

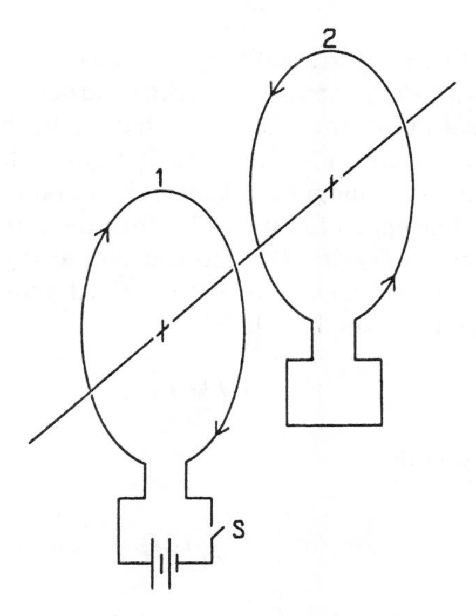

Figure 32.4 Two parallel coils, coil 1 and coil 2. When switch S is closed or opened, there is a change in flux through coil 2 resulting from the change in current in coil 1. The induced current in coil 2 is always such that the change in flux is opposed.

4. *Inductance*. In most applications involving emf, we find fixed geometry, as in Figure 32.4, in which a changing current in coil 1 induces an emf in coil 2. The induced emf in coil 2 is given by

$$\mathcal{E}_2 = -\frac{d\Phi_{B_1}}{dt}$$

where the subscript 1 indicates that the flux through coil 2 is due to the magnetic flux of coil 1. But this flux Φ_{B_1} is proportional to the current I_1 in coil 1, so that

$$\Phi_{B_1} = L_{21}I_1$$

The constant of proportionality, L_{21}, depends on the size of the coils and their relative separation and orientation—that is, it depends on the geometry. L_{21} is called the *mutual inductance* of the two coils. We can now write the induced emf as

$$\mathcal{E}_2 = -\frac{d}{dt}(L_{21}I_1) = -L_{21}\frac{dI_1}{dt}$$

If in Figure 32.4 we were to move the battery and switch from coil 1 to coil 2, we would find that a changing current in coil 2 induces an emf in coil 1. We would find that again we could write, for the emf induced in coil 1, that

$$\mathcal{E}_1 = -L_{21}\frac{dI_2}{dt}$$

where the constant of proportionality L_{21} is exactly the same as before.

The concept of *self-inductance* is a slightly harder concept to appreciate than that of mutual inductance. Suppose that we have a coil all by itself, and we change the current in it. Since the flux through the coil changes as the current changes, this changing flux must, by Faraday's Law, produce an emf $\mathcal{E} = -d\Phi_B/dt$. The net emf in the coil is then the sum of the external emf and the induced emf, $-d\Phi_B/dt$. This induced emf always acts to oppose the change of current (Lenz' Law). Again, for fixed geometry, the magnetic flux through the coil is given by

$$\Phi_B \propto I = LI$$

so that the induced emf

$$\mathcal{E} = -d\Phi_B/dt = -\frac{d}{dt}(LI) = -L\,dI/dt$$

L is defined to be the *self-inductance* of the coil, and it depends on such geometric factors as the size of the coil and the number of turns. For example, consider a solenoid of cross-sectional area A and with n turns per meter of length. The magnetic field inside the solenoid is $B = \mu_0 nI$. The flux through each turn is $BA = \mu_0 nIA$. The magnetic flux through all n turns per meter of length is $nBA = \mu_0 n^2 IA$, and since $\Phi_B = LI$, the self-inductance of the solenoid is $L = \Phi_B/I = \mu_0 n^2 A$ H per meter of length.

5. *Magnetic energy*. Most conductors, whether or not they are coils, have some self-inductance, as defined by the relationship $L = \Phi_B/I$, and so act as inductors to oppose a sudden change of current. When the current increases at a rate dI/dt in an inductor, the induced emf (sometimes called the *back emf*) is $\mathcal{E} = -LdI/dt$. The power delivered to the inductor is then $\mathcal{E}I = LIdI/dt$. The energy stored in the inductor in time dt is then $dU = LIdI$, so that the total energy stored in an inductor is given by

$$U = \int_0^I LIdI = \tfrac{1}{2}LI^2$$

This magnetic energy is capable of doing external work, as is the electric energy stored in a capacitor.

If we again consider a solenoid with n turns per meter of length, cross-sectional area A, and a current I flowing through the windings, we find that the stored magnetic energy is

$$U = \tfrac{1}{2}LI^2 = \tfrac{1}{2}\mu_0 n^2 I^2 Al$$

where l is the length of the solenoid. Since the magnetic field strength $B = \mu_0 nI$, we can rewrite

$$U = \frac{1}{2}\frac{(\mu_0 n^2 I^2)}{\mu_0} lA = \frac{1}{2}\frac{B^2}{\mu_0} lA$$

But lA is just the volume of the magnetic field of the solenoid (the total volume inside the solenoid), so that $U/\text{volume} = u = \dfrac{1}{2}\dfrac{B^2}{\mu_0}$. u is defined as the magnetic energy per unit volume of the magnetic field, so that we can consider the energy as stored either in the current I or in the magnetic field B produced by the current. (In Chapter 26, on electric energy, we showed that the energy can be considered as stored either in the charges on the capacitor plates or in the electric field between the plates.) The expression $u = \dfrac{1}{2}\dfrac{B^2}{\mu_0}$ is the general expression for magnetic field energy per unit volume, just as $u = \frac{1}{2}\varepsilon_0 E^2$ is the expression for the energy density of an electrical field E.

6. *RL Circuit*. The RL circuit consists of an inductor and a resistor connected in series to a source of emf, as shown in Figure 32.5.

Assume that the battery is suddenly connected at time $t = 0$ and that the current starts to increase. The self-inductance generates an emf across the inductor that opposes the increase of the current I (Lenz' Law). We can use Kirchhoff's rule to obtain an equation giving us the time dependence of the current in the circuit. From Kirchhoff's rule, the circuit equation is $\varepsilon - L\, dI/dt - IR = 0$. This equation has the same mathematical form as the equation for the charge in an RC circuit, and its solution, by comparison, is

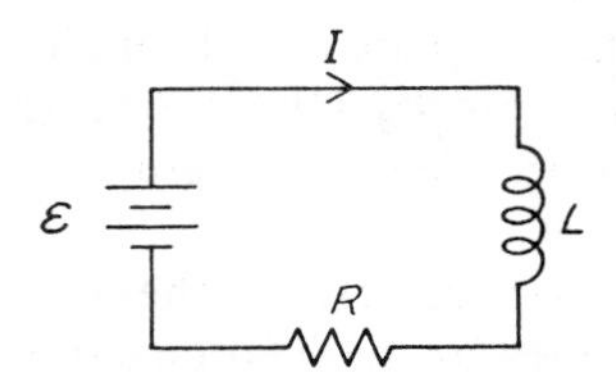

Figure 32.5 An RC circuit.

$$I = (\varepsilon/R)[1 - e^{-Rt/L}]$$

A plot of this current as a function of time is shown in Figure 32.6(a); we see that the current in a circuit containing an inductor cannot increase suddenly. If, after the current has reached its final, steady value of ε/R, we suddenly switch the battery out of the circuit and "discharge" the inductor through the resistor, the current gradually decays to zero according to

$$I = (\varepsilon/R)e^{-Rt/L}$$

A plot of current versus time is shown in Figure 32.6(b).

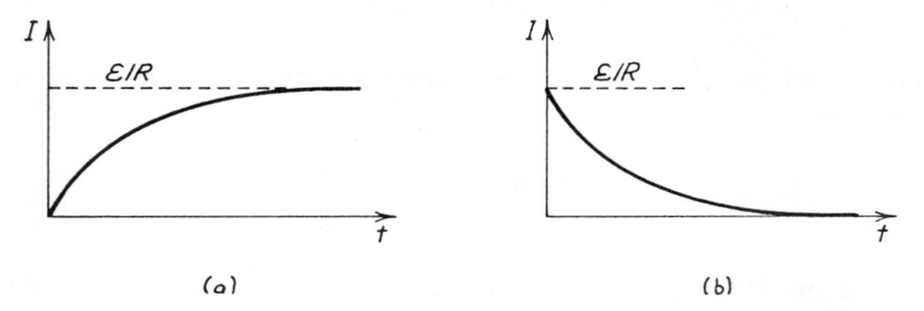

Figure 32.6 Current versus time in an RL circuit: (a) growth of current and (b) decay of current.

Sample Problems

1. *Worked Problem.* A copper rod 1 m long slides on a pair of frictionless rails, assumed to have zero resistance, at a speed of 10 m/s, as shown in Figure 32.7. There is a 0.2 T magnetic field into the paper, as indicated.

a. What is the induced emf between the ends of the rod, and which end is positive and which negative? What force is needed to keep the rod at constant speed?

b. A 4.0-ohm resistor is now connected between points *c* and *d*, forming a closed circuit *abcd*. What now is the emf between points *a* and *b*? What is the current in the circuit *abcd*, and in which direction does it flow? How much force is now needed to keep the speed υ constant?

c. Compare the power needed to keep the rod in motion and at constant speed with the power dissipated in the resistor.

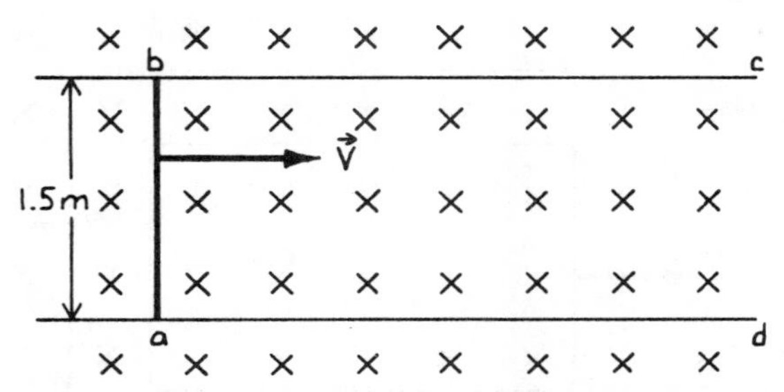

Figure 32.7 Problem 1. A copper rod sliding along metal rails in a uniform magnetic field.

Solution

a. The emf is given by $\mathcal{E} = VBl = 10 \times 0.2 \times 1.5 = 3.0$ volts. The direction of the force on the free electrons in the copper, given by $\mathbf{F} = q\mathbf{v} \times \mathbf{B}$, is from top to bottom, so that electrons are forced toward the bottom of the rod. The top of the rod, therefore, is positive with respect to the bottom of the rod. Since no current flows in an external circuit, equilibrium between the magnetic force on the electrons and the electric field in the rod is quickly set up, and no further redistribution of electrons occurs. Therefore, no force is needed to keep the rod moving at constant speed.

b. With the resistor connected between c and d, we have a closed circuit with total resistance of 4 ohms. The sliding rod acts as a source of emf, with the positive terminal at b. A current $i = \mathcal{E}/R = 0.75$ A flows in a clockwise direction around the loop. The emf between a and b, 3.0 V, is the same as in part a, but because there is now a current flowing between a and b, there is now a force on the rod, given by $F = Bil = (0.2) \times 0.75 \times 1.5 = 0.225$ N. The direction of the force, given by $\mathbf{F} = i\mathbf{l} \times \mathbf{B}$, is from right to left (the direction opposite to υ). An equal and opposite force, 0.225 N from left to right, is needed to keep the rod moving at constant speed.

c. Power is given by [force] × [velocity] $= 0.225 \times 10 = 2.25$ watts. The power dissipated in the resistor is $i^2R = (0.75^2) \times 4 = 2.25$ watts.

2. *Guided Problem.* A rectangular loop of copper wire 60 cm × 50 cm, as shown in Figure 32.8, is pulled through a uniform 0.5 T magnetic field, 1.0 meter long, at a constant speed of 1.0 m/s. The resistance of the loop is 1.0 ohm. Plot the current generated in the loop as a function of time, from $t = 0$ to $t = 2$, with $t = 0$ being the time when the left side of the loop first enters the magnetic field. Be sure to indicate the direction of the current in the loop.

Solution scheme

a. With only one side of the loop in the magnetic field, there is an emf generated between the top and the bottom of that side of the loop, just as in the preceding problem. Can you calculate both the magnitude of the emf and its direction?

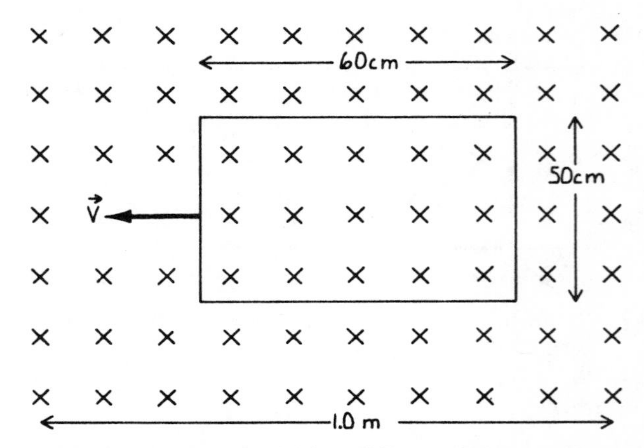

Figure 32.8 Problem 2. A rectangular loop being pulled through a uniform magnetic field.

b. Knowing the emf and the resistance of the loop, you can determine the current flowing around the loop (and the direction of current flow).
c. When the whole loop is within the magnetic field, there is an emf generated between the top and the bottom of *both* ends of the loop. The situation is just like that in which two batteries *oppose* each other. Do we expect any current to flow around the loop in this situation?
d. With the loop entirely out of the field, is any emf generated?
e. You should now be able to graph the current i versus time from $t = 0$ to $t = 2.0$ s. Be sure to indicate the direction of current flow on your graph.

3. *Worked Problem.* It is possible, by connecting an alternating emf source to a solenoid, to cause a metal ring placed on top of the solenoid to "levitate" above the solenoid, or at least to jump violently off the end of the solenoid. Suppose that an aluminum ring is placed on top of a long solenoid, as shown in Figure 32.9. The solenoid field at the ring is about 1/2 that at the midpoint of the solenoid. If the current in the solenoid is increased

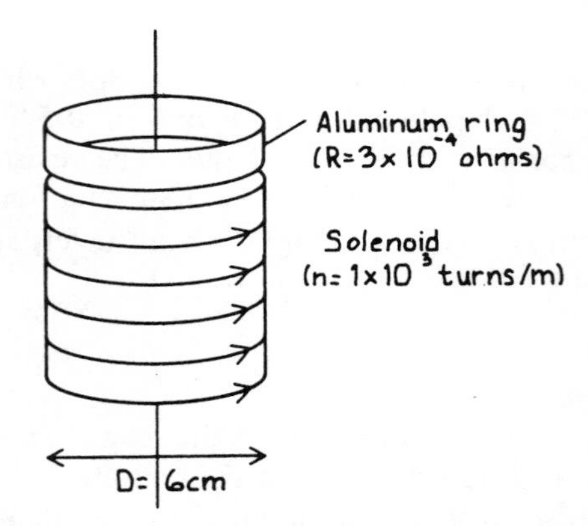

Figure 32.9 Problem 1. A thin aluminum ring resting on top of a solenoid 6 cm in diameter.

suddenly at the rate of 100 A/s, find the magnitude and the direction of the induced current in the ring. Will there be a net force on the ring and, if so, in what direction? What if the current is *decreased* suddenly at the same rate; what is the direction of the force on the ring in this case?

Solution: From Faraday's Law, we know that induced emf in the ring is given by

$$\mathcal{E} = -\frac{\Delta\Phi}{\Delta t} = -\frac{\Delta(BA)}{\Delta t} = -\frac{\Delta}{\Delta t}\left(\frac{\mu_0 niA}{2}\right)$$

$$= -\frac{\mu_0 n\pi R^2}{2}\frac{\Delta i}{\Delta t}$$

$$= -\frac{1.26(10^{-6})10^3\pi(.03)^2}{2}(100) = 3.6 \times 10^{-4}\ \text{V}$$

Since the resistance of the ring is 3×10^{-4} ohms, the induced current is $I = \mathcal{E}/R = 1.2$ A.

When the current in the solenoid is increased in the direction indicated by the arrows, the upward magnetic flux through the aluminum ring increases. The direction of the induced current is such as to oppose this flux, and therefore this current flows clockwise, looking down on the ring from above the solenoid. Since the magnetic field of the ring is in the direction opposite that of the solenoid, there is a *repulsive* force between the ring and the solenoid.

When the current in the solenoid is *decreased* suddenly, there is again an induced emf of the same magnitude in the ring, if the rate of decrease is 100/s. Again, the direction of the induced current opposes the change in flux, and therefore this current flows counterclockwise, looking down on the ring. Again, there must be a repulsive force between the ring and the solenoid.

4. *Guided Problem.* A heavy, flat copper pendulum is swung between the poles of an electromagnet, as shown in Figure 32.10. The pendulum decelerates abruptly as the copper plate enters the magnetic field. A second

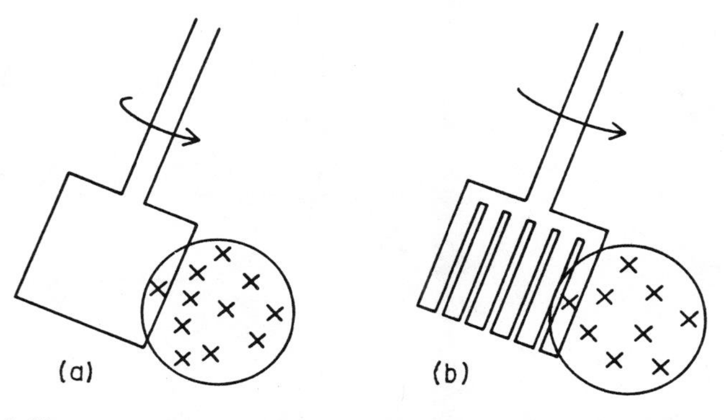

Figure 32.10 Problem 4. (a) A copper-plate pendulum swung between the poles of an electromagnet. (b) The copper plate has slots cut through it and experiences very little deceleration.

pendulum, identical to the first except that the pendulum bob has slots cut through the copper plate, experiences very little deceleration as it enters the magnetic field. Explain, using diagrams, why the first pendulum decelerates and the second one does not.

Solution scheme

a. Draw a diagram of a copper plate entering a magnetic field with speed v. In what direction will the magnetic force $q\mathbf{v} \times \mathbf{B}$ cause the free electrons along the edge of the plate to move? Will this movement set up a circulating current in the plate? What is the direction of this circulating current in the magnetic field?

b. The magnetic force on a conductor with current I is given by $\mathbf{F} = I\mathbf{l} \times \mathbf{B}$. Show that the direction of this force is such as to oppose the motion of the plate through the liquid, so that the plate is decelerated.

c. If we apply the same arguments to the slotted plate, we see that because the plate is slotted, currents cannot flow freely through the plate, so the magnetic force on the conductor is minimized.
The deceleration of the copper plate by the magnetic field is called magnetic braking. Magnetic brakes have important industrial applications.

5. *Worked Problem.* (a) A piece of copper wire of length R is rotated at a frequency f, as shown in Figure 32.11(a), such that the plane of rotation is perpendicular to a uniform magnetic field. Find the emf generated between the axis of rotation and the end of the wire. Which end of the wire will be at a positive potential? (b) Figure 32.11(b) shows a "homopolar generator," a device with a solid conducting disk as a rotor. This machine can produce a greater emf than wire loop rotors, since it can spin at a much higher angular speed before centrifugal forces destroy the rotor. Show that the emf produced is given by $\varepsilon = \pi f B R^2$, where f is the frequency of the rotor. What torque must be provided by the motor spinning the rotor when the output current is I?

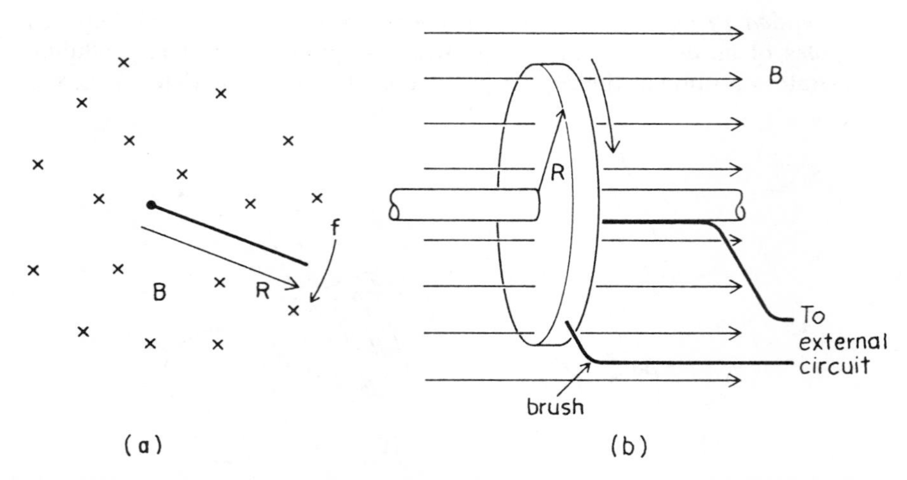

Figure 32.11 Problem 5. (a) A piece of copper wire of length R rotating perpendicular to a uniform magnetic field, B. (b) A "homopolar generator" spinning in a uniform magnetic field, B. The brushes provide connections to an external circuit.

Solution

a. Since the expression for the magnetic force on a moving charge is given by $\mathbf{F} = q\mathbf{v} \times \mathbf{B}$, the free electrons in the rod will move to the left (that is, to the pivot) in Figure 32.11. The outer end of the wire will be positive. The emf generated is given by $\varepsilon = -d\varphi_B/dt = -d(BA)/dt$. This is calculated most easily by considering that the area swept out by the wire in $1/f$ seconds is the area πR^2, so that

$$d(BA)/dt = B(\pi R^2)/(1/f) = \pi f B R^2 = \varepsilon$$

b. For the homopolar generator, the answer is clearly the same. The real difference is that higher values of f are possible and that the large volume provided by the conducting disk will allow a much smaller resistance (and therefore greater current and power output) than a single wire. That is,

$$\varepsilon = \pi f B R^2$$

The output power is just $\varepsilon I = \pi f B R^2 I$, and since the angular frequency $\omega = 2\pi f$, $\varepsilon I = BR^2\omega I/2 = \tau\omega$, where τ is the torque, so that the torque is given by

$$\tau = BR^2 I/2$$

6. *Guided Problem.* When we change the current through a solenoid, we induce an emf in a coil concentric with the solenoid (Figure 32.12). For a long solenoid with 400 turns per meter, suppose that the current changes at the rate of 100 A/s. What is the emf induced in a single concentric loop of wire of radius r? Plot the induced emf and the mutual inductance of the loop and the solenoid as a function of r from $r = 0$ to $r \gg R$.

Solution scheme

a. Write down the expression for the emf induced in the loop by the changing magnetic flux in the solenoid. (This expression is given by Faraday's Law.)

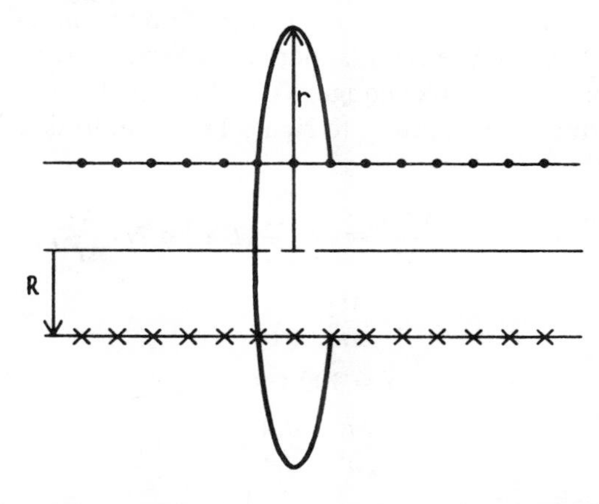

Figure 32.12 Problem 6. A single loop of radius r concentric with a long solenoid of radius R.

b. Now evaluate the induced emf in terms of the field of the solenoid and the area linking the loop to the solenoid. You should now be able to obtain an explicit value for the induced emf in terms of the radius of the loop, for both $r < R$ and $r > R$. For $r \geqslant R$, does the size of the loop matter?

c. Write the corresponding expressions for the induced emf in terms of the mutual inductance L_{12}. You should now be able to graph both the induced emf and the mutual inductance as a function of r.

7. *Worked Problem.* When switch S in Figure 32.13 is closed, the current in the long solenoid increases suddenly from 0 to 0.4 A in 0.1 s. The direction of the field in the solenoid is from left to right. What is the direction of

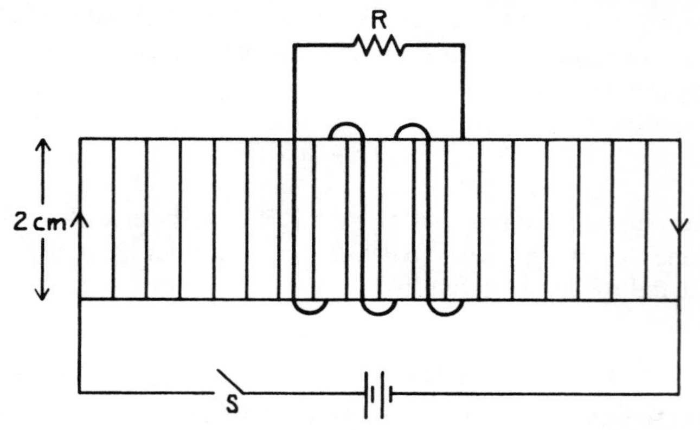

Figure 32.13 Problem 7. A second four-turn winding in series with a resistor R wound over a solenoid.

the current flow through the resistor, the current in the resistor during this time period, and the voltage across the resistor? When switch S is opened, the current in the solenoid drops abruptly to 0 in 0.001 s. Again, find the direction of the current in R, the average current through R, and the voltage across R. $R = 20$ ohms; the long solenoid has 300 turns per meter and has a radius of 2 cm.

Solution: We can easily use Lenz' Law to find the direction of the induced current in R. Since the field is increasing from left to right, the induced flux will oppose this flux change. To do so, the direction of the current in the resistor must be from left to right. The magnitude of the induced emf is given by

$$\mathcal{E}_2 = N_2 \frac{\Delta\Phi_{21}}{\Delta t} = N_2 \frac{\Delta}{\Delta t}(BA) = N_2 \frac{\Delta}{\Delta t}(\mu_0 n i A)$$

$$= N_2 \mu_0 n A \frac{\Delta i}{\Delta t}$$

$$N_2 = 4 \quad n = 300 \text{ turns/m} \quad A = \pi(.01)^2$$

$$\frac{\Delta i}{\Delta t} = \frac{0.4}{0.1} = 4 \text{ A/s}$$

Then

$$\mathcal{E}_2 = 1.9 \times 10^{-6}\ \text{V}$$

Assuming that the four turns of the coil have negligible resistance compared with that of the 20-ohm resistor, this is the average emf across the resistor during this 0.1 s period; the average current, which flows from left to right, is $\mathcal{E}_2/R = 9.5 \times 10^{-8}$ A. When the switch is opened, the current drops more suddenly, so that the induced emf and current are both 100 times greater. The emf is 1.9×10^{-4} volt, and the average current is 9.5×10^{-6} A. Since the induced flux is now opposing the *decrease* in current, the direction of the induced current in the resistor R is now from right to left.

8. *Guided Problem.* A coaxial cable consists of two thin, concentric copper cylinders with respective diameters of 2 mm and 1.0 cm, as shown in Figure 32.14. The cylinders carry equal currents of 0.5 A in opposite directions. Does the cable have a self-inductance L, and, if so, what is the self-inductance per meter of length of the cable?

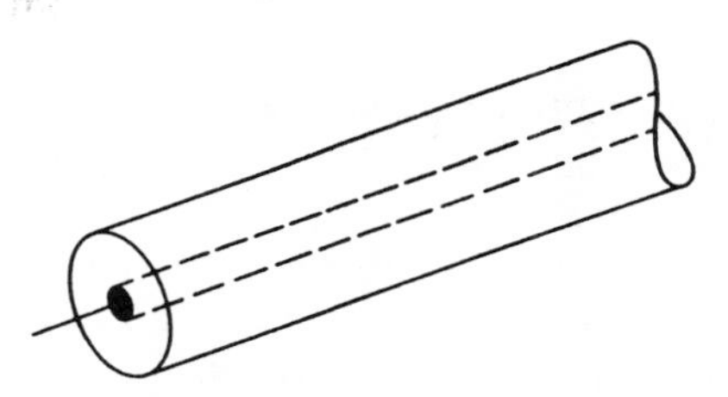

Figure 32.14 Problem 8. A coaxial cable consisting of two concentric copper cylinders.

Solution scheme

a. Write down the expression for the self-inductance L in terms of the magnetic flux Φ_m and the current I in the two cylinders.
b. Use Ampère's Law to find the magnetic field B in the regions (a) inside the inner cylinder, (b) between the two cylinders, and (c) outside the outer cylinder. In which of these three regions is there a nonzero magnetic field?
c. Since the magnetic flux $\Phi_m = \int \mathbf{B} \cdot d\mathbf{A}$, choose the appropriate area over which to evaluate Φ_m. This is the hardest part of the problem; since the magnetic field lines are concentric circles whose axes are parallel to the cylinder axes, Φ_m must be evaluated over a rectangle of width $R_2 - R_1$ (the radii of the two cylinders). Subdivide this rectangle into strips, each of width dr.
d. You should now be able to evaluate Φ_m and determine the self-inductance of the cable, L, per meter of length.

9. *Worked Problem.* The large superconducting toroid shown in Figure 32.15 has a rectangular 20 cm $\times$ 15 cm cross section and an inner radius of 1.0 m. The toroid has 2000 turns, with a current of 100 A in each turn. What is the self-inductance of the toroid and the magnetic energy stored in

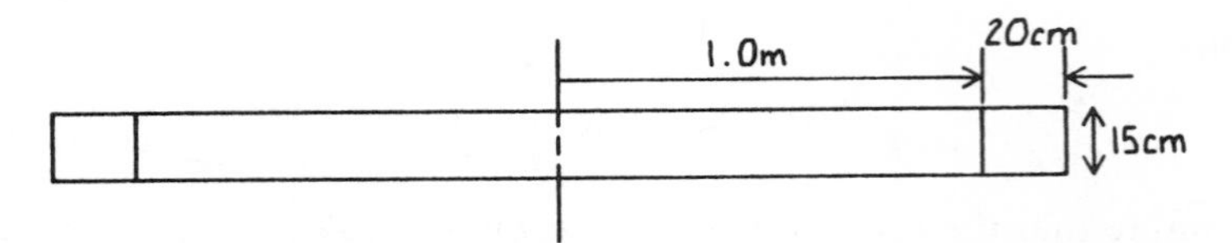

Figure 32.15 Problem 9. A large superconducting toroid.

the current in the winding of the toroid? Compare your answer with the amount of energy stored in the magnetic field of the toroid.

Solution: The self-inductance is defined by the expression $LI = N\Phi_m$, where the magnetic flux

$$\Phi_m = \int \mathbf{B} \cdot d\mathbf{A}$$

From Ampère's Law, the field of a toroid can be given by

$$2\pi rB = \mu_0 NI \quad \text{or} \quad B = \frac{\mu_0 NI}{2\pi r}$$

Since the magnetic field B is a function of r, we must subdivide the cross-sectional area of the toroid into vertical strips of width dr, as shown in Figure 32.16. Then, since

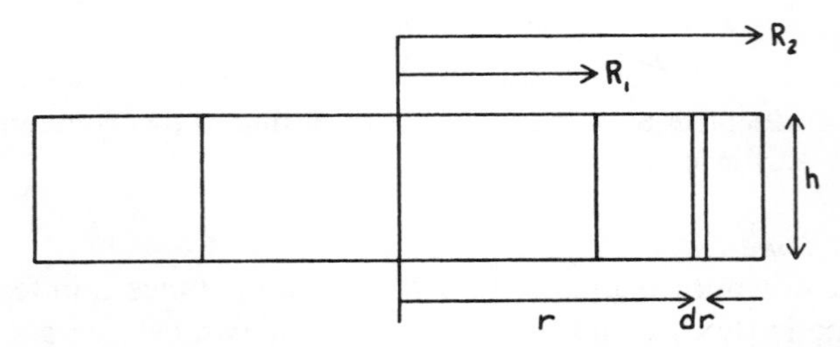

Figure 32.16 Problem 9. The cross-sectional area of the toroid subdivided into vertical strips, each of width *dr*.

$$L = N\Phi_m / I = \frac{N}{I}\int_{R_1}^{R_2} \mathbf{B} \cdot d\mathbf{A}$$

$$L = \frac{N}{I}\int_{R_1}^{R_2}\left(\frac{\mu_0 Ni}{2\pi r}\right)h\,dr = \frac{\mu_0 N^2 h}{2\pi}\ln\frac{R_2}{R_1}$$

Since $N = 2000$, $h = 0.15$ m, $R_2 = 1.2$ m, and $R_1 = 1.0$ m,

$$L = \frac{1.26(10^{-6})(2000)^2(.15)}{2\pi}\ln 1.2 = 0.022 \text{ H}$$

The magnetic energy $U = \frac{1}{2}LI^2 = \frac{1}{2}(0.022)(10^4) = 110$ J. The energy density of the magnetic field $u = \frac{1}{2}B^2/\mu_0$. The total magnetic field energy is arrived at by integrating over the volume of the field: $U = \int u d\upsilon$.

$$U = \int_{R_1}^{R_2} \left(\frac{1}{2}\frac{B^2}{\mu_0}\right) d\upsilon$$

Since here we need to integrate over the whole volume of the field, we subdivide the field into thin, concentric toroids of radius r, thickness dr, and volume $2\pi rhdr$. Then

$$U = \int_{R_1}^{R_2} \frac{1}{2\mu_0}\left(\frac{\mu_0 NI}{2\pi r}\right)^2 (2\pi rhdr)$$

$$= \frac{\mu_0}{4\pi} N^2I^2h \int_{R_1}^{R_2} \frac{dr}{r} = \frac{\mu_0}{4\pi} N^2I^2h \ln \frac{R_2}{R_1}$$

If we compare this expression with our expression for L, we see that our expression for U is exactly $\frac{1}{2}LI^2$, so that again we can consider the magnetic energy to be stored either in the magnetic field or in the current in the windings.

10. *Guided Problem*. A battery of emf 4.0 V and internal resistance 0.6 ohm is connected across a large inductor at $t = 0$. At $t = 0.06$ s, the current in the circuit is 0.2 A; at $t = 6.0$ s, it is 0.5 A. What is the self-inductance? What is its internal resistance?

Solution scheme

a. Write down the expression for the growth of the current in an RL circuit. Note that the value of R in the equation is the *total* resistance of the circuit.
b. At $t = 0.06$ s, the current is 0.2 A; by $t = 6.0$ s (100× as large as 0.06 s), the current has risen only to 0.5 A. What does this tell you about the maximum value of current in the circuit? Note that the limiting value of current is given by ε/R. Can you now calculate the value of R? What is the internal resistance of the self-inductance?
c. Now go back to the equation you wrote in part a. Can you solve the equation for L at time $t = 0.06$ s? Was your assumption in part b that at time $t = 6.0$ s the current had reached its limiting value (if you made this assumption) a good one?

Answers to Guided Problems

2. With one side of the loop in the field, $I = 0.25$ A; with both sides of the loop in the field, $I = 0$. The graph of I versus t is shown in Figure 32.17.
4. The explanation is given in the problem.
6. For $r < R$, the emf is given by $0.158\ r^2$ V.
   For $r \geqslant R$, the emf is given by $0.158\ R^2$ V.
   For $r < R$, the inductance L is $1.58 \times 10^{-3}\ r^2$ H.
   For $r \geqslant R$, the inductance L is $1.58 \times 10^{-3}\ R^2$ H.

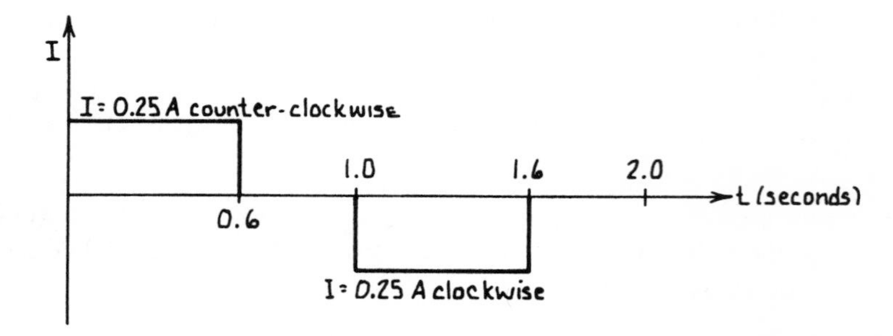

Figure 32.17 Problem 2. I versus time t for a loop pulled through a magnetic field.

Graphs of emf and L_{12} versus r are presented in Figure 32.18.

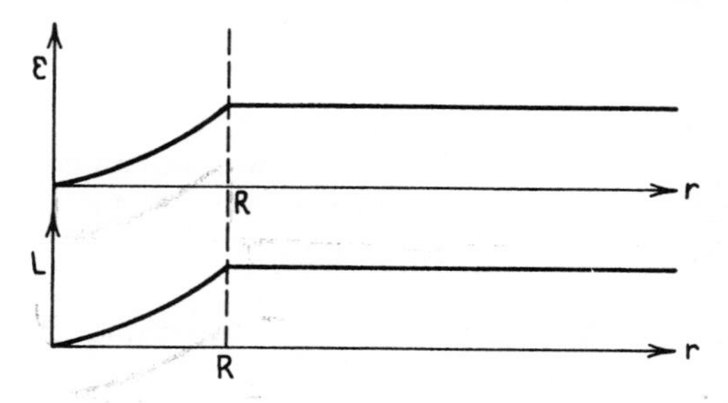

Figure 32.18 Problem 6. The induced emf and the mutual inductance plotted as a function of the coil radius r.

8. The self-inductance is given by

$$L = \frac{\Phi_m}{I} = \frac{\int \mathbf{B} \cdot d\mathbf{A}}{I} = \int_{R_1}^{R_2} \left(\frac{\mu_0 I}{2\pi r I}\right)(l dr)$$

The area of integration is shown in Figure 32.19. Then

$$L = \frac{\mu_0}{2\pi} \ln \frac{R_2}{R_1} \text{ H}$$

per meter of length.

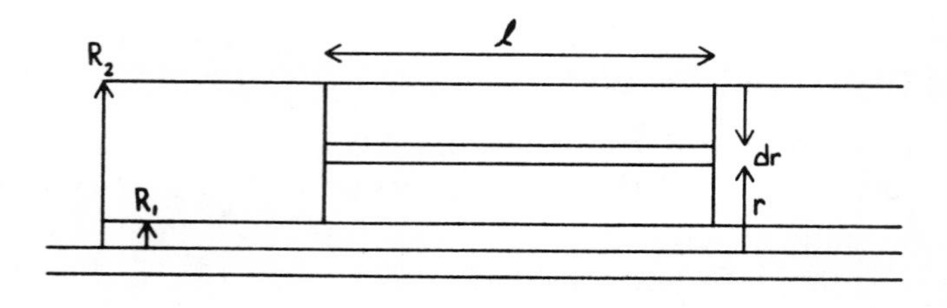

Figure 32.19 Problem 8. The area of integration.

10. $R = 8.0\ \Omega$; $\therefore$ $R_L = 7.4\ \Omega$ for the internal resistance of the self-inductance.
$L = 0.94$ H.

At $t = 6.0$ s, $I = \frac{\varepsilon}{R}$.

Chapter 33

Magnetic Materials

Overview

Apart from a brief mention of the magnetic field generated by a moving point charge, we have discussed only the magnetic field produced by the flow of current through conductors. When we think of atoms, we realize that the orbital motion of electrons around a nucleus may be considered a flow of electric current. Electric currents produce magnetic fields, so that each atom in a bulk sample of material may be thought of as a tiny magnetic dipole. The response of these atomic dipoles to an external magnetic field gives rise to the bulk magnetic behavior of the sample. In this chapter we study the three general types of bulk magnetic behavior exhibited by different magnetic materials.

Essential Terms

atomic dipole moment
nuclear dipole moment
paramagnetic material
relative permeability constant
ferromagnetic material
domain
Curie temperature
diamagnetic material
Làrmor frequency

Key Concepts

1. *Atomic dipole moment.* An electron revolving in a circular orbit with speed υ at a radius r about a nucleus, as shown in Figure 33.1, generates an average current given by $I = e \times$ [no. of revolutions per second] $= e\upsilon/2\pi r$. This atomic current loop has a magnetic moment μ, where, as usual,

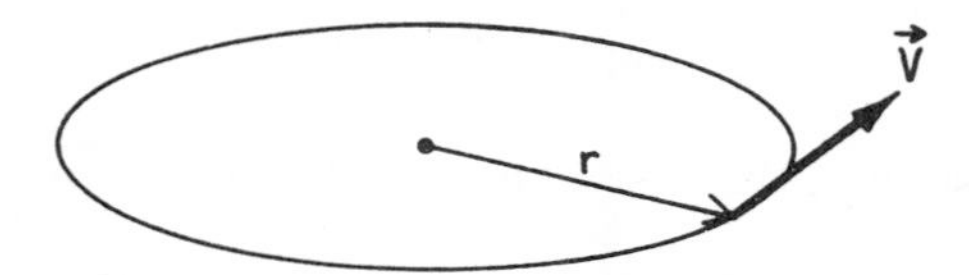

Figure 33.1 An electron revolving in a circular orbit with speed υ at radius r about a nucleus.

$$\mu = I(\pi r^2) = e\upsilon r/2$$

But the orbital angular momentum of the rotating electron, L, is given by $L = \mu_e \upsilon r$, so that the magnetic moment $\mu = eL/2m_e$. The orbital angular momentum of any atom is quantized; that is, it is always some integral multiple of Planck's constant, $\hbar$, so that L has possible values of 0, $\hbar$, $2\hbar$, and so on. Then the atomic magnetic moment associated with orbital angular momentum has possible values $\mu = en\hbar/2m_e$, where $n = 0, 1, 2, \ldots$. If $n = 1$, then $\mu = e\hbar/2m = 9.27 \times 10^{-24}\ \text{A} \cdot \text{m}^2$.

In addition to the orbital motion, every electron has an *intrinsic* angular momentum or spin, as if it rotated about its own axis at a fixed rate. The magnitude of this spin is $\hbar/2$, giving the electron an intrinsic magnetic moment $\mu = e\hbar/2m_e = 9.27 \times 10^{-24}\ \text{A} \cdot \text{m}^2$. The net magnetic moment of an atom is obtained by adding vectorially the orbital and the spin moments of all the electrons. For hydrogen, with only one electron, for which the orbital angular momentum is zero, the magnetic moment $\mu = 9.27 \times 10^{-24}\ \text{A} \cdot \text{m}^2$, because of the spin of the one electron. For oxygen, with eight electrons, $\mu = 13.9 \times 10^{-24}\ \text{A} \cdot \text{m}^2$. For most of the electrons, the magnetic moments (and angular momenta) act to cancel one another out.

The nucleus of an atom also has a magnetic moment, but because the mass of a proton or neutron is very much greater than the mass of an electron, the corresponding nuclear magnetic moments are much smaller than atomic magnetic moments. (If you substitute the mass of a proton or neutron in the equation we have presented for magnetic moment, you will see that the value of μ becomes *much* less!)

2. *Paramagnetic material.* All materials are at least somewhat magnetic. The magnetic behaviors of bulk samples are divided into three general classes—paramagnetic, ferromagnetic, and diamagnetic. Paramagnetism occurs in materials, such as oxygen, whose molecules have a permanent magnetic dipole moment. In the absence of an external magnetic field, the molecules are randomly oriented, and no magnetic effects are seen. When an external magnetic field is applied, it exerts a torque on the magnetic dipoles, tending to align them parallel to the field. The total magnetic field in the sample, resulting from the external field plus the field of the aligned dipoles, is greater than the external field, because the axial field of a dipole is parallel to the direction of its magnetic moment. The thermal motion of the molecules tends to reduce the alignment of the dipoles, so that for a paramagnetic material the field of the sample, $\boldsymbol{B}$, is only slightly greater than the external field, $\boldsymbol{B}_{\text{free}}$. The ratio between the field in the sample and the external field $B/B_{\text{free}} = \kappa_m$, where κ_m is called the *relative permeability constant* and is slightly greater than unity. Ampère's Law,

$$\oint \mathbf{B}_{\text{free}} \cdot d\mathbf{l} = \mu_0 I_{\text{free}}$$

still holds in a magnetic material and is usually written as

$$\oint \frac{1}{\kappa_m} \mathbf{B} \cdot d\mathbf{l} = \mu_0 I_{\text{free}}$$

Written this way, Ampère's Law permits us to express the magnetic field in the sample in terms of the true, or free, current.

3. *Ferromagnetic material.* In ferromagnetic materials, of which iron is the most common example, the molecules also have a permanent dipole moment but the magnetic field in the sample is very large compared with the external magnetic field. There is very strong alignment of the atomic diples parallel to the external field. A ferromagnetic material is made up of tiny regions called *domains*; in a domain the magnetic moments of the atoms are perfectly aligned, so that each domain behaves like a small magnet. The domains are randomly oriented relative to one another, so that in the absence of an external magnetic field, the sample does not behave like a magnet. When the sample is placed is an external magnetic field, those domains that are already aligned parallel to the field grow at the expense of those that are not. Eventually, as the external magnetic field is increased, nearly all the atomic magnetic moments become aligned parallel to the field. The field in a sample of iron can be about 5000 times the external field. An iron magnet can retain its magnetism for a long time even after the external field is removed; it is called a *permanent magnet*. Dropping or hitting a magnet will jar and reorient some of the domains, decreasing the magnetism. Above a certain temperature, known as the *Curie temperature* (1043°K for iron), the thermal motion of the atoms is so large that the domains are randomized and the material loses all its magnetism.

4. *Diamagnetic material.* Diamagnetic materials, such as bismuth, are made up of molecules that have no permanent dipole moment. When an external magnetic field is applied, magnetic dipoles are induced, with the induced dipole moment being in the direction opposite to that of the external field. The total field in the sample is therefore slightly less than the external field, and the relative permeability constant κ_m is slightly less than unity.

The external field has the effect of increasing the orbital speed of electrons revolving in one direction and decreasing the speed of electrons revolving in the other direction. This change causes a net dipole moment opposing the external field. Diamagnetic effects are present in all materials, but such effects are so small that they are noticed only in materials that have no permanent atomic dipole moments.

Sample Problems

1. *Worked Problem.* An atom has a single electron revolving about a nucleus. The orbital angular momentum of the electron is $\hbar$. The spin angular momentum of the electron may be either parallel to the orbital angular

momentum, in which case the angular momentum of the atom is $\frac{3}{2}\hbar$, or antiparallel to the orbital angular momentum, in which case the total angular momentum of the atom is $\frac{1}{2}\hbar$.

a. What is the magnetic moment of the atom if the orbital and spin angular momenta are parallel to each other?
b. What is the magnetic moment if the orbital and spin angular momenta are antiparallel?
c. Is the magnetic moment parallel or antiparallel to the orbital angular momentum in parts a and b?

Solution: The orbital and spin angular momenta and magnetic moments are depicted in Figure 33.2. For both orbital angular momentum and spin angular momentum, the magnetic moment is oriented opposite to the angular momentum. The magnetic moment associated with orbital angular momentum, μ_L, is given by $\mu_L = eL/2m_e$, and since $L = \hbar$, $\mu_L = 9.27 \times 10^{-24}$ A • m², the magnetic moment associated with spin angular momentum is $\mu_s = 9.27 \times 10^{-24}$ A • m² also.

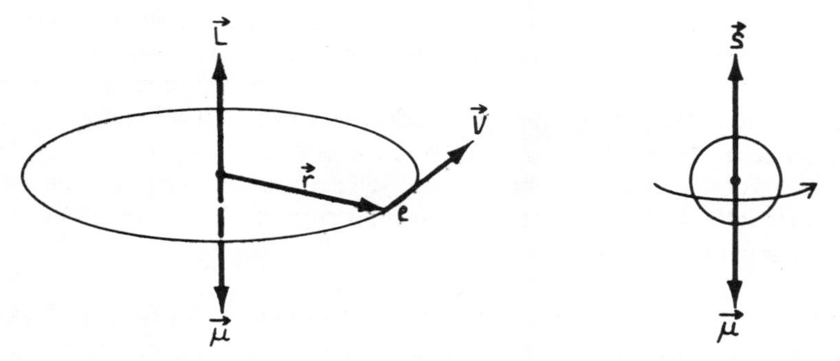

Figure 33.2 Problem 1. (a) The orbital angular momentum of a rotating electron is opposite to its magnetic moment. (b) The intrinsic spin momentum is opposite to the spin magnetic moment of an electron.

a. When the orbital and spin angular momenta are parallel, the total magnetic moment

$$\mu = \mu_L + \mu_s = 2\ (9.27 \times 10^{-24}\ \text{A} \cdot \text{m}^2)$$

or

$$\mu = 1.85 \times 10^{-23}\ \text{A} \cdot \text{m}^2$$

and is oriented antiparallel to the direction of total angular momentum.

b. If the orbital and spin angular momenta are antiparallel, the magnetic moments μ_L and μ_s are also antiparallel, and because they are equal in magnitude, they cancel exactly, so that the total magnetic moment $\mu = 0$.
c. The magnetic moment in part a is antiparallel to the orbital angular momentum.

2. *Guided Problem*

a. The magnetic moment μ of a current loop is parallel to a magnetic field B. What is the change in potential energy if the loop is flipped into a position antiparallel to the field? How much work is needed to flip the loop?

b. The electron in a hydrogen atom orbits around the proton in a circular orbit of radius 0.53×10^{-10} m. When the magnetic moment of the proton is flipped from antiparallel to parallel with the magnetic field of the electron, 9.47×10^{-25} J of energy is released. What is the magnetic field experienced by the proton? The magnetic moment of the proton is 1.41×10^{-26} A • m^2.

Solution scheme

a. The potential energy of a magnetic dipole in a magnetic field is given by $U = -\boldsymbol{\mu} \cdot \mathbf{B}$, so that $U = -\mu B$ if the magnetic moment is parallel to the field. Deduce the change in energy if the magnetic moment is flipped from antiparallel to parallel with the field. The work needed is the increase in potential energy, so if you can find the change in potential energy, it is easy to find the work needed.

b. The potential energy is a minimum when the magnetic moment is parallel to the magnetic field, so that the energy release when the dipole is flipped is equal to the decrease in potential energy of the system. Write down, from step a, the change in potential energy in terms of the magnetic moment of the proton and the magnetic field strength. From this information you should be able to obtain the magnetic field strength.

3. *Worked Problem.* The relative permeability constant κ_m for liquid oxygen, a paramagnetic substance, is 1.00327. If the magnetic moment of the oxygen atom is 13.9×10^{-24} A • m^2, what fraction of the atoms of a liquid oxygen sample will tend to align themselves parallel to the field of a long solenoid having 2000 turns per meter and a current $i = 1.5$ A? The specific gravity of liquid oxygen is 1.14.

Solution: Imagine that we have a solenoid filled with liquid oxygen, as indicated in Figure 33.3. The solenoid length is L, and its cross-sectional area is A.

The magnetic moment of the liquid oxygen $M_T = MV$, where M is the moment per unit volume of the liquid oxygen and V is the volume, given by LA. But the magnetic dipole moment $M_T = I_T A$, where I_T is the total effective, or equivalent, current produced by the dipole moment. But

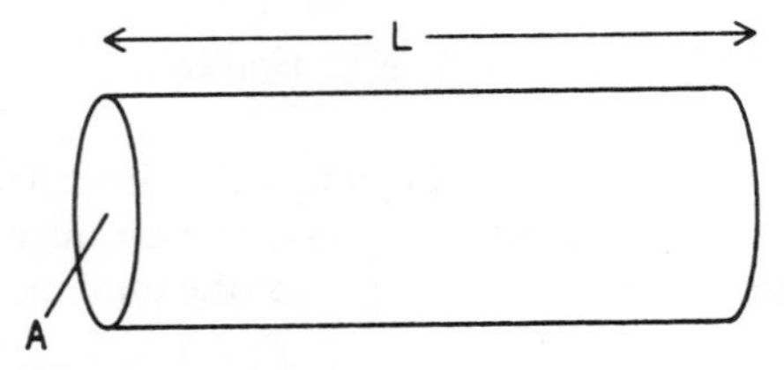

Figure 33.3 Problem 3. A cylinder of liquid oxygen inside a solenoid of length L and cross-sectional area A.

$M_T = I_m LA$, where I_m is the equivalent magnetization current per unit length produced by the liquid oxygen. Then $MV = I_m LA = I_m V$, so that $I_m = M$, the dipole moment per unit volume.

The liquid oxygen produces an effective surface current of I_m amperes per unit length, so that the expression

$$\oint \mathbf{B} \cdot d\mathbf{l} = \mu_0 I_{\text{free}}$$

for Ampère's Law for the solenoid can be written

$$\oint \mathbf{B} \cdot d\mathbf{l} = \mu_0 [I_{\text{free}} + I_m] = \mu_0 [I_{\text{free}} + M]$$

The field inside the solenoid is therefore given by

$$B = \mu_0 [n i_{\text{free}} + M]$$

where $i_{\text{free}} = I_{\text{free}}/n$ is the true current in the solenoid winding. This expression can be rewritten as

$$B = B_0 + \mu_0 M$$

where B_0 is the field in the solenoid in the absence of any magnetic material and $B = \kappa_m B_0$.

$$\kappa_m B_0 = B_0 + \mu_0 M$$

$$M = \frac{(\kappa_m - 1)}{\mu_0} B_0$$

B_0 is given by $B_0 = \mu_0 n i_{\text{free}} = 1.256 \times 10^{-6}(2000)(1.5) = 3.77 \times 10^{-3}$ T, giving $M = 9.81$ A/m for the liquid oxygen. But

$$M = [\text{no. of atoms per m}^3][\text{fraction aligned}][\text{magnetic moment/atom}]$$
$$= (6.022 \times 10^{23} \times 1.14 \times 10^6/16)(F)(13.9 \times 10^{-24})$$

Then $F = 1.88 \times 10^{-5}$ for the fraction of oxygen atoms with their magnetic moments aligned parallel to the magnetic field.

4. *Guided Problem.* A long (70 cm) solenoid has a cross-sectional area of 2.0 cm^2 and has 12,000 turns. The current in the winding is 2.00 A. (a) What is the magnetic field in the solenoid if it is in a vacuum? By how much will the magnetic field change (and will it increase or decrease) if it is filled with (b) air, (c) manganese chloride, or (d) bismuth?

Solution scheme

a. Write down the expression for the magnetic field B_0 for the field in a solenoid in the absence of magnetic materials. From this you should be able to calculate B_0.
b. Using Ampère's Law for a magnetic material, you should be able to relate B, the magnetic field in the presence of magnetic material, to B_0 in terms of the relative permeability of the various materials.

c. Calculate the change in the magnetic field ΔB, caused by the different magnetic materials. You should be able to obtain the values of κ_m from the tables in the text. Remember that the field will increase for paramagnetic materials and decrease for diamagnetic materials.

5. *Worked Problem.* A solenoid has an iron core inserted into it, so that the core fills the space inside the solenoid. The magnetic properties of each iron atom may be considered to be caused by the intrinsic magnetic moment of two of the electrons, with their spins aligned parallel. Suppose that the magnetic moments of all these electrons in the iron core are perfectly aligned parallel to the field of the solenoid. By what ratio is the field inside the solenoid increased by the addition of the iron core? Does your answer depend on the size of the solenoid? Let n equal 1000 turns per meter and the current, i, equal 1.0 A. The density of iron is 7.9 g/cm^3, and the gram atomic weight (GAW) is 56.

Solution: This problem is similar to problem 3, in which we showed that the field inside a solenoid with a core of magnetic material could be expressed by

$$B = B_0 + \mu_0 M$$

where B_0 is the field in the solenoid in the absence of any magnetic material and M is the dipole moment per unit volume of the core material. But M = [no. of iron atoms/m^3][dipole moment of each atom]. The number of iron atoms per cubic meter = [density][Avogadro's number]/GAW = $7.9\,(6.02 \times 10^{23})(1 \times 10^6)/56 = 8.49 \times 10^{28}$ atoms/m^3. (The number 1×10^6 is used because there are 10^6 cm^3 per cubic meter.) Then $M = (8.49 \times 10^{28})(2 \times 9.27 \times 10^{-24}) = 1.57 \times 10^6$ A/m. (The factor of 2 comes from the fact that there are *two* aligned electron spin magnetic moments per atom.) Then, since $B = B_0 + \mu_0 M$,

$$B = \mu_0 ni + \mu_0 M = 1.26\,(10^{-3}) + 1.98\ \text{T}$$

(Nearly all the magnetic field is due to the ferromagnetic iron in the solenoid core.) Since $B = \kappa_m B_0$, $\kappa_m = B/B_0 = 1.98/1.26\,(10^{-3}) = 1570$—that is, the field is increased by the ratio of 1570, independent of the size of the solenoid.

6. *Guided Problem*

a. You have available to you a superconducting cable of diameter 0.7 mm. This cable, when superconducting, will carry a current of 1900 A. If this cable is wound tightly in a single layer to form a long solenoid, approximately what magnetic field would you expect inside the solenoid?

b. A long solenoid has a core of annealed iron. Two of the electrons in each iron atom (the maximum number) have their spins aligned parallel to the solenoid axis. What is the magnetic field in the solenoid, ignoring any field contribution due to current in the solenoid windings? Compare this field to the field you obtained in part a for the superconducting winding. The density of iron is 7.9 g/cm^3, and the gram atomic weight is 56.

Solution scheme

a. What is the expression for the magnetic field inside a long solenoid? Remember that n is the number of turns per meter in length. What should this be, given that the cable diameter is 0.7 mm and that the turns are wound tightly? Having deduced n, you should be able to calculate the magnetic field strength, B. Does your answer surprise you?

b. Remember that in worked problem 5 we arrived at the expression $B = B_0 + \mu_0 M$ for the field inside a solenoid with a core of magnetic material, where M is the dipole moment per unit volume of the core material. What is B_0 here? M = [no. of iron atoms/m^3][dipole moment of each atom]. Remember that there are two aligned electron spin magnetic moments per atom. Can you now calculate first the no. of iron atoms/m^3, and then find M? What do your results tell you—if you want a really high magnetic field, should you use an iron core, with conventional windings, or just superconducting windings, and never mind about the core?

7. *Worked Problem.* A solenoid 40 cm long and 1 cm in diameter has a winding of 400 turns.

a. What current in the windings is required to produce a field of 8×10^{-4} T in the solenoid?
b. The current is held constant at this value, and a core of annealed iron is inserted into the solenoid. What is the new value of the field in the solenoid? (Use Figure 33.7 of your textbook to obtain your answer.)
c. Compare the energy stored in the magnetic field of the solenoid before and after the core is inserted.

Solution: The field in the solenoid is given by $B = \mu_0 n i$, where $n = 400/0.4 = 1000$ turns/m. Then

$$i = B/\mu_0 n = 0.63 \text{ A}$$

If we refer to Figure 33.7, we see that if $B_0 = 8 \times 10^{-4}$ T, then $B = 1.5$ T. That is, the field is increased by a factor of 1875. The magnetic energy density is given by

$$u = B^2/2\mu_0 = 0.254 \text{ J/m}^3$$

The field volume is $AL = [0.785 \times 10^{-4}][0.4] = 0.314 \times 10^{-4}$ m^3, so that the magnetic energy without the core is $[0.314 \times 10^{-4}][0.254] = 7.97 \times 10^{-6}$ J. With the core, $u = 8.93 \times 10^5$ J/m^3, so that the magnetic energy = 28.0 J. That is, the magnetic energy is increased by a factor of 3.5×10^6.

8. *Guided Problem.* If we apply a bunsen flame to a nail clinging to a mag-

net, after some time the nail will drop off the magnet. Explain why. Let us suppose that at the Curie temperature, which for iron is 1043°K, the mean thermal energy of the iron atoms, $kT/2$, is about that of the difference in the magnetic potential energy U between the alignment of a very small domain parallel and antiparallel to an external magnetic field. If the magnetic field strength is 0.1 T, estimate the number of iron atoms in this small domain. (Take the magnetic moment of the iron atom to be 1.8×10^{-23} A • m^2.)

Solution scheme

a. Calculate the mean thermal energy of the iron atom at the Curie temperature of 1043°K. The Curie temperature is that temperature above which a ferromagnetic material behaves like a paramagnetic material. Can you explain what this implies about the alignment of the domains, or even about the existence of the domains, in a ferromagnetic material above the Curie temperature?
b. The expression for the change in potential energy of a magnetic dipole between alignment parallel and alignment antiparallel to a magnetic field can be used here. This expression relates the change in potential energy to the product of magnetic field strength and magnetic dipole moment. If you equate ΔU with the mean thermal energy you obtained in step a, you will obtain the dipole moment of the very small domain.
c. Since the magnetic moments of all the atoms in a domain are perfectly aligned, you should be able to find the number of atoms in this small domain.

9. *Worked Problem.* It is clear that a ferromagnetic material, such as a nail, will become magnetized when placed in an external magnetic field. It is not so clear just why there is a strong force of attraction between a magnet and that same nail. Can you explain why there is such an attraction?

Solution: When we say that the iron nail becomes magnetized, we are saying that it acquires a dipole moment and so itself behaves like a magnet. One way of dealing with magnets is to treat them as having separate north (N) and south (S) poles, instead of as behaving as current loops. We will do so for the purposes of this problem, because it is then easier to compare the forces on electric and magnetic dipoles. Figure 33.4(a) shows a small magnet (an iron nail) being attracted by the field of a large bar magnet. Figure 33.4(b) shows a small electric dipole being attracted by the field of a large one. In the case of the electric dipole, there is an attractive force because the positive charge of the small dipole is in a stronger electric field than is the negative charge. The net force on the small dipole is $F = F_+ - F_- = q(E_+ - E_-)$, where E_+ is the electric field due to the large dipole at the location of the positive charge of the small dipole and E_- is the electric field at the negative charge. $E_+ - E_- = l\Delta E/\Delta y$, where l is the distance between the two charges, so that $F = p dE/dy$ with $p = ql$.

Even though a magnetic dipole does not have isolated N and S poles, qualitatively the attractive force on the N pole of the iron nail is greater than the repulsive force on the S pole because the N pole, being closer to the large magnet, is in a stronger magnetic field. Again, the net attractive

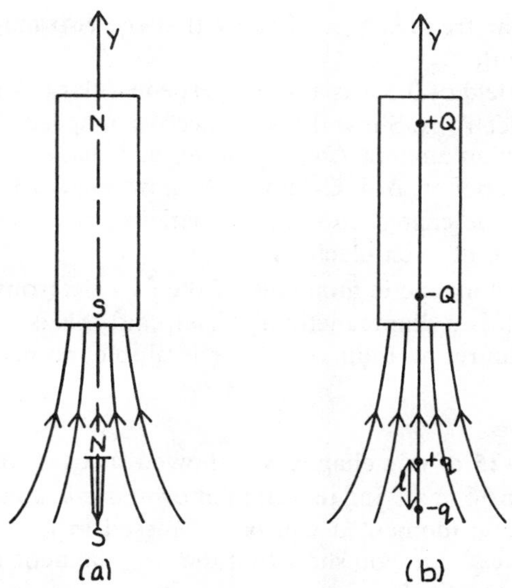

Figure 33.4 Problem 9. (a) A small magnetic dipole (an iron nail) in the field of a large permanent magnet. The field lines near the nail are indicated. (b) A small electric dipole in the electric field of a large electric dipole. The (nonuniform) field lines near the dipole are indicated.

force on the iron nail is given by $F = \mu dB/dy$, where μ is the magnetic moment of the nail and dB/dy is the spatial variation of the magnetic field along the y axis. What is crucial is that the small dipole magnet find itself in a *nonuniform* magnetic field, and because in fact the magnetic fields near most magnets are nonuniform, we usually find a net force on a small magnetic dipole (or on any ferromagnetic object) placed near a magnet.

10. *Guided Problem.* Two electrons in circular orbits of radius r in an atom of bismuth are rotating in opposite directions, as shown in Figure 33.5.

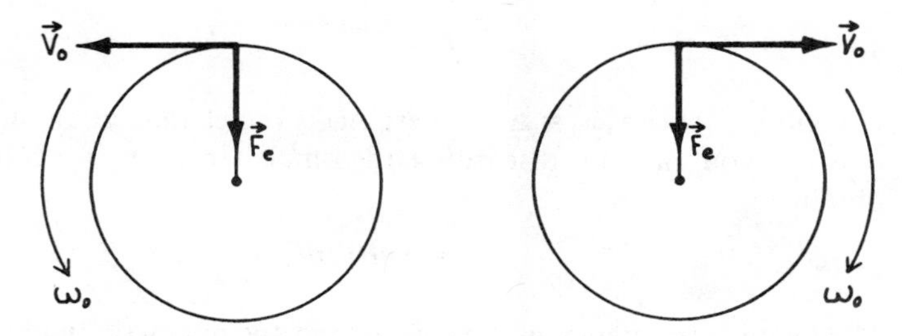

Figure 33.5 Problem 10. Two electrons in an atom of bismuth rotating at an angular frequency ω_0. The orbit radius is r, and the centripetal force is F, due to the electrical force of attraction between the electron and the nucleus.

a. Find an expression for the magnetic moment of the electrons in terms of the angular frequency, and show that the two magnetic moments cancel each other.

b. A magnetic field of 0.5 T is applied perpendicular to the plane of rotation of the electrons. Show that one electron is speeded up by the magnetic field by an amount $\Delta\omega = eB/2m_e$ and that one is slowed down by the same amount, $\Delta\omega$. Calculate the new angular frequency of each electron and the change in total magnetic moment resulting from the orbital motion of each electron.

c. Show that the magnetic moments of the two electrons no longer cancel. (Bismuth is a diamagnetic material, and this is qualitatively how the atom acquires an induced magnetic dipole moment.)

Solution scheme

a. On page 425 of this chapter we showed that the magnetic moment of an electron revolving in a circular orbit is $evr/2$, and since $v = r\omega$, the magnetic moment μ can be expressed in terms of the angular frequency ω_0. Can you show that the two magnetic moments should cancel each other?

b. With the 0.5 T magnetic field applied, find the direction of the magnetic force on the rotating electron in each of the two cases. Redraw Figure 33.5, showing the magnetic force on the electrons in addition to the electric force that provides the centripetal force alone in the absence of the magnetic field.

 Show that in one case the centripetal force on the electron is increased by the magnetic force ($F_B = evB$) and that in the other it is decreased by the same amount. You should now have the following equation relating the centripetal force and the angular frequency:

$$F_e \pm F_B = m_e\omega^2 r$$

but you already know that $F_e = m_e\omega_0^2 r$ and that $F_B = evB$. Now substitute for F_e and F_B, and obtain the following expression relating ω and ω_0:

$$\omega^2 \pm \frac{eB}{m_e} - \omega_0^2 = 0$$

If you now let $\omega = \omega_0 = \Delta\omega$, where $\Delta\omega$ is the change in angular frequency, you can, by discarding the small terms in the equation, obtain

$$\Delta\omega = \pm eB/2m_e$$

c. The angular frequency in the presence of the magnetic field is given by $\omega = \omega_0 + \Delta\omega$, where we know both ω_0 and $\Delta\omega$; now write down the new angular frequency and magnetic moment of each electron.

d. Do the magnetic moments now cancel each other? Write down the net magnetic moment of the two electrons?

Answers to Guided Problems

2. a. The change in potential energy is 2 μB, and this is also the work needed to flip the loop. (We are flipping the coil from stable to unstable equilibrium, from a lower to a higher potential energy, so that a net input of work is needed.)
   b. The magnetic field is 33.6 T.
4. a. $B_0 = 0.04306$ T
   b. $(\Delta B)_{\text{air}} = 1.31 \times 10^{-5}$ T increase
   c. $(\Delta B)_{\text{mn. chloride}} = 5.77 \times 10^{-5}$ T increase
   d. $(\Delta B)_{\text{bismuth}} = 8.18 \times 10^{-7}$ T decrease
6. a. 3.4 T
   b. 1.98 T
   For a very high magnetic field, use superconducting windings.
8. $kT/2 = 7.2 \times 10^{-21}$ J. The number of atoms in the very small domain is 2000.

10. a. For each electron $\mu = er^2/2\ \omega_0$, and since the directions of rotation are opposite, the magnetic moments are in opposite directions and so cancel.
    b. Figure 33.5 should be redrawn as shown in Figure 33.6.

$$\omega = \omega_0 \pm \Delta\omega = \omega_0 \pm \frac{eB}{2\mu_e}$$

The change in the magnetic moment is therefore

$$\Delta\mu = \frac{er^2}{2}\Delta\omega$$

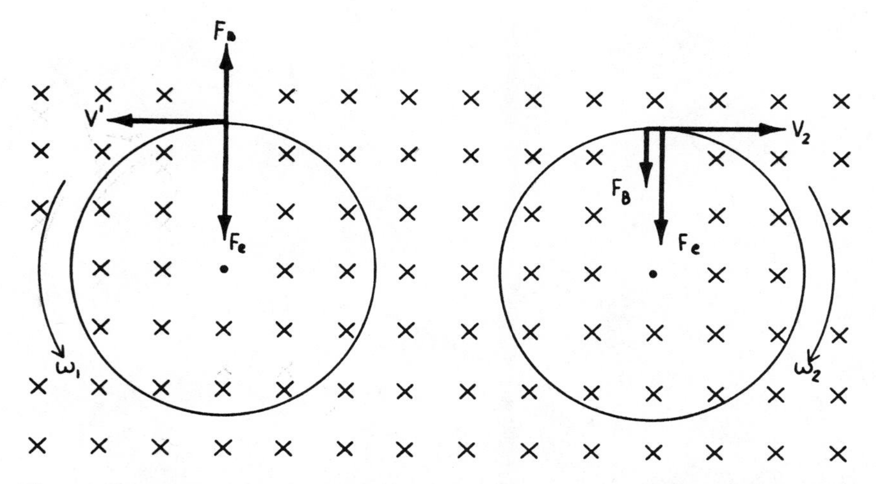

Figure 33.6 Problem 10. The two electrons shown in Figure 33.5 are now rotating perpendicular to a magnetic field; the centripetal force is now $F_e \pm F_B$. The direction of the magnetic field is into the paper.

The magnetic moments for the two electrons are therefore

$$\mu_1 = \frac{er^2}{2}(\omega_0 + \Delta\omega) \qquad \text{(for which } F_{\text{cent}} = F_e + F_B)$$

$$\mu_2 = \frac{er^2}{2}(\omega_0 - \Delta\omega) \qquad \text{(for which } F_{\text{cent}} = F_e - F_B)$$

$\mu_1 + \mu_2 = er^2\Delta\omega^2$ so that the two magnetic moments no longer cancel. (The net magnetic moment is in the direction opposite to the applied field.)

c. $\mu_1 + \mu_2 = er^2\Delta\omega$.

Chapter 34

AC Circuits

Overview

We now return to circuit theory. We developed the ideas of DC, or direct current, circuits in Chapter 29. Now we will consider AC, or alternating current, circuits, in which the voltage and current vary sinusoidally with time. Virtually all our electric power comes to us as AC power, and all our appliances involve circuits with oscillating currents, so AC circuits have vast practical application. We have delayed our study of AC circuits because AC circuit theory involves both electric and magnetic effects. In fact, for the first time we are entering the arena of electromagnetic effects—effects that require us to understand the interplay between electricity and magnetism.

Essential Terms

AC circuit
instantaneous electric power
root-mean-square voltage
capacitive reactance; inductive reactance
freely oscillating LC circuit
"Q"
phasor diagram; phasor
transformer

Key Concepts

1. *AC circuit*. The essential feature of an AC circuit is an oscillating source of emf. Figure 34.1 shows an AC circuit, consisting of an oscillating source of emf connected across a resistor. The emf in an AC circuit is usually taken to be of the form

$$\mathcal{E} = \mathcal{E}_{max} \sin \omega t$$

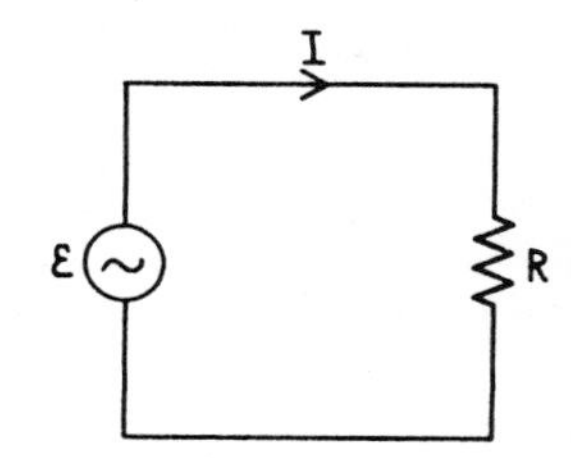

Figure 34.1 A simple AC circuit.

Kirchhoff's rules of circuit analysis hold for any circuit, so that

$$\mathcal{E} - IR = 0$$

This gives us the following relation:

$$I = \mathcal{E}/R = \mathcal{E}_{max} \sin \omega t/R = I_{max} \sin \omega t$$

where I_{max} is the maximum value of current at any time in the circuit.

2. The *instantaneous power* dissipated in the resistor is the product of the voltage drop across the resistor and the current in the resistor:

$$P = \mathcal{E}I = \frac{\mathcal{E}_{max}^2}{R} \sin^2 \omega t$$

The average power, P_{av}, over a complete cycle of the emf is obtained by integrating over the complete cycle. The time period of the cycle $T = 2\pi/\omega$, where ω is the angular frequency.

$$P_{av} = \frac{1}{T}\int_0^T \frac{\mathcal{E}_{max}^2}{R} \sin^2 \omega t dt = \frac{\mathcal{E}_{max}^2}{R}\int_0^T \frac{1}{T} \sin^2 \omega t dt$$

since

$$\frac{1}{T}\int_0^T \sin^2 \omega t dt = \frac{1}{2}$$

$$P_{av} = \frac{\mathcal{E}_{max}^2}{2R} = \frac{\mathcal{E}_{rms}^2}{R}$$

where

$$\mathcal{E}_{rms} = \frac{\mathcal{E}_{max}}{\sqrt{2}}$$

In the United States the oscillating voltage available in normal wall outlets has an amplitude $\mathcal{E}_{max} = 156$ volts, so that $\mathcal{E}_{rms} = 110$ volts. The power dissipated in the circuit of Figure 34.1, $\mathcal{E}_{rms}^2/R$, would be the same if the oscillating emf source were replaced by a DC voltage source $\mathcal{E}_{rms}$.

3. *Capacitive reactance.* For a circuit consisting of a capacitor hooked across an oscillating source of emf, as shown in Figure 34.2, the circuit equation is

$$\mathcal{E}_m \sin \omega t - \frac{Q}{C} = 0$$

or

$$Q = C\mathcal{E}_m \sin \omega t$$

Since $I = dQ/dt$,

$$I = \omega C\mathcal{E}_m \cos \omega t = \frac{\mathcal{E}_m \cos \omega t}{X_C}$$

where $X_C = \dfrac{1}{\omega C}$ and is called the *capacitive reactance* of the circuit. The units of X_C are ohms, as for a resistor. The maximum value of current (a *cosine* function) occurs one-quarter cycle before the maximum value of voltage (a *sine* function), so we say that the current leads the voltage in the circuit.

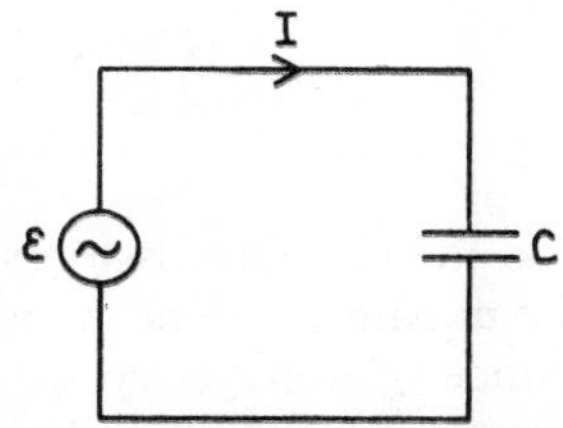

Figure 34.2 An AC circuit consisting of a capacitor hooked across an oscillating source of emf.

The instantaneous power is given by

$$P = I\mathcal{E} = \omega C\mathcal{E}_m^2 \cos \omega t \sin \omega t$$

The *average*, as before, is given by $P_{av} = \frac{1}{T}\int_0^T I\mathcal{E} dt$

$$P_{av} = \omega C\mathcal{E}_m^2 \frac{1}{T}\int_0^T \cos \omega t \sin \omega t dt = 0$$

This result is to be expected, because although a capacitor can store electric energy, it cannot dissipate it as a resistor can.

4. *Inductive reactance.* Suppose that we look at a circuit consisting of an inductor hooked across an oscillating source of emf, as shown in Figure 34.3. From Kirchhoff's second rule, the circuit equation is

$$\mathcal{E} - L\frac{dI}{dt} = 0$$

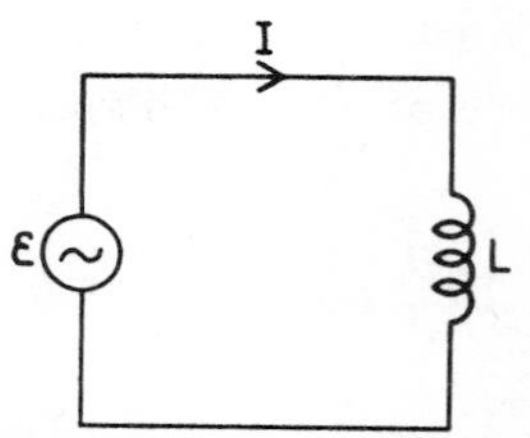

Figure 34.3 An AC circuit consisting of an inductor hooked across an oscillating source of emf.

so that

$$\frac{dI}{dt} = \frac{\mathcal{E}}{L} = \frac{\mathcal{E}_{\text{max}}}{L} \sin \omega t$$

$$dI = \frac{\mathcal{E}_{\text{max}}}{L} \sin \omega t dt$$

Integrating, we obtain

$$I = -\frac{\mathcal{E}_{\text{m}}}{\omega L} \cos \omega t$$

or

$$I = -\frac{\mathcal{E}_{\text{m}}}{X_{\text{L}}} \cos \omega t$$

where $X_{\text{L}} = \omega L$ and is called the *inductive reactance*. The units of X_{L}, like those of X_{C}, are ohms. The maximum value of current occurs one-quarter cycle after the maximum value of voltage, so we say that the current *lags* the voltage in the circuit. Again, we find that the average power dissipated in the circuit is zero, even though the instantaneous power is not.

5. *Freely oscillating LC circuit*. This circuit, shown in Figure 34.4(a), consists of a capacitor C, which at time zero is taken to have a charge $+Q$ on the top plate (and $-Q$ on the lower plate), and an inductor L. At time zero cur-

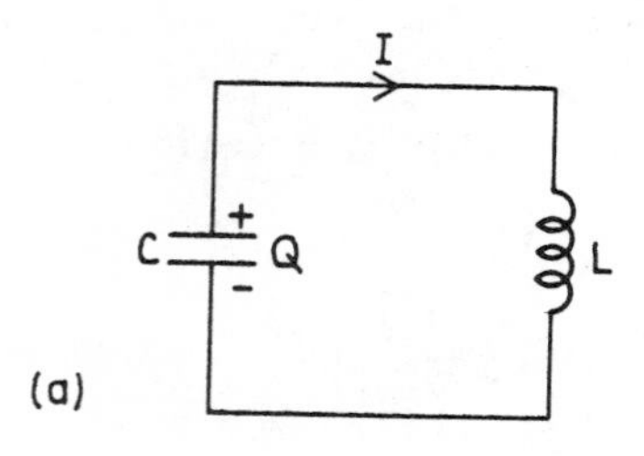

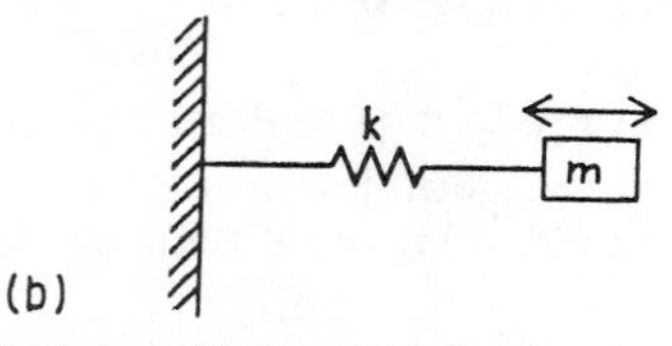

Figure 34.4 (a) A freely oscillating LC circuit. (b) A mass m oscillating at the end of a spring with spring constant k.

rent starts to flow through the inductor, discharging the capacitor but also causing a back emf Ldi/dt across the inductor. By the time that the capacitor is discharged ($Q = 0$), all the energy of the system has been transferred into the magnetic energy $\frac{1}{2}LI^2$ of the inductor and the current has reached a maximum. The current then decreases until the capacitor is fully charged again, but with a charge of $+Q$ on the *bottom* plate. Now all the energy of the system is back in the electric energy $Q^2/2C$ of the capacitor. The system oscillates back and forth in this way, exactly as does a mass on the end of a spring, as shown in Figure 34.4(b). In fact, there is a one-to-one correspondence between the two systems, with the inductance L being equivalent to the mass and $(1/C)$ being equivalent to the spring constant k of the mechanical system. The corresponding equations for the two systems are given in Table 34.1.

Table 34.1 Corresponding equations for a freely oscillating LC circuit and for a mass oscillating at the end of a spring

LC circuit	Mass on spring
Electrical energy $U_E = Q^2/2C$	Kinetic energy $U_k = \frac{1}{2}mv^2$
Magnetic energy $U_B = \frac{1}{2}LI^2$	Potential energy $U_P = \frac{1}{2}kx^2$
Circuit equation:	Equation of motion:
$Ld^2Q/dt^2 + Q/C = 0$	$md^2x/dt^2 + kx = 0$
Charge on the capacitor:	Displacement x:
$Q = Q_0 \cos\left(\frac{1}{\sqrt{LC}}t\right)$	$x = x_0 \cos\left(\sqrt{\frac{k}{m}}t\right)$
where the frequency $\omega = 1/\sqrt{LC}$	where the frequency $\omega = \sqrt{k/m}$

Now suppose that the LC circuit contains a resistor R, as shown in Figure 34.5. The resistor dissipates electrical energy as heat, so that the total energy of the system decreases with time. The amplitude of the oscillations decreases with time, or is *damped*, and in addition the frequency is also reduced. These effects are indicated in Figure 34.6.

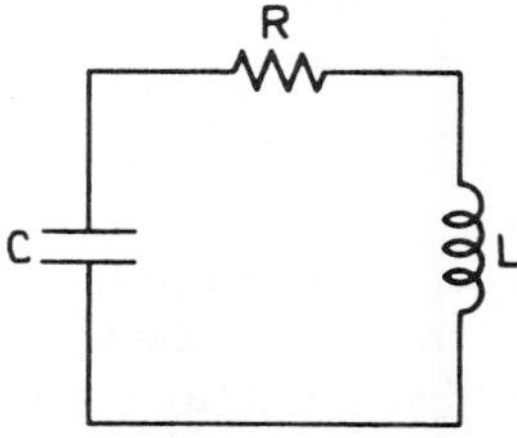

Figure 34.5 A freely oscillating LC circuit with a resistor R.

The charge on the capacitor is given by the expression

$$Q = Q_0 e^{-\gamma t/2} \cos \omega_0 t$$

where $e^{-\gamma t/2}$ is the damping factor and

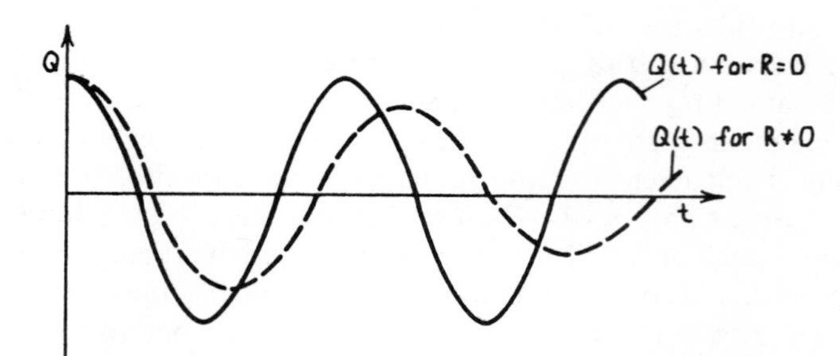

Figure 34.6 The charge Q on a capacitor in an LC circuit as a function of time (a) without and (b) with a resistance R. The resistance damps the oscillatory motion.

$$\omega_0 = \frac{1}{\sqrt{LC}}\left(\sqrt{1 - \frac{CR^2}{4L}}\right)$$

so that $\sqrt{1 - \dfrac{CR^2}{2L}}$ is the factor by which the frequency is decreased with respect to an LC circuit without resistance.

$$\gamma = \frac{R}{L}$$

6. ***Phasor; phasor diagrams.*** Suppose we look at an RLC circuit driven by a source of alternating emf, $\mathcal{E} = \mathcal{E}_{max} \sin \omega t$, as shown in Figure 34.7. The current I in the circuit is expected to oscillate with the same frequency as the

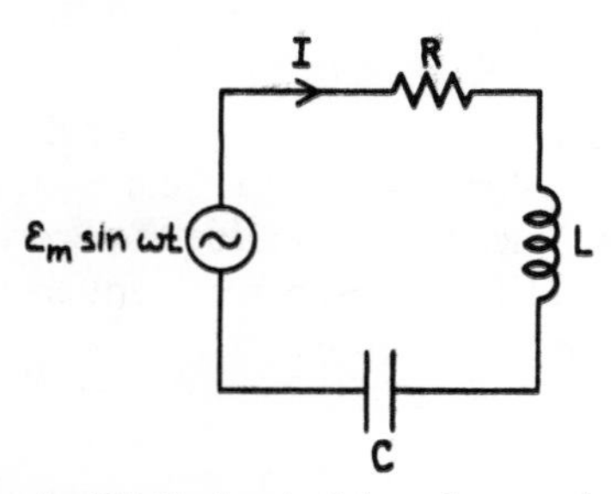

Figure 34.7 An RLC circuit driven by an alternating voltage source.

voltage—that is, $I = I_{max} \sin(\omega t + \phi)$, where I_{max} and the phase angle ϕ are unknown. We represent the current I by means of a rotating vector diagram, or *phasor diagram*, shown in Figure 34.8(a). $I = I_{max} \sin(\omega t + \phi)$ is given by the projection of the phasor I_{max} on the vertical axis. From Kirchhoff's first rule we know that the applied emf is equal to the vector sum of the voltages across the circuit elements:

$$\mathcal{E} = \Delta V_R + \Delta V_L + \Delta V_C$$

We can also represent the voltages across the circuit elements as phasors, as shown in Figure 34.8(b).

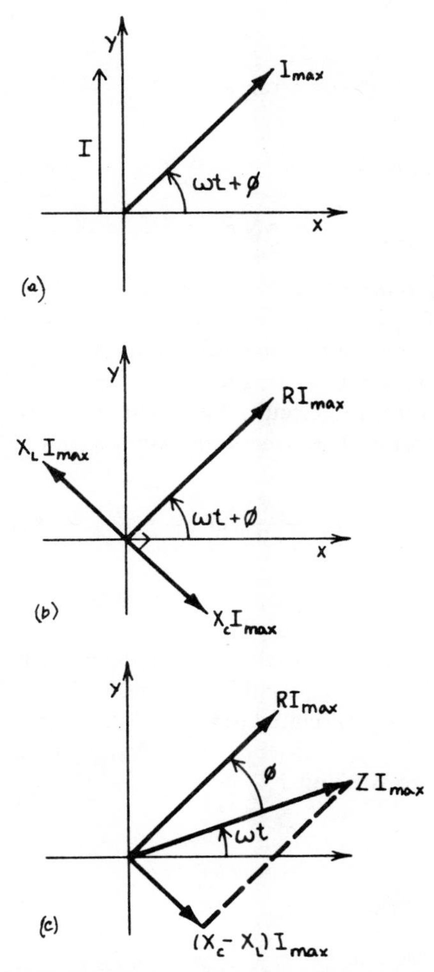

Figure 34.8 (a) Phasor representation of the current in an RLC circuit. (b) Phasor representations of the voltages across the resistor, inductance, and capacitance. (c) The impedance Z of the circuit represented in the phasor diagram.

The amplitude of the voltage across the resistor R is $\Delta V_R = RI_{max}$ and is in phase with the current. ΔV_L leads I by 90° and is given by $X_L I_{max}$. ΔV_C lags I by 90° and is given by $X_C I_{max}$. The vector sum of the three phasors, $\Delta V_R + \Delta V_L + \Delta V_C$, shown in Figure 34.8(c), is given by

$$\mathcal{E}_{max} = \sqrt{R^2 + (X_C - X_L)^2}\, I_{max}$$

Therefore

$$I_{max} = \frac{\mathcal{E}_{max}}{\sqrt{R^2 + (X_C - X_L)^2}} = \frac{\mathcal{E}_{max}}{Z}$$

and

$$I = I_{max} \sin(\omega t + \phi)$$

where

$$\tan \phi = \frac{X_C - X_L}{R}$$

The quantity Z, given by

$$Z = \sqrt{R^2 + (X_C - X_L)^2}$$

is defined as the *impedance* of the circuit.

If $X_L = X_C$, $Z = R$ and $I_{max} = \mathcal{E}_m / R$. This value of current is the maximum value of current in the circuit and occurs at the frequency given by $\omega_0 L = 1/\omega_0 C$, so that $\omega_0 = 1/\sqrt{LC}$, the very same frequency as the frequency of oscillation of a pure LC circuit. This value of ω_0 is called the *resonant frequency* of the circuit. The average power dissipated in the circuit is

$$P_{av} = \frac{\mathcal{E}_m^2}{2Z} \cos \phi = \frac{\mathcal{E}_{rms}^2}{Z} \cos \phi$$

The factor $\cos \phi$ is called the *power factor* of the circuit.

7. *The transformer*. A transformer is a device, consisting of two coils, for changing the voltage of an oscillating source of emf. The emf in the second, or secondary, coil, $\mathcal{E}_2$, is related to the emf in the first coil (the primary coil), $\mathcal{E}_1$, according to $\mathcal{E}_2 = (N_2/N_1)\mathcal{E}_1$, where N_2 and N_1 are the number of turns in, respectively, coils 2 and 1.

Sample Problems

1. *Worked Problem*

a. For the circuit shown in Figure 34.9, plot the power P versus time t for a complete cycle.

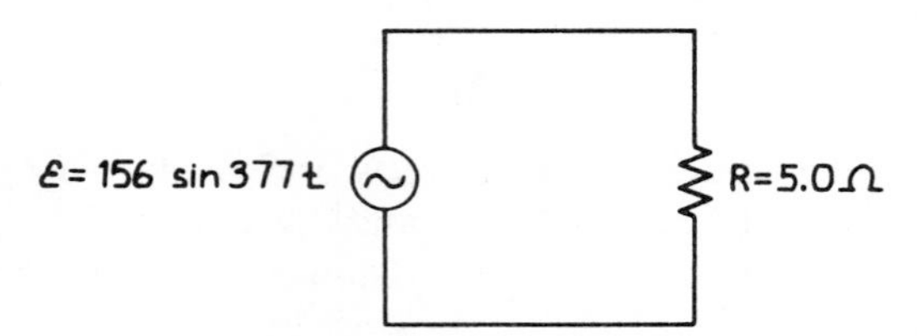

Figure 34.9 Problem 1. An AC circuit consisting of an oscillating source of emf and a resistor R.

b. If the oscillating source of emf is replaced by a battery, what must the emf of the battery be if the average power consumption is to remain the same as it was?

c. If the resistor R is replaced by a 20 μF capacitor, again plot P versus t for a complete cycle. What is the *average* power over the cycle?

Solution: The instantaneous power P is given by

$$P = \mathcal{E}I = \frac{\mathcal{E}_{max}^2}{R} \sin^2 \omega t = 4867 \sin^2 377t$$

The length of a complete cycle, T, is given by $T = 1/f = 2\pi/377 = 1/60$ s. The plot of the power versus time for the time interval between $t = 0$ and $t = 1/60$ s is shown in Figure 34.10(a).

The average power, P_{av}, is given by

$$P_{av} = \frac{\mathcal{E}_{max}^2}{2R} = \frac{\mathcal{E}_{rms}^2}{2}$$

so that

$$\mathcal{E}_{rms} = \frac{\mathcal{E}_{max}}{\sqrt{2}} = 110 \text{ V}$$

This is the emf needed by the battery for the average power consumption to remain the same.

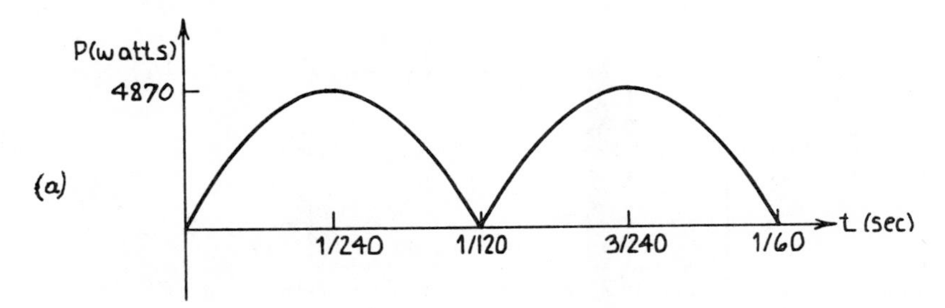

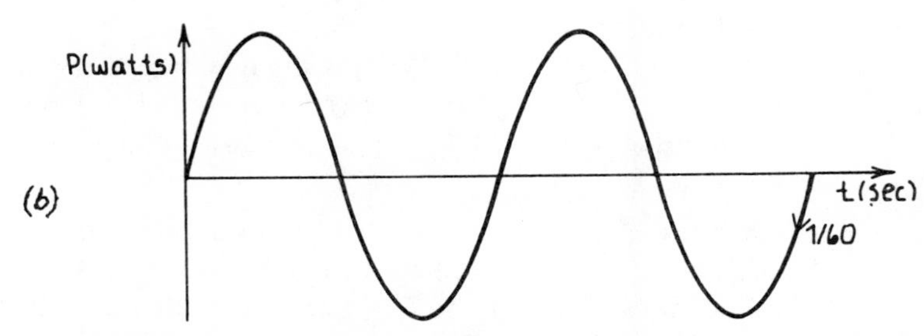

Figure 34.10 Problem 1. (a) Power over a cycle for the circuit shown in Figure 34.9. (b) Power over a cycle when R is replaced by a capacitor.

If we now replace the resistor by a 20 μF capacitor, the reactance of the capacitor is $X_C = 1/\omega C = 1/377\ (20 \times 10^{-6}) = 133\ \Omega$. The instantaneous power is then

$$P = \mathcal{E}I = (\mathcal{E}_{max} \sin \omega t) \frac{\mathcal{E}_{max}}{X_C} \cos \omega t$$

$$= 184\ (\sin 377t)(\cos 377t)$$

This relationship is shown plotted over a complete cycle in Figure 34.10(b). The average power over the cycle is *zero*.

2. *Guided Problem.* A sinusoidal signal generator whose output is 10 sin (2πft) is connected across a 20 mH inductance. Plot the change in the maximum current I_{max} through the inductance as the frequency of the signal generator is increased from 100 to 1000 Hz. Repeat the plot of current versus frequency if the inductance is replaced by a 5 μF capacitor.

Solution scheme

a. Write down the expression for the current I in terms of the applied voltage and the reactance of the inductance. What, then, is the maximum current, I_{max}? Plot the graph of I_{max} versus frequency over the specified frequency range.
b. Repeat the process with the capacitor replacing the inductance. Do you notice that the two curves behave very differently as the frequency is increased?

3. *Worked Problem.* At $t = 0$, switch S is moved from A to B in the circuit of Figure 34.11. The system is observed to oscillate at a frequency of 1000 Hz.

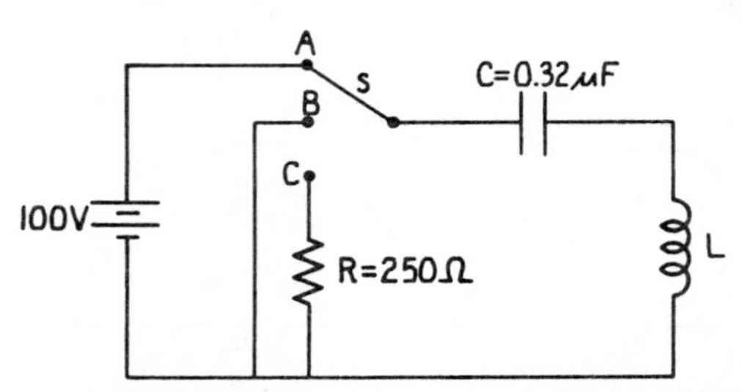

Figure 34.11 Problem 3. An LC circuit with a three-position switch S.

a. Find the value of the inductance L.
b. The switch is now moved from B to C. What is the new frequency of oscillation of the circuit? How long does it take, after we switch from B to C, for the amplitude of the oscillations to die down by a factor of 2?

Solution: In position A the capacitor is charged to 100 V; when we switch to position B, the circuit will begin to oscillate with a frequency given by $\omega = 2\pi f = 1/\sqrt{LC}$. Since we are given f and C, we can calculate the value of L from $L = 1/4\pi^2 f^2 C = 0.079$ H. When we switch from B to C, a damping resistor $R = 250$ ohms is introduced into the circuit, so that the new frequency is given by

$$\omega_0 = \omega\left(\sqrt{1 - \frac{CR^2}{4L}}\right) \quad \text{or} \quad f_0 = f\left(\sqrt{1 - \frac{CR^2}{4L}}\right) = 1000\ (0.968)$$

The new frequency is therefore 968 Hz.

The damping factor is

$$e^{-\gamma t/2} = e^{-Rt/2L} = e^{-1580t}$$

The amplitude of the oscillations is damped by this factor, so that $Q/Q_0 = e^{-\gamma t/2}$. The time needed for Q to equal $Q_0/2$ is then given by $Q/Q_0 = 0.5 = e^{-1580t}$. Solving for t, we obtain $t = 4.38 \times 10^{-4}$ s.

4. *Guided Problem.* A 0.0022 μF capacitor is charged to a potential difference of 150 V across its plates and at $t = 0$ is connected across a 50 mH inductance. Find the maximum current in the inductance, the current at $t = 2 \times 10^{-5}$ s through the inductance, the frequency of oscillations, and the total energy stored in the system.

Solution scheme

a. It is easiest to solve first for the total energy stored in the system. At $t = 0$ there is no current, so that all the energy is stored in the capacitor. Write down the energy stored in a capacitor in terms of the voltage across its plates, and the capacitance. You should be able to determine the total energy of the system easily.

b. The maximum current occurs when there is zero charge on the capacitor, so that all the energy is stored in the magnetic energy of the inductance. What is the expression for the energy of an inductance in terms of L and the current I? This should give you the maximum current.

c. What is the expression for the frequency of oscillation of an LC circuit in terms of L and C? Use this to find the frequency of oscillation.

d. What is the expression for the charge on the capacitor as a function of time? Since $I = dQ/dt$, you can obtain the time dependence of the current by differentiating your expression for Q. (You will find that it has a sinusoidal dependence.) Since you know the frequency f and the time, you should be able to determine $\sin \omega t$ for $t = 2 \times 10^{-5}$ s, and so to find I at 2×10^{-5} s.

5. *Worked Problem.* A phase shift circuit is shown in Figure 34.12. R is a variable resistor; by varying R the phase shift between the input voltage V (equal to $3 \sin 600\, t$) and the output voltage V_0 can be made to vary. Find V_0 and the phase difference ϕ between V and V_0 when R = 10, 20, 30, and 40 ohms. Does V_0 lag or lead V?

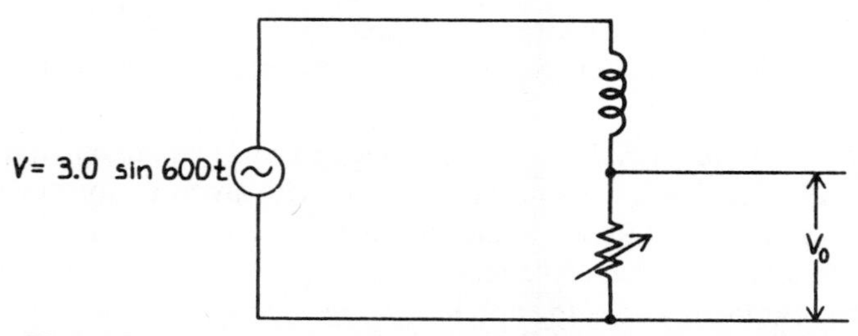

Figure 34.12 Problem 5. A simple L–R phase shift circuit.

Solution: First draw a phasor diagram, as shown in Figure 34.13. The output voltage, V_0, is given by $V_0 = IR$. V_0 *lags* V by the phase angle ϕ. The maximum current, I_m, is given by

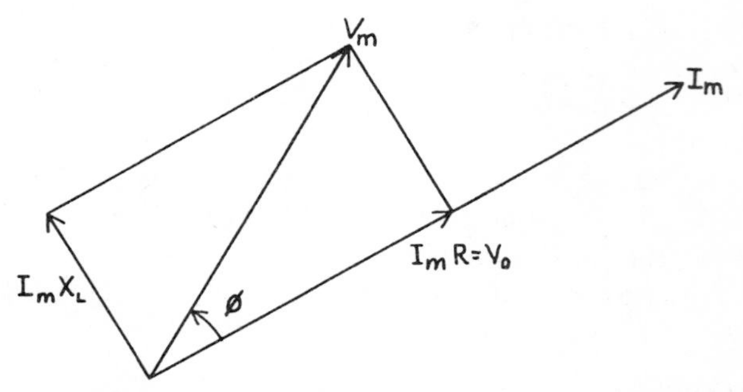

Figure 34.13 Problem 5. A phasor diagram for the phase shift circuit.

$$I_m = \frac{V_m}{Z} = \frac{V_m}{\sqrt{R^2 + X_L^2}}$$

where $X_L = \omega L = 12\ \Omega$.

$$\cos \phi = \frac{R}{Z}$$

so that

$$\phi = \cos^{-1}\left(\frac{R}{Z}\right)$$

Since we need to calculate ϕ at four different values of R, it is easiest to construct a table of values:

R	Z	I_m	V_0	$\left(\frac{R}{Z}\right)$	ϕ
10	15.6	.19	1.92	0.64	50.2°
20	23.3	.13	2.58	0.86	30.9°
30	32.3	.093	2.78	0.93	21.9°
40	41.8	.072	2.87	0.96	16.6°

Notice that as R becomes larger, the phase angle decreases. The output voltage V_0 remains fairly constant as R is varied.

6. *Guided Problem.* A coil of self-inductance 88 mH and unknown resistance and a 0.94 μF capacitor are connected in series with an oscillator of frequency 600 Hz. If the phase angle between the applied voltage from the oscillator and the current is 25°, what is the resistance of the inductor? Show in a phasor diagram the relation between the applied voltage and the current in the circuit.

Solution scheme

a. Calculate the reactance of the inductor and of the capacitor.
b. Draw a phasor diagram showing the voltages across the three circuit elements and their vector sum.
c. What is the relation between the phase angle ϕ, the resistance R, and the reactances X_C and X_L? Can you now calculate the resistance R?

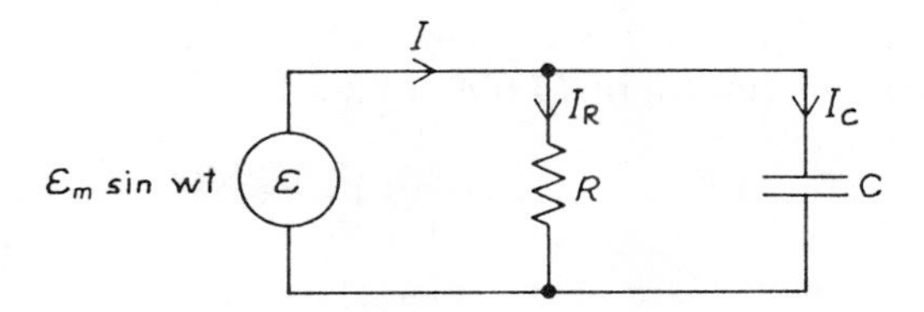

Figure 34.14 Problem 7. A parallel RC circuit.

7. *Worked Problem.* In the parallel RC circuit shown in Figure 34.14, the oscillator has a frequency of 800 Hz and an amplitude of 20 V. $R = 100\ \Omega$, and $C = 1.2\ \mu F$. What is the maximum current in the circuit, and what is the phase angle between current and voltage?

Solution: Since the oscillator voltage output can be written as $\varepsilon = \varepsilon_m \sin \omega t$, the current through the resistor is

$$I_R = \varepsilon/R = \varepsilon_m \sin \omega t/R$$

Similarly, the current through the capacitor is

$$I_C = \varepsilon_m \cos \omega t/X_C$$

(note that the two currents are 90° out of phase!). We can add these two currents by means of a phasor diagram, as shown in Figure 34.15.

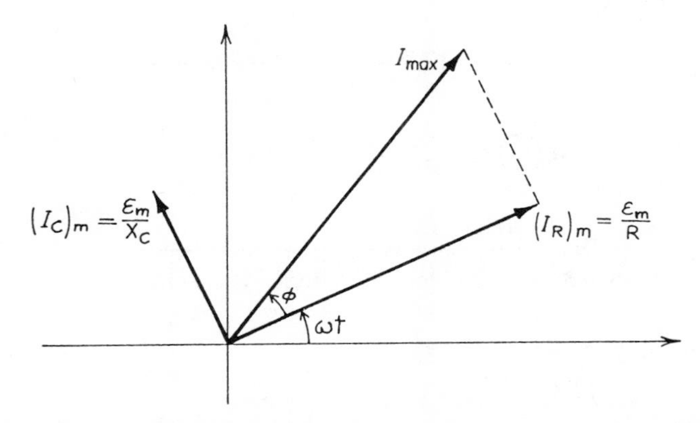

Figure 34.15 Problem 7. A phasor diagram of the currents in a parallel RC circuit.

$$(I_C)_m = \varepsilon_m/X_C;\ (I_R)_m = \varepsilon_m/R$$

Therefore

$$I_{max} = \varepsilon_m\sqrt{(1/X_C)^2 + (1/R^2)} = I_m \sin(\omega t + \phi)$$

Since $f = 800$ Hz, $X_C = 166\ \Omega$ and $R = 100\ \Omega$. We obtain $I_m = 0.0117\ \varepsilon_m$, and since $\varepsilon_m = 2.0$ V, $I_m = 0.0234$ A. From the phasor diagram, the phase angle between current and voltage is given by $\tan \phi = (1/X_C)/(1/R) = 0.602$, so that $\phi = 31.1°$.

8. *Guided Problem.* An industrial user finds that an inductive load is drawing an average rms current of 73.0 A from a 110 V_{rms}, 60-cycle supply. The voltage leads the current by 12°.

a. Find the impedance of the load and the average power consumption.
b. The user is required to keep the current in phase with the supply voltage. To do this he decides to put a capacitor in series with the load. What capacitance does he need, and what is the current through the load after the capacitor is added to the circuit?

Solution scheme

a. What is the relationship between I_{rms}, V_{rms}, and the circuit impedance, Z? Use this to find Z.
b. What is the average power consumption in terms of the rms voltage, the circuit impedance, and the power factor of the circuit? Use this to find the average power consumption.
c. Knowing that current lags voltage by 12°, and knowing the total impedance of the circuit, can you find the inductance, as well as the resistance, of the circuit? (Remember that $\cos\phi = R/Z$.)
d. Knowing the inductance and the frequency, you can find the inductance reactance. What value of capacitive reactance is needed to make the circuit purely resistive? Knowing X_C, you should be able to find the value of capacitance C needed.
e. What is the new current, I_{rms}, with the capacitor in the circuit?

9. *Worked Problem.* A 40 mH coil with a resistance of 50 ohms is connected in series with a 5500 pf (1 pf = 10^{-12} F) capacitor, as shown in Figure 34.16.

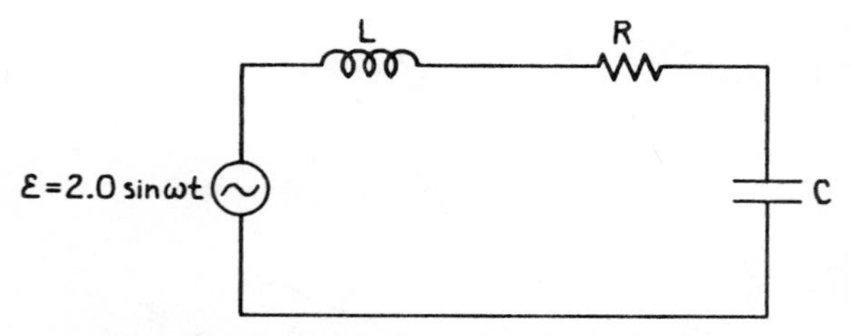

Figure 34.16 Problem 9. A series RLC circuit.

a. What is the resonant frequency of the circuit?
b. Plot the maximum current I_{max} in the circuit as the frequency of the emf is varied from 8 kHz to 13 kHz. What is the maximum value of current, and at what frequency does it occur? At what approximate values of frequency is the current 1/2 of its maximum value?
c. A 150-ohm resistor is added in series to the circuit. Again plot the current as the frequency is varied from 8 to 13 kHz.

Solution: The resonant frequency is given by $\omega_0 = 2\pi f_0 = 1/\sqrt{LC}$.

$$f_0 = \frac{1}{2\pi}\frac{1}{\sqrt{40(10^{-3})5500(10^{-12})}} = 10.7 \text{ kHz}$$

At any frequency the maximum current is given by $I_{max} = \mathcal{E}_{max}/Z$, where $Z = \sqrt{R^2 + (X_C - X_L)^2}$. We therefore need to calculate X_C, X_L, Z, and from these I_m for every frequency; the easiest approach is to tabulate the values:

f (kHz)	X_C	X_L	$(X_C - X_L)$	Z ($R = 50$)	I_{max} (mA)	Z ($R = 150$)	I_{max} (mA)
8	3617	2008	1609	1609	1.24	1621	1.23
9	3216	2259	957	958	2.09	978	2.04
10	2894	2510	384	387	5.17	436	4.59
11	2631	2761	130	139	14.4	238	8.40
12	2412	3012	598	600	3.33	631	3.17
13	2226	3263	1037	1037	1.93	1056	1.89
10.7	2689	2689	0	50	40	200	10

The last row in the table gives values for the resonant frequency of the circuit. Note that when we add a 150-ohm resistor in series, the total circuit resistance is 200 ohms, not 150 ohms!

The plots of maximum current I_{max} versus frequency are given in Figure 34.17. For $R = 50$ ohms, the curve is much narrower than at $R = 200$ ohms, and I_{max} is about 1/2 of its value at resonance at the frequencies of 10.2 and 10.9 kHz.

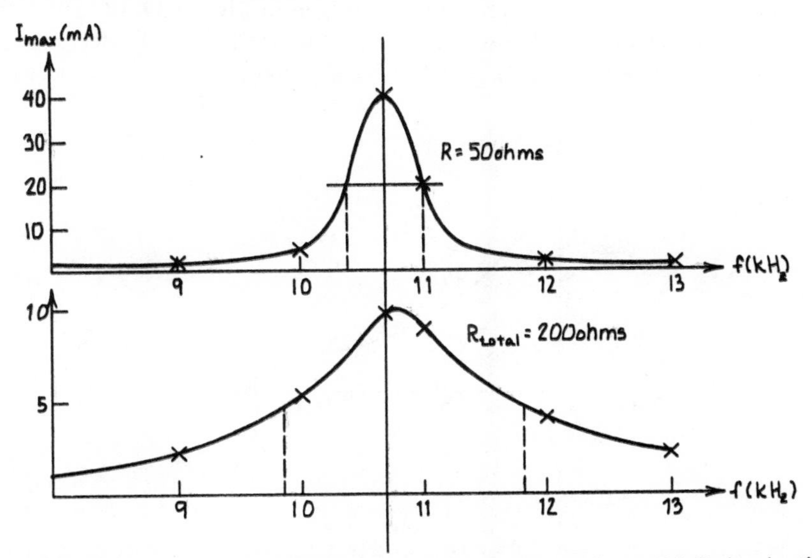

Figure 34.17 Problem 9. Plots of I_{max} versus frequency for an RLC circuit with two different values of resistance. For $R = 50$ ohms, the curve is much narrower than at $R = 200$ ohms.

10. *Guided Problem.* Fifty kilowatts of power is to be transmitted over a 50-mile distance from a power plant to a small town. Given a transmission line, you are to decide on whether to transmit the power at 120 VAC or at some higher voltage. What is the ratio of the power loss in the transmission

line at 2000 V to that at 120 V? What is the corresponding ratio at 20,000 V? If you decide to transmit the power at 20,000 V, what ratio of secondary to primary turns should your step-up transformer have, and what is the current in the 50-mile-long transmission line at this voltage?

Solution scheme

a. Write down the expression for power in terms of current and voltage, and for power loss in a resistance in terms of the current and the resistance. What is the ratio of the transmission line currents for 2000 V versus 120 V transmission? What is the corresponding ratio of the resistive losses?
b. Evaluate the same ratios for 20,000 V versus 120 V transmission. Which voltage will minimize your transmission losses?
c. Write down the ratio of output to input voltage of a transformer in terms of the turns ratio. Since you know the input and output voltages, you can calculate the needed turns ratio for your transformer.

Answers to Guided Problems

2. For the inductance, $I = \dfrac{\mathcal{E}_m \cos \omega t}{X_L}$; $I_{max} = \dfrac{10}{.126f}$.

   For the capacitor, $I_{max} = \dfrac{\mathcal{E}_m}{X_C} = \dfrac{f}{3180}$.

   The plots of I_{max} versus frequency are given in Figure 34.18. Since the inductive reactance increases with increasing frequency, I_m decreases as f increases; conversely, X_C decreases as f increases, so that I_m increases as f increases.

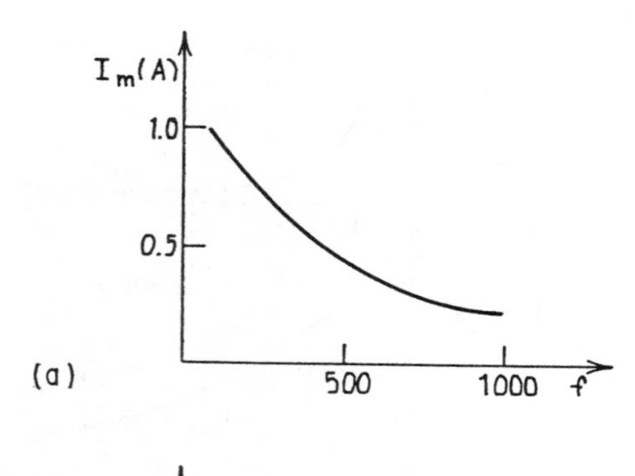

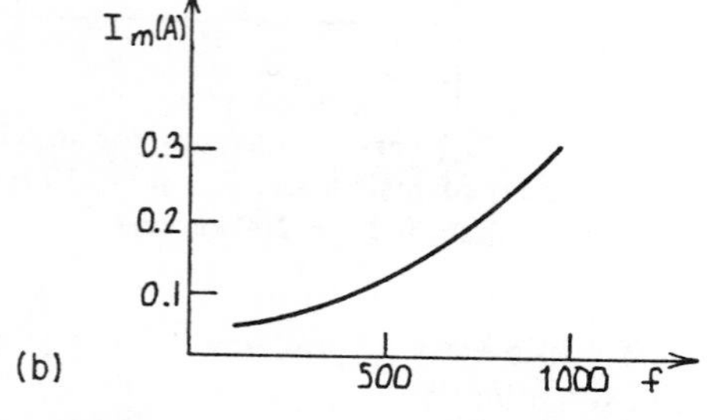

Figure 34.18 Problem 2. (a) I_m versus frequency for an inductive load. (b) I_m versus frequency for a capacitive load.

4. $I_m = 0.0315$ A, $f_0 = 1.52 \times 10^4$ Hz; the total energy $U = 2.48 \times 10^{-5}$ J. At $t = 2 \times 10^{-5}$ s, $I = 0.03$ A.

6. $X_L = 332\ \Omega$
   $X_C = 282\ \Omega$
   $R = 107\ \Omega$
   The relation between voltage and current is shown in Figure 34.19.

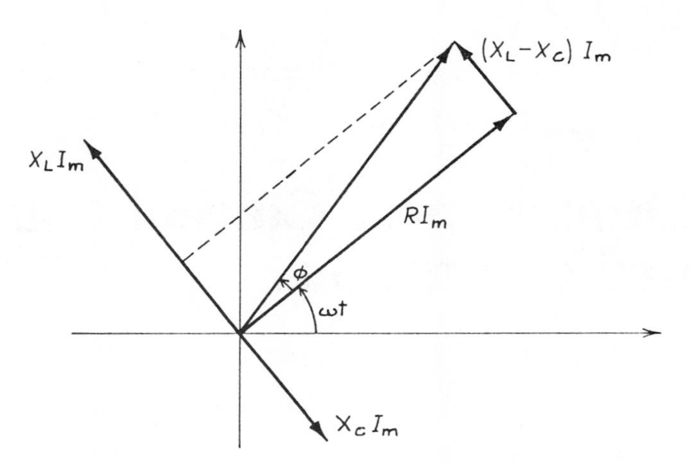

Figure 34.19 Problem 6. Phasor diagram for the circuit of Problem 6.

8. $Z = 1.51$ ohms; $P_{av} = 7.58$ kW.
   The capacitance needed is $C = 8.44 \times 10^{-3}$ F; the new current is $I_{rms} = 74.5$ A.

10. $(\text{Loss})_{2000}/(\text{Loss})_{120} = 3.6 \times 10^{-3}$
    $(\text{Loss})_{20,000}/(\text{Loss})_{120} = 3.6 \times 10^{-5}$
    $N_2/N_1 = 20{,}000/120 = 167$; $I_{2000} = 2.5$ A

Chapter 35

The Displacement Current and Maxwell's Equations

Overview

At last, having firmly established the fundamental equations of electricity and magnetism, we are ready to express these equations, with one addition, in the form known as Maxwell's equations. This one addition, formulated by Maxwell, will show that a changing electric field induces a magnetic field, just as Faraday's Law tells us that a changing magnetic field induces an electric field. Maxwell's equations link and interrelate electric and magnetic effects. In addition, they show how electromagnetic waves can be generated and propagated through free space. In short, Maxwell's equations form the basis of classic electromagnetic theory and allow us to understand an enormous number of everyday phenomena.

Essential Terms

displacement current
Maxwell's equations
cavity oscillations
waveguide
radiation field

Key Concepts

1. *Displacement current*. There is a flaw in Ampère's Law, which says that the integral of **B** around a closed path is proportional to the total current passing through an arbitrary surface enclosed by the path of integration; that is,

$$\int \mathbf{B} \cdot d\mathbf{l} = \mu_0 I$$

Suppose that we evaluate the field a distance r from a wire carrying a current I that is charging a capacitor. (I is then dQ/dt, where Q is the charge on the capacitor.) Using Ampère's Law, we easily obtain the result that the field a distance r from the wire is given by

$$\int \mathbf{B} \cdot d\mathbf{l} = B\int dr = B(2\pi r) = \mu_0 I$$

or

$$B = \frac{\mu_0 I}{2\pi r}$$

I is the current enclosed by a circle of radius r around the wire; that is, I is the current through the surface S_1 in Figure 35.1.

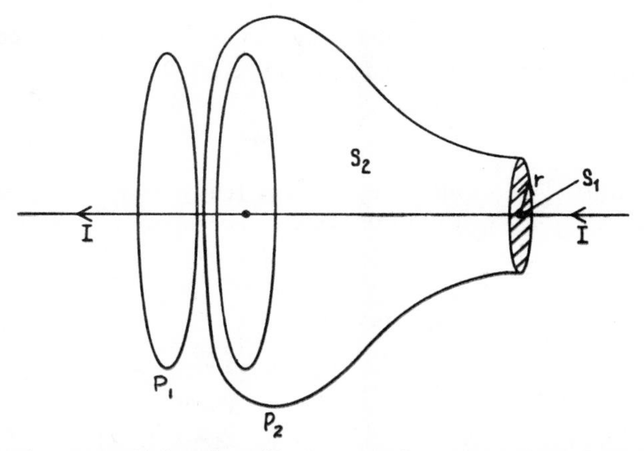

Figure 35.1 A current I charging circular capacitor plates P_1 and P_2. S_1 and S_2 are two surfaces enclosed by the circle of radius r around the wire carrying the current I.

But S_2 is another possible surface enclosed by the circle of radius r, and since no current passes through S_2, which extends in between the capacitor plates, Ampère's Law fails. If we add a term $\mu_0 \varepsilon_0 d\Phi_E/dt$, where Φ_E is the electric flux $\int \mathbf{E} \cdot d\mathbf{A}$, to Ampère's Law, however, Ampère's Law works, because if we evaluate $\mu_0 \varepsilon_0 d\Phi_E/dt$ over the surface S_2 in Figure 35.1, we find that

$$\mu_0 \varepsilon_0 \frac{d\Phi_E}{dt} = \mu_0 I$$

It therefore does not matter whether we evaluate Ampère's Law over surface S_1 or S_2; both give the result that

$$\oint \mathbf{B} \cdot d\mathbf{l} = \mu_0 I$$

The term $\varepsilon_0 \frac{d\Phi_E}{dt}$ is called the *displacement current*, and Ampère's Law, with the displacement current term, is now written

$$\oint \mathbf{B} \cdot d\mathbf{l} = \mu_0 I + \mu_0 \varepsilon_0 \frac{d\Phi_E}{dt}$$

In this form Ampère's Law is perfectly general; we cannot find any more situations where Ampère's Law breaks down. It is important to notice the symmetry between Faraday's Law,

$$\oint \mathbf{E} \cdot d\mathbf{l} = -\frac{d\Phi_B}{dt}$$

and Ampère's Law,

$$\oint \mathbf{B} \cdot d\mathbf{l} = \mu_0 I + \mu_0 \varepsilon_0 \frac{d\Phi_E}{dt}$$

Faraday's Law says that a changing magnetic field produces an electric field; Ampère's Law says that a changing electric field produces a magnetic field.

2. *Maxwell's equations.* We are now ready to write the four Maxwell equations. They are as follows:

$$\oint \mathbf{E} \cdot d\mathbf{S} = Q/\varepsilon_0$$
$$\oint \mathbf{B} \cdot d\mathbf{S} = 0$$
$$\oint \mathbf{E} \cdot d\mathbf{l} = -d\Phi_B/dt$$
$$\oint \mathbf{B} \cdot d\mathbf{l} = \mu_0 I + \mu_0 \varepsilon_0 d\Phi_E/dt$$

The first equation is the familiar Gauss' Law for electricity; the second equation, Gauss' Law for magnetism, is really an assertion that there are no magnetic monopoles. The last two equations are, respectively, Faraday's Law of induction and Ampère's Law with the displacement current term. With the addition of one equation, that for the Lorentz force on a charged particle, Maxwell's equations give a complete description of electric and magnetic fields in a vacuum. That last equation is:

$$\mathbf{F} = q\mathbf{E} + q\mathbf{v} \times \mathbf{B}$$

3. *Cavity oscillations.* Suppose a parallel-plate capacitor with circular plates of radius R is connected to a source of alternating voltage, so that the electric field between the plates is given by $E_{(1)} = E_0 \sin \omega t$. The modified Ampère's Law equation tells us that this oscillating electric field will give rise to a magnetic field in the region between the plates. To find the magnetic field at a radius r from the center of the plate, we need to evaluate the rate of change of the electric flux through a circle of radius r, indicated in Figure 35.2. The electric flux $\Phi_E = EA = E(\pi r^2)$. Then

$$\mu_0 \varepsilon_0 d\Phi_E/dt = \mu_0 \varepsilon_0 \frac{d}{dt}(\pi r^2 E) = \mu_0 \varepsilon_0 \pi r^2 \frac{dE}{dt}$$

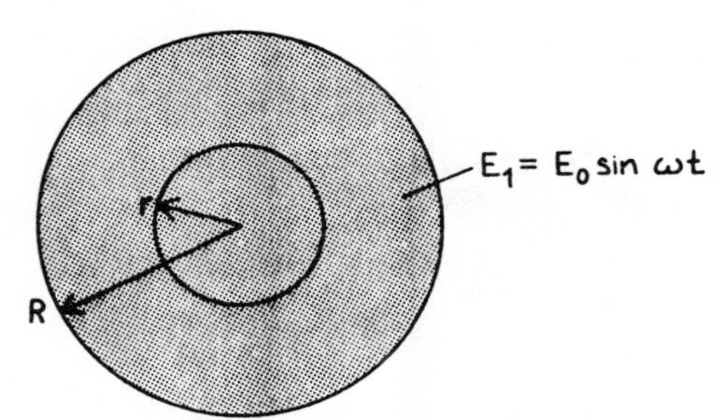

Figure 35.2 A uniform electric field $E_{(1)}$ between capacitor plates of radius R. The direction of the field is out of the paper. The magnetic field at radius r is being evaluated.

where

$$E = E_{(1)} = E_0 \sin \omega t$$

Then

$$\int \mathbf{B} \cdot d\mathbf{l} = \mu_0 \mathcal{E}_0 \pi r^2 \omega E_0 \cos \omega t$$

From symmetry we expect the magnetic field lines to be closed concentric circles, so that

$$\int \mathbf{B} \cdot d\mathbf{l} = \int B dl = B \int dl = 2\pi r B$$

$$B = \frac{\mu_0 \mathcal{E}_0}{2} r \omega E_0 \cos \omega t$$

This means the magnetic field increases linearly with r. The direction of the field is indicated by the arrows; at this instant, current is flowing toward the capacitor plate from below the paper, and the direction of the magnetic field between the plates is consistent with the field direction around the wire that is carrying current toward the capacitor plate.

According to Faraday's Law of induction, this time-dependent magnetic field must induce a second electric field $E_{(2)}$ that will have to be added to $E_{(1)}$. By applying Faraday's Law, we find that

$$E_{(2)} = -\frac{\mu_0 \mathcal{E}_0}{4} r^2 \omega^2 E_0 \sin \omega t$$

where the negative sign indicates that $E_{(2)}$ is antiparallel to $E_{(1)}$. The total electric field

$$E = E_{(1)} + E_{(2)} = E_0 \sin \omega t - \frac{\mu_0 \mathcal{E}_0}{4} r^2 \omega^2 E_0 \sin \omega t$$

$$= \left(1 - \frac{\mu_0 \mathcal{E}_0 \omega^2}{4} r^2\right) E_0 \sin \omega t$$

Actually, $E_{(2)}$ induces another time-dependent magnetic field, which in turn induces a third time-dependent electric field $E_{(3)}$, and so on, but the total magnetic and electric fields are fairly close to those that we have already calculated. The electric field vanishes at a radius R given by

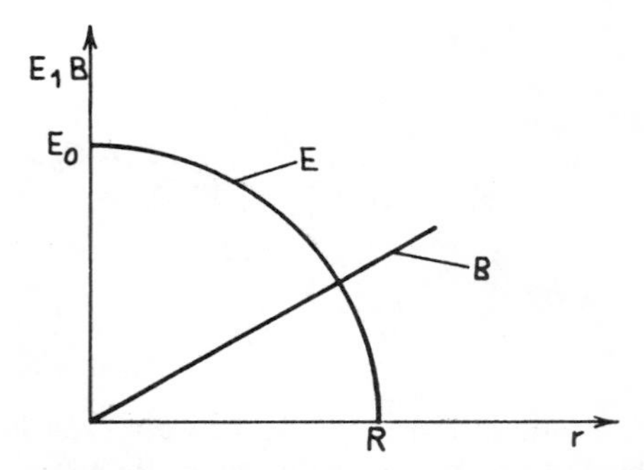

Figure 35.3 Plots of the amplitude of the electric and the magnetic fields in a parallel-plate capacitor driven by a source of alternating voltage.

$$\left(1 - \frac{\mu_0 \mathcal{E}_0 \omega^2}{4} R^2\right) = 0$$

$$R = \frac{2}{\sqrt{\mu_0 \mathcal{E}_0}} \frac{1}{\omega}$$

Plots of E and B as a function of the radius r are shown in Figure 35.3.

If we insert a conducting cylindrical wall of radius R between the capacitor plates, we find that the cylinder wall does not interfere with the electric field, because the field is zero at $r = R$. Current can flow from one plate to the other along the cylinder wall, and we find that we have a resonant cavity that will sustain oscillations once charge is placed on a capacitor plate in much the same way that an LC circuit will oscillate. The energy of the system is alternately stored as electric and then magnetic energy as the charge flows back and forth along the cylinder wall.

A resonant cavity of practical size will usually have a radius R of a few centimeters. If $R = 2$ cm, the resonant frequency is of the order of 5×10^9 Hz, as given by the expression

$$f = \frac{1}{\pi} \frac{1}{R\sqrt{\mu_0 \mathcal{E}_0}}$$

By most standards of measurement, this is a very high frequency.

A resonant electromagnetic cavity has many practical applications, one of which is the generation of high-frequency electromagnetic radiation. Such devices are called *klystrons*. A *waveguide* is a long, open-ended conducting pipe or cavity used to carry high-frequency electromagnetic radiation from one place to another. The electromagnetic oscillations in a waveguide are traveling, rather than standing, waves.

4. *Radiation field*. A charge q that experiences an acceleration a emits a transverse electric field, called a radiation field, which, in vacuum, moves outward from the charge with speed c. The magnitude of this field is given by

$$E_\theta = \frac{1}{4\pi\mathcal{E}_0} \frac{qa \sin\theta}{c^2 r}$$

where θ is the angle between the direction of the acceleration and the point at which we measure the field. Suppose that at $t = 0$ a charge is accelerated for a short time along the x axis and that at time t we are interested in the radiation field at point P, a distance r away from the charge, as indicated in Figure 35.4.

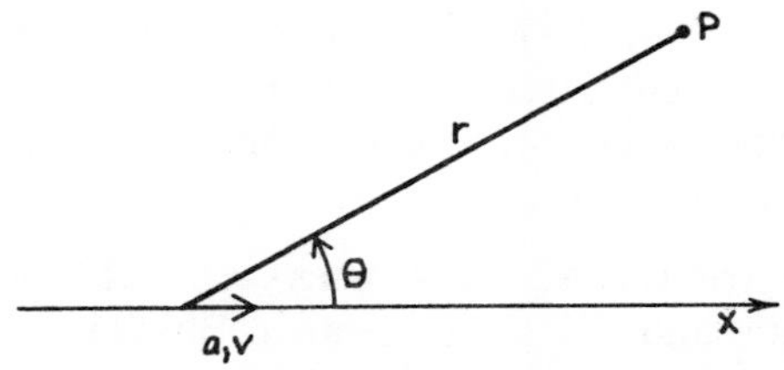

Figure 35.4 A charge is accelerated parallel to the x axis. We need to calculate the radiation electric field at point P, a distance r from the charge.

At $t = 0$ the radiation field at point P is zero, and in fact it takes a time r/c for the field to reach point P, so that the field at time t is produced by an acceleration at time $(t - r/c)$. We therefore express the radiation field as

$$E_\theta(t,r) = \frac{1}{4\pi\varepsilon_0}\frac{q \sin\theta}{c^2 r}\, a\left(t - \frac{r}{c}\right)$$

where E_θ is a function of time t and the distance away from the charge q. If the charge happens to oscillate back and forth with simple harmonic motion, so that $a = a_0 \sin \omega t$, then the radiation field is given by

$$E_\theta(t,r) = \frac{1}{4\pi\varepsilon_0}\frac{q \sin\theta}{c^2 r}\, a_0 \sin \omega\left(t - \frac{r}{c}\right)$$

This is the equation of a harmonic wave of angular frequency ω traveling in a radial direction with speed c.

In addition to a transverse electric field, an accelerated charge also emits a magnetic field, which can be considered as induced by the electric field according to

$$\oint \mathbf{B} \cdot d\mathbf{l} = \mu_0 \varepsilon_0 d\phi/dt$$

We can show that the magnitude of $\mathbf{B}$ is given by

$$B = c\varepsilon_0\mu_0 E_\theta$$

$\mathbf{B}$, like the electric field, is transverse to the direction of propagation, but is *also* perpendicular to $\mathbf{E}$. It can be shown, too, that the speed of propagation, c, can be expressed by $c = 1/\sqrt{(\mu_0\varepsilon_0)}$, so that $\mathbf{B} = E_\theta/c$. Thus, the radio wave emitted from an antenna or the light wave emitted from an atom consists of a transverse electric field E_θ and a magnetic field B, also transverse to the direction of propagation, but perpendicular to E_θ, and given by $B = E_\theta/c$.

Sample Problems

1. *Worked Problem.* A 10 mA current is charging square capacitor plates, 20 cm on a side, with a plate separation of 3 mm. Find the electric flux Φ_E and the displacement current I_d between the plates. Do the same for round capacitor plates of radius 10 cm, also with a 3 mm plate separation. For the round capacitor plates, find the magnetic field midway between the plates at respective distances of 2 cm, 10 cm, and 20 cm from the plate axis. Compare your answers with the magnetic field at distances of 2 cm, 10 cm, and 20 cm from a wire carrying a current of 10 mA.

Solution: The electric flux $\Phi_E = EA$, where E is the electric field between the capacitor plates and A is the plate area. But E is given by

$$C = \frac{Q}{V} = \frac{Q}{Ed} = \frac{\varepsilon_0 A}{d}$$

so that $E = Q/\varepsilon_0 A$. The electric flux

$$\Phi_E = Q/\varepsilon_0$$

We don't know the total charge Q on the capacitor plates, and so we cannot evaluate the absolute value of Φ_E. The displacement current $I_d = \varepsilon_0 d\Phi_E/dt$, however, and the time rate of change of electric flux is $\frac{1}{\varepsilon_0}\frac{dQ}{dt} = \frac{1}{\varepsilon_0} I$. Then $I_d = I$, the real current into the capacitor plates, which is 10 mA for both the square and the round capacitor plates.

To find the magnetic field between the capacitor plates, we apply Ampère's Law with the displacement current, which says that

$$\oint \mathbf{B} \cdot d\mathbf{l} = \mu_0 \varepsilon_0 \frac{d\Phi_E}{dt}$$

Because we have circular symmetry, B is everywhere the same at a given radius from the axis and everywhere tangent to the circle, so that the left side of the equation can be written $(2\pi r)B$. The right side of the equation is

$$\mu_0 \varepsilon_0 (\pi r^2) \frac{dE}{dt}$$

where $dE/dt = I/\varepsilon_0 A$ and A is the total plate area = 0.0314 m^2. The right side of the equation then becomes $\mu_0 \pi I r^2/0.0314$. By equating the left and the right sides of the equation, we get

$$2\pi r B = \frac{\mu_0 \pi I r^2}{.0314}$$

$$B = \frac{\mu_0 I r}{.0628}$$

for values of r less than the plate radius.

For $r = 0.02$ m, $B = 4 \times 10^{-9}$ T. For $r = 0.10$ m, $B = 2 \times 10^{-8}$ T. For $r = 20$ m, the flux is given by AdE/dt, where A is the total plate area, so that the right side of the equation is $\mu_0 I$ and $B = \mu_0 I/2\pi r$, just as it is outside a wire carrying a current I. Then for $r = 0.20$ m, $B = 1 \times 10^{-8}$ T, both outside the capacitor and 20 cm away from a wire carrying a current of 10 mA. At 0.02 m away from the wire, $B = 10 \times 10^{-8}$ T, and at 0.10 m away from the wire, $B = 2 \times 10^{-8}$ T. Note that at $r = 2$ cm, well inside the plates, the field in the plates is much less than the field 2 cm away from a wire; at $r = 10$ cm, just at the outer edge of the plates, the field at the outer edge of the plates is the same as it would be outside the wire and remains so for all values of r greater than 10 cm.

2. *Guided Problem.* Point P is 30 cm from the midpoint of the capacitor plates shown in Figure 35.5. Using Ampère's Law, find the magnetic field at point P. Use two different surfaces, one in the region between the capacitor plates and one entirely outside the capacitor plates, in your evaluation of the right side of Ampère's Law. Show that your answer is independent of the surface you choose.

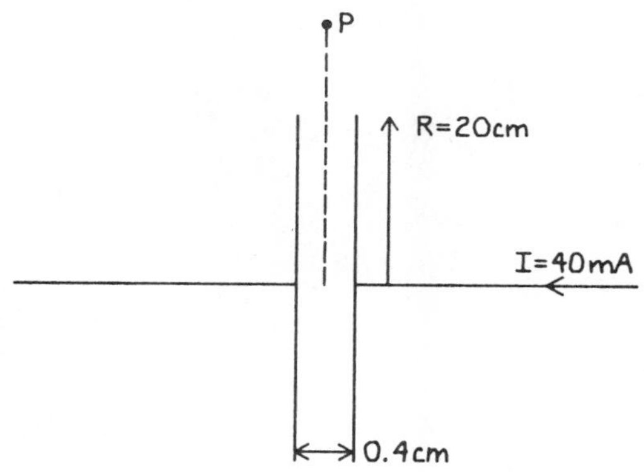

Figure 35.5 Problem 2. Point P is 30 cm from the midpoint of the capacitor plates of 20 cm radius.

Solution scheme

a. Write down Ampère's Law with the displacement current term. For the *left* side of the equation, the easiest path of integration is a circle of radius 30 cm centered on the axis of the capacitor. Using the symmetry, you can express the left side of Ampère's Law as $2\pi rB$.

b. Now sketch two surfaces enclosed by the path of integration, one just a circle of radius 30 cm centered on the plate axis and the other a bag extending outside the capacitor plates. The wire carrying the current to the plates must pass through the surface of this bag.

c. For the first surface, the circle, the true current passing through the surface is zero. You can evaluate the right side of Ampère's Law in terms of the displacement current $I_d = \varepsilon_0 \dfrac{d\Phi_E}{dt}$ passing through this surface. Assume that the electric field, and so the electric flux, is zero outside the capacitor plates, that is, for r greater than 20 cm.

You should be able to express your result in terms of the true current flowing onto the plates.

d. Now evaluate the right side of Ampère's Law in terms of the current passing through the baglike surface outside the capacitor plates. The electric flux is everywhere zero on this surface, but the true current does pass through it. You should find that the right side of Ampère's Law is the same for both surfaces, and you should now be able to find the magnetic field at point P.

3. *Worked Problem.* A long solenoid with n turns per meter and radius R m has a time-dependent current $I = I_0 \sin \omega t$ flowing through its coil. Will there be an electric field in the solenoid? If so, obtain an expression for the electric field and sketch the lines of force. Is the electric field conservative?

Solution: A cross-sectional view of the solenoid is presented in Figure 35.6. The magnetic field in the solenoid, shown coming up out of the paper, is given by

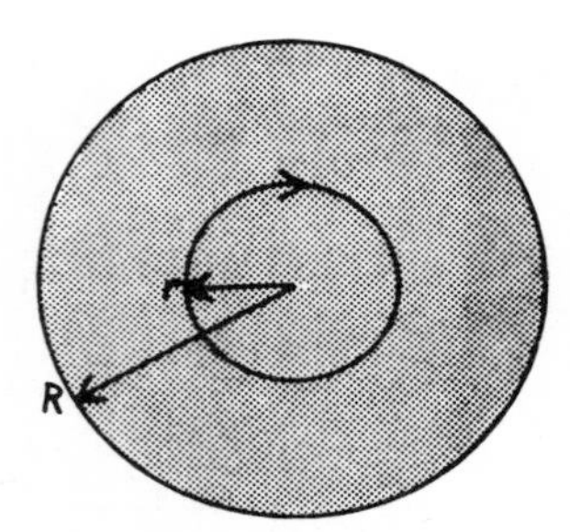

Figure 35.6 Problem 3. A cross section through a solenoid of radius R. The magnetic field is out of the paper. We are evaluating the electric field at a distance r from the solenoid axis.

$$B = \mu_0 n I = \mu_0 n I_0 \sin \omega t$$

From Faraday's Law we know that the induced emf is given by

$$\int \mathbf{E} \cdot d\mathbf{l} = -\frac{d\Phi_B}{dt}$$

where the enclosed magnetic flux, Φ_B, is given by

$$\Phi_B = \int \mathbf{B} \cdot d\mathbf{A}$$

For a circle of radius r, $\Phi_B = BA$, since B is independent of r. Then

$$\Phi_B = (\mu_0 n I_0 \sin \omega t)\pi r^2$$

and

$$\frac{d\Phi_B}{dt} = (\mu_0 n I_0 \omega \cos \omega t)\pi r^2$$

From symmetry we expect that E is everywhere the same at a circle of radius r and is everywhere tangent to the circle; that is, we expect that

$$\oint \mathbf{E} \cdot d\mathbf{l} = 2\pi r E$$

$$2\pi r E = -(\mu_0 n I_0 \omega \cos \omega t)\pi r^2$$

$$E = -\frac{\mu_0 n I_0}{2} \omega r \cos \omega t$$

There is an induced electric field inside the solenoid, just as there is an induced magnetic field between capacitor plates when the electric field between the plates is changing. The direction of the induced electric field is given most easily by Lenz' Law; the direction of E should be so as to oppose the change in flux. When the magnetic field is increasing in the direction out of the paper, the induced electric field will be counterclockwise in Figure 35.6. (It is the direction of the change in flux that counts here!) The electric field is *not* conservative, because the integral of $\mathbf{E} \cdot d\mathbf{l}$ around a closed loop is not zero.

In this example we find uniform, straight magnetic field lines and electric field lines that are concentric circles everywhere perpendicular to the direction of the magnetic field lines. Compare this to the field lines between capacitor plates, where we have straight, uniform electric field lines and magnetic field lines that are concentric circles perpendicular to the electric field lines. In this example the electric field strength increases with r, just as the magnetic field strength increases with r inside capacitor plates.

4. *Guided Problem.* A cylindrical cavity has a radius $R = 2.0$ cm; the distance d between the end plates is 1.0 cm. The cavity is oscillating such that the maximum voltage between the centers of the end plates is 500 volts. Find the frequency of oscillation in hertz, the amplitude of the electric field at the center of the cavity and at $r = R$, and the amplitude of the magnetic field at $r = 0$ and $r = R$.

Solution scheme

a. What is the expression for the frequency of a cylindrical cavity in terms of its radius, R? Using this, you should be able to determine the angular frequency, ω, as well as the frequency, f, in hertz.

b. What are the expressions for the total electric field, $E = E_{(1)} + E_{(2)}$, and for the magnetic field in terms of the radius r? What does the expression for E become if $r = 0$? Since the field is uniform along the axis of the cylinder, the maximum voltage difference between the two ends of the cylinder $V = Ed$; because you know V and d, you should be able to find the amplitude of the electric field. What is the value of E at $r = R$? What is therefore the amplitude of E at $r = R$?

c. From the expression for the magnetic field, you should be able to see that its behavior is opposite to that of the electric field; that is, the amplitude *increases* with increasing r. Calculate the amplitudes at $r = 0$ and $r = R$.

5. *Worked Problem.* In the cylindrical cavity of the preceding problem ($R = 2.0$ cm, $d = 1.0$ cm, $V_{max} = 500$ volts between the end plates), find the maximum charge stored on the end plates (*inside* the cavity), the maximum current flowing along the cavity walls, and the total energy stored in the system.

Solution: The electric field at the surface of a conductor is given by $E = \sigma/\varepsilon_0$, so that $\sigma = \varepsilon_0 E$. The electric field

$$E = E_0\left(1 - \frac{\mu_0\varepsilon_0\omega^2 r^2}{4}\right)\sin\omega t$$

so that the maximum surface charge density is given by

$$\sigma_{max} = \varepsilon_0 E_{max} = \varepsilon_0 E_0\left(1 - \frac{\mu_0\varepsilon_0\omega^2 r^2}{4}\right)$$

Since the surface charge density is a function of radius r, we need to integrate over the area of the plate to obtain the total charge,

$$Q_{max} = \int \sigma_{max} da$$

and we divide the plate into concentric rings of radius r and width dr, so that $da = 2\pi r dr$. Then

$$\begin{aligned} Q_{max} &= \int_0^R \sigma_{max}(2\pi r dr) \\ &= \varepsilon_0 E_0 \int_0^R \left(1 - \frac{\mu_0\varepsilon_0\omega^2 r^2}{4}\right)(2\pi r dr) \\ &= \varepsilon_0 E_0 (2\pi)\left[\frac{R^2}{2} - \frac{\mu_0\varepsilon_0\omega^2}{4}\frac{R^4}{4}\right] \end{aligned}$$

From the preceding problem, we know $E_0 = 5(10^4)$ V/m, $\omega = 3 \times 10^{10}$ radians/s, and $R = 0.02$ m. Therefore, $Q_{max} = 2.78 \times 10^{-10}$ C. This is the maximum charge on the end plates of the cavity; since we obtained Q from

$$\begin{aligned} Q &= \int \sigma da = \int \varepsilon_0 E da \\ &= \int \varepsilon_0 \left(1 - \frac{\mu_0\varepsilon_0\omega^2 r^2}{4}\right) E_0 \sin\omega t da \\ &= Q_{max}\sin\omega t \end{aligned}$$

Then the charge Q, like the electric field, must have a sinusoidal time dependence. The current I along the walls of the cavity must therefore be given by

$$\begin{aligned} I &= dQ/dt = \omega Q_{max}\cos\omega t = 3(10^{10})2.78(10^{-10})\cos\omega t \\ &= 8.34\cos\omega t \text{ A} \end{aligned}$$

Even though the total charge is small, the current is not, because of the very high frequency of oscillation.

We can obtain the total energy by integrating the expression $u_E = \frac{1}{2}\varepsilon_0 E^2$ over the cavity volume; we choose $E = E_{max}$ because when $E = E_{max}$, $B = 0$, and there is then no contribution from magnetic energy. The total energy is therefore

$$U = \int u_{E_{max}} dV$$

where

$$u_{E_{max}} = \tfrac{1}{2}\varepsilon_0 E^2_{max}$$

$$U = \tfrac{1}{2}\varepsilon_0 \int E^2_{max} dV$$

$$= \tfrac{1}{2}\varepsilon_0 \int E_0^2 \left(1 - \frac{\mu_0\varepsilon_0\omega^2 r^2}{4}\right)^2 dV$$

We subdivide the cavity volume into concentric cylinders, each of radius r, thickness dr, and height $d = 0.01$ m. Then $dV = 0.01(2\pi r dr)$. The total energy is then given by

$$U = 0.01\pi\varepsilon_0 E_0^2 \int_0^R \left[1 - \frac{\mu_0\varepsilon_0\omega^2 r^2}{4}\right]^2 rdr$$

$$= 0.01\pi\varepsilon_0 E_0^2 \int_0^R \left(1 - \frac{\mu_0\varepsilon_0\omega^2 r^2}{2} + \frac{\mu_0\varepsilon_0\omega^4 r^4}{16}\right) rdr$$

$$= 0.01\pi\varepsilon_0 E_0^2 \left[\frac{R^2}{2} - \frac{\mu_0\varepsilon_0\omega^2 R^4}{9} + \frac{\mu_0^2\varepsilon_0^2\omega^4}{(16)}\frac{R^6}{6}\right]$$

where $E_0 = 5 \times 10^4$ V/m, $R = 0.02$, $\omega = 3 \times 10^{10}$ radians/s. Then $U = 4.64 \times 10^{-8}$ J.

6. *Guided Problem.* In the LC circuit shown in Figure 35.7, the capacitor is connected across a 100 V battery. At $t = 0$ the capacitor is suddenly switched to the 5 mH inductance shown, so that the system starts to oscillate. Starting with the expression for the charge on the capacitor, $Q = Q_0 \cos \omega t$, where $\omega = 1/\sqrt{LC}$, calculate the displacement current in the gap between the capacitor plates as a function of time, and compare your answer with the current I in the inductance.

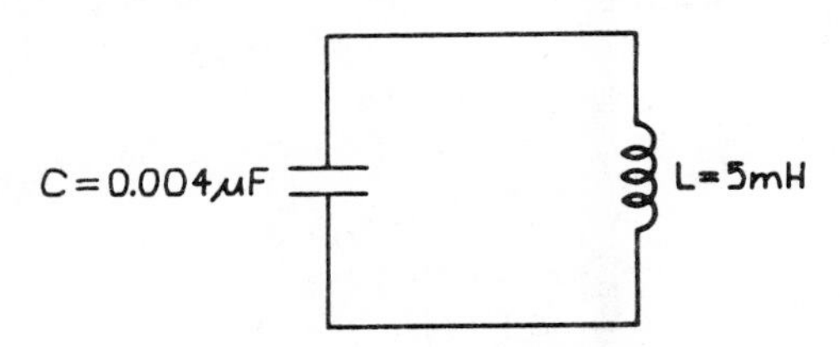

Figure 35.7 Problem 6. A freely oscillating LC circuit.

Solution scheme

a. What is the expression for the displacement current I_d in terms of the time rate of change in electric flux? Express this in terms of dQ/dt, where $Q = Q_0 \cos \omega t$.
b. Is there any difference between your expression for I_d and your expression $I = dQ/dt$ for the current I in the inductance?
c. Calculate explicit values for the frequency and Q_0 to obtain an explicit expression for the displacement current I_d.

7. *Worked Problem.* An electron beam is accelerated through a potential difference of 20,000 volts and strikes a copper block. Electrons, on the average, travel 10 nanometers (1 nm = 10^{-9} m) into the copper before stopping. Assuming uniform deceleration, find the amplitude of the electric field seen by a detector 1 m away from the copper block at angles of 10°, 30°, 60°, and 90°, as indicated in Figure 35.8. If the beam were to consist of a single electron that strikes the block at $t = 0$, sketch the approximate response of the detector at $\theta = 90°$ between $t = 0$ and $t = 5$ nanoseconds. Assume that the electron is nonrelativistic.

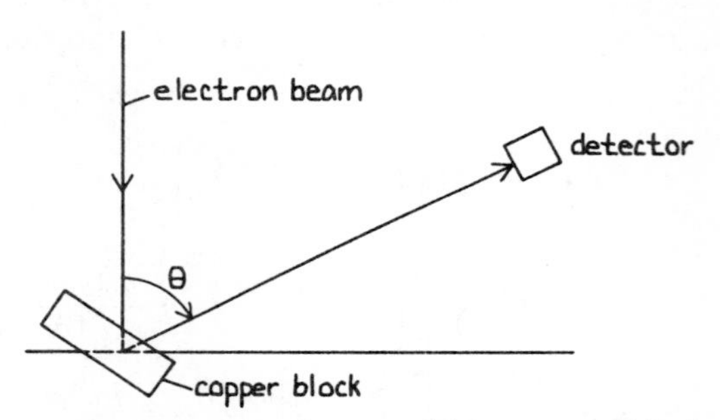

Figure 35.8 Problem 7. An electron beam striking a copper block. The detector is 1 m away from the block.

Solution: The electric field is given by

$$E_\theta(t,r) = \frac{1}{4\pi\varepsilon_0} \frac{q \sin \theta a\left(t - \frac{r}{c}\right)}{c^2 r}$$

where $q = 1.6 \times 10^{-19}$ C, $c = 3 \times 10^8$ m/s, and $r = 1.0$ m. The magnitude of the acceleration a is given from $v_f^2 = v_0^2 + 2\,as$, where $s = 10 \times 10^{-9}$ m, and we obtain v_0 from $\frac{1}{2}mv_0^2 = 20{,}000$ eV $= 20{,}000 \times 1.6 \times 10^{-19}$ J. Since the mass of the electron $m = 9.1 \times 10^{-31}$ kg, we obtain for v_0 the value 8.39×10^7 m/s. By substituting this value into our equation of motion (note that the final velocity of the electron $v_f = 0$), we find that the acceleration is 3.52×10^{23} m/s^2. Using this value for $a(t - r/c)$, we find that $E_\theta(t,r) = 5.63 \times 10^{-3} \sin \theta$ V/m. Then

$$E_{10}(r) = 9.78\,(10^{-4})\ \text{V/m}$$
$$E_{30}(r) = 2.82\,(10^{-3})\ \text{V/m}$$
$$E_{60}(r) = 4.88\,(10^{-3})\ \text{V/m}$$
$$E_{90}(r) = 5.63\,(10^{-3})\ \text{V/m}$$

The time for the pulse to travel 1 m is $1/3 \times 10^8 = 3.33$ nanoseconds. We can estimate the acceleration time period from $a = (\upsilon_f - \upsilon_0)/t$, or $t = 8.39 \times 10^7/3.52 \times 10^{23} = 2.4 \times 10^{-16}$ s; the acceleration time period is negligible. The detector therefore sees nothing until t = 3.33 nanoseconds; we then see a pulse as the electric field passes the detector. [The electric field can be expressed as $E_\theta(3.33 \text{ ns}, 1.0)$.] After the pulse at t = 3.33 nanoseconds, there is nothing further in the detector, assuming that the beam consists of a single electron. The response is shown in Figure 35.9.

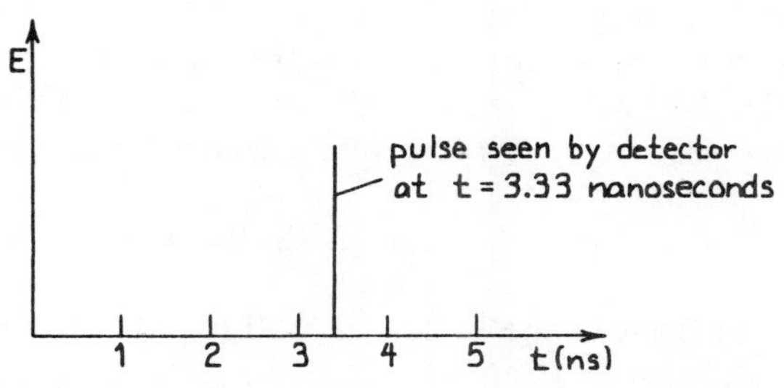

Figure 35.9 Problem 7. The response seen by the detector shown in Figure 35.8 to a single electron striking the block.

8. *Guided Problem.* A transmitter applies an electric field of E equal to $1 \times 10^{-4} \cos(3 \times 10^6\, t)$ V/m along a vertical copper antenna. The antenna is 3.0 m long and 3 mm in diameter.

a. What is the acceleration of the free electrons in the antenna?
b. If there is one free electron per copper atom and each free electron contributes equally to the electric field radiated by the antenna, what is the amplitude of the electric field at a horizontal distance of 1.0 km from the antenna?
c. If your receiver has a sensitivity of 1 V/m, within what range (in kilometers) can you expect to pick up the signal from the antenna?

Solution scheme

a. What is the acceleration of the electrons in terms of the electric field, the charge of the electron, and the mass of the electron? Use this information to obtain the explicit expression for a.
b. To find the number of free electrons in the antenna, we need to find the total number of copper atoms, which is given by [mass] $\times$ [Avogadro's number] / [atomic number] $= \dfrac{V[\text{density}](6.02 \times 10^{23})}{63}$. The volume $\pi r^2 L$ = 21.2 cm^3, and the density of copper is 8.96 g/cm^3. The number of atoms of copper in the antenna, which is also the number of free electrons, is therefore 1.82×10^{24}.
c. Write down the expression $E_\theta(t,r)$ for the electric field resulting from the accelerated electrons in the antenna. What is the value of θ in this case?
d. To find the range, given $E_\theta(t,r)$, solve for r in the expression for the amplitude of the electric field.

9. *Worked Problem.* An electron in a TV tube has a kinetic energy of 3.5×10^{-15} J. It hits the screen and stops in a distance of 1 mm. What are the magnitudes of the electric and the magnetic radiation fields that the electron generates at a distance of 30 cm (a) directly in front of the TV screen and (b) to one side of the screen, that is, at an angle of 90° to the direction of motion of the electron? (c) Draw a diagram showing the direction of the acceleration of the electron, and the direction of the electric and magnetic radiation fields in parts a and b.

Solution: Kinetic energy $= 1/2\, m\upsilon^2 = 3.5 \times 10^{-15}$ J. Since $m = 9.1 \times 10^{-31}$ kg, the initial speed of the electron is 8.77×10^7 m/s. We obtain the deceleration of the electron through the equation $\upsilon_f^2 = \upsilon_0^2 + 2\,as$, where $s = 1 \times 10^{-3}$ m. Then $a = -3.846 \times 10^{18}$ m/s^2. We know that the electric and the magnetic radiation fields are given by

$$E_\theta = (1/4\pi\varepsilon_0)[\,qa \sin\theta/c^2 r]$$

a. Directly in front of the screen, $\theta = 0$ (or 180°), and therefore $E = B = 0$.
b. To one side of the screen, $\theta = 90°$, and $\sin\theta = 1$, so that

$$E = (9 \times 10^9)(1.6 \times 10^{-19})(3.846 \times 10^{18})/(9 \times 10^{16})(0.3)$$
$$= 2.05 \times 10^{-7} \text{ V/m}$$

and since $B = E/c$, $B = 6.83 \times 10^{-16}$ T. The fields are sketched in Figure 35.10.

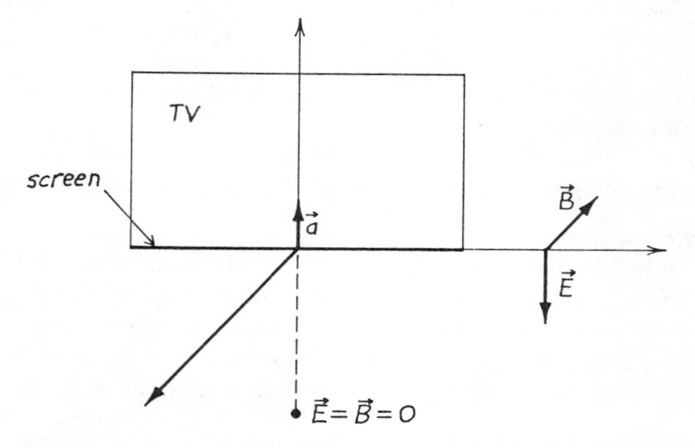

Figure 35.10 Problem 9. The electric and the magnetic radiation fields created by an electron accelerated in a TV tube.

10. *Guided Problem.* How would you expect Maxwell's equations to be modified for the case of a material medium? Write down Maxwell's equations for a medium with dielectric constant κ and relative permeability constant κ_m.

Solution scheme

a. The electric field in a material with dielectric constant κ is modified by $\mathbf{E} = \mathbf{E}_{\text{free}}/\kappa$, so that $\oint \kappa\mathbf{E} \cdot d\mathbf{S} = Q/\varepsilon_0$. We thus need to replace $\mathbf{E}_{\text{free}}$ in Gauss' Law by $\kappa\mathbf{E}$.

b. Gauss' Law for magnetism, $\oint \mathbf{B} \cdot d\mathbf{S} = 0$ is really a statement that magnetic monopoles do not exist. Would you expect this law to be changed in a magnetic material?

c. Faraday's Law of induction holds true both for vacuum and for material media.

d. Remember that Ampère's Law (without displacement current), $\oint \mathbf{B} \cdot d\mathbf{l} = \mu_0 I$ had to be modified for a magnetic material, since the free current and the current in the magnetic material produce a total field $\mathbf{B} = \kappa_m \mathbf{B}_{\text{free}}$. Therefore, Ampère's Law becomes

$$(1/\kappa_m)\oint \mathbf{B} \cdot d\mathbf{l} = \mu_0 I_{\text{free}}$$

However, the displacement current $\varepsilon_0 d\phi_E/dt$ must also be modified. Since $\kappa E = E_{\text{free}}$, the modified displacement current is $\kappa\varepsilon_0 d\phi_E/dt$. Can you now write the modified Ampère's Law, with displacement current?

Answers to Guided Problems

2. The magnetic field at point P is 2.66×10^{-8} T. The answer is independent of which of the two surfaces over which you choose to integrate the right side of Ampère's Law.
4. The angular frequency, $\omega = 3.0 \times 10^{10}$ radians/s. The frequency is 4.77×10^9 Hz.
   At $r = 0$, $E_0 = 5 \times 10^4$ V/m, where E_0 is the amplitude of the field.
   At $r = R$, $E_0 = 0$.
   At $r = 0$, the amplitude of the magnetic field is zero.
   At $r = R$, the amplitude of the magnetic field is 1.67×10^{-4} T.
6. The displacement current $I_d = 0.09 \sin 22.4(10^4)t$ amperes. This is the same as the real current I in the inductance.
8. a. The acceleration of the free electrons is given by $a = 1.76 \times 10^7 \cos 3 \times 10^6 t$ m/s^2.
   b. At 1.0 km from the antenna, the amplitude of the electric field is 512 V/m.
   c. The range at which the receiver can pick up the signal is 512 km.

10. $$\oint \kappa\mathbf{E} \cdot d\mathbf{s} = \frac{\phi}{\varepsilon_0}$$

$$\oint \mathbf{B} \cdot d\mathbf{s} = 0$$

$$\oint \mathbf{E} \cdot d\mathbf{l} = -d\phi_B/dt$$

$$\frac{1}{\kappa_m}\oint \mathbf{B} \cdot d\mathbf{l} = \mu_0 I + \mu_0 \kappa \varepsilon_0 \frac{d\phi_E}{dt}$$

Chapter 36

Light and Radio Waves

Overview

We saw in the last chapter that when a charge is accelerated, it emits a transverse electric field, or radiation field, that moves outward from the accelerated charge with speed c. In this chapter we study radiation fields in more detail, and we find that they have a magnetic component as well as an electric component, with the two mutually perpendicular and in phase. The overall field is called an electromagnetic wave. Radio waves and light waves are electromagnetic waves of this kind, and their behavior can be predicted by Maxwell's equations. In this chapter we investigate some of the general properties of electromagnetic waves.

Essential Terms

electromagnetic wave
plane wave
polarization
energy flux
Poynting vector
radiation pressure; momentum of a wave
Doppler shift of light

Key Concepts

1. *Electromagnetic wave.* In the last chapter we found that the field of an accelerated charge is given by

$$E_\theta(r,t) = \frac{1}{4\pi\varepsilon_0}\frac{q\sin\theta}{c^2 r}\,a\,(t - r/c)$$

so that, at a fixed angle θ, $E(r,t) = [\text{const}]\ a(t - r/c)/r$, where the direction of E is perpendicular to the direction of propagation. The magnitude of the magnetic field, which is perpendicular to both E and the direction of propagation, is $B(r,t) = E(r,t)/c$. The electric and magnetic field vectors at a distance r from an accelerated charge are indicated in Figure 36.1.

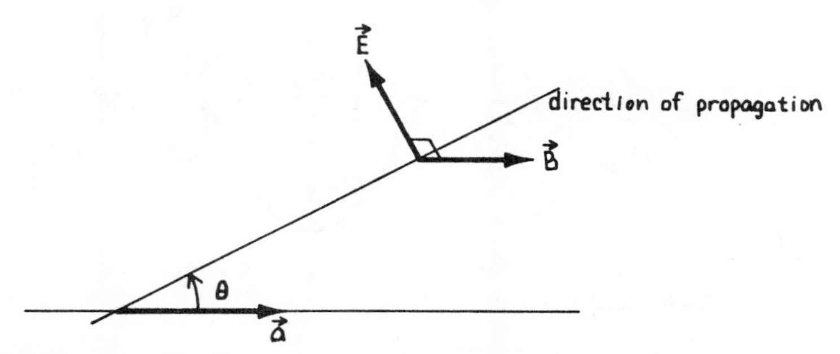

Figure 36.1 The electric and magnetic field vectors at a distance r from an accelerated charge.

If the acceleration of the charge is simple harmonic, as is often true in practice, then the acceleration a = [constant] sin ωt, so that

$$E(r,t) = [\text{constant}]\ \frac{\sin \omega(t - r/c)}{r}$$

For a range of distances small compared with $1/r$, we can treat $1/r$ as a constant also, so that

$$E(r,t) = [\text{constant}]\ \sin \omega\left(t - \frac{r}{c}\right) = E_0 \sin \left(\omega t - \frac{\omega r}{c}\right)$$

Let us choose the x direction as the direction of propagation; then, if the electric field is along the y axis,

$$E_y(x,t) = E_0 \sin \left(\omega t - \frac{\omega x}{c}\right)$$

$$B_z(x,t) = B_0 \sin \left(\omega t - \frac{\omega x}{c}\right)$$

with B_0 equal to E_0/c.

These are the equations of a *plane electromagnetic wave* traveling along the x axis. The waveform is shown graphically in Figure 36.2. We could also have chosen the direction of the electric field to be along the z axis, in which case the magnetic field would be parallel to the y axis.

In Figure 36.2 the electric field is always either parallel or antiparallel to the y axis, and consequently the electromagnetic wave is said to be *polarized* along the y axis. The magnetic field is always either parallel or antiparallel to the z axis. The cross product $\mathbf{E} \times \mathbf{B}$ is parallel to the direction of propagation, so that if we had chosen the direction of the electric field to be along the z axis, the magnetic field would be pointed along the *negative* y axis. The wavelength is λ, and the time needed for a complete wavelength to pass

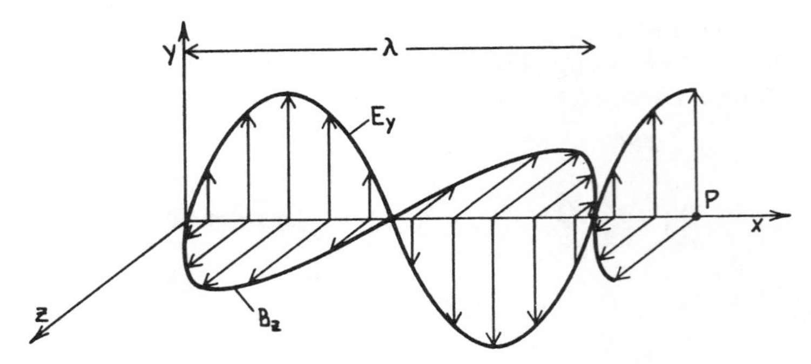

Figure 36.2 A plane electromagnetic wave traveling along the x axis. P is any point on the x axis.

some point P is T. Then $T = \lambda / c = 1/f$, where f is the number of wavelengths passing point P per second. f is usually called the *frequency* of the radiation. Then $1/f = \lambda / c$, or, as it is usually written, $c = f\lambda$. It is usual to define a *wave number, k,* where $k = 2\pi / \lambda$, so that

$$\omega x / c = 2\pi f x / c = kx$$

Then the plane wave equations can be written as

$$E_y(x,t) = E_0 \sin(\omega t - kx)$$
$$B_z(x,t) = B_0 \sin(\omega t - kx)$$

Although we have stated without proof that the amplitude of the magnetic component of an electromagnetic wave $B_0 = E_0 / c$ and that the velocity of propagation of the wave in vacuum is c, the speed of light, it is fairly easy to show that both of these statements are consistent with, and in fact predicted by, Maxwell's equations. If $E_y(x,t)$ and $B_z(x,t)$ are electromagnetic waves, then necessarily $B_0 = E_0 / c$ and the velocity of propagation in vacuum is $c = 1/\sqrt{\mu_0 \varepsilon_0}$.

Electromagnetic radiation has an enormous range of wavelength, varying from radio waves hundreds of kilometers in length down to high-energy gamma rays of wavelength of the order of 10^{-15} m or less. Visible light, whose wavelength is about 5×10^{-7} m, is approximately in the middle of the electromagnetic spectrum.

2. *Energy flux.* The energy densities of electric and magnetic fields are, respectively,

$$u_E = \frac{1}{2} \varepsilon_0 E^2 \quad \text{and} \quad u_B = \frac{1}{2} \frac{B^2}{\mu_0}$$

so that we expect the energy density of an electromagnetic wave to be

$$u = u_E + u_B = \frac{1}{2} \varepsilon_0 E^2 + \frac{1}{2} \frac{B^2}{\mu_0}$$

or, since $B = E/c$,

$$u = u_E + u_B = \tfrac{1}{2}\varepsilon_0 E^2 + \tfrac{1}{2}\varepsilon_0 E^2 = \varepsilon_0 E^2 \ \mathrm{J/m^3}$$

Thus, field energy is carried along as the electromagnetic wave propagates.

Suppose that we have a plane electromagnetic wave moving parallel to the x axis. Let us calculate the energy flux through a thin, rectangular slab of thickness Δx and area A lying in the y–z plane, as shown in Figure 36.3. In

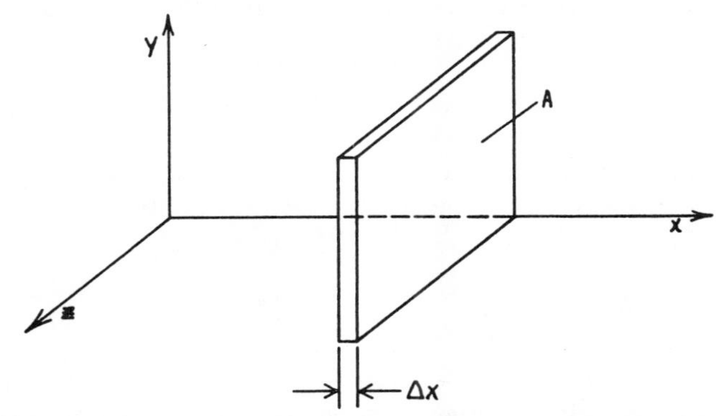

Figure 36.3 A thin, rectangular slab through which an electromagnetic wave is passing. SA is the power in watts through the slab.

time Δt the wave front moves a distance $\Delta x = c\Delta t$, so that all energy in a volume $A\Delta x$ passes through A. If S is the energy flux in watts per square meter past any point on the x axis, the energy flux through the slab is $SA = uA\dfrac{\Delta x}{\Delta t} = uAc$. But $u = \varepsilon_0 E^2$, so that $S = \varepsilon_0 cE^2 = \varepsilon_0 cE(cB) = EB/\mu_0$. Written in vector form,

$$\mathbf{S} = \frac{1}{\mu_0}\mathbf{E} \times \mathbf{B}$$

S is called the *Poynting vector* and is parallel to the direction of propagation, which it must be, because energy is carried along by the wave. If we write S in the form

$$S = \varepsilon_0 cE^2 = \varepsilon_0 cE_0^2 \sin^2(\omega t - kz)$$

we see that S oscillates between zero and $\varepsilon_0 cE_0^2$. The time-average energy flux is

$$\frac{\varepsilon_0 cE_0^2}{2} = \frac{E_0^2}{2\mu_0 c}$$

3. *Momentum of a wave.* Suppose that an electromagnetic wave moving along the x axis interacts with a charged particle, giving the particle an energy dW. We can show that the particle also acquires a momentum $dp_x = dW/c$. If the wave were to be completely absorbed by a cloud of particles, the particles would acquire an energy U and momentum $p_x = U/c$. Therefore, an

electromagnetic wave carries momentum U/c as well as energy U. Since force equals the rate of change of momentum, that is, $F = dp/dt$, it must be true that if an electromagnetic wave strikes a piece of material of cross-sectional area A and is absorbed by it, the wave exerts a force on the material given by

$$F_x = \frac{dp_x}{dt} = \frac{1}{c}\frac{dU}{dt}$$

where $dU/dt = SA$, the energy flux passing through the material. Then

$$F_x = 1/c(SA) = \frac{1}{c}\frac{EBA}{\mu_0}$$

so that the *pressure* on the material $F_x/A = p_x = S/c = EB/\mu_0 c$. p_x is called the radiation pressure. The time average of the radiation pressure is

$$\bar{p}_x = \frac{\bar{S}}{c} = \frac{E_0^2}{2\mu_0 c^2}$$

If an electromagnetic wave is completely reflected by an object, the force and pressure are twice as great, because the motion of the wave is completely reversed.

4. *Doppler shifts of light*. If a wave source and a receiver are moving toward each other, the observed frequency is greater than the source frequency; when they are moving away from each other, the observed frequency is less than the source frequency. This effect is called the *Doppler effect*. For electromagnetic radiation, if the source is receding from the receiver, the frequency seen by the receiver, ν, is given by

$$\nu = \sqrt{\frac{1 - u/c}{1 + u/c}}\,\nu_0$$

where ν_0 is the frequency emitted by the source and u is the speed of the source relative to the receiver. If the source is approaching the receiver, then

$$\nu = \sqrt{\frac{1 + u/c}{1 - u/c}}\,\nu_0$$

For $u \ll c$, these formulas reduce to

$$\nu = \left(1 \pm \frac{u}{c}\right)\nu_0$$

The derivation of these formulas is a straightforward extension of the formulas for the Doppler shift of sound waves. For sound waves in *air*, if the source is receding from the receiver, the apparent frequency is

$$\nu = \left(\frac{1}{1 + \frac{u}{v}}\right)\nu_0$$

where υ is the speed of the sound wave. For light waves, when $\upsilon = c$, the source is itself subject to time dilation effects, however. The frequency of the source in the frame of reference of the receiver is not ν_0 but $\nu_0\sqrt{1 - u^2/c^2}$, since moving clocks tick more slowly than fixed clocks. Therefore, if the source is receding from the receiver, the resulting frequency seen by the receiver is

$$\nu = \frac{\sqrt{1 - u^2/c^2}}{(1 + u/c)}\,\nu_0$$

which reduces to

$$\nu = \sqrt{\frac{1 - u/c}{1 + u/c}}\,\nu_0$$

Since most stars are receding from us, their light reaching us is lower in frequency, or "red-shifted."

Sample Problems

1. *Worked Problem.* Sunlight is normally unpolarized; that is, the electric vector may have any orientation with respect to the axis of propagation. Suppose that electromagnetic radiation has the following electric vector: $\mathbf{E}(x,t) = \hat{\mathbf{y}}E_0 \cos(\omega t - kx) + \hat{\mathbf{z}}E_0 \cos(\omega t - kx)$. What is the direction of propagation? Is this radiation polarized? Explain your answer. What is the magnetic component of the radiation? Suppose that a filter, called a polarizer, is able selectively to absorb radiation whose electric vector is parallel to the y axis without affecting components parallel to the z axis. Write down and sketch the electric and the magnetic vectors of such radiation after it passes through the filter.

Solution: The direction of propagation is the x axis. The electric vector can be rewritten as $\mathbf{E}(x,t) = (\hat{\mathbf{y}} + \hat{\mathbf{z}})E_0 \cos(\omega t - kx)$—that is, the electric vector has one component parallel to the y axis and one parallel to the z axis, as shown in Figure 36.4. The sum of the two components is a single

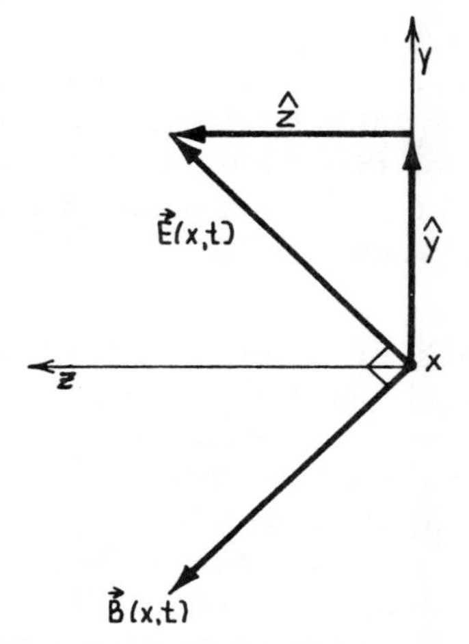

Figure 36.4 Problem 1. An electromagnetic wave polarized at 45° with respect to the y axis. The direction of propagation is along the x axis.

vector oriented at 45° with respect to the y axis, so that the radiation *is* polarized, with the direction of polarization being 45° with respect to the y axis. Since the magnetic component is perpendicular to both **E** and the direction of propagation, the magnetic vector is at an angle of 45° with respect to the *negative* y axis, as shown in Figure 36.4.

The magnetic field is given by $\mathbf{B}(x,t) = (-\hat{\mathbf{y}} + \hat{\mathbf{z}})B_0 \cos(\omega t - kx)$, where $B_0 = E_0/c$. If the filter removes the component of the electric field parallel to the y axis, the remaining electric field is given by

$$\mathbf{E}(x,t) = \hat{\mathbf{z}}E_0 \cos(\omega t - kx)$$
$$\mathbf{B}(x,t) = -\hat{\mathbf{y}}B_0 \cos(\omega t - kx)$$

with B_0 equal to E_0/c. The electric and the magnetic waves are depicted in Figure 36.5.

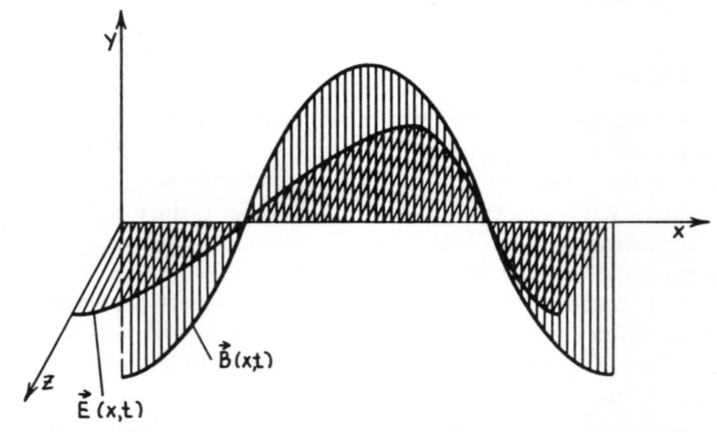

Figure 36.5 Problem 1. An electromagnetic wave, polarized in the z direction, propagating along the x axis. The magnetic field is parallel to the y axis.

2. *Guided Problem.* Write down expressions for the electric and the magnetic fields of 460 nm wavelength blue light polarized along the z axis. The amplitude of the electric vector is 1.0 V/m. Assume that the light can be represented by a plane wave. Show that at least two different propagation directions are possible, and write down expressions for **E** and for **B** for each. Calculate explicit values of amplitude, frequency, and wave number.

Solution scheme

a. In what direction is the electric vector? Choosing a direction of propagation perpendicular to **E**, write down a general expression for the electric vector **E**. For convenience choose this direction of propagation along a coordinate axis.
b. Since $\mathbf{E} \times \mathbf{B}$ is parallel to the direction of propagation, can you find the direction along which **B** must point? Knowing this, you should be able to write a general expression for **B**.
c. Suppose that you had chosen the other coordinate axis as the direction of propagation of **E**. Along which direction would **B** then

point? Be careful about the sign of the vector **B**! Now write down general expressions for **E** and for **B** for this new propagation direction.

d. Knowing the amplitude E_0, find B_0. What is the expression relating frequency and wavelength for electromagnetic radiation? Knowing λ, find both the frequency ν and the angular frequency ω. What about the wave number, k?

3. *Worked Problem.* A very large, flat copper sheet 1 mm thick carries a surface current density of 200 A/m of width parallel to the z axis, as shown in Figure 36.6. The resistivity of copper is 1.7×10^{-8} A • m. Find the electric field parallel to the current flow and the magnetic field at the surface of the copper sheet. Calculate the Poynting vector, and show that the power flowing into 1 m^2 of the sheet from the space around it is equal to the Joule heating in the sheet.

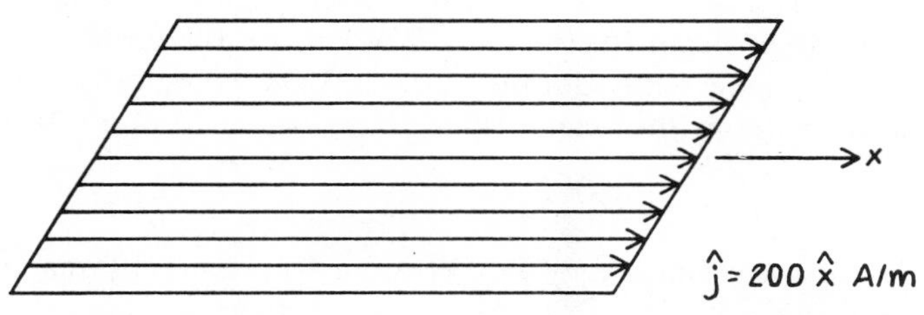

Figure 36.6 Problem 3. A very large, flat copper sheet carrying a current of 200 A/m of width parallel to the x axis.

Solution: To find the electric field parallel to the current flow, we need to find the voltage drop along the sheet and determine the field from the relation $(E)(L) = V$. Consider a piece of the sheet 1 m long and 1 m wide. Its cross-sectional area A is 1×10^{-3} m^2. Then the resistance R is given by $R = \rho L/A = 1.7 \times 10^{-8}(1)/10^{-3}$, or $R = 1.7 \times 10^{-5}$ ohm. Then the voltage drop per meter length of the sheet (which is numerically equal to the field strength) is $V = IR = 200\,(1.7 \times 10^{-5}) = 3.4 \times 10^{-3}$ V, so that $E = 3.4 \times 10^{-3}$ V/m. To find the magnetic field, we need to apply Ampère's Law, as shown in Figure 36.7. From Ampère's Law, $\int \mathbf{B} \cdot d\mathbf{l} = \mu_0 i = \mu_0(200)(w)$, where the line integral is to be evaluated around the narrow closed rectangle of width w. From the right-hand rule, we expect the field above the

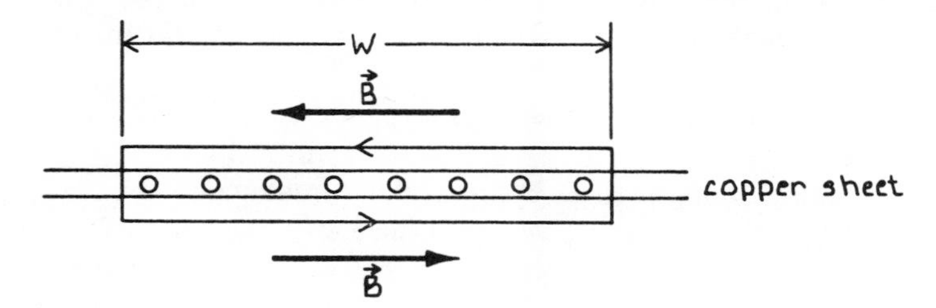

Figure 36.7 Problem 3. The application of Ampère's Law to find the magnetic field at the surface of a sheet of current. The direction of the current is out of the page. Ampère's Law is evaluated around the rectangular loop shown.

sheet to be parallel to the sheet and pointing to the left; below the sheet the field should be pointing to the right. The left side of the equation then becomes $2\ wB$, so that $2\ wB = \mu_0\ 200\ w$, or $B = 100\ \mu_0 = 1.26 \times 10^{-4}$ T. Note that B does not depend on the distance from the sheet.

The Poynting vector, **S**, is given by $\mathbf{S} = (\mathbf{E} \times \mathbf{B})/\mu_0$, so that above the copper sheet **S** points downward (into the sheet) and below the sheet **S** points upward, again into the sheet. The total power per square meter radiated into the sheet is therefore $2\ S$, where $S = EB/\mu_0$; numerically, $2S = 2(3.4 \times 10^{-3})(1.26 \times 10^{-4})/1.26 \times 10^{-6} = 0.68$ W. The Joule heating in the sheet is $I^2R = 200^2(1.7 \times 10^{-5}) = 0.68\ \text{W/m}^2$, so that the power flowing into the sheet as calculated by the Poynting vector is exactly equal to the Joule heating of the sheet.

4. *Guided Problem.* A pulsed laser has a power output, during pulses, of 1 MW, where 1 MW = 10^6 W. The pulse length is 2 μs, with 1 pulse per second on the average. What is the time-average energy flux (Poynting vector) of the laser beam *during pulses*? What are the amplitudes of the electric and magnetic fields during pulses? What pressure and force are exerted on the pellet by the beam? What is the *average* power output of the laser?

Solution scheme

a. The time-average energy flux is defined as [power]/[area] = S. Use this formula to find S.
b. What is the relation between the time-average energy flux and the amplitude of the electric vector? Deduce both the electric and the magnetic field amplitudes.
c. What is the relation between the time-average energy flux and the radiation pressure? Use this to obtain both the pressure and the force on the pellet.
d. The output of the laser is zero except for 2 μs per second, so that the average power output is reduced by the factor 2×10^{-6}. What, then, is the average power output?

5. *Worked Problem.* About 1350 W/m^2 of radiant energy reaches the upper atmosphere of the Earth. The Earth-to-Sun distance is 1.5×10^{11} m, and the Sun's radius is 7×10^8 m.

a. Estimate the luminosity of the Sun, that is, its total power output.
b. Assuming that the Earth absorbs all of the sunlight, what average force does the radiation from the Sun exert on the Earth? (The Earth's radius is 6.4×10^6 m.)
c. Does this force affect the motion of the Earth significantly? Justify your answer.
d. Assuming sunlight to be a single electromagnetic wave, calculate the amplitude of the electric vector at the Earth's surface. What would the amplitude be 5 million miles from the surface of the Sun?

Solution

a. Assuming that the Sun emits radiation uniformly in all directions, the total radiation is distributed uniformly over a sphere of radius R and surface area $4\pi R^2$, where R is the Earth-to-Sun distance, so that the Sun's luminosity, or total power output, is given by $L = 4R^2S = 4\pi(1.5 \times 10^{11})^2(1350) = 3.82 \times 10^{26}$ W.

b. The radiation pressure is given by $p = S/c = 1350/3 \times 10^8 = 4.5 \times 10^{-6}$ N/m^2. The effective area of the Earth seen by direct sunlight is the area of a disk of radius r, where r is the Earth's radius, so that the area $A = \pi r^2 = \pi(6.4 \times 10^6)^2 = 1.29 \times 10^{14}$ m^2. Then the total force exerted on the Earth, which represents a repulsive force away from the Sun, is $F = pA = 5.79 \times 10^8$ N.

c. Yes, this force is bound to affect the motion of the Earth. If we compare this force with the gravitational force of attraction between the Earth and the Sun, $F_G = GM_SM_E/R^2$, however, we find $F_G = 3.6 \times 10^{22}$ N, so that the radiation force is negligible in comparison.

d. The time-average Poynting vector $S = E_0^2/2\ \mu_0 c = 1350$ W/m^2. Therefore, we obtain $E_0 = 1008$ V/m. Five million miles, which is 8×10^9 m, must be compared with the Earth-to-Sun distance of 1.5×10^{11} m. Since the electric field strength falls off as $1/r$ as we move away from the source (the Sun's surface), we expect the electric vector at 5 million miles to be given by $E_{5\ \mathrm{million}}/E_{\mathrm{Earth}} = (1.5 \times 10^{11}/8 \times 10^9)$. Since $(E_0)_{\mathrm{Earth}} = 1008$ V/m, $(E_0)_{5\ \mathrm{million}} = 18{,}900$ V/m.

6. *Guided Problem*. A 50 cm × 40 cm aluminum sheet that is 0.1 mm thick is placed in outer space. A light beam whose electric vector $E_0 = 400$ V/m at the sheet is directed perpendicular to the surface of the sheet from a nearby space station. Assuming that the light is totally reflected from the face of the aluminum sheet, what force does the light beam exert on the sheet and what is the acceleration of the sheet? The density of aluminum is 2.7 g/cm^3. How far will the sheet, initially at rest with respect to the space station, move in the first hour after the light beam is turned on it?

Solution scheme

a. What is the expression for the time-average radiation pressure in terms of the amplitude of the electric vector? How does this change if the light wave is totally reflected? Use this expression to obtain the radiation pressure on the aluminum sheet?

b. Calculate the force exerted on the sheet, knowing the pressure and the area of the sheet.

c. What is the expression for acceleration in terms of force and mass? Calculate the mass of the aluminum sheet. You should now be able to find the acceleration of the sheet.

d. From the equation of motion, what is the expression for distance traveled in terms of acceleration and the total time? You should now be able to find the distance traveled in 1 h (3600 s).

7. *Worked Problem*. Radar using microwaves with a wavelength of 4.0 cm is used to measure the speed of a car by reflecting the waves from the car and picking up the reflected signal. A car traveling at 140 km/h is approaching the radar transmitter. By how much does the reflected signal received by the radar set differ from the transmitted signal? Is the signal that is received of higher or lower frequency than the transmitted signal?

Solution: Since u, the speed of the source relative to the observer, is much less than c, the speed of light, we can use for the Doppler shift the expression

$$\nu = \left(1 + \frac{u}{c}\right)\nu_0$$

As seen by the moving car, the microwave signal frequency ν' is

$$\nu' = \left(1 + \frac{u}{c}\right)\nu_0$$

But this same signal is reflected from the car, and so its frequency, as seen by the receiver, must also be corrected:

$$\nu = \left(1 + \frac{u}{c}\right)\nu' = \left(1 + \frac{u}{c}\right)\left(1 + \frac{u}{c}\right)\nu_0 = \left(1 + \frac{u}{c}\right)^2\nu_0$$

$$\nu \simeq \left(1 + \frac{2u}{c}\right)\nu_0$$

$$\nu - \nu_0 = \frac{2u}{c}\nu_0 = \frac{2u}{\lambda}$$

since $c = \nu_0\lambda$. The problem is indicated schematically in Figure 36.8. Because the car is approaching the receiver, the frequency is *increased*. $u = 140$ km/h $= 140(10^3)/3600 = 38.9$ m/s, and $\lambda = 0.04$ m. Therefore, $\nu - \nu_0 = 2(38.9)/0.04 = 1945$ Hz.

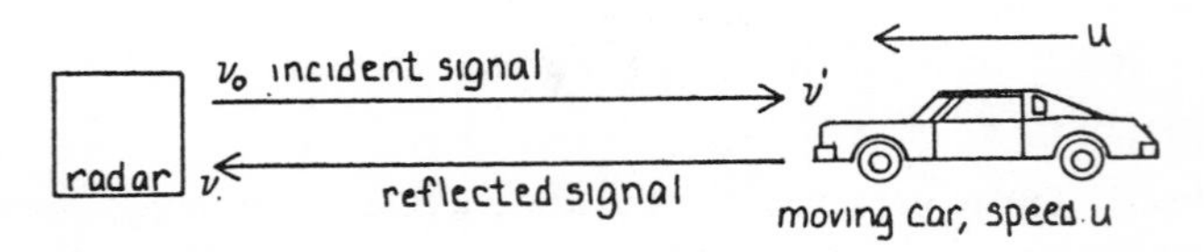

Figure 36.8 Problem 7. A radar set sends a signal (frequency ν_0) that is reflected back to the radar set by a moving car. The change in frequency is $\nu - \nu_0$.

8. *Guided Problem.* A rocket ship receding from the Earth emits a powerful flash of light of wavelength 400 nm (1 nm = 10^{-9} m). Observers back on Earth record the flash but measure its wavelength to be 650 nm. With what speed is the rocket ship receding from the Earth? What color is the light flash as seen (a) in the rocket ship and (b) back on Earth?

Solution scheme

a. Since the source is receding from the receiver, what expression for the Doppler shift of light do you use?
b. We are given the wavelength of the emitted light; convert this to the frequency ν_0.
c. We are also given the wavelength of the light seen by the receiver (observers back on Earth); can you convert this to the frequency ν?

d. Knowing both ν and ν_0 in the expression derived in step a for the Doppler shift, solve the equation for the speed of the source relative to the observer, u. It is simpler if you express your answer in terms of u/c.

e. The range of visible light is approximately 400 to 700 nm; the colors of the rainbow vary from red (longest wavelength) to violet (shortest wavelength). Guess at the colors seen in the situation in this problem.

9. *Worked Problem.* A 0.001 mm diameter dust particle of density 1.0 gm/cm^3 is out in interplanetary space. Calculate the ratio of the gravitational force of the Sun on the particle to the force due to radiation pressure exerted on the same particle. The luminosity of the Sun is 3.82×10^{26} W, and the Sun's mass is 1.99×10^{30} kg. What will happen to this dust particle?

Solution: Let the particle be a distance R from the Sun. Then the gravitational force, $F_g = GM_s m/R^2$, where m, the mass of the particle, is given by $(4/3)\pi r^3 \rho$. Then

$$F_g = GM_s 4\pi r^3 \rho / 3R^2$$

The force due to radiation pressure is given by $F_{rad} = \overline{p}_x A$, where A = area of the dust particle $= \pi r^2$, and $\overline{p}_x = \overline{S}/c$. At a distance R from the Sun (measured from the center of the Sun), $4\pi R^2 \overline{S}$ = total luminosity $= 3.82 \times 10^{26}$ W. Therefore

$$\overline{S} = 3.82 \times 10^{26}/4\pi R^2$$

We can therefore write for the ratio

$$\frac{F_g}{F_{rad}} = \frac{(GM_s\ 4\pi r^3 \rho)(4\pi R^2 c)}{(3.82 \times 10^{26})(\pi r^2)(3R^2)} = \frac{GM_s\ 16\pi\ \rho r c}{(3.82 \times 10^{26})(3)}$$

$$= \frac{(6.67 \times 10^{-11})(1.99 \times 10^{30})\ 16\pi\ (1 \times 10^{+3})(.5 \times 10^{-6})(3 \times 10^{8})}{3\ (3.82 \times 10^{26})}$$

$$= 0.87$$

Since the value of this ratio is <1 and since this ratio is independent of R, the force on the particle due to solar radiation is less than the attractive force of gravity, and the dust particle will be blown out of the Solar System.

Answers to Guided Problems

2. Choosing the direction of propagation to be along the x axis,

$$\mathbf{E}(x,t) = \hat{\mathbf{z}} E_0 \sin(\omega t - kx)$$
$$\mathbf{B}(x,t) = -\hat{\mathbf{y}} B_0 \sin(\omega t - kx)$$

Since $\hat{\mathbf{z}} \times (-\hat{\mathbf{y}}) = \hat{\mathbf{x}}$, $\mathbf{E} \times \mathbf{B}$ is a vector parallel to the x axis. Choosing the direction of propagation to be along the y axis,

$$\mathbf{E}(y,t) = \hat{\mathbf{z}}E_0 \sin(\omega t - ky)$$
$$\mathbf{B}(y,t) = \hat{\mathbf{x}}B_0 \sin(\omega t - ky)$$

Since $\hat{\mathbf{z}} \times \hat{\mathbf{x}} = \hat{\mathbf{y}}$, $\mathbf{E} \times \mathbf{B}$ is a vector parallel to the y axis. $E_0 = 1.0$ V/m $= cB_0$, so that $B_0 = 3.33 \times 10^{-9}$ T.
$k = 2\pi/\lambda = 1.36 \times 10^7$ m^{-1}; $c = \nu\lambda$; $\nu = 6.52 \times 10^{14}$ s^{-1}; and $\omega = 4.1 \times 10^{15}$ s^{-1}.
Then $\mathbf{E}(x,t) = \hat{\mathbf{z}} \sin(4.1 \times 10^{15}\, t - 1.36 \times 10^7\, x)$.

4. $\bar{S} = 2 \times 10^{12}$ W/m^2; $E_0 = 3.88 \times 10^7$ V/m; $B_0 = 0.13$ T.
   The pressure $p = 6.67 \times 10^5$ N/m^2; the force $F = 0.33$ N; the average power $P = 2$ watts.
6. Force $F = 2.83 \times 10^{-7}$ N; the acceleration $a = 5.24(10^{-6})$ m/s^2; the distance $s = 35$ m.
8. $u = 0.45\, c$
   The light flash seen on the space ship is violet. The light flash seen back on Earth is orange.

Chapter 37

Reflection, Refraction, and Polarization

Overview

In vacuum an electromagnetic wave will propagate in a fixed direction with speed c. If the wave encounters a region filled with matter, it will, as a result of the interaction with the matter, experience changes in direction, intensity, and polarization. Although these changes can be calculated from Maxwell's equations, such calculations are usually complicated. To find the change of direction, we usually resort to a simple geometric construction, called Huygens' Construction, developed for light waves but applicable to any kind of wave. In this chapter we therefore leave, for the present, the study of electromagnetic theory and begin the study of a subtopic, the wave propagation of light. This subject is often called geometric, or ray, optics.

Essential Terms

Huygens' Construction
reflection of a light wave
refraction of a light wave
index of refraction
Snell's Law
total internal reflection
dispersion
polarization
polaroid
polarization by reflection
Brewster's angle
birefringence
Law of Malus

Key Concepts

1. *Huygens' Construction*. When we stand on a shore and watch waves roll in, it is easy to see the advancing wave crests, or wave fronts, as they are often called. Huygens' Construction, which is as true for electromagnetic waves

as for water waves, tells us that the change in position of a wave front in a time Δt can be found by drawing many small spheres of radius $\upsilon\Delta t$ with centers on the old wave front, where υ is the wave speed. The new wave front is the surface tangent to the spheres. A plane wave front moving parallel to the x axis is shown in Figure 37.1. The vertical line on the left shows the position of the wave front at time t; the line a distance $c\Delta t$ to its right shows the position of the wave front at time $t + \Delta t$. A *ray* is a line drawn perpendicular to the wave front; it is used to indicate the direction of propagation of the wave front. In Figure 37.1 the x axis is a ray.

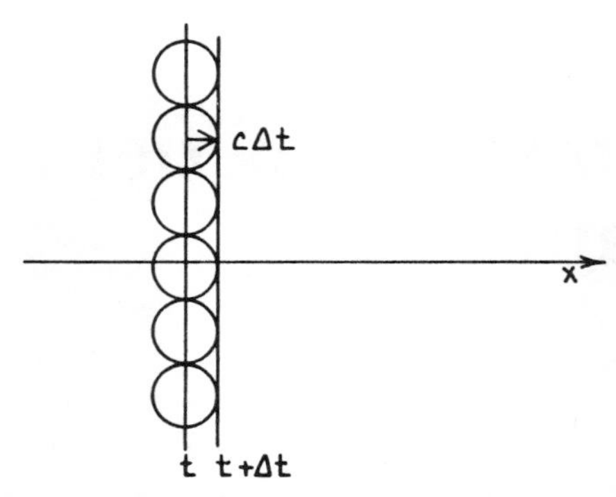

Figure 37.1 A wave front moving parallel to the x axis. Huygens' Construction is used to show the new position of the wave front at time $t + \Delta t$.

2. *Reflection of a light wave*. When a light wave strikes a surface, part of the wave is reflected; the law of reflection tells us that the angle of incidence equals the angle of reflection. Figure 37.2 shows a wave front approaching a reflecting surface. Again we apply Huygens' Construction to the wave front to find the new position of the wave front at time $t + \Delta t$. Point A of the wave front strikes the surface at time t and is reflected. Point B of the wave front strikes the surface at time $t + \Delta t$; the total wave front at time $t + \Delta t$ is partly an incident wave front (BQ) and partly a reflected wave front (PB).

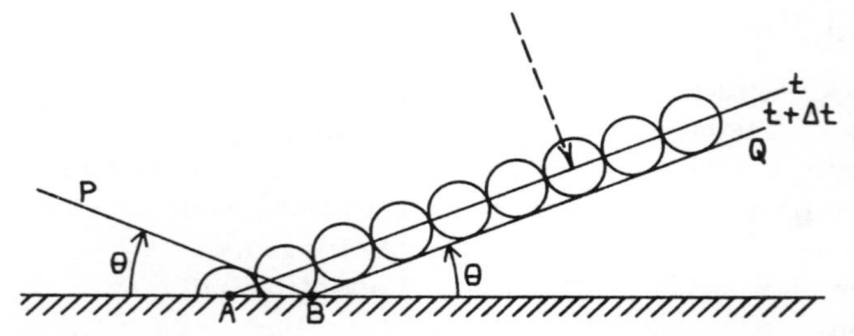

Figure 37.2 Huygens' Construction used to show that the angle of incidence of a light wave equals the angle of reflection. The dashed arrow is used to indicate the direction of the incident wave front.

From the geometry the angle θ of the incident wave front can be seen to equal the angle θ of the reflected part. The rays associated with the incident and the reflected wave fronts are shown in Figure 37.3.

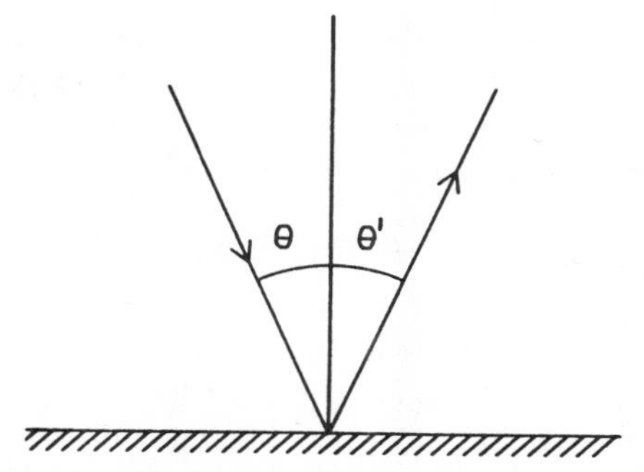

Figure 37.3 The angle of incidence, θ, is equal to the angle of reflection, θ'.

When light from a source strikes a plane mirror, the reflection of the light leads to the formation of an image of the source, as is indicated in Figure 37.4. The object is a small candle. A ray from the candle flame is reflected from the mirror into the observer's eye. A second ray, from the base of the candle, is also reflected into the eye of the observer. To the observer the rays appear to originate from behind the mirror, even though they do not. The image appears the same distance behind the mirror that the object is in front of it. The image is said to be *virtual* (the light rays do not actually pass through the image; they just appear to do so), *erect* (as opposed to inverted), and *reversed*. To see what reversed means, look at the image of some large printing in a mirror!

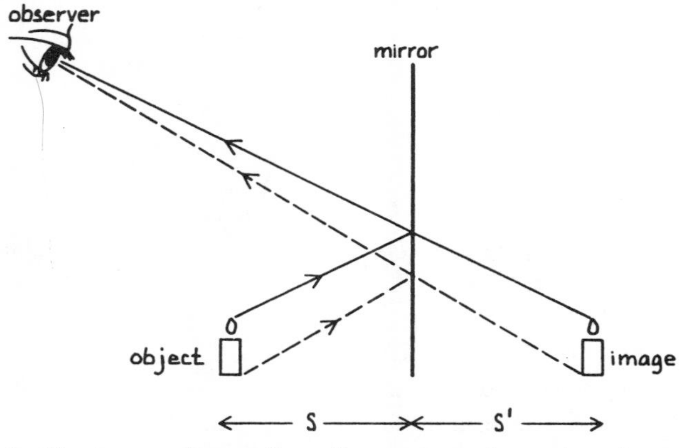

Figure 37.4 The image of a small candle seen in a plane mirror by an observer. Even though the light rays seem to come directly from the image, they originate from the object and are reflected at the mirror surface into the eye of the observer.

3. *Refraction of a light wave.* Maxwell's equations tell us that the speed of light in a substance (such as glass) is less than the speed of light in a vacuum. For a vacuum, Maxwell's equations tell us that $c = 1/\sqrt{\mu_0\varepsilon_0}$. For a material medium, the speed of light is given by

$$\upsilon = 1/\sqrt{\kappa\kappa_m\mu_0\varepsilon_0} = (1/\sqrt{\kappa\kappa_0})(1/\sqrt{\mu_0\varepsilon_0}) = (1/\sqrt{\kappa\kappa_m})c = c/n$$

where $n = \sqrt{\kappa\kappa_m}$, and is called the index of refraction. For most materials, $\kappa_m \simeq 1$, so that $n \simeq \sqrt{\kappa}$, although we cannot use the value of κ we used in our study of electrostatics, because we are now interested in the value of κ for the high-frequency light waves. The relation between frequency and wavelength for a light wave now becomes $\upsilon = \nu\lambda = c/n$, so that $\lambda = \dfrac{c}{n}\dfrac{1}{\nu}$.

Although the *wavelength* of light changes when it enters a material medium from vacuum, its *frequency* does not. We already know that when a light wave strikes a surface, part of it is reflected. The rest of the wave is *refracted*, which means that the wave changes direction as it enters the medium. Again, we can use Huygens' Construction to show that the reduction in the speed of propagation as the wave enters the medium causes the wave front to swing around, changing its direction. We find that the angle of incidence and the angle of refraction are related according to $\sin\theta = n \sin\theta'$, as indicated in Figure 37.5. This relation is called *Snell's Law*.

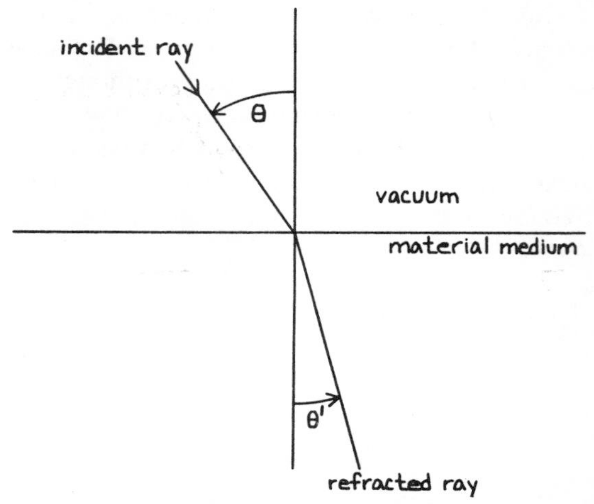

Figure 37.5 The refraction of a ray of light as it enters a material medium from vacuum.

In the general case, when two dielectrics are involved with respective indices of refraction n_1 and n_2, Snell's Law becomes

$$n_1 \sin\theta_1 = n_2 \sin\theta_2$$

where θ_1 and θ_2 are the angles between the rays and the normal to the interface between the two dielectrics. This relation is indicated in Figure 37.6. The two parts of the figure are identical, except that in the second the direction of the arrows is reversed; if light can go in one direction through a medium, it can also go in the opposite direction.

It is also a general rule that when light enters a medium with a higher index of refraction, it is bent toward the normal [Figure 37.6(a)]; when it enters a medium with a lower index of refraction, it is bent away from the normal [Figure 37.6(b)]. For the special case in which the angle θ_1 in Figure

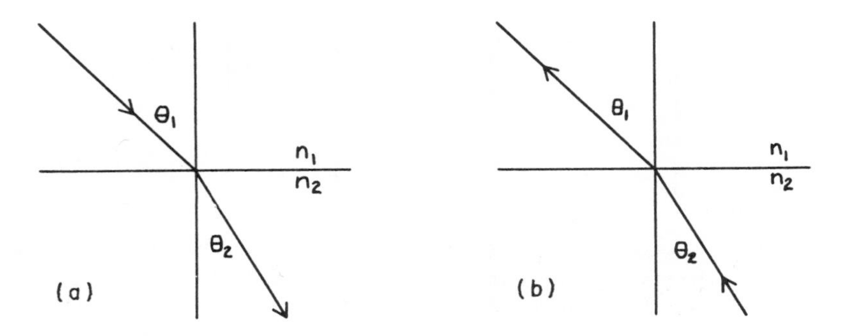

Figure 37.6 (a) The refraction of a light ray at the interface of two materials with respective indices of refraction n_1 and n_2. n_2 is larger than n_1. (b) The light ray is traveling in the opposite direction, so that it is entering a material with a smaller index of refraction.

37.6(b) is equal to or greater than 90°, the light cannot enter the second medium and is said to be *totally internally reflected* (Figure 37.7). Total internal reflection occurs for values of θ_2 equal to or greater than that given

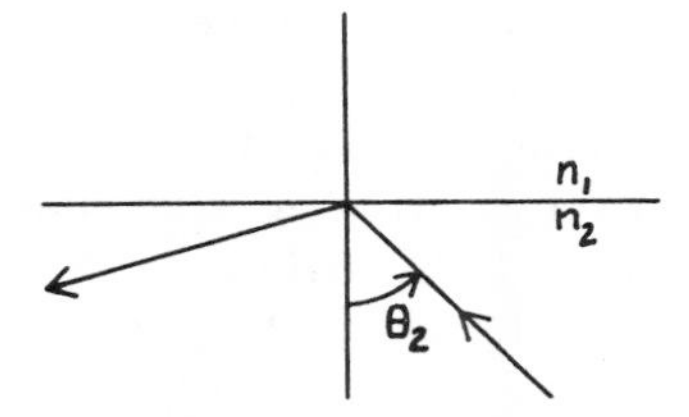

Figure 37.7 Total internal reflection of a ray of light.

by $\sin \theta_2 = n_1 / n_2$. For an air–water interface $n_2 = 1.33$ and $n_1 = 1.0$, so that $\theta_2 = 49°$.

To observe the effect of total internal reflection, look up at the water surface through the bottom of a jar of water, or, while swimming underwater (with goggles), look up at the surface. There is only a fairly narrow cone through which you can see above the surface of the water.

4. *Polarization*. The direction of *polarization* of an electromagnetic wave is defined to be the direction of the electric field. We saw in Chapter 35 that the direction of the electric field of an accelerated charge lies in the same plane as the direction of motion of the accelerated charge itself. For example, the radio waves emitted at right angles from a straight antenna, in which the currents, and charges, move up and down the antenna, are polarized in the direction parallel to the antenna (Figure 37.8).

Any individual light wave is always polarized in a definite direction, but the directions of vibrations (accelerations) of the atoms of most light sources, such as the Sun or an electric light bulb, are randomly oriented over all directions. Consequently, the directions of the electric field vectors are also randomly oriented, and the light beam, which consists of a large number of light

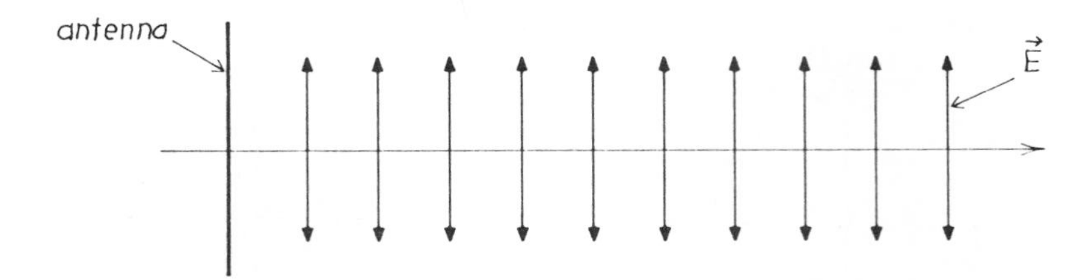

Figure 37.8 The wave emitted at right angles from a straight antenna is polarized in the direction parallel to the antenna. Since the charges are accelerated in both the "up" and the "down" directions, the electric field oscillates between positive and negative values.

waves, is said to be *unpolarized*. That is, there is no preferred direction for the electric vectors.

To polarize this kind of light, we must pass it through a *polarizer*, which selects only those components of the electric vectors parallel to a preferential direction of polarization. Components perpendicular to this direction are blocked.

5. *Polarizers*. A sheet of *polaroid* contains long chains of organic molecules aligned parallel to each other over the sheet. Electric field vectors parallel to the direction of these chains are absorbed; field vectors perpendicular to these molecular chains can pass through the polaroid sheet.

We can also polarize light by reflection. When light is incident on a transparent material, some portion of the light is reflected; the remainder is refracted. The fraction of light reflected depends on the polarization of the light. The reflection is always stronger for light rays polarized at right angles to the plane of incidence (the plane defined by the incident and reflected rays). At a critical angle, called *Brewster's angle*, the surface of the transparent material does not reflect any of the light polarized in the plane of incidence. The reflected light is therefore completely polarized at right angles to the plane of incidence, as is indicated in Figure 37.9. The value of the critical angle, θ_B, is given by *Brewster's Law*: $\tan \theta_B = n$. Thus, for glass with $n = 1.5$, Brewster's angle is $\theta_B = 56°$.

The phenomenon of double refraction, or *birefringence*, displayed by calcite, quartz, and some other crystals, can also be used to polarize light. For these crystals, the indices of refraction are different for light polarized in the plane of incidence than for light polarized perpendicular to the plane of incidence. Therefore, an unpolarized light ray incident on the crystal is split into two rays inside the crystal whose polarizations are at right angles to each other, as is indicated in Figure 37.10.

Two polarizers in tandem can be used to demonstrate clearly the blocking of light by a polarizer.

Figure 37.11 shows the two polarizers. The lines drawn on the polarizers are used to indicate the preferred axis of each polarizer, namely, the preferential direction of polarization for which waves will pass freely through the polarizer. Unpolarized light is incident on the first polarizer, which selects waves of vertical polarization and allows them to pass; the light emerging from this polarizer is vertically polarized. The electric field vector of light incident on the second polarizer, E_0, makes an angle ϕ with the preferential

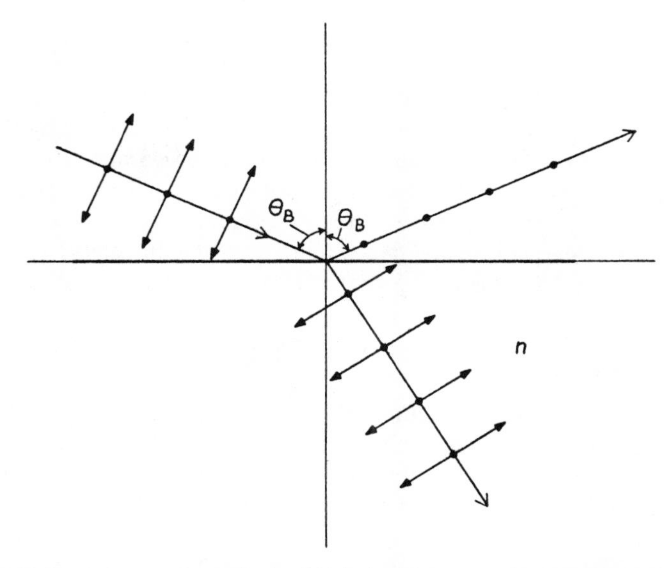

Figure 37.9 A beam of unpolarized light incident on a glass surface at Brewster's angle. The reflected light is polarized at right angles to the plane of incidence.

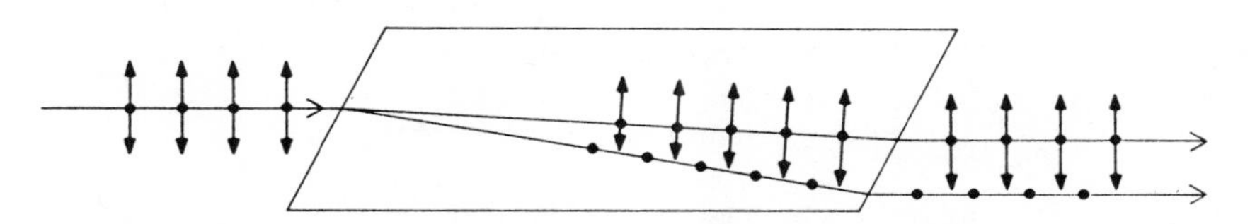

Figure 37.10 Refraction inside a calcite crystal of an unpolarized incident ray of light. The crystal splits the incident light into two refracted rays, with different polarizations.

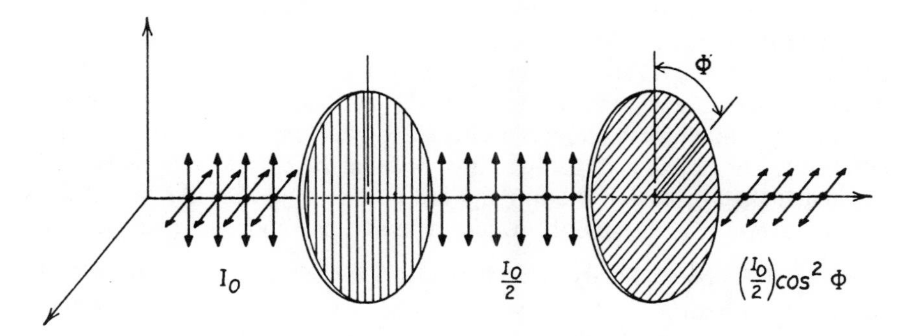

Figure 37.11 Two polarizers in tandem. Unpolarized light of intensity I_0 is incident on the left polarizer. The intensity of the light transmitted through the second polarizer is $(I_0/2)\cos^2\phi$.

direction. The amplitude of the wave parallel to the preferred axis is then $E_0 \cos \phi$. Since the intensity is proportional to the square of the amplitude, the intensity of the wave transmitted by the second polarizer is then proportional to $E_0^2 \cos^2 \phi$. That is, the transmitted intensity is reduced by the factor $\cos^2 \phi$. This relation is called the Law of Malus. Note that if $\phi = 90°$, all of the incident light is blocked. A single polarizer will reduce the intensity of incident unpolarized light by a factor of 2, since unpolarized light can be regarded as a superposition of two waves of equal intensity, respectively parallel and perpendicular to the preferred direction of a polarizer.

Sample Problems

1. *Worked Problem.* A small object is located near a pair of mirrors placed at 90° relative to each other, as indicated in Figure 37.12. An observer looks at the images of the object in the mirror. If the object is 4 cm from one mirror and 5 cm from the other, where do the images appear to the observer?

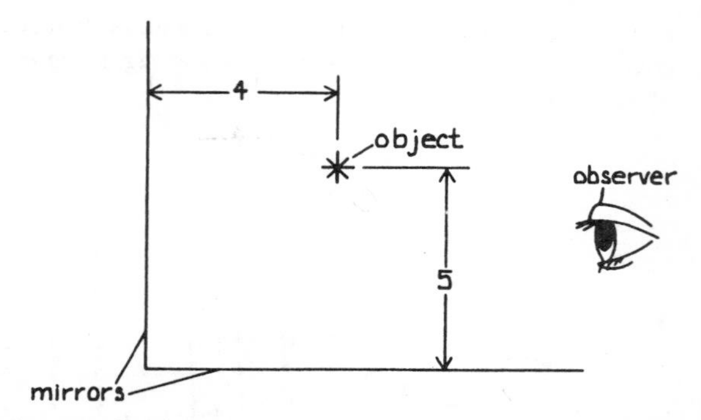

Figure 37.12 Problem 1. An object located near two flat mirrors placed at 90° to each other.

Solution: The observer will see three images, one each from the single reflections from the vertical mirror and from the horizontal mirror, and one from the double reflection in both mirrors. The third image is caused by the reflection of the image of the object in the second mirror. The image in the vertical mirror is located 4 cm to the left of the mirror, as is shown in Figure 37.13(a). In the horizontal mirror, the image is 5 cm below the mirror, as shown in Figure 37.13(b). The position of the image caused by double reflection is indicated by the ray diagram in Figure 37.13(c). The object and the three images form the corners of a rectangle, so that if the corners of the mirrors are at the origin and the object has coordinates (4,5), the doubly reflected image has coordinates (−4, −5).

2. *Guided Problem.* Sandy, who is 5 feet 8 inches tall, wants to order a "full-length mirror" for her new apartment. How long does the mirror need to be, and where should the top of the mirror be placed on her wall?

Solution scheme

a. Make a sketch of a person looking at a long mirror.
b. Now use ray tracing. If a person wants to see the top of her head in

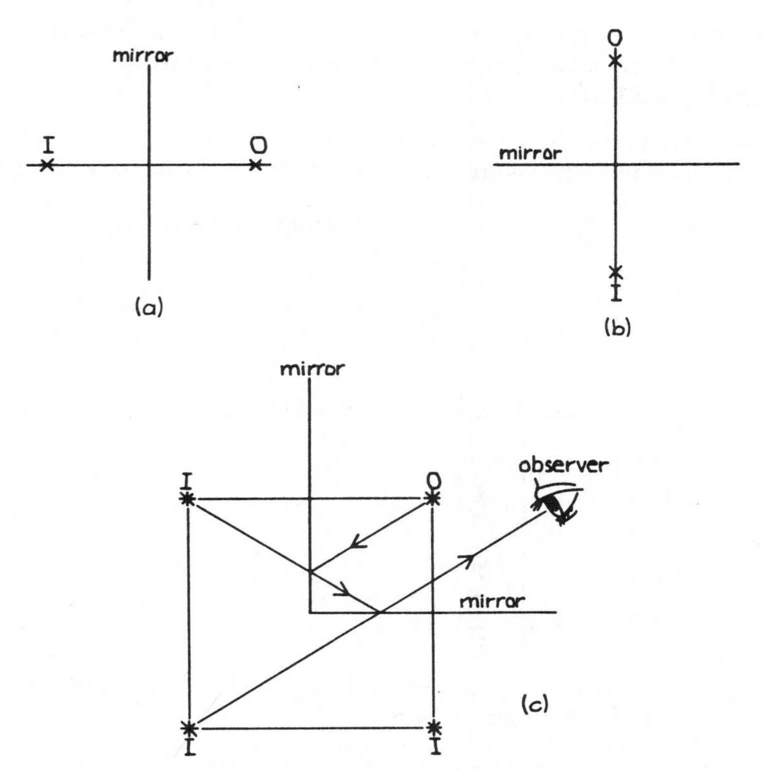

Figure 37.13 Problem 1. The three images formed of an object located near two plane mirrors at right angles to each other. In (c) the doubly reflected ray from the mirror is parallel to the ray incident on the first mirror.

the mirror, where do the light rays from the top of her head strike the mirror before being reflected back into her eyes? Remember that the angle of incidence equals the angle of reflection. You should now be able to decide on the minimum height of the *top* of the mirror.

c. If a person wants to see her feet in the mirror, where do the light rays from her feet need to strike the mirror before being reflected back upward into her eyes? Again use ray tracing on your sketch. Does the mirror have to extend any lower than this point you have just located? You should now be able to decide where the *bottom* of the mirror must be.

d. Knowing the positions of the top and the bottom of the mirror, you can decide on its length. To answer the question *exactly*, you would need to know the distance between a person's eyes and the top of her head, but you can make a good guess of this distance. Does your answer depend on how far the person stands in front of her mirror?

3. ***Worked Problem.*** A hawk flying over the surface of a lake sees a fish almost directly below it. The fish appears to be swimming at a depth of 50 cm. How far below the surface is the fish actually swimming? To the fish

the hawk appears to be flying at a height of 10 m above the surface of the lake. What is the actual height of the hawk above the surface? (The index of refraction $n = 1.33$ for water.)

Solution: To answer the first question, we sketch the light rays reaching the hawk from the fish (Figure 37.14). The arrows indicate the actual path

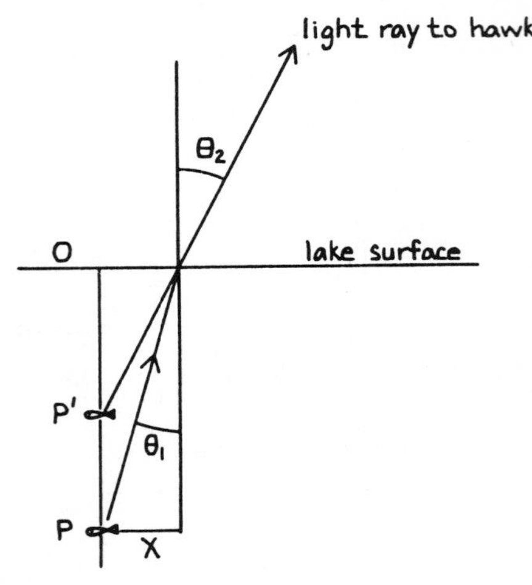

Figure 37.14 Problem 3. The true (P) and apparent (P') position of a fish to a hawk above the surface.

of the light rays from the fish to the hawk, so that although the hawk thinks the fish is at a depth OP', the fish is actually at a depth OP. The fish is almost directly below the hawk, which means that the angles θ_1 and θ_2 are small. We apply Snell's Law, $n_1 \sin \theta_1 = n_2 \sin \theta_2$, where $n_1 = 1.33$ and $n_2 = 1.0$, so that $1.33 \sin \theta_1 = \sin \theta_2$, or $\sin \theta_1 / \sin \theta_2 = 1/1.33$. But $\sin \theta_1 \simeq \tan \theta_1$ and $\sin \theta_2 \simeq \tan \theta_2$. We then have

$$\frac{\tan \theta_1}{\tan \theta_2} = \frac{x/OP}{x/OP'} = \frac{OP'}{OP} \cong \frac{1}{1.33}$$

Then, since $OP' = 50$ cm, $OP = (1.33)(50) = 66.5$ cm.

To find the actual height of the hawk we sketch the ray diagram of light rays reaching the fish from the hawk (Figure 37.15). The arrows are again used to indicate the actual path of light rays from the hawk to the fish, so that although the fish thinks the hawk is at a height OQ', the actual height is OQ. Again, from Snell's Law, $\sin \theta_1 = 1.33 \sin \theta_2$, or $1.33 = \sin \theta_1 / \sin \theta_2$. Since the angles are small,

$$1.33 = \tan \theta_1 / \tan \theta_2 = \frac{x/OQ}{x/OQ'} = OQ/OQ'$$

Then $OQ = OQ'/1.33 = 7.52$ m. So the actual height of the hawk is 7.52 m above the lake.

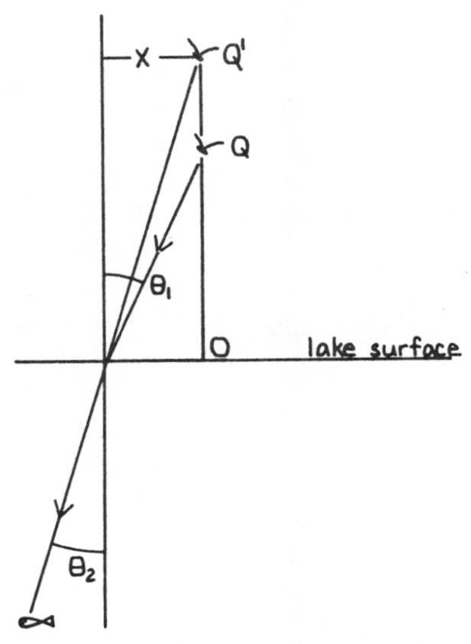

Figure 37.15 Problem 3. The true (Q) and apparent (Q') position of a hawk to a fish swimming below the surface of a lake.

4. *Guided Problem.* A laser beam is shone at normal incidence into a small prism with an apex angle of 20°, as indicated in Figure 37.16. What is the angle of deviation of the ray emerging from the far side of the prism with respect to the incident beam if the prism is made of (a) glass, with an

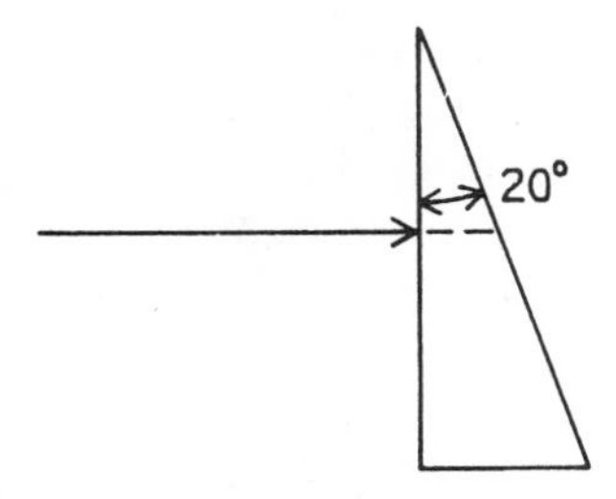

Figure 37.16 Problem 4. A beam of light shone at normal incidence into a prism with apex angle 20°.

index of refraction of 1.50, and (b) diamond, with an index of refraction of 2.42. (c) Answer the same questions if the apex angle of the prism is 25°.

Solution scheme

a. Since the light beam enters the prism at normal incidence, is it bent at that point?
b. The light beam hits the right side of the prism at some angle θ_1 relative to the normal to the surface. Draw a diagram showing θ_1 and θ_2, where θ_2 is the angle of the beam relative to the normal to the surface on *leaving* the prism. You can now apply Snell's Law (knowing n_1 and n_2) to find θ_2. From θ_2 find the angle of deviation from the incident direction.

c. You should be able to find θ_2 and the angle of deviation for both glass and diamond. Do you find a large difference between your results for the two materials?

d. Now repeat the problem for an apex angle of 25°. What do you find when you try to find θ_2 for the diamond? Physically, what does the light ray do when it hits the far side of the diamond prism?

5. *Worked Problem.* In a primary rainbow, the angle of deviation for red light is 42.3°, as indicated in Figure 37.17. For violet light it is 40.4°. A per-

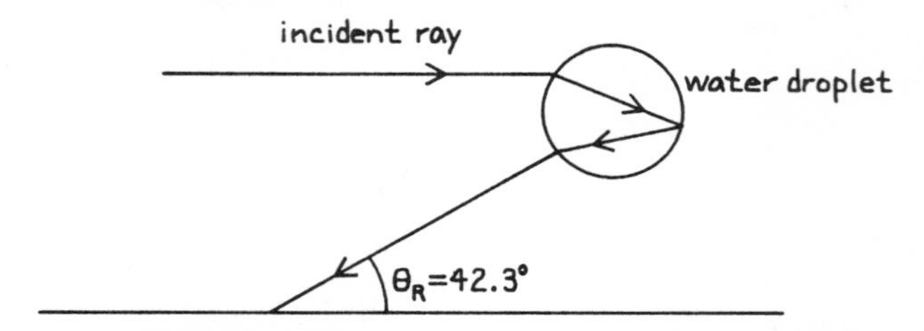

Figure 37.17 Problem 5. The refraction of red light in a water droplet to cause part of a primary rainbow.

son standing on a level plain near sunset sees a primary rainbow caused by a shower in the air 5 km away (person–to–water droplet distance). What fraction of the circular arc of the rainbow does the observer see, and what is the radius of the arc of red light in the rainbow? What is the radius of the violet arc of the rainbow?

Solution: Near sunset the sunlight incident on the water droplets is approximately parallel to the plain. The person sees the rainbow by looking back at an angle of 42.3° away from the direction of the approaching light for the red light, and at an angle of 40.4° for violet. (The other colors lie in between.) The situation is indicated schematically in Figure 37.18.

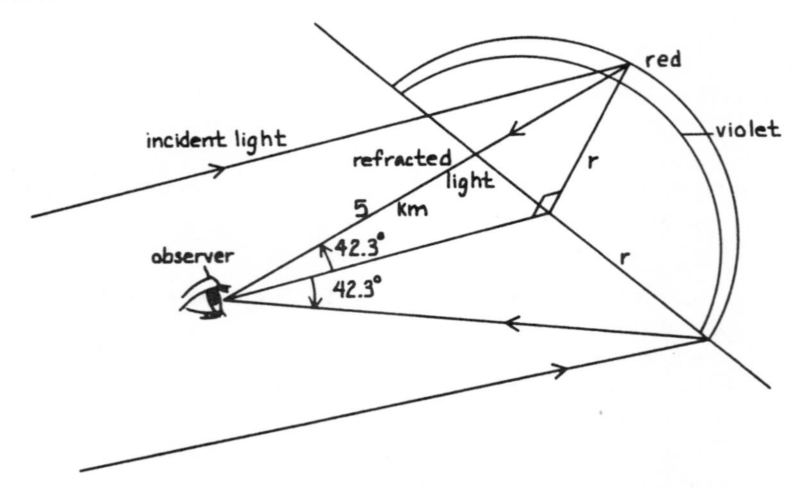

Figure 37.18 Problem 5. An observer looking at a primary rainbow. The water droplets are 5 km away from the observer.

It is clear from the figure that for a person on a level plain, the rainbow will be a semicircle, the cutoff being caused by the incident light rays striking the plain surface. If the person were *above* the surface of the plain, he or she could see *more* than a semicircular rainbow. (How high above the surface would a person need to be to see a complete circle?)

From the geometry, the rainbow radius r can be derived from $r/5 = \sin 42.3°$ for red, so that $r_{red} = 3.36$ km. For violet $r/5 = \sin 40.4°$, so that $r_{violet} = 3.24$ km.

6. *Guided Problem.* The very high index of refraction of diamond ($n = 2.42$) gives the stone its extraordinary brilliance when properly cut and polished. What is the critical angle of total internal reflection of light incident on the surface of diamond from within? Figure 37.19 shows a typical cut diamond. Show that the two light rays shown entering the top of the diamond will be reflected back out through the top surfaces of the diamond as a result of total internal reflection.

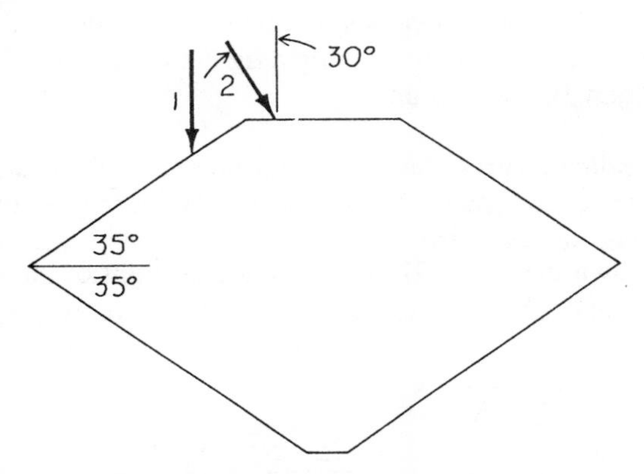

Figure 37.19 Problem 6. A typical cut diamond. The two light rays will be reflected back out the top surfaces of the diamond, contributing to the brilliance of the stone.

Solution scheme

a. Write down the criterion for total internal reflection. What value do you obtain for a diamond–air interface?
b. Trace ray 1 through the diamond by means of Snell's Law and the law of reflection. Show, using a diagram, where it leaves the diamond.
c. Do the same for ray 2. Do the rays leave the diamond through the top surfaces and travel in more or less the direction opposite to their incident direction?

7. *Worked Problem.* A glass plate ($n = 1.53$) is submerged in water ($n = 1.33$) as is shown in Figure 37.20. What should be the angle of incidence of light at the air–water interface if the light reflected at the surface of the glass plate is to be linearly polarized?

Solution: Brewster's angle is given by $\tan \theta_B = n_g/n_w = 1.53/1.33$, so that $\theta_B = 49°$. Now apply Snell's Law to the air–water interface: $\sin \theta_1$

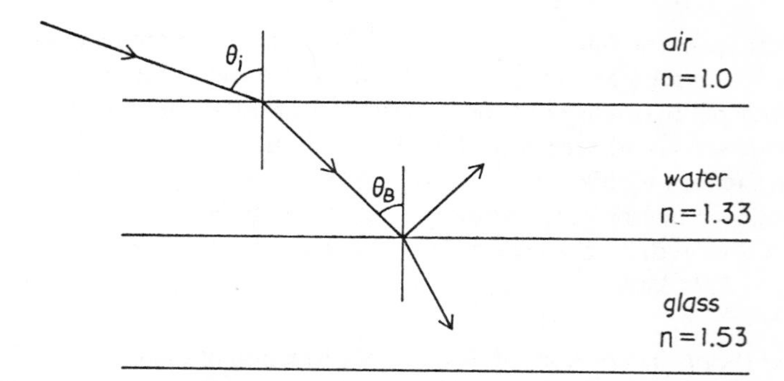

Figure 37.20 Problem 7. A glass plate submerged in water. The light reflected at the water–glass interface is linearly polarized.

$= 1.33 \sin 49°$. We obtain $\sin \theta_1 = 1.00$; that is, the incident light ray must be parallel to the surface of the water and therefore will not enter the surface of the water. All values of angle of incidence at the water–glass interface that we can obtain by shining a light source from above the water are smaller than Brewster's angle!

8. *Guided Problem.* Three sheets of polaroid are stacked such that the preferred axis of the second makes an angle of 45° with respect to the first, and the preferred axis of the third makes an angle of 45° with respect to the second, as is shown in Figure 37.21. Unpolarized light of intensity I_0 is incident on the first sheet. What is the intensity of light emerging from the third sheet?

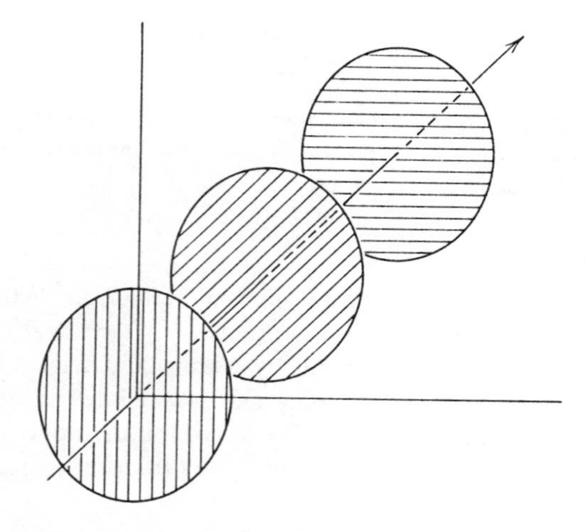

Figure 37.21 Problem 8. Three sheets of polaroid. The preferred axis of each one is at an angle of 45° with respect to the preferred axis of an adjacent sheet.

Solution scheme

a. The incident intensity of unpolarized light is I_0. Remember that the intensity of light transmitted through a polarizer is proportional to

$\cos^2 \phi$. What is the polarization of light emerging from the first sheet, and what is its intensity?

b. What is the direction of polarization, and what is the intensity, between the second and the third polarizers?

c. Can you answer these questions for the light emerging from the third polarizer?

9. *Worked Problem.* Calcite has an index of refraction of 1.658 for light waves polarized perpendicular to the optic axis (o rays) and an index of refraction for light waves polarized parallel to the optic axis (e rays) of 1.486. A prism made by cementing together two pieces of calcite with their optic axes mutually perpendicular can be used for producing beams of polarized light. In the prism shown in Figure 37.22 find the angular separation $\delta = \theta_{o2} + \theta_{e2}$ between the emergent beams for incident unpolarized light.

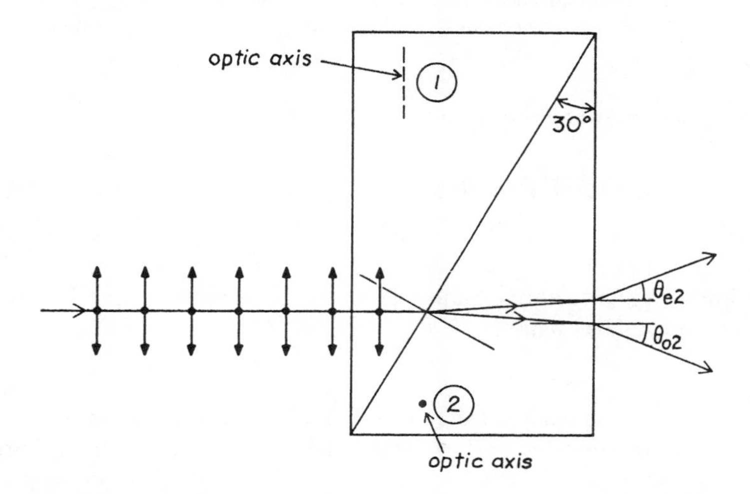

Figure 37.22 A prism made by cementing together two pieces of calcite with their optic axes mutually perpendicular.

Solution: Apply Snell's Law for both e and o rays at the interface between the two pieces of calcite. [Note that the o ray in (1) becomes an e ray in (2) and vice versa.]

$$n_o \sin 30° = n_e \sin \theta_{e1}$$
$$1.658 \sin 30° = 1.486 \sin \theta_{e1} \qquad \theta_{e1} = 33.91°$$

Similarly, $n_e \sin 30° = n_o \sin \theta_{o1}$

$$1.486 \sin 30° = 1.658 \sin \theta_{o1} \qquad \theta_{o1} = 26.62°$$

Now apply Snell's Law at the calcite–air interface of (2), as shown in Figure 37.23.

$$1.486 \sin 3.91° = \sin \theta_{e2} \qquad \theta_{e2} = 5.82°$$
$$1.658 \sin 3.38° = \sin \theta_{o2} \qquad \theta_{o2} = 5.61°$$

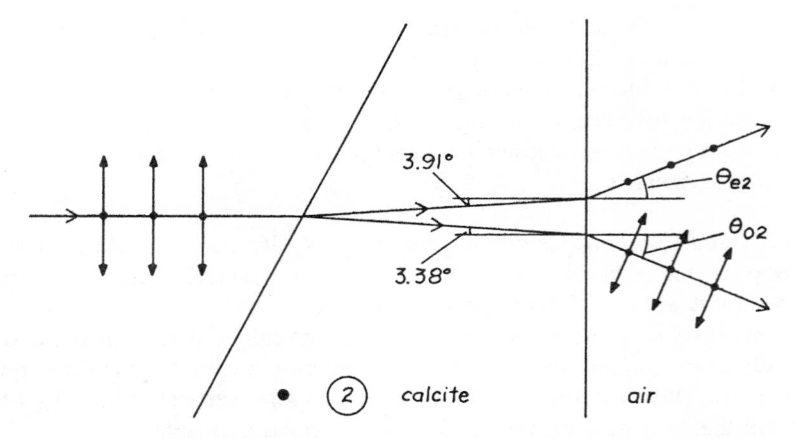

Figure 37.23 Problem 9. Snell's Law applied to e and o rays in a prism made of two pieces of calcite with their optic axes mutually perpendicular.

Therefore the total angular separation $\delta = 11.43°$.

Answers to Guided Problems

2. The mirror has to be only slightly (about 2 inches) longer than one-half of Sandy's height, or about 36 inches long. The top of the mirror must be about 2 inches above Sandy's eye level; the bottom, one-half of the floor-to-eye level height.
4. a. 10.9°
   b. 35.9°
   c. 19.3° for the glass prism. For the diamond prism the light is totally internally reflected and will strike the bottom of the diamond prism. It is harder for light to *get out* of diamond, and this is why diamond sparkles so brightly.
6. For total internal reflection, $n \sin \theta = 1$, where $n = 2.42$.
   Then $\theta = 24.4°$.
   The path of ray 1 through the diamond is shown in Figure 37.24.

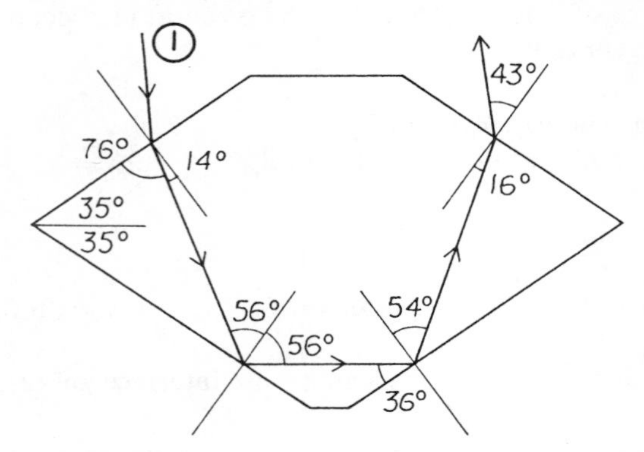

Figure 37.24 Problem 6. The path of a light ray inside the diamond of Figure 37.19. There are two refractions and two total internal reflections.

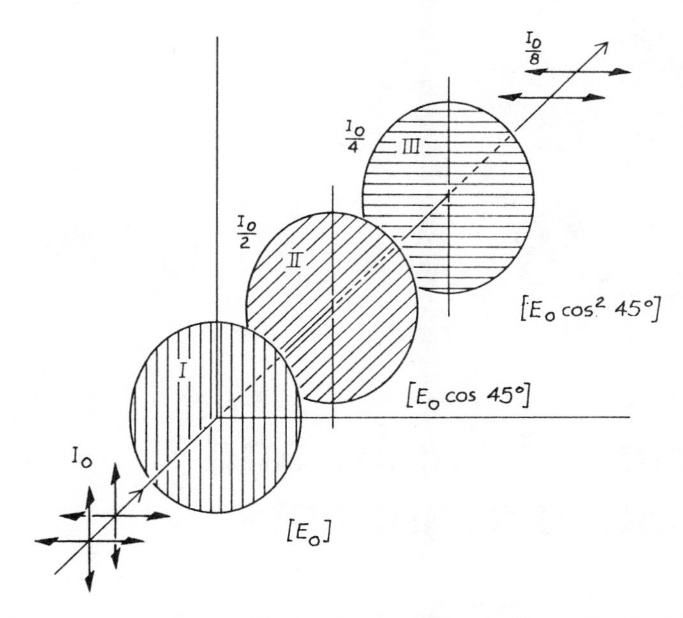

Figure 37.25 Problem 8. Three sheets of polaroid, with the axis of each succeeding sheet rotated by 45° with respect to the preferred axis of the sheet in front of it.

By a similar procedure we can show that ray 2 also emerges from the upper-right slanted face of the diamond, and approximately parallel to ray 1.

8. The three sheets of polaroid are indicated schematically in Figure 37.25. The incident intensity is I_0; the intensity after the first polaroid is $I_0/2$. Since the preferred axis of the second polaroid is at 45° with respect to that of the first, the intensity after the second polaroid is $(1/2)(I_0/2) = I_0/4$. After the third polarizer the intensity is $\frac{1}{2}(I_0/4) = I_0/8$.

Chapter 38

Mirrors, Lenses, and Optical Instruments

Overview

In this chapter we continue and extend our study of geometric, or ray, optics by calculating the trajectories of rays through mirrors, lenses, and prisms. These calculations, called ray tracing, involve the application of the laws of reflection and refraction, which we studied in the preceding chapter. We then extend our study to include a discussion of optical instruments, such as the camera, telescope, microscope, and human eye.

Essential Terms

concave spherical mirror
spherical aberration
convex spherical mirror
focal length
mirror equation
thin lens
lens-maker's formula
lens equation
photographic camera
f-number
pinhole camera
near point
magnifier
angular magnification
microscope
eyepiece
telescope

Key Concepts

1. *Spherical mirrors.* We have talked about reflection from a plane mirror; now we extend the discussion to spherical reflecting surfaces, which may be either *concave* or *convex*. If rays from a point source of light are

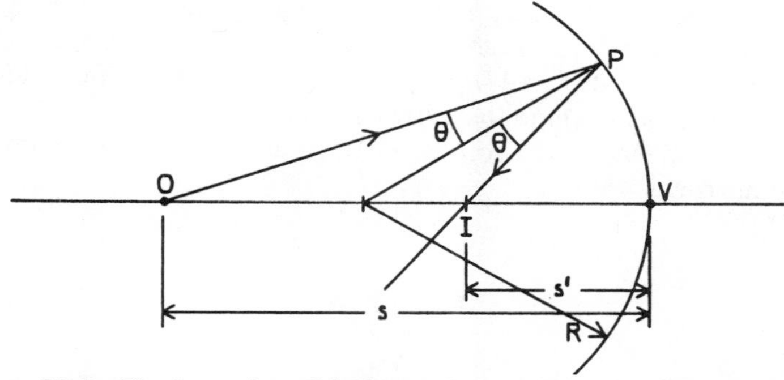

Figure 38.1 The formation of the image I of a point source O by reflection from a concave spherical mirror.

brought together at a common point by reflection at a surface, that point is the *image* of the object point. Figure 38.1 shows a concave spherical mirror with radius of curvature R. The line joining the point source at O and the center of the mirror at C defines the *mirror axis*, which intersects the mirror at the *vertex* V. Ray OP from the point source is reflected from the mirror surface just as it would be for a flat mirror tangent to the spherical surface at P. The reflected ray, and all other rays from O, intersect the axis at I, the image point of the object O. The object distance is s and the image distance is s'.

It is fairly easy to show that $1/s + 1/s' = 2/R$. There is, however, another definition to learn. The point at which the mirror causes incident parallel rays to converge is called the *focal point* of the mirror. When the source O is far away from the mirror, so that $1/s = 0$, the incident rays become parallel. Then $1/s' = 2/R = 1/f$, so that the *focal length* $f = R/2$. The *mirror equation* may then be written $1/s + 1/s' = 1/f$.

For an extended object we can use ray tracing to identify the image, as is indicated in Figure 38.2. A ray drawn from the object parallel to the axis is reflected back through the focal point. A second ray from the tip of the object drawn through C is reflected straight back, since it is normal to the mirror surface. The two rays intersect at the image point of the arrow tip, identifying the location of the image. The image is *real*, since light rays actually pass through the image. The image is also upside down, or inverted, and smaller than the object, since the image distance s' is less than the object distance s.

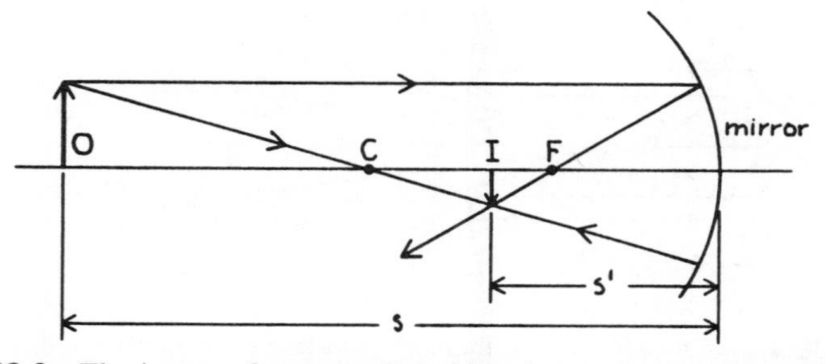

Figure 38.2 The image of an extended object in a concave mirror. The image is real, inverted, and smaller than the object.

For a convex mirror the focal point and the image position are located on the far side of the mirror, as is shown in Figure 38.3. We can again use the mirror equation $1/s + 1/s' = 1/f$ to find the position of the image algebraically, but all distances behind the mirror are taken to be negative. Thus, for example, the focal length of a convex mirror is always negative and that for a concave mirror is positive.

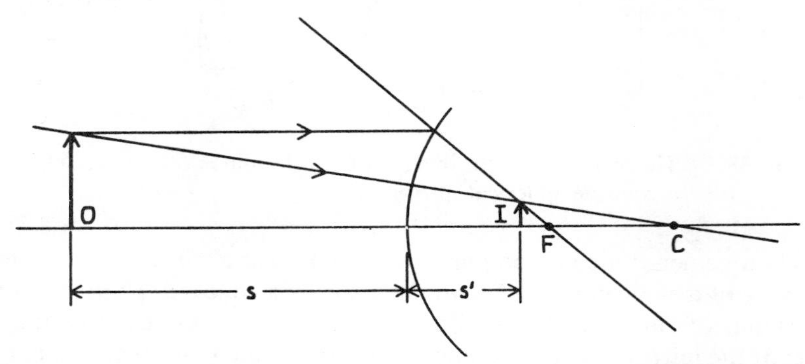

Figure 38.3 The image of an object in a convex mirror. The image is virtual, erect, and smaller than the object.

2. *Thin lenses*. A lens is usually a thin, circular piece of glass (or other transparent substance) with spherical surfaces. Such a lens can, through refraction, focus a beam of light to a point. The focal distance f is given by the *lens-maker's formula*, $1/f = (n - 1)(1/R_1 + 1/R_2)$, where n is the index of refraction of the lens material; R_1 and R_2 are the radii of the two spherical surfaces. A lens can be convex (both surfaces curved outward), concave (both surfaces curved inward), or concave–convex, in which case one of the surfaces is concave and the other convex.

Figure 38.4 shows the refraction of incident parallel rays by a convex (a)

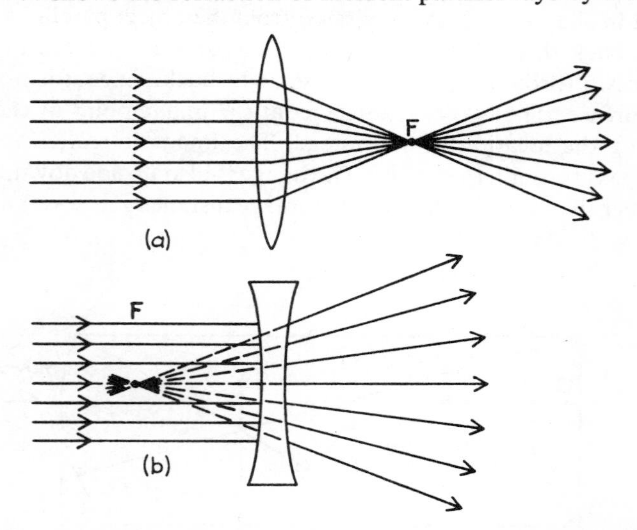

Figure 38.4 (a) A convex lens. Incident parallel rays converge to the focal point F. (b) A concave lens. Incident parallel rays diverge but appear to be coming from the focal point F.

and a concave (b) lens. For a convex lens, R_1, R_2, and f are taken to be positive; for a concave lens, R_1, R_2, and f are negative. For the concave lens, the focal point is that point from which incident parallel rays appear to diverge. In general, any thin lens is called converging or diverging depending on whether incident parallel rays of light converge or diverge after passing through the lens. The word *thin* means that the thickness of the lens is small in comparison to R_1 and R_2. Just as in the case of spherical mirrors, we can use ray tracing to find the image of an object placed near a thin lens. We do so in Figure 38.5(a) for a converging lens and in Figure 38.5(b) for a diverging lens. One ray from the top of the object is drawn parallel to the lens axis. For the converging lens it passes through the focal point F; for the diverging lens it passes through the lens and diverges, as if from the focal point. For both lenses a second ray passes undeflected through the center of

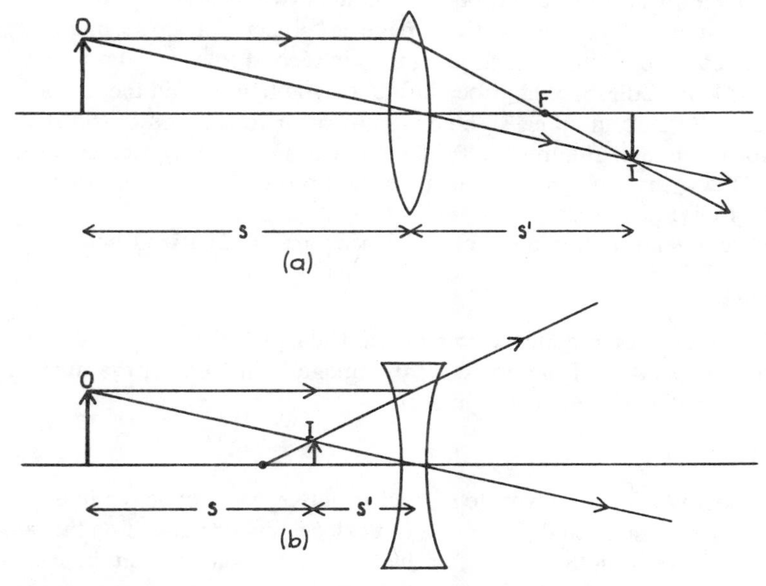

Figure 38.5 (a) The use of ray tracing to find the image of an object placed near a converging lens. (b) The use of ray tracing to find the image of an object placed near a diverging lens.

the lens (this is an approximation). The point of intersection of the two rays is the image point. We can also use the *thin lens formula*, $1/s + 1/s' = 1/f$, to find the position of the image algebraically, just as we did for the spherical mirror. For a converging lens, f is positive; for a diverging lens, f is negative. If the image is on the side of the lens opposite from the object, the image distance s' is positive; if the image is on the same side of the lens as the object [as it is in Figure 38.5(b)], the image distance is negative.

Note that a lens has two focal points at equal distances right and left of the lens. The point on the right of a converging lens is the focus for a parallel beam coming from the left. The point on the left is the focus for a parallel beam coming from the right. For a diverging lens, the two focal points at equal distances left and right of the lens are the points from which parallel beams coming from the left and right, respectively, appear to diverge.

The sizes of the images, I, in Figure 38.5 are clearly different from those of the objects, O. For a concave lens, or a convex mirror, we find that the size of the image is always smaller than that of the object. For a convex lens, or a concave mirror, the image may be either smaller or larger than the object. From Figure 38.5, it is easy to see that the ratio of image to object size is given by $I/O = M = -s'/s$. M is called the magnification produced by a lens or a mirror. We use the minus sign since in Figure 38.5(a) we measure the image size downward (the image is inverted). In Figure 38.5(b), the image distance s' is negative, and the image is upright. That is, a minus sign for M tells us that the image is inverted (and real). If M is positive, the image is upright and virtual.

Lenses may be, and often are, used in tandem, and in these cases it is necessary to trace the image through the system to find the position of the final image of an object. To begin, calculate where the first lens by itself forms an image; then use this first image as "object" for the second lens, and find the corresponding second image. This second image is the "object" for the next lens. Continue the procedure until you have found the image of the last lens. Be careful about sign conventions for image and object distances and focal lengths. Figure 38.20 of your text shows such a calculation. To obtain the overall magnification, take the product of the magnifications of all the individual lenses.

Sign conventions for mirrors and lenses are summarized below:

MIRRORS

f is positive for a concave mirror, and negative for a convex mirror.
s or s' is positive if the object or the image is in front of the mirror, and negative if it is behind the mirror.

LENSES

f is positive for a convex lens, and negative for a concave lens.
s is positive if it is on the near side of the lens, and negative if on the far side.
s' is positive if on the far side of the lens, and negative if on the near side.

There is a close correspondence between a convex lens and a concave mirror, and between a concave lens and a convex mirror. Note that the focal lengths of concave lenses and of convex mirrors are negative and that the images produced are always virtual.

3. *The photographic camera and the eye*. The simple lenses that we have described suffer from various aberrations, and for this reason most optical instruments use "compound" lenses, made by combining several simple lenses into a single lens. For the purposes of ray tracing, a compound lens can usually be represented by a single thin lens of the appropriate focal length.

In a camera (Figure 38.6) the distance between the lens and the film is adjustable so that the image, which is real, can always be made to fall on the film. A shutter controls the exposure time during which light is admitted to the camera. The size of the camera lens is labeled by the *f-number*, defined as the ratio of the focal length of the lens to its diameter. Good cameras have an adjustable iris diaphragm that blocks part of the lens and so allows the effective f-number to be changed.

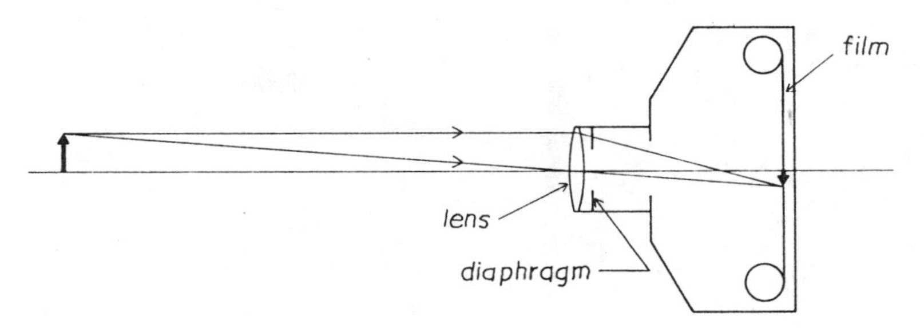

Figure 38.6 Schematic diagram of a photographic camera.

A *pinhole camera* has no lens. The iris diaphragm is, in effect, closed down to a single point, through which the rays pass. Since the rays suffer no deflection, no lens is needed, but we must either have a very bright light or a very long exposure time to expose the film.

The eye is similar to a camera in that the lens of the eye forms a real image on the retina, a membrane containing light-sensitive cells at the back of the eye socket. The lens of the eye is flexible. If the eye is viewing a distant object, the muscles that adjust the lens are relaxed and the lens is fairly flat, with a long focal length. To view nearby objects, the lens becomes more rounded, with a shorter focal length. The *near point* (Figure 38.7) is the shortest distance at which an object can be placed from the eye and still be seen sharply. For a normal young adult, the near point is typically 25 cm. As we grow older, the lens of the eye loses its flexibility and the near point recedes.

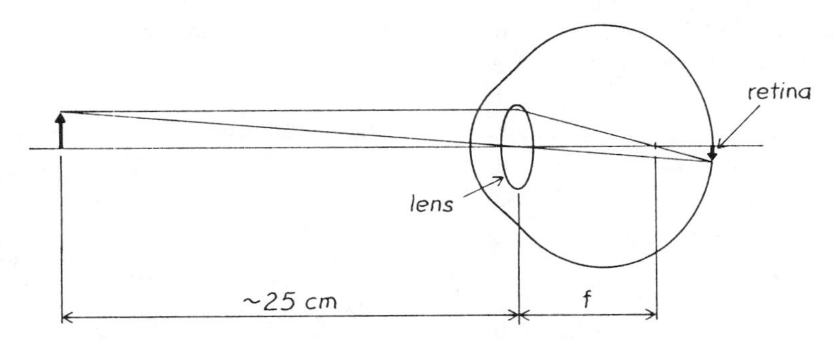

Figure 38.7 The object being viewed is at the near point of the eye. The focal length is a minimum. If the image is brought closer to the eye, the image will be formed beyond the retina.

Nearsightedness, a condition in which the focal length of the eye is too short, can be corrected for by eyeglasses with divergent lenses. In a *farsighted* eye, the focal length is too long, so that rays from a nearby object converge toward an image beyond the retina. Farsightedness can be corrected for by eyeglasses with converging lenses.

The *magnifier* consists of a strongly convergent lens placed adjacent to the eye. It allows us to bring the object closer to the eye than the near point and so increases the size of the image on the retina. The magnifier is shown in Figure 38.8; the object to be viewed is placed just inside the focal length of the lens. The eye sees the image produced by the lens, much enlarged,

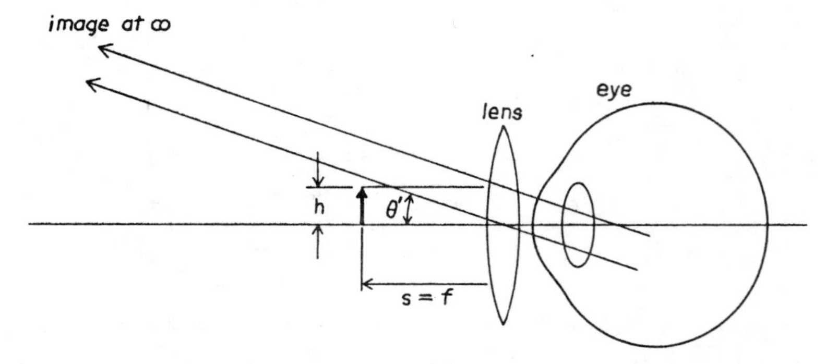

Figure 38.8 The magnifier. The object to be viewed is placed at f, the focal point of the lens. The eye sees the image at infinity. The angular size of the image is θ'.

at infinity. The angular size of the object seen by the eye is θ'. $\tan \theta' \doteq h/f$, where f is the focal length of the lens. Without the magnifier, the angular size of the object (see Figure 38.8) is θ, where $\tan \theta = h/25$ cm, where 25 cm is taken to be the near point of the eye. The angular magnification of the magnifier is defined to be the ratio $\theta'/\theta \cong \tan \theta'/\tan \theta = [h/f]/[h/25] = 25\text{ cm}/f$.

The *microscope*, shown in Figure 38.9, consists of two lenses: the objective and the eyepiece. Both have very short focal lengths. The object is placed just beyond the focal distance of the objective. The real, enlarged image of the objective serves as the object for the eyepiece, which acts as a magnifier and forms a virtual image at infinity. The magnification of the objective is just s'/s, and that of the eyepiece, as shown above, is 25 cm/f. The overall angular magnification of the microscope is therefore $[25\text{ cm}/f][s'/s]$, where f is the focal length of the eyepiece.

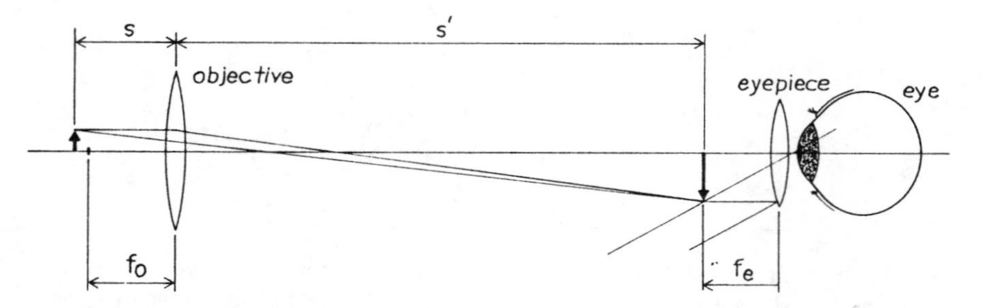

Figure 38.9 Arrangement of lenses in a microscope. The image of the objective is the object for the eyepiece. The eye sees the virtual image at infinity formed by the eyepiece.

The *telescope*, shown in Figure 38.10, consists of an objective lens of long focal length and an eyepiece of short focal length. The two lenses are separated by a distance nearly equal to the sum of the two focal lengths. The objective forms a real image of a distant object, and this image is the object for the eyepiece, which forms a magnified virtual image at infinity. Note that for an infinite object distance, the lens equation $(1/\infty) + (1/s') = 1/f_{\text{obj}}$, so that $s' = f_{\text{obj}}$, the focal length of the objective. The angular magnification of the

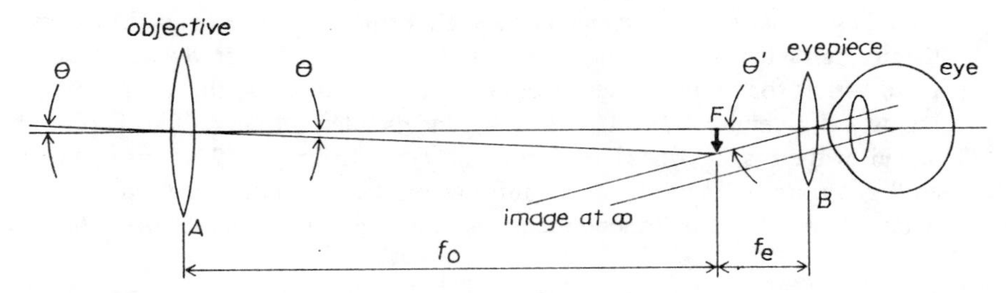

Figure 38.10 A telescope. The angular magnification is the ratio of the two focal lengths, f_{obj}/f_{eye}.

telescope is the ratio θ'/θ. Since both angles are small, $\theta'/\theta \approx \tan\theta'/\tan\theta = FA/FB = f_{obj}/f_{eye}$. That is, the angular magnification is equal to the ratio of the focal lengths of the two lenses.

Because it is easier to manufacture a large mirror than a good large lens, many astronomical telescopes use concave mirrors in place of the objective lens. The mirror forms a real image that serves as the object for the eyepiece. Modern large telescopes tend to use several mirrors in parallel, in order to collect more light.

Sample Problems

1. *Worked Problem.* A thin concave lens has a focal length of 15 cm. If a thin convex lens of focal length 9 cm is placed immediately behind it, where will rays of sunlight be brought to a focus? Where will they focus if the convex lens is placed right in front of the concave lens?

Solution: The arrangement of the two lenses is shown in Figure 38.11.

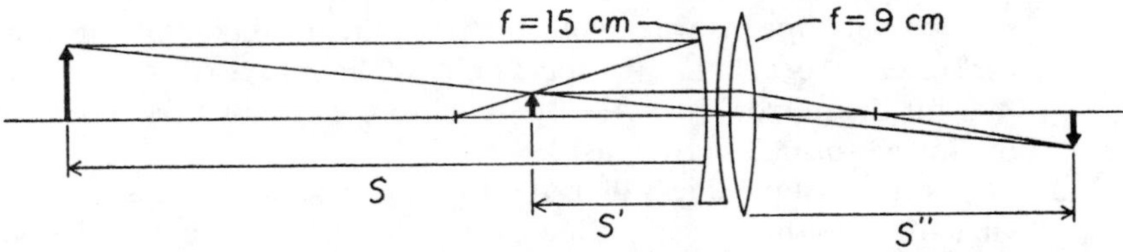

Figure 38.11 Problem 1. A concave lens with a convex lens right behind it. The image formed by the concave lens forms the object for the convex lens.

Apply the thin lens formula, $(1/s) + (1/s') = (1/f)$, to the diverging lens. The image formed by this lens is the object for the converging lens, with the object distance given by $-[(1/f_d) - (1/s)]$. The image distance for the diverging lens is negative, since the image is on the near side of the lens, but the object distance of the converging lens is positive. Now apply the thin lens formula to the converging lens: $-[(1/f_d) - (1/s)] + (1/s'') = (1/f_c)$, where s'' is the new image distance. Rearranging, $(1/s) + (1/s'') = (1/f_d) + (1/f_c)$, which we can write as $(1/s) + (1/s'') = 1/f$, where $1/f = (1/f_d) + (1/f_c)$. If we had placed the converging lens first, its image, given by $1/s' = (1/f_c) - (1/s)$, would be the object for the diverging lens, and we would again use a negative sign because this object distance would be *behind* the diverging lens and, therefore, negative. We have thus arrived at a general result—that for two thin lenses adjacent to one another, the overall focal length f is given by $1/f = (1/f_1) + (1/f_2)$, independent of which lens is placed ahead of the other. Since $f_c = 9$ cm, and $f_d = -15$ cm, $f = 22.5$ cm, so that sunlight (parallel rays from an infinite distance) would be focused at a distance of 22.5 cm, independent of which lens is ahead of the other.

2. *Guided Problem.* A person walks slowly toward a large concave mirror with a radius of curvature of 1.0 m. Where is the image of the person's face when he is at distances of 5.0, 3.0, 1.0, 0.6, 0.5, and 0.4 m away, respectively, from the mirror? Does the person see any sudden or drastic changes in his image as he approaches the mirror?

Solution scheme

a. Write down the mirror equation relating object distance, s, image distance, s', and the focal length, f. What is the focal length of the mirror, and is it positive or negative?

b. Make a table of image distances s' versus object distances s. Be sure to note whether the image distance, s', is positive or negative. A negative s' means that the image is virtual, and a positive s' means that the image is real. Remember also that a real image is inverted and is on the same side of the mirror as the object. A virtual image is erect and is on the side of the mirror opposite to the subject. Draw one or two ray diagrams to make sure that you understand the image formation.

c. By looking at your table of s' values, determine whether, as the person approaches the mirror from 5.0 m away, the image becomes larger or smaller. (The image-to-object size is given by the ratio s'/s.) What happens when your face is at the focal point of the mirror?) Remember that at the focal point all light rays from your face that hit the mirror will reflect parallel to the mirror axis. Where is the image formed in this case?

d. For object distances less than the focal length, is the image real or virtual? Does the image size increase or decrease as you get closer to the mirror surface from inside the focus?

3. *Worked Problem.* A cylinder 10 cm long is placed with its near end 20 cm from a convex mirror, as shown in Figure 38.12. The radius of curvature of the mirror is 15 cm. Describe the image of the cylinder in the mirror—where is it, and how long is it?

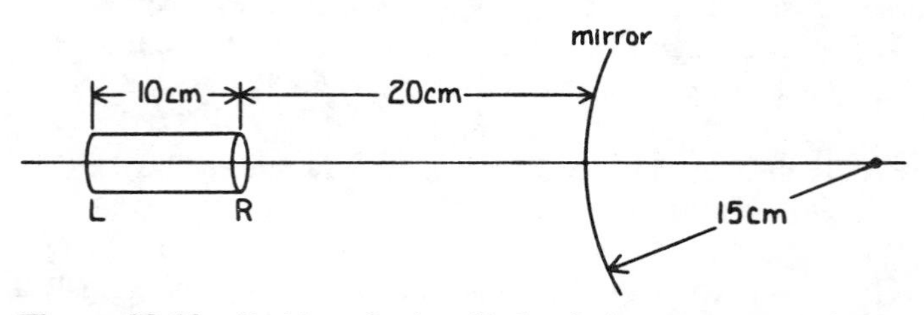

Figure 38.12 Problem 3. A cylinder facing a convex mirror.

Solution: We can use the mirror equation $1/s + 1/s' = 1/f$ to find the location of the image of the ends of the cylinder. The focal length $f = -7.5$ cm, since the mirror is convex. If $s = 20$ cm, $s' = -5.45$ cm, and if $s = 30$ cm, $s' = -6.0$ cm, so that the image of the right side of the cylinder is 5.45 cm behind the mirror and that of the left side of the cylinder is 6.0 cm behind the mirror. Thus, the right and left ends of the cylinder are reversed in the image, which is shown schematically in Figure 38.13. The length of the image is therefore 0.55 cm, so that the image is much smaller than the object. The image is virtual and erect.

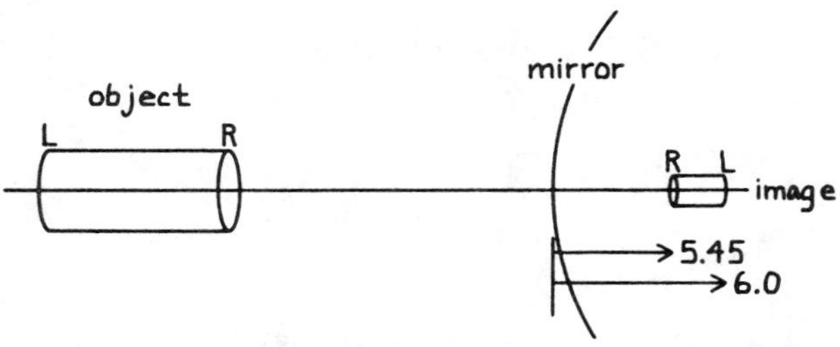

Figure 38.13 Problem 3. The image of the cylinder shown in Figure 38.12 in a convex mirror.

4. *Guided Problem.* A camera has a converging lens of focal length 4.0 cm. What must be the film-to-lens distance to focus on the film
a. a distant mountain?
b. a flower 50 cm away from the lens?

Solution scheme
a. Write down the lens equation for this problem. What is the focal length, and is it positive or negative?
b. Is the image distance just the film-to-lens distance? If so, apply the lens equation to solve for the image distance s' for the two cases. What object distance s should you use when the object is a distant mountain?

5. *Worked Problem.* Two converging lenses, the first with focal length 18 cm and the second with focal length 2.0 cm, are 20 cm apart, as shown in Figure 38.14. Where does the first lens (18 cm focal length) form the image of a tree 10 m distant? Where does the second lens form an image of this image? Is this final image erect or inverted?

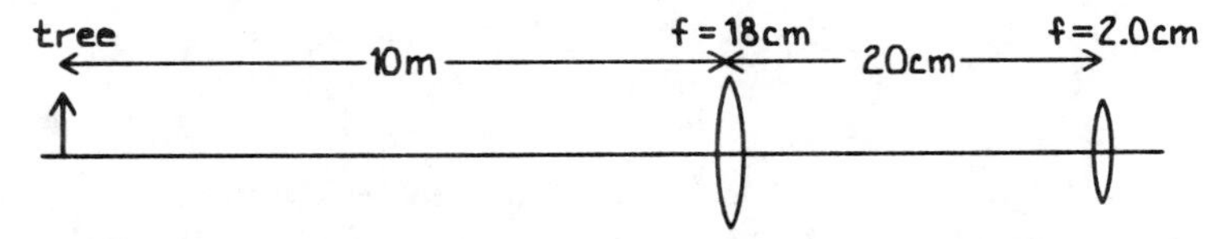

Figure 38.14 Problem 5. Two converging lenses 20 cm apart. Where is the image of the tree?

Solution: We can solve this problem by first finding the image of the tree produced by the lens with 18 cm focal length. This image then becomes the object seen by the second lens. For the first lens—that is, the lens with $f = 18$ cm—the object distance $s = 10 \times 10^2$ cm and $f = 18$ cm (converging lens), so that the lens equation gives an image distance $s' = 18.33$ cm. (The image is to the right of the lens; the image is real, inverted, and much smaller than the object.) This image now becomes the object for the $f = 2.0$ cm lens, so that the new object distance for this lens is $20 - 18.33 = 1.67$ cm. By applying the lens equation, with $f = 2.0$ cm, we obtain $s' = -10$ cm. The minus sign indicates that the image is on the same side of the lens as the object, so that the image is midway between the two lenses. The image is inverted, since the object was inverted also. We show the formation of the images with ray tracing in Figure 38.15.

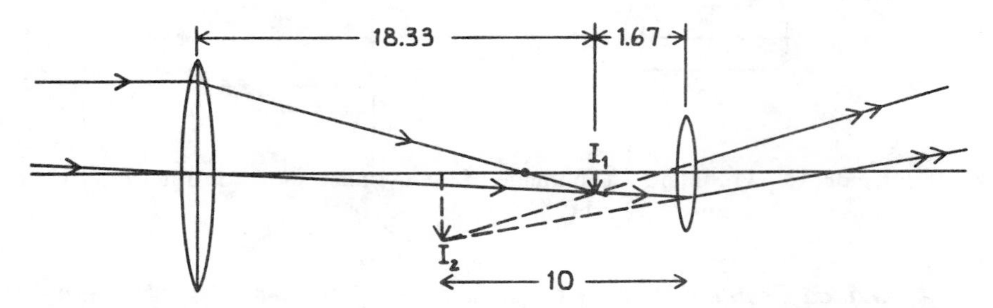

Figure 38.15 Problem 5. The images of the tree in the two lenses shown in Figure 38.14. The image I_1 produced by the first lens becomes the object for the second lens. The final image is I_2. It is a virtual image—the light rays do not actually pass through I_2.

6. *Guided Problem.* Which of the thin lenses shown in Figure 38.16 are converging and which diverging, and what are their respective focal lengths? In all cases the index of refraction $n = 1.50$. In the figure legend, R_1 is the radius of curvature of the left face of the lens and R_2 is the radius of curvature of the right face. Distances are in centimeters. Note that a flat surface has an infinite radius of curvature.

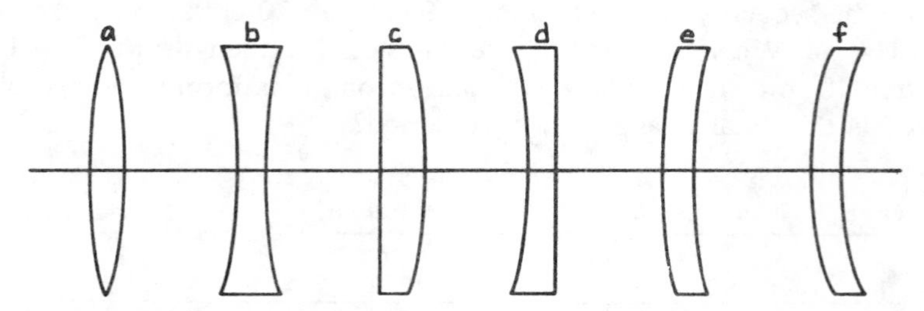

Figure 38.16 Problem 6. Some thin lenses. The radii of the curvature are as follows:

	a	b	c	d	e	f
R_1	10	10	∞	15	30	15
R_2	15	15	20	∞	15	30

Solution scheme

a. Write down the lens-maker's equation for a thin lens. What are the sign conventions to follow for R_1 and R_2? For example, you should know that for lens *a*, R_1 and R_2 are both positive; for *b*, both R_1 and R_2 are negative.

b. For a flat surface ($R = \infty$), does it matter whether R is positive or negative? Solve for the effective focal length f of each lens. What does it mean if f is positive? negative?

c. Make a table of your results.

7. *Worked Problem*. A nearsighted eye has a near point of 12 cm and can focus clearly on an object only for distances out to 20 cm. (Vision is blurred for objects beyond 20 cm.)

a. What kind of corrective lens is required, and what is its focal length, to allow the eye to focus on objects at infinity?

b. With this corrective lens in place, what is the new near point of the eye?

Solution: It is worth using a diagram to indicate the properties of the lens of the eye in question, and this is done in Figure 38.17. Note that the focal length of the eye changes from its minimum value, f_{12}, to its maximum value (when the eye is relaxed), f_{20}, in such a way so as to keep the image distance, s', constant.

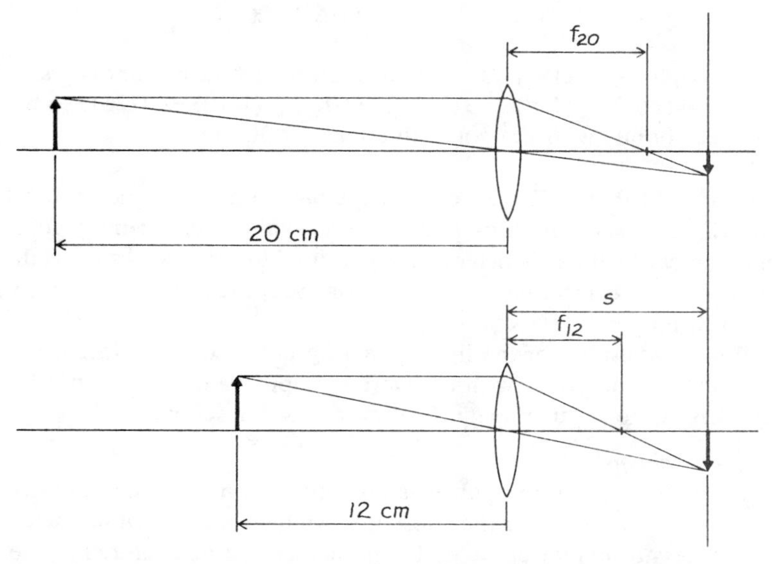

Figure 38.17 Problem 7. The focal properties of a shortsighted eye. The near point is 12 cm, and the eye cannot focus objects farther away than 20 cm.

We need to increase the focal length of the eye and therefore expect to use a concave, or diverging, corrective lens, to be placed close to the eye. From worked problem 1, for thin adjacent lenses, we know that the focal length of the lens combination is given by $1/f = (1/f_1) + (1/f_2)$. For an object at 20 cm the lens equation is

$$(1/20) + (1/s') = 1/f_{20} \tag{1}$$

With a corrective lens, objects at infinity can be seen, and the lens equation is $(1/\infty) + (1/s') = 1/f$, where we can now express the new focal length f as $1/f = (1/f_L) + (1/f_{20})$, where f_L is the focal length of the corrective lens. We therefore have $(1/\infty) + (1/s') = (1/f_L) + (1/f_{20})$ or

$$(1/s') = (1/f_L) + (1/f_{20})$$

But from Eq. (1), above, $(1/s') = (1/f_{20}) - (1/20)$, so that $(1/f_L + (1/f_{20}) = (1/f_{20}) - (1/20)$, or $f_L = -20$ cm. That is, the corrective lens has a focal length of -20 cm (the minus sign indicates that the lens is concave).

To find the new near point (np), the lens equation without the corrective lens is

$$(1/12) + (1/s') = (1/f_{12})$$

Again, with the corrective lens, s' remains the same, so that with the corrective lens,

$$(1/np) + (1/s') = (1/f_{12}) - (1/20)$$

Since $(1/f_{12}) = (1/12) + (1/s')$, the equation for the near point is just $(1/np) + (1/s') = (1/12) + (1/s') - (1/20)$, or $(1/np) = (1/12) - (1/20)$, so that the new near point, with the corrective lens, is 30 cm.

8. *Guided Problem.* The Hale telescope has a mirror diameter of 5.08 m and a primary focal length of 17 m. For viewing at the primary focus, an observation cage with a diameter of 1.8 m is used to hold the observer, thereby blocking out the central part of the mirror. Suppose that the eyepiece used has a focal length of 3.8 cm.

a. What is the ratio of the light-gathering power of the telescope to that of the unaided eye, with an average pupil diameter of 2 mm?
b. What is the angular magnification of the telescope?

Solution scheme

a. The light-gathering power is just proportional to the total area used to collect light. For the telescope, we have a circle of diameter 5.08 m less a smaller circle of 1.8 m diameter at its center. For the eye, we have a circle of diameter 2 mm.
b. What is the expression for the angular magnification of a telescope? Note the large difference in the two numbers you have just calculated—a large telescope has a very large light-gathering power!

9. *Worked Problem.* When we discussed the magnifier, we said that the object to be viewed is placed so that the eye sees the image produced by the lens, much enlarged, at infinity. The resulting angular magnification is 25 cm/f. You have a magnifier with a focal length of 5 cm. Where must the

object to be magnified be placed with respect to the lens, and what is the resulting angular magnification? Suppose that you want to view the image at the near point of your eye, 25 cm. Where must you now place the object, and what is the new angular magnification? Can you think of why you might prefer to use the magnifier with the image at infinity?

Solution: From the lens equation, $(1/s) + (1/s') = 1/f$, $s' = -\infty$, and $f = 5$ cm, to that $(1/s) - (1/\infty) = 1/5$, or $s = 5$ cm. The object distance must be at the focal point of the lens, 5 cm. The angular magnification of a magnifier is given by $\theta'/\theta = 25$ cm$/f$. Here $f = 5$ cm, so that $\theta'/\theta = 5$.

Suppose that $s' = -25$ cm (minus sign since the image is virtual). Then, from the lens equation, $(1/s) - (1/25) = 1/5$, from which we obtain $s = 4.17$ cm.

The new angular magnification can be seen from Figure 38.18.

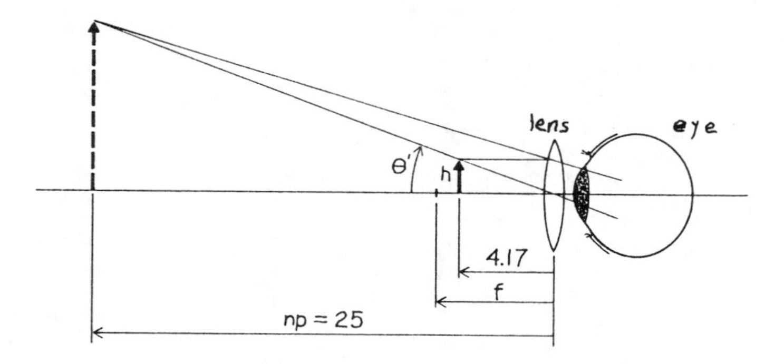

Figure 38.18 Problem 9. The angular magnification of a magnifier with the image at the near point of the eye.

The ratio $\theta'/\theta = [h/4.17]/[h/25] = 25/4.17 = 6$, so that by moving the image point to the near point of the eye, we have increased the angular magnification from 5 to 6. However, the eye is relaxed when focused at infinity (the lens has its longest focal length). Focusing the eyes for long at the near point, where the eye muscles must adjust the lens shape to give the shortest possible focal length, can result in eyestrain.

Answers to Guided Problems

2. In the following table the object distance is s; the image distance is s'.

s (m)	s' (m)
5.0	0.56
3.0	0.60
1.0	1.0
0.6	3.0
0.5	
0.4	−2.0

The image becomes larger as the person approaches the mirror. As the person goes through the focal point (0.5 m), the image "blows up," since the image distance becomes infinite. For object distances less than the focal length, the image is virtual; the image size decreases as the object gets closer to the mirror face.

4. For the mountain, $s' = 4.0$ cm. For the flower, $s' = 4.35$ cm.

6.

Lens	f
a	12 cm, converging
b	−12 cm, diverging
c	40 cm, converging
d	−30 cm, diverging
e	−30 cm, diverging
f	60 cm, converging

8. a. $R = 5.64 \times 10^6$

   b. ang. mag. $= f_{obj}/f_{eye} \; \dfrac{1700}{3.8} = 4.47$

Chapter 39

Interference

Overview

In the last two chapters we studied the changes in the directions of light waves as a result of interactions with matter, as predicted by Huygens' Construction. We now begin a more complex study, an attempt to answer the question of what happens when electromagnetic waves meet. The resulting electromagnetic wave is obtained by adding the individual fields at every point (superposition). The phenomenon is called *interference* and may be destructive, in which case the resulting waveform is smaller than the original waves, or constructive, in which case the resulting waveform is larger than the original waves. Our discussion holds for any electromagnetic wave, but we will again largely restrict ourselves to light waves, for which the interference effects can usually be seen by eye.

Essential Terms

interference
standing electromagnetic wave
constructive interference
destructive interference
Michelson interferometer
interference fringes
interference from two slits
interference from multiple slits
resolving power
diffraction grating

Key Concepts

1. *Interference*. As our first example of interference, we add the electric field of a light wave incident on a mirror to the field of the reflected light wave. If the electromagnetic wave moving in the $+x$ direction strikes a mirror lying in the y–z plane at $x = 0$, the electric field of the incident wave is

given by $E_{in} = E_0 \cos(\omega t - kx)$, as shown in Figure 39.1. The equation of a wave of the same amplitude reflected directly backward is $E_0 \cos(\omega t + kx)$. Although we haven't shown it, a perfect reflector is also a perfect conductor, so that the total electric field at the reflecting surface (that is, at $x = 0$) must equal zero. For $x = 0$, $E_{tot} = E_{in} + E_{refl} = 0$. There must therefore be a 180° phase change upon reflection, so that the reflected wave is given by $-E_0 \cos(\omega t + kx)$ and the resulting total electric field, obtained by adding the incident and the reflected waves, is

$$\begin{aligned} E_{tot} &= E_0 \cos(\omega t - kx) - E_0 \cos(\omega t + kx) \\ &= 2E_0 \sin \omega t \sin kx \end{aligned}$$

The incident and the reflected waves are shown in Figure 39.1.

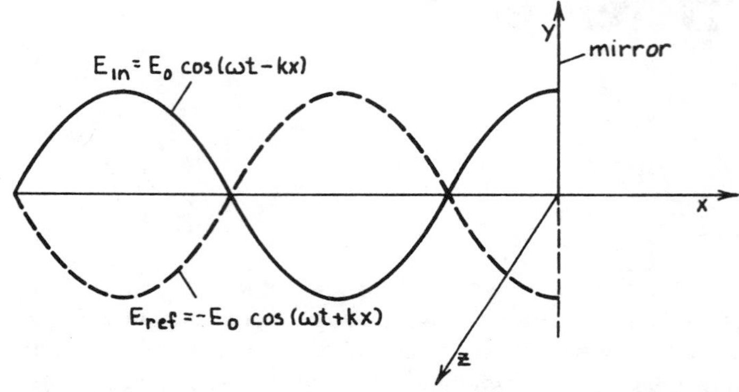

Figure 39.1 The electric field of the incident and the reflected light wave on a mirror at $x = 0$.

The equation $E_{tot} = 2E_0 \sin \omega t \sin kx$ is the equation of a *standing electromagnetic wave*, a wave that remains fixed in space but that pulsates up and down. For those places where $\sin kx = 0$ at all times, $E_{tot} = 0$ and the incident and reflected waves are said to *interfere destructively*. These points occur when $kx = 2\pi x/\lambda = 0, \pi, 2\pi, 3\pi, \ldots$—that is, they occur at values of x given by $x = 0, \lambda/2, \lambda, 3\lambda/2, \ldots$. For those places where $\sin kx = 1$ at all times, $E_{tot} = 2E_0 \sin \omega t$ and the incident and reflected waves are said to interfere *constructively*. These points occur at $kx = 2\pi x/\lambda = \pi/2, 3\pi/2, 5\pi/2, \ldots$ or at values of x given by $x = \lambda/4, 3\lambda/4, 5\lambda/4, \ldots$.

The magnetic fields associated with the incident and the reflected electric fields are in the z direction. Again, $B_{tot} = B_{in} + B_{refl}$, and it is easy to show that

$$B_{tot} = 2E_0/c \cos \omega t \cos kx$$

This is the equation for a standing wave displaced in phase one-quarter wavelength from the standing electric wave and with its maxima and minima also displaced one-quarter wavelength from those of the electric field.

Because the wavelength of visible light is very small, we cannot see the maxima and minima associated with the reflection of light backward from a mirror. We can, however, easily see interference effects caused by two or more reflections from a thin film—most of us have seen colored fringes on oil slicks on water or on soap bubbles. The reflection for nearly perpendicular incident light from a thin film of thickness d is indicated in Figure 39.2. The wave reflected from the lower surface has to travel approximately $2d$ farther than the wave reflected from the upper surface.

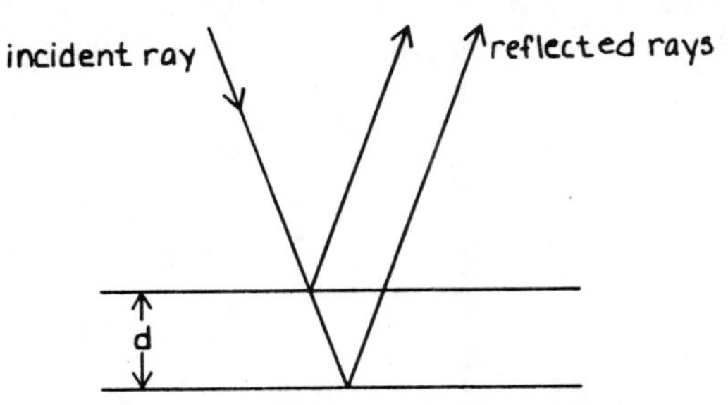

Figure 39.2 Incident and reflected rays at approximately normal incidence on a thin film of thickness d.

If this extra distance is exactly equal to one or more wavelengths, the wave reflected from the lower surface will meet crest to crest with waves reflected from the upper surface and constructive interference will occur. Thus, constructive interference occurs when $2d = \lambda, 2\lambda, 3\lambda, \ldots$. (Note that λ is the wavelength *inside* the film.) If the extra distance $2d$ is exactly one-half a wavelength, the two reflected waves will cancel each other and destructive interference will occur. Thus, destructive interference occurs when $2d = \lambda/2, 3\lambda/2, 5\lambda/2, \ldots$. The two reflected waves do not set up a standing wave with closely spaced maxima and minima; the interference is the same along the whole reflected wave train, so that we can see these interference effects with our eyes. What we see are bright or dark *interference fringes* on the thin film, corresponding, respectively, to constructive or destructive interference.

When the index of refraction is increasing (as at an air–water interface), a wave suffers a 180° phase change on reflection; when the index of refraction decreases, no phase change occurs on reflection. Thus, the equations for constructive and destructive interference that we have developed hold for an oil slick on water, since the index of refraction for the oil is less than that for water and a 180° phase change occurs for *both* reflected waves. For reflection from both sides of a soap bubble in air, shown in Figure 39.3, the lower

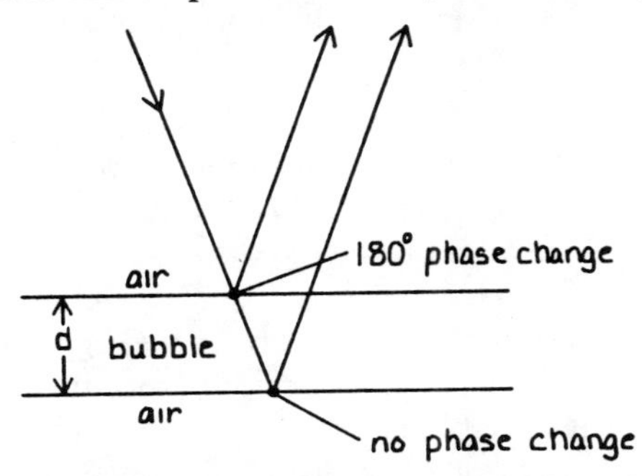

Figure 39.3 Incident and reflected light rays from the surface of a soap bubble. Because of the phase change, constructive interference will occur when $2d = \lambda/2, 3\lambda/2, \ldots$.

reflected wave suffers no phase change on reflection and the equations for interference must be reversed.

2. *Michelson interferometer*. The Michelson interferometer uses the idea of light interference to allow distance measurements to be made with very high precision. Figure 39.4 indicates how the interferometer works. Mono-

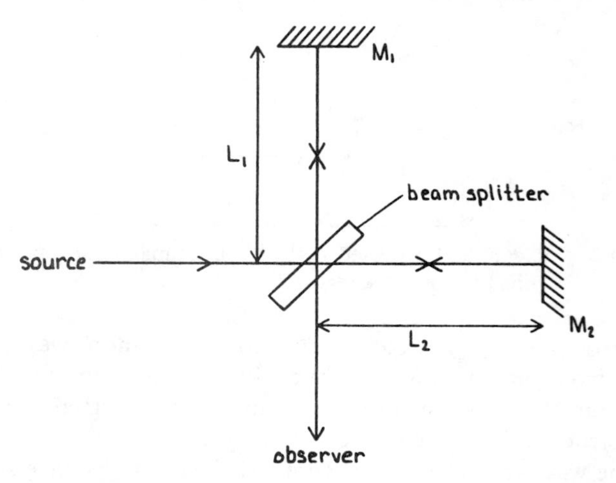

Figure 39.4 A Michelson interferometer.

chromatic light (light of a single wavelength) from a source passes through a beam splitter, where part of it is reflected to mirror M_1 and part is transmitted directly to mirror M_2. The light from the mirrors is reflected back to the beam splitter, where it is directed to the observer. The two wave trains interfere in the region between the beam splitter and the observer. The interference is constructive if the path difference $2(L_1 - L_2) = 2d$ is an integral number of complete wavelengths, that is, if $2d = 0, \lambda, 2\lambda, 3\lambda, \ldots$. The interference is destructive if the path difference is an odd integral number of half wavelengths, that is, if $2d = \lambda/2, 3\lambda/2, 5\lambda/2, \ldots$. The device is made so that mirror M_1 (or M_2) can be moved very precisely parallel to the path of the light beam. As the mirror is moved, the observer sees bright (constructive) and dark (destructive) interference fringes. It is possible to observe a shift as small as 1/40 of fringe spacing, corresponding to a distance of about 75 Å, where 1 Å = 10^{-8} cm.

3. *Interference from two slits*. Suppose that we pass a beam of monochromatic light through narrow parallel slits S_1 and S_2 a distance d apart, as shown in Figure 39.5. A screen is located a distance D from the slits. S_1 and S_2 become sources of secondary waves of equal amplitude and phase that radiate outward and interfere in the region to the right of the slits. At a point P on the screen, the path length for light from the lower slit, S_2P, is longer than the path length for light from the upper slit, S_1P, as shown in Figure 39.6. For D very much greater than d, the rays S_1P and S_2P are very nearly parallel and the path difference $S_2P - S_1P = \Delta l$ is approximately equal to $d \sin \theta$. We find maximum intensity on the screen (constructive interference) when

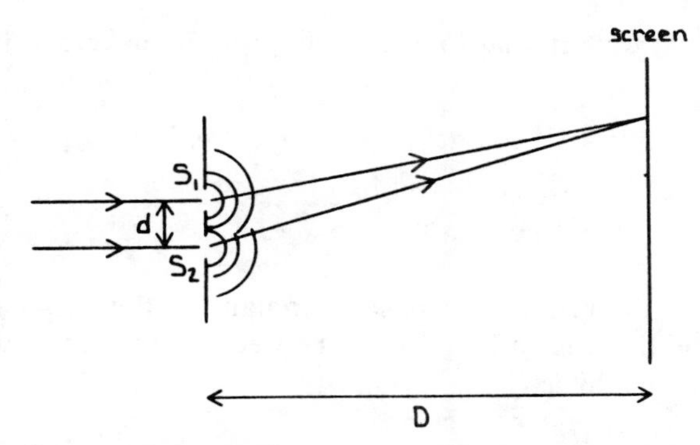

Figure 39.5 A light beam strikes two parallel slits S_1 and S_2. The slits act as two sources of spherical light wave fronts, so that waves from the two slits interfere.

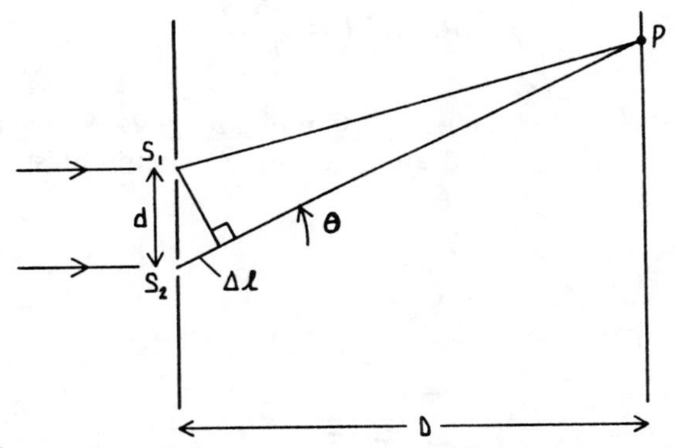

Figure 39.6 The path difference, Δl, for the two rays from slits S_1 and S_2 is approximately $d \sin \theta$.

$$d \sin \theta = 0, \lambda, 2\lambda, 3\lambda, \ldots$$

and minimum intensity (destructive interference) when

$$d \sin \theta = \lambda/2, 3\lambda/2, 5\lambda/2, \ldots$$

On the screen, parallel to the slits, we see bright and dark bands. At the center there is a bright band corresponding to $d \sin \theta = 0$, followed by a dark band on either side, corresponding to $d \sin \theta = \lambda/2$, followed by a bright band, and so on. To calculate the intensity distribution as a function of the angle θ, consider two electromagnetic waves propagating along the x axis, given respectively by $E_1 = E_0 \cos(\omega t - kx)$ and $E_2 = E_0 \cos(\omega t - kx - \delta)$. At a screen placed at $x = x_0$, superposition of the two waves gives $E = E_1 + E_2$, so that

$$\begin{aligned} E &= E_0[\cos(\omega t - kx) + \cos(\omega t - kx - \delta)] \\ &= 2\,E_0 \cos\left(\omega t - kx - \frac{\delta}{2}\right) \cos\frac{\delta}{2} \end{aligned}$$

The intensity I is proportional to E^2, so that the time-average intensity is given by

$$\begin{aligned}\bar{I} &= 4\, E_0^2 \cos^2 (\omega t - kx - \delta/2) \cos^2 \delta/2 \\ &= 4\, E_0^2 (1/2 \cos^2 \delta/2) \\ &= 2\, E_0^2 \cos^2 \delta/2\end{aligned}$$

The quantity δ is the phase difference in radians of the two waves at the screen. We found that the path difference between the two waves was $d \sin \theta$; expressed in radians this is

$$\left(\frac{d \sin \theta}{\lambda}\right) 2\pi$$

Therefore

$$\bar{I} = 2\, E_0^2 \cos^2 \left(\frac{\pi d \sin \theta}{\lambda}\right)$$

The intensity is plotted versus the angle θ in Figure 39.7. We see that there is an intensity maximum at $\theta = 0$ and successive maxima at $\sin \theta = \pm\lambda/d$, $\pm 2\lambda/d$, $\pm 3\lambda/d$, ...

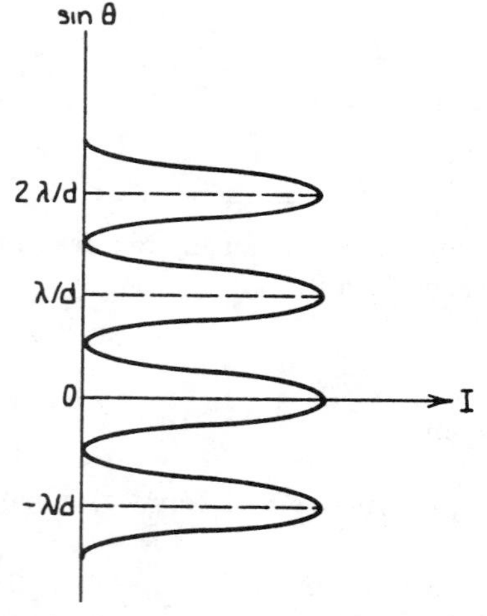

Figure 39.7 The intensity pattern versus sin θ for interference from two slits.

4. *Interference from multiple slits.* Suppose that we have an arbitrary number of evenly spaced slits. As Figure 39.8 shows, maximum intensity will still occur for $d \sin \theta = n\lambda$, where n has possible value 0, 1, 2, . . . , since if rays from adjacent slits interfere constructively, rays from all slits will also interfere constructively. The maximum of intensity corresponding to $n = 1$ is called the first principal maximum, and so on.

Minimum intensities are not, in general, given when $d \sin \theta = \lambda/2, 3\lambda/2$, . . . , since this condition ensures that waves from *only* adjacent slits will cancel. We can show that for N slits, the minima are given by $d \sin \theta = m\lambda/N$, where $m = 1, 2, 3, \ldots N - 1$. For a three-slit system, there are

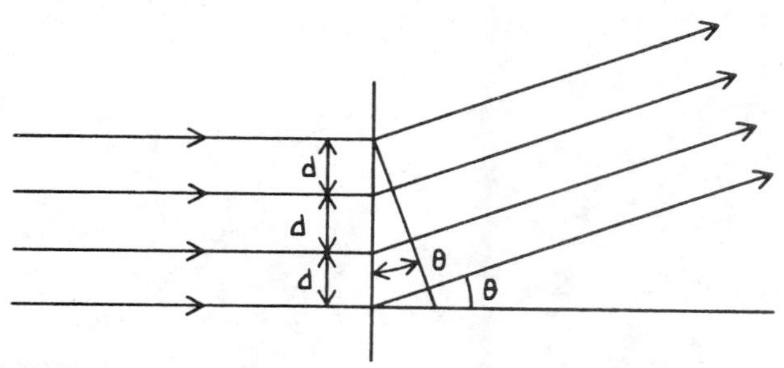

Figure 39.8 Constructive interference ($d \sin \theta = n\lambda$) for a plate with four evenly spaced slits.

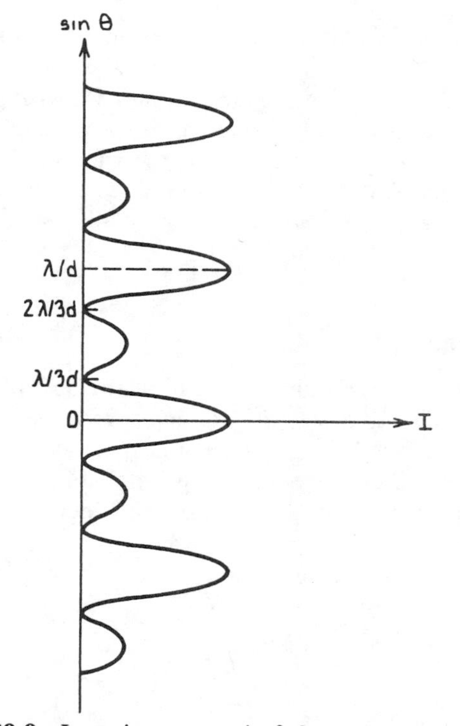

Figure 39.9 Intensity versus sin θ for a three-slit system.

two minima between each maximum of intensity, with a small "secondary" maximum between them, as shown in Figure 39.9. For a four-slit system, there are *three* minima and two small secondary maxima between each principal maximum, and so on.

A multislit system can be used to analyze precisely light's colors, or wavelengths, since the maxima (except for the central maximum at $\theta = 0$) for different wavelengths occur at different angles. For a system with a very large number of slits, N, the principal maxima are very narrow. We can show this as follows: if we let $n = 1$ and, since θ is small, assume that $\sin \theta = \theta$, the first principal maximum is given by $d\theta_{max} = \lambda$, or $\theta_{max} = \lambda/d$. The adjacent minimum is given by

$$d\theta_{min} = (N - 1)\lambda/N$$

so that

$$\theta_{min} = (N - 1)\lambda/Nd$$

Then

$$\theta_{max} - \theta_{min} = \frac{\lambda}{d}\left[1 - \frac{(N-1)}{N}\right] = \frac{\lambda}{Nd}$$

or

$$\Delta\theta = \frac{\lambda}{Nd}$$

The larger the number of slits, N, the narrower the width of the maxima. *The resolving power* of a slit system is the smallest wavelength difference that can be detected and is defined by the ratio $\lambda/\Delta\lambda$, where $\Delta\lambda$ is the smallest wavelength difference that can be detected for a wavelength λ. We can show that $\lambda/\Delta\lambda = Nn$, so that to detect a small wavelength difference, we need a large number of slits, N. A system with a very large number of slits is called a *diffraction grating*. Diffraction gratings usually have several thousand slits per centimeter of width.

Sample Problems

1. *Worked Problem*. Two rather thick, uniform glass plates are separated at one end by a needle of diameter h, forming a wedge-shaped air gap. Interference occurs between light rays reflected at the glass–air interfaces at the top and the bottom of the air wedge, as shown in Figure 39.10. Sunlight is incident on the upper plate at nearly normal incidence.

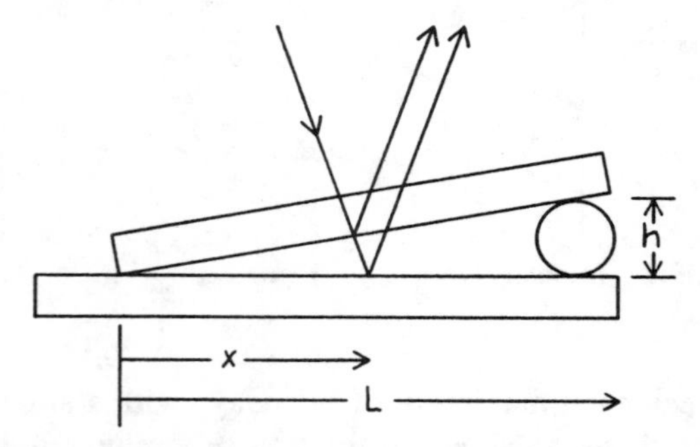

Figure 39.10 Interference in a wedge-shaped air gap.

a. Will you expect a dark or a bright fringe at $x = 0$? Explain your answer.
b. If $L = 10$ cm and $h = 0.1$ mm, find the positions of the first and the second bright red ($\lambda = 700$ nm) and violet ($\lambda = 400$ nm) fringes (1 nm $= 10^{-9}$ m).

Solution

a. At the upper interface, since light is passing from glass into air and therefore the index of refraction is decreasing, there is no phase change on reflection. At the lower interface there is a 180° phase change on reflection. At $x = 0$ the path difference between the upper and the lower reflected rays is zero; because of the 180° phase change at the lower surface, however, the two reflected waves are 180° out of phase and destructive interference occurs. We therefore expect a dark fringe at $x = 0$.

b. If $L = 10$ cm and $h = 0.1$ mm, the wedge thickness d a distance x cm from the edge of the wedge is given by $d = xh/L$. There will be bright fringes at $2d = \lambda/2$, $3\lambda/2$ or $d = \lambda/4$, $d = 3\lambda/4$. The first and second bright red fringes therefore require wedge thicknesses d of 1750×10^{-8} cm and 5250×10^{-8} cm, respectively. The corresponding distances x are 1.75 and 5.25 mm, respectively. For green light the distances x are 1.0 mm and 3.0 mm for the first and the second bright fringes, respectively. The human eye can resolve such distances, so that we will see rainbowlike interference patterns along the wedge.

2. *Guided Problem.* A standing wave is set up by reflecting radio waves from a large, flat metal dish. It is found that maxima in the electric field of the standing waves occur at distances of 20, 60, and 100 cm from the dish.

a. What is the frequency of the radio waves?
b. Where are the corresponding maxima in the magnetic field?

Solution scheme

a. Where do the maxima in the electric field occur in terms of the wavelength? Using this information, solve for the wavelength of the radio waves.
b. Now that you know the wavelength of the wave, find the frequency.
c. Where do the maxima in the magnetic field occur in terms of the wavelength? Now that you know the wavelength, you can predict where the maxima will occur.

3. *Worked Problem.* A glass tube with parallel glass end plates and with an inside length of 8.0 cm is placed in one arm of a Michelson interferometer that is illuminated with green light (wavelength of 5300 Å). The tube is filled with air and has a rubber squeeze bulb attached to it that allows the pressure inside the tube to be increased above atmospheric. If the pressure in the tube is increased slowly until it is 10% above atmospheric, how many bright fringes will have passed the cross hair of the viewing telescope? Assume that the index of refraction of air varies linearly with the pressure; the index of refraction of air at atmospheric pressure is 1.000293. Causing fringes to move by increasing the pressure is a common laboratory demonstration.

Solution: The index of refraction for a vacuum is 1.0000; for air at atmospheric pressure it is 1.000293, so that the change caused by one atmosphere of air is 0.000293. The additional change caused by a pressure 10% above atmospheric should be (0.000293)(.1) = 0.000029. The index of refraction, n', for a pressure 10% above atmospheric is then 1.000322. Since the wavelength is inversely proportional to the index of refraction, the ratio

$$\frac{\lambda n}{\lambda n'} = \frac{n'}{n} = \frac{1.000322}{1.000293} = 1.000029$$

This is the ratio of the wavelength at atmospheric pressure to the wavelength at 10% above atmospheric.

In the tube 8 cm long, there are $8/5300(10^{-8}) = 150943.4$ wavelengths at atmospheric pressure; at 10% above atmospheric there are $(8)(1.000029)/5300(10^{-8}) = 150947.8$ wavelengths. The path length has therefore been changed by 4.4 wavelengths by increasing the pressure by 10% in the glass tube 8 cm long. Four bright fringes will therefore pass the cross hair of the viewing telescope as the pressure is increased to 10% above atmospheric.

4. *Guided Problem.* Sunlight, striking the wall of a soap bubble at nearly normal incidence, makes it appear a mixture of red and green colors ($\lambda_{red} = 700$ nm; $\lambda_{green} = 500$ nm). How can this be so? What is the thickness of the bubble wall if the index of refraction of the bubble is 1.35?

Solution scheme

a. Are the bright colors the result of constructive or destructive interference? What are the conditions for this kind of interference in terms of the bubble thickness? (Remember that this is a soap bubble!)

b. Can we perhaps have *simultaneously* constructive interference for red and green? Remember that the path difference for the two reflected waves does not have to be the same number of wavelengths for both the colors. Write down an expression for the thickness d of the bubble wall in terms of the wavelength of the red light, and do the same for the green light. Since the thickness d is the same for both colors, you can now equate the two equations. This will allow you to solve for the path difference in terms of the wavelength for both the red and the green.

c. You should now be able to solve for the thickness of the bubble wall.

5. *Worked Problem.* A double-slit apparatus with a slit separation $d = 0.09$ mm is placed in oil ($n = 1.40$). At what angle θ do the rays from the slits form the first minimum and the first maximum away from the central ray when light from a ruby laser ($\lambda = 693$ nm) is used to illuminate the slit apparatus? The apparatus is now removed from the oil. Where now do the first minimum and maximum occur? If a *very thin* piece of transparent plastic ($n = 1.50$) is used to cover only one of the two slits, we find that the central point on the screen, instead of being a maximum of intensity, is now a minimum. What is the thinnest possible plastic thickness that can cause this effect?

Solution: In oil the wavelength is $693/n = 693/1.40 = 495$ nm. The first minimum occurs when $d \sin \theta = \lambda/2$. Then

$$\theta_{min} = \sin^{-1} [\lambda/2d] = \sin^{-1} \left[\frac{495\ (10^{-9})}{2\ (.09 \times 10^{-3})}\right]$$

$$\theta = 0.16°$$

The first maximum occurs when $d \sin \theta = \lambda$. Then

$$\theta_{max} = \sin^{-1} [\lambda / d] = \sin^{-1} \left[\frac{495\ (10^{-9})}{0.09\ (10^{-3})} \right]$$

$$\theta_{max} = 0.32°$$

When the apparatus is removed from the oil, the wavelength is 693 nm. Again

$$\theta_{min} = \sin^{-1} [\lambda / 2d] = \sin^{-1} \left[\frac{693\ (10^{-9})}{(2)(.09 \times 10^{-3})} \right]$$

from which

$$\theta_{min} = 0.22°$$

$$\theta_{max} = \sin^{-1} [\lambda / d]$$

from which

$$\theta_{max} = 0.44°$$

When the apparatus is immersed in oil, the interference minima and maxima occur at smaller angles θ away from the central ray, since the effective wavelength is decreased by the factor $1/n$.

With the plastic sheet over one slit, the central point on the screen is now a minimum, indicating that the path difference for the two rays is one-half wavelength. Let the thickness of the plastic be t nm. The wavelength of light in the plastic is $\lambda' = 693/1.50 = 462$ nm. The wavelength of light from the slit without the plastic $= \lambda = 693$ nm. The number of wavelengths in the plastic is t/λ'. The number of wavelengths for the same distance t without plastic is t/λ. But since the path difference for the two rays is $\lambda/2$,

$$t/\lambda' - t/\lambda = 1/2$$

$$t \left[\frac{1}{\lambda'} - \frac{1}{\lambda} \right] = \frac{1}{2}$$

$$t \left[\frac{\lambda - \lambda'}{\lambda \lambda'} \right] = \frac{1}{2}$$

Substituting in values for λ and λ', we obtain $t = 693$ nm.

6. *Guided Problem.* When a student illuminates two narrow slits separated by 0.12 mm with laser light, she measures the distance between the central beam and the second lateral maximum on a screen 2.0 m away from the slits to be 1.6 cm.

a. What is the wavelength of the light?
b. At what lateral displacement is the first lateral maximum?
c. If she replaces the double slit with three slits with the same spacing between slits, where does she now see the first lateral principal maximum?
d. Where is the first lateral maximum if there are four slits, again with the same spacing between slits?

Solution scheme

a. What is the expression relating wavelength, the slit separation, and the angular position of the second maximum? From the information given in the problem, you can find the angle at which the second maximum is seen. Using this, find the wavelength of the light.
b. Knowing the wavelength of the light and the slit separation, find the angular position of the first maximum. Now that you know the slit-to-screen distance, find the lateral displacement of the maximum.
c. Is the expression for the position of the maxima any different if there are three, rather than two, slits? You can now find the lateral displacement of the first maximum if there are three slits.
d. Answer the same question if there are four slits. Just how does the interference pattern change as the number of slits is increased?

7. *Worked Problem.* A typical diffraction grating has 7000 lines per centimeter. If sunlight (400–700 nm wavelength) is shone at normal incidence on the grating, what angular ranges are spanned by the first-, second-, and third-order maxima, respectively, in the grating spectra? Will there be overlapping of the colors? Could you avoid any overlapping of the spectra by using a different grating? Explain.

Solution: The angular positions of the maxima are given by the expression $n\lambda = d \sin \theta$, where n is the order of the maximum. d, the slit separation, is here (1/7000) cm. Then $(\sin \theta)_{max} = n\lambda/d$. To get the angular range we must find θ_{max} for the wavelength extremes 700 nm (red) and 400 nm (violet).

Then for $n = 1$, $(\theta_{max})_V = 16.3°$; $(\theta_{max})_R = 29.3°$; for $n = 2$, $(\theta_{max})_V = 34.1°$; $(\theta_{max})_R = 78.5°$; for $n = 3$, $(\theta_{max})_V = 57.1°$; $(\theta_{max})_R$ would appear, if it were possible, at an angle greater than 90°. Light of 476 nm will have its third-order maximum at 90°, so that most of the third-order spectrum cannot be obtained with this grating. There is no overlapping of colors between the first- and second-order maxima, but there is overlapping between the second- and third-order maxima; the second-order maximum spans the angular range 34.1 to 78.5°, and the third-order maximum extends from 57.1 to 90°.

Suppose that we decide to use a grating with only 3000 lines per centimeter, to see if this will prevent overlapping of the colors in the second- and third-order maxima.

For $n = 1$, the angular range of the maximum extends from 6.9 to 12.1°. For $n = 2$, the angular range is from 13.9 to 24.8°. For $n = 3$, the angular range is from 21.1 to 39.1°. Again, we have overlapping of the colors between the second- and third-order maxima.

If we look at the expression we used to calculate the positions of the maxima, $(\theta_{max}) = \sin^{-1} (n\lambda/d)$, we see that for $n = 2$, the product of $2(\lambda_R)$ is always greater than the product of $3(\lambda_V)$ for $n = 3$. It is therefore not possible to avoid this overlapping of the colors by any choice of slit separation.

8. *Guided Problem.* The yellow doublet in the emission spectrum of sodium vapor has respective wavelengths $\lambda_1 = 589.0$ nm and $\lambda_2 = 589.6$ nm.

A diffraction grating 2 cm wide is used in an attempt to resolve the two spectral lines in first order ($n = 1$). What is the minimum number of lines per centimeter needed in the grating? At what angle does the doublet appear in first order? At what angle does it appear in second order ($n = 2$)?

Solution scheme

a. What is the expression for the resolving power of a diffraction grating in terms of the number of slits? Use this expression to solve for the total number of slits required. What is the smallest wavelength difference that we need to detect? Does it matter greatly which value of wavelength (λ_1 or λ_2) we use in the equation?
b. Knowing the total number of slits, find the number of lines per centimeter required.
c. What is the expression for the angular position of the first maximum in terms of the slit separation and the wavelength? Use this (remember that you now know the slit separation) to find the angular position of the doublet in both first and second orders. (It is usual to use the *average* wavelength of the doublet.)

9. *Worked Problem.* A large radio telescope has twelve parabolic antennas for picking up radio waves from a distant source arranged in a straight line. The separation between adjacent antennas is 0.5 km. The antennas are connected to a single radio receiver by waveguides of equal length. If the telescope is used to detect 2 cm wavelengths from a distant galaxy, what is the angular separation between the central maximum and the first principal maximum? What is the angular distance from the center of a principal maximum to the adjacent minimum?

Solution: The telescope, which we can consider a 12-element diffraction grating in which parallel radio waves from a distant source enter the antennas and interfere within the receiver, is shown schematically in Figure 39.11.

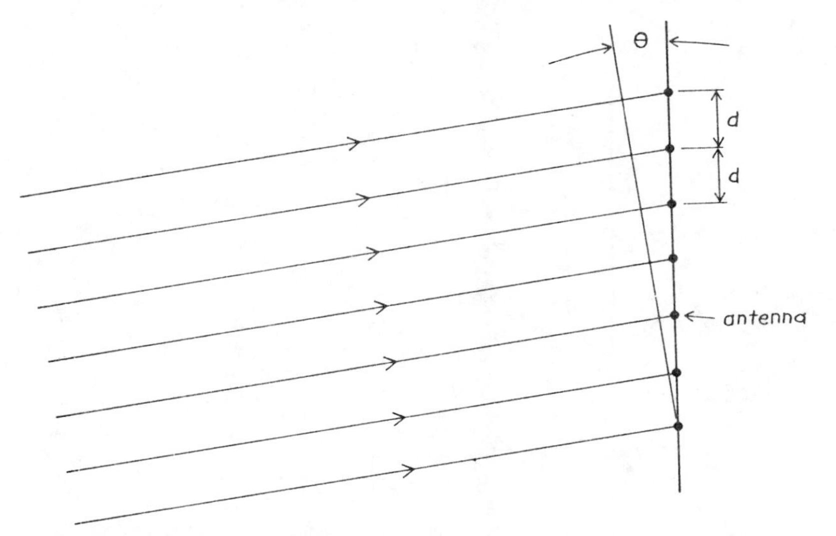

Figure 39.11 Problem 9. Part of a 12-element radio telescope. Constructive interference within the receiver will occur for values of θ such that $n\lambda = d \sin \theta$.

For the central maximum, $\theta = 0$. At the first principal maximum, $\theta \approx \sin\theta = \lambda/d = 0.02/500 = 4 \times 10^{-5}$ radian, or 8.25 seconds of arc. The separation of an adjacent minimum from a principal maximum is given by $\Delta\theta = \lambda/Nd$, where $N = 12$ and $d = 500$ m. Then $\Delta\theta = 3.33 \times 10^{-7} = 0.69$ second of arc.

Answers to Guided Problems

2. a. The frequency is 3.75×10^8 Hz.
   b. Maxima in the magnetic field occur at distances of 0, 40 cm, 80 cm, and 120 cm from the metal dish.
4. The bright colors are the result of simultaneous constructive interference for the two colors red and green. The expression for the bubble thickness d in terms of the respective wavelengths is

$$d = \frac{p}{4}\lambda_R \quad d = \frac{(p + 2)}{4}\lambda_G$$

   where p has possible values 1, 3, 5, and so forth. Since we know both of the wavelengths, however, we can solve explicitly for p, which gives $p = 5.0$. The bubble thickness is therefore $d = 5/4(700/1.35) = 648$ nm.
6. a. The wavelength is 480 nm.
   b. If $n = 1$, the first maximum is 0.8 cm from the central beam.
   c. For either three slits or four slits (or for any number), the first maximum is still 0.8 cm from the central beam. The maxima become narrower as more slits are added.
8. 494 lines per centimeter are needed in the grating.
   The doublet appears at an angle of 1.67° in first order and at an angle of 3.34° in second order.

Chapter 40

Diffraction

Overview

Diffraction refers to the bending of waves around obstacles and the spreading out of waves after their passage through narrow slits. Diffraction effects become noticeable if the size of the obstruction, or the width of the slit, is of the same order of magnitude as the wavelength, which for light is very small. Interference effects are due to the addition of two or more waves; diffraction is caused by the interaction of a single wave front with an obstacle. For example, a single wave front will spread out in passing through a narrow slit. We can, however, explain diffraction of light waves through the interference of rays that travel different distances, so that diffraction is in fact a very special case of interference. Although diffraction is a characteristic feature of waves, as discussed in Chapter 17 of your text, we will here restrict ourselves to the special case of the diffraction of light waves.

Essential Terms

diffraction by a single slit	Rayleigh's criterion
diffraction by a circular aperture	Babinet's Principle

Key Concepts

1. *Diffraction by a single slit*. Consider parallel rays of light passing through a narrow slit of width a, as shown in Figure 40.1(a). As we know from Huygens' Construction, the waves passing through the slit spread out in all directions. We will study the interference of the waves passing through different parts of the slit and hitting a screen a very long distance away from the slit; in such a case the rays heading toward any point on the screen are

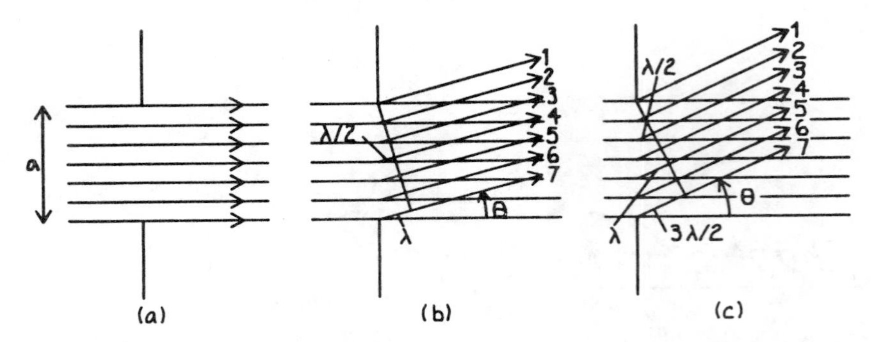

Figure 40.1 The interference of light rays passing through a single slit of width a causes diffraction maxima and minima.

parallel. Rays that pass straight through the slit undeflected all travel the same distance and so arrive at the screen *in phase*. There will be a central bright spot on the screen.

In Figure 40.1(b) we show rays moving at an angle θ such that the ray from the bottom edge of the slit (ray 7) moves exactly one wavelength farther than the ray from the top edge (ray 1). The ray passing through the center of the slit (ray 4) travels one-half wavelength farther than ray 1. These two rays, being one-half wavelength out of phase, will interfere destructively. Similarly, rays 2 and 5 will interfere destructively, as will rays 3 and 6. Rays between 3 and 4 will interfere destructively with rays between 6 and 7. Thus, all rays will interfere destructively in pairs, so that there will be a minimum of intensity on the screen at this angle θ, which is given by $\sin\theta = \lambda / a$.

Now consider Figure 40.1(c). The ray from the bottom of the slit, ray 7, travels $3\lambda/2$ farther than ray 1 from the top of the slit. Ray 5 travels a distance λ farther than ray 1, and ray 3, $\lambda/2$ farther than ray 1. Ray 1 will interfere destructively with ray 3; ray 2 will similarly cancel ray 4. Thus, rays from the top one-third of the slit will cancel rays in the middle third of the slit, leaving rays in the bottom one-third of the slit to reach the viewing screen; there thus will be a maximum of intensity on the screen but not nearly as bright a maximum as the central maximum. In general, for a single slit we find a bright central maximum, minima defined by $a \sin\theta = m\lambda$, where $m = 1, 2, 3, \ldots$, with small maxima in between the minima (the first small maximum occurring at $a \sin\theta = 3\lambda/2$). The intensity pattern is indicated in Figure 40.2. Note that the width of the central maximum is inversely proportional to the slit width; that is, the narrower the slit, the wider the central maximum becomes.

The calculation of the complete intensity distribution as a function of angle θ is similar to the calculation for the double-slit interference described in the last chapter. Here we must integrate over the amplitude of the waves originating from the slit, as indicated in Figure 40.3. An element dy of the slit contributes an amplitude

$$dE = (A/r)\cos(\omega t - kr)(dy)$$

where $r = r_0 - y\sin\theta$. Then

$$dE = (A/r_0)\cos(\omega t - kr_0 + ky\sin\theta)(dy)$$

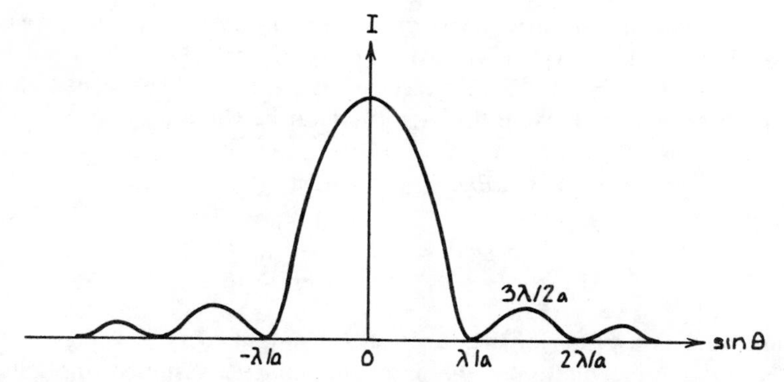

Figure 40.2 The intensity pattern in single-slit diffraction as a function of sin θ.

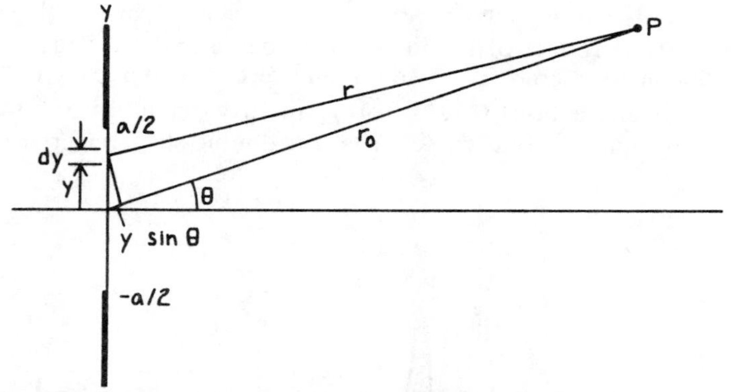

Figure 40.3 The calculation of the wave amplitude at point P due to a short element dy of the slit. The total slit width is a.

Since the denominator is not very sensitive to r, we have replaced r (in the denominator) by r_0. We now integrate over the width of the slit.

$$E = \int_{-a/2}^{a/2} dE = \int_{-a/2}^{a/2} (A/r_0)\cos(\omega t - kr_0 + ky\sin\theta)dy$$

We obtain the intensity by integrating and squaring this expression for E:

$$I = E^2 \propto \frac{1}{r_0^2}\,\frac{\sin^2\left(\frac{\pi a}{\lambda}\sin\theta\right)}{\sin^2\theta}$$

By defining the parameter $\alpha = \pi a \sin\theta/\lambda$ we can express the intensity as $I \propto (\sin\alpha/\alpha)^2$ or, as it is usually written,

$$I = I_0\left(\frac{\sin\alpha}{\alpha}\right)^2$$

At $\theta = 0$, $\alpha = 0$, and since $\lim_{\alpha\to 0}\dfrac{\sin\alpha}{\alpha} = 1$, the central position corresponds to a *maximum* of intensity. There are secondary *maxima* given closely, but not exactly, by $\sin \alpha = \pm 1$—that is, $\pi a \sin \theta / \lambda = \pm 3\pi/2, \pm 5\pi/2, \pm 7\pi/2,$ . . . , so that $\sin \theta = \pm 3\lambda/2a, \pm 5\lambda/2a, \pm 7\lambda/2a,$ . . . . The intensity minima occur when $\sin \alpha = 0$ (excluding $\alpha = 0$). That is, when

$$\alpha = \pi a \sin \theta/\lambda = \pm\pi, \ \pm 2\pi, \ \pm 3\pi, \ . . .$$

so that

$$\sin \theta = \pm\lambda/a, \ \pm 2\lambda/a, \ \pm 3\lambda/a, \ . . .$$

Figure 40.2 is a graph of the function $I = I_0 (\sin \alpha/\alpha)^2$.

The two-slit calculation of the preceding chapter assumed implicitly that each slit was very narrow, so that the diffraction pattern was wide compared with the interference pattern and so could be ignored. Therefore, we considered only the interference pattern. The actual pattern that we see, however, is the product of the interference curve and a diffraction curve. Figure 40.4 shows a graph of a two-slit interference pattern; the dotted line, the single-slit diffraction pattern, forms an envelope and acts to decrease the secondary maxima that we see from the double-slit interference.

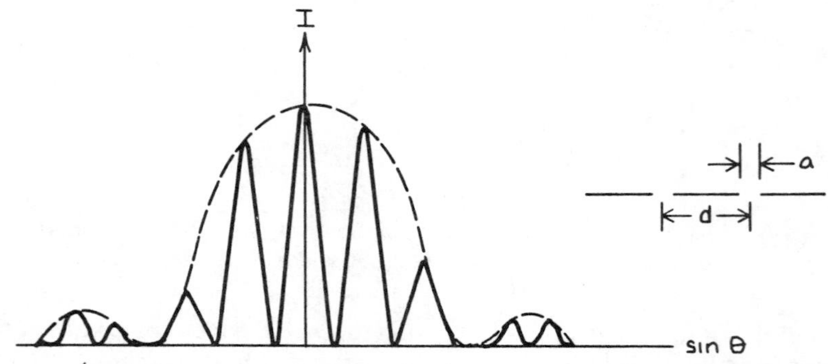

Figure 40.4 Intensity as a function of sin θ for interference from two slits. The dotted curve, which represents the intensity versus sin θ for a single slit, indicates how the two-slit interference pattern is modulated by single-slit diffraction. As indicated, the slit separation is d and the width of each slit is a.

2. *Diffraction by a circular aperture*. The diffraction of light by a circular aperture is similar to diffraction by a single slit; there is a strong central maximum and a sequence of minima and secondary maxima, just as for a single slit. The angular position of the first minimum is given by $\sin \theta = 1.22 \lambda/a$, where a is the diameter of the circular aperture. Many optical instruments, such as telescopes, microscopes, cameras, and even the human eye, have circular or nearly circular apertures, and these will diffract light. Parallel rays of light from a distant object arriving at such an aperture will be diffracted and produce an intensity distribution similar to that shown in Figure 40.2. The image of the distant object is thus "spread out" by the diffraction, putting a limit on the detail of the object that can be measured. Thus, the image of a distant point light source, such as a star, as viewed

through a telescope lens is a diffraction pattern whose central maximum has an angular half width given by $\theta = 1.22\ \lambda/a$. When two point objects are very close to each other, the diffraction patterns of their images will overlap. As the two objects are moved closer, a point is reached at which it is impossible to tell if there are two overlapping images or just a single image. *Rayleigh's criterion* says that two images can just be resolved when the center of the diffraction disk of one is directly over the first minimum in the diffraction pattern of the other, as in Figure 40.5. The angular separation of two images that can just be resolved is therefore given by $\theta = 1.22\ \lambda/a$.

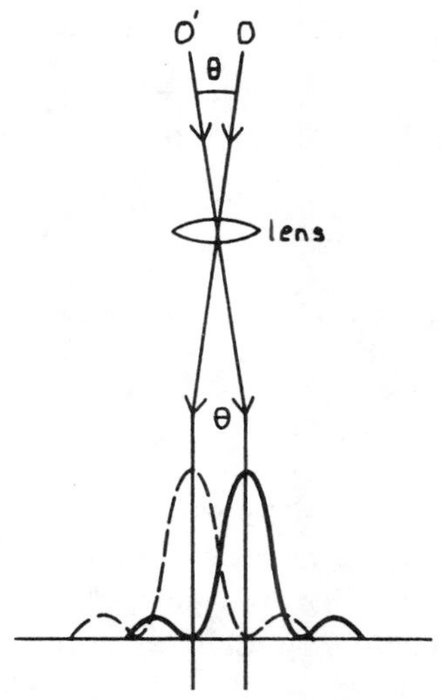

Figure 40.5 The center of the diffraction peak from O is just over the first minimum of the diffraction peak from O'. The two images can just be resolved.

3. *Babinet's Principle.* In our study of single-slit diffraction, we have assumed that the amplitude of the light in the region beyond the slit can be calculated by assuming that all points in the slit act as point sources of light. Babinet's Principle, which in part is a justification of this assumption, states that the field beyond the plate with a single slit in it, which we call $\mathbf{E}_{\text{beyond plate}}$, is the negative of the field resulting from a plug, or opaque object just the size of the slit itself, which we call $\mathbf{E}_{\text{rad,plug}}$. That is, $\mathbf{E}_{\text{beyond plate}} = -\mathbf{E}_{\text{rad,plug}}$. Thus, apart from a minus sign, the amplitude beyond the plate is just what we would calculate if we were to pretend that the points in the aperture radiate as point sources. A very interesting consequence of Babinet's Principle is that a plate with a single slit (or with any apertures in it) and its complement, an opaque plug, produce the same diffraction pattern (outside of the path of the original light wave).

Sample Problems

1. *Worked Problem.* In a laboratory demonstration, a He–Ne laser (wavelength 632.8 nm) is used to illuminate a single precision adjustable slit. A viewing screen is placed 2.0 m behind the slit. Sketch the central maximum

on the screen when the slit width is (a) 4 mm, (b) 1.0 mm, (c) 0.4 mm, (d) 0.15 mm, and (e) 0.04 mm. What is the slit width when the central maximum has a full width of 10.0 cm? Could you use this as a technique for measuring the width of a narrow slit?

Solution: The half width of the central maximum is given by $\sin\theta = \lambda/a$, and to a good approximation $\sin\theta = \theta = \tan\theta = w/200$, where w is the half width of the central maximum in centimeters. We can make a table of values:

a	$\sin\theta$	$2w$
4.0	$0.158(10^{-3})$	0.063
1.0	$0.633(10^{-3})$	0.25
0.4	$1.58\ (10^{-3})$	0.64
0.15	$4.22\ (10^{-3})$	1.68
0.04	$1.58\ (10^{-2})$	6.32

$2w$ is the full width of the central maximum in centimeters. When $a = 4.0$ mm, we obtain a value of 0.063 (or 0.63 mm) for the width of the central maximum; because the slit width is so much larger than the wavelength, the diffraction effect is small and will produce only a fuzzy, smeared edge to the 4 mm wide central maximum—it is *not* true that the width of the central maximum is *less* than the slit width. The central maxima are depicted in Figure 40.6. When the central maximum has a full width of 10.0 cm, we can obtain the width a from $\sin\theta = \lambda/a = 0.025$, so that $a = 0.025$ mm. This technique could be used to measure the width of a narrow slit, although for a very narrow slit the intensity pattern becomes extremely faint.

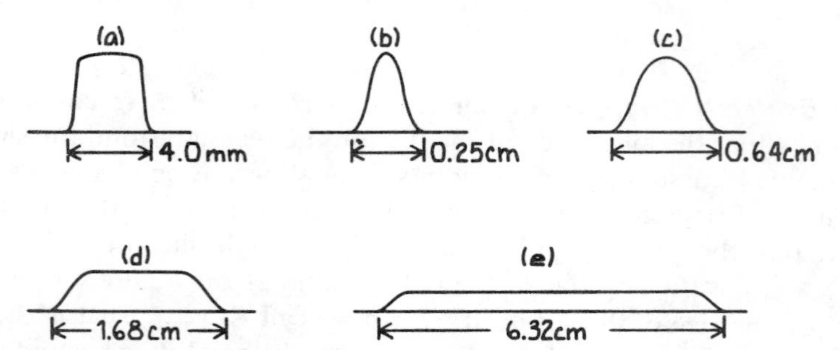

Figure 40.6 Problem 1. The central maximum for single-slit diffraction for varying slit widths. In (a) there is only a small diffraction effect.

2. *Guided Problem.* Two narrow slits are separated by a distance of 0.18 mm. When light is shone on the slits, the central diffraction peak contains nine double-slit maxima. What is the width of each slit? What is the ratio of slit width to slit separation?

Solution scheme

a. Write down where the maxima and minima are found for double slits. If the pattern is to contain nine double-slit maxima, can you express the limits of the pattern in terms of the slit separation? (It is worthwhile to sketch the observed intensity pattern to deduce just where the limits are.)

b. What is the half width of a single-slit diffraction pattern? Equate this to the limit you obtained for the two-slit pattern. When you do this, the wavelength should cancel, so that you do not need to know the wavelength of the light to solve the problem. You should now be able to solve for the slit width, a.

c. Since you already know the slit separation, d, you should be able to find the ratio a/d.

3. *Worked Problem.* Light of 600 nm is shown through a slit 0.03 mm wide. Plot the intensity distribution on a screen for an angular range of 5° on either side of the central maximum.

Solution: The expression for the intensity pattern from a single slit is given by

$$I = I_0 \left(\frac{\sin \alpha}{\alpha}\right)^2$$

with

$$\alpha = \frac{\pi a \sin \theta}{\lambda}$$

There are minima at $\sin \theta = \pm\lambda/a, \pm 3\lambda/a, \pm 4\lambda/a$ and since $\lambda/a = 600(10^{-6})/0.03 = 0.02$, the first four minima occur at angles of 1.15°, 2.29°, 3.44°, and 4.59°, respectively.

The maxima are given approximately by $\sin \theta = \pm 3\lambda/2a, \pm 5\lambda/2a, \pm 7\lambda/2a$, corresponding to $\sin \theta = \pm 0.03, \pm 0.05$, and ± 0.07, giving values of $\theta = \pm 1.72°, \pm 2.86°$, and $\pm 4.01°$.

To evaluate the relative intensities at these maxima we need to calculate the quantity $(\sin \alpha/\alpha)^2$. We tabulate the values as follows:

θ	1.72	2.86	4.01
α	4.5	7.85	11.0
$\sin \alpha$	0.98	1.0	1.0
$\left(\frac{\sin \alpha}{\alpha}\right)^2$	0.05	0.016	0.008

The observed intensity distribution is plotted in Figure 40.7. Note how quickly the intensity falls off away from the central maximum.

4. *Guided Problem.* Comment on the single-slit diffraction effect in a diffraction grating with 5000 lines per centimeter. If the slit width is one-third of the slit separation, what is the angular width of the central diffraction pattern in relation to the positions of the first-, second-, and third-order maxima for 632.8 nm light from a He–Ne laser?

Solution scheme

a. Write down the expression for the half width of the central maximum for single-slit diffraction in terms of the wavelength and the slit width. Knowing this, calculate the half width of the central maximum for the grating.

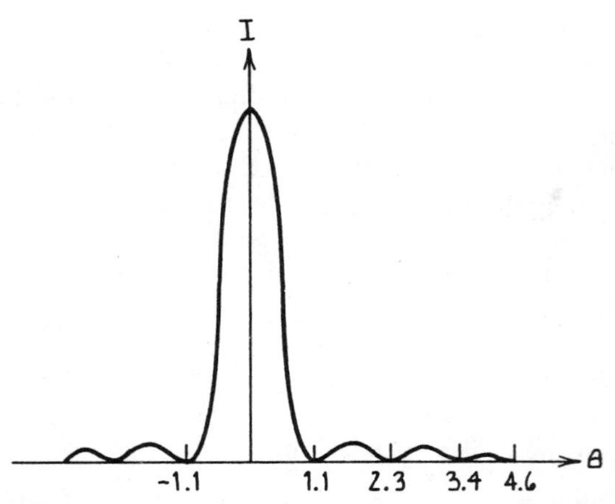

Figure 40.7 Problem 3. The intensity versus θ for a single slit of width 0.03 mm. The values of θ are in degrees.

b. Write down the expression for the position of the nth-order maximum caused by interference for the grating in terms of the distance between slits. Where do the first-, second-, and third-order maxima occur?

c. Does the minimum in the central maximum for diffraction coincide with any of the interference maxima? In other words, will we be able to see the first-, second-, and third-order maxima in the grating spectra? In most cases the slit width of a grating is less than $d/3$, so that the width of the central diffraction maximum is even greater than you calculate for this example.

5. *Worked Problem.* In a certain experiment, laser light is bounced off the surface of the Moon (the Earth-to-Moon distance is 3.84×10^5 km) and reflected back to Earth. If a He–Ne laser ($\lambda = 632.8$ nm) with an aperture diameter of 5.0 cm is used, what is the minimum central spot diameter of the beam when it his the Moon's surface? What *maximum* fraction of the light can be reflected back to Earth if the reflector diameter on the Moon is 2 m?

Solution: The half angle of the central diffraction spot on the Moon's surface is given by $\theta = 1.22\ \lambda/a = R/D$, where R is the radius of the central spot and D is the Earth-to-Moon distance. Therefore

$$R = (D)\ 1.22\ \lambda/a = 3.84(10^8)1.22(632.8 \times 10^{-9})/0.05$$
$$R = 5930 \text{ m or } 5.93 \text{ km}$$

The diameter of the central diffraction spot on the Moon is then 11.86 km.

To calculate the maximum fraction that can be reflected back exactly, we would need to assume that the reflector is located at the very center of the diffraction spot, where the intensity is greatest. Then if we knew the *average* intensity over the central maximum, $\bar{I}$, where $\bar{I}(\pi R^2) = \int I(\theta)dA$, we could calculate the fraction from

$$I_0[\text{reflector area}]/\bar{I}[\text{total spot area}]$$

We can get an estimate of the maximum fraction by assuming uniform intensity over the central spot, so that the fraction reflected back is just [reflector area]/[spot area] = $r^2/R^2 = 1/(5.93 \times 10^3)^2 = 2.8 \times 10^{-8}$.

6. *Guided Problem.* Considering only diffraction, what is the maximum distance at which the human eye can distinguish the two headlights on an automobile? The headlights are separated by 1.60 m, the pupil of the eye is 5.0 mm in diameter, and the wavelength of the light is 500 nm. At what distance can the headlights be distinguished if the wavelength is increased to 650 nm?

Solution scheme

a. Write down the expression for the angular separation of two images that can just be resolved.
b. Express the angle in terms of the headlight separation and the distance D of the automobile. Remember that since θ is very small, $\sin\theta = \theta = \tan\theta$ to a very good approximation.
c. Now solve for the distance D, and then solve for D when the wavelength is increased to 650 nm. By increasing the wavelength do you increase or decrease the resolving power of the eye?

7. *Worked Problem.* A spy satellite carries a camera with a lens 20 cm in diameter. It also carries ultraviolet and infrared scanners with effective lens diameters of 8 cm and 40 cm, respectively. Assuming the mean wavelength for visible light to be 5×10^{-5} cm, that for ultraviolet light to be 10^{-5} cm, and that for infrared light to be 10^{-3} cm, which detector should be able to detect the finest detail on Earth 100 km below the satellite? What are the approximate limits of linear resolution attainable with each device for objects on Earth?

Solution: The limiting resolution for each device is given by $\Delta\theta = 1.22\,\lambda/a$. Since we know a for all three devices, and the respective wavelengths, we can calculate $\Delta\theta$, and we obtain the following:

$$(\Delta\theta)_{\text{visible}} = 3.05 \times 10^{-6}$$
$$(\Delta\theta)_{\text{uv}} = 1.52 \times 10^{-6}$$
$$(\Delta\theta)_{\text{IR}} = 3.05 \times 10^{-5}$$

The highest resolution is therefore attainable with the ultraviolet device, namely, 1.52×10^{-6}.

To find the respective linear resolutions, since

$$(\Delta\theta) = [\text{linear resolution}]/[\text{distance from device}] = [\text{linear resolution}]/10^7\ \text{cm}$$

the respective linear resolutions (LR) are as follows:

$$(LR)_{\text{visible}} = 30.5\ \text{cm}$$
$$(LR)_{\text{uv}} = 15.2\ \text{cm}$$
$$(LR)_{\text{IR}} = 305\ \text{cm}$$

8. *Guided Problem.* Sketch the diffraction pattern produced when 693 nm light from a ruby laser is shone on a very narrow opaque strip 0.04 mm across.

Solution scheme

a. Refer back to Babinet's Principle, which says that a plate with a single slit and its complement, an opaque plug, produce the same diffraction pattern.

b. What will the diffraction pattern from the narrow opaque strip look like? Sketch it to determine the positions of the maxima and the minima.

9. *Worked Problem.* A circular opaque disk of diameter 0.15 mm is placed centrally in a laser beam with a diameter of 0.5 mm and a wavelength of 458 nm. Sketch the diffraction pattern made by the disk on a screen 5.8 m away. What is the diameter of the first dark circular ring?

Solution: From Babinet's Principle, we know that the diffraction pattern is the same as that made by a circular hole of diameter 0.15 mm in an opaque screen, and consists of a bright central maximum, followed by a dark circular ring and then by weak secondary maxima and minima. The diameter of the first dark circular ring, whose half angle is given by $\sin\theta = 1.22\lambda/d = 1.22(458 \times 10^{-9})/[0.15 \times 10^{-3}] = 3.725 \times 10^{-3} \approx \theta$, is therefore $3.725 \times 10^{-3} \times 2 \times 5.8 = 0.044$ m, or 4.4 cm. Note that the width of the central maximum, due to diffraction from the disk, is much broader than the 0.5 mm diameter of the laser beam. The diffraction pattern is sketched in Figure 40.8.

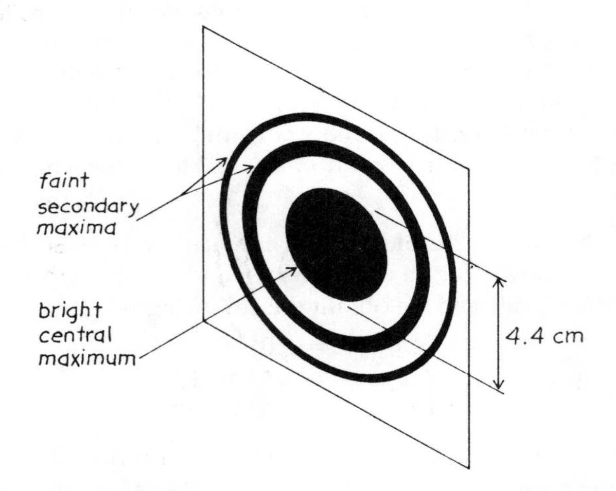

Figure 40.8 Problem 9. The diffraction pattern due to a laser beam shining on an opaque disk of diameter 0.15 mm.

10. *Guided Problem.* With a telescope lens of diameter 12 cm, what is the minimum separation at which we can resolve two objects on the central visible surface of the Moon? The objects are viewed in light of wavelength 500 nm. (The Earth–Moon distance is 3.84×10^8 m.)

Solution scheme

a. What is the expression for the angular separation, θ, of two images that can just be resolved by means of a telescope lens of diameter a? Can you calculate this for the lens of 12 cm diameter?

b. Using the known Earth–Moon distance, you should be able to find the minimum separation distance of two objects that are just resolved.

11. *Worked Problem.* Two slits with slit width $D=0.04$ mm and slit separation $d=0.10$ mm are illuminated with 450 nm light, and the resulting pattern is viewed on a screen 2.0 m away.

a. What is the full width of the central diffraction maximum on the screen?
b. How many interference maxima are contained in this diffraction maximum?
c. Sketch the interference pattern in the central diffraction maximum, and also sketch what you would observe in this region if one of the slits were covered with an opaque material.

Solution

a. The full width is given by $2\lambda/D = 2 \sin \theta \approx 2w/s$, where s = slit-screen distance. Therefore, the full width

$$2w = 2\lambda s/D = 2(450 \times 10^{-9})(2)/(0.04 \times 10^{-3}) = 0.045 \text{ m} = 4.5 \text{ cm}$$

b. Interference maxima are found at $\sin \theta = n\lambda/d = n\lambda/2.5D$, since $d = 2.5D$. Minima occur at $\sin \theta = (n + 0.5)\lambda/2.5D = \lambda/5D$, $3\lambda/5D$, λ/D, . . ., so that there are five interference maxima in the first diffraction maximum. The overall interference pattern is sketched in Figure 40.9.

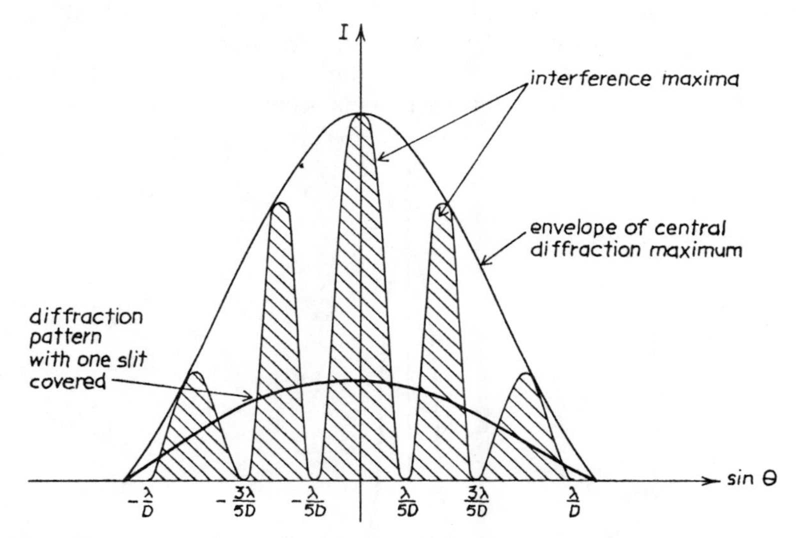

Figure 40.9 Problem 11. The two-slit interference pattern in the central single-slit diffraction maximum, with slit width 0.04 mm and slit separation 0.10 mm.

c. If one of the slits were covered with an opaque material the interference pattern would disappear, and we would see only the single-slit diffraction pattern. The intensity of the central maximum would be only one-fourth that of the central maximum of the two-slit interference pattern.

Answers to Guided Problems

2. The observed pattern is shown in Figure 40.10. The edges of the central *diffraction* peak are given by $\sin\theta = \pm 9\,\lambda/2d$. Therefore, $9\,\lambda/2d = \lambda/a$, so that $a = 2d/9 = 0.04$ mm; $a/d = 0.222$.

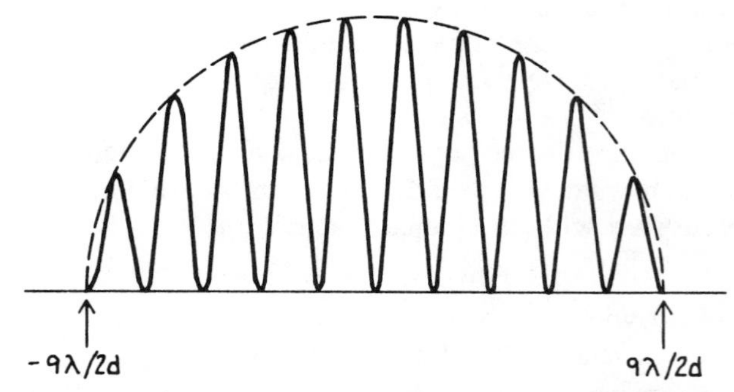

Figure 40.10 Problem 2. The central single-slit diffraction peak of this double-slit interference pattern contains nine double-slit maxima.

4. The angular width of the central diffraction pattern has a half width of $\theta = 71.6°$.

   The first-, second-, and third-order maxima occur at angles $\theta_1 = 18.4°$, $\theta_2 = 39.3°$, and $\theta_3 = 71.6°$, so that the *third-order* maximum will not be visible.
6. For 500 nm light, $D = 13.1$ km; for 650 nm light, $D = 10.0$ km. The resolution decreases as the wavelength increases.
8. The diffraction pattern is that of a single slit 0.04 mm across, as shown in Figure 40.11. The first three minima occur at 1.0°, 1.95°, and 2.92°, respectively.

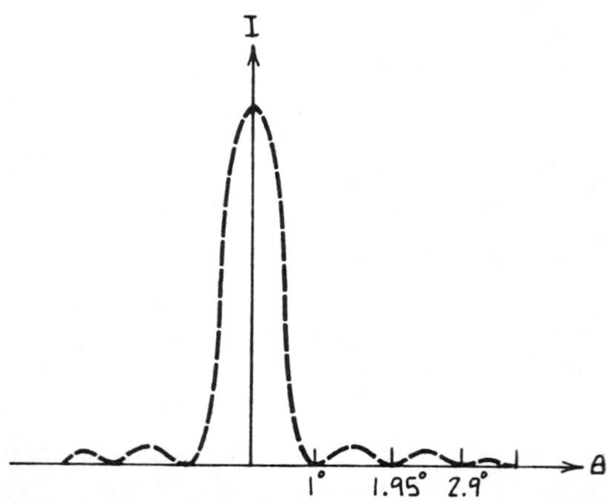

Figure 40.11 Problem 8. Diffraction pattern produced by 693 nm light on an opaque strip 0.04 mm wide.

10. Separation $s = R\,[1.22\lambda/a] = 1952$ m, or ~2 km.

Chapter 41

The Theory of Special Relativity

Overview

Although the theory of relativity has become almost a household phrase, it is understood by only a minute percentage of the population. The theory of special relativity runs counter to one's intuitive feelings about space and time. But intuition is based largely on experience, and Newtonian physics, which matches our intuition closely, is adequate for everyday situations. It is when speeds approach that of light that strange things begin to happen. In a sense, Einstein's theory of special relativity corrected classical mechanics to explain these strange happenings. In this chapter we deal with the basic postulates of relativity and some of the consequences of these postulates.

Essential Terms

ether
Principle of Relativity
time dilation
length contraction
spacetime
spacetime diagram
light-second
Lorentz transformation
addition of velocities
relativistic momentum
relativistic kinetic energy
relativistic total energy

Key Concepts

1. *Einstein's Principle of Relativity.* Having failed to detect the presence of the ether and, consequently, an absolute reference frame, scientists were at a loss to explain how the speed of light could be the same in all reference frames and the laws of electricity and magnetism remain valid for all inertial reference frames. Einstein proposed that one simply accept these as givens

and see what they implied for the basic relations developed in the Newtonian framework. His two postulates regarding the constancy of the speed of light and the equivalence of reference frames are statements based on evidence (or lack of it). Taking these as the basic operating principles, he then set out to see what, if any, changes had to be made in the Galilean relativistic relations and the dynamical laws of Newtonian physics. The most difficult to accept, perhaps, is the idea that the speed of light is the same to all observers, regardless of the particular inertial reference frames the observers happen to be in. Einstein made no attempt to explain why this should be so—it is an experimental fact, so we have to determine its physical consequences.

The most surprising result is that *simultaneity is relative*, that is, two events that are simultaneous in one inertial frame will not be simultaneous in another frame moving at constant velocity relative to the first. For example, suppose a flash of light is sent from station B to stations A and C located equally distant and on opposite sides of B. In the reference frame for which these stations are at rest, the signals (flashes of light) arrive at A and C simultaneously. The situation is shown in Figure 41.1, in which the worldlines for A, B, and C are vertical lines in the reference frame S, which is the rest frame of the three stations. The worldlines are vertical because the stations move in time but not in space. The worldlines for the light signals sent out from B cross the worldlines for stations A and C at the same time, as indicated by the horizontal dashed line representing equal times in S.

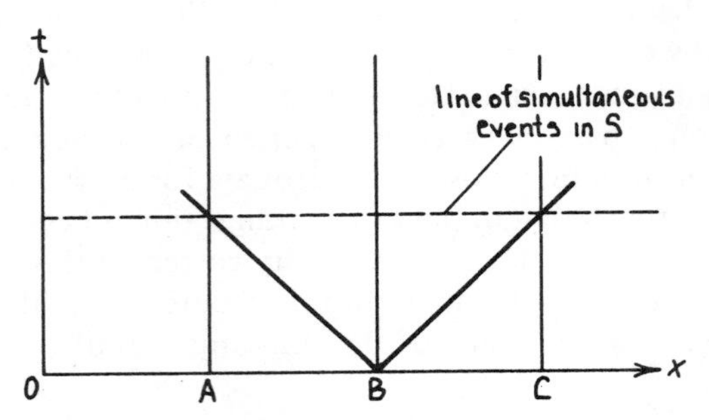

Figure 41.1 Worldlines for stations A, B, and C at rest in frame S. The worldlines for light signals sent out simultaneously from B arrive at A and C at the same time.

But now consider a reference frame S' moving at some velocity υ relative to S. Because the velocity of light must be the same in all inertial reference frames, the worldlines for the light signals remain unchanged. In S', however, the stations are moving with a velocity υ (both time *and* space motion); the worldlines for these stations therefore become oblique in S', as shown in Figure 41.2. As a result, the light-signal worldlines cross the worldlines for A and C at different times (measured in S'). In fact, according to S', the light signal reaches A before it reaches C. But in S the stations are at rest and are equidistant from B, so here these light signals must reach A and C simultaneously. Therefore, the dashed line in Figure 41.2 connecting the points at which the worldlines intersect is the locus of events that are simultaneous in S'. We see that events that are simultaneous in one frame are not simultaneous in another frame moving relative to the first.

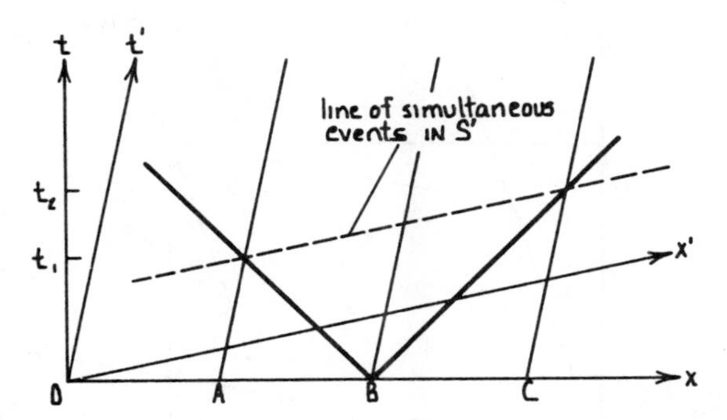

Figure 41.2 If stations A, B, and C are at rest in frame S moving at speed relative to S', the worldlines for these stations in frame S' are slanted (both x and t motion) but the worldlines for the light signal remain unchanged. Although the signals reach A and C simultaneously in S, they do not in S'.

Suppose S uses the light signals to synchronize all the clocks in that frame. For example, because all these clocks are at rest in S, the arrival of the light signal at a clock can be used to set the clock if the distance from the clock to the light source is known. Now suppose A and C have been given instructions to set their clocks to read 6:00:00 when they receive the light signal. (Recall that A and C are equidistant from B, so they should receive their signals simultaneously.) An observer in S' will note that C receives the light signal later than A, however, so that when C sets his clock to read 6:00:00, A's clock has already moved ahead because it was set to read 6:00:00 earlier. Therefore, when an observer in S asserts that all the clocks are synchronized, an observer in S' will argue that they are not. For example, if the S observer claims that the clocks at A and C read 6:00:00, the S' observer may claim that C reads 6:00:00 but that A reads, say, 6:00:05. In other words, time is not absolute. It is this lack of agreement about synchronization that results in the symmetry between frames of such phenomena as time dilation and length contraction.

2. *Lorentz transformation*. Using the information developed in the preceding section, let's see what form the coordinate transformations take when the postulates of relativity are used as constraints. As we know, the Galilean transformations will not suffice, because they are based on the assumption that simultaneity is the same in all inertial frames. But we also know that the Galilean transformations work very well for ordinary phenomena, so whatever new set of transformations we devise, they must reduce to the Galilean form when the relative velocity between frames is such that $v \ll c$. Figure 41.3 shows the space–time coordinates of some event, E, in two different inertial frames, S and S'. In the preceding section we saw that the events simultaneous in S' are not simultaneous in S, and vice versa. In Figure 41.2 the line connecting the intersections of the worldlines is the line of simultaneity for the system S', that is, the line of constant t'. By definition, the x' axis must be parallel to this line, so that the coordinate axes for the S' system are as shown in Figure 41.3. The (x,t) and (x',t') coordinates are shown also in the figure. Now we wish to show how

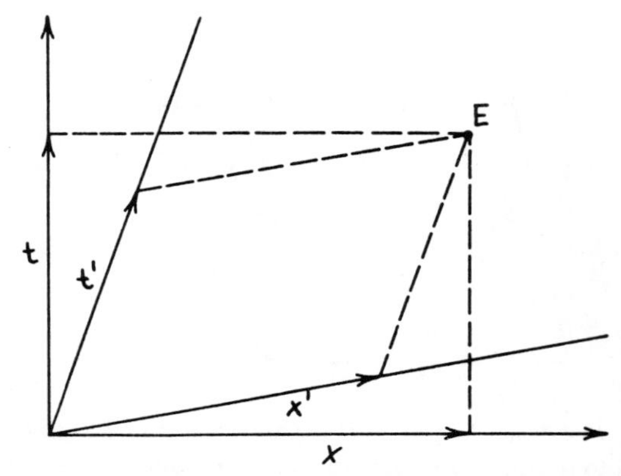

Figure 41.3 The coordinates of a single event, E, in both the S and the S' frames. The x' and t' coordinates are drawn according to Figure 41.2, in which the x' coordinate is parallel to the line of simultaneity in S'.

these two sets of coordinates are related. First, each set of coordinates must be a *linear* combination of the other if the Galilean law of inertia and the equivalence of reference frames is to hold. Therefore, we know that because of the symmetry of the transformations,

$$x = mx' + nt'$$

and

$$x' = mx - nt$$

Given these expressions, for $\upsilon \ll c$, the Galilean transformation laws hold —for example, for $\upsilon \ll c$, $m = 1$ and $n = \upsilon$. Now the motion of the *origin* of S' with respect to S is found by setting x' equal to zero in the second equation. Similarly, by setting x equal to zero in the first equation, we get the motion of the origin of S relative to S'. Thus

$$O = mx' + nt' = x' + \frac{n}{m} t'$$

and

$$O = mx - nt = x - \frac{n}{m} t$$

But since the velocity of each frame relative to the other is υ, we know that

$$\frac{n}{m} = \upsilon$$

Now consider a light signal traveling along the positive x direction, starting at the origin O in Figure 41.4. Because the speed of this signal, c, is the same in both frames, we have simply

$$x = ct$$

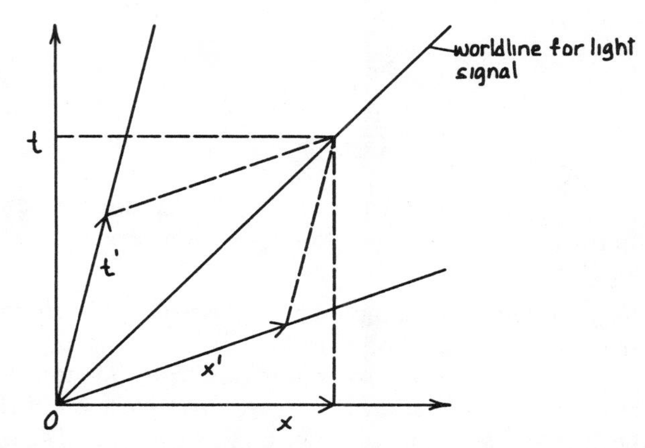

Figure 41.4 The worldline for a light ray emitted from the origin of the frames S and S'. Because the measured value of c must be the same in each frame, $x/t = x'/t'$.

and

$$x' = ct'$$

Note what this implies. Because c is postulated to be the same in both frames, if $x \neq x'$, then $t \neq t'$. This is the significant departure from the Galilean transformation. Now we get

$$ct = mct' + nt' = (mc + n)t'$$
$$ct' = mct - nt = (mc - n)t$$

Dividing these two equations gives us

$$\frac{c}{mc - n} = \frac{mc + n}{c}$$

or

$$c^2 = (m^2c^2 - n^2) = m^2\left(c^2 - \frac{n^2}{m^2}\right)$$

which, since $n/m = \upsilon$, gives

$$c^2 = m^2(c_2 - \upsilon^2)$$

or

$$m^2 = \frac{c^2}{c^2 - \upsilon^2} = \frac{1}{1 - \upsilon^2/c^2}$$
$$m = \frac{1}{\sqrt{1 - \upsilon^2/c^2}}$$

Note that when $\upsilon \ll c$, $m = 1$ as required and $n = \upsilon$. Our transformation equations now may be written as

$$x = \frac{1}{\sqrt{1 - \upsilon^2/c^2}} (x' + \upsilon t')$$

$$x' = \frac{1}{\sqrt{1 - \upsilon^2/c^2}} (x - \upsilon t)$$

These are the Lorentz transformations, and they are the same as those developed in the text, with the exception that in the text, the transformation for t is given also.

3. *The time dilation*. Although the mathematical development of the equations regarding time dilation is rather straightforward, the conceptual aspects of this phenomenon are sometimes difficult to reconcile with one's intuition. How can observers in each of two different inertial frames claim that the other's clocks are running slow? First, we must realize that the measurements are not made using only one clock in each frame. For example, for S to make measurements of the clock in S', S cannot carry his clock alongside S', for then both clocks would be moving at the same speed and there would be no disagreement. S must have a series or chain of clocks set up so that as the clock in S' passes by each of these clocks, comparisons can be made. But because S is moving relative to S', S' must also have a chain of clocks to make comparisons with *one* of the clocks in S as it passes by each of these clocks. Now suppose that when one of S's clocks, say, S_1, passes clock S_1', they both read 6:00:00. Later, when S_1 passes S_3', which now reads 6:00:10, an observer in S' notes that S_1 reads only 6:00:05. Now S will agree that his clock reads only 6:00:05 while S_3' reads 6:00:10, but S will say that S_3' is a different clock from S_1' and that these clocks were not synchronized properly in the first place. S is monitoring only one clock S', say S_1', and it will appear to be running slower than the clocks in S. The key to the difficulty is that each thinks his own clocks are synchronized and the other's are not.

4. *Length construction*. The arguments here are similar to those used in the preceding section, namely, that the observed length contraction is based on lack of simultaneity between frames. Suppose you wish to measure the length of a meter stick that is moving with some velocity υ. To measure the length of this stick in the stationary frame, one can note the *positions* of the front and the rear of the meter stick at some specified instant of time. For example, at an instant when all of S's clocks read (according to S) 6:00:00, measurements are made of the positions of the ends of the meter stick. Observers in S' will find that the meter stick is less than one meter in length. Now the observer in S will agree that S''s measurements showed a length less than one meter but will argue that the position measurements were *not made simultaneously* in S', because the clocks in S' are not properly synchronized. In fact, the S observer will note that the S' measurement of the position of the front end was made at one instant, whereas the measurement of the position of the other end was made at a later time. The end of the meter stick will have moved forward during that time interval and therefore will appear to be closer to the front end than it actually is. The arguments work both ways. Each claims the other made the position measurements at different times, not the same time. Again the problem is with the inability to synchronize clocks between frames.

5. *Combination of velocities*. The essential difference between the Galilean transformations and the Lorentz transformations is that the speed of light is a constant value, regardless of the reference frame, in the Lorentz transformations. Let's see what this means. In the Galilean formulation the velocities add in a very simple way. Suppose a bullet is fired from a rifle at a speed (relative to the rifle) of 500 m/s. If the rifle is traveling at a speed of 500 m/s relative to the ground, then an observer on the ground will measure the speed of the bullet to be 1000 m/s, and so on. If the bullet is replaced with a photon (particle) of light, then the ground observer would see the photon traveling at $c + 500$ m/s. Experimentally, however, light is never measured to travel at a velocity greater than c in any reference frame. Therefore, our velocity addition rules must be modified to account for this behavior of light. The velocity addition formulas derived from the Lorentz transformation laws do not allow the measured velocity of *any* object, from *any* inertial reference frame, to exceed c. But, as we have come to expect by now, the formulas reduce to the Galilean formulas when $v \ll c$.

6. *Momentum and energy*. We have seen that the conservation of momentum depends on the absence of any net external force acting on the system. Because collision processes involve internal action–reaction pairs of forces, momentum is conserved over the collision time if the collision time is very short or if the external forces are zero or small compared with the interaction forces. Consider the following thought experiment. A person standing on a stationary cart has a ball of mass m, identical to one held by another person standing on the ground. At a given instant they throw the balls directly toward each other with identical speeds. The balls collide elastically and return at the same speeds, as shown in Figure 41.5. Now let the

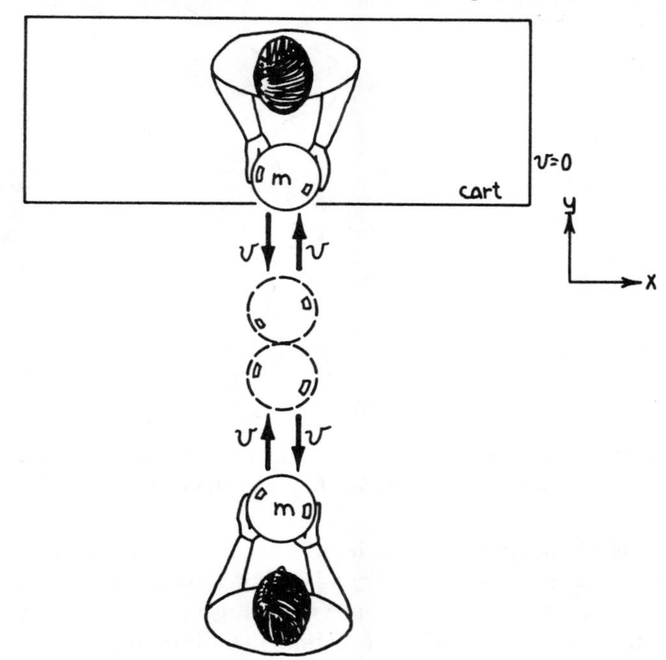

Figure 41.5 Completely elastic collision with both objects having zero velocity in the x direction.

cart move by with a constant speed υ relative to the person on the ground. For the balls to collide in such a way that each person in his own frame sees the collision the same way, they must be thrown before the two persons are directly across from each other (Figure 41.6). Each person, however, will see his ball come back the same way as before. Therefore, there has been no change from a momentum standpoint in the collision process. But because the cart is moving, the person on the ground sees all processes on the cart running more slowly, so the component of velocity of the ball transverse to the motion of the cart will also be measured as slower. But if the momentum is not different, then the mass must have increased! In fact, because time dilation is the responsible factor, we should not be surprised that the relationship for the comparison of momentum in the two frames is just

$$p(\upsilon) = \frac{m\upsilon_0}{\sqrt{1 - \upsilon^2/c^2}}$$

where $m\upsilon_0$ is the momentum of the ball in the rest frame. Note that as υ approaches c, the momentum increases indefinitely, indicating that c is the upper limit of the velocity.

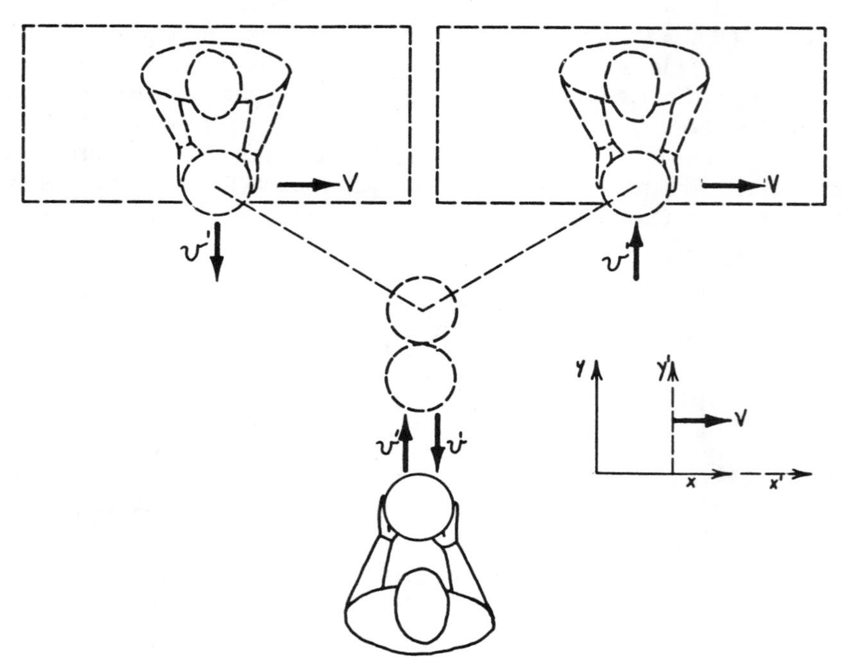

Figure 41.6 The same situation as in Figure 41.5, except that one of the balls has a velocity V along the x axis in the rest frame of the other ball.

But what about energy? First, let us refer to the many experiments involving subatomic particles and those involving radiation pressure. In these experiments there is ample evidence that photons behave like particles with momentum given by $p = E/c$. Now imagine a box of length L and mass M completely enclosed. A burst of radiant energy E is emitted at one end, as in Figure 41.7. If the radiation *does* carry momentum E/c, then the box must

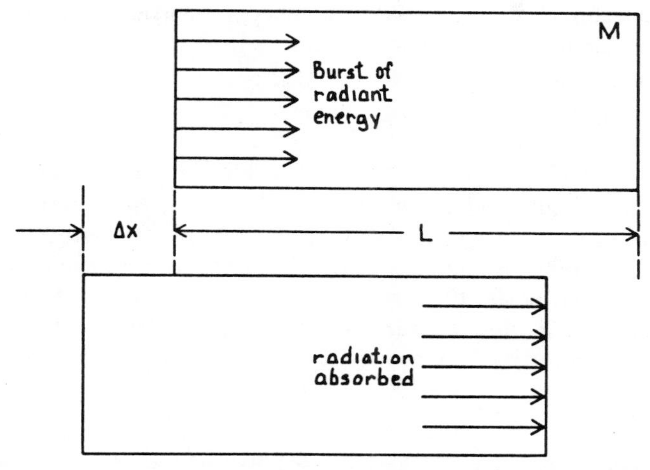

Figure 41.7 A burst of radiant energy causes the box to recoil and move a distance Δx before the energy is absorbed at the other end.

acquire an equal and opposite momentum $-E/c$, because we have an isolated system. The box, then, recoils with a velocity given by

$$\upsilon = \frac{E}{Mc}$$

After traveling the length of the box for a time $\Delta t = L/c$ (assuming $\upsilon \ll c$), the radiation is absorbed at the other end and the box comes to rest once more, because there is no *net* change in momentum. During this time the box has traveled a distance

$$\Delta x = \upsilon \Delta t = \left(-\frac{E}{Mc}\right) \times \frac{L}{c} = -\frac{EL}{Mc^2}$$

But because the system is isolated, the center of mass of the box plus its "contents" cannot have moved. If so, then the radiation must have associated with it a mass m such that

$$mL + M\Delta x = 0$$

Then

$$m = -\frac{M}{L}\Delta x = -\frac{M}{L} \times \left(-\frac{EL}{Mc^2}\right)$$
$$m = E/c^2$$
$$E = mc^2$$

This equation says that the energy of the photons can be expressed as the photon mass multiplied by c^2. But what about the box? It surely must have suffered a decrease in mass of E/c^2 when the radiation was emitted. Thus we can see that $E = mc^2$ applies to all matter, not just to photons. Actually, we should be more explicit and write the energy relation as $E(\upsilon) = m(\upsilon)c^2$ to indicate the velocity dependence, with

$$m(\upsilon) = \frac{m_0}{\sqrt{1 - \upsilon^2/c^2}}$$

where m_0 is the so-called *rest mass* of the particle and is the ordinary inertial mass we have been referring to in Newtonian dynamics. Whether one actually thinks of the mass as changing with velocity is not important. The pertinent dynamical variables are momentum and energy, given by $\mathbf{p}(\upsilon) = m(\upsilon)\mathbf{v}$ and $E(\upsilon) = m(\upsilon)c^2$. *In these expressions* m(υ) *is just a construct that allows us to preserve the form of momentum as mass × velocity.* In other words, the momentum is given by $\mathbf{p}(\upsilon) = \dfrac{m_0\upsilon}{\sqrt{1 \times + \upsilon^2/c^2}}$. If we wish to think of momentum as mass × velocity, then we can say that it is *as if* the inertial mass increased with velocity. The other point of view is to state simply that the (rest) mass of a body is an invariant property of that body and that **p** is velocity-dependent according to the above expression.

In the relativistic formulation the kinetic energy takes on no special role. It is merely the difference between the total energy of the particle and its rest energy, that is

$$K(\upsilon) = E(\upsilon) - E_0 = m_0c^2 \quad \frac{1}{\sqrt{1 - \upsilon^2/c^2}} - 1$$

and is the extra energy given to a particle through the work done by external forces.

One caution: Because we were able to derive a relativistic formula for momentum by making a simple replacement $m \rightarrow m(\upsilon)$, we might be tempted to express the relativistic kinetic energy in terms of the Newtonian expression $\frac{1}{2}m\upsilon^2$ by the same procedure. This will not work, however, as a direct application of $W = \int F \cdot dS$ will show.

Sample Problems

1. *Worked Problem.* Consider an inertial system S_2 moving at constant velocity V_1 relative to an inertial system S_1. Another inertial system S_3 moves at constant velocity V_2 relative to S_2. Show that the transformation of velocities from S_1 to S_3 is given by

$$\upsilon_{x1} = \frac{\upsilon_{x3} + V}{1 + \dfrac{\upsilon_{x3}V}{c^2}}$$

where

$$V = \frac{V_1 + V_2}{1 + \dfrac{V_1V_2}{c^2}}$$

Solution: Because V_1 is given relative to S_1, we can use the Lorentz transformation to go from S_1 to S_2. Then, because V_2 is specified relative to S_2, we can effect another transformation from S_2 to S_3. The velocity transformation has the general form

$$\upsilon_x = \frac{\upsilon_x' + V}{1 + \dfrac{\upsilon_x' V}{c^2}}$$

Then

$$\upsilon_{x1} = \frac{\upsilon_{x2} + V_1}{1 + \dfrac{\upsilon_{x2} V_1}{c^2}} \tag{1}$$

and

$$\upsilon_{x2} = \frac{\upsilon_{x3} + V_2}{1 + \dfrac{\upsilon_{x3} V_2}{c^2}} \tag{2}$$

Substituting Eq. (2) into Eq. (1), we get

$$\upsilon_{x1} = \frac{\left[\dfrac{\upsilon_{x3} + V_2}{1 + \dfrac{\upsilon_{x3} V_2}{c^2}}\right] + V_1}{1 + \dfrac{\upsilon_{x3}}{c^2}\left[\dfrac{\upsilon_{x3} + V_2}{1 + \dfrac{\upsilon_{x3} V_2}{c^2}}\right]}$$

where the term in the brackets is υ_{x_2}. This can be further simplified to

$$\upsilon_{x1} = \frac{\upsilon_{x3} + V_2 + V_1 + V_1 V_2 \upsilon_{x3}/c^2}{1 + \dfrac{V_2 \upsilon_{x3}}{c^2} + \dfrac{V_1 \upsilon_{x3}}{c^2} + \dfrac{V_1 V_2}{c^2}}$$

$$\upsilon_{x1} = \frac{\upsilon_{x3}\left(1 + \dfrac{V_1 V_2}{c^2}\right) + V_1 + V_2}{\left(1 + \dfrac{V_1 V_2}{c^2}\right) + \upsilon_{x3}\left(\dfrac{V_1 + V_2}{c^2}\right)}$$

Factoring out $\left(1 + \dfrac{V_1 V_2}{c^2}\right)$, we get

$$\upsilon_{x1} = \frac{\upsilon_{x3} + \left(\dfrac{V_1 + V_2}{1 + \dfrac{V_1 V_2}{c^2}}\right)}{1 + \dfrac{\upsilon_{x3}}{c^2}\left(\dfrac{V_1 + V_2}{1 + \dfrac{V_1 V_2}{c^2}}\right)}$$

which has the form

$$v_{x_1} = \frac{v_{x3} + V}{1 + \dfrac{v_{x3} V}{c^2}}$$

2. *Guided Problem.* Two inertial coordinate frames, S and S', have their origins coincident at $t = t'$. At that instant a flash of light is emitted from a point source located at the origin. Show that an observer in each frame sees an expanding spherical wave front traveling at speed c. Assume that the relative velocity between frames is V along the x direction.

Solution scheme

a. In frame S the equation for the wave front is $x^2 + y^2 + z^2 = (ct)^2$.
b. What are the equivalent expressions for x', y', z', and t' in frame S'?
c. What must be the observed value for c in each frame?
d. Show that the expression for the wave front in S' has the same form as in S.

3. *Worked Problem.* Consider once again two inertial frames, S and S', having coincident x axes. Frame S' moves relative to S with a speed of $0.5\,c$ in the x direction. A rocket is moving along the y axis with a speed of $0.2\,c$, measured in S. With what speed and in what direction does the rocket move as measured by an observer in S'?

Solution: Because the S' frame moves relative to S in the x direction, we know that $y = y'$. Since $dy = dy'$, however, we have

$$v'_y = \frac{dy'}{dt'} = \frac{dy}{dt'} = \frac{dy}{dt} \cdot \frac{dt}{dt'} = v_y \frac{dt}{dt'}$$

But

$$t = \sqrt{1 - v^2/c^2}\, t' + \frac{V_x}{c^2}$$

so that

$$\frac{dt}{dt'} = \sqrt{1 - v^2/c^2} + \frac{V}{c^2}\frac{dx}{dt} \cdot \frac{dt}{dt'}$$

$$\frac{dt}{dt'} = \sqrt{1 - v^2/c^2} + \frac{V}{c^2} v_x \frac{dt}{dt'}$$

$$\frac{dt}{dt'}\left(1 - \frac{Vv_x}{c^2}\right) = \sqrt{1 - v^2/c^2}$$

$$\frac{dt}{dt'} = \frac{\sqrt{1 - v^2/c^2}}{\left(1 - \dfrac{Vv_x}{c^2}\right)}$$

Then

$$v'_y = \frac{v_y \sqrt{1 - v^2/c^2}}{\left(1 - \dfrac{Vv_x}{c^2}\right)}$$

So we see that although $y = y'$, $\upsilon_y \neq \upsilon_y'$. This result stems from the time dilation between frames.

From the text, Eq. 41.36, we have

$$\upsilon_x' = \frac{\upsilon_x - V}{1 - \frac{\upsilon_x V}{c^2}}$$

Therefore, for the data given

$$\upsilon_x' = \frac{0 - 0.5\ c}{1 - 0} = -0.5\ c$$

and

$$\upsilon_y' = \frac{0.2\ c\sqrt{1 - \frac{(0.5\ c)^2}{c^2}}}{(1 - 0)} = 0.2\ c \times \sqrt{1 - 0.25}$$

$$\upsilon_y' = 0.17\ c$$

Thus, in the S' frame, the rocket has components $\upsilon_x' = -0.5\ c$ and $\upsilon_y' = 0.17\ c$, so that

$$v' = \sqrt{\upsilon_x'^2 + \upsilon_y'^2} = \sqrt{(-0.5\ c)^2 + (0.17\ c)^2}$$

$$\upsilon' = 0.53\ c$$

The angle made by the rocket's path with the x' axis is

$$\theta = \tan^{-1} \frac{0.17\ c}{-0.5\ c} = \tan^{-1}(-0.34)$$

$$\theta = 18.7^\circ \text{ with respect to the negative } x' \text{ axis}$$

4. *Guided Problem*. In the preceding problem, let the rocket ship be replaced by a beam of light traveling along the same path.

a. Calculate the speed and direction of this light beam as measured in S'.

b. Show that the calculated value for the speed υ' (but not the angle) holds for any value of V.

Solution scheme

a. For the υ_y' expression, substitute c for $0.2\ c$.

b. Calculate υ' as before. What do you expect to get?

c. Use V in the expressions for υ_x' and υ_y' to get the general result.

5. ***Worked Problem***. **The Milky Way galaxy is approximately 10^5 light-years in diameter. A proton is traveling at 0.99999 c relative to the rest frame of the galaxy. How long will it take the proton to traverse the galaxy as measured in**

a. **the rest frame of the galaxy?**

b. **the rest frame of the proton?**

Solution

a. In the rest frame of the galaxy, the time would be approximately 10^5 yr, because the proton is traveling at very nearly the speed of light.
b. In the rest frame of the proton, we have

$$\Delta t' = \Delta t \sqrt{1 - v^2/c^2}$$
$$\Delta t' = 10^5 \text{ yr} \sqrt{1 - \frac{(0.99999\, c)^2}{c^2}}$$

Although this expression can be evaluated directly, there are situations in which V may differ from c no more than, say, one part in 10^{10}. In such a case the limitations of the rounding errors on the calculator will prevent a solution. Therefore, we will take a slightly different tack, which can be used here as well as in the more extreme case. We know that

$$1 - \frac{v^2}{c^2} = \left(1 + \frac{v}{c}\right)\left(1 - \frac{v}{c}\right)$$

Then

$$\left(1 - \frac{v}{c}\right) = \frac{c - v}{c} = \frac{c - 0.99999\, c}{c} = 10^{-5}$$

and

$$\left(1 + \frac{v}{c}\right) = 2 - \left(1 - \frac{v}{c}\right) = 2 - 10^{-5} \approx 2$$

from which we get

$$\left(1 - \frac{v^2}{c^2}\right) = 2 \times 10^{-5}$$
$$\left(1 - \frac{v^2}{c^2}\right)^{1/2} = 4.5 \times 10^{-3}$$

Finally

$$\Delta t' = 10^5 \text{ yr} \times 4.5 \times 10^{-3}$$
$$\Delta t' = 4.5 \times 10^2 \text{ yr}$$

Comment: How can the proton cover a distance of 10^5 ly in only 450 yr? In the rest frame of the proton, the diameter of the galaxy is *not* 10^5 ly but, rather,

$$\Delta l = \Delta l' \sqrt{1 - v^2/c^2} = 10^5 \text{ ly} \times 4.5 \times 10^{-3} = 450 \text{ ly}$$

Thus, the proton sees a galaxy 450 ly in diameter moving at 0.99999 c relative to the rest frame of the proton.

6. *Guided Problem.* A space ship is traveling at 10^5 km/s relative to the Earth.
   a. If the ship travels at this speed for 2 hr of its own time, how much Earth time has elapsed?
   b. If the spaceship were to travel at this speed to a star 50 ly away (measured in the Earth frame) and return, how much time will have elapsed in each frame (the Earth and the spaceship)?

Solution scheme

a. What is the ratio v^2/c^2 for these data?
b. Using the time dilation formula, calculate Δt when $\Delta t' = 2$ hr.
c. What interval Δt is required for the ship to travel 2×50 ly at the given v/c?
d. What is $\Delta t'$ if Δt is the value calculated in part c?

7. *Worked Problem.* A particle of mass m at rest is subjected to a constant force F.
   a. Compare the Newtonian and the relativistic expressions for the speed attained by the particle after a time t.
   b. Compare the answers obtained for the final speed of a rocket of mass m experiencing a constant F/m equal to g ($g = 9.80$ m/s^2) for 1 yr.

Solution

a. For the Newtonian case we know that

$$v = at = \frac{F}{m}\, t$$

For the relativistic situation, we have, because F is a constant,

$$p = Ft$$

But we also know that

$$p = \frac{mv}{\sqrt{1 - v^2/c^2}}$$

Then

$$Ft = \frac{mv}{\sqrt{1 - v^2/c^2}}$$

which can be written

$$\frac{Ft}{mc} = \frac{v/c}{\sqrt{1 - v^2/c^2}}$$

$$\left(1 - \frac{v^2}{c^2}\right)\left(\frac{Ft}{mc}\right)^2 = \frac{v^2}{c^2}$$

Solving for υ^2/c^2:

$$\frac{\upsilon^2}{c^2} = \frac{(Ft/mc)^2}{1 + \left(\dfrac{Ft}{mc}\right)^2}$$

$$\frac{\upsilon}{c} = \frac{Ft/mc}{\left[1 + \left(\dfrac{Ft}{mc}\right)^2\right]^{1/2}}$$

$$\upsilon = \frac{Ft/m}{\left[1 + \left(\dfrac{Ft}{mc}\right)^2\right]^{1/2}}$$

which gives for small values of t

$$\upsilon = \frac{F}{m}\,t$$

which is the Newtonian result.

b. If the rocket has a constant $F/m = g$, then for the Newtonian case

$$\upsilon = gt = 9.80 \text{ m/s}^2 \times 3.16 \times 10^7 \text{ s} = 3.09 \times 10^8 \text{ m/s}$$
$$\upsilon = 1.03\ c$$

Thus the Newtonian formula predicts a speed greater than c. The relativistic formulation gives us

$$\upsilon = \frac{gt}{\left[1 + \left(\dfrac{gt}{c}\right)^2\right]^{1/2}} = \frac{9.80 \text{ m/s}^2 \times 3.16 \times 10^7 \text{ s}}{\left[1 + \left(\dfrac{9.80 \text{ m/s}^2 \times 3.16 \times 10^7 \text{ s}}{3 \times 10^8 \text{ m/s}}\right)^2\right]^{1/2}}$$
$$\upsilon = 2.15 \times 10^8 \text{ m/s} = 0.72\ c$$

8. *Guided Problem.* Suppose you wish to send a rocket probe to the nearest star, Alpha Centauri, which is 4.3 ly away. If you are able to give a constant $F/m = g$ to the rocket starting from rest on Earth, in how many years (Earth time) will the rocket reach the star?

Solution scheme

a. Start with the expression for υ in problem 7, then set υ equal to dx/dt.

b. Integrate $\int_0^x dx = \int_0^t \upsilon dt$. You will have an integral of the form

$$\int \frac{u\,du}{(1 + u^2)^{1/2}} \text{ with } u = Ft/mc.$$

c. Carry out the integration.

d. Note that $F/m = g$ does *not* mean that the force is constant. It means only that there is an external force on the rocket equal to mg, where m is the *rest mass*.

9. *Worked Problem.* A box at rest in the laboratory has dimensions given by $L_0 \times W_0 \times H_0$. The box is completely filled with a fluid of density $\rho_0 = 1.5 \times 10^3$ kg/m³. If the box is given a velocity 0.6 c along its L_0 dimension, what will be the measured value of ρ in the laboratory frame?

Solution: Let us call the density ρ equal to m/V in whatever frame we happen to be measuring it. In the rest frame of the box,

$$\rho_0 = \frac{m_0}{V_0} = \frac{m_0}{L_0 \times W_0 \times H_0}$$

In the rest frame of the laboratory,

$$\rho = \frac{m}{V} = \frac{m_0/\sqrt{1 - \upsilon^2/c^2}}{L_0\sqrt{1 - \upsilon^2/c^2} \times W_0 \times H_0}$$

Then

$$\frac{\rho}{\rho_0} = \frac{m_0(1 - \upsilon^2/c^2)^{-1/2}}{L_0(1 - \upsilon^2/c^2)^{1/2} \times W_0 \times H_0} \times \frac{L_0 \times W_0 \times H_0}{m_0} = (1 - \upsilon^2/c^2)^{-1}$$

$$\rho = \frac{\rho_0}{1 - \upsilon^2/c^2} = \frac{1.5 \times 10^3 \text{ kg/m}^3}{1 - \dfrac{(0.6\ c)^2}{c^2}} = \frac{1.5 \times 10^3 \text{ kg/m}^3}{0.64}$$

$$\rho = 2.34 \times 10^3 \text{ kg/m}^3$$

10. *Guided Problem.* The front of a building facing a street is half as high as it is wide. A high-speed spacecraft traveling parallel with the street sees the front of the building as a square. How fast is the spacecraft traveling relative to the building?

Solution scheme

a. If the height of the building is L, what is its width (in the building's rest frame)?
b. What is the relation between the widths measured in the two frames?
c. What width does the spacecraft observer measure?
d. Use the Lorentz transformations to calculate V.

11. *Worked Problem.* At what speed is the kinetic energy of a particle equal to 1.5 times its rest energy?

Solution: The kinetic energy is given by

$$K = \frac{m_0c^2}{(1 - \upsilon^2/c^2)^{1/2}} - m_0c^2 = E(\upsilon) - E_0$$

Thus when $K = 1.5\ E_0$

$$\frac{m_0c^2}{(1 - \upsilon^2/c^2)^{1/2}} - m_0c^2 = 1.5\ m_0c^2$$

$$\frac{m_0 c^2}{(1 - \upsilon^2/c^2)^{1/2}} = 2.5\ m_0 c^2$$

$$1 - \upsilon^2/c^2 = \left(\frac{1}{2.5}\right)^2$$

$$\upsilon^2/c^2 = 1 - \left(\frac{1}{2.5}\right)^2 = 0.84$$

$$\upsilon = 0.92\ c$$

In general, if $K = fE_0$, then

$$\upsilon^2/c^2 = 1 - \frac{1}{(1+f)^2}$$

For example, if we wish to know what f is for a given υ, we can write

$$f = \frac{1}{(1 - \upsilon^2/c^2)^{1/2}} - 1$$

which is just the term left in the parentheses if we factor $m_0 c^2$ out of the kinetic energy expression. For the electrons in the Stanford accelerator traveling at 0.99999999967 c, we have

$$(1 - \upsilon^2/c^2)^{1/2} = 2.57 \times 10^{-5}$$

$$f = 3.9 \times 10^4$$

that is, the electron kinetic energy is approximately 40,000 times its rest energy.

12. *Guided Problem.* Two identical bodies, each having rest mass m_0, are moving toward each other with equal speeds in the laboratory frame. They suffer a perfectly inelastic collision, after which the combined mass M_0 is at rest. Show that rest mass is not conserved in this collision.

Solution scheme

a. What is the total relativistic energy of the two masses before the collision?
b. What is the total energy of the mass M_0 after the collision?
c. Because total relativistic energy must be conserved, equate these two expressions and solve for M_0 in terms of m_0.

Answers to Guided Problems

2. $x'^2 + y'^2 + z'^2 = (ct)'^2$
4. a. $\upsilon' = c$ (as it must, according to the postulates of relativity)

   b. $\theta = \tan^{-1} \dfrac{0.87\ c}{-0.5\ c} = 60°$ with respect to the $-x$ axis

6. a. $\Delta t = 2.12$ h
   b. $\Delta t = 300$ yr
      $\Delta t' = 283$ yr

8. $x = \dfrac{mc^2}{F}\left\{\left[1 + \left(\dfrac{Ft}{mc}\right)^2\right]^{1/2} - 1\right\}$

   $t = 10$ yr
   Let $F/m = g$. Then

$$x = \frac{c^2}{g}\left\{\left[1 + \left(\frac{gt}{c}\right)^2\right]^{1/2} - 1\right\}$$

   and

$$g = \frac{2x}{t^2 - x^2/c^2}$$

   and

$$t = \sqrt{\frac{2x}{g} + \frac{x^2}{c^2}}$$

10. $0.87\,c$

12. $M_0 = \dfrac{2m_0}{(1 - v^2/c^2)^{1/2}}$; then $M_0 > 2m_0$

Chapter 42

Quanta of Light

Overview

The wave properties of light, as we have discovered, are all consistent with the predictions of Maxwell's equations. As physicists discovered in approximately 1900, however, light sometimes behaves not as a wave but as a particle, particularly at the atomic level. An entire new theory, called quantum physics, is required to fit the puzzling dual behavior of light.

Essential Terms

quantum physics
blackbody radiation
spectral emittance
ultraviolet catastrophe
Planck's constant
energy quanta
Wien's Law
Stefan–Boltzmann Law
photon
photoelectric effect
stopping potential
threshold frequency
work function
Compton effect
Heisenberg's uncertainty principle

Key Concepts

1. *Spectral emittance.* Any hot object, such as a glowing fireplace log, emits heat in the form of radiation. The wavelength of the emitted radiation has a continuous, smooth distribution of wavelengths over a wide range. The *spectral emittance,* S_λ, is defined as the power per unit area emitted by a hot surface per unit wavelength interval, so that $S_\lambda d_\lambda$ is the energy (or flux) emitted for wavelengths between λ and $(\lambda + d\lambda)$. The flux of thermal radiation emerging from the surface of a body depends on the characteristics of the surface.

2. *Blackbody radiation*. A body with a perfectly absorbing (and emitting) surface is called a *blackbody*. For a blackbody the spectral emittance depends *only* on the temperature of the body. S is a function only of the wavelength, λ, and of the temperature, T. The term *blackbody radiation* refers to the radiation from a blackbody. The spectral emittance of a blackbody at a temperature of 1500°K is shown in Figure 42.1.

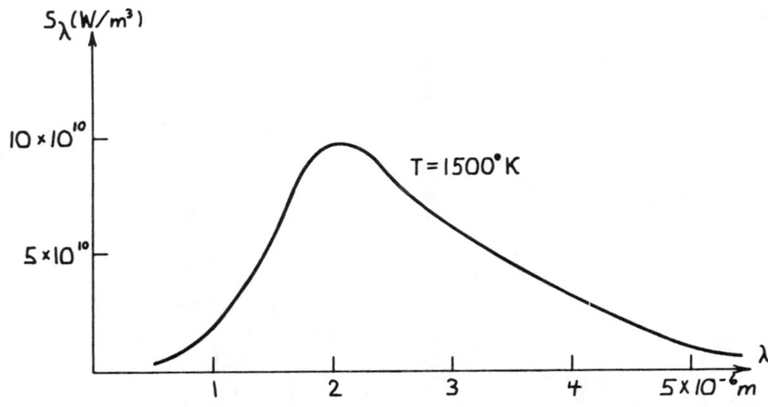

Figure 42.1 The spectral emittance of a blackbody whose surface temperature is 1500°K.

3. *Ultraviolet catastrophe*. An early theoretical calculation of the spectral emittance was made by Lord Rayleigh, who was able to show that a blackbody should contain an *infinite* amount of short-wavelength radiation and therefore should radiate an infinite amount of energy. This calculation, which was based on classical theory, was clearly unrealistic and became known as the *ultraviolet* (short-wavelength) *catastrophe*. This was the first instance of a total breakdown of classical theory.

4. *Planck's formula*. In 1900 Planck arrived at a formula that agreed precisely with the experimentally measured spectral emittance of a blackbody:

$$S_\lambda = \frac{2\pi c^2 h}{\lambda^5} \frac{1}{e^{hc/kT\lambda} - 1}$$

h is a new constant called *Planck's constant* and has the value 6.626×10^{-34} J · s. Although Planck knew that his formula was correct (that is, that it agreed with experimental findings), it was not consistent with classical theory.

5. *Energy quanta*. Planck could derive his formula only if he postulated that the energy states of the oscillating atoms inside a blackbody were *quantized*, so that the only values of energy possible were the discrete values $E = 0, h\nu, 2h\nu, 3h\nu, 4h\nu, \ldots$. Neither Planck's formula nor the quantization of energy states could be justified on the basis of classical theory.

6. *Wien's Law*. Two additional formulas describing the spectral emittance of a blackbody were discovered empirically, well before Planck arrived at his

formula. Both of these formulas are called "laws." The first, *Wien's Law*, says that the wavelength λ_{max} at which the spectral emittance of a blackbody at temperature T has a maximum is given by

$$\lambda_{max} T = 2.898 \times 10^6 \text{ nm} \cdot {}^\circ\text{K}$$

(1 nm = 10^{-9} m).

7. *Stefan–Boltzmann Law.* The second law, the *Stefan–Boltzmann Law*, says that the total flux emitted by a blackbody is given by

$$S = \sigma T^4$$

where $\sigma = 5.67 \times 10^{-8}$ W/(m$^2 \cdot {}^\circ$K^4). This formula can be obtained by integrating Planck's formula, that is, $S = \int S_\lambda d\lambda$.

8. *Photoelectric effect.* In 1905 Einstein, in seeking a rigorous derivation of Planck's formula, proposed that, instead of the energy of the vibrating atoms, the electromagnetic energy emitted by a blackbody is quantized according to $E = nh\nu$—that is, the electromagnetic radiation itself consists of packets of energy such that if an electromagnetic wave has frequency ν, each packet, or *photon*, has an energy $h\nu$. A wave can therefore be considered to be composed of particle-like photons, each with energy $h\nu$. Using statistical mechanics, Einstein, like Planck before him, was able to arrive at the correct expression for the spectral emittance of a blackbody. The main triumph of Einstein's theory, however, is his explanation of the *photoelectric* effect. When a light beam is shone on a metal surface inside a vacuum, electrons are emitted from the surface of the metal. Classical electromagnetic theory holds that the kinetic energy of the electrons should depend on the intensity of the light beam, but this is not observed in experiments. Instead, the kinetic energy of the emitted electrons increases linearly with the frequency of the light, as denoted by the straight line in Figure 42.2.

The equation of the straight line is $K = h\nu - \phi$, and is Einstein's photoelectric equation. The quantity, $\phi = h\nu_0$, is called the *work function* of the metal, because this is the energy needed to overcome the force binding the

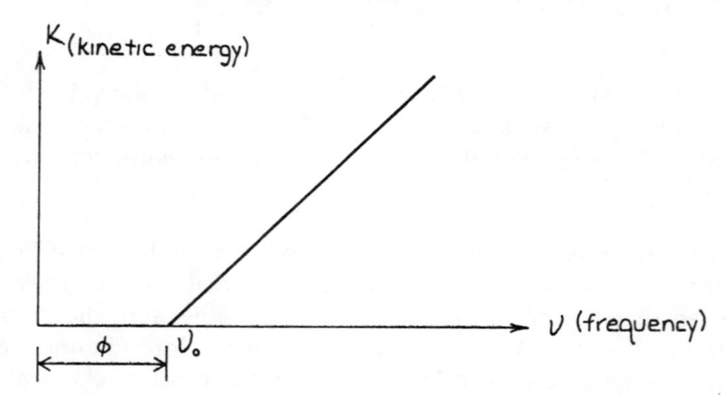

Figure 42.2 Kinetic energy of photoelectrons versus the frequency of the incident light in photoelectric emission. ϕ is called the work function of the metal.

electrons to the metal. The frequency ν_0 is called the threshold frequency, since photons of frequency less than ν_0 do not have enough energy, $h\nu_0$, to overcome the work function ϕ; if the light has a frequency less than ν_0, no electrons will be emitted from the surface of the metal. For frequencies greater than ν_0, the kinetic energy of the electrons is given by $K = h\nu - \phi = h(\nu - \nu_0)$, since this is the surplus energy of each photon after overcoming the binding energy of the electron to the metal surface. Increasing the intensity of the light beam increases the number of electrons emitted but does not change their kinetic energy.

9. *Compton effect*. In 1924 Compton, in measurements of the scattering of X rays from electrons, found that the scattered X rays had a wavelength *larger* than the incident X rays (that is, a *lower* energy). Again, this result contradicted classical theory. However, by treating the scatterings as collisions between individual photons and electrons (as indicated in Figure 42.3),

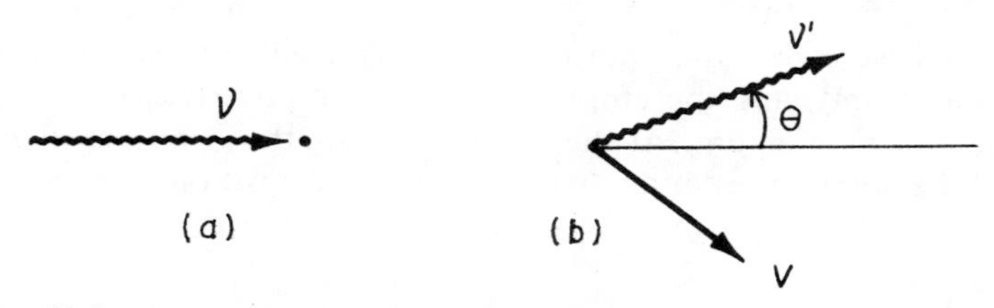

Figure 42.3 (a) A photon of frequency ν striking an electron. (b) The photon, with smaller frequency ν', is scattered through the angle θ, and the electron recoils with velocity υ.

Compton showed that the experimental results were exactly what the principles of energy and momentum conservation would predict. (The energy of a photon is $h\nu$; its momentum is $h\nu/c$, where c is the speed of light.) We require that the initial and the final energies and momenta are the same; this leads directly to the result that the increase in wavelength of the photon is given by

$$\Delta\lambda = \frac{h}{m_e c}(1 - \cos\theta)$$

It is clear that here the X ray behaves just like a particle. The Compton effect thus established definitively the existence of light quanta.

10. *Heisenberg's uncertainty principle*. We are left in a dilemma: In most everyday phenomena light, and all other electromagnetic radiation, behaves like a wave, but we now know that sometimes we must regard light as particle-like photons. Therefore, we must face the reality that neither model of light is adequate. In fact, if we go a little further, we will find that "particles" such as electrons and even neutrons and protons sometimes behave not like particles, but like waves, so that there is in nature a wave–particle duality that led, in the 1920s, to the development of what we call *quantum physics*. With respect to light (and to other particles in general), we find that a single photon always acts like a particle. When we deal with large numbers of photons (or of any particle), as we usually do in actual situations, the probability of find-

ing a photon at some point is proportional to the intensity of the wave at that point, as obtained from the wave theory of light. We used the word *probability*; this means that although we may know the likelihood of finding a particle at a certain point, because it has a wavelike character, we do not know the exact position of a given photon or particle. *Heisenberg's uncertainty principle* says that the product of the uncertainty in the position of a photon (Δy) and in its momentum (Δp_y) must necessarily be equal to or greater than Planck's constant, h. Thus, if y and p_y are, respectively, the y components of position, and momentum, then

$$\Delta y \Delta p_y \geqslant h$$

We can show how this relation comes about by passing a photon through a narrow slit. The slit's width determines an uncertainty Δy in the y position of the photon. The photon is diffracted by the slit, however; the narrower the slit, the greater the angular width of the central maximum (and all we know is that we expect the photon to be diffracted somewhere inside this central maximum), and therefore the greater the uncertainty Δp_y in the y component of momentum. At the atomic level we thus find that there exist inescapable uncertainties in any measurements of position and energy.

Sample Problems

1. *Worked Problem.* Plot a graph of the wavelength of the intensity maximum of the radiation of a blackbody versus the temperature of the blackbody for temperatures between 500°K and 8000°K. The maximum intensity of sunlight occurs at a wavelength of 480 nm (1 nm = 10^{-9} m). (The human eye has its maximum sensitivity at about this wavelength.) Assuming that the Sun is a blackbody, use your graph to find the approximate surface temperature of the Sun. As you heat an object, at about what temperature would you first expect the object to glow brightly, and why? The response of the human eye extends approximately from 400 to 800 nm.

Solution: This problem involves the application of Wien's Law,

$$\lambda_{max} T = 2.898(10^6) \text{ nm} \cdot {}^\circ\text{K}$$

For $T = 1000$°K, $\lambda_{max} = 2900$ nm, and for $T = 8000$°K, $\lambda_{max} = 362$ nm. A graph of λ_{max} versus T is shown in Figure 42.4.

For a λ_{max} value of 480 nm, we find a corresponding temperature on the graph to be about 6000°K, roughly what we would expect the surface temperature of the Sun to be. We would expect an object to glow very brightly when the wavelength of maximum intensity is within the visible region, that is, when the value of λ_{max} reaches about 800 nm. This occurs at a temperature of about 3600°K, so that we would expect an object to be glowing very brightly by the time it reaches this temperature.

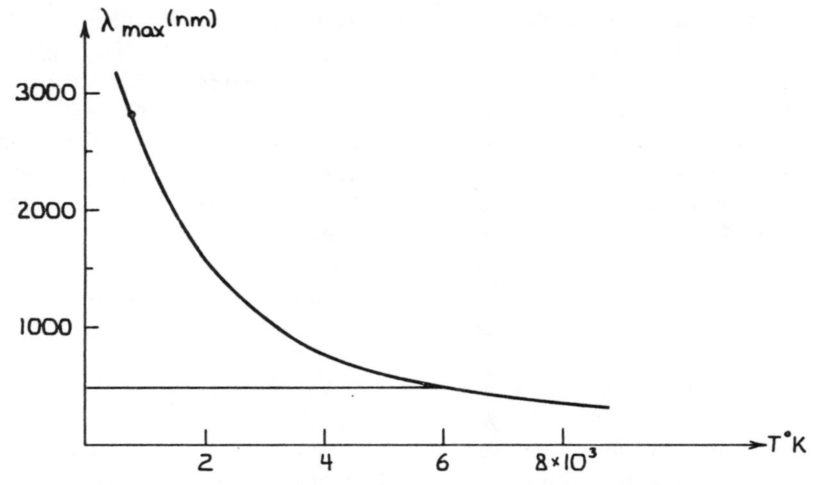

Figure 42.4 Problem 1. λ_{max} versus T for a blackbody.

2. *Guided Problem.* The flux of sunlight per unit area facing the Sun is 1.34×10^3 W/m^2 at the Earth's surface. Assuming that the Sun is a blackbody radiator, estimate a value for the surface temperature of the Sun. The Sun's radius is 7×10^8 m, and the Earth-to-Sun distance is 1.496×10^{11} m.

Solution scheme

a. What is the relationship between the total flux emitted by a blackbody and the temperature of its surface?

b. Write down the expression for the surface area of a sphere. From this, the Earth-to-Sun distance, and the flux of sunlight at the Earth's surface, you can determine the total power output of the Sun in watts per second.

c. From the result in step b you can find the flux emitted by the Sun in watts per square meter. Use the surface area of the Sun to find this value.

d. You are now ready to use the relation you wrote down in step a to find the surface temperature of the Sun. Don't be surprised if your answer does not agree closely with the answer in problem 1—we have made loose assumptions and are interested only in approximate results.

3. *Worked Problem.* If 5% of the energy supplied to an incandescent light is radiated as visible light, how many visible quanta are emitted per second by a small 10-watt bulb? Assume the wavelength of the light to be 520 nm. How many would enter a human eye (diameter of the pupil is 0.7 cm) 1000 m distant from the light bulb? Supposing that it takes at least 1000 photons per second for the eye to be able to register a light signal, could the eye, in principle, see the light from the bulb at a distance of 1000 m?

Solution: The energy of a photon is given by the expression $E = h\nu$, and since the wavelength and frequency are related through $c = \nu\lambda$, the energy is

$$\begin{aligned} E &= hc/\lambda \\ &= (6.63 \times 10^{-34})(3 \times 10^8)/(520 \times 10^{-9}) \\ &= 3.825 \times 10^{-19} \text{ J} \end{aligned}$$

Since the power emitted as visible light is $10 \times 0.05 = 0.5$ J/s, the number of 520 mm photons emitted per second is given by $0.5/(3.825 \times 10^{-19}) = 1.31 \times 10^{18}$. We assume that these are emitted uniformly in all directions, so that the fraction entering the eye at a distance of 1000 m is given by [area of eye]/[area of sphere at 1000 m radius] = F. Then $F = (0.35 \times 10^{-2})^2/(1000)^2 = 1.225 \times 10^{-11}$. The number per second entering the eye is just $F \times$ [no. emitted], or $(1.225 \times 10^{-11}) \times (1.31 \times 10^{18}) = 1.6 \times 10^7$ per second. Since the eye can respond to a signal of only 1000 photons per second, the eye can in principle see the signal from the light bulb very easily at a distance of 1000 m. This does assume that there is little "background" signal from other light sources.

4. *Guided Problem.* The minimum sensitivity of the human eye is about 1.5×10^{-11} J/m$^2 \cdot$ s. In other words, the power input from a light source must be at least this much for the eye–brain combination to "see" the light. If the diameter of the pupil of the eye is 0.6 cm, what is the power level (in watts) actually entering the eye at the given power level? How many photons per second of wavelength 500 nm enter the eye at this intensity level? Could the eye possibly see a single photon? Planck's constant $h = 6.63 \times 10^{-34}$ J • s.

Solution scheme

a. Calculate the area of the eye (in square meters), and so find the power level in watts actually entering through the pupil of the eye.
b. What is the expression for the energy of a photon in terms of its wavelength? (Remember that frequency and wavelength are related through $c = \nu\lambda$.) Use this expression to calculate the energy of each photon in joules.
c. From your answers to steps a and b, you should be able to find out how many photons per second are entering the eye. Does your answer surprise you? It partly explains why modern technology is needed to study quantum phenomena—the eye by itself is not an adequate sensor.

5. *Worked Problem.* Most photographic films contain silver bromide, AgBr. The absorption of light dissociates the molecules, exposing the film. If 1×10^5 J are needed to dissociate one mole of AgBr, find the energy and the wavelength of a photon just barely able to dissociate a single molecule. In what part of the electromagnetic spectrum does this lie? (The visible region extends approximately from 400 to 800 nm.) Is the film likely to be exposed by a high-power, 200 MHz radio beam?

Solution: Since 1 mole contains 6.022×10^{23} molecules, the energy per molecule to dissociate AgBr is $(1 \times 10^5)/(6.022 \times 10^{23}) = 1.66 \times 10^{-19}$ J. Since the photon energy is $E = h\nu$, this corresponds to a frequency, $\nu = E/h = (1.66 \times 10^{-19})/(6.63 \times 10^{-34}) = 2.5 \times 10^{14}$ Hz, or a wavelength $\lambda = c/\nu = (3 \times 10^8)/(2.5 \times 10^{14}) = 1.2 \times 10^{-6}$ m = 1200 nm. This wavelength is clearly well beyond the visible region and lies in the infrared part of the electromagnetic spectrum. Any visible photon will have enough energy to dissociate the AgBr molecule.

In a 200 MHz radio beam, the wavelength is given by $\lambda = c/\nu$, or $\lambda = (3 \times 10^8)/(200 \times 10^6) = 1.5$ m. This wavelength is too long by a factor of 10^6 to dissociate the AgBr molecule and so cannot expose the film.

6. *Guided Problem*. You are given the equipment shown in Figure 42.5 to study the photoelectric effect, with a sample of clean metal inside the evacuated tube. When the wavelength of the radiation is 236 nm (this is in the ultraviolet range), you find that the "stopping potential", V, is 1.04 volts—that is, for all voltages greater than 1.04 volts, zero current flows through the meter G.

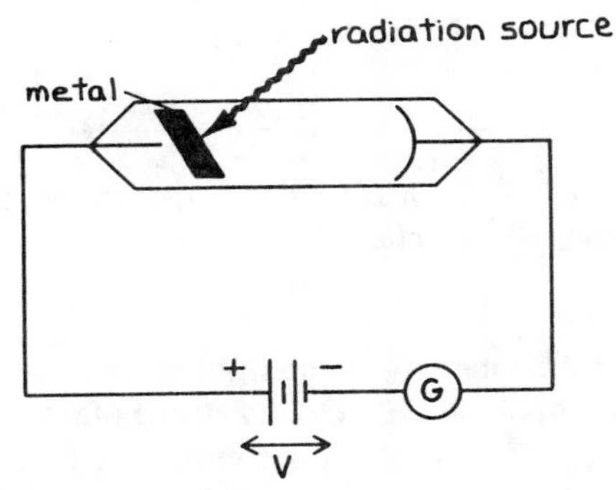

Figure 42.5 Problem 6. Equipment for the study of the photoelectric effect.

What is the work function of the metal in electron-volts? Given that potassium, chromium, and zinc have respective work functions of 2.26, 4.37, and 4.24 eV, identify the metal used in your study. If the metal were potassium, what "stopping potential" would be needed for the 236 nm radiation?

Solution scheme

a. If you look at the polarity of V, you will see that the electrons emitted from the metal must have enough kinetic energy to overcome V to reach the collector plate. The stopping potential is a measure of the maximum kinetic energy of the electrons, so a stopping potential of 1.04 volts means that the kinetic energy of the electrons emitted from the metal surface is 1.04. With this in mind, write down Einstein's photoelectric equation, and identify the terms.
b. It is easier to work with the photoelectric equation if all terms are expressed in electron-volts. Find the energy of the radiation, first in joules and then in electron-volts. (Remember that 1 eV = 1.6×10^{-19} J.)
c. What is the only unknown in the equation? Can you now find the work function of the metal? Can you identify the metal?
d. What is the work function for the potassium? You should now be able to calculate the kinetic energy of the emitted electrons.

7. *Worked Problem*. A sample of potassium metal (in a vacuum) is illuminated by a light source as indicated in Figure 42.6. The photoelectrons are collected on plate P, and the electron current is measured by meter G. The work function of potassium is 2.0 eV. Initially, red light is shone on the metal, but the wavelength of the light is slowly decreased. At what wavelength should a person first read a current on the meter? As the wavelength is slowly decreased until the light is violet, will the current continue to in-

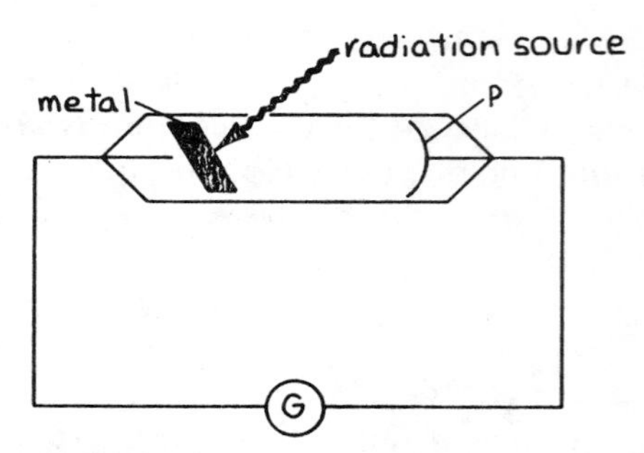

Figure 42.6 Problem 7. Equipment for the study of the photoelectric effect. Here the stopping potential used is zero.

crease? Assume that the intensity of the light stays constant. Find the kinetic energy (in electron-volts) and the velocity of the electrons emitted when the light source is violet, of wavelength 400 nm. The electron mass $m_e = 9.1 \times 10^{-31}$ kg.

Solution: Einstein's photoelectric equation is $K = h\nu - \phi$, where K is the kinetic energy of the photoelectrons. We should first see a current in the meter when the photon energy becomes equal to the work function of the metal, 2.0 eV. At this wavelength the photoelectrons will be ejected from the metal with zero kinetic energy, so that $h\nu = \phi = hc/\lambda$. The wavelength is given by $\lambda = hc/\phi$, so that $\lambda = (6.63 \times 10^{-34})(3 \times 10^8)/(2.0 \times 1.6 \times 10^{-19})$ (where we are converting 2.0 eV into joules), so that

$$\lambda = 6.24 \times 10^{-7} \text{ m} = 624 \text{ nm}$$

As the wavelength is slowly decreased, the current will *not* increase. The kinetic energy with which the electrons are ejected from the metal will increase.

When $\lambda = 400$ nm, the photon energy is given by $E = hc/\lambda$ or

$$E = h\nu = (6.63 \times 10^{-34})(3 \times 10^8)/(400 \times 10^{-9}) = 5 \times 10^{-19} \text{ J}$$
$$= (5 \times 10^{-19})/(1.6 \times 10^{-19}) = 3.12 \text{ eV}$$

Again applying Einstein's photoelectric equation, $K = h\nu - \phi$, we obtain $K = 3.12 - 2.0 = 1.12$ eV for the kinetic energy of the photoelectrons.

The velocity of the electrons is obtained from $\frac{1}{2}mv^2 = K = 5 \times 10^{-19}$ J, and since $m = 9.1 \times 10^{-31}$ kg, we obtain the value $v = 6.27 \times 10^5$ m/s.

8. *Guided Problem.* A 50 KeV photon collides with a free electron at rest. What is the wavelength of the photon before the collision? For which respective scattering angles will the photon experience its smallest and its greatest changes of wavelength? What are these wavelength changes? What are the respective energies of the scattered photon in each case?

Solution scheme

a. Express 50 KeV in joules (1 eV $= 1.6 \times 10^{-19}$ J). What is the relation between the energy and the wavelength of a photon?

b. Write down the expression giving the change in wavelength of a photon, $\Delta\lambda$, as a function of scattering angle θ, during Compton scattering. For what angle is $\Delta\lambda$ smallest (or zero)? For what angle is $\Delta\lambda$ largest? Substitute in your values of θ to find the minimum and the maximum values of $\Delta\lambda$.

c. Knowing the new wavelengths of the scattered photon, can you find the respective energies?

9. ***Worked Problem.*** A 0.51 MeV photon collides head-on with an electron at rest. (By "head-on" we mean that the photon is scattered through 180°.) What are the respective energies of the scattered photon and of the recoiling electron? Express your answer in MeV. At what angle relative to the incident photon does the electron recoil? Why doesn't the photon give up all its energy to the recoiling electron?

Solution: It is useful to illustrate this problem schematically, treating the problem as an instance of scattering between two particles. Figure 42.7

Figure 42.7 Problem 9. (a) A photon collides head-on with a free electron at rest. (b) The photon is scattered through 180°. The electron recoils directly backward.

provides such an illustration. The incident photon scatters directly backward (at 180°) at a lower frequency (and energy) ν'. To conserve momentum, the electron must recoil along the same direction as the incident photon—that is, the initial momentum $h\nu/c$ must equal the final momentum $m_e\upsilon - h\nu'/c$. To conserve energy the energy lost by the photon must equal the energy gained by the electron—that is, $h\Delta\nu = \frac{1}{2}m_e\upsilon^2$, where $\Delta\nu = \nu - \nu'$. These equations lead directly to the expression

$$\Delta\lambda = \frac{h}{m_e c}(1 - \cos\theta)$$

The change in wavelength of the photon after scattering is then

$$\Delta\lambda = \frac{h}{m_e c}[1 - (-1)] = \frac{2h}{m_e c} = 0.0486 \times 10^{-10}\ \text{m}$$

It is useful, but not necessary, to know that the quantity $h/m_e c = 0.0243 \times 10^{-10}$ m. This quantity comes up so often in atomic physics that it is called the "Compton wavelength" of the electron.

If we calculate the initial photon wavelength, before scattering, from $E = hc/\lambda$, we find that $\lambda = 2.437 \times 10^{-12}$ m, so that the wavelength after scattering $\lambda' = 7.24 \times 10^{-12}$ m. The energy of the scattered photon is therefore given by

$$E' = hc/\lambda' = (6.63 \times 10^{-34})(3 \times 10^8)/(7.24 \times 10^{-12}) = 2.75 \times 10^{14}\ \text{J}$$
$$E' = 0.172\ \text{MeV}$$

The energy of the scattered electron, through conservation of energy, is just $0.51 - 0.172 = 0.338$ MeV.

We find that in the collision process we have conserved both energy and momentum. If the photon were to give up all its energy to the recoiling electron, we could not *simultaneously* conserve energy and momentum. Therefore, the process cannot occur. You may remember that in the photoelectric effect, the electron disappears—that is, it gives up all its energy to the surface of the metal, and an electron is emitted. Momentum is conserved only if the electron interacts with the atom in the metal—photoelectric emission *cannot* occur in the scattering of a photon from a free electron, but only in the scattering of a photon with an electron bound in an atomic lattice.

10. *Guided Problem.* Suppose that we need to measure, at a given instant of time, the y component of both position and momentum of a 500 nm photon. Our equipment allows an accuracy of 0.01 nm in position. Estimate the minimum uncertainty in the frequency of the photon. What is the minimum uncertainty in the wavelength?

Solution scheme

a. Write down Heisenberg's uncertainty principle, relating the uncertainty in position and momentum. Do you know Δy?
b. Express the momentum in terms of the frequency, and the uncertainty in momentum, Δp, in terms of the uncertainty in frequency, $\Delta\nu$. (Remember that you are really just differentiating the momentum p in terms of the frequency ν.)
c. You can now find explicitly the uncertainty in frequency, $\Delta\nu$.
d. To find the uncertainty in wavelength knowing the uncertainty in frequency, we must differentiate the expression $\nu = c/\lambda$—that is, we need to express $\Delta\nu$ in terms of $\Delta\lambda$. Doing this, you should now be able to find $\Delta\lambda$, since you know both $\Delta\nu$ and λ.

11. *Worked Problem.* In the situation of the preceding problem, suppose that instead of equipment that allowed an uncertainty of 0.01 mm in position, we use equipment that allows a measurement of the wavelength of the 500 nm light to an uncertainty of 1 part in 10^{-3}. What now is the minimum uncertainty in the *position* of the photon?

Solution: According to Heisenberg's uncertainty principle, $\Delta y \Delta p_y \geqslant h$ where here $\Delta\lambda/\lambda = 10^{-3}$. Thus, the uncertainty in wavelength $\Delta\lambda = 500 \times 10^{-9} \times 10^{-3} = 0.5 \times 10^{-9}$ m. Then since $\nu = c/\lambda$,

$$\Delta\nu = c\Delta\lambda/\lambda^2$$

$$\Delta\nu = \frac{3(10^8)0.5(10^{-9})}{[500 \times 10^{-9}]^2} = 6 \times 10^{11} \text{ Hz}$$

Since the momentum $p = h\nu/c$, $\Delta p = (h/c)\Delta\nu$, so that $\Delta p = (6.63 \times 10^{-34}) \times (6 \times 10^{11})/(3 \times 10^8) = 1.326 \times 10^{-30}$ and the minimum uncertainty in position $\Delta y = h/\Delta p$. Therefore, $\Delta y = (6.63 \times 10^{-34})/(1.326 \times 10^{-30}) = 0.0005 \text{ m} = 0.5 \text{ mm}$.

If we measure the wavelength more accurately, we increase the uncertainty in position, just as, as the preceding problem demonstrated, if we increase the precision with which we measure the position, we increase the uncertainty in the wavelength.

12. *Guided Problem*. For Earth and its two nearest neighbors, Venus and Mars, we have the following data: Venus: 108×10^6 km from the Sun; radius 6052 km. Earth: 150×10^6 km from the Sun; radius 6378 km. Mars: 228×10^6 km from the Sun; radius 3397 km. The luminosity of the Sun is 3.9×10^{26} W. Each planet absorbs heat from the Sun and reradiates the same amount of heat as thermal radiation, behaving more or less like a blackbody. Estimate the average surface temperatures of the three planets in degrees Celsius.

Solution scheme

a. Knowing the distance of each planet from the Sun, and the total exposed area offered by each planet to solar radiation, calculate the total power absorbed by each planet from the Sun.
b. Your results give the total power each planet then reradiates. The Stefan–Boltzmann Law gives the total flux emitted by a blackbody as $S = \sigma T^4$ per m^2 of total surface area. For each planet, SA = the total power absorbed from the Sun.
c. You should now be able to calculate T (in degrees Kelvin) for each planet.

Answers to Guided Problems

2. $T = 5730°K$.
4. The power entering the eye is 4.24×10^{-16} watts, corresponding to 1060 photons per second. The eye cannot possibly respond to (see) a single photon of light.
6. The work function of the metal is 4.24 eV, so the metal is zinc. If the metal were potassium, a stopping potential of 3.02 volts would be needed for the 236 nm radiation.
8. Before collision the wavelength $\lambda = 0.245 \times 10^{-10}$ m. The photon will experience no change in wavelength and no change in energy for a scattering angle of 0° (no real scattering).

   The photon will experience a *maximum* change in wavelength, $\Delta\lambda = 2h/m_e c$, for a scattering angle of 180°. At 180° the change in wavelength $\Delta\lambda = 4.86 \times 10^{-12}$ m, and the energy of the scattered photon is $E' = 4.24$ KeV.
10. The uncertainty in frequency $\Delta\nu = 3 \times 10^{13}$ H.
    The uncertainty in wavelength $\Delta\lambda = 2.5 \times 10^{-8}$ m = 25 nm.
12. $T_{\text{Venus}} = 56°C$
    $T_{\text{Earth}} = 6°C$
    $T_{\text{Mars}} = -46°C$

Chapter 43

Atomic Structure and Spectral Lines

Overview

At about the same time, during the early part of the twentieth century, that some physicists were studying the puzzling particle-like behavior of light, others were concerned with the structure of atoms that emitted light and other electromagnetic radiation. It was soon discovered that atoms consist of a tiny but massive central positive nucleus, surrounded by electrons in roughly circular orbits. The electrons do not obey Newton's laws of motion, and in addition they have wavelike properties. Their behavior, like the behavior of light discussed in the last chapter, is described in terms of a new theory called quantum physics. The development of quantum physics is perhaps the greatest triumph of physics during the twentieth century.

Essential Terms

spectral lines
absorption spectra; Fraunhofer lines
Balmer series
Rydberg constant
"plum-pudding" model of the atom
Rutherford scattering
Rutherford atom
impact parameter
Bohr orbits
angular-momentum quantum number
correspondence principle
quantum mechanics
de Broglie wavelength
Schrödinger equation

Key Concepts

1. *Spectral lines*. Spectroscopy with optical prisms, as well as with diffraction gratings, became a well-developed science by the end of the nineteenth century. It was found that the light from any element could be resolved into a unique set of spectral lines, called emission lines, and that a wavelength could be measured for each line. Each element therefore had its own "signature," and its presence could therefore be distinguished from that of every other element. Moreover, it was found that an atom capable of emitting light of a given wavelength can also absorb light of the same wavelength. For example, if white light, a mixture of all wavelengths, is passed through hydrogen gas, the hydrogen atoms will absorb light of their characteristic wavelengths. The white light, after passing through the gas, therefore has dark lines, called *absorption lines*, in the white background. These lines are also called *Fraunhofer lines*.

2. *Balmer series*. Part of the hydrogen spectrum is shown in Figure 43.1. In 1885 Balmer, trying to find an interrelation in the wavelengths of the spectral lines, hit upon the following formula:

$$\lambda = 911.76\ \text{Å}\ 4n^2/(n^2 - 4)$$

where $n = 3, 4, 5, 6, \ldots$. This formula reproduces the wavelengths of the observed lines, which became known as the *Balmer series*. Balmer's formula can be rewritten in terms of the frequency:

$$\nu = cR_{\text{H}}\left(\frac{1}{2^2} - \frac{1}{n^2}\right) \quad n = 3, 4, 5, 6, \ldots$$

where c is the speed of light and $R_{\text{H}} = 1/911.76\ \text{Å} = 109678\ \text{cm}^{-1}$ and is called the *Rydberg constant*.

In 1908 another series of hydrogen spectral lines, later called the Paschen series, was found, with frequencies given by

$$\nu = cR_{\text{H}}\left(\frac{1}{3^2} - \frac{1}{n^2}\right) \quad n = 4, 5, 6, \ldots$$

It was soon realized that the spectrum of hydrogen really consisted of a large number of sets of spectral lines but that the observed frequencies could all be described by the single relation

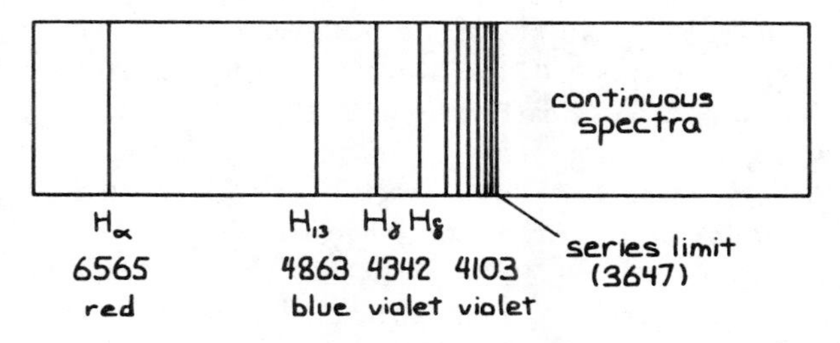

Figure 43.1 Part of the Balmer series of hydrogen. Wavelengths are in angstroms.

$$\nu = cR_{\mathrm{H}}\left(\frac{1}{n_2^2} - \frac{1}{n_1^2}\right)$$

where n_1 and n_2 are positive integers and $n_1 > n_2$. Most of the lines are in either the ultraviolet or the infrared part of the spectrum and so cannot be seen by the eye.

The observed frequencies of the spectral lines from other atoms could not be fitted by a single simple relation, but as with hydrogen, the observed frequencies could, in all cases, be expressed as the difference between two terms.

3. *Models of the atom.* The studies of spectral lines led scientists to wonder about the structure of the atoms emitting the radiation of such well-defined frequencies. Since it was soon clear that atoms are too small to be seen by even the most powerful microscope, models of the atom were postulated. An atomic model makes it possible to predict how the atom should respond to a given experiment, so that the model can be tested and either rejected or modified on the basis of experimental findings. One of the earliest atomic models, developed by Thompson, was called the "plum-pudding" model; it postulated that an atom consists of electrons embedded in a heavy cloud of positive charge. The model had the correct atomic size and mass and equal numbers of positive and negative charges, and it predicted that atoms could, when disturbed, emit light of various frequencies. It could not predict the observed frequencies, however.

The correct atomic model was developed by Rutherford in 1911, after he carried out what is now called a *Rutherford scattering experiment.* The experiment consisted of passing a beam of 5 MeV alpha particles, which are helium nuclei, through a very thin gold foil and measuring the angles through which the particles are scattered as they pass close to the atoms in the foil, as indicated in Figure 43.2.

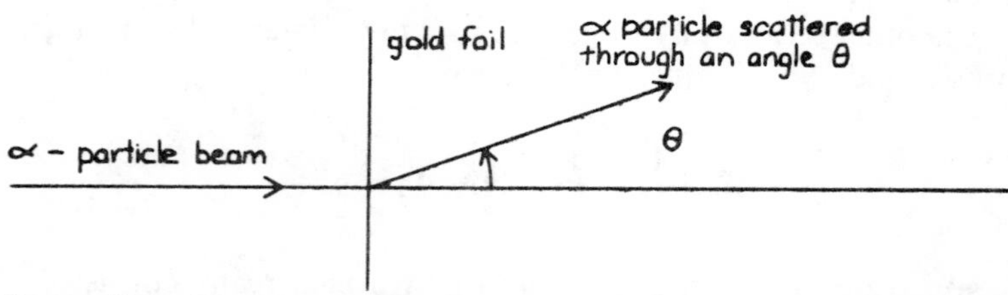

Figure 43.2 The scattering of an alpha particle through an angle θ by a thin gold foil.

If the "plum-pudding" model of the atom were correct, the angle θ would always be very small; instead, the observed scattering angles θ were often very large. The model was clearly wrong. Rutherford showed that his results agreed with an atomic model consisting of a very small, massive, positive "nucleus," with charge $+Ze$, surrounded by a negative "cloud" of Z electrons. This model of the atom soon became known as the *Rutherford model.* The scattering of the alpha particles is caused by the strong electric repulsive force between the alpha particle and the positive nucleus. The scattering is indicated schematically in Figure 43.3.

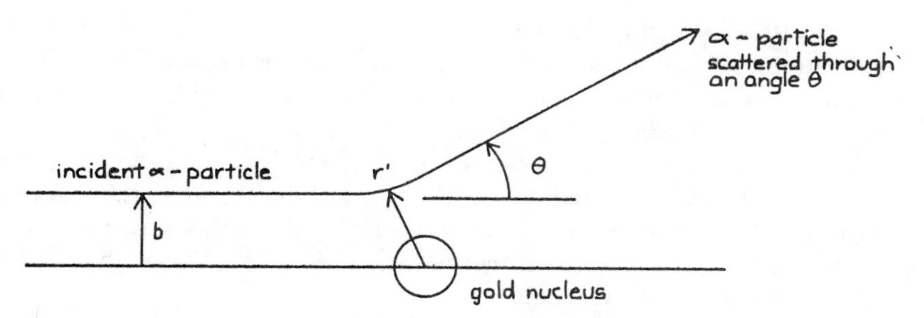

Figure 43.3 The Rutherford scattering of an alpha particle from a gold nucleus. r' is the distance of closest approach, and b is the impact parameter.

The smaller the *impact parameter, b,* the smaller the distance of closest approach, r', and the larger the scattering angle θ. If the impact parameter is zero, the alpha particle is scattered directly backward, through an angle of 180°. Suppose that $b = 0$. Then the incident particle slows down as it approaches the gold nucleus, and at some point, r' from the center of the gold nucleus, instantaneously stops and is repelled directly backward. From the principle of the conservation of energy, we know that the kinetic energy of the particle a long distance away, $\frac{1}{2}mv^2$, is equal to the potential energy $2Ze^2/4\pi\varepsilon_0 r'$ at the instant that the particle comes to a stop. Then

$$\frac{1}{2}mv^2 = \frac{2Ze^2}{4\pi\varepsilon_0 r'}$$

or

$$r' = \frac{Ze^2}{\pi\varepsilon_0 m v^2}$$

When the impact parameter is not zero, the situation is more difficult, but we can show that the distance of closest approach and the impact parameter are related as follows:

$$r' = \frac{Eb^2}{-\left(\frac{Ze^2}{4\pi\varepsilon_0}\right) + \sqrt{\left(\frac{Ze^2}{4\pi\varepsilon_0}\right)^2 + E^2b^2}}$$

where $E = \frac{1}{2}mv^2$ and the atomic number $Z = 79$ for gold.

4. *Bohr orbits.* Rutherford's model did not exlain the observed atomic spectral lines. The simplest atomic model, based on the need for Rutherford's massive, positive nucleus, is to allow the electrons to rotate in circular orbits around the nucleus. The electric force of attraction will then "bind" the electrons to the nucleus, just as the planets are bound to the Sun by gravitational attraction. It was quickly realized that the centripetal acceleration of an electron in a circular orbit should, according to classical theory, cause the electron to radiate energy. As the electron radiated, it would quickly collapse in toward the nucleus. Atoms would have a lifetime of about 10^{-10} s!

In 1913 Bohr, realizing that the laws of classical physics could not describe atomic structure, suggested that although the electrons revolve around the nucleus in circular orbits, only those orbits occur for which the angular momentum of the electron, L, is given by $nh/2\pi$, where $n = 1, 2, 3, 4, \ldots$ — that is, the angular momentum of an orbital electron is quantized. He also postulated that no electron radiates while in one of the allowed orbital states. Radiation occurs when an electron goes from a higher energy state to a lower one, in which case the radiated energy $h\nu = \Delta E$, where ΔE is the energy difference between the two states.

We can now, by applying the laws of classical mechanics, easily calculate the energies of the allowed states for hydrogen, with orbital angular momentum, $L = m_e v r = n\hbar = nh/2\pi$. n is called the *angular-momentum quantum number* of the state. The radii, r, of the *Bohr orbits* are given by

$$r = \frac{4\pi\varepsilon_0 n^2\hbar^2}{m_e e^2}$$

and the energies, E_n, of the states with these radii are given by

$$E_n = -\frac{m_e e^4}{2(4\pi\varepsilon_0)^2\hbar^2}\frac{1}{n^2}$$

For the state with the least energy ($n = 1$), $E_1 = -13.6$ eV, and for the other states, $E_n = -13.6\ \text{eV}/n^2$. The minus sign indicates that the electrons are bound to the nucleus; 13.6 eV of energy is needed to remove an electron in the lowest ($n = 1$) state completely from the proton. This energy is called the *binding energy* of the electron in hydrogen. The radiation of energy is illustrated in Figure 43.4.

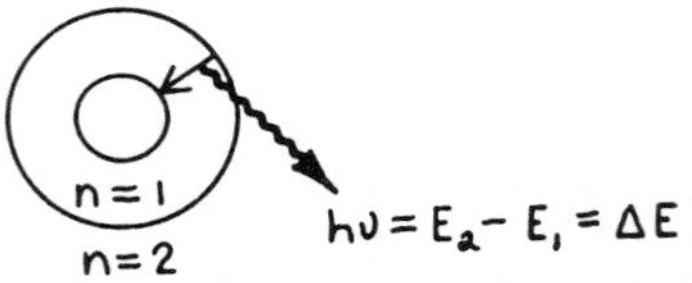

n=1
n=2
$h\nu = E_2 - E_1 = \Delta E$

Figure 43.4 An electron in the second Bohr orbit of hydrogen ($n=2$) drops back to the ground state ($n = 1$). Energy $h\nu = E_2 - E_1$ is radiated in the process.

Usually, an electron will remain in the lowest possible energy state ($n = 1$), which is called the "ground state." If the electron is kicked into a higher orbit (for example, the $n = 2$ state) by absorbing some energy, it is then said to be in an "excited state" and will deexcite by dropping back into a lower state, here the ground state. The energy radiated in this example is $E = (-13.6/4 + 13.6/1) = 10.2$ eV, so that the frequency of the radiation is given by

$$\nu = \frac{\Delta E}{h} = \frac{13.6(1.6 \times 10^{-19})}{6.63 \times 10^{-34}}\left[\frac{1}{2^2} - \frac{1}{1}\right]$$

$$= 3.282(10^{15})\left[\frac{1}{2^2} - \frac{1}{1^2}\right]$$

in agreement with our earlier expression for the frequency

$$\nu = cR_H\left[\frac{1}{n_2^2} - \frac{1}{n_1^2}\right]$$

Although our formulas have been derived explicitly for hydrogen, they can also be applied to hydrogen-like systems. For example, a singly ionized helium atom, that is, a helium atom with one missing electron, consists of one electron orbiting around a nucleus of charge $2e$. The energies E_n are given with the same formula that gives them for hydrogen, but with the factor e^4 = [electron charge]^2[nuclear charge]^2 replaced by $e^2(2e)^2$. In general, for an atom with several electrons, the behavior of only those electrons nearest the nucleus can be considered "hydrogen-like," because of the mutual interactions of the electrons. That is, most of the electrons are "shielded" from the nucleus. For the electrons very near the nucleus, the attractive force between the electron and the nucleus overshadows the electron–electron forces, and the energies E_n have the factor e^4 replaced by $e^2(Ze)^2$. These energies are rather large, and transitions between the innermost orbits of complex atoms give rise to high-energy photons, called X rays.

5. *Bohr's correspondence principle* states that in the limiting case of large quantum numbers, the results obtained from quantum theory must agree with those obtained from classical theory. For example, for hydrogen, the transition frequency between states n_2 and n_1 is given by

$$\nu = kR[1/n_2^2 - 1/n_1^2]$$

In the limiting case of large n, we can show that this expression gives, for a transition from the state n to the state $n-1$, $\nu = kR[2/n^3]$ for the frequency according to quantum theory. In classical theory, for an accelerated charge, the frequency of the emitted light is that of the frequency of the motion. For an electron in a circular orbit, the frequency of the motion is $v/2\pi r$, which can be shown to be the same as $kR[2/n^3]$. That is, for large values of n, the classical calculation agrees with the quantum-mechanical calculation.

6. *Quantum mechanics*. Although the Bohr theory agreed well with early experimentally observed spectral lines, more details of the spectra, such as a finer splitting of the spectral lines, were soon observed that the theory could not explain. In addition, it was discovered that electrons, as well as all other particles found in nature, have wave properties. For example, when a beam of electrons passes through a narrow slit, the electrons exhibit diffraction. All the above results were explained by means of quantum physics, or quantum mechanics, which replaced Newton's classical mechanics in describing the motion of particles on the atomic level. The wavelength associated with any particle is inversely proportional to its momentum: $\lambda = h/p$, where λ is called the *de Broglie wavelength*. The heart of quantum mechanics is a wave equation called the *Schrödinger equation*, which plays the same role for electrons (and for other particles) that Maxwell's equations play for photons. The amplitude of the wave associated with the particle is represented by ϕ, called the wave function. As a consequence of the Schrödinger wave equation, electrons, and in fact all particles, obey the uncertainty principle just as photons do. The behaviors of particles are predicted by the laws of probability, and there is a fundamental uncertainty in nature. For example, there is always an uncertainty in the position of an electron in an atom; furthermore, an electron follows no definite orbit. There is quantum uncertainty in everything, but

when objects have masses much larger than atomic masses, the uncertainties soon become very small and, in fact, completely negligible.

In wave mechanics, the quantization of the energy in the hydrogen atom (and in other atoms) is an automatic consequence of the wave properties of the electron. The attractive electric force of the nucleus confines the electron wave to some region near the nucleus and causes the wave to reflect back and forth across the region, forming a standing wave. Different stationary states of the atom correspond to different standing-wave modes. The standing electron waves in the atom have a discrete set of wavelengths and, therefore, a discrete set of energies.

Sample Problems

1. *Worked Problem.* The four series of spectral lines of hydrogen with the shortest wavelengths are, respectively, the Lyman, Balmer, Paschen, and Brackett series. Write down the expressions for the frequencies for each series, and calculate the shortest wavelength seen for each series (that is, the highest frequency). Although Balmer discovered the formula for the Balmer series in 1885, not until 1908, 23 years later, was the next series, the Paschen series, even observed. Can you explain why this might have been so?

Solution

$$\nu = cR_H(1/1^2 - 1/n_1^2) \quad n_1 = 2, 3, 4, \ldots \quad \text{Lyman series}$$
$$\nu = cR_H(1/2^2 - 1/n_1^2) \quad n_1 = 3, 4, 5, \ldots \quad \text{Balmer series}$$
$$\nu = cR_H(1/3^2 - 1/n_1^2) \quad n_1 = 4, 5, 6, \ldots \quad \text{Paschen series}$$
$$\nu = cR_H(1/4^2 - 1/n_1^2) \quad n_1 = 5, 6, 7, \ldots \quad \text{Brackett series}$$

These are the expressions for the four series with the shortest wavelengths (highest frequencies). For each series the transition with the highest frequency is obtained by setting n_1 to its minimum value. We obtain the following $[cR_H = 3.29(10^{15})]$:

$$(\nu_1)_L = 2.468 \times 10^{15} \text{ Hz} \quad (n_1 = 2)$$
$$(\nu_1)_B = 4.570 \times 10^{14} \text{ Hz} \quad (n_1 = 3)$$
$$(\nu_1)_P = 1.599 \times 10^{14} \text{ Hz} \quad (n_1 = 4)$$
$$(\nu_1)_B = 7.403 \times 10^{13} \text{ Hz} \quad (n_1 = 5)$$

We can obtain the wavelengths through $c = \nu\lambda$, from which

$$(\lambda_1)_L = 121.6 \text{ nm} \quad \text{ultraviolet}$$
$$(\lambda_1)_B = 656.4 \text{ nm} \quad \text{visible (red)}$$
$$(\lambda_1)_P = 1876 \text{ nm} \quad \text{infrared}$$
$$(\lambda_1)_B = 4052 \text{ nm} \quad \text{infrared}$$

Only the Balmer series lies in the visible region; the others are in either the ultraviolet or the infrared and so are not visible to the unaided eye. These had to wait for more advanced technology to be discovered and measured.

2. *Guided Problem*. A continuous spectrum of visible and ultraviolet light is passed through gaseous hydrogen atoms at room temperature. Find the energy and the wavelength of the lowest-energy photons that can be absorbed by the radiation. Are these photons in the visible region? Suppose that the hydrogen is subjected to an electric discharge, so that many of the hydrogen atoms are excited to their $n = 2$ level. What now is the wavelength of the lowest-energy photons that can be absorbed?

Solution scheme

a. What is the expression for the energy, E_n, for the state in hydrogen with angular-momentum quantum number n? What is the value of n for the atoms at room temperature?
b. What is the least amount of energy that will excite an atom to the next highest level? What is the wavelength of a photon with this much energy?
c. Is this photon in the visible region, which is approximately 400 to 800 nm? If not, in what region of the spectrum is it?
d. Can you answer the same questions, now supposing that many of the atoms are *initially* in the $n = 2$ state?
e. You should now be able to understand the statement "Hydrogen gas at room temperature is transparent to visible light"!

3. *Worked Problem*. Alpha particles of 4 MeV are scattered from ^{40}Ca ($Z = 20$). Calculate the distance of closest approach, and sketch the approximate trajectory of the alpha particle as it passes along the calcium nucleus for respective impact parameters b of 10^{-12}, 10^{-13}, 10^{-14}, and 0 m. Which of the impact parameters has the smallest distance of closest approach, and to what scattering angle θ does this correspond?

Solution: The distance of closest approach r' and the impact parameter are related through

$$r' = \frac{Eb^2}{-\left(\dfrac{Ze^2}{4\pi\varepsilon_0}\right) + \sqrt{\left(\dfrac{Ze^2}{4\pi\varepsilon_0}\right)^2 + E^2b^2}}$$

$E = \frac{1}{2}mv^2 = 4(1.6 \times 10^{-19})(1 \times 10^6) = 6.4 \times 10^{-13}$ J and $Ze^2/4\pi\varepsilon_0 = 4.61 \times 10^{-27}$. It is easiest here to set up a table of values:

b (m)	Eb^2	E^2b^2	r' (m)
10^{-12}	6.4×10^{-37}	4.1×10^{-49}	1.007×10^{-12}
10^{-13}	6.4×10^{-39}	4.1×10^{-51}	1.075×10^{-13}
10^{-14}	6.4×10^{-41}	4.1×10^{-53}	1.95×10^{-14}

For $b = 0$, we can calculate r' from $r' = Ze^2/\pi\varepsilon_0 mv^2$ which gives a value for r' of 1.44×10^{-14} m. It is clear that the *smallest* distance of closest approach is obtained when the impact parameter $b = 0$, that is, when the alpha particle is aimed directly at the nucleus. In this case the scattering angle is 180°, so that the particle is scattered directly backward. The respective trajectories are depicted in Figure 43.5.

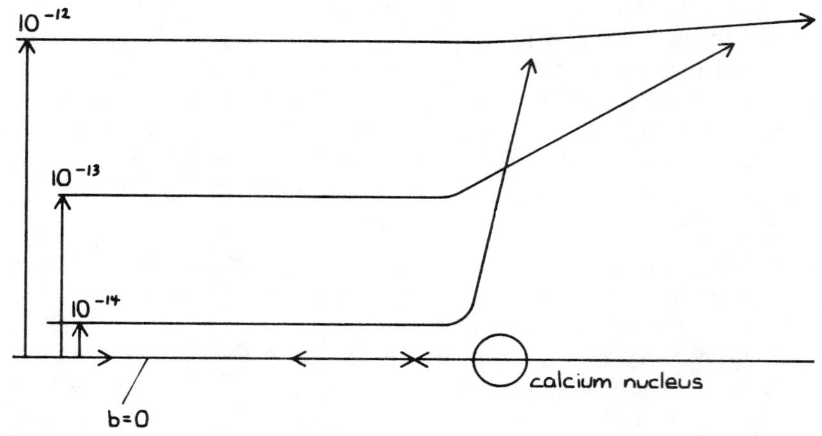

Figure 43.5 Problem 3. Trajectories of alpha particles scattered from ^{40}Ca. The impact parameter of the incident particle is indicated at the left.

4. *Guided Problem.* An 8 MeV alpha particle experiences a "head-on" (zero impact parameter) collision with a gold nucleus ($Z = 79$). What is the distance of closest approach? If instead of an alpha particle an 8 MeV proton experiences a head-on collision with gold, what is the distance of closest approach of the proton to the gold nucleus? If the gold nucleus has a radius of 7×10^{-15} m, will either of the particles actually come into contact with the nucleus? By contact, we mean will the particles experience the powerful short-range nuclear force?

Solution scheme

a. What is the expression for the distance of closest approach when the impact parameter is zero? Using this, find r' for the alpha particle.
b. How does the expression in step a need to be changed when the incident particle is a proton? Find the distance of closest approach for the proton.
c. Compare your answers to the nuclear radius. Does either of the particles contact the nuclear surface? Which comes closer?

5. *Worked Problem.* Calculate the radii of the first Bohr orbits of hydrogen. Calculate the radius and the energy of the 100th Bohr orbit. Find r_n and E_n as n approaches infinity, and explain the physical significance of your answers.

Solution: The radius of the nth Bohr orbit is given by

$$r_n = \frac{4\pi\varepsilon_0 n^2\hbar^2}{m_e e^2} = n^2\left(\frac{4\pi\varepsilon_0\hbar^2}{m_e e^2}\right) = 0.529 \times 10^{-10}\, n^2 \text{ m}$$

Therefore

$$\begin{aligned} r_1 &= 0.529 \times 10^{-10}\ \text{m} \\ r_2 &= 2.12 \times 10^{-10}\ \text{m} \\ r_3 &= 4.76 \times 10^{-10}\ \text{m} \\ r_4 &= 8.46 \times 10^{-10}\ \text{m} \\ r_{100} &= 5.29 \times 10^{-7}\ \text{m} \end{aligned}$$

The only *stable* state is the ground state, n_1, for which $r_1 = 0.529 \times 10^{-10}$ m. For progressively higher excitation, the electron moves farther and farther away from the nucleus. As n approaches infinity, r_n approaches infinity and E_n approaches zero. Thus, an electron infinitely far away from the nucleus is totally unbound and therefore is a free electron. The energy of any state is given by $E_n = -13.76\ \text{eV}/n^2$, so that the energy of the ground state is -13.6 eV, whereas $E_{100} = -0.0014$ eV; that is, for $n = 100$, the electron is very nearly free with respect to the ground state.

6. *Guided Problem.* A hydrogen atom is in an excited state for which the binding energy of the electron to the proton is 1.51 eV.

a. The atom makes a transition to the ground state. What is the energy of the photon emitted?
b. Instead of the ground state, the atom makes a transition to a state for which the excitation energy is 10.18 eV. What is the energy of the photon emitted? Sketch the energy levels involved, giving their angular-momentum quantum numbers.

Solution scheme

a. What is the binding energy of the ground state of hydrogen?
b. If a state has a binding energy of 1.51 eV, what is its excitation energy (relative to the ground state)?
c. What is the energy difference between the excited state and the state with an excitation energy of 10.18 eV?
d. You should now be able to calculate the energies of the photons emitted in parts a and b by calculating the energy differences between the states involved. You should also now be able to sketch the energy levels involved and to see whether their energies agree with those expected from the Bohr formula.

7. *Worked Problem.* Assume that 3000 hydrogen atoms are initially in the $n = 4$ state. The atoms then make transitions to lower energy states.

a. How many distinct spectral lines will be emitted?
b. Assuming that for any given excited state all possible downward transitions are equally probable, what is the total number of photons emitted?

Solution: The easiest approach to this problem is to draw an energy-level diagram of the states involved. This we do in Figure 43.6. The wavy lines are used to indicate the transition, with the dots indicating the ends of the lines (transitions). Since all transitions are equally probable, we find 1000 transitions each between the $n = 4$ and the $n = 3$ state, between the $n = 4$ and the $n = 2$ state, and between the $n = 4$ and the $n = 1$ (ground) state.

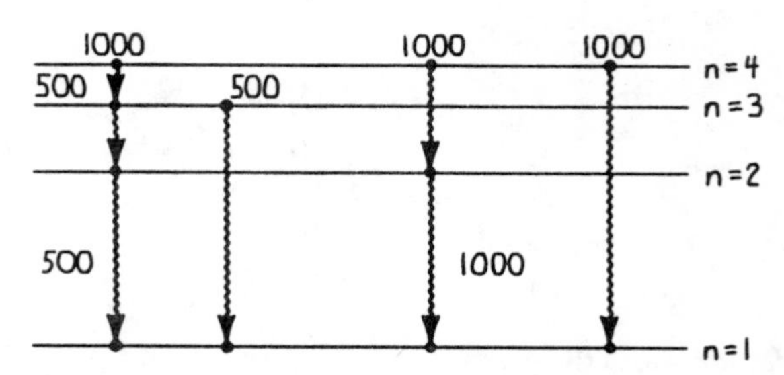

Figure 43.6 Problem 7. There are 3000 atoms initially in the $n = 4$ state of hydrogen. The transitions down to the ground state are shown.

Similarly, there are 500 transitions between the $n = 3$ and the $n = 2$, and also between the $n = 3$ and the $n = 1$ state. There are, however, a total of 1500 transitions from the $n = 2$ to the $n = 1$ state! We tabulate the transitions as follows:

$n = 4$ to 3: 1000
$n = 4$ to 2: 1000
$n = 4$ to 1: 1000
$n = 3$ to 2: 500
$n = 3$ to 1: 500
$n = 2$ to 1: 1500

There are, therefore, six distinct spectral lines; the total number of photons emitted is 5500.

8. *Guided Problem.* Singly ionized helium, He^+, is hydrogen-like in that it has a single electron. Since the Coulomb force of attraction between a single electron and the nucleus is now $2e^2/4\pi\varepsilon_0 r^2$ instead of $e^2/4\pi\varepsilon_0 r^2$ for the hydrogen atom, the energies of the states in He^+ are given by

$$E_n = \frac{-4\, m_e e^4}{2(4\pi\varepsilon_0)^2\hbar^2 n^2}$$

Make a sketch of the energy levels for He^+. Which states up to $n = 4$ are identical in energy to states in the hydrogen atom? What is the ionization energy for He^+?

Solution scheme

a. Do you have enough information to calculate the ratio $(E_n)_{He^+}/(E_n)_H$? Use this to obtain a simple expression for E_n for He^+.
b. Now calculate the energies of the first four states of He^+, and make a diagram of the energy levels.
c. On the same diagram sketch in the energies of the lowest levels of hydrogen. You should now be able to identify which, if any, of the states of He^+ have the same energies as states of H. From the definition of ionization energy, you should be able to deduce the ionization energy for He^+.

9. *Worked Problem.* The correspondence principle states that classical mechanics works well when the quantum uncertainties are small compared with the magnitudes of positions and moments. In a television tube the accelerating voltage is 15,000 volts and the electron beam passes through an aperture 0.4 mm in diameter to a screen 0.35 m away. Find the approximate uncertainty in position of the point where the electrons strike the screen. Does this uncertainty significantly affect the clarity of the picture?

Solution: According to the uncertainty principle, $\Delta y \Delta p_y \simeq h$, so that if $\Delta y = 0.4 \times 10^{-3}$ m, then the uncertainty in momentum $\Delta p_y \simeq h/0.4 \times 10^{-3}$ $= 6.63 \times 10^{-34}/0.3 \times 10^{-3} = 1.66 \times 10^{-30}$ kg • m/s. But $\Delta p_y = \Delta(m\upsilon_y)$ $= m\Delta\upsilon_y$ so that $\Delta\upsilon_y = 1.66 \times 10^{-30}/9.1 \times 10^{-31} = 1.82$ m/s. But since the kinetic energy of the electron beam $\frac{1}{2}m\upsilon^2$ is given by the voltage

$$KE = 15000 \text{ eV} = 15000\ (1.6 \times 10^{-19}) \text{ J}$$

the x component of velocity of the electrons is $\upsilon_x = 7.26 \times 10^7$ m/s. The time for the beam to reach the screen after passing through the aperture is therefore only $s/\upsilon_x = 0.35/7.26 \times 10^7 = 4.82 \times 10^{-9}$ s. The uncertainty in position of the electrons at the screen resulting from the uncertainty principle is therefore only $\Delta\upsilon_y$ [time to reach screen] $= 1.82(4.82 \times 10^{-9}) = 8.8$ $= 10^{-9}$ m, or 8.8 nm. This uncertainty is totally insignificant.

10. *Guided Problem.* A 40 kg satellite at a height of 300 km above the Earth's surface has an orbit period of 1.51 h.

a. Assuming that Bohr's angular-momentum postulate applies to satellites just as it does to an electron in the hydrogen atom, find the quantum number of the orbit of the satellite.
b. Show from the Bohr postulate and from Newton's Law of gravitation that the radius of the satellite orbit is proportional to n^2, and, using this, find the separation between the satellite orbit and the next quantum mechanically "allowed" orbit. Do you see why quantum mechanics does not play a significant role in the calculation of satellite orbits? The Earth's radius $R_E = 6380$ km, its mass $M_E = 5.98 \times 10^{24}$ kg, and the universal gravitational constant $G = 6.67 \times 10^{-11}$ Nt • m^2/kg^2.

Solution scheme

a. What is Bohr's angular-momentum postulate? Write it down, and calculate the quantum number n for this case. Would you expect n to be a very large number here? For your calculation of n you will need to calculate first some characteristics of the orbit, such as its radius and the speed of the satellite.
b. Remember that for an Earth satellite the gravitational force provides the centripetal force required for the orbit. Equate these two. Then eliminate the satellite velocity, using the Bohr postulate. This should provide you with an expression for the radius of the Bohr orbit of the satellite in terms of n^2, where n is the quantum number of the orbit.
c. By differentiation you should be able to obtain the dependence of the orbit radius r on the quantum number n.
d. You should now be able to obtain the change in r as n changes by 1 unit. Does your answer surprise you?

11. *Worked Problem.* The negative muon is a particle with mass $m_\mu = 206.8\ m_e$ and charge $-e$. In a muonic atom one of the atomic electrons is replaced by a muon.

a. Calculate the radius of the first Bohr orbit ($n = 1$) for a gold ($Z = 79$) muonic atom, and compare it with that of the "normal" first Bohr orbit (i.e., with an electron instead of a muon).

b. Compare the energies emitted if an electron falls from the $n = 2$ to the $n = 1$ state in gold with the energy emitted if a muon falls from the $n = 2$ to the $n = 1$ state in muonic gold.

Solution

a. For hydrogen, the radius of the $n = 1$ orbit is expressed as $r_1 \sim (1/m_e e^2)$, where e^2 is the product of the charge of the electron and that of the proton. For a gold atom, we replace e^2 by $e(79e) = 79e^2$, so that $[r_1]_{Au}/[r_1]_H = 1/79$. Since $[r_1]_H = 0.529 \times 10^{-10}$ m, $[r_1]_{Au} = 6.70 \times 10^{-13}$ m. For muonic gold, we must divide by the ratio $m_\mu/m_e = 206.8$, so that $r_1 = 3.24 \times 10^{-15}$ m. Essentially, the muon is inside the nucleus!

b. The energies of the states of hydrogen are given by $E_n = -13.6/n^2$ eV, and the expression for E is of the form $E \sim -m_e e^4$, so that for gold we must multiply E by $79^2 = 6241$. Then $E_1 = -84.86$ KeV, and $E_2 = E_1/4 = -21.22$ KeV. The energy emitted in falling from $n = 2$ to $n = 1$ is then 63.66 KeV. For muonic gold, we must multiply the energies further by the ratio $m_\mu/m_e = 206.8$, so that $E_1 = -17.553$ MeV, and $E_2 = -4.388$ MeV. The energy emitted in falling from $n = 2$ to $n = 1$ is then 13.165 MeV!

Answers to Guided Problems

2. 10.2 eV. The wavelength = 121.9 nm (in the ultraviolet). These photons are not in the visible region, so that hydrogen gas at room temperature is transparent to visible light. Atoms in the $n = 2$ level can be excited up to the $n = 3$ level, so that photons of energy 1.855 eV can be absorbed. This is in the red region of the spectrum (wavelength of 672 mm).
4. For the alpha particle, $r' = 2.84 \times 10^{-14}$ m. For the proton, $r' = 1.42 \times 10^{-14}$ m. Both these values are well outside the nuclear radius.
6. The energy levels are shown in Figure 43.7.

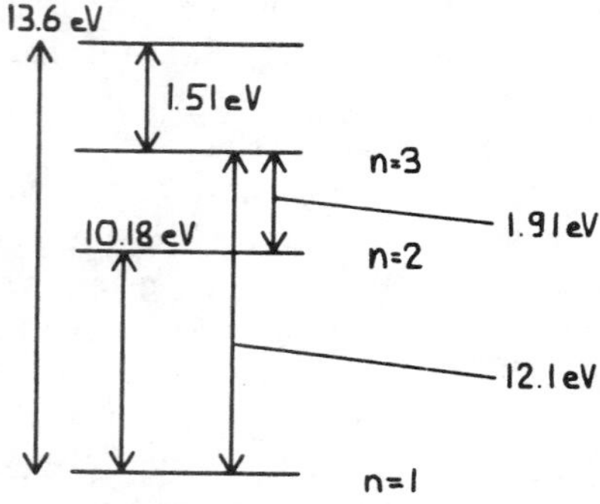

Figure 43.7 Problem 6. Involved energy levels. The transitions are shown as double arrows.

a. The photon energy is 12.1 eV.
b. 1.91 eV

8. For He^+, $E_n = -54.4\ \text{eV}/n^2$, so that the ionization energy for He^+ is 54.4 eV. The energy levels are shown in Figure 43.8. The energies of the $n = 2$ and $n = 4$ states of He^+ are the same as those of the $n = 1$ and $n = 2$ states, respectively, of hydrogen.

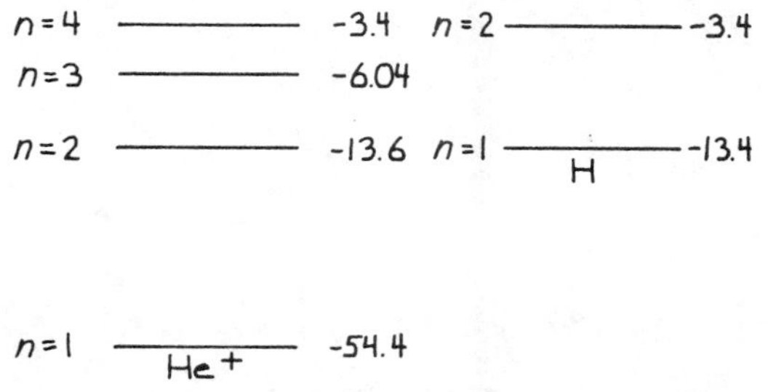

$n=4$ ——— -3.4 $n=2$ ——— -3.4
$n=3$ ——— -6.04
$n=2$ ——— -13.6 $n=1$ ——— -13.4
H
$n=1$ ——— -54.4
He^+

Figure 43.8 Problem 8. A comparison of energy levels of He^+ with those of hydrogen.

10. a. $m\upsilon r = nh/2\pi$, where $\upsilon = 2\pi r/T = 7730$ m/s, so that $n = 1.957 \times 10^{46}$.

b.
$$\frac{GM_E m}{r^2} = \frac{m\upsilon^2}{r} = \frac{m}{r}\left(\frac{n^2h^2}{4\pi^2m^2r^2}\right)$$

$$r = \frac{n^2h^2}{4\pi^2 GM_E m^2}$$

$$\Delta r = \frac{h^2}{4\pi^2 GM_E m^2}(2n\Delta n)$$

and if $\Delta n = 1$

$$\Delta r = 6.83 \times 10^{-40}\ \text{m}$$

Chapter 44

Quantum Structure of Atoms, Molecules, and Solids

Overview

Nearly all the properties of atoms and of molecules that we can observe or measure depend on the quantum behavior of their electrons. Almost the entire volume of the atom is occupied by the electrons, and consequently their orbits determine such things as the spectrum of light emitted and absorbed and the chemical bonds the atom forms, in addition to the physical, magnetic, electrical, and mechanical properties. The arrangement of the electrons in an atom or a solid is subject to an important restriction called the exclusion principle, which says that no more than two electrons can occupy the same orbital state. We will see that the exclusion principle is closely related to the spin, or intrinsic angular momentum, of the electron.

Essential Terms

elliptical orbits
principal quantum number
orbital quantum number
magnetic quantum number
spin quantum number
Zeeman effect
exclusion principle
vibrational energy of a molecule
rotational energy of a molecule
energy bands in solids
valence band
conduction band
acceptor and donor impurities
semiconductor
n- and *p*-type semiconductors
forward and reverse bias
transistor
light-emitting diode (LED)
solar cell

Key Concepts

1. *Elliptical orbits*. In Bohr's theory, the electron in the hydrogen moves in circular orbits, each allowed orbit being characterized by a quantum number n. Soon after Bohr formulated his theory, Sommerfeld extended the model to allow for *elliptical orbits*. For an elliptical orbit, both the radial distance of the electron from the nucleus and the angular position vary. There can thus be a second quantum number, l. Elliptical orbits are thus characterized by the two quantum numbers, n, the *principal quantum number*, and l, the *orbital quantum number*. For a given value of n, l can take any value from 0 to $n-1$, where the greater the l value, the greater the elongation, or eccentricity, of the elliptical orbit. For $n=1$, the lowest, or ground, state, $l=0$, and the orbit is circular. For $n=2$, l can be 0 (circular orbit) or 1 (elliptical orbit). Sommerfeld recognized also that the possible orbits need not all be in the same plane, but can be tilted at different angles. This leads to a quantization of the orientation of the orbit in space, called space quantization. For an orbit with orbital quantum number l, there are $2l+1$ possible orientations, which can be characterized by a quantum number m with integer values from $-l$ to $+l$. m is called the *magnetic quantum number*. Although we have introduced the quantum numbers n, l, and m on the basis of semiclassical considerations, a more rigorous analysis, based on quantum mechanics, confirms that the states of an electron in an atom are characterized by the above three quantum numbers.

In addition to n, l, and m, one more quantum number is needed to characterize completely the states of an electron in the hydrogen atom. This is the *spin quantum number*, m_s, which describes the spin, or intrinsic, angular momentum of the electron. When the spectral lines of hydrogen (and of other atoms) are examined with high resolution, they are often found to consist of pairs of very closely spaced lines. This "splitting" of the spectral lines is called "hyperfine structure." It can be explained only by a similar splitting of the energy levels due to another quantum number. Imagine the electron as a small ball of charge spinning about its axis, with an angular momentum of magnitude $\hbar/2$, as shown in Figure 44.1.

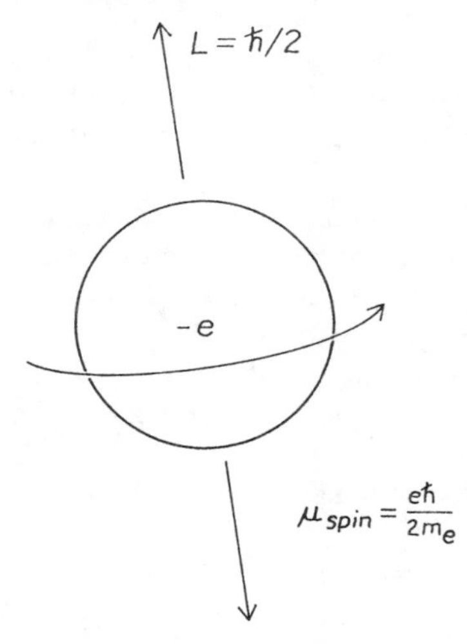

Figure 44.1 Picture of the electron as a rotating ball of charge.

There are two possible orientations of the spin, one characterized by $m_s = +1/2$ and the other by $m_s = -1/2$, and therefore two possible energy states. If the electron is represented as a small ball of spinning charge, it should behave like a current loop and should have a magnetic moment. Experiment shows that the electron has an intrinsic magnetic moment $\mu_{spin} = e\hbar/2m_e = 9.27 \times 10^{-24}$ A • m^2. The direction of μ_{spin} is opposite to the direction of the spin itself, as we would expect.

When an atom is placed in an external magnetic field, both the intrinsic magnetic moment (μ_{spin}) and the orbital magnetic moment interact with the magnetic field and change the energies of the atomic states. The result is a splitting of the energy levels into closely spaced multiplets, and a corresponding splitting of the spectral lines. The splitting is called the *Zeeman effect*.

The quantum numbers n, l, m, and m_s completely characterize the states of the hydrogen atom. The energy of the state depends mainly on the principal quantum number, n. In the Bohr model the energy is given by $E = -13.6/n^2$ eV. However, we now find that E also depends slightly on l as well as on m, so that the above equation for E is not quite accurate. We summarize the quantum numbers and their possible values below:

Quantum Number	Symbol	Values
Principal	n	1, 2, 3, . . .
Orbital	l	0, 1, 2, . . . , $n-1$
Magnetic	m	$-l, -l+1, -l+2, \ldots, l-1, l$
Spin	m_s	$-1/2, +1/2$

2. *The structure of atoms*. The arrangement, or *configuration*, of electrons in their orbits around the nucleus of an atom determines all the physical and chemical properties of the atom. For example, similarities of chemical properties among groups of elements are caused by similarities in their electronic configurations. We can easily determine the electron configurations for hydrogen. For the ground state $n = 1$, $l = 0$, $m = 0$, and $m_s = \pm 1/2$. For atoms with several electrons, the situation is more complicated. We might think that for the ground state, at least, we should simply put all of the electrons into the lowest state, with $n = 1$, $l = 0$, $m = 0$, and $m_s = \pm 1/2$ again. If this were possible, all atoms would have a spectrum similar to that of hydrogen, and in addition atoms of large Z would be very small; neither of the above is true.

According to the *exclusion principle*, each state of quantum numbers n, l, m, and m_s can be occupied by no more than one electron. Since $m_s = \pm 1/2$ only, the exclusion principle, which is predicted by quantum theory for particles of half-integer spin, also says that each orbital state of the quantum numbers n, l, and m can be occupied by no more than two electrons. We can construct a list of the quantum numbers of the available states in order of increasing energy.

The lowest energy states have $n = 1$. There are only two of these states:

$$n = 1,\ l = 0,\ m = 0,\ m_s = -1/2$$
$$n = 1,\ l = 0,\ m = 0,\ m_s = +1/2$$

There are 8 states with $n = 2$:

$$n = 2,\ l = 0,\ m = 0,\ \ [m_s = -1/2, \text{ or } m_s = +1/2]$$
$$n = 2,\ l = 1,\ m = -1,\ [m_s = -1/2, \text{ or } m_s = +1/2]$$
$$n = 2,\ l = 1,\ m = 0,\ \ [m_s = -1/2, \text{ or } m_s = +1/2]$$
$$n = 2,\ l = 1,\ m = +1,\ [m_s = -1/2, \text{ or } m_s = +1/2]$$

For $n = 3$, there are 18 available states. The energy levels are shown schematically in Figure 44.2.

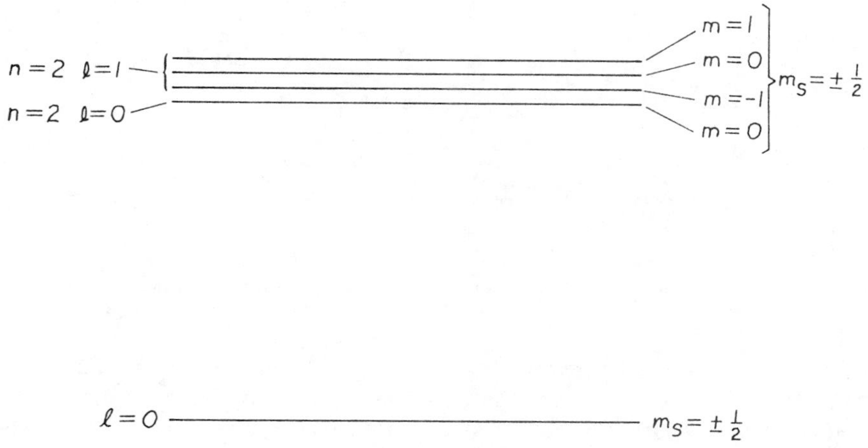

Figure 44.2 The lowest energy levels for hydrogen. Only the largest energy splittings are indicated in the diagram.

States with a given value of n are called shells. The two states with $n = 1$ form the K shell, the eight with $n = 2$ form the L shell, and those with $n = 3$ form the M shell. Since each state cannot have more than one electron, if an atom with Z electrons is in its lowest, or ground, state, the electrons will occupy the first of the above states. We can build up the ground-state configurations for the atoms in the Periodic Table of elements, beginning with hydrogen and adding electrons one by one. Thus, helium, with two electrons, fills up the K shell. Lithium has three electrons; the third electron is in the L shell, with quantum numbers $n = 2$, $l = 0$, $m = 0$, and $m_s = -1/2$.

We find that helium completely fills the K shell; neon, with ten electrons, fills the K and L shells. The two elements have similar chemical behavior and rather similar spectra. Hydrogen (one electron), lithium (three electrons), and sodium (eleven electrons) all have a single electron outside a full electron shell. This outer single electron is fairly weakly bound and dominates the chemical behavior of the atom. These three atoms have fairly similar chemical and spectroscopic properties. Thus, the exclusion principle accounts for the broad, qualitative features of the Periodic Table of elements.

3. *Vibrational energies*. The chemical bonds that bind two or more atoms together in a molecule, such as HCl, are due to a rearrangement of the outer, or valence, electrons of the atoms. Sometimes, as in O_2, the outer electrons

are shared between the atoms, producing an attractive, or covalent, bond. In other cases (HCl), one atom loses an electron to the other atom, and the atom with the missing electron is then electrically attracted by the atom with the extra electron (ionic bond). The chemical bonds are elastic and behave like springs tying the atoms together, allowing the atoms to move back and forth over an average equilibrium distance.

Let us represent a diatomic molecule as two pointlike masses connected by a massless spring, as shown in Figure 44.3. The system is an oscillator,

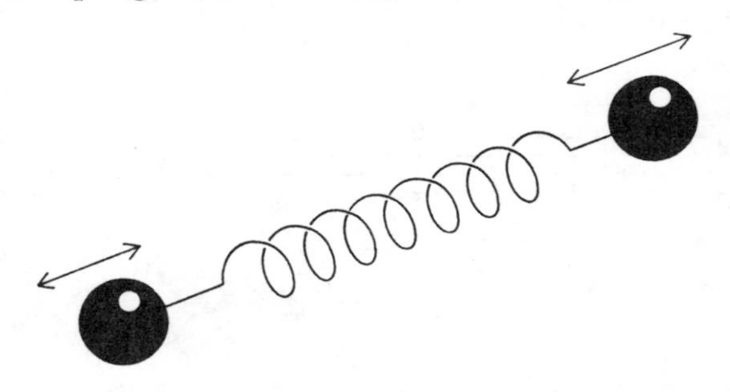

Figure 44.3 Schematic diagram of an oscillating diatomic molecule.

with the atoms oscillating together relative to the center of mass. The energy of the oscillator is subject to Planck's quantization condition. If the oscillation frequency is ν, the energy is $E = nh\nu$, where $n = 0, 1, 2, \ldots$. The energy-level diagram is shown in Figure 44.4. It turns out that transitions can proceed only from one level to the next; the arrows show the allowed transitions. The frequencies of the transitions are all the same, typically about 10^{13} Hz, corresponding to infrared radiation.

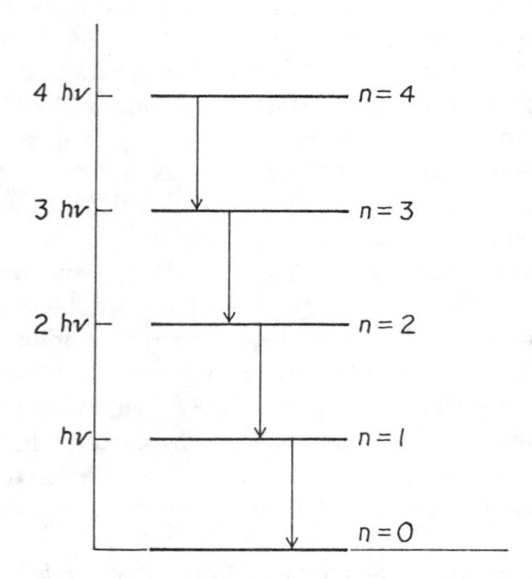

Figure 44.4 Energy-level diagram for the oscillating molecule. The arrows indicate possible transitions.

4. *Rotational energies*. The molecule can also rotate; in doing so, it behaves like two pointlike masses connected by a massless rod, that is, a dumbbell with a moment of inertia I. The kinetic energy of rotation is $E = (1/2)I\omega^2$, where ω is the angular frequency of the rotation. In terms of angular momentum, $L = I\omega$, and $E = L^2/2I$, and with angular momentum quantized according to $L = n\hbar$, the rotational energy is quantized according to $E = n^2\hbar^2/2I$, with $n = 0, 1, 2, \ldots$. The corresponding energy-level diagram is shown in Figure 44.5. Transitions again can proceed only from one level to the next, as the arrows show.

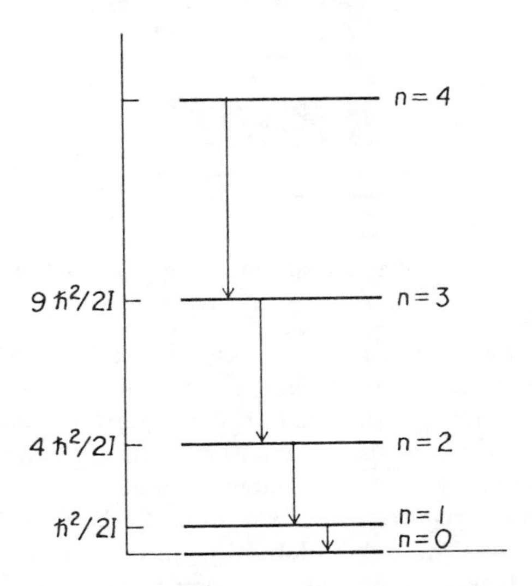

Figure 44.5 Energy-level diagram for a rotating molecule. The arrows indicate the possible transitions.

5. *Energy bands in solids*. In a metal, the valence electrons are detached from their atoms and are free to move all over the volume of the metal. However, when such a "free" electron passes by an atom, it experiences an attractive force. If the distance between atoms in the crystal lattice of the metal, a, satisfies the condition $2a = \lambda, 2\lambda, 3\lambda, \ldots$, where λ is the de Broglie wavelength for the free electrons, the electrons can form a standing wave. Since a standing wave does not travel, the electrons with the wavelengths above are not free to move through the metal. The corresponding electron momenta values are $p = h/2a, h/a, 3h/2a, \ldots$. These momenta are "forbidden," and corresponding to them there are forbidden energy values. These forbidden energies are really forbidden energy gaps, or intervals, and are shown in Figure 44.6. The allowed ranges of energy, shown shaded in the figure, are called *energy bands*.

The electron configuration of crystals can be deduced by means of the exclusion principle. In the ground state, the free electrons fill all the available states of lowest energy. The lowest energy bands will be completely filled, but the uppermost energy band will be either filled or partly filled, depending on the number of free electrons and the number of available states. In a conductor, the highest energy band is only partly filled. When the elec-

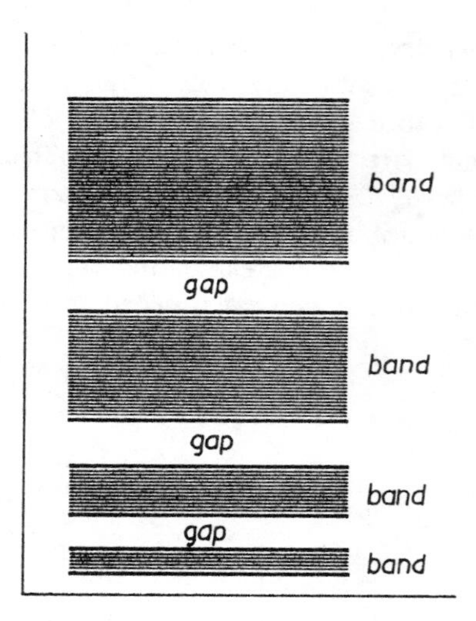

Figure 44.6 Energy-level diagram for an electron moving in a crystal lattice.

trons in this band experience an electric field, they absorb energy and make transitions to some of the slightly higher, empty states of the band and begin to carry an electric current. In an insulator, the uppermost band is completely filled. When the electrons in such a band experience an electric field, they can make transitions only to the next higher, empty energy band, but this is too difficult, since it requires the absorption of a large amount of energy from the electric field. Thus, they do not begin to carry a current. Figure 44.7 indicates the difference between a conductor and an insulator.

In a *semiconductor*, such as silicon or germanium, the uppermost band is completely filled, as it is in an insulator, but the energy gap between this band the next band (~1 eV) is much smaller here than in an insulator, where it is about 6 eV. The electrons can jump this smaller gap and can carry a

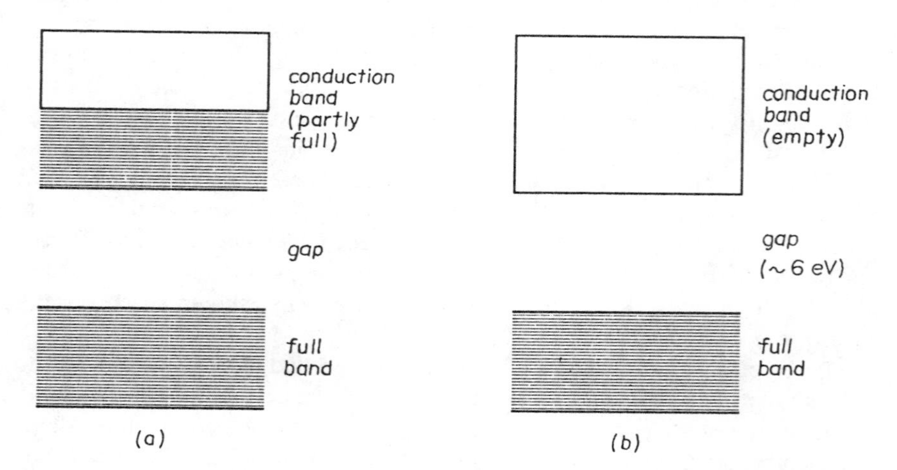

Figure 44.7 In a conductor (a), the uppermost energy band is only partly filled with electrons. In an insulator (b), the uppermost band is filled.

current. In an n-type semiconductor, the carriers of current are free electrons that have reached the conduction band. In a p-type semiconductor, the carriers of current are "holes" of positive charge. The valence band is almost, but not quite, filled with electrons, so that there are holes in the electron distribution, and if they move, they will transport charge. The current flow is really due to electrons jumping from one atom to the next in the direction opposite to that of the apparent motion of the holes, but it can be conveniently described in terms of moving holes.

An n-type semiconductor is produced by "doping" a pure semiconductor (silicon) with donor impurities, atoms such as those of arsenic that release their valence electrons when placed in the semiconductor. A p-type semiconductor has been doped with acceptor impurities such as boron that trap electrons when placed in the semiconductor and so generate holes.

A semiconductor rectifier consists of a piece of n-type semiconductor adjoining a piece of p type. At the p–n junction, some of the free electrons wander from the n region into the p region, and some of the holes wander from the p region into the n region. Wherever the electrons and holes meet, they annihilate each other, so that both the electron and the hole disappear (the electron has simply fallen into the hole and filled it). The loss of some electrons and holes leaves residual positive and negative ions near the interface of the two regions, and the electric charges of these ions generate an electric field across the interface; this acts to inhibit further movement of holes or of electrons across the interfaces.

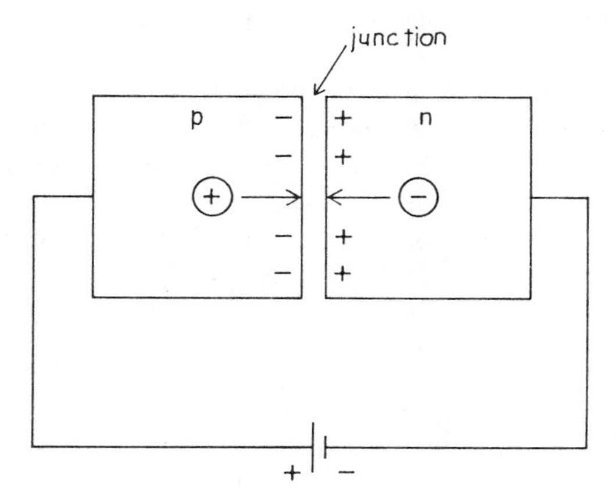

Figure 44.8 A p–n junction connected to a battery. The p region is at high potential and the n region at low potential (forward bias).

When such a p–n junction is connected to a battery, it allows the flow of current from the p region into the n region, but not in the opposite direction. Figure 44.8 shows *forward biasing*, with the p region at high potential and the n region at low potential. The battery removes electrons from the p region, which is equivalent to adding holes, and adds electrons into the n region. The electrons and holes meet at the junction and annihilate. The process can continue indefinitely.

If we reverse the battery leads, as shown in Figure 44.9, we have *reverse bias*. The free electrons in the n region flow away from the junction, as do

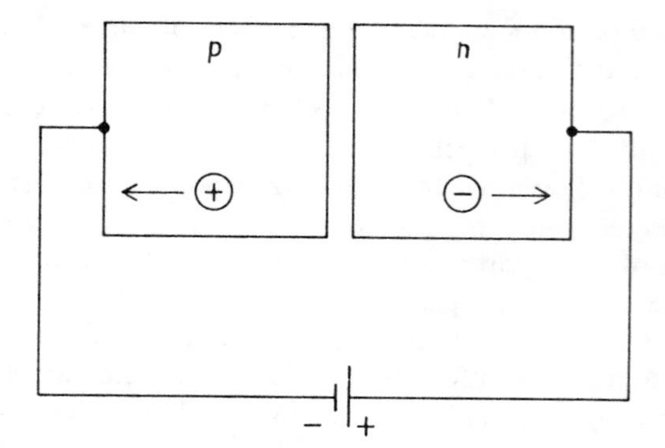

Figure 44.9 A *p–n* junction connected to a battery (reverse bias).

the free holes in the *p* region, so that each region is depleted of its charge carriers and the flow of current stops. The *p–n* junction can be used to convert an alternating current into a direct current, since if it is connected to a source of alternating current, it will pass current only during the "forward" part of the cycle.

A transistor, shown in Figure 44.10, consists of a thin piece of semiconductor of one type (the base) sandwiched between two pieces of semiconductor of the other type (the emitter and the collector). The transistor has three terminals that are connected to two sources of emf, V_B and V_C, so that the emitter-base junction has a forward bias and the base-collector junction a reverse bias.

The emitter-base junction acts as a diode with forward bias, allowing electrons to flow from the emitter into the base. Since the *p* region is thin, most of the electrons pass through it and into the collector. A small increase in the base voltage V_B causes a large increase in the flow of electrons entering the base from the emitter, but since most of these flow right through the base to the collector, the result is a large increase of collector current and only

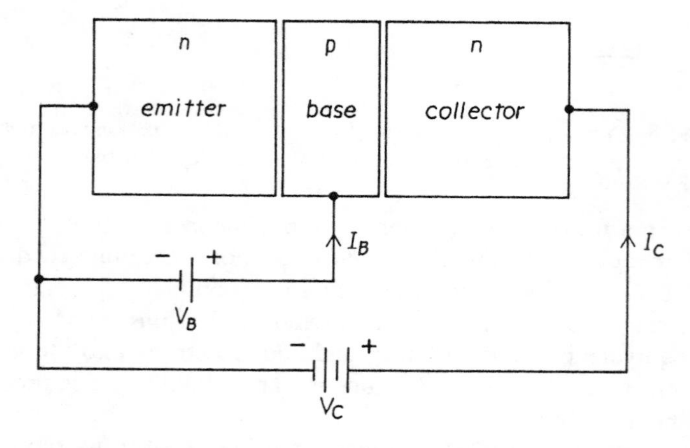

Figure 44.10 An *n–p–n* junction transistor.

a small increase of base current. The ratio of these current increments is called the gain factor:

$$[\text{gain factor}] = \Delta I_C/\Delta I_B$$

It is typically of the order of 100 to 200.

A *light-emitting diode (LED)* is a special *p–n* junction operated with forward bias. The annihilation of electrons and holes at the junction is a transition from the conduction band to the valence band, and it thus releases energy. In an LED the transition energy is in the visible region, so that the *p–n* junction emits light when an electric current passes through it.

A *solar cell* is an LED diode operating in reverse. When sunlight is absorbed at the *p–n* junction, it excites an electron from the valence band to the conduction band, creating a free hole and a free electron. The electric field at the junction pulls the electron toward the *n* region and the hole toward the *p* region. Thus, the sunlight striking the junction generates an electric current in the external circuit.

Sample Problems

1. *Worked Problem*. A hydrogen electron is excited into the state with quantum numbers $n = 3$, $l = 2$, $m = -2$, $m_s = -1/2$.

a. What are the values of the orbital and the spin angular momenta of the electron?

b. What are the values of the orbital and the spin magnetic moments?

Solution

a. The orbital quantum number $l = 2$. It is *almost* true that the magnitude of the orbital angular momentum $L = l\hbar = 2\hbar$. According to the Schrödinger equation, the magnitude of the orbital angular momentum, L, is given by $L = \sqrt{l(l+1)}\hbar = \sqrt{2(3)}\hbar$. Note, however, that the projection of L along any defined axis (which we define as the z axis) is given by $L_z = m\hbar = -2\hbar$. This means that the orbital-angular-momentum vector is restricted to a discrete set of components $m\hbar$ along the z axis. Similarly, it is *almost* true that the magnitude of the spin angular momentum $S = \hbar/2$. Actually, the magnitude of the spin angular momentum, S, is given by $S = \sqrt{s(s+1)}\hbar = \sqrt{1/2(3/2)}\hbar$, where $s = 1/2$, and, again, the projection of S along the z axis is restricted to $m_s\hbar = \pm 1/2\hbar$. Thus, $S_z = -\hbar/2$.

b. The magnitude of the orbital magnetic moment is given by $\mu_l = -(e/2m)L$, and that of the spin magnetic moment is given by $\mu_s = -(e/m)S$. Their projections along the z axis are respectively $-(e\hbar/2m)m$ and $-(e\hbar/m)m_s$. That is, $\mu_{\ell,z} = (e\hbar/2m)(2)$ and $\mu_{s,z} = (e\hbar/m)(1/2)$. The projection of the total magnetic moment along the z axis is just the sum of these, $(e\hbar/m) + (e\hbar/2m) = 3e\hbar/2m$.

2. *Guided Problem*. In the absence of a magnetic field, the quantum numbers for the ground and the first excited states of the hydrogen atom are $n = 1$, $l = 0$, $m = 0$, and $n = 2$, $l = 0$, $m = 0$, respectively. The intrinsic electron

spin has the possible values $m_s = \pm 1/2\hbar$ for each of these states. In the presence of an external magnetic field, these two states are split by what is called the Zeeman effect, the interaction of the spin magnetic moment with the magnetic field, given by $U_B = \mu_s B \cos\theta$, where θ is the angle between the spin orientation and the magnetic field. Since m_s is restricted to values of $\pm 1/2\hbar$ parallel to any magnetic field, possible values of U_B are also restricted. Calculate and plot the approximate energies of the ground and the first excited states, above, and the values of the Zeeman splittings of these energy levels in an external magnetic field of 1 T. Sketch an energy-level diagram of the original energies and of the new energies in the presence of the magnetic field. How many lines is the original transition from the first excited state back to the ground state split into?

Solution scheme

a. In the Bohr model, the energy of a state is given by $E = -13.6/n^2$ eV. Use this relation to calculate the approximate energies of the states in the absence of a magnetic field.
b. What are the possible values of magnetic moment of each of the above states? What are the possible projections of magnetic moment along the magnetic field? What is the numerical value of μ? Can you use the expression $U_B = -\mu_s B \cos\theta = -[\pm \mu_s B]$ (since the spin is aligned either parallel or antiparallel to the magnetic field) to calculate the Zeeman splitting? (Remember to convert your answer to eV.)
c. To obtain the splitting, you need to add the magnetic energy to the original energy of the state. Draw an energy-level diagram, showing the original, "unperturbed" levels, and the splitting of the levels in the magnetic field.
d. How many transition energies are now possible between the states? Note that although each state has been split into two new states, so that four transitions are now possible, two of these transitions have the same energy.

3. *Worked Problem.* List the quantum numbers of a neon atom in its ground state. What would you expect the values of the orbital and the spin angular momenta of this atom to be?

Solution: The shell with $n = 1$, the K shell, is filled by two electrons. The shell with $n = 2$, the L shell, has eight possible states. Helium, with ten electrons, exactly fills the K and the L shells. The quantum numbers are therefore $n = 1$, $l = 0$, $m = 0$, $m_s = \pm 1/2$ for the K shell, and for the L shell,

$$\begin{aligned} n = 2,\ l = 0,\ m = 0,\ \ & m_s = \pm 1/2 \\ n = 2,\ l = 1,\ m = -1,\ & m_s = \pm 1/2 \\ n = 2,\ l = 1,\ m = 0,\ \ & m_s = \pm 1/2 \\ n = 2,\ l = 1,\ m = +1,\ & m_s = \pm 1/2 \end{aligned}$$

For each electron the orbital-angular-momentum vector is restricted to a discrete set of components, given by m, along the z axis. The net value of the orbital-angular-momentum vector along the z axis is the sum of the values of m for each electron. Clearly, this is 0 for the K shell; it is also 0 for the L shell. Similarly, the component of spin angular momentum for a given electron along the z axis is either +1/2 or –1/2. By adding these, we find that the net spin angular momentum along the z axis = 0 for each shell. It can be shown that for a closed shell, the electrons tend to "cancel their spins"; that is, both the net orbital angular momentum and the net spin angular momentum due to all the electrons in the atom are zero.

4. *Guided Problem.* An electron in a hydrogen atom is known to have the magnetic quantum number $m = 3$. What can you say about the rest of its quantum numbers? What are the quantum numbers of the state of lowest energy that the electron could be in?

Solution scheme

a. If $m = 3$, what values of orbital quantum number l are possible?
b. Given the above values of orbital quantum numbers, what values of principal quantum number n are possible?
c. What values of spin quantum number m_s are always possible?
d. Which of these quantum numbers would have the lowest possible energy?

5. *Worked Problem.* The effective spring constant, k, for the CO molecule is 187 N/m. Calculate the frequency of vibration of the CO molecule and the spacing between its vibrational energy levels.

Solution: The frequency of a vibrating body of mass m connected to a spring of force constant k is $\nu_0 = (1/2\pi)\sqrt{(k/m)}$. For a two-body oscillator, the frequency of oscillation is given by the same formula, but with the reduced mass m' substituted for m, where $m' = m_1 m_2/(m_1 + m_2)$. The mass of the carbon atom is $12(1.67 \times 10^{-27}) = 2.00 \times 10^{-26}$ kg, and that of the oxygen atom is $16(1.67 \times 10^{-27}) = 2.67 \times 10^{-26}$ kg, so that $m' = 1.14 \times 10^{-26}$ kg. Then $\nu_0 = (1/2\pi)\sqrt{(k/m')} = 2.04 \times 10^{13}$ cycles/s. The spacing between the vibrational levels of CO is therefore given by

$$\Delta E = h\nu_0 = 6.63 \times 10^{-34} \text{ J} \cdot \text{s} \times 2.04 \times 10^{13} \text{ s}^{-1} = 8.4 \times 10^{-2} \text{ eV}$$

Notice that this energy is considerably less than the energy gap between the first few levels of the hydrogen atom, as given from the relation $E_n = -13.6/n^2$ eV.

6. *Guided Problem.* On the basis of the result of worked problem 3, what can you say about the orbital angular momentum, the spin angular momentum, and the magnetic moment of sodium in its ground state?

Solution scheme

a. Problem 3 showed that the net orbital and spin angular momenta of the electrons in the closed-shell atom neon, with ten electrons, were zero. How many electrons does sodium have? Does this imply that the outermost electron in sodium will dominate the chemical behavior of the atom?
b. Write down the quantum numbers for the lowest state with $n = 3$. From this can you deduce the values of the orbital and the spin angular momenta for sodium?

c. What value do you expect for the magnetic moment of sodium?

7. *Worked Problem.* Find the rotational inertia of the $H^{35}Cl$ molecule if the distance between the atoms is 1.25×10^{-10} m. What are the energies of the two lowest energy levels? What transition frequencies would you see from these levels?

Solution: To a good approximation, we can consider the Cl atom as fixed and the H atom to be rotating around it at a distance of 1.25×10^{-10} m. Then

$$I = mR^2 = 1.67 \times 10^{-27} \times (1.25 \times 10^{-10})^2 = 2.61 \times 10^{-47} \text{ kgm}^2$$

Then for the lowest excited, $n = 1$, level,

$$E = \hbar^2/21 = (1.05 \times 10^{-34})^2/21 = 2.11 \times 10^{-22} \text{ J} = 1.32 \times 10^{-3} \text{ eV}$$

For $n = 2$,

$$E = 4\hbar^2/21 = 8.45 \times 10^{-22} \text{ J} = 5.28 \times 10^{-3} \text{ eV}$$

The two transition energies would be $E_2 - E_1 = 6.34 \times 10^{-22}$ J, and $E_1 - 0 = 2.11 \times 10^{-22}$ J, with corresponding transition frequencies given by $E = h\nu$, with $h = 6.63 \times 10^{-34}$ J • s. Then $\nu_{2-1} = 9.56 \times 10^{11}$ Hz and $\nu_{1-0} = 3.18 \times 10^{-11}$ Hz.

8. *Guided Problem.* The energy gap in silicon is 1.1 eV, and in diamond it is 6 eV. The eye is sensitive to light in the range 400–700 nm. Do you expect either silicon or diamond, or both, to be transparent to visible light?

Solution scheme

a. Suppose that the energy gap of one of these substances falls within the energy of visible light. This would mean that visible light of energy at least equal to that of the energy gap could cause electrons to "jump" the energy gap from the valence to the conduction band. Photons of light that caused the electron to jump the gap would be absorbed; consequently, the material would appear opaque. What is the energy range, in eV, of visible light? Will visible-light photons cause electrons to jump the energy gap in either silicon or diamond?

b. What about the other material? If light cannot be absorbed by the material, there is a good chance that the material will be "transparent," or nearly so. Is this true?

9. *Worked Problem.* In the molecule CO, the $n = 0 \rightarrow n = 1$ rotational absorption line occurs at a frequency of 5.72×10^{10} cycles/sec. Assuming that the CO molecule acts as a rigid rotating dumbbell, what is the moment of inertia of the dumbbell? What is the bond length between the two molecules?

Solution: The rotational energy is quantized according to $E = n^2\hbar^2/2I$, so that for the transition from the lowest, $n = 0$, state, to the $n = 1$ level,

$E = \hbar^2/2I$, where $E = h\nu$. Then $\hbar^2/2I = h\nu$, or

$$I = \hbar/4\pi\nu = 1.46 \times 10^{-46} \text{ kgm}^2$$

Then, since $I = m'R^2$, where m' is the reduced mass, and is given by $m' = m_1m_2/(m_1 + m_2)$, where $m_1 = 12(1.67 \times 10^{-27})$ kg, and $m_2 = 16(1.67 \times 10^{-27})$ kg, $m' = 1.14 \times 10^{-26}$ kg. Therefore, the bond length

$$R = \sqrt{1/m'} = (1.46 \times 10^{-46})/(1.14 \times 10^{-26}) = 1.13 \times 10^{-10} \text{ m}$$

10. *Guided Problem.* Figure 44.11(a) shows a "half-wave rectifier," in which a *p–n* junction diode is connected as a rectifier. The circuit passes the positive half of the input waveform but suppresses the negative half. The average output waveform [Figure 44.11(b)] is positive. Design a circuit to act as a "full-wave rectifier," in which more than one *p–n* junction diode is used to produce a waveform of one polarity, using both halves of the input waveform. Explain how your circuit works.

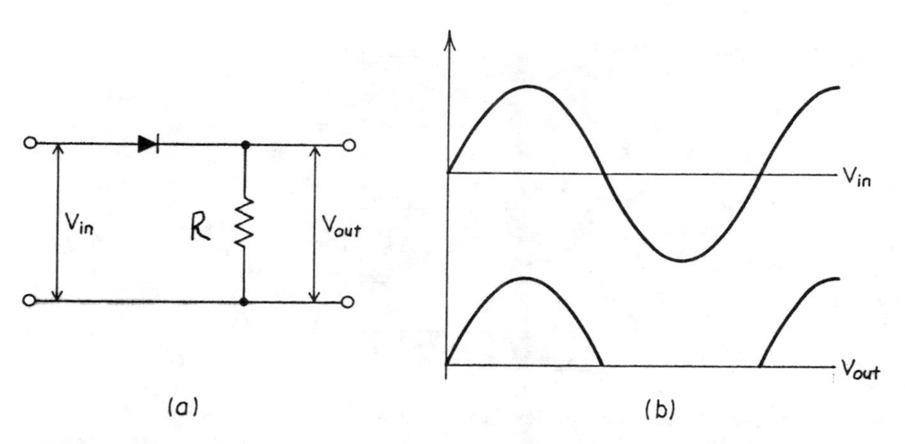

Figure 44.11 Problem 10. (a) A half-wave rectifier, using a *p–n* junction diode. (b) The input and output waveforms of the circuit.

Solution scheme: Suppose we connect four diodes as shown in Figure 44.12. Note that the output waveform is taken across the resistor *R*. Which diodes will conduct during the positive half of the cycle? Which during the negative half of the cycle? Will the direction of the current through *R* change during the cycle? Can you sketch the output waveform as measured across *R*?

11. *Worked Problem.* In the ground state of a metal, the free, or conduction, electrons (those occupying the highest, partly filled conduction band) fill all the available states of lowest energy (Figure 44.13). The highest occupied level in this band, at $T = 0°$K, is called the Fermi level. The energy corresponding to the Fermi level is called the Fermi energy. For metals, the Fermi energy is about 5 eV. Calculate the number of conduction electrons in a copper wire 1 m long and 1 mm in diameter. If the Fermi energy is 5

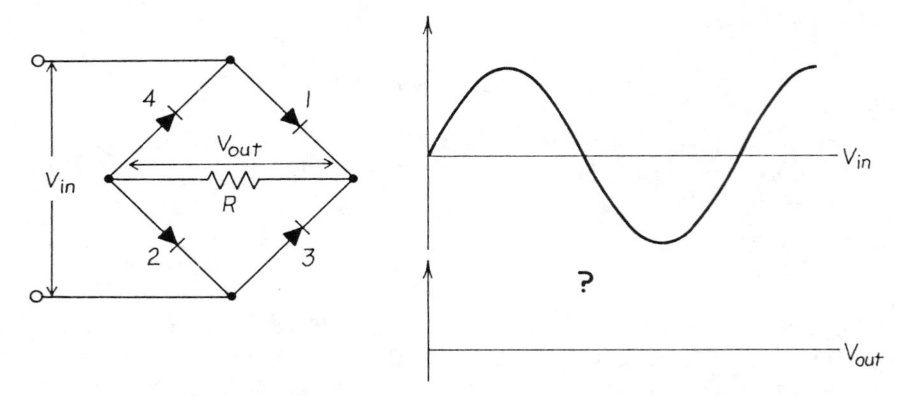

Figure 44.12 A full-wave rectifier circuit. Why is it called a "full-wave" rectifier?

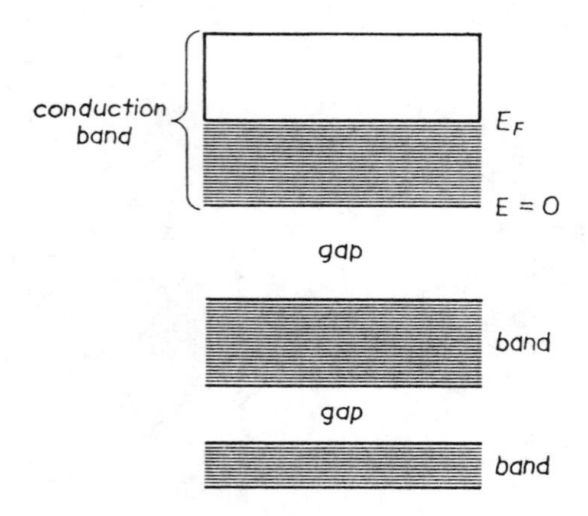

Figure 44.13 Energy levels in a metal. E_f = Fermi energy.

eV, and if the probability of an electron's having any energy up to the Fermi energy is uniform, find the average energy and the average speed of a conduction electron at 0°K. Contrast your answers with those you would get for a perfect gas at 0°K. What is the density, $n(E)$, of occupied states in no. of electrons per m^3 per eV at 0°K?

Solution: In copper, there is one conduction electron per atom. For copper, $d = 8.9$ gm/cm^3, the atomic number $A = 63.6$ gm/mole, and Avogadro's number, $N_A = 6.02 \times 10^{23}$ atoms/gm-mole. The volume of the wire $= [\pi d^2/4]L = 0.785$ cm^3, corresponding to a mass of 6.99 gm. The number of atoms (and conduction electrons) is therefore $N_A(6.99/63.6) = 6.62 \times 10^{22}$. The average energy of the conduction electrons is just 2.5 eV. The average speed, given by

$$(1/2)m\upsilon^2 = E = 2.5 \times 1.6 \times 10^{-19}\text{ J} = (1/2)(9.1 \times 10^{-31})\upsilon^2$$

so that $\upsilon_{av} = 9.4 \times 10^5$ m/s. (For a perfect gas at 0°K, $E_{av} = \upsilon_{av} = 0$, corresponding to a state of 0 energy.) For the conduction electrons, the density of occupied states is given by

$$[\text{no.}]/[\text{vol-}E_f] = 6.62 \times 10^{22}/(0.785 \times 10^{-6})(5) = 1.7 \times 10^{28}/\text{m}^3\text{-eV}$$

Answers to Guided Problems

2.

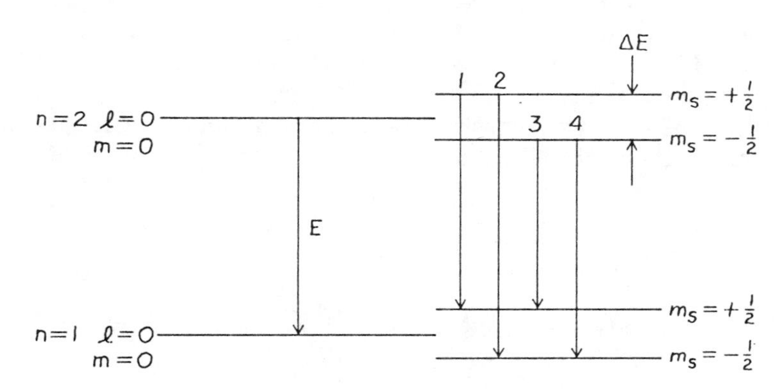

Figure 44.14 Energies of the ground and the first excited states of hydrogen.

$$E_1 = -13.16 \text{ eV} \qquad E_2 = -\frac{13.6}{4} = -3.4 \text{ eV}$$

$$E_2 - E_1 = 10.2 \text{ eV}$$

$$\mu_s = \pm \frac{e\hbar}{2m} = \pm 9.27 \times 10^{-24} \text{ A} \cdot \text{m}^2$$

The potential energy

$$U_B = \pm \mu B = \pm (9.27 \times 10^{-24})(1) = \pm 9.27 \times 10^{-24} \text{ J}$$
$$= \pm 5.8 \times 10^{-5} \text{ eV} = \pm \Delta E$$

Note that for $m_s = +1/2$, the spin angular momentum is aligned *parallel* to B, so that the spin magnetic moment is aligned *antiparallel* to B. Thus, U_B is *higher* for $m_s = +1/2$ than for $m_s = -1/2$.
The new transition energies are E [for (1)], $E + 2\Delta E$ [for (2)], $E - 2\Delta E$ [for (3)], and E [for (4)].
Thus, the original spectral line is split into three lines—one with the same energy as the original, one with larger energy, and one with less energy. Note that the splitting is very small.

4. $m = 3$. Therefore l can have the values 3, 4, 5, . . .
   n can have the values 4, 5, 6, . . .
   m_s has the possible values $\pm 1/2$.
   For the state of minimum energy, $n = 4$, $l = 3$, $m = -3$, $m_s = -1/2$.

6. On the basis of problem 3, the entire angular momentum and magnetic moment of the atom is that of the single electron outside the closed L shell. This electron is in the lowest energy state in the $n = 3$ shell. Its quantum numbers are $n = 3$, $l = 0$, $m = 0$, $m_s = -1/2$.
   The orbital angular momentum = 0, the magnetic quantum number $l = 0$. Since $m_s = -1/2$, the magnetic moment of sodium is $e\hbar/2m = 9.27 \times 10^{-24}$ A • m².
8. The frequency range 400–700 nm corresponds to an energy range, given by $E = h\nu$, of $E_{400} = 3.11$ eV, and $E_{700} = 1.78$ eV. Visible light of *all colors* can cause transitions from the valence band to the conduction band and therefore can be *absorbed* by the atom. Since silicon can absorb light, it will appear opaque.
   Visible light cannot cause similar transitions in diamond. The light therefore cannot be absorbed, and diamond will appear transparent.

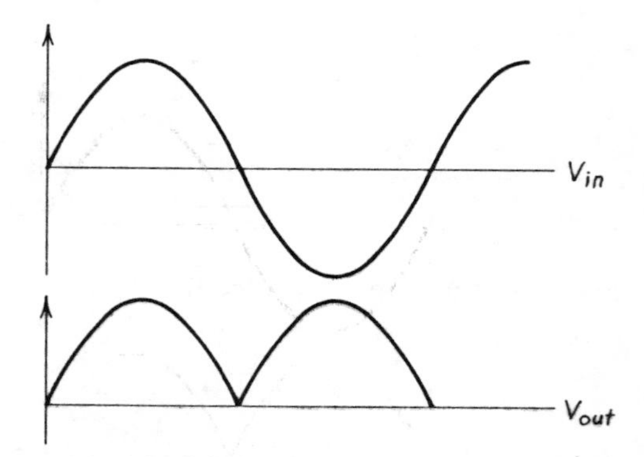

Figure 44.15 Input and output waveforms for the full-wave rectifier of Figure 44.12. The output voltage is measured across R.

10. Figure 44.15 shows the output waveform for the full-wave rectifier. During the positive half cycle, diodes 1 and 2 will conduct, while diodes 4 and 3 will not. During the negative half cycle, the reverse occurs. Diodes 3 and 4 conduct, while diodes 2 and 1 do not. The direction of current flow through R is always from right to left, during both halves of the cycle. We therefore have a *full-wave* rectifier.

Chapter 45

Nuclei

Overview

Before 1911, when Rutherford carried out the first of his famous scattering experiments, very little was known about the structure of atoms, except that they are electrically neutral and therefore must contain positive charge in addition to electrons. Most people believed in the "plum-pudding" model of the atom, proposed by J. J. Thomson, in which positive charge is distributed over the volume of the atom. Rutherford's experiments showed that the positive charge of the atom is contained within a tiny but massive "nucleus." We now know that the nucleus consists of two kinds of "nucleon," protons (charge $+e$) and neutrons (no charge), bound together by the nuclear, or "strong," force. Protons and neutrons have masses about 1800 times that of the electron and, like the electron, have a spin of $\hbar/2$. Most nuclei have approximately equal numbers of protons and neutrons. A given atom, such as carbon, always has a specific number of protons (six), but may have different numbers of neutrons in the nucleus. Atoms such as ^{12}C and ^{13}C, with six and seven neutrons, respectively, are called "isotopes" of carbon.

We learn about nuclear densities and binding energies, and we study radioactivity, in which a nucleus ejects a particle and in doing so "decays." Finally, we investigate the conservation laws of nuclear reactions, through which we obtain most of our knowledge of nuclei.

Essential Terms

proton and neutron; nucleon
nuclear or strong force
isotopes; mass number
nuclear binding energy
mass defect
radioactivity
alpha decay
parent nucleus; daughter nucleus
radioactive series
beta decay; weak interaction
gamma emission
half-life; mean life
accelerators
reaction Q value
threshold energy

Key Concepts

1. *Nucleons*. The proton and neutron, or nucleon, has a mass about 1800 times that of the electron, a radius of about 10^{-15} m, a spin of $\hbar/2$, the same as that of the electron, and a magnetic moment about 1000 times smaller than that of the electron. The magnetic moment of the proton is in the same direction as its spin, as one would expect from classical physics. The magnetic moment of the neutron is in the direction opposite to that of the spin, indicating that the neutron contains both negative and positive rotating charges.

2. The *mass number*, A, of a nucleus is the sum of the number of protons, Z, and the number of neutrons, N. That is, $A = Z + N$. Atoms with the same number of protons, but with different numbers of neutrons, are called *isotopes*. Most elements, such as carbon, have several isotopes. For example, ^{12}C has six protons and six neutrons, while ^{13}C has six protons and seven neutrons. Most isotopes are unstable and decay into a different element by emitting radiation. The carbon atom is used as a mass standard for the measurement of atomic masses. The atomic mass unit, u, is exactly equal to [(1/12)(mass of the ^{12}C atom)]. $u = 1.66 \times 10^{-27}$ kg. The mass of the proton, $m_p = 1.00728$ u, and that of the neutron, $m_n = 1.00866$ u. Experiments have shown that the size of the nucleus is proportional to $A^{1/3}$. The specific formula for the nuclear radius is $R = 1.2 \times 10^{-15} A^{1/3}$ fm (fermi), where 1 fm $= 10^{-15}$ m.

3. *The strong force*. Since the protons inside a nucleus are very close together, they experience a strong electric repulsive force. The *nuclear*, or *strong*, force is a powerful attractive force between nucleons and acts to hold the nucleus together in spite of the electrical repulsion. The strong force is complicated; it is attractive only over the range from about one to two fermis. In this range it is nearly 100 times as strong as the electrical, or Coulomb, force. For nucleon separations of less than 1 fm, the force is repulsive; for distances greater than 2 fm, the strong force decreases abruptly. A nucleon inside a nucleus is pulled in all directions by those nucleons immediately around it; the net force on it is zero. A nucleon at the nuclear surface is pulled inward by the nucleons inside it, and so cannot escape through the nuclear surface, in much the same way that molecules of a water drop are pulled toward the inside of the drop.

4. *Nuclear binding energy*. It is clear that the mass of the ^{12}C atom, 12 u, is much less than the mass of 6 protons + 6 neutrons + 6 electrons. That is, energy must be supplied to separate ^{12}C or any other nucleus into its constituent nucleons. The negative of this "stored" energy is called the *binding energy* (B.E.). For most nuclei, the average binding energy of a nucleus is 8 MeV per nucleon. Note that the rest-mass energy of a nucleon is about $1\ u \times c^2 = 931$ MeV, so that the binding energy is about 1% of the rest-mass energy. The *mass defect* of a nucleus = [mass of N neutrons + Z protons] − [mass of nucleus] = B.E./c^2.

5. *Radioactivity*. Uranium and many other unstable isotopes are found to emit three kinds of radiation, which are called alpha, beta, and gamma rays. Alpha rays are helium nuclei, beta rays are high-speed electrons, and gamma

rays are energetic photons. When a nucleus emits an alpha particle, Z decreases by 2, and A decreases by 4. Thus, the alpha decay of ^{238}U can be written as

$$^{238}U \rightarrow {}^{234}Th + \alpha$$

^{238}U is called the *parent nucleus* and ^{234}Th the *daughter*. Alpha decay is an example of a *fission* process, in which a heavy nucleus, because of the strong electrical repulsive force between its protons, splits up into two or more lighter nuclei.

6. *Beta decay*. An example of beta decay is the decay of the neutron into a proton, an electron, and an antineutrino: $n \rightarrow p + e^- + \bar{\nu}$. Neutrinos and their antiparticles, antineutrinos, have spin $\hbar/2$ and, like photons, have zero mass. Beta decay of a nucleus can be considered the equivalent of the decay of one of the neutrons inside the nucleus into a proton. Thus

$$^{60}Co \rightarrow {}^{60}Ni + e^- + \bar{\nu}$$

As always, energy conservation applies to the decay process. Since the energy of the decay is shared between the electron and the neutrino, the energy of the electron can vary between zero and the maximum amount that corresponds to the electron carrying away all the energy of the decay. Beta decay is governed by a new kind of force, called the *weak interaction force*, one of the four fundamental forces in nature, together with the strong, electromagnetic, and gravitational forces.

7. *Gamma emission*, caused by nucleons making transitions from one nuclear state down to another, is similar to the emission of light due to transitions of the atomic electrons. When a nucleus emits an alpha or a beta ray, it is often left in an excited state, and deexcites via gamma emission. Thus, alpha or beta emission often precedes gamma emission, such as in the case

$$\begin{array}{l} ^{60}Co \rightarrow {}^{60}Ni + e^- + \bar{\nu} \\ \qquad\qquad \searrow {}^{60}Ni + \gamma \end{array}$$

8. *The law of radioactive decay*. When a nucleus decays by alpha or beta emission, the original, or parent, nucleus disappears and is replaced by the daughter nuclei. Radioactive decay obeys an exponential law, such that if the original sample contains n_0 radioactive nuclei, the number n remaining after time t is given by $n = n_0 e^{-\lambda t}$, where λ is called the *decay constant*. The *half-life*, $t_{1/2}$, is the time after which only 1/2 of the original number of radioactive nuclei remain, and is given by

$$n_0/2 = n_0 e^{-\lambda(t/2)}$$

It is easy to show that $t_{1/2} = 0.693/\lambda$. The *mean life*, the average lifetime of radioactive nuclei, is just $1/\lambda$. The *decay rate*, the number of decays per second, is just $-dn/dt = \lambda n$. The usual unit for radioactive decay is the curie, where 1 curie $= 3.7 \times 10^{10}$ disintegrations per second.

9. *Nuclear reactions*. The first studies of nuclear reactions were carried out by Rutherford, using alpha particles from radioactive sources. Later, *accelerators*, which produce beams of high-speed ions, were used to bombard targets to induce nuclear reactions, an example of which is

$$p + {}^{7}Li \rightarrow n + {}^{7}Be$$

Here protons strike a lithium target, producing beryllium and neutrons. Nuclear reactions obey the conservation laws for rest-mass energy, momentum, angular momentum (including spin), charge, and a conservation law for mass, or "baryon" number. Essentially, the total number of nucleons present must remain constant. In the example above, there are eight nucleons before and after the reaction. Note also that charge is conserved. We can express the conservation of rest-mass energy as follows: the initial kinetic energy, K, + the change in rest-mass energy, Q, = the final kinetic energy, K'. If the initial masses are m_1 and m_2 and the final masses are $m_{1'}$ and $m_{2'}$, Q is defined to be $(m_1 + m_2 - m_{1'} - m_{2'})c^2$, so that $K + Q = K'$. Note that if Q is positive, the initial masses are greater than the final masses, and some of the initial rest-mass energy is converted into kinetic energy. If Q is negative, some of the initial kinetic energy must be converted into rest mass in order for the reaction to proceed. Assuming that the target is at rest, the available part of the initial kinetic energy, the center-of-mass energy, is given by $K_{cm} = (1/2)m_1m_2/(m_1 + m_2)[v_1^2]$. This must be greater than $-Q$. The *threshold energy*, the minimum energy at which the reaction can occur, is given by $K_{cm} = -Q$.

Sample Problems

1. *Worked Problem*. Tritium decays to form helium 3. The mass of a tritium atom is 3.016049 u and the mass of the helium 3 atom is 3.016029 u. What particle or particles are emitted in the decay process, and with what energy? (Note that the electron mass is 5.49×10^{-4} u.)

Solution: The decay can be written

$${}^{3}H \rightarrow {}^{3}He + e^{-} + \bar{\nu}$$

The energy corresponding to 1 u = 931.5 MeV, so that in terms of rest-mass energy, the initial energy = 3.016049(931.5) = 2809.4496 MeV. The final rest-mass energy = (3.016029)(931.5) = 2809.4310 MeV. Note that we do not add the electron rest mass to the final rest mass, even though an electron is emitted, since the ^{3}He resulting from the ^{3}H decay is singly ionized. We just need to use the difference in atomic masses. Then the energy released in the decay is 2809.4496 − 2809.4310 = 0.0186 MeV, or 18.6 keV. This is the maximum kinetic energy of the emitted electron, since the energy is shared between the electron and the neutrino.

2. *Guided Problem*. The radius of the Earth is 1.74×10^6 m, and its mass is 7.35×10^{22} kg. Its mean density is 3340 kg/m^3. If the entire mass of the Earth were confined to nuclear matter, what would be its radius?

Solution scheme

a. What is the expression for the radius of a nucleus?

b. What is the volume of a nucleus of mass number A? Note that the nucleus can be regarded as spherical and that the expression for the volume of a sphere is $(4/3)\pi R^3$.

c. Can you arrive at an expression for nuclear density, where density, $d = \text{mass/volume}$? You should find that this is independent of A, so that the composition of the Earth does not matter. What is your explicit expression for the mass density of nuclear matter?

d. Since $d = m/V$, $V = m/d$. Knowing m for the Earth, and d for nuclear matter, you should now be able to deduce a radius for the "nuclear matter" Earth.

3. *Worked Problem.* Compute the total binding energy and the binding energy per nucleon for (a) deuterium, (b) tritium, and (c) helium 4 (^{4}He). The mass of deuterium is 2.014102 u; that of tritium, 3.016049 u; and that of helium 4, 4.002603 u.

Solution: The binding energy (B.E.) of deuterium is given by

$$\begin{aligned}[\mathrm{u_p} + \mathrm{u_n} + \mathrm{u_e} - \mathrm{u_d}]c^2 &= [1.00728 + 1.00866 + 5.49 \times 10^{-4} - 2.014102]931.5 \\ &= 2.223 \text{ MeV}\end{aligned}$$

The binding energy per nucleon = 1.112 MeV/nucleon. For tritium,

$$\begin{aligned}\text{B.E.} &= [1.00728 + 2(1.00866) + 5.49 \times 10^{-4} - 3.016049](931.5) \\ &= 8.478 \text{ MeV, or } 2.826 \text{ MeV/nucleon}\end{aligned}$$

For helium,

$$\begin{aligned}\text{B.E.} &= [2(1.00728) + 2(1.00866) + 2(5.49 \times 10^{-4}) - 4.002603](931.5) \\ &= 28.294 \text{ MeV, or } 7.074 \text{ MeV/nucleon}\end{aligned}$$

Note how rapidly the binding energy per nucleon increases!

4. *Guided Problem.* Protons from a 5 MeV Van de Graaff accelerator are used to bombard a thin gold foil. Do you expect that the protons have enough energy to reach the surface of a gold nucleus, even in a head-on collision?

Solution scheme

a. Gold has a mass number A of 197 and a Z number of 79. What is its nuclear radius?

b. At the distance of closest approach, the kinetic energy of the protons is momentarily zero. The Coulomb energy between the proton and the nucleus, given by $(1/4\pi\varepsilon_0)[e(Ze)/R]$, is at this point equal to the kinetic energy that the proton had, namely, 5 MeV, at a very large distance from the nucleus. Can you now calculate R? Is R greater than the nuclear radius?

5. *Worked Problem.* Carbon 14 beta-decays with a half-life of 5730 years.

^{14}C dating makes use of the fact that living matter is slightly radioactive because of its ^{14}C content. When an organism dies, it stops acquiring new ^{14}C, and its content of ^{14}C decreases according to the radioactive decay law. A sample of wood from an ancient ship gives 9.4 decays per minute. A recent sample of the same size and the same kind of wood gives 16.3 decays per minute. How old is the ship?

Solution: The decay rate is given by $-dn/dt = \lambda n$, and since $t_{1/2} = 0.6793/\lambda$, $-dn/dt = 0.693n/t_{1/2}$. For the ancient ship, $-dn/dt = 9.4/\text{minute} = 4.94 \times 10^6$ per year. For the new sample, $-dn/dt = 8.57 \times 10^6$ per year. Then the ratio of radioactive ^{14}C atoms in the old sample to that in the new sample,

$$n_{old}/n_{new} = 4.94 \times 10^6/8.57 \times 10^6 = 0.576$$

Writing the decay law in the form

$$n = n_0 e^{-.693t/t_{1/2}}$$

we can solve for t.

$$n/n_0 = 0.576 = e^{-.693/t_{1/2}}, \text{ or } \ln(1.734) = 0.693t/(5730 \text{ years})$$

Therefore, $t = 4550$ years, so the ancient ship is 4550 years old.

6. *Guided Problem.* The nucleus fermium 251 (^{251}Fm) decays by alpha emission to several states of californium 247 (^{247}Cf), as indicated in Figure 45.1. The reaction Q value, defined to be $(m_x - m_{x'} - m_\alpha)c^2 = 7423$ keV for alpha decay to the ground state of ^{247}Cf. What are the energies of the alpha particles and gamma rays that you would observe if all the possible delays occur?

Solution scheme

a. Since we can assume that the parent nucleus decays at rest, the reaction Q is also equal to the total kinetic energy given to the decay

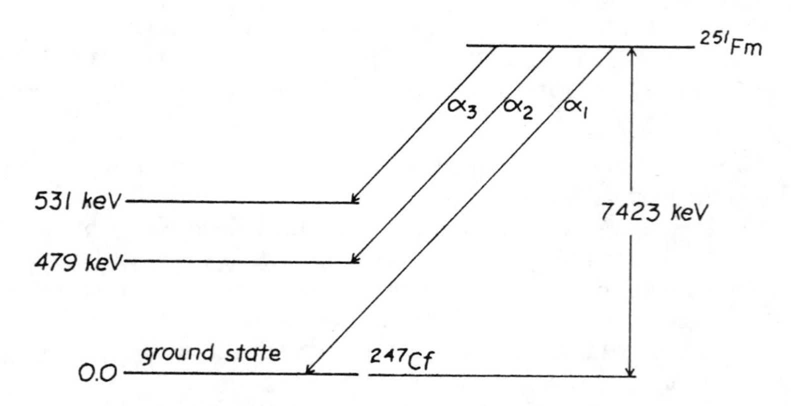

Figure 45.1 Problem 6. Alpha decays at ^{251}Fm to the ground and several excited states of ^{247}Cf.

fragments. That is, $Q = T_{x'} + T_{\alpha'}$. From conservation of momentum, $p_{x'} = p_{\alpha'}$, allowing us to eliminate $T_{x'}$, so that the kinetic energy of the alpha particle, $T_\alpha = Q/(1 + m_\alpha/m_{x'})$. Can you now solve for T_α for each of the allowed alpha decays? Note that the Q value is different for alpha decay to each different excited state.

b. Assuming that each excited state of ^{247}Cf can decay to any state lower in energy, what gamma ray energies do you expect?

7. *Worked Problem.* What is the Q value for the reaction

$$p + {}^{14}C \rightarrow {}^{14}N + n$$

(protons strike a ^{14}C target, producing ^{14}N + neutrons)? What is the threshold energy, that is, the energy at which we will first observe the production of neutrons from the reaction? The mass of ^{14}C is 14.003242 u and that of ^{14}N is 14.003074 u.

Solution: The reaction

$$\begin{aligned} Q &= (m_H + m_C - m_n - m_N)c^2 \\ &= (1.007825 + 14.003242 - 1.008665 - 14.003074)931.5 \\ &= (-6.62 \times 10^{-4})931.5 = -0.626 \text{ MeV} \end{aligned}$$

Note that, as always, we must be very careful to conserve electron number when using atomic mass units. We use m_H, the atomic mass of hydrogen, rather than the proton mass, in determining the Q value; otherwise, we would need to subtract 1 electron mass from the mass of ^{14}N, since the reaction does not add the electron, but leaves ^{14}N singly ionized, that is, with one electron short! The reaction Q is negative, so that the threshold energy is given by $K_{cm} = -Q = 0.625$ MeV.

$$K_{cm} = (1/2)m_1 m_2/(m_1 + m_2)(v_1^2) = (14/15)T_1 = 0.625 \text{ MeV}$$

Therefore, $T_1 = 0.670$ MeV. This is the incident proton energy at which neutrons will be first seen.

8. *Guided Problem.* Complete the following *fission* reaction, in which the heavy nucleus ^{235}U absorbs a very low energy (thermal) neutron and splits up into two or more lighter nuclei, with a net energy release. Also calculate the total energy release, namely, the Q value:

$${}^{235}U + n \rightarrow {}^{141}Cs + {}^{93}Rb + ?$$

The atomic mass values are as follows: ^{235}U: 235.043924 u; ^{141}Cs: 140.91949 u; ^{93}Rb: 92.92172 u.

Solution scheme

a. What are the Z numbers for U, Cs, and Rb? Note that the reaction must conserve charge. Is charge conserved, or do we need to add a reaction product in order to conserve charge?

b. The reaction must also conserve baryon number. What do we need to add in order to conserve baryon number? Can you now complete the reaction?

c. Write down the expression for the reaction Q. Note that the energy of the absorbed neutron may be taken to be zero. What is the total energy release, in MeV?

Answers to Guided Problems

2. $d = 2.3 \times 10^{17}$ kg/m^3; $V = m/d = (4/3)\pi r^3$, from which $r = 42.4$ m
4. The distance of closest approach for a head-on collision,

$$R_{\min} = 2.275 \times 10^{-14} \text{ m}$$

The radius of the gold nucleus, from

$$r = 1.2 \times 10^{-15}(197)^{1/3} = 6.98 \times 10^{-15} \text{ m}$$

The proton does not get close enough to reach the nuclear surface. Note that we have neglected the size of the proton, a reasonable assumption, since the gold nucleus is much larger than the proton.

6. $T_{\alpha 1} = 7423/(1 + 4/247) = 7305$ keV
$T_{\alpha 2} = [7423 - 479]/(1 + 4/247) = 6833$ keV
$T_{\alpha 3} = [7423 - 531]/(1 + 4/247) = 6782$ keV
The energies of the three possible gamma rays, shown in Figure 45.2, are $E_{\gamma 1} = 52$ keV, $E_{\gamma 2} = 531$ keV, $E_{\gamma 3} = 479$ keV.

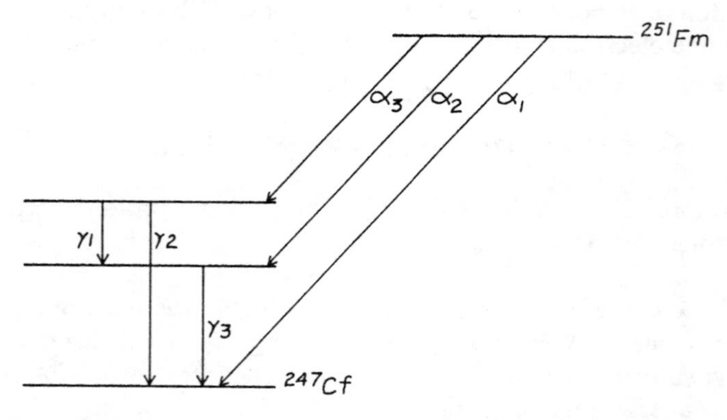

Figure 45.2 Problem 6. Alpha and gamma rays resulting from the decay of ^{251}Fm.

8. $Z_U = 92$; $Z_{Cs} = 55$; $Z_{Rb} = 37$. Since $37 + 55 = 92$, Z number (charge) is conserved. $N_U = 143$; $N_{Cl} = 86$; $N_{Rb} = 56$. Since a total of 144 neutrons are on the left side of the equation, in order to conserve baryon number there must be 2 extra neutrons on the right side of the equation. The complete equation is

$$^{235}\text{U} + \text{n} \rightarrow {}^{141}\text{Cs} + {}^{93}\text{Rb} + 2\text{n}$$

The Q value is given by

$$Q = [m_U + m_n - m_{Cs} - m_{Rb} - 2m_n](931.5)\ \text{MeV}$$

$Q = 0.1940(931.5) = 180.7$, or 181 MeV.

Chapter 46

Elementary Particles

Overview

Much of physics is concerned with the question "What is matter really made of—what are the elementary, or fundamental, building blocks?" Until we found that atoms were really composed of electrons orbiting a nucleus, we thought that atoms were "fundamental." Soon, helped by particle accelerators, we were able to study nuclei, and found not only that nuclei could be split into their constituent protons and neutrons but also that the protons and neutrons themselves were not "fundamental." High-energy collisions of neutrons and protons result in the formation of many new "elementary particles." However, most of these particles are not fundamental; many are more massive than nucleons, live for only a short time period, and decay into a variety of smaller particles. Recently, we have come to believe that there is, in fact, a rather small number of fundamental particles. The electron is a member of a family of particles called leptons. Neutrons, protons, and most other "elementary particles" are really "built" from subunits called quarks, of which there are six "flavors." Quarks have never been directly observed by experiment, yet the evidence for their existence is almost overwhelming.

Essential Terms

Energy units: eV, keV, MeV, GeV, TeV
Fermilab, CERN, SLAC, SSC
detectors: bubble chambers
leptons, baryons, mesons, hadrons
strong force, electromagnetic force, weak force
baryon number, lepton number, isospin, strangeness, parity
field quanta; pions; W and Z particles; photons
Feynman diagram; quantum electrodynamics
electroweak force
quarks: up, down, charmed, strange, top, bottom
color force, gluon, quantum chromodynamics

Key Concepts

1. We need high-energy accelerators because the energies required to create new particles are often huge. At a more fundamental level, in order to study an object effectively, we need a wavelength that is as short as, or shorter than, the object we are probing. The de Broglie wavelength of a particle is given by $\lambda = h/p$. The higher the energy, the shorter the wavelength. A proton or neutron has a radius of the order of 10^{-15} m. To obtain such short wavelengths, we need very high energies, of the order of hundreds of MeV, or even GeV, where 1 GeV = 10^3 MeV. The world's largest accelerator, as of 1988, is the Fermilab's Tevatron, which produces beams of protons and antiprotons, whirling in opposite directions in a circular tunnel 6 km in circumference. The maximum energy for each particle is 1 TeV = 10^3 GeV. The *superconducting supercollider*, or SSC, if it is funded and built, will have beams of 20 TeV particles spinning in orbits 20 km in diameter.

2. The basic tools of elementary-particle physics are accelerators and detectors. There are many kinds of detectors. A *bubble chamber* is a tank filled with superheated liquid (usually hydrogen). A charged particle leaves in its wake a fine stream of bubbles, which can be photographed by means of high-speed cameras. Nowadays, bubble chambers have for the most part been replaced by huge, complex *electronic bubble chambers*, in which the particles produced in high-energy collisions are tracked and measured by many different kinds of detectors, coupled to high-speed computers. These detector systems, which are not bubble chambers at all, are more versatile and can process events much faster than the older bubble chambers. The UA1 detector at CERN, used to detect the W and Z^0 particles in 1982, has a weight of 2000 tons; even larger detectors are now being built.

3. The known particles fall into three families: *leptons, baryons,* and *mesons*. In addition, for every particle there is a corresponding *antiparticle*. The lepton family is made up of the electron, e, and its neutrino, ν_e; the muon, μ, and its neutrino, ν_μ; and the massive tau, τ, and its neutrino, ν_τ. The neutrinos have zero mass and zero charge and are very difficult to detect, since even objects as massive as the Earth are nearly transparent to neutrinos. The electron, muon, and tau all have a charge of -1. The antiparticles have the same mass and spin as the particle, but they have the opposite charge. The antielectron is the positron, e^+, and the antimuon is the μ^+.

The strong force acts on *baryons* and on *mesons*, both of which are called *hadrons*, but not on leptons. The nucleon (proton and neutron), with a mass of 939 MeV, is the lightest member of the baryon family. The *pion* (mass 140 MeV) is the lightest member of the meson family. Baryons are distinguished from mesons in that baryons are usually composed of three quarks, and mesons of only two. A natural consequence of this is that sometimes an uncharged meson, such as the π^0, can be either a particle or an antiparticle. Two π^0 mesons can annihilate.

4. The four fundamental forces are the *strong, electromagnetic, weak,* and *gravitational*. Of these, the strong and the weak have very short ranges; the electromagnetic and the gravitational have infinite ranges. The strong, electromagnetic, and weak forces, or interactions, can also be characterized by

typical reaction times, ranging from 10^{-23} s for the strong to 10^{-10} s for the weak interaction. In the language of elementary-particle physics, a "stable" particle is one *stable against decay through the strong interaction* and is therefore one with a lifetime long compared with 10^{-23} s. We tabulate the differences between the interactions below:

Force	**Acts On**	**Relative Strength**	**Typical Reaction Time**
Strong	hadrons	1	$\sim 10^{-23}$ s
Electromagnetic	all charged particles	10^{-2}	$\sim 10^{-14}$ s
Weak	leptons, hadrons	10^{-6}	$\sim 10^{-10}$ s
Gravitational	matter and energy	10^{-38}	

The electromagnetic force acts on *charge*. Thus, it acts not only on charged particles but also on neutral particles, such as the neutron and the π^0, which contain an *internal distribution* of electric charge.

5. The strong, electromagnetic, and weak interactions are characterized by the conservation laws that they obey. All three conserve energy-momentum, charge, and two quantities we call *baryon number* and *lepton number*. Each baryon has a baryon number of +1; each antibaryon, a baryon number of −1. In any reaction, the net baryon number remains constant. Similarly, each lepton has a lepton number of +1; each antilepton, a lepton number of −1. Again, in any reaction, the net lepton number remains constant. A strong (and very fast) decay is that of the delta: $\Delta^{++} \rightarrow p + \pi^+$ ($\sim 10^{-23}$ s). A weak decay is that of the lambda particle: $\Lambda^0 \rightarrow p + \pi^-$ ($\sim 10^{-10}$ s). What is the difference between the above two decays? Both conserve mass-energy and momentum, charge, baryon number, and lepton number. In the decay of the Δ^{++}, both the Δ^{++} and the proton have baryon number +1. The π^+, a meson, has baryon number 0. In the decay of the Λ^0, both the Λ^0 and the proton have baryon number +1, and the π^- has baryon number 0.

There are additional quantities that may be conserved. The strong interaction conserves *strangeness, parity,* and *isospin*. Of these, the electromagnetic interaction does not conserve isospin. The weak interaction conserves *none* of these three quantities. *Isospin* is a very special quantum number that characterizes the strong interaction. *Strangeness* is a number that we will later relate to the strange quark. Each hadron has a strangeness number; that of the proton, neutron, pion, or delta particle is 0, and that of the lambda is −1. The decay $\Delta^{++} \rightarrow p + \pi^+$ conserves strangeness; the decay $\Lambda^0 \rightarrow p + \pi^-$ does not. The lambda decay is therefore a weak decay. Conservation of *parity* can be best described by saying that if parity is conserved, the mirror image of the quantum-mechanical wave describing the system is the same as the original wave.

6. Forces are mediated by *fields*. Thus, there are *gravitational, electromagnetic, weak,* and *strong* fields. According to quantum theory, the energy stored in fields is found in *quanta*, or lumps of energy. Photons are the quanta of the electromagnetic field; pions, of the strong field; and $W^\pm$ and Z^0 particles, of the weak field. The quanta of the weak and strong field are therefore particles with mass. Forces between any two particles are generated by

the exchange of the field quanta; for example, photons in the case of the electromagnetic force. These exchanged field quanta are *virtual*, because they live for only a very short time, being emitted by one particle, then propagating through the distance between the two particles, and finally being absorbed by the second particle. A *Feynman diagram*, showing the exchange of a photon in an electron–proton interaction, is shown in Figure 46.1. The larger the mass of the exchange quanta, the shorter the range of the force. Photons are massless—the range of the electromagnetic force is infinite. W and Z particles have a large mass, larger even than that of the pion, and the range of the weak force is very short.

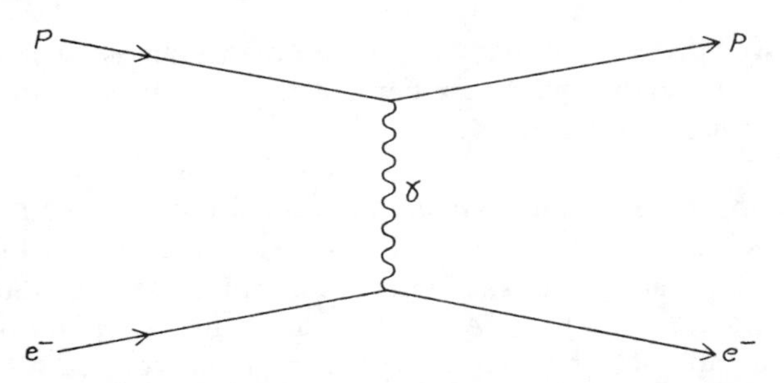

Figure 46.1 The exchange of a photon in an electron–proton interaction.

7. The weak and the electromagnetic forces have been shown to be two aspects of a single *electroweak force*. The unified theory predicted that the field quanta of the weak force consist of two kinds of particles, called $W^{\pm}$ and Z^0 particles. At a fundamental level, the weak and the electromagnetic forces can be regarded as having the same strength (and so as being essentially one and the same force) if the W and the Z particles have masses about 80 and 90 times, respectively, that of the proton. The W and Z particles were detected in 1982, confirming the unified theory of the weak and the electromagnetic forces.

8. All experiments have shown the electron and, in fact, all members of the lepton family are "pointlike" and have no internal structure. However, none of the nearly 300 hadrons seen in experiments appear to be "elementary particles"—that is, none have emerged as elementary or indivisible building blocks of matter. We now believe that the real building blocks of matter are particles called *quarks*. Quarks come in six *"flavors"*: *down* (d), *up* (u), *strange* (s), *charmed* (c), *bottom* (b), and *top* (t). All have charge of either −1/3 or +2/3, and all have a spin-angular-momentum number of 1/2. The quarks, and their quantum numbers, are tabulated below:

	d	**u**	**s**	**c**	**b**	**t**
Q (charge)	−1/3	+2/3	−1/3	+2/3	−1/3	+2/3
Z (component of isospin)	−1/2	+1/2	0	0	0	0
S (strangeness)	0	0	−1	0	0	0
C (charm)	0	0	0	1	0	0
B (bottomness)	0	0	0	0	−1	0
T (topness)	0	0	0	0	0	1

Most of the known hadrons are made up of the d, u, and s quarks; for example, the proton consists of (u, u, d) and therefore has a total charge of $+2/3+2/3-1/3=1$. The neutron is (u, d, d) and therefore has a total charge of $+2/3-1/3-1/3=0$. For each quark there is a corresponding antiquark, of opposite electric charge. In general, baryons comprise three quarks; mesons, two quarks. Thus, the π^+ is (u, $\bar{d}$), where the bar indicates an anti-d quark, and so has charge $+2/3+1/3=+1$. The π^- is ($\bar{u}$, d), with charge $-2/3-1/3=-1$. Thus, the π^+ is the antiparticle in the π^-, and vice versa. We can build up all known hadrons from the quarks and, in addition, make reasonable predictions about some of their properties, such as magnetic moments.

However, we have never directly observed isolated quarks in experiments, and it is now believed that quarks are permanently "*confined*" within hadrons by the very strong "*color force*."

9. Several apparently identical quarks are often found in the same quantum state, which would violate the exclusion principle. To avoid this violation, we postulate that the quarks are not identical but can have the different *colors* of red, green, and blue. Antiquarks have the *anticolors* of antired, antigreen, and antiblue. The three quarks found in baryons each have a different color, and just as a mixture of red, green, and blue colors is white, or colorless, each baryon is colorless. Similarly, the mesons, which constitute a quark–antiquark pair, are colorless. The strong attractive *color force* that confines the quarks within hadrons is closely related to the strong force, so that the strong force between nucleons is caused by the color force acting within each nucleon. The color force remains constant as the distance between quarks increases. The quarks behave as though linked by rubber bands. During a collision, the bands stretch, but they pull the quark back again after the collision.

The color force between quarks is due to an exchange of virtual particles, called *gluons*, between the quarks. Thus, the exchange of a gluon between two colored quarks is analogous to the exchange of a photon between two charged particles. However, unlike the photon, which has no charge, the gluon has color, and gluon exchange alters the color of the quarks. The theory of the color force is called *quantum chromodynamics* (QCD). QCD predicts that in violent particle–antiparticle collisions, a quark, an antiquark, and a gluon may be produced. Each will decay, mostly into pions, which should be experimentally observable as "pion jets" spurting out from the collision. Experiments have confirmed such "triple jets," lending support to the theory of QCD.

The charmed (c), bottom (b), and top (t) quarks are much more massive than the up (u), down (d), or strange (s) quarks, so that particles containing these quarks are more massive than particles composed of the lighter u, d, or s quarks. The study of hadrons containing, c, b, or t quarks has therefore required increasingly higher-energy accelerators. Evidence for the c quark was found at the 30 GeV accelerator at Brookhaven National Laboratory; particles containing the b quark were found at Fermilab in 1977. At the present time (1988) we are still searching for experimental evidence for the t quark.

Sample Problems

1. *Worked Problem.* The Tevatron, at Fermilab, has a circumference of 6 km and produces beams of up to 1 TeV (10^{12} eV) of protons and antiprotons, which rotate in opposite directions. What average magnetic field is required to cause a 1 TeV proton to rotate in a circular orbit 6 km in circumference? How many revolutions per second does the proton make around the circular ring?

Solution: The centripetal force is given by $m\upsilon^2/R$ and the magnetic force by $Bq\upsilon$, so that, for a circular orbit, $m\upsilon^2/R = Bq\upsilon$, or $BqR = p$, where p is the relativistic momentum of the particle. The total energy is given by

$$E = \sqrt{p^2c^2 + m^2c^4}$$

where $E = 10^{12}$ eV. Since $mc^2 = 938(10^6)$ eV, we can neglect the mass term, m^2c^4, and $E = pc$, or $p = E/c$. Therefore

$$p = 10^{12}\ \text{eV}/c = 10^{12}(1.6 \times 10^{-19})/(3 \times 10^8) = 5.33 \times 10^{-16}\ \text{kg} \cdot \text{m/s}$$

Therefore

$$B = p/qR = (5.33 \times 10^{-16})(2\pi)/(1.6 \times 10^{-16})(6 \times 10^3) = 3.49\ \text{T}$$

For a 1 TeV proton, $\upsilon = 0.9999\ c$, or to a very good approximation, $\upsilon = c$. The period $T = 6 \times 10^3/3 \times 10^8 = 2 \times 10^{-5}$ s, or $(1/T) = 50{,}000$ turns/s.

2. *Guided Problem.* Much experimental effort has gone into looking for the decay $\mu^- \rightarrow e^- + \gamma$. It has not been found, down to a "branching ratio" limit of about 10^{-11}. Instead, the muon decays by

$$\mu^- \rightarrow e^- + \nu_\mu + \bar{\nu}_e$$

What conservation law allows the second, but not the first, decay?

Solution scheme

a. The mass of the muon is 106 MeV, that of the electron is 0.51 MeV, and the neutrino has zero mass. Are both reactions energetically possible? Can both conserve momentum? Note that each of the three particles involved has a spin of 1/2.
b. Is charge conserved in both decays?
c. Are any of the particles baryons?
d. The muon is a lepton, as is the electron. The first decay would appear to conserve lepton number. The lepton family is made up of three kinds of leptons—the electron, the muon, and the tau particle—each with its neutrino. What we have *not* yet mentioned is that total lepton number is separately conserved for *each* type of lepton. Can you now

see that the allowed (observed) decay obeys lepton conservation, whereas the decay $\mu \rightarrow e + \gamma$ fails to conserve both muon and electron lepton number?

3. *Worked Problem.* The following "stable" particle decays are observed; their mean lifetimes are indicated in parentheses. A "stable" particle is defined as being stable against strong decay, so that none of these particles decay through the strong interaction.

$$\begin{array}{lll} \pi^0 & \rightarrow \gamma + \gamma & (10^{-16}\ \text{s}) \\ \Lambda^0 & \rightarrow p + \pi^- & (2.5 \times 10^{-10}\ \text{s}) \\ \Lambda^0 & \rightarrow p + e^- + \bar{\nu} & (2.5 \times 10^{-10}\ \text{s}) \\ K^0 & \rightarrow \pi^+ + \pi^- & (\sim 10^{-10}\ \text{s}) \\ \Sigma^+ & \rightarrow p + \pi^0 & (\sim 10^{-10}\ \text{s}) \end{array}$$

Which, if any, of the above decays do not conserve strangeness? Indicate for each decay whether it is an electromagnetic or a weak decay.

Solution: Of the decay reaction products, photons, electrons, and neutrinos do not have strangeness. For a hadron to have strangeness different from zero, it must contain one or more strange quarks. Protons and pions are made from the u, d quarks, and their antiquarks, and therefore have $S = 0$. Thus, in no case is S different from zero for the reaction products. The π^0 also has $S = 0$. Thus, the first decay conserves strangeness and is an electromagnetic decay.

The quark composition of the Λ^0 is (u, d, s), that of the K^0 is (d, $\bar{s}$), and that of the Σ^+ is (u, u, s). Thus, each of the particles decaying in the last four decays has $S = \pm 1$; S is conserved in none of the decays, and, therefore, each of the decays is a weak decay. Note that the time scales of the final four decays are all about 10^{-10} s, and that of the first decay (an electromagnetic decay) is faster, 10^{-16} s.

4. *Guided Problem.* Note the following reactions:

$$\begin{array}{l} \Lambda^0 \rightarrow \Sigma^0 + \gamma \\ n + p \rightarrow d + \gamma \\ p + p \rightarrow n + p + \pi^+ \\ K^+ \rightarrow \mu^+ + \nu \end{array}$$

Do each of the above reactions conserve mass-energy, charge, baryon number, and lepton number? Are any of the reactions "forbidden"? The quark composition and masses of the hadrons are given as follows:

Λ^0 = (u, d, s), 1116 MeV Σ^0 = (u, d, s), 1192 MeV
n = (u, d, d), 939 MeV p = (u, u, d), 938 MeV
π^+ = (u, $\bar{d}$), 139.6 MeV K^+ = (u, $\bar{s}$), 494 MeV

Solution scheme

a. Do any of the reactions violate mass–energy conservation? Note that you need only look at the decays here (the first and the last), since for the middle two, one of the particles may have the energy (say, from an accelerator) to make the reaction occur. The letter d in the second reaction refers to the deuteron.

b. Is charge conserved in all cases?
c. Is baryon number conserved in all cases?
d. How about lepton number? Can you categorize any of the reactions as being "strong," "weak," or "electromagnetic"?

5. *Worked Problem.* No proton decay experiment has found firm evidence of the decay of a proton, in spite of several years of observation with very large detectors. Assume that in order to be sure of seeing a proton decay, we need to devise a detector to see at least one *event per year*. How large a detector, in kg and in m^3, do we need to observe ~1 event per year, assuming a water detector, if the mean life of the proton is 10^{33} years? If the mean life of the proton is even longer, say, ~10^{34} years, are we likely to be able to observe the decay? Assume that the decay of *any* proton in the water molecule is possible.

Solution: At any time t, the number of particles remaining is given by $N = N_0 e^{-\lambda t} = N_0 e^{-t/\tau}$, where τ is the mean life. Therefore, $-(dN/dt) = -N/\tau$. Let $(dN/dt) = 1/\text{yr}$, and let $\tau = 10^{33}$ years. Then $(1/\text{yr})(10^{33}$ years$) = 10^{33} = N$, so that we need 10^{33} protons. Assume that H_2O has 10 protons per molecule, so that we need 10^{32} molecules, or $[10^{32} \times 18]/[6.02 \times 10^{23}] = 2.99 \times 10^9$ gm, or 2.99×10^6 kg. The density of water = 1000 kg/m^3, so that we need 2.99×10^3 m^3. The detector thus needs to be ~15 m on each side. If $t \sim 10^{34}$ years, we need a detector with ten times as many protons in it! The expense and complexity of a detector this large is likely to be prohibitive.

6. *Guided Problem.* Quarks are believes to exist in six "flavors": down, up, strange, charmed, bottom, and top (d, u, s, c, b, t). We have experimental evidence for the existence of all but the top (t) quark (and searches are now under way for the t quark). All known baryons and mesons shown in Tables 46.3 and 46.4 in your text can be "constructed" from the quarks. The baryons are three quark states, mesons are generally quark–antiquark pairs.

a. Write down the quark composition of the proton (p), neutron (n), and the delta (Δ^{++}), where the ++ indicates a charge of +2, the Λ^0 (mass of 1116), the π^+, π^-, and K^0. Make sure that your particles have the right charge.
b. Write down the quark compositions of the antiproton, antineutron, anti π^-, and anti K^0.

Solution scheme

a. None of the above particles have the exotic c, b, or t quarks. Only the Λ^0 and the K^0 contain s quarks (that is, have strangeness different from 0).
b. Remember that baryons are three-quark states, mesons, a quark–antiquark pair. Which of the above are mesons, and which are baryons? Can you now construct the quark constituents of each particle, remembering to conserve charge?

7. *Worked Problem.* What is the threshold energy for antiproton production by means of a proton beam from a high-energy accelerator incident on a liquid hydrogen target? The masses of the proton and of the antiproton are 938 MeV.

Solution: The reaction is $p + p \to p + \bar{p} + p + p$. Note that since the antiproton has baryon number −1, we must produce proton–antiproton pairs in order to conserve baryon number (and charge). The Q of this reaction, the initial mass-energy minus the final mass-energy, is given by

$$Q = [m_p + m_p - 4m_p] = 2(938) - 4(938) = -1876 \text{ MeV}$$

This is the center-of-mass energy that the incident proton must supply. For a particle incident on a particle of the same mass at rest, the *total* center-of-mass energy is given by

$$\sqrt{S} = \sqrt{2mc^2[2mc^2 + K]}$$

where K is the laboratory kinetic energy of the incident particle. Then $\sqrt{S} = 2(1876)$ MeV, since the total center-of-mass energy must be *twice* 1876 MeV because the rest-mass energy of the two incident particles themselves (two protons) is 1876 MeV. Therefore,

$$2(1876) = \sqrt{2(938)[2 \times 938 + K]}$$

from which $K = 5628$ MeV $= 5.6$ GeV. This is the observed threshold for antiproton production.

8. *Guided Problem.* Find the threshold energy for pion production in the following reactions:

(a) $p + p \to p + n + \pi^+$ $[m_p = 938 \text{ MeV},\ m_m = 939.6 \text{ MeV},\ m_{\pi^\pm} = 139.6 \text{ MeV}]$
(b) $p + p \to d + \pi^0$ $[m_d = 1876 \text{ MeV},\ m_{\pi^0} = 135 \text{ MeV}]$
(c) $p + p \to d + \pi^+ + \pi^-$

Solution scheme

a. Can you calculate the reaction Q for each of the above reactions?
b. Knowing the reaction Q, can you apply the equation we used in worked problem 7, above, to find the reaction threshold energy, K? Note that in each case the equation used in problem 7 for $\sqrt{S}$ gives the *total* center-of-mass energy needed.

9. *Worked Problem.* It is instructive to draw quark diagrams for some reactions, especially for strong interactions. For example, the reaction $p + \pi^+ \to K^+ + \Sigma^+$ can be represented as in Figure 46.2. Note that the worldline of an antiquark (here d) is represented as that of a quark (d) traveling backward in time. The strong interaction occurs within the dotted circle.

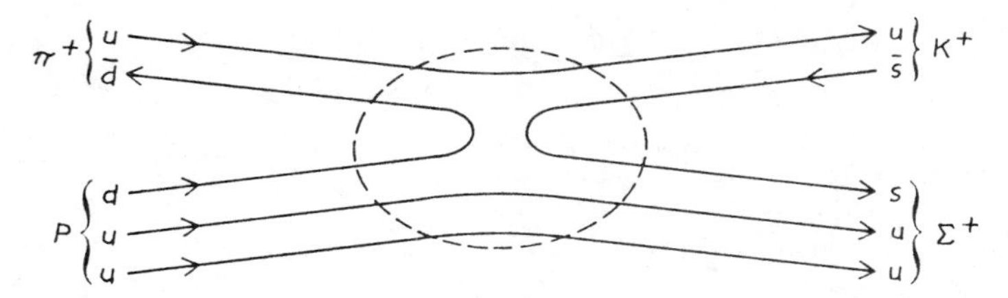

Figure 46.2 Problem 9. The quark diagram for the reaction $p + \pi^+ \rightarrow K^+ + \Sigma^+$.

Note that strangeness is conserved in the diagram since we have simultaneously created an s quark–antiquark pair. Draw corresponding quark diagrams for the following:

$$\Delta^{++} \rightarrow p + \pi^+ \qquad [\Delta^{++} = uuu] \quad [\pi^- = \bar{u}d]$$
$$\pi^- + p \rightarrow K^0 + \Lambda^0 \qquad [K^0 = d\bar{s}] \quad [\Lambda^0 = uds]$$
$$p + K^- \rightarrow K^0 + \Xi^0 \qquad [K^- = \bar{u}s] \quad [\Xi^0 = uss]$$

Solution: The quark diagrams are drawn in Figure 46.3(a), (b), and (c), respectively, where the circles are meant to represent the regions where the strong interaction occurs. Note that in all cases quarks are created, or annihilated, only in quark–antiquark pairs. In all cases charge, baryon number, and strangeness are conserved.

Answers to Guided Problems

2. Both reactions are energetically possible, and both can conserve momentum. Both also conserve charge. The particles are all leptons, or field quanta (gamma rays), so that baryon number is conserved. The observed decay conserves *both* muon and electron lepton number; the decay $\mu \rightarrow e + \gamma$ would violate *both* of these, while conserving *overall* lepton number. Conservation of lepton number for *each family* of leptons appears to be a rather good conservation number.
4. The decay $\Lambda^0 \rightarrow \Sigma^0 + \gamma$ is energetically forbidden, since the mass of the Σ^0 is greater than that of the Λ^0. Each of the remaining reactions conserves mass-energy, charge, baryon number, and lepton number. Note that in the decay $K^+ \rightarrow \mu^+ + \nu$, the neutrino is ν_μ, in order to conserve muon lepton number. (K^+ is a meson, with lepton number 0, and the μ^+ has lepton number -1.) Of the above interactions, the $p + p \rightarrow n + p + \pi^+$ is a strong interaction.
6. $p = (u, u, d)$ $\quad [q = 2/3 + 2/3 - 1/3 = +1]$
   $n = (u, d, d)$ $\quad [q = 2/3 - 1/3 - 1/3 = 0]$
   $\Delta^{++} = (u, u, u)$ $\quad [q = 2/3 + 2/3 + 2/3 = +2]$
   $\Lambda^0 = (u, d, s)$ $\quad [q = 2/3 - 1/3 - 1/3 = 0]$
   $\pi^+ = (u, \bar{d})$ $\quad [q = 2/3 + 1/3 = +1]$
   $\pi^- = (\bar{u}, d)$ $\quad [q = -2/3 - 1/3 = -1]$
   $K^0 = (d, \bar{s})$ $\quad [q = -1/3 + 1/3 = 0]$

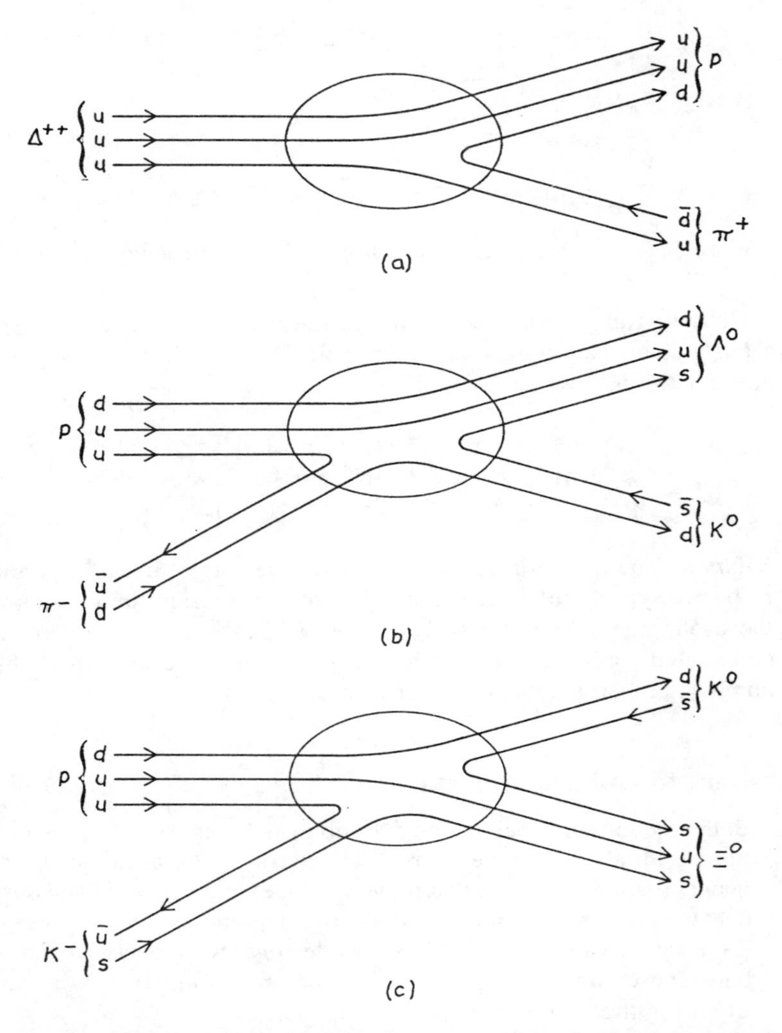

Figure 46.3 Problem 9. Quark diagrams for three strong interactions. The circles are meant to represent the regions where the strong interaction occurs.

$\bar{p} = (\bar{u}, \bar{u}, \bar{d})$ $[q = -2/3 - 2/3 + 1/3 = -1]$
$\bar{n} = (\bar{u}, \bar{d}, \bar{d})$ $[q = -2/3 + 1/3 + 1/3 = 0]$
$\bar{\pi}^- = (u, \bar{d}) = \pi^+$ $[q = 2/3 + 1/3 = +1]$
$\bar{K}^0 = (\bar{d}, s)$ $[q = 1/3 - 1/3 = 0]$

8. a. $Q = -141.2$ MeV; $K_{thres} = 295$ MeV
   b. $Q = -134.4$ MeV; $K_{thres} = 280$ MeV
   c. $Q = -278.6$ MeV; $K_{thres} = 596$ MeV